Handbuch der experimentellen Pharmakologie

Handbook of Experimental Pharmacology

Heffter-Heubner New Series

Volume XXVIII/3

Concepts in Biochemical Pharmacology

Part 3

Contributors

I. M. Arias, M. Black, B. Chabner, S. N. Cohen, D. S. Davies, G. Digenis, C. T. Dollery, S. Garattini, M. Gibaldi, J. R. Gillette, L. L. Goldberg, J. A. Hinson, E. Jawetz, G. Levy, F. Marcucci, J. R. Mitchell, E. Mussini, J. A. Oates, V. T. Oliverio, G. L. Plaa, W. Z. Potter, L. F. Prescott, D. G. Shand, W. R. Snodgrass, J. V. Swintosky, J. A. Timbrell, E. S. Vesell, W. W. Weber, R. M. Welch

Editors

J. R. Gillette and J. R. Mitchell

Assistant Editor

Patricia S. Randall

With 87 Figures

Springer-Verlag New York · Heidelberg · Berlin 1975

James R. Gillette—Jerry R. Mitchell
Laboratory of Chemical Pharmacology
National Heart and Lung Institute, National Institutes
of Health, Bethesda, Maryland 20014/USA

ISBN 0-387-07001-X Springer-Verlag New York Heidelberg Berlin
ISBN 3-540-07001-X Springer-Verlag Berlin Heidelberg New York

Library of Congress Cataloging in Publication Data (Revised) Main entry under title: Concepts in biochemical pharmacology. Handbuch der experimentellen Pharmakologie. Handbook of experimental pharmacology. New series, v. XXVIII, 1, etc. Includes bibliographies. 1. Drug metabolism-Collected works. I. Argy, W. P. II. Brodie, B. B., ed. III. Gillette, James R., 1928- ed. IV. Ackerman, Helen S., ed. V. Series: Handbuch der experimentellen Pharmakologie, v. XXVIII, 1 QP905.H3 Bd. 28, t. 1, etc. [RM301] 615'. 7 79-135957 ISBN 0-387-05134-1.

Typesetting, printing, and bookbinding: Brühlsche Universitätsdruckerei, Gießen

Preface

Part 3 of the Handbook of Experimental Pharmacology (Concepts in Biochemical Pharmacology) applies the principles enunciated in Parts 1 and 2 to clinical pharmacology and toxicology. The major objective is to elucidate the many factors that determine the relationships between pharmacokinetic aspects of the disposition and metabolism of drugs and their therapeutic or toxic actions in man. Because of the more restricted information obtainable in human studies, this volume reflects the editors' bias that an understanding of pharmacokinetics is fundamental for assessing pharmacologic or toxicologic effects of drugs in humans. The first chapter is a unique primer on when to apply and how to use pharmacokinetic tools in human pharmacology. The second chapter explains the general assumptions underlying pharmacokinetic approaches both in simple terms for the novice and in mathematical form for the more sophisticated reader.

Several chapters on determinants of drug concentration and activity discuss drug absorption, drug latentiation, drugs acting through metabolites, enterohepatic drug circulation, influence of route of drug administration on response, genetic variations in drug disposition and response, age differences in absorption, distribution and excretion of drugs, and pathologic and physiologic factors affecting absorption, distribution and excretion of drugs and drug response. The focus of these chapters is data obtained in human, rather than animal, studies. Most of the chapters contain new material never summarized previously.

The section on drug interactions opens with a chapter in which all drug interactions are viewed as resulting either from pharmacokinetic alterations or from pharmacodynamic alterations. Previously, drug interactions have been classified according to multiple proposed mechanisms. The present chapter critically reviews reported pharmacokinetic drug interactions, assesses their significance in humans, and provides guidelines for evaluating the significance of possible pharmacokinetic drug interactions in patients. The accompanying chapters discuss special aspects of drug interactions as they relate to cardiovascular, cancer chemotherapeutic and antimicrobial drugs. These chapters demonstrate that drug interactions are not always adverse but can be used to therapeutic advantage under certain circumstances.

The final section provides theoretical and clinical perspectives on the importance of drug metabolism and disposition for an understanding of toxic drug reactions.

The editors are indebted to Dr. Bernard B. Brodie, who inspired this trilogy on "Concepts in Biochemical Pharmacology" and served as coeditor of the first two parts. His numerous pioneering contributions have led to the development of many of the concepts enunciated in this series. The editors also wish to thank Mrs. Bonnie J. Chambers for her devotion in typing many of the more difficult manuscripts.

Bethesda, Maryland/USA — J. R. Gillette

Autumn 1974 — J. R. Mitchell

Contents

Foreword. J. R. Gillette and J. R. Mitchell XXI

Section Six: Determinants of Drug Concentration and Activity

Chapter 59:

Pharmacokinetics. G. Levy and M. Gibaldi. With 18 Figures

I. Introduction 1
II. Absorption and Elimination; Linear, One-Compartment Systems 3
A. Elimination of Single Intravenous Doses 3
B. Apparent Volume of Distribution 6
C. Urinary Excretion of Drug and Metabolites 7
D. Drug Metabolite Levels in the Body 9
E. Bioavailability and Absorption Kinetics. 11
F. Intravenous Infusions 14
G. Repetitive Drug Administration 15
III. Absorption, Distribution, and Elimination; Linear, Multicompartment Systems . . 17
A. Distribution and Elimination of Single Intravenous Doses 17
B. Apparent Volume of Distribution 20
C. The Meaning of β 20
D. Bioavailability and Absorption Kinetics. 20
E. Intravenous Infusion 21
F. Repetitive Drug Administration 22
G. The "First-Pass" Effect. 23
H. Other Multicompartment Models. 24
IV. Pharmacokinetics of Non-Linear Systems 24
A. Elimination of Single Intravenous Doses 24
B. Various Processes Resulting in Non-Linear Elimination Kinetics 27
C. Non-Linear Absorption Kinetics 29
D. Repetitive Drug Administration 29
V. Kinetics of Reversible Pharmacologic Effects 30
A. One-Compartment Systems 30
B. Multicompartment Systems 31
VI. Some Useful Suggestions for the Experimentalist 32
References 33

Chapter 60:

Other Aspects of Pharmacokinetics. J. R. Gillette. With 7 Figures

I. Introduction 35
II. Factors that Affect the Rate of Distribution of Drugs in Tissues. 35
A. Permeability of Capillaries and Cells 35
B. Lipid Solubility and Ionization of Drugs 36
C. Differences in pH between Cells and Blood 36
D. Reversible Binding of Drugs to Proteins and Other Components 36
1. General Considerations 36

2. Dissociation Rate Constants and Transit Times 37
3. Diffusion into Extracellular Spaces 38
4. Effect of Reversible Binding on the Clearance of Drugs by Liver 38
E. Diffusion Barriers in the Extracellular Space 39
F. Active Transport of Drugs 39
G. Time-Delay Factors 39
III. Some Problems in the Development of Pharmacokinetic Models for Tissues and Organs 40
A. General Aspects 40
B. Closed Model for a Tissue 41
C. Perfusion Systems (Nonrecirculatory) 44
1. General 44
2. Model 1 45
3. Model 2 49
4. Model 3 49
5. Model 4 50
6. Emptying of Tissue Compartments 50
D. Pharmacokinetics of Drug Disposition and Metabolism after Intravenous Administration of Drugs 51
1. General 51
2. *In vivo* Kinetic Models 52
a) Model I — Caternary 3-Pool System 52
b) Model Ia (a Mammillary, Degenerate Diffusion Limited System) 54
c) Model II (a Mammillary, Caternary System) 56
d) Model III (2-Pool System) 58
e) Model IV (Mammillary, 3-Pool Flow-Limited System) 60
3. General Comments on the Central Compartment 62
4. Concepts of Volume of Distribution 62
a) Equations for $V_d(eq)$, and $V_d(\omega)$ in 2 and 3 Compartment Systems . . . 63
b) Volume of Distribution by the Extrapolation Method 65
c) The Partition of Areas Method 66
5. Area under the Curve Method Applied to Metabolites 67
6. Effects of Reversible Binding of Drugs to Macromolecules on Pharmacokinetic Parameters 68
a) General 68
b) Effects of Reversible Binding on the Clearance of Drugs by a Single Organ 69
c) Effects of Reversible Binding on the Volume of Distribution of Drugs . 70
d) Effects of Reversible Binding on Rate Constants of Elimination in Linear, First Order Models 71
e) Non-Linear Binding 78
f) Area Under the Curves (AUC) of Total Drug Concentration in Plasma . 79
Appendix 1 80
References 84

Chapter 61:

Drug Latentiation. G. A. Digenis and J. V. Swintosky

I. Introduction 86
II. Chemical and Structural Consideration 87
III. Prodrug Design 92
IV. Carboxylic Acid Drugs 92
V. Alcohol Drugs 97
VI. Amine Drugs 106
References 109

Chapter 62:

Biotransformation of Drugs to Pharmacologically Active Metabolites. S. GARATTINI, F. MARCUCCI, and E. MUSSINI

I. Introduction . . . 113

II. Drugs Acting on the Central Nervous System . . . 114

A. Hypnotics and Sedatives . . . 114
a) Chloralhydrate . . . 114

B. Anticonvulsants . . . 114
a) Mephobarbital . . . 114
b) Primidone . . . 114
c) Trimethadione . . . 115
d) Methsuximide . . . 115

C. Centrally Acting Muscle Relaxant . . . 115
a) Zoxazolamine . . . 115

D. Narcotic Analgesics . . . 115
a) Codeine . . . 115
b) Meperidine . . . 116
c) Acetylmethadol . . . 116
d) Diphenoxylate . . . 116

E. Analgesic-Antipyretics, Anti-Inflammatory Agents, and Inhibitors of Uric Acid Synthesis . . . 116
a) Salicylates . . . 116
b) Phenacetin . . . 116
c) Aminopyrine . . . 117
d) Phenylbutazone . . . 117
e) Allopurinol . . . 117
f) (4-phenylthioethyl) . . . 118

F. Drugs Used in the Treatment of Psychiatric Disorders . . . 118

1. Drugs for Treatment of Psychoses . . . 118
a) Tetrabenazine . . . 118

2. Drugs for Anxiety . . . 118
a) Chlordiazepoxide . . . 118
b) Diazepam . . . 119
c) Medazepam . . . 120
d) Prazepam . . . 121

3. Psychotropic Drugs for Affective Disorders . . . 121
a) Tricyclic Compounds . . . 121
b) Monoaminoxidase (MAO) Inhibitors . . . 121

III. Drugs Acting at Synaptic and Neuroeffector Functional Sites . . . 122

A. Anticholinesterase Agents . . . 122
a) Parathion . . . 122

B. Drugs Acting on Postganglionic Adrenergic Nerve Endings and Structures Innervated by them (Sympathomimetic Drugs) . . . 122
a) *N*-isopropylmethoxamine . . . 122
b) Fenfluramine . . . 122
c) Fenproporex . . . 122

C. Drugs Inhibiting Adrenergic Nerves and Structures Innervated by them . . . 123

1. Adrenergic Neuron Blocking Agents . . . 123
a) γ-methyldopa . . . 123
b) γ-methyl-*m*-tyrosine . . . 123

IV. Cardiovascular Drugs . . . 123

A. Antiarrhythmic Drugs . . . 123
a) Lidocaine . . . 123
b) Propranolol . . . 123

B. Vasodilator Drugs . . . 124
a) Diallylmelamine . . . 124
b) Prenylamine . . . 124

C. Drugs Lowering Blood Lipids or Glucose . . . 124
a) Nicotinic Acid . . . 124
b) Clofibrate . . . 125
c) 3,5-dimethylpyrazole . . . 125
d) 3,5-dimethylisoxazole . . . 125
e) Acetohexamide . . . 125

V. Chemotherapy of Parasitic Diseases . . . 125

A. Drugs Used in the Chemotherapy of Helminthiasis . . . 125
a) Lucanthone . . . 125

VI. Chemotherapy of Neoplastic Diseases . . . 126

A. Alkylating Agents . . . 126
a) Cyclophosphamide . . . 126

VII. Conclusions . . . 126

References . . . 126

Chapter 63:

The Enterohepatic Circulation. G. L. Plaa

I. Introduction . . . 130

II. Methods for Studying the Enterohepatic Circulation . . . 131

III. The Enterohepatic Circulation of Bile Salts . . . 131
A. Excretion of Bile Salts . . . 132
B. Absorption of Bile Salts . . . 132

IV. Enterohepatic Circulation of Drugs . . . 134
A. Morphine . . . 134
B. Methadone . . . 135
C. Etorphine . . . 135
D. Digitoxin . . . 136
E. Diethylstilbestrol . . . 138
F. Steroids . . . 138
G. Indomethacin . . . 139
H. Glutethimide . . . 140
I. Amphetamine . . . 141
J. Butylated Hydroxytoluene . . . 141
K. Pentaerythritol Trinitrate . . . 142
L. Fenamates . . . 142
M. Phenothiazines . . . 143
N. Antibiotics . . . 143

V. Enhanced Biliary Excretion of Drugs . . . 143
A. Enhanced Biliary Flow . . . 143
B. Enhanced Formation of Metabolites . . . 144
C. Formation of Complexes . . . 145

References . . . 145

Chapter 64:

Routes of Administration and Drug Response. C. T. Dollery and D. S. Davies. With 10 Figures

I. Introduction . . . 150

II. Enteral Administration of Drugs . . . 151
A. The Oral Route . . . 151
1. Advantages of Oral Dosing . . . 151
2. Disadvantages of Oral Dosing . . . 152
a) Slow Onset of Action . . . 152
b) Nonavailability of Oral Route . . . 152
c) Poor Drug Availability . . . 153
d) Selective Local Action . . . 155
B. Sublingual and Rectal Administration . . . 155
III. Parenteral Administration . . . 155
A. Intravascular Injection . . . 155
B. Intramuscular Injection . . . 156
C. Subcutaneous and Percutaneous Administration . . . 157
D. Inhalation of Drugs . . . 157
IV. Influence of Route of Administration on Drug Response . . . 158
A. Isoproterenol . . . 158
B. Chlorpromazine . . . 162
C. Lidocaine . . . 163
D. Propranolol . . . 164
V. Conclusions . . . 165
References . . . 166

Chapter 65:

Genetically Determined Variations in Drug Disposition and Response in Man. E. S. VESELL. With 6 Figures

I. Introduction . . . 169
II. Hereditary Conditions Affecting Drug Response Transmitted as Simple Single Factors . . . 170
A. Genetic Conditions Transmitted as Single Factors Affecting the Manner in which the Body Acts on Drugs . . . 173
1. Acatalasia . . . 173
2. Slow Inactivation of Isoniazid . . . 176
3. Succinylcholine Sensitivity or Atypical Pseudocholinesterase . . . 178
4. Deficient Parahydroxylation of Diphenylhydantoin . . . 182
5. Bishydroxycoumarin Sensitivity . . . 184
6. Acetophenetidin-Induced Methemoglobinemia . . . 184
B. Genetic Conditions Probably Transmitted as Single Factors Altering the Way Drugs Act on the Body . . . 185
1. Warfarin Resistance . . . 185
2. G6PD Deficiency, Primaquine Sensitivity or Favism . . . 187
3. Drug-Sensitive Hemoglobins . . . 191
4. Taste of Phenylthiourea or Phenylthiocarbamide (PTC) . . . 192
5. Responses of Intraocular Pressure to Steroids: Relationship to Glaucoma . . . 193
6. Malignant Hyperthermia with Muscular Rigidity . . . 194
III. Atypical Liver Alcohol Dehydrogenase . . . 195
IV. Ethanol Metabolism in Various Racial Groups . . . 195
V. Correlation of Certain Genetic Factors with Adverse Reactions to Various Drugs . . . 196
VI. Reduced Drug Binding Capacity in Fetal and Newborn Blood . . . 197

VII. Variation Among Individuals in Rate of Drug Elimination . . . 197
A. Genetic Control of Variations in Drug Clearance . . . 197
B. Environmental Effects on Drug Action and Genetic Control of their Expression 201
References . . . 203

Chapter 66:

Aging Effects and Drugs in Man. W. W. WEBER and S. N. COHEN. With 3 Figures

I. Introduction . . . 213
II. Absorption . . . 214
III. Distribution and Elimination . . . 217
A. Body Water Compartments . . . 218
B. Binding . . . 219
1. Serum Proteins . . . 220
2. Plasma Lipids . . . 221
3. Other Binding Components . . . 222
C. Renal Excretion . . . 222
D. Relations Affecting Distribution and Elimination . . . 225
IV. Concluding Remarks . . . 228
References . . . 230

Chapter 67:

Pathological and Physiological Factors Affecting Drug Absorption, Distribution, Elimination, and Response in Man. L. F. PRESCOTT. With 8 Figures

I. Introduction . . . 234
II. Drug Absorption . . . 234
A. Parenteral Administration . . . 234
B. Oral Administration . . . 235
1. Effects of Food . . . 235
2. Gastrointestinal pH . . . 236
3. Gastric Emptying Rate . . . 236
4. Malabsorption Syndromes . . . 238
5. Gastrointestinal Surgery . . . 239
6. Mesenteric Blood Flow and Biliary Tract Disease . . . 239
III. Drug Distribution . . . 240
A. Regional Tissue Distribution . . . 240
B. Volume of Drug Distribution . . . 240
C. Plasma Protein Binding . . . 241
IV. Drug Metabolism . . . 242
A. Liver Disease . . . 243
B. Acetaminophen-Induced Acute Hepatic Necrosis . . . 244
C. Drug Metabolism in Other Pathological Conditions . . . 245
D. Extrahepatic Drug Metabolism . . . 246
V. Renal Excretion of Drugs . . . 247
A. Renal Failure and Drug Excretion . . . 247
B. Urine Flow and pH . . . 248
C. Increased Hepatic Drug Metabolism in Renal Failure . . . 249
VI. Receptor Sensitivity . . . 250
A. Diminished Response to Drugs . . . 250
B. Enhanced Response to Drugs . . . 250
C. Acid-Base and Electrolyte Balance . . . 251

VII. Conclusions . . . 252
Acknowledgement . . . 252
References . . . 252

Chapter 68:

Absorption, Distribution, Excretion, and Response to the Drug in the Presence of Chronic Renal Failure. M. BLACK and I. M. ARIAS. With 4 Figures

I. Introduction . . . 258
II. Absorption . . . 258
III. Distribution: Protein Binding . . . 259
IV. Metabolism . . . 261
V. Renal Elimination of Drugs in Patients with CRF . . . 265
VI. Response to Drugs . . . 267
VII. Conclusions . . . 268
References . . . 268

Section Seven: Drug Interactions and Adverse Drug Reactions

Chapter 69:

Pharmacokinetic Drug Interactions. D. G. SHAND, J. R. MITCHELL, and J. A. OATES. With 4 Figures

I. Introduction . . . 272
II. Biological Determinants of Kinetic Parameters . . . 274
A. Drug Half-Life . . . 275
B. Drug Clearance . . . 276
C. Volume of Distribution . . . 277
D. Drug Concentration . . . 277
E. Summary . . . 278
III. Mechanisms of Drug Interactions . . . 278
A. Altered Absorption . . . 278
1. Alterations in Gastrointestinal pH . . . 279
2. Gut Motility . . . 279
3. Sequestration or Metabolism in the Gut Lumen . . . 280
4. Alteration in the Absorptive Process . . . 281
5. Summary . . . 282
B. Altered Elimination . . . 282
1. Drug Metabolism . . . 282
a) Stimulation of Drug Metabolism . . . 285
b) Inhibition of Metabolism . . . 288
c) Altered Formation of Active Drug Metabolites . . . 290
2. Renal Excretion . . . 290
a) Altered Tubular Reabsorption . . . 291
b) Tubular Secretion . . . 291
3. Hemodynamic Drug Interactions . . . 292
C. Redistribution . . . 294
1. Drug Binding in the Blood . . . 294
a) Restrictive Elimination . . . 294
b) Non-Restrictive Elimination . . . 297
2. Tissue Binding or Uptake . . . 299
D. Multiple Mechanisms . . . 300
IV. Investigation of Drug Interactions . . . 302

V. Clinical Relevance . . . 304
A. Factors Determining the Clinical Significance of Pharmacokinetic Drug Interactions . . . 305
1. Inherent Properties of the Drug or Disease State . . . 305
2. Pharmacokinetic Factors . . . 305
3. Pharmacogenetic Factors . . . 306
4. Disease-Induced Pharmacokinetic Factors . . . 307
B. Summary . . . 307
References . . . 307

Chapter 70:

Interactions of Cardiovascular Drugs at the Receptor Level. L. I. GOLDBERG

I. Introduction . . . 315
II. Drugs that Increase Myocardial Contractility . . . 315
A. Cardiac Glycosides . . . 316
1. Influence of Cations . . . 316
a) Potassium . . . 316
b) Calcium . . . 316
c) Magnesium . . . 316
2. Specific Digitalis Blocking Agents . . . 316
3. Relationship of Digitalis Blood Levels to Cardiac Activity . . . 317
B. Sympathomimetic Amines . . . 317
C. Methylxanthines . . . 318
D. Glucagon . . . 318
E. Interaction: Therapeutic and Adverse . . . 318
III. Drugs Acting on Blood Vessels . . . 319
A. Vasoconstricting Agents . . . 319
1. Sympathomimetic Amines . . . 319
2. Angiotensin . . . 320
3. Ergot Derivatives . . . 320
B. Vasodilating Agents . . . 320
1. Sympathomimetic Amines . . . 321
2. Dopamine . . . 321
3. Cholinergic Drugs . . . 322
4. Histamine . . . 322
IV. Summary . . . 322
References . . . 322

Chapter 71:

Drug Interactions in Cancer Chemotherapy. B. A. CHABNER and V. T. OLIVERIO. With 3 Figures

I. Introduction . . . 325
II. Use of Drug Combinations for Enhanced Antitumor Effect . . . 325
A. Definition of Terms . . . 325
B. Experimental Assessment of Drug Interaction . . . 326
C. Biochemical Rationale for Combination Chemotherapy . . . 328
III. Clinical Experience with Combination Chemotherapy . . . 331
A. Acute Leukemia . . . 331
B. Lymphomas . . . 334
C. Therapy of Solid Tumors . . . 335
IV. Antagonistic Drug Interactions . . . 337
References . . . 338

Chapter 72:

Combined Actions of Antimicrobial Drugs. E. JAWETZ. With 5 Figures

I. Introduction . . . 343

II. Possible Clinical Indications for Combined Antimicrobial Drug Treatment 343
a) Overwhelming, Life-Threatening Infections 343
b) Rapid Emergence of Drug-Resistant Microbial Mutants 344
c) Microbial Infections that May Require Drug Synergism for Eradication of an Infectious Process 344
d) Mixed Infections 344
e) Possible Reduction in Toxicity 344

III. Clinical Disadvantages of Combined Antimicrobial Drug Treatment 344

IV. Problems in Defining and Measuring Effects of Antimicrobial Drugs in Combination 345

V. Dynamics of Combined Antimicrobial Action 347
A. Antagonism 347
1. Mechanism of Antimicrobial Antagonism 350
2. Antimicrobial Antagonism in the Treatment of Clinical Disease 350
B. Synergism 351
1. Mechanism of Antimicrobial Synergism. 351
a) Blocking Successive Steps in a Metabolic Sequence 351
b) Inhibition by one Drug of an Enzyme Capable of Destroying the Second Drug 351
c) Antimicrobial Synergism Manifested by Marked Enhancement of Early Bactericidal Rate 352
2. Antimicrobial Synergism in Clinical Disease. 353

VI. Conclusion 355

Acknowledgements 355

References 355

Section Eight: Perspectives on the Importance of Drug Disposition in Drug Therapy and Toxicology

Chapter 73:

Drug Actions and Interactions: Theoretical Considerations. J. R. Gillette and J. R. Mitchell. With 6 Figures

I. Pharmacologic and Toxicologic Effects are Mediated Solely by the Parent Drug . . 359
A. What Proportion of the Drug is Excreted Unchanged? 361
B. Does Increasing the Dose of a Drug Affect its Total Body Clearance?. 362
C. What is the Relative Importance of Drug Metabolizing Enzymes in Different Tissues? 362
D. Is the Rate of Drug Metabolism Limited Mainly by the Blood Flow Rate through the Tissues? 363
E. What is the Relative Importance of the Pathways of Drug Metabolism in the Tissues? 363
F. Do Substances that Deplete the Body of Cosubstrates for Conjugation Limit the Rate of Metabolism of the Drug? 363
G. Are Changes in the Plasma Level of a Drug Therapeutically or Toxicologically Significant? 364

II. Pharmacologic and Toxicologic Effects are Mediated Solely by Reversibly Acting Metabolites of the Drug 364
A. Do Inducers or Inhibitors Alter the Proportion of the Dose that is Converted to the Active Metabolite? 365
B. Do Inducers or Inhibitors Alter the Rate of Formation of the Active Metabolite without Significantly Changing the Proportion of the Dose that is Converted to the Active Metabolite? 366
C. Do Inducers or Inhibitors Alter the Rate of Elimination of the Active Metabolite? 366
D. Do Substances Change the Tissue Levels of Cosubstrates Used in the Conjugation Reactions of the Active Metabolite? 367
E. Do High Doses of the Drug Alter the Pattern of Metabolites Derived from the Active Metabolite by Depleting the Cosubstrate? 367

F. Do Inducers or Inhibitors Alter the Rates of Both the Formation and Inactivation of the Active Metabolite? . . . 367
G. Is the Clearance of the Active Metabolite Greater than the Clearance of the Parent Drug? . . . 368
III. Pharmacologic and Toxicologic Effects are Mediated Solely by Chemically Reactive Metabolites . . . 368
A. Do Inducers or Inhibitors Alter the Relative Proportion of the Dose that is Converted to the Reactive Metabolite? . . . 370
B. Do Inducers or Inhibitors Alter the Relative Proportion of the Reactive Metabolite that Becomes Covalently Bound? . . . 371
C. Do Substances Change the Tissue Levels of Cosubstrates Used in Conjugation Reactions? . . . 371
D. Do High Doses of the Drug Lead to Depletion of Cosubstrates Used in Conjugation Reactions? . . . 373
E. Are Chemically Reactive Metabolites Formed in Different Tissues? . . . 374
F. Do Reactive Metabolites Leave the Tissues in which They are Formed? . . . 375
G. Examples of the Effects of Inducers, Inhibitors and Depleting Substances on Covalent Binding and Drug Toxicity . . . 375
Acknowledgement . . . 381
References . . . 381

Chapter 74:

Toxic Drug Reactions. J. R. MITCHELL, W. Z. POTTER, J. A. HINSON, W. R. SNODGRASS, J. A. TIMBRELL, and J. R. GILLETTE. With 13 Figures

I. Exaggerated or Unwanted Drug Actions . . . 383
1. Pharmacokinetic Drug Reactions . . . 383
2. Pharmacodynamic Drug Reactions . . . 384
II. Toxic Drug Reactions . . . 384
A. Cell Necrosis . . . 384
1. Acetaminophen (Paracetamol) . . . 384
2. Phenacetin . . . 391
3. Furosemide (Frusemide) . . . 393
4. Cephaloridine . . . 397
5. Isoniazid and Iproniazid . . . 397
6. Fluroxene . . . 399
7. Acetanilide . . . 400
8. Halobenzenes . . . 401
9. Spironolactone . . . 402
10. Porphyria . . . 403
B. Drug-Induced Neoplasia . . . 404
1. Antineoplastic Drugs . . . 405
2. Estrogens . . . 406
3. Phenacetin . . . 406
4. Drugs Used Clinically Not Known to be Carcinogenic in Humans but Known to Induce Cancer in Animals . . . 406
a) Isoniazid . . . 407
b) Nitrofuran Derivatives . . . 407
c) Miscellaneous Drugs . . . 407
C. Drug Allergy . . . 407
III. Drug Reactions of Unknown Etiology . . . 409
IV. Perspective . . . 410
References . . . 412
Authors Index . . . 420
Subject Index . . . 461

List of Contributors

ARIAS, IRWIN, M., Professor of Medicine, Department of Medicine, Albert Einstein College of Medicine of Yeshiva University, 1300 Morris Park Avenue, Bronx, New York 10461/USA

BLACK, MARTIN, Department of Medicine, Albert Einstein College of Medicine of Yeshiva University, 1300 Morris Park Avenue, Bronx, New York 10461/USA

CHABNER, BRUCE, Laboratory of Chemical Pharmacology, National Cancer Institute, National Institutes of Health, Bethesda, Maryland 20014/USA

COHEN, SANFORD N., Department of Pharmacology, New York University Medical Center, New York, N. Y. 10016/USA

DAVIES, DONALD S., Royal Postgraduate Medical School, Hammersmith Hospital, London, W12 OHS/England

DIGENIS, GEORGE, College of Pharmacy, University of Kentucky, Lexington, Kentucky 40506/USA

DOLLERY, COLIN T., Royal Postgraduate Medical School, Hammersmith Hospital, London, W12 OHS/England

GARATTINI, SILVIO, Istituto di Ricerche Farmacologiche, "Mario Negri", Via Eritrea, 62, 20157 Milan/Italy

GIBALDI, M., Department of Pharmaceutics, School of Pharmacy, State University of New York at Buffalo, Buffalo, New York 14214/USA

GILLETTE, JAMES R., Chief, Laboratory of Chemical Pharmacology, National Heart and Lung Institute, National Institutes of Health, Bethesda, Maryland 20014/USA

GOLDBERG, LEON L., Director, Division of Clinical Pharmacology, Emory University School of Medicine, Atlanta, Georgia 30303/USA

HINSON, J.A., Laboratory of Chemical Pharmacology, National Heart and Lung Institute, Bethesda, Maryland 20014/USA

JAWETZ, ERNEST, Professor and Chairman, Department of Microbiology, School of Medicine, University of California, San Francisco, California 94122/USA

LEVY, GERHARD, Professor of Biopharmaceutics, Department of Pharmaceutics, School of Pharmacy, State University of New York at Buffalo, Buffalo, New York 14214 /USA

MARCUCCI, F., Istituto di Ricerche Farmacologische, "Mario Negri" Via Eritrea, 62, 20157 Milan/Italy

MITCHELL, JERRY R., Laboratory of Chemical Pharmacology, National Heart and Lung Institute, National Institutes of Health, Bethesda, Maryland 20014/USA

MUSSINI, E., Istituto di Ricerche Farmacologische, "Mario Negri", Via Eritrea, 62, 20157 Milan/Italy

OATES, JOHN A., Professor of Pharmacology and Medicine, Vanderbilt University, Nashville, Tennessee 37203/USA

OLIVERIO, VINCENT T., Chief, Laboratory of Chemical Pharmacology, National Cancer Institute, National Institutes of Health, Bethesda, Maryland 20014/USA

PLAA, GABRIEL L., Professor and Chairman, Department of Pharmacology, University of Montreal Faculty of Medicine, Montreal, Quebec/Canada

POTTER, W. Z., National Heart and Lung Institute, Laboratory of Chemical Pharmacology, Bethesda, Maryland 20014/USA

PRESCOTT, L. F., Department of Therapeutics, The Royal Infirmary, Edinburgh, EH3 9YW/Scotland

SHAND, DAVID G., Department of Pharmacology, Vanderbilt University School of Medicine, Nashville, Tennessee 37203/USA

SNODGRASS, W. R., Laboratory of Chemical Pharmacology, National Heart and Lung Institute, Bethesda, Maryland 20014/USA

SWINTOSKI, JOSEPH V., Dean, College of Pharmacy, University of Kentucky, Lexington, Kentucky 40506/USA

TIMBRELL, J. A., Laboratory of Chemical Pharmacology, National Heart and Lung Institute, Bethesda, Maryland 20014/USA

VESELL, ELLIOT S., Professor and Chairman, Department of Pharmacology, College of Medicine, The Pennsylvania State University, Hershey, Pennsylvania 17033/USA

WEBER, WENDELL W., Department of Pharmacology, New York University Medical Center, New York, N. Y. 10016/USA

WELCH, RICHARD M., Senior Biochemist, Wellcome Research Laboratories, Burroughs Wellcome & Company, Tuckahoe, New York 10707/USA

Contents of Part 1

Section One: Routes of Drug Administration

Biological Membranes and Their Passage by Drugs. C.A.M. Hogben
Absorption of Drugs from the Gastrointestinal Tract. L. S. Schanker
Buccal Absorption of Drugs. A.H. Beckett and R.D. Hossie
Subcutaneous and Intramuscular Injection of Drugs. J. Schou
Absorption, Distribution and Excretion of Gaseous Anesthetics. H. Rackow
Absorption of Drugs through the Skin. M. Katz and B. J. Poulsen

Section Two: Sites of Drug Transport and Disposition

The Nature of Drug-Protein Interaction. W. Settle, S. Hegeman, and R.M. Featherstone
Physical Methods for Studying Drug-Protein Binding. C.F. Chignell
Effect of Binding to Plasma Proteins on the Distribution, Activity and Elimination of Drugs. P. Keen
Competition between Drugs and Normal Substrates for Plasma and Tissue Binding Sites. H.M. Solomon
Drug Entry into Brain and Cerebrospinal Fluid, D.P. Rall
Translocation of Drugs into Bone. H. Foreman
Translocation of Drugs and Other Exogenous Chemicals into Adipose Tissue. L.C. Mark
Placental Transfer of Drugs and their Distribution in Fetal Tissues. M. Finster and L.C. Mark
The Use of Autoradiography in Experimental Pharmacology, L.J. Roth
Accumulation of Drugs at Sympathetic Nerve Endings. I.J. Kopin

Section Three: Sites of Drug Excretion

Excretion of Drugs by the Kidney. I.M. Weiner
Excretion of Drugs in Bile. R.L. Smith
Excretion of Drugs by Milk. F. Rasmussen
Extracorporeal and Peritoneal Dialysis of Drugs. G.E. Schreiner, J.F. Maher, W.P. Argy, Jr., and L. Siegel

Contents of Part 2

Section Four: Methods of Studying the Metabolism of Drugs

Subsection A. Assay of Drugs and Their Metabolites

Basic Principles in Development of Methods for Drug Assay. B.B. Brodie
Absorption Spectrophotometry. R.P. Maickel and T.R. Bosin
Fluorometry. H.S. Ackerman and S. Udenfriend
Radioactive Techniques: The Use of Labeled Drugs. P.P. Maickel, W.R. Snodgrass, and R. Kuntzman
Radioactive Techniques: Radioactive Isotope Derivatives of Nonlabeled Drugs. R. Kuntzman, R.H. Cox, Jr., and R.P. Maickel
Gas Chromatography. M.W. Anders
Enzymatic Assays in Pharmacology. A.K. Cho
Bioassay. M. Vogt
Immunoassay. S. Spector

Subsection B. Isolation and Identification of Drug Metabolites

Paper, Column and Thin Layer Chromatography, Counter-Current Distribution and Electrophoresis. E.O. Titus
Isotope Dilution Analysis. V.T. Oliverio and A.M. Guarino

Gas Chromatography-Mass Spectrometry. A.M. GUARINO and H.M. FALES
The Application of Various Spectroscopies to the Identification of Drug Metabolites. P. BOMMER and F. M. VANE

Section Five: Sites of Drug Metabolism

Introduction: Pathways of Drug Metabolism. R. T. WILLIAMS

Subsection A. Microsomal Enzymes

Some Morphological Characteristics of Hepatocyte Endoplasmic Reticulum and Some Relationships between Endoplasmic Reticulum, Microsomes, and Drug Metabolism. J.R. FOUTS
Model Systems in Studies of the Chemistry and the Enzymatic Activation of Oxygen. V. ULLRICH and H.J. STAUDINGER
Cytochrome P-450—Its Function in the Oxidative Metabolism of Drugs. R.W. ESTABROOK
Enzymatic Oxidation at Carbon. J. DALY
N-Oxidation Enzymes. J.H. WEISBURGER and ELIZABETH K. WEISBURGER
Enzymatic *N*-, *O*-, and *S*-Dealkylation of Foreign Compounds by Hepatic Microsomes. Th. E. GRAM
Reductive Enzymes. J.R. GILLETTE
Oxidative Desulfuration and Dearylation of Selected Organophosphate Insecticides. P. A. DAHM
Metabolism of Halogenated Compounds. E.A. SMUCKLER
Glucuronide-Forming Enzymes. G.J. DUTTON
Metabolism of Normal Body Constituents by Drug-Metabolizing Enzymes in Liver Microsomes A.H. CONNEY and R. KUNTZMAN
Tissue Distribution Studies of Polycyclic Hydrocarbon Hydroxylase Activity. L. W. WATTENBERG and J.L. LEONG
Mechanisms of Introduction of Drug Metabolism Enzymes. H.V. GELBOIN
Inhibition of Drug Metabolism. G.J. MANNERING

Subsection B. Nonmicrosomal Enzymes

Esterases of Human Tissues. B.N. LA DU and H. SNADY
Enzymatic Oxidation and Reduction of Alcohols, Aldehydes, and Ketones. R.E. MCMAHON
Amine Oxidases. E.A. ZELLER
Sulphate Conjugation Enzymes. A.B. ROY
Acetylating, Deacetylating and Amino Acid Conjugating Enzymes. W.W. WEBER
Mercapturic Acid Conjugation. E. BOYLAND
Methyltransferase Enzymes in the Metabolism of Physiologically Active Compounds and Drugs. J. AXELROD
Enzymes that Inactivate Vasoactive Reptides. E.G. ERDÖS
The Metabolism of Analogs of Endogenous Substrates: Wider Application of a Limited Concept. H.G. MANDEL

Foreword

Since the blood serves as the medium of exchange among all tissues of the body, measurements of drug concentrations in blood plasma provide a unique clinical tool for separating those factors that control the disposition of the drug in the body from those that limit the passage of the drug from the blood to its site of action or those that alter the responsiveness of the drug-action site complexes. Thus a recurrent theme in this book is the usefulness of correlating drug concentrations in plasma with drug action. However, attempts to relate the plasma levels of drugs with their therapeutic response should be put in perspective and carefully interpreted. Unfortunately, many studies in comparative pharmacology and clinical pharmacology frequently are not sufficiently complete to permit proper interpretation of the data or to confirm the conclusions suggested by the investigators.

It has become increasingly obvious that much of the variability in drug responses among different pharmaceutical preparations, different individuals and various animal species results from differences in the availability or disposition of the drug in the body rather than to variations in the responsiveness of tissue action sites. For example, some of the variability in the effects of different preparations of a drug may be due to differences in either the amount of drug that is absorbed from the administration site or to the rate at which it is absorbed. Some of the differences may be due to chemical instability of the drug in the stomach or to metabolism of the drug by bacterial flora or enzymes in the intestinal mucosa. Some of the variability may also be due to individual differences in drug metabolism by enzymes in liver and other tissues of the body. These variations in drug response due to alterations in the bioavailability and disposition of the drug may be assessed by studying changes in the plasma levels of the drug after the administration of either single or multiple doses of the drug. Provided that the binding sites on serum proteins are not saturated at the doses used and that the drug is metabolized or eliminated by first order processes, the apparent bioavailability of the drug may be determined by comparing the area under a rectangular plot of the plasma level against time after the administration of the drug orally and intravenously in the same individual. If the area under the curve (AUC) after oral administration is less than that after intravenous administration and if the biological half-life of the drug and the bound/free ratio of the drug in plasma are not changed, then it may be concluded 1) that a portion of the dose is not absorbed or 2) that a portion of the dose is converted to other substances, either nonenzymatically (chemical instability), or enzymatically by digestive enzymes, bacterial flora or enzymes in intestinal mucosa or 3) that the drug is metabolized so rapidly by enzymes in the liver that only a portion of the dose enters the systemic blood or 4) that a combination of these kinetic factors occur. The investigator may confirm that only a fraction of a drug is absorbed by measuring how much of the drug is excreted unchanged in the feces or how much of the dose is excreted into urine both as unchanged drug and as metabolites. However, the interpretation of such data is clouded when the drug undergoes enterohepatic circulation and thus these measurements should be compared with the pattern of elimination after intravenous administration of the drug. If the total body clearance of the drug after its intravenous administration approaches the plasma flow rate through the liver, then the investigator might suspect that mechanism 3 accounts for the decreased AUC after the oral dose. In animals this may be confirmed by comparing the AUC after intraperitoneal doses with those after oral and intravenous doses, but this procedure would be too dangerous in man. Alternatively, valuable information concerning "first pass effects" is sometimes obtained by comparing the clearance of the drug after intravenous infusion with the hepatic blood flow rate of the patient.

The AUC after various oral doses of a drug is also useful in determining whether saturable processes are involved in the metabolism and distribution of drugs. In some instances, the AUC/dose may increase as the dose is increased. Such findings suggest: 1) that the concentration of the drug in the gastrointestinal tract or in the portal blood may reach levels that saturate drug metabolizing enzymes during the absorption phase, or 2) that the excretory mechanisms in the kidney or liver may become saturated or 3) that noncompetitive product inhibition of drug metabolizing enzymes may have occurred. On the other hand, the AUC/dose may decrease as the dose is increased, suggesting a decrease in bioavailability owing to the insolubility of the drug or to saturation of binding sites in blood plasma. Obviously, the saturation of binding sites in blood plasma may be confirmed by studying the binding of the drug to the patient's blood *in vitro*.

The effects of various treatments and drug interactions on drug disposition may be assessed by use of both AUC and the kinetic constants of drug disposition. With these values, the investigator may determine whether the treatment or drug interaction resulted in a change in the biological half-life of the drug and whether the change in half-life is due primarily to changes in the activity of the enzyme systems, changes in the volume of drug distribution, or changes in blood flow rate.

In order to assess such data properly, however, it is important to determine not only the proportion of the drug in plasma that is unbound but also the concentrations of the drug and the proteins in the system and whether the proportion changes as the drug concentration decreases. From this kind of data, together with the kinetic constants that describe the disposition of drug, it is frequently possible to determine whether decreases in the binding of drugs to plasma proteins caused by displacement of one drug by another or by changes in the conformation or concentration of plasma proteins are sufficiently large to account for the observed changes in the clearance of the drug from the body or changes in biological half-life. From some of the fallacious reasoning appearing in the literature, it becomes obvious that many investigators fail to realize that when the increase in the unbound fraction is not due to saturation of binding sites of plasma proteins, the unbound fraction of drug in plasma and the apparent clearance of the drug from the body may be increased two or three-fold without significantly affecting either the biological half-life of the drug or the unbound concentration of the drug in plasma. Apparently these investigators have forgotten that the blood constitutes only about eight percent of the body weight, that the unbound form of the drug frequently distributes with body water and that drug may be bound in extravascular tissues, all of which would tend to decrease the concentration of unbound drug both in plasma and in other tissues of the body.

Studies on the disposition of single doses of drugs are also useful in predicting the plasma levels of drug during the administration of multiple doses. When all of the processes of absorption, distribution metabolism and elimination of a drug are first order and their rate constants do not change during therapy, the equations for predicting the steady-state plasma levels of drugs are relatively simple. Indeed, these equations even provide estimates of the loading and maintenance doses and the dosage intervals required to maintain the plasma levels within predetermined limits. Nevertheless, these simple equations do not take into account daily and diurnal variations in the rate of drug absorption, blood flow rates, pH changes or drug metabolism. In addition they do not take into account the possibility that different foods may affect absorption or that drugs may increase their own metabolism by inducing the synthesis of drug metabolizing enzymes.Furthermore, when the dosage interval is much shorter than the biological half-life, the drug levels achieved after the steady-state has been established may be high enough to saturate binding sites, active transport systems and drug elimination mechanisms, which would not have been predicted by a single dose study. In this situation the levels of unbound drug may reach dangerously higher levels than those predicted by the equations. Thus, the investigator should confirm the validity of the equation by measuring the plasma levels and the proportion of the bound drug as the steady-state levels are approached.

Occasionally a drug may be metabolized so rapidly during the early phases of drug distribution that most of the drug is eliminated before the drug in the "deep pools" becomes equilibrated with that in the plasma and thus the presence of such pools may be missed by measuring the plasma levels alone. Under these conditions, high concentrations of the drug would reach action sites in the central and intermediate compartments but only low concentrations of the drug would reach the deep compartments. Indeed, this kind of distribution has led to the concept of membrane barriers such as the blood-brain barrier and the placental barrier that prevent the entrance of drugs into the brain and fetuses. However, when drugs are infused for many hours or days or are administered repeatedly at short dosage intervals, theoretically the drug in the "deep pool" should accumulate until the average concentration of unbound drug in the deep pool reaches that in the plasma; indeed the tissue representing the deep pool would have to possess a fluid comparable to the cerebral spinal fluid (which continuously sweeps drugs out of the brain), or an active transport system, or a drug metabolizing system in order to prevent the accumulation of the drug, Thus, the accumulation of polar drugs in "deep tissues" depends not only on their rates of diffusion but also on their rates of elimination. This concept of the kinetics of drug accumulation in deep compartments has been used in designing dosage schedules for cancer chemotherapy that do not expose healthy tissue (central and intermediate compartments) to unduly high concentrations of drug but permit the drug to reach therapeutic levels in tumors (the deep compartment).

The concept that the activity of a drug after the administration of a single dose depends on its plasma concentration is based on the assumption 1) that the unbound concentration in the immediate environment of the action site is similar to that in blood, 2) that the drug equilibrates instantaneously with that bound to the action site, 3) that the magnitude of the response is directly proportional to the concentration of the drug-action site complex and 4) that rapid changes in the concentration of the complex are reflected instantaneously by changes in the magnitude of the response. According to this concept the *duration of action* of a drug may vary markedly because of differences in metabolism and elimination of the drug and the response disappears when the plasma level of the drug declines to a threshold concentration. For example, the metabolism of barbiturates, such as hexobarbital and pentobarbital in animals may be altered markedly by prior administration of inhibitors or inducers of cytochrome P-450 enzymes, but the plasma levels at which the animals awake are almost identical to those at which nonpretreated animals awake.

When evaluating the relationship of the *intensity of response*, however, the relationship between the plasma level of the drug and the response is more complex. According to the law of mass action, the proportion of the receptor sites occupied by the drug is directly proportional to the unbound drug concentration until about 10% of the sites are occupied. In the range of 10 to 90%, the proportion of the receptor sites occupied by the drug is virtually proportional to the log of the concentration of unbound drug. Although the relationship between the response and the drug concentration docs not always obey the law of mass action exactly (that is the slope of the log dose-response does not always have the theoretical value of 0.576), still the magnitude of the response between 10 to 90% is more closely proportional to the log of the concentration than to the concentration. As pointed out by LEVY and GIBALDI (see Chapter 59), the log concentration values can be substituted for by the first order kinetic equation for the disappearance of the drug. Thus, 1) Response/Response (max) = log C; 2) log $C = -kt/2.3 + \log Co$; and 3) Response/Response (max) = $(-kt/2.3) + \log Co$. The decrease in the intensity of the response, therefore, should be directly proportional to time and the rate of decrease is determined by the value $k/2.3$ when the intensity is between 10 and 90% of the maximal response. But the relationship should not be expected to hold when the response is initially greater than 90% or less than 10% of the maximal response.

The time at which the initial response appears depends on whether the receptor sites are present in the central or peripheral kinetic compartments. If they are located in the central compartment the response should appear shortly after the drug is administered intravenously. On the other hand, if the receptor sites are in the peripheral compartment, the response will appear slowly and reach a maximum when the concentration of the drug at the receptor site reaches a maximum and then the response will decrease as the plasma level decreases. In this case, the plasma levels at which animals recover from the effects of the drug would be virtually identical regardless of the rate of metabolism of the drug, and the intensity of the response may be related to the log of the plasma concentration during the terminal phase of drug disposition. Obviously, there would be no relationship between plasma levels of the drug and the response during the early distribution phases that occur shortly after the administration of the drug.

Because of the vast differences in the mechanisms of action of drugs and the many factors that affect the distribution of drugs and the responsiveness of action sites, it should not be surprising that many investigators fail to find direct relationships between the plasma level and drug response, especially after the administration of single doses of the drug. In some instances, the lack of correlation is due to inadequate methodology or to a misunderstanding of the fundamental precepts on which the concept is based. In other instances, the kinetic factors that relate the drug concentration to the response are so complex that there is little relationship between them.

Some of the problems that have arisen in relating plasma concentration to response include:

1) The analytical method is not specific; thus the plasma levels appear to be high, especially during the terminal phases of drug elimination.

2) Some investigators have assumed that the relationship between the plasma level and the response should be identical for all therapeutically active analogues of a drug.

3) The drug is administered as a racemic mixture of optically active forms but the responsiveness and the metabolism of the forms differ.

4) One or more of the metabolites of the drug may combine with the receptor sites; the total response thus depends on the complex interrelationships between the plasma concentrations of the drug and its metabolites, the relative affinities of the receptor sites for the drug and its active metabolites and their relative intrinsic activities.

5) Either the drug or its metabolites evoke several different responses, some of which may be antagonistic, by combining with either the same or different kinds of receptor sites.

6) The drug enters its site of action by active transport systems, and thus drug at the receptor site may equilibrate slowly with that in the blood.

7) The pharmacologically inert metabolites of a drug may compete for binding sites on plasma proteins and thereby alter the distribution of the active drug.

8) The presence of other drugs or endogenous substances may displace the drug from plasma proteins.

9) The drug combines with receptor sites irreversibly and thus the response may appear slowly but persist long after the drug has disappeared from the body. Alternatively the drug causes irreversible changes in receptor sites without actually combining with them.

10) Variations occur in either the affinity or the intrinsic response of receptor sites owing to genetic differences, age differences, disease, other drugs or diet.

11) The receptor sites are located in a small compartment whose filling and emptying is not reflected by the kinetics of the major compartments of the drug.

12) The drug may evoke several responses some of which are mediated by reversible complexes of the drug with receptor sites and others of which are mediated by irreversible complexes.

13) The measured response may be imprecise or may be affected by the action of a variety of receptors and physiological feed-back control mechanisms.

14) The measured response results indirectly from the primary pharmacologic response and thus its appearance may be delayed or may result in tachyphylaxis. For example, a) A drug such as tyramine may evoke its action by causing the release of neurotransmitters such as norepinephrine and hence its response persists only as long as the neurotransmitters are present in storage sites, regardless of the blood level of the drug. b) A drug, such as a decarboxylase inhibitor that inhibits a non-rate-limiting step may initially cause a decrease in the concentration of the product of the enzyme, but the rate of synthesis of the product will increase as the concentration of the endogenous substrate increases until the rate of product formation returns to that occurring before the administration of the drug. c) The decrease in product formation may activate feed-back mechanisms that stimulate the synthesis of the endogenous substrate. d) A drug may evoke its action by combining with vitamins or hormones and thus the intensity of action may be modified by the amount of the vitamin in the diet. e) A drug may block the synthesis of a substance that has a slow turnover rate and thus the response is delayed until the substance declines. f) A drug may block the catabolism of a substance that has a slow turnover rate and thus the response is delayed until the substance accumulates to a new steady-state concentration.

Many of these problems in relating plasma levels to drug response after the administration of single doses of the drug diminish in importance when the drug is administered repeatedly. For example, when the plasma concentration of the drug is maintained between specified limits for extended periods of time, the average concentration of the drug at the receptor site is likely to be proportional to the average concentration of the drug in plasma, whether the receptor sites are located in the central compartment or in a deep compartment or whether the drug gains access to the receptor sites by active transport systems. Moreover, the average plasma concentration of the drug and its metabolites should eventually become relatively constant after repeated administration of the drug even though the drug may induce its own metabolism. Similarly the ratio of the average concentrations of the *d*- and *l*-forms of racemic mixtures of a drug should eventually reach relatively constant values. The effects of physiological control mechanisms on drug response should also become relatively stable and thus the relationship between the plasma level and the response should become relatively constant. Moreover, relationships between the direct and indirect responses of the drug should become approximately proportional to the average drug concentration in plasma as the administration of the drug is continued. Even the effects of drugs that act irreversibly on receptor sites should be a mathematical function of the average concentration of the drug in plasma, although the relationship will obviously depend on the rate at which the drug is eliminated from the body, the rate and extent it becomes irreversibly bound to the receptor sites during the dosage interval and the rate at which the receptor sites are regenerated either by the removal of the bound drug or by resynthesis. For these reasons the investigator is more apt to find better correlations between the average plasma concentration of drugs and their pharmacological response after the administration of repeated doses than after single doses.

Despite the vast variability in drug responses in different animal species, suitable animal models will continue to be needed to evaluate various compounds as potential therapeutic agents in the hope that the experimental results obtained with animals may be applied to man. Since such studies frequently demand large numbers of animals for screening purposes, it is not always possible nor necessarily warranted to use rare or unusually large animal species, such as old world apes and monkeys.

How then can we obtain useful data with common laboratory animal species? It should be evident that species differences may be due to either pharmacokinetic differences in drug metabolism and distribution or to pharmacodynamic differences in drug responsiveness. Indeed a number of studies have shown that many of the species differences in drug responses can be attributed to differences in drug metabolism rather than to differences in receptor sites or their responsiveness. For example, despite a 50-fold species difference in the duration of action of hexobarbital in animals receiving the same doses, the plasma levels of barbiturate were similar when the animals recovered. Similarly, the sedative action of carisoprodol varied in four species from 0.1 hr in mice to 10 hrs in cats, but plasma concentrations were almost identical on recovery.

In addition to showing that there is little species difference in the intrinsic responsiveness of the target organs to the drugs, the studies demonstrated several other important principles. Since the pharmacologic actions of these drugs are caused by the drugs themselves and not by their metabolites, the duration of their actions is dependent on their rate of metabolism, but *not* on their pattern of metabolism. For this reason, virtually any mammalian species might serve as a suitable model for studying the pharmacology of these drugs, even though the species metabolized the drugs by pathways which markedly differ from those found in man.

Moreover, the species selected as the model need not metabolize the drug at rates that approximate those occurring in man. In animal species that metabolize drugs considerably more rapidly than man, the plasma levels can be maintained within prescribed limits by adjusting either the size of the dose or the time interval between repeated doses. Indeed if the rate of metabolism of the drug is proportional to the plasma concentration (i.e., the rate follows first-order kinetics) and if the half-life of the drug is not changed during repeated administration, the time between doses may be calculated from the following formula:

$$T = 3.3\, t\text{-}1/2 \log Cm/Ca$$

in which Cm is effective plasma concentration and Ca is the maximum desired plasma concentration. The size of the priming dose required to give Ca is Ca times Vd (the volume of distribution of the drug) and the size of the maintenance dose is $(Ca - Cm)\ Vd$.

In carrying out these studies, of course, the investigator must be constantly aware that significant amounts of an active metabolite may accumulate in one animal species but not in another, that drug response should be related to the concentration of unbound drug rather than total concentration of the drug and that the formation of cosubstrates such as PAPS (phosphoadenosylphosphosulfate) or glutathione may become rate-limiting when drugs are rapidly metabolized in animals along conjugative pathways. Indeed, whenever a response is clearly unrelated to the plasma concentration of the drug or clearly differs among animal species, the investigator should determine whether differences are due to variations in protein binding of the drug or to species variations in the formation of active metabolites. Searches for active metabolites may be especially rewarding because the investigator may discover a more selective and safer drug.

Even when a lack of correlation between drug response and drug levels in plasma is not due to the formation of pharmacologically active drug metabolites, the maintenance of relatively constant plasma levels of drug in different animal species permits one to exclude pharmacokinetic variables from experiments in order that the mechanism of action of the drug and the reasons for the species differences in responsiveness of drug receptors might better be understood. By such studies, one can then determine whether the results obtained with given animal species are relevant to man.

In addition, however, studies on drug disposition and metabolism may play a valuable role in the differentiation of disease processes. Many clinical syndromes may have different etiologies and a drug that is designed to act on one abnormal physiological control mechanism, such as the sympathetic nervous system in depressive illness, may not improve patients manifesting the syndrome caused by the abnormal function of another control system, such as the serotonin system. Thus by discovering patients that are refractory to the drug even when it is known that the blood levels are adequate, the clinical pharmacologist may be able to identify subgroups that have different variations of the disease and thus should be treated by a different kind of drug.

Studies of drug disposition and metabolism, therefore, will undoubtedly continue to play an important role in the discovery and development of new drugs, in the determination of proper drug formulations and dosages, in the elucidation of mechanisms underlying drug interactions and adverse drug reactions, and perhaps in understanding new variants of disease syndromes. The usefulness of such studies for the practicing physician, however, is less clear because most drugs have such a wide therapeutic range between effective and toxic drug levels. It is increasingly apparent that the routine monitoring of drug blood levels is not warranted. Nevertheless, monitoring of plasma levels of certain drugs that have a small therapeutic index has led to better management of individual patients. In addition, routine monitoring of drugs used prophylactically to control intermittent disease processes such as seizures or cardiac arrhythmias seems justifiable on clinical grounds. For these compounds, the therapeutic effect of the drug cannot safely be used as an index of drug concentration, that is, an insufficient drug concentration would only become apparent by the occurrence of a life-threatening seizure or arrhythmia.

J. R. Gillette
J. R. Mitchell

Section Six: Determinants of Drug Concentration and Activity

Chapter 59

Pharmacokinetics

G. Levy and M. Gibaldi

With 18 Figures

I. Introduction

Almost all aspects of pharmacology involve dynamic processes. This is true whether one is concerned with drug absorption, distribution, metabolism and excretion, or whether one focuses on pharmacologic effects. It is necessary therefore to describe and to analyze these processes or effects in terms of their rates, rate constants or time course. A qualitative description is usually quite insufficient and seldom leads to adequate characterization of the effects of the body on drugs or of the effects of drugs on the body.

Pharmacokinetics is a scientific discipline that deals with the mathematical description of biologic processes affecting drugs and affected by drugs. The purposes of pharmacokinetics are, among others, to develop mathematical expressions and to determine constants that describe succinctly the temporal patterns of drug-related processes, to make at least tentative predictions and extrapolations based on these mathematical expressions, and to help establish the mechanisms of the phenomena being described. Almost every pharmacologist requires at least an orientation in pharmacokinetics and preferably a working knowledge of the elements of pharmacokinetics. Such knowledge is also desirable for the physician because it is he who has to determine the dosage regimen of drugs for individuals who may differ in their intrinsic responses and in their abilities to absorb and eliminate such drugs. Adjustment of dosage regimens to account for these individual differences is in essence an exercise in clinical pharmacokinetics.

The purpose of this chapter is to provide the pharmacologist, and those in disciplines and professions related to pharmacology, an orientation to and an operative understanding of basic pharmacokinetics. Emphasis will be on how to carry out pharmacokinetic analyses of data and how to use the results of such analyses for predictive purposes. The mathematical equations that are presented in this chapter have been chosen because of their general usefulness; detailed derivations are not presented in most cases, but references to such derivations are given.

The most commonly used approach to pharmacokinetic characterizations is to depict the body as a system of compartments even though these compartments often have no apparent physiologic or anatomic reality. Three frequently used compartment models are shown in Fig. 1. The one-compartment model depicts the body as a single, homogeneous unit. This, the most simple model, is particularly useful for the pharmacokinetic analysis of blood plasma* concentrations and urinary excretion data for drugs that are very rapidly distributed in the body. However, it is also useful under certain conditions for some types of pharmacokinetic analyses involving drugs that are distributed in the body relatively slowly. The two-compartment model consists of a so-called central compartment,

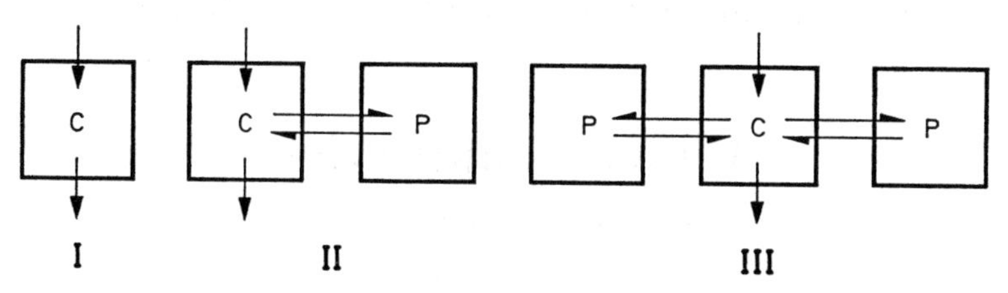

Fig. 1. Schematic representation of a one-compartment (I), two-compartment (II), and three-compartment (III) model commonly used in pharmacokinetic analyses. The arrows originating from a compartment represent apparent first-order processes. The central compartment is designated C; the peripheral compartments are designated as P

which includes the plasma, connected to a so-called peripheral or "tissue" compartment. It is usually assumed, in the absence of data to the contrary, that the sites of drug metabolism and excretion are part of the central compartment. Each of the compartments can be considered to include a group of fluids, tissues, organs or parts of organs. These groups differ in their relative accessibility to drugs or in the ease with which a drug can leave them. A somewhat more complex model consists of a central compartment connected to two peripheral compartments that differ in their relative accessibility to a drug. Many other pharmacokinetic models can be or have been developed, depending upon the available experimental data and physiologic considerations. There is no such thing as a unique model for a particular drug. The choice of model depends on the number of fluids and/or tissues being sampled, the frequency of sampling and the purpose of the pharmacokinetic analysis, among others.

Many processes in pharmacokinetics can be described satisfactorily by first-order (or "linear") kinetics. This means, for example, that the rate of biotransformation of a drug is proportional to the amount of drug in the apparent compartment in which the site of biotransformation is located. Also, the rate of transfer of drug between compartments, in any one direction, is proportional to the amount of drug in the originating compartment. It is recognized that drug biotransformation and many drug transfer processes (which include biliary and renal secretion) are not strictly first-order because they are usually specialized processes with limited capacity, involving enzymes or carriers. However, drug concentrations in the body are often so low that the limited capacity of these processes is not evident.

The next four sections of this chapter deal with linear single-compartment systems, multicompartment systems, non-linear systems and with the application of the pharmacokinetic principles developed in the foregoing sections to the kinetics of reversible pharmacologic effects. A final section is devoted to some useful suggestions for the experimentalist.

---

* This applies equally to serum. The same holds true whenever the word plasma is used subsequently in this chapter.

II. Absorption and Elimination; Linear, One-Compartment Systems

A. Elimination of Single Intravenous Doses

If a drug is distributed very rapidly in the body after intravenous injection and thereby confers upon the body the pharmacokinetic characteristics of a one-compartment system, and if drug elimination from the body can be described by first-order kinetics, then a plot of the logarithm of drug concentration in the

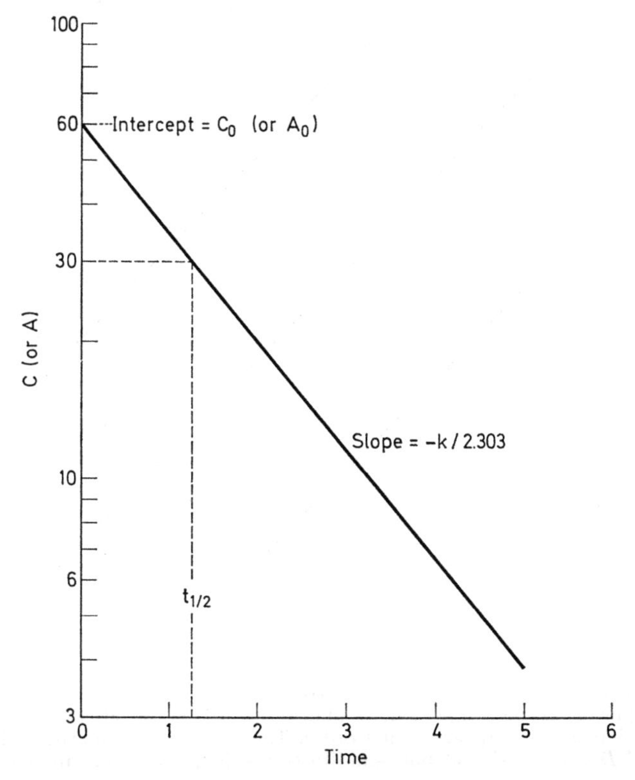

Fig. 2. Schematic representation of the concentration (C) or amount (A) of drug in the body as a function of time after rapid intravenous injection if the body acts as a one-compartment system and drug elimination is apparent first-order with a rate constant (k)

plasma as a function of time after injection yields a straight line as shown in Fig. 2. The time required for any given drug concentration to decrease by one-half is known as the biologic half-life or elimination half-life of the drug ($t_{1/2}$). The word elimination includes all biotransformation and excretion routes. A drug is considered eliminated even if its metabolites are still in the body. The half-life of a drug with first-order elimination characteristics is independent of dose; in the ideal case a plot of drug concentration/dose as a function of time yields a single curve independent of dose. This is known as the principle of superposition. Implied in this principle is the existence of a direct proportionality between drug concentration at any specified time and the size of the dose (Fig. 3). Whether or not the plasma concentrations of a drug reflect one-compartment or multi-compartment characteristics will depend on how soon the first blood sample is taken.

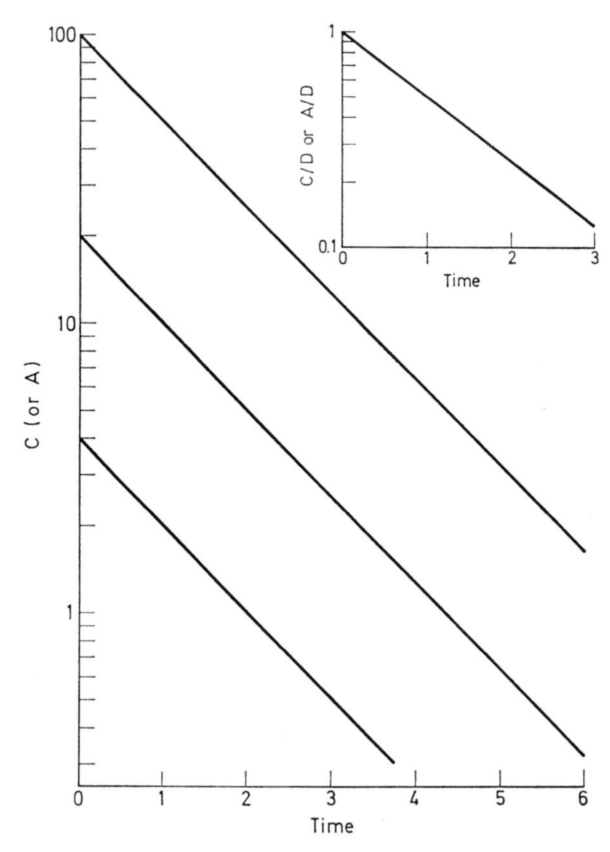

Fig. 3. Schematic representation of the time course of elimination of different doses (D) of a drug after rapid intravenous injection, assuming that the body acts as a one-compartment system and elimination is apparent first-order. The inset is a semilogarithmic plot of C/D or A/D as a function of time and illustrates the principle of superposition

In the case of a first-order process in a single compartment, the rate at any time is proportional to the concentration or amount of drug in the system at that time:

$$\text{Rate} \propto \text{Concentration} \tag{1}$$

or, in differential form

$$dC/dt = -kC \tag{2}$$

in which C is drug concentration, t is time and the proportionality constant, k, is the first-order rate constant of the process with units of reciprocal time (t^{-1}). In pharmacokinetics, this constant is usually referred to as an *apparent* first-order rate constant to convey the fact that the kinetics are only apparent first-order.

When input of a certain amount of drug into the system is instantaneous (i. e., by intravenous injection), the time course of drug concentration after injection may be described by integrating Eq. (2). This yields

$$C = C_0 \exp(-kt) \tag{3}$$

or

$$\ln C = (\ln C_0) - kt \tag{4}$$

or

$$\log C = (\log C_0) - kt/2.303 \tag{5}$$

In the above three equations, C is the drug concentration at time (t) after injection, C_0 is the drug concentration immediately after injection and k is the apparent first-order elimination rate constant. C_0 may be obtained by extrapolation of a log C versus t plot to $t = 0$. The term *exp* $(-kt)$ in Eq. (3) is often depicted as e^{-kt}; e represents the base of the natural logarithm (ln). Equation (4) was obtained by taking the natural logarithm of both sides of Eq. (3), and Eq. (5) was obtained by converting the logarithm in Eq. (4) to the base of 10 (log). This conversion is made on the basis of the relationship

$$2.303 \log x = \ln x \tag{6}$$

Tables to convert a given number x to ln x or log x values, or to determine exp $(-x)$, are found in the Handbook of Chemistry and Physics (Chemical Rubber Publishing Co., Cleveland, Ohio), in Scientific Tables (J. R. GEIGY S. A., Basle, Switzerland), and in other reference books.

Equation (5) shows that a plot of log C versus t will be linear under the stated conditions. The rate constant k can be determined from the slope of the line. It is much easier, however, to determine k by making use of the relationship

$$k = 0.693/t_{1/2} \tag{7}$$

since $t_{1/2}$ is determined readily by plotting drug concentrations directly on the logarithmic scale of semilogarithmic graph paper versus time on the linear scale (Fig. 2). It is often possible to fit the best straight line to the data graphically, (by eye). A more objective method is to convert all concentration values to logarithms and then to determine the best fitting line by the method of least squares, which is described in elementary textbooks of statistics. Computer programs are available for non-linear least squares fitting of data that does not require logarithmic conversions. "Weighting" of individual data points is a problem in all pharmacokinetic analyses. This problem is still subject to considerable debate and uncertainty. Perhaps the best way of reducing these statistical problems is to utilize precise and specific analytical methods, and to adhere rigorously to well defined protocols in order to obtain the best possible data.

The derivation of Eq. (7) from Eq. (5) provides an example of calculations involving these equations. Let $C_0 = 100$ mg/L. At $t = t_{1/2}$, C_0 is therefore equal to $C_0/2$ or 50 mg/L. Substituting in Eq. 5 for C and C_0, and rearranging, yields

$$kt_{1/2}/2.303 = \log (100/50) = 0.301.$$

Therefore

$$kt_{1/2} = 2.303 \cdot 0.301 = 0.693$$

or

$$k = 0.693/t_{1/2}$$

and

$$t_{1/2} = 0.693/k \tag{8}$$

If, for example, $t_{1/2}$ is 3 hours, $k = 0.231$ h^{-1}.

B. Apparent Volume of Distribution

The body is obviously not homogeneous even if plasma concentration and urinary excretion data can be described by representing the body as a one-compartment system. Drug concentrations in the liver, kidneys, heart, muscle, fat, and other tissues will therefore differ from one another as well as from the concentration in the plasma. If the relative binding of a drug to components of these tissues and fluids is essentially independent of drug concentration, then the ratio of drug concentrations in the various tissues and fluids is constant. Consequently, a constant relationship will exist between drug concentration in the plasma and the amount of drug (A) in the body:

$$A = VC \tag{9}$$

The proportionality constant V in this equation is known as the apparent volume of distribution. Despite its name, this constant has no direct physiologic meaning and does not refer to a real volume. For example, the apparent volume of distribution of a drug in a 70 kg man may be larger than 500 L or as small as 4 L.

Since A equals the intravenously injected dose at $t = 0$, V may be estimated from the relationship

$$V = i.\ v.\ \text{dose}/C_0 \tag{10}$$

A better method for determining V is to use the relationship (DOST, 1968),

$$V = D/(\text{Area} \cdot k) \tag{11}$$

in which *Area* is the *total* area under the drug concentration in the plasma *versus* time curve plotted on rectilinear graph paper. D is the amount of drug that actually enters the body. In the case of intravenous administration, D is equal to the amount of drug administered; in the case of oral administration this may or may not be the case depending on whether or not the drug is completely absorbed. The area method for determining V has the advantage of being applicable not only to one-compartment but also to some multicompartment linear systems. Another advantage is that the drug does not have to be injected intravenously. For determining the area under the curve, it is essential to obtain a sufficient number of blood samples to characterize the curve adequately. The usual experiment does not yield plasma concentration data all the way to zero concentration and it is then necessary to extrapolate the terminal portion of the plasma concentration versus time curve. This can be done by the use of Eq. (5). Another way to determine the area beyond the last concentration data point (C_{t*} at time $t*$) is to apply the relationship

$$\text{Area from } t^* \text{ to } \infty = C_{t*}/k \tag{12}$$

This area must be added to the area calculated from time zero to time $t*$ to obtain the total area under the curve. The area under a plasma concentration versus time curve has the units of concentration · time and can be determined by several methods. One method is to use a planimeter, an instrument for measuring mechanically the area of plane figures. It can be obtained at most university bookstores. Another procedure, known as the "cut and weigh" method, is to cut out the area under the entire curve on rectilinear graph paper and to weigh it on an analytical balance. The weight thus obtained is converted to the proper units by dividing it by the weight of a unit area of the same paper. A third and preferred method to determine the area under the curve is to estimate it by means of the trapezoidal rule. This method is described in many textbooks of college mathematics (such as BERS, 1969) and may be programmed readily on many calculators.

C. Urinary Excretion of Drug and Metabolites

The elimination kinetics of a drug often can be determined from urinary excretion data. This requires that at least some of the drug is excreted unchanged. If a drug is eliminated from the body partly by renal excretion and partly by biotransformation to two metabolites, X and Y, as shown in the following scheme:

$$\text{Drug in body} \begin{array}{l} \xrightarrow{k_{ex}} \text{Drug in Urine} \\ \xrightarrow{k_{mX}} \text{Metabolite } X \text{ in body} \rightarrow \text{Metabolite } X \text{ in urine} \\ \xrightarrow[k_{mY}]{} \text{Metabolite } Y \text{ in body} \rightarrow \text{Metabolite } Y \text{ in urine} \end{array}$$

(Scheme I)

then the elimination rate constant (k) is the sum of the individual rate constants that characterize the three parallel elimination processes.
Thus,

$$k = k_{ex} + k_{mX} + k_{mY} \tag{13}$$

Since, in first-order kinetics, the rate of a process is proportional to the amount of drug in the body (in the case of one-compartment systems),

$$\text{Rate of Excretion} = dA_u/dt = k_{ex}A \tag{14}$$

$$\text{Rate of Formation of Metabolite } X = k_{mX}A \tag{15}$$

$$\text{Rate of Formation of Metabolite } Y = k_{mY}A \tag{16}$$

in which A_u is the amount of unchanged drug excreted in the urine, and A is the amount of drug in the body.

Substitution for A in Eq. (14) by VC according to Eq. (9) yields

$$dA_u/dt = k_{ex}VC \tag{17}$$

Further substitution for C according to Eq. (3) results in

$$dA_u/dt = k_{ex}VC_0 \exp(-kt) \tag{18}$$

and

$$\log(dA_u/dt) = \log(k_{ex}\, VC_0) - kt/2.303 \tag{19}$$

Therefore, a semilogarithmic plot of excretion rate of unmetabolized drug *versus* time is linear, with a slope of $-k/2.303$. This is the same slope as is obtained from a semilogarithmic plot of plasma concentration versus time. Thus, the elimination rate constant of a drug can be obtained from either plasma concentration or urinary excretion data. The slope of the log excretion rate *versus* time curve is a function of the elimination rate constant (k) and *not* of the urinary excretion rate constant (k_{ex}). The urinary excretion rates determined experimentally are obviously not instantaneous rates (i.e., dA_u/dt) but average rates over a finite time period (i.e., $\Delta A_u/\Delta t$). However, the average excretion rate closely approximates the instantaneous rate at the midpoint of the urine collection period as long as the latter is no longer than the half-life of the drug. It is important to remember that urinary excretion rates must be plotted at the midpoints of the urine collection periods and not at the beginning or end of these periods. For example, if a subject excretes 30 mg of a drug during the first hour after administration and 20 mg during the next two hours, the average urinary excretion rates are 30 mg/h at one-half hour and 10 mg/h at two hours after drug administration.

It is sometimes difficult to fit a straight line to a semilogarithmic plot of urinary excretion rates *versus* time because the data may show considerable scatter. Some investigators prefer therefore to plot the logarithm of the amount of unchanged drug remaining to be excreted *versus* time. This type of plot is based on the integrated form of Eq. (18), expressed in logarithmic form:

$$\log (A_u^\infty - A_u) = \log A_u^\infty - kt/2.303 \qquad (20)$$

In this equation, A_u^∞ is the total amount of unchanged drug ultimately excreted in the urine, while A_u is the cumulative amount of unchanged drug excreted to time t. The slope of this type of plot is $-k/2.303$, i.e., the same slope as a plot of log C *versus* t or of a plot of log dA_u/dt *versus* t. To determine A_u^∞, urine must be collected until no more unchanged drug can be detected in the urine. It is incorrect to plot log (Dose $- A_u$) rather than log $(A_u^\infty - A_u)$ *versus* time. Urine must be collected for a period of time equal to about seven half-lives of the drug in order to determine A_u^∞. This presents considerable practical difficulties if the drug has a long half-life. The problem does not arise in plots of log excretion rate *versus* time since excretion rates can be determined intermittently and over a more limited period of time.

Nelson and O'Reilly (1960) showed that

$$k_{ex} = A_u^\infty \cdot k/D \qquad (21)$$

in which D is the intravenous dose or the actual amount of drug that enters the body following administration by any other route.
Also

$$k_{mX} = M_{Xu}^\infty \cdot k/D \qquad (22)$$

and

$$k_{mY} = M_{Yu}^\infty \cdot k/D \qquad (23)$$

in which k_{mX} and k_{mY} are the *formation* rate constants for metabolites X and Y, and M_{Xu}^∞ and M_{Yu}^∞ are the total amounts of metabolites X and Y eventually excreted. If a primary metabolite, X, is further metabolized totally or in part to a secondary metabolite X', then

$$k_{mX} = (M_{Xu}^\infty + M_{X'u}^\infty)\, k/D \qquad (24)$$

It is often not possible to account for all of the administered drug in the urine even when the drug is administered intravenously. This may be due to excretion of some of the drug by an extra-renal route (lung, bile, etc.), excretion of metabolite by an extra-renal route, further biotransformation of primary metabolites to chemical forms that are not identified by the analytical methods used, or formation of unknown and unidentified primary metabolites. This may introduce considerable uncertainty in the determination of rate constants for individual elimination processes, but not in the determination of k.

The kinetics of urinary excretion of a drug may be characterized not only by the renal excretion rate constant (k_{ex}) but also by a renal clearance value. Renal clearance (Cl_r) is the urinary excretion rate divided by the plasma concentration at the midpoint of the urine collection period:

$$Cl_r = (dA_u/dt)/C \qquad (25)$$

Substitution for dA_u/dt from Eq. (17) yields a second method for calculating renal clearance:

$$Cl_r = k_{ex}\, VC/C = k_{ex}\, V \qquad (26)$$

The clearance ratio of a drug is the ratio of its renal clearance to that of inulin in the same subject and is useful for determining the mechanism of renal clearance of a drug. Since inulin clearance is a measure of glomerular filtration rate, clearance ratios greater than one indicate that the drug is excreted in part by renal tubular secretion. However, if glomerular filtration is not complete, and/or if renal tubule reabsorption is appreciable, then the clearance ratio may be less than unity, even though the drug is subject to renal tubular secretion. Glomerular filtration will be incomplete if the drug is extensively bound to plasma proteins.

Body clearance (Cl_b) is a concept and a parameter that is analogous to renal clearance and may be defined as

$$Cl_b = kV \tag{27}$$

This constant reflects the sum of the contributions of all drug elimination processes and is particularly useful in comparing data obtained by using different compartment models.

Changes of k_{ex} (due to renal disease or urine pH), and of k_{mX} or k_{mY} (due to stimulation or inhibition of drug metabolizing enzymes), affect the elimination rate constant and half-life of a drug and the composition of urinary metabolites. Let it be assumed that $k_{ex} = 0.2\ \text{h}^{-1}$, $k_{mX} = 0.2\ \text{h}^{-1}$ and $k_{mY} = 0.1\ \text{h}^{-1}$. The elimination rate constant (k) is therefore $0.5\ \text{h}^{-1}$ and the half-life is 0.693/0.5 or 1.39 h. The fraction of the dose excreted unchanged, A_u^∞/D, may be determined from Eq. (21), i.e., $A_u^\infty/D = k_{ex}/k = 0.4$. Similarly, the fraction of the dose, M_{Xu}^∞/D, eliminated as metabolite X is 0.4 or 40 percent of D and $M_{Yu}^\infty/D = 20$ percent of D. If impaired renal function reduces k_{ex} to $0.1\ \text{h}^{-1}$ without affecting k_{mX} and k_{mY}, then k is decreased from $0.5\ \text{h}^{-1}$ to $0.4\ \text{h}^{-1}$, the latter corresponding to a half-life of 1.73 h. The metabolic fate of the drug is then changed from 40 percent unchanged drug, 40 percent metabolite X and 20 percent metabolite Y, to 25 percent unchanged drug, 50 percent metabolite X and 25 percent metabolite Y. The increase of the fractions X and Y is not due to any compensatory increase in the rate constants for the formation of these metabolites but is solely the result of the reduced contribution of the parallel and competing urinary excretion process to the overall elimination of the drug. The same mathematical approach may be used to determine the effect of changes in k_{mX} or k_{mY} on k and on metabolite composition.

The renal excretion rate constant (k_{ex}) of some drugs seems to be proportional to kidney function as reflected by creatinine or inulin clearance. If the biotransformation rate constants are not significantly affected by the renal disease, or if the drug is largely excreted unchanged, one may find an essentially linear relationship between k and creatinine or inulin clearance (DETTLI, 1970).

D. Drug Metabolite Levels in the Body

Scheme 1, showing the parallel routes of elimination of a drug, may be extended into Scheme 2, which includes the elimination of the metabolites by routes designated by the rate constants k_X and k_Y. These constants may represent excretion, further biotransformation, or both.

$$\text{Drug in body} \begin{cases} \xrightarrow{k_{ex}} \text{Drug in urine} \\ \xrightarrow{k_{mX}} \text{Metabolite } X \text{ in body} \xrightarrow{k_X} \\ \xrightarrow{k_{mY}} \text{Metabolite } Y \text{ in body} \xrightarrow{k_Y} \end{cases} \qquad \text{(Scheme 2)}$$

The time course of metabolite X (or Y) levels in the body is a function of the rates of formation and elimination of this metabolite,

$$dM_X/dt = k_{mX}\,A - k_X\,M_X \tag{28}$$

in which M_X is the amount of metabolite in the body at a given time. Substituting $D \exp(-kt)$ for A and integration of Eq. (28) yields

$$M_X = \frac{k_{mX} D}{k_X - k} [\exp(-kt) - \exp(-k_X t)] \tag{29}$$

This equation permits calculation of the amount of metabolite in the body at any time after intravenous injection of a dose (D) of a drug. It is informative to consider two different cases, one in which $k_X > k$ and the other in which $k > k_X$.

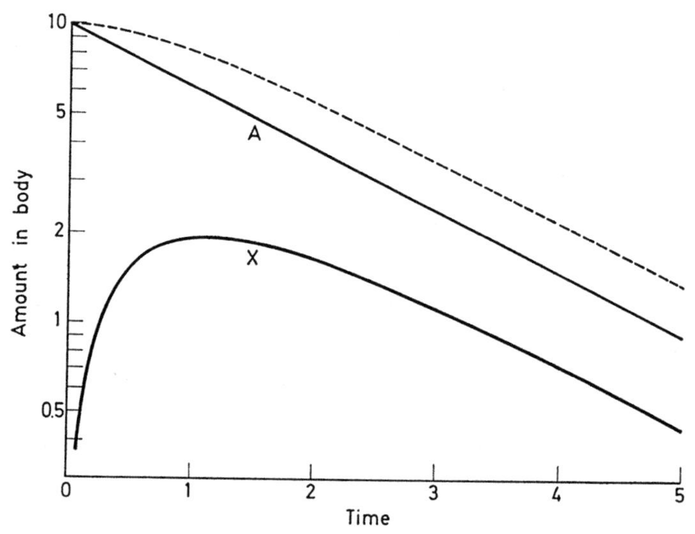

Fig. 4. Time course of drug (A) and metabolite (X) levels in the body after intravenous injection of the drug, if drug elimination is first-order and the body behaves like a one-compartment system. The rate constant for metabolite elimination in this simulation is three times higher than the rate constant for drug elimination. The stippled line represents the sum of A and X in the body, i.e., the type of curve that might be obtained by determinations of total radioactivity in the body after injections of labelled drug

If k_X is larger than k, then the curve of log M_X or log plasma concentration of X versus time will eventually become parallel to the curve of log plasma concentration of unchanged drug versus time as shown in Fig. 4. This will be obvious only when k_X is several times larger than k. Conversely, if k is larger than k_X, metabolite levels in the body will decline more slowly than the levels of unchanged drug (Fig. 5).

The mathematical basis for these effects can be ascertained from Eq. (29). If $k_X > k$, then at some time after drug administration $\exp(-k_X t)$ approaches zero while $\exp(-kt)$ still has a finite value. Consequently,

$$M_X \cong \frac{k_{mX} D}{k_X - k} \exp(-kt)$$

Therefore,

$$\log M_X \cong \log \frac{k_{mX} D}{k_X - k} - kt/2.303$$

and a plot of log M_X or log plasma concentration of X versus time will eventually be linear with a slope of $-k/2.303$. Similarly, if $k > k_X$, Eq. (29) reduces eventually to

$$\log M_X \cong \log \frac{k_{mx} D}{k - k_X} - k_X t/2.303$$

so that the terminal slope of a plot of the logarithm of metabolite levels versus time is $-k_X/2.303$.

Figures 4 and 5, which show the time course of drug and metabolite concentrations, also depict the time course of total drug levels (i.e., drug plus metabolite). The latter is of interest particularly in the interpretation of experimental

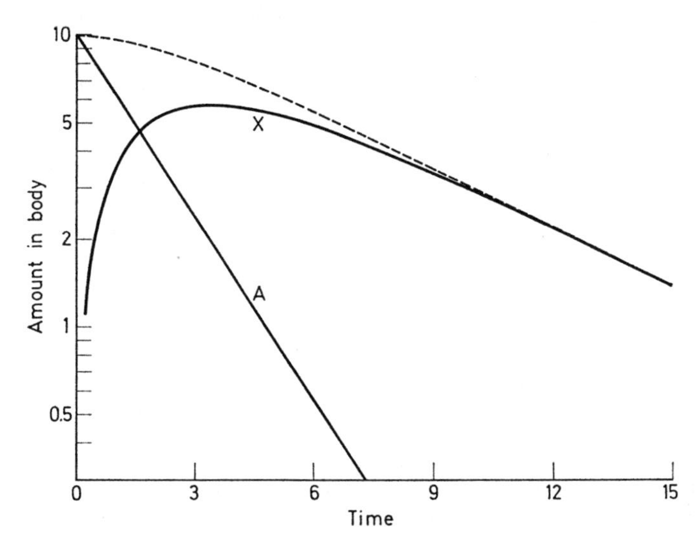

Fig. 5. As in Fig. 4 except that the rate constant for metabolite elimination is one-third as large as the rate constant for drug elimination

data obtained from measurements of either total radioactivity or from some other nonspecific assay. The figures show that data obtained by a nonspecific assay will indicate the true half-life of the drug only if the rate constant for metabolite elimination (k_X) is several times greater than the rate constant for drug elimination. Even then, the apparent volume of distribution of the drug cannot be determined accurately because the area under the combined drug and metabolite concentration curve is larger and the apparent C_0 is higher than the respective values derived from data obtained by specific analytical methods.

E. Bioavailability and Absorption Kinetics

An important application of pharmacokinetics is the determination of the extent of absorption, known as bioavailability or physiologic availability, of a drug. Bioavailability can be determined from plasma concentration or urinary excretion data. To determine bioavailability from plasma concentrations, a rearranged form of Eq. (11) may be utilized

$$D = k \cdot V \cdot \text{Area} \tag{30}$$

D is the intravenously injected amount of drug or the amount of drug that actually enters the body (exclusive of the gastrointestinal tract, intramuscular injection sites, etc.) following administration by other than the intravenous route. For example, in the cases of oral and i.v. administration of the same amount of a drug, bioavailability may be defined as

$$\text{Bioavailability} = \frac{D_{\text{oral}}}{D_{\text{i.v.}}} \times 100 \tag{31}$$

Substitution for D from Eq. (30) yields

$$\text{Bioavailability} = \frac{(\text{Area})_{\text{oral}}}{(\text{Area})_{\text{i.v.}}} \times 100 \tag{32}$$

Implicit in Eq. (32) is the assumption that k and V have not changed from one experiment to another.

Determinations of bioavailability from urinary excretion of unchanged drug can be made by using the rearranged form of Eq. (21)

$$D = \frac{A_u^\infty k}{k_{ex}} \tag{33}$$

Similar to the development of Eq. (32) from Eq. (31),

$$\text{Bioavailability} = \frac{A_{u,\text{ oral}}^\infty}{A_{u,\text{ i.v.}}^\infty} \times 100 \tag{34}$$

Bioavailability also can be determined from the amount of total drug (drug and metabolites) excreted in urine if no chemical or metabolic conversion of the drug occurred prior to its entry to the systemic circulation.

The rate of absorption of a drug can be determined from the plasma concentrations. Absorption of most drugs under the usual clinical conditions can be described by first-order kinetics, which is sometimes preceded by an apparent lag time. Under these conditions, the absorption and elimination of a drug involve two consecutive apparent first-order processes:

$$\begin{matrix}\text{Drug at} \\ \text{absorption site}\end{matrix} \xrightarrow{k_{ab}} \text{Drug in body} \xrightarrow{k} \begin{matrix}\text{Excreted and} \\ \text{metabolized drug}\end{matrix} \qquad \text{(Scheme 3)}$$

Mathematically, the time course of drug levels is described by the same kind of equation as Eq. (29)

$$A = \frac{k_{ab} D}{k_{ab} - k} [\exp(-kt) - \exp(-k_{ab} t)] \tag{35}$$

which can be converted to concentration terms by substitution for A from Eq. (9)

$$C = \frac{k_{ab} D}{(k_{ab} - k) V} [\exp(-kt) - \exp(-k_{ab} t)] \tag{36}$$

if $k_{ab} > k$, a plot of log C versus t will eventually have a slope of $-k/2.303$, i.e., the same slope as is obtained after intravenous administration. Figure 6 is a simulation by computer of drug concentrations as a function of time when the same dose of a drug is absorbed at different rates, with all other variables (k, V and D) being constant. Slower absorption results in decreased and delayed concentration peaks. This is the reason why the onset and duration of drug action and the intensity of pharmacologic effects, including lethality as reflected by the LD_{50}, may be affected by the route and form of administration of a drug.

The apparent first-order absorption rate constant depicted in Scheme 3 may often be estimated by the method of residuals. This method consists of back-extrapolating the terminal linear portion of the log plasma concentration versus time curve to time zero and determining the difference between the experimentally determined plasma concentrations and the corresponding concentrations on the extrapolated line (Fig. 7). These differences are called residuals. A semilogarithmic plot of the residuals versus time has a slope of $-k_{ab}/2.303$ and thereby permits determination of the absorption rate constant. If $k > k_{ab}$, then the terminal linear portion of a plot of log C versus time will have a slope of $-k_{ab}/2.303$ rather than $-k/2.303$, and the method of residuals yields a measure of k rather than of k_{ab}. This is sometimes called a "flip-flop" model.

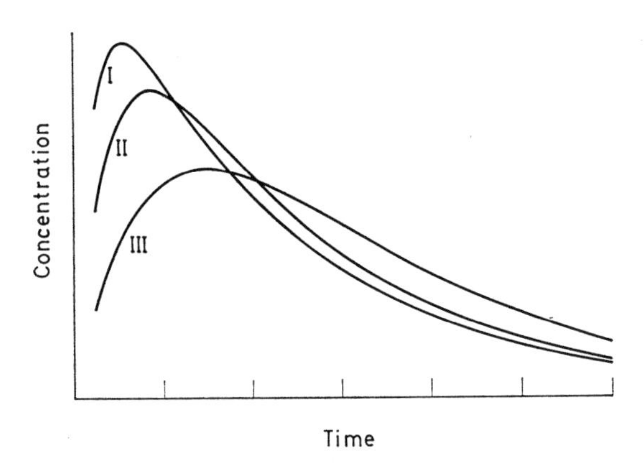

Fig. 6. Simulation of drug concentrations in the body according to Eq. (36) for a drug with $k = 0.23\ h^{-1}$ and $k_{ab} = 2.3$ (I), $1.15\ h^{-1}$ (II), and $0.46\ h^{-1}$ (III)

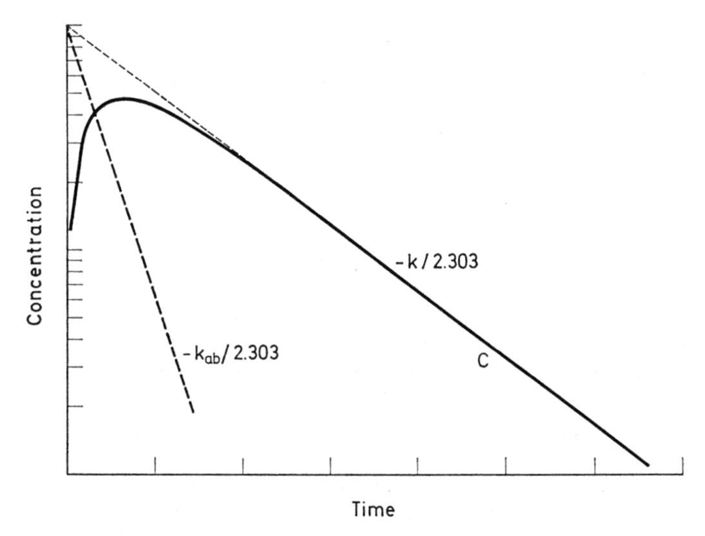

Fig. 7. Determination of k_{ab} by the method of residuals. The terminal linear portion of a log plasma concentration-time curve (C) is extrapolated to the concentration axis (....). The difference between the concentration represented on the dotted line and the actual values of C is plotted semilogarithmically (- - - -) to obtain k_{ab}

Another method of determining the kinetics of drug absorption is by means of percent absorbed versus time plots (WAGNER and NELSON, 1963) in which the percent absorbed is calculated by determining the relative amounts absorbed as a function of time and expressing these values relative to the total amount of drug ultimately absorbed. In the first step of this procedure

$$A_T/V = C_T + k \int_0^T C\,dt \quad (37)$$

in which A_T is the amount of drug absorbed from zero time to time T. C_T is the drug concentration in the plasma at time T and $\int_0^T C\,dt$ is the area under the drug concentration in the plasma versus time curve from time zero to time T. A_T/V will increase with time until it reaches a constant value (A_T^∞/V), which is equal to the total amount of drug absorbed from the dose divided by V. The ratio of $(A_T/A_T^\infty) \cdot 100$ at any time is the percent of drug absorbed at that time, relative to the total amount of drug ultimately absorbed (which may or may not be equal to the administered dose):

$$\text{Percent absorbed} = \frac{A_T/V}{A_T^\infty/V} \cdot 100 = \frac{A_T}{A_T^\infty} \cdot 100 \quad (38)$$

and therefore,

$$\text{Percent remaining to be absorbed} = 100\left(1 - \frac{A_T}{A_T^\infty}\right) \quad (39)$$

If absorption is apparent first-order then a plot of log (percent remaining to be absorbed) versus time will be linear with a slope of $-k_{ab}/2.303$.

The methods described here for determining the absorption rate constant apply only if the drug is completely absorbed. If absorption is incomplete, then these calculations may not yield correct values. However, even the incorrect values are useful in that they permit a quantitative description and simulations of the time course of drug levels in the body under certain conditions.

F. Intravenous Infusions

If a drug is administered intravenously at a constant rate, the plasma concentration at any time (t) will be (Dost, 1968):

$$C = \frac{k_0}{k\,V}\,[1 - \exp(-kt)] \quad (40)$$

in which k_0 is the rate of drug infusion, expressed in amount per unit time. Drug concentrations will rise with time after the start of the intravenous infusion and slowly approach a constant level. At that level the rate of elimination of the drug equals the rate of infusion. After an infusion time equal to about 4 times the biologic half-life of the drug, the plasma concentration is within ten percent of its plateau level. After a period of time equal to 7 half-lives, the drug concentration is within one percent of the plateau level. The plateau concentration, C_∞, can be determined from Eq. (40) by recognizing that the term $\exp(-kt)$ approaches zero with time. Thus,

$$C_\infty = k_0/k\,V \quad (41)$$

so that the plateau concentration is directly proportional to the rate of infusion. The decline of plasma concentrations after the infusion is stopped can be calculated from Eq. (5) by substituting C_∞ for C_0 and time *after* stopping the infusion for t. Equation (41) also shows a convenient way of determining the apparent volume of distribution by means of intravenous infusion experiments. If k_0, C_∞, k are known, it is possible to calculate V.

Since the time required to obtain C_∞ will be quite long for drugs with a long half-life, it is often convenient in such cases to administer an intravenous loading dose to attain immediately the desired drug concentration and then to maintain this concentration by continuous infusion. The loading dose required to attain concentration C is $C \cdot V$ and the rate of infusion required to maintain this concentration is $k \cdot C \cdot V$.

G. Repetitive Drug Administration

If a fixed intravenous dose of drug is administered repeatedly, at constant time intervals (τ), then the concentration of drug at any time may be calculated by the following expression (Dost, 1968).

$$C = \frac{D\,[1 - \exp(-nk\tau)]\exp(-kt)}{V\,[1 - \exp(-k\tau)]} \tag{42}$$

in which n is the number of doses that have been administered, and t is the elapsed time since the *last* dose.
At the plateau

$$C_\infty = \frac{D\exp(-kt)}{V\,[1 - \exp(-k\tau)]} \tag{43}$$

The maximum concentration at the plateau $C_{\infty,\max}$ and the minimum concentration $C_{\infty,\min}$ may be obtained by setting t equal to zero and to τ, respectively. Then

$$C_{\infty,\max} = \frac{D}{V\,[1 - \exp(-k\tau)]} \tag{44}$$

and

$$C_{\infty,\min} = \frac{D\exp(-k\tau)}{V\,[1 - \exp(-k\tau)]} \tag{45}$$

It is often useful to determine the "average" concentration at the plateau, $\bar{C}_\infty$. This is given by (Wagner et al., 1965)

$$\bar{C}_\infty = \frac{D}{Vk\tau} \tag{46}$$

$\bar{C}_\infty$ is not the arithmetic or geometric mean of $C_{\infty,\max}$ and $C_{\infty,\min}$. Rather, it is the area under the plasma concentration curve during a dosing interval at the plateau, divided by the dosing interval, τ. Equation (46) applies not only to intravenous injections but also to drug administration by other routes (Fig. 8).

The mathematical equation for calculating C_∞ after oral administration is considerably more complicated than the analogous equation for intravenous dosing (Dost, 1968):

$$C_\infty = \frac{k_{ab}D}{(k_{ab} - k)\,V}\left[\frac{\exp(-kt)}{1 - \exp(k\tau)} - \frac{\exp(-k_{ab}t)}{1 - \exp(-k_{ab}\tau)}\right] \tag{47}$$

in which k_{ab} is the apparent first-order absorption rate constant and t is the elapsed time since the last dose. Equation (47) reduces to

$$C_{\infty,\min} = \frac{k_{ab}D}{(k_{ab} - k)\,V}\left[\frac{\exp(-k\tau)}{1 - \exp(-k\tau)}\right] \tag{48}$$

if each successive dose is administered after the previous dose has been completely absorbed.

The bioavailability of a drug can be determined from the area under the drug concentration curve during a dosing interval, at the plateau (also known as the steady-state). Rearrangement of Eq. (46) yields

$$D = \overline{C}_\infty \cdot V \cdot k \cdot \tau$$

and since $\overline{C}_\infty \cdot \tau$ is equal to the area (Area_τ) under the plasma concentration curve during a dosing interval at the steady state, then

$$D = V \cdot k \cdot \text{Area}_\tau$$

Consequently, to determine the bioavailability of an orally administered dosage form of a drug relative to an intravenously administered dose of the drug,

$$\text{Bioavailability} = \frac{D_{\text{oral}}}{D_{\text{i.v.}}} = \frac{V k \,\text{Area}_{\tau,\,\text{oral}}}{V k \,\text{Area}_{\tau,\,\text{i.v.}}} = \frac{\text{Area}_{\tau,\,\text{oral}}}{\text{Area}_{\tau,\,\text{i.v.}}} \tag{49}$$

The same approach may be used to determine the bioavailability of a test product relative to an orally administered standard.

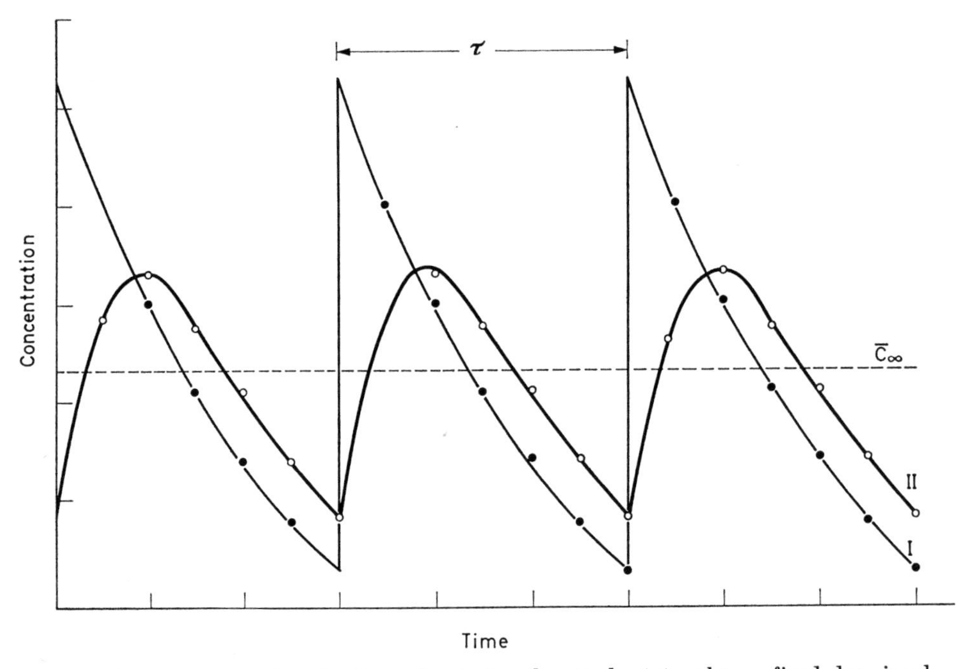

Fig. 8. Drug concentrations in the body during the steady-state when a fixed dose is administered repeatedly at constant time intervals (τ) after intravenous (I) or oral (II) administration. It is assumed that the drug is completely absorbed after oral administration. The dashed line represents $\overline{C}_\infty$ for both I and II [see Eq. (46)]

Repeated administration of fixed doses of a drug at constant time intervals results in a gradual increase of drug levels in the body until the steady state is reached. It is important therefore to be able to predict the degree of accumulation of a drug under defined conditions. If each side of Eq. (46) is multiplied by V and divided by D, one obtains

$$\frac{\overline{A}_\infty}{D} = \frac{1}{k\tau} = \frac{1.44\, t_{1/2}}{\tau} \tag{50}$$

$\bar{A}_{\infty}$ in this equation is the "average" amount of drug in the body at the steady state. The ratio $\bar{A}_{\infty}/D$ may be called the accumulation ratio of a drug. If, for example, a drug with a half-life of 12 hours is administered every 6 hours, $\bar{A}_{\infty}/D$ is 2.9. This means that repeated administration of a fixed dose of this drug every 6 hours will eventually result in a steady state at which the average amount of drug in the body is 2.9 times the amount administered in a single dose. The accumulation ratio is directly proportional to the half-life of a drug and inversely proportional to the dosing interval (τ).

Since considerable time may be required in order to attain steady state conditions during repeated drug administration, it is often desirable to administer a larger dose initially (the loading dose), in order to achieve the desired drug levels immediately. Equation (3), which describes the time course of drug concentrations after a single intravenous dose, may be written as

$$C_{1,\min} = \frac{D}{V} \exp(-k\tau) \tag{51}$$

in which $C_{1,\min}$ is the drug concentration immediately before administration of the next dose, i.e., the minimum drug concentration that occurs at $t = \tau$ following administration of the first dose. The minimum drug concentration during the steady state, $C_{\infty,\min}$, is given by Eq. (45). Thus, the ratio $C_{\infty,\min}/C_{1,\min}$ is another measure of drug accumulation. This ratio may be calculated by means of the expression

$$\frac{C_{\infty,\min}}{C_{1,\min}} = \frac{1}{1 - \exp(-k\tau)} \tag{52}$$

This ratio of *minimum* drug concentrations is not numerically equal to the ratio of $\bar{A}_{\infty}/D$. If the investigator wants to administer a loading dose (D^*) that produces a minimum concentration equal to $C_{\infty,\min}$, in which

$$C_{1,\min} = C_{\infty,\min} = \frac{D^* \exp(-k\tau)}{V} \tag{53}$$

then he can divide Eq. (45) by Eq. (53) with the result

$$1 = \frac{D}{D^*[1 - \exp(-k\tau)]}$$

which upon rearrangement yields an expression to calculate D^*:

$$\frac{D^*}{D} = \frac{1}{1 - \exp(-k\tau)} \tag{54}$$

This equation, which is true not only for intravenous drug administration but also for other routes of administration if each maintenance dose is given after complete absorption of the previous dose, permits calculation of the loading dose (D^*) for any chosen maintenance dose (D) and dosing interval (τ). Conversely, it also permits determination of the maintenance dose required to maintain the minimum drug level produced by an initial dose (D^*) for any chosen dosing interval (τ).

III. Absorption, Distribution, and Elimination; Linear, Multicompartment Systems

A. Distribution and Elimination of Single Intravenous Doses

If a drug is injected intravenously, it usually takes some time before it is distributed in the body. During this distribution phase, drug concentrations in the plasma will decrease more rapidly than in the post-distributive phase (Fig. 9). Whether or not this distributive phase is apparent will depend on the timing of blood samples. A distributive phase may last for only a few minutes or for hours or even for days. A semi-logarithmic plot of plasma concentrations as a function of time after rapid intravenous injection of a drug can often be resolved into two

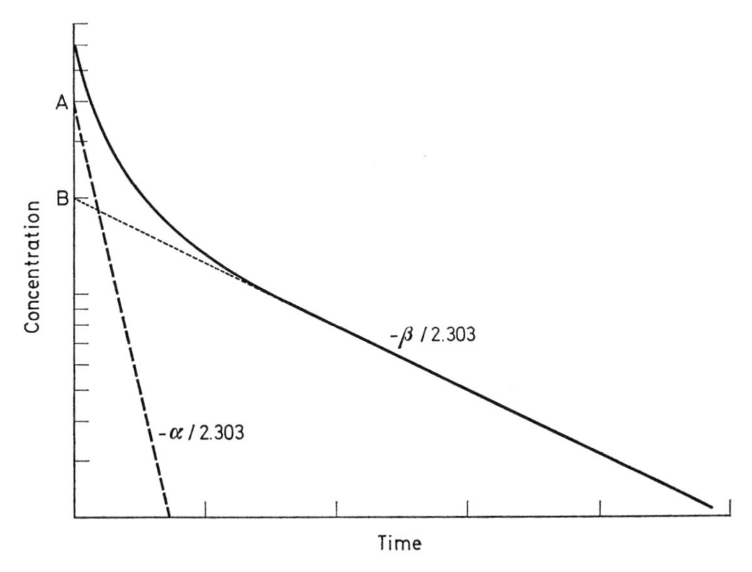

Fig. 9. Semilogarithmic plot of drug concentrations in the plasma after rapid intravenous injection when the body may be represented as a two-compartment, open, linear system. The dashed line is obtained by "feathering" the curve

linear components. This can be done graphically, by the method of residuals (also called "feathering"), as seen in Fig. 9. The slopes of the rapid and slow exponential components are designated as $-\alpha/2.303$ and $-\beta/2.303$, respectively. The intercepts on the concentration axis are designated A and B. Hence the entire curve may be described by the following bi-exponential equation:

$$C = A \exp(-\alpha t) + B \exp(-\beta t) \tag{55}$$

A bi-exponential decline of drug concentrations in the plasma following rapid intravenous injection usually justifies, from a mathematical point of view, the representation of the body as an open, two-compartment, linear system. There are three possible types of two compartment systems and they are depicted in Fig. 10. They differ in that elimination occurs either from the central compartment, from the peripheral compartment, or from both of these compartments. These three types of two-compartment models are mathematically indistinguishable on the basis of the usually available experimental data (drug concentrations in the plasma and/or urinary excretion rates). In the absence of information to the contrary, it is usually assumed that drug elimination from two-compartment systems takes place exclusively in the central compartment. All subsequent equations in this section are based on this assumption. The reason for this assumption is that the major sites of biotransformation and excretion, i.e., the liver and kidneys, are well perfused with blood and are therefore, presumably, rapidly accessible to drugs in the systemic circulation. The rate constants k_{12}, k_{21} and k_{10} for the model can be determined from α, β, A and B. This is done by the following steps (MAYERSOHN and GIBALDI, 1971)

$$k_{21} = \frac{A\beta + B\alpha}{A + B} \tag{56}$$

$$k_{10} = \alpha\beta / k_{21} \tag{57}$$

$$k_{12} = \alpha + \beta - k_{21} - k_{10} \tag{58}$$

Determination of these rate constants permits an assessment of the relative contribution of distribution and elimination processes to the drug concentration

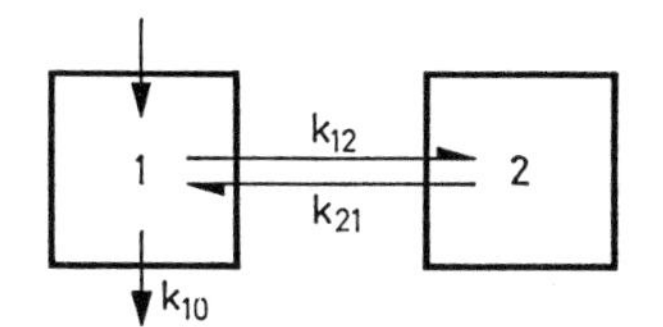

Fig. 10. Schematic representation of three types of two-compartment systems consisting of a central compartment (1) and a peripheral compartment (2). Note that the numbering of the rate constants (k) indicates the originating compartment (first numeral) and the receiving compartment (second numeral); 0 denotes "outside the body"

versus time profile of a drug. It also aids in elucidating the mechanism of drug interactions, and of the effects of disease, age, genetic influences, etc., on drug disposition. One of the rate constants, k_{12}, is also required to calculate the amount of drug in the peripheral compartment (A_p) as a function of time after intravenous administration (MAYERSOHN and GIBALDI, 1971)

$$A_p = \frac{D k_{12}}{\beta - \alpha} [\exp(-\alpha t) - \exp(-\beta t)] \tag{59}$$

This equation may be useful in determining the relationship between pharmacologic effect and "tissue" levels of a drug. However, the time course of drug levels in a hypothetical peripheral compartment, as inferred from the mathematical analysis of plasma concentration data, may not correspond to the actual drug levels in any real tissue. The peripheral compartments of pharmacokinetic models are, at best, hybrids of several much more complex functional physiologic units. Also, the time course of inferred drug levels in the peripheral compartment depends upon the type of two-compartment model used.

B. Apparent Volume of Distribution

The apparent volume of the central compartment with respect to a given drug (V_c) may be calculated from the equation

$$V_c = \frac{\text{i.v. dose}}{A + B} \tag{60}$$

The apparent volume of distribution, V (also called V_{area} or V_β) of a drug in a two-compartment system, with elimination occurring only from the central compartment, may be determined from the equation (GIBALDI et al., 1969)

$$V = \frac{D}{(\text{Area})\,\beta} \tag{61}$$

or

$$V = \frac{\text{i.v. dose}}{\left[\frac{A}{\alpha} + \frac{B}{\beta}\right]\beta} \tag{62}$$

Equation (61) may be used regardless of route of administration while Eq. (62) applies only to data obtained after rapid intravenous injections. As in the one-compartment model, V is a proportionality constant relating concentration in the plasma to the amount of drug in the body, except that it applies only to drug concentration in the post-distributive phase. Body clearance is then defined as $V\beta$, and as is seen in Eq. (61), is equal to D/Area.

C. The Meaning of β

A clear distinction must be made between k_{10}, the elimination rate constant, and β, which may be termed the disposition rate constant of a drug. The two constants are related in the following manner (GIBALDI et al., 1969):

$$\beta = f_c\, k_{10} \tag{63}$$

in which f_c is the fraction of drug in the central compartment under post-distributive conditions, relative to the total amount of drug in the body. The biologic half-life of a drug in a two-compartment system is

$$t_{1/2} = 0.693/\beta \tag{64}$$

If, because of insufficient data, the plasma concentrations of a drug with multi-compartment characteristics after rapid intravenous injection show only the terminal exponential phase, then what is actually the β value will appear to be the elimination rate constant (k) of a one-compartment system.

D. Bioavailability and Absorption Kinetics

Rearrangement of Eq. (61) yields

$$D = V \cdot \text{Area} \cdot \beta \tag{65}$$

This permits determination of the extent of absorption (bioavailability) of dosage forms of drugs with two-compartment pharmacokinetic characteristics when elimination occurs only from the central compartment:

$$\text{Bioavailability} = \frac{D_{\text{oral}}}{D_{\text{i.v.}}} \cdot 100 = \frac{\text{Area}_{\text{oral}}}{\text{Area}_{\text{i.v.}}} \cdot 100 \tag{66}$$

in which $\text{Area}_{\text{i.v.}}$ is the area under the C versus t curve for an intravenously injected dose. Equation (66) implies that V and β remain constant so that these terms may be cancelled in the numerator and denominator. Equation 66 also can be used to determine the *relative* bioavailability of a drug if the standard is other than an intravenous injection.

Determination of absorption kinetics for drugs with two-compartment characteristics is quite complicated and is usually done by computer. A method to determine the amount absorbed as a function of time has been described (Loo and RIEGELMAN, 1968) but a detailed discussion of this and other methods (WAGNER et al., 1968) to determine the kinetics of absorption in multicompartment systems is beyond the scope of this chapter.

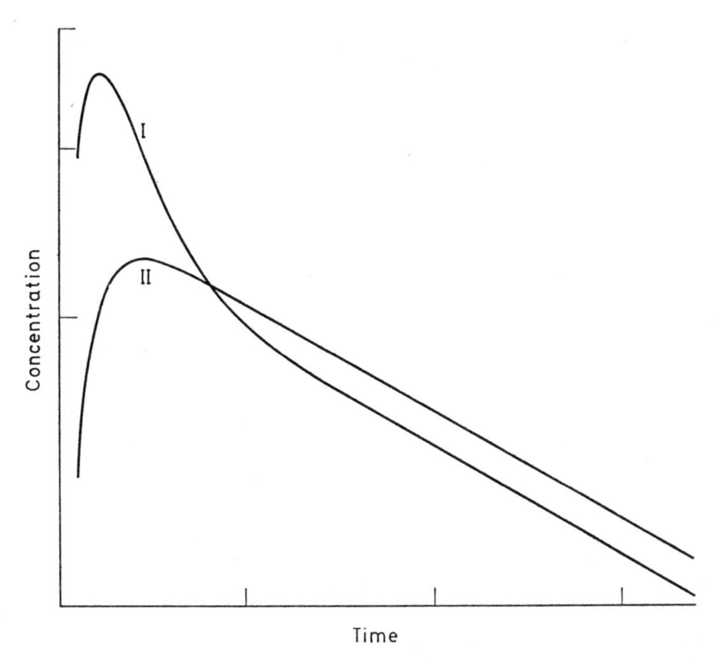

Fig. 11. Effect of the absorption rate on the time course of drug concentrations in the plasma when the body has the characteristics of a two-compartment system. $k_{ab} = 1.0\ hr^{-1}$ (I) or $k_{ab} = 0.4\ \mathrm{h}^{-1}$ (II); $k_{12} = 1.0\ \mathrm{h}^{-1}$; $k_{21} = 0.4\ \mathrm{h}^{-1}$; $k_{10} = 0.2\ \mathrm{h}^{-1}$

Plasma concentration versus time curves following drug administration by routes other than intravenous injection may or may not show evidence of a distributive phase (see Fig. 11), depending on the rate of absorption. If a distributive phase is not apparent and if adequate intravenous data are not available, one may come to the mistaken conclusion that the drug has one-compartment characteristics. Determination of absorption kinetics on the basis of this assumption will give incorrect results.

E. Intravenous Infusion

Intravenous infusion at a constant rate (k_0) results in the eventual attainment of steady state conditions in which

$$C_\infty = \frac{k_0}{k_{10}\, V_c} = \frac{k_0}{\beta\, V} \tag{67}$$

A simulation of the time course of drug levels in the central and peripheral compartments during and after intravenous infusion, shown in Fig. 12, reveals some interesting characteristics. Under certain conditions, the ratio of the amount of drug in the central compartment to that in the peripheral compartment will be considerably higher during the steady state phase of the infusion than in the post-infusion phase (GIBALDI, 1969). Consequently, the relationship between plasma concentration and pharmacologic effect may be different during and after infusion, if the site of action is in the peripheral compartment.

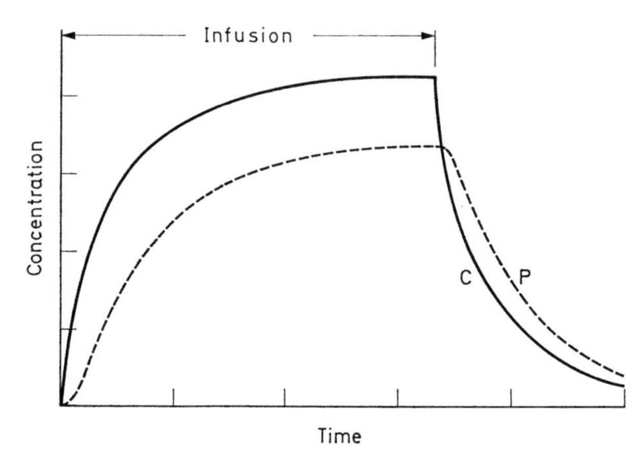

Fig. 12. Drug concentrations in the central (C) and peripheral (P) compartments of a two-compartment open system during and after constant rate intravenous infusion

Administration of a drug by rapid intravenous injection is frequently difficult or impossible because of limited solubility of the drug (requiring a large injection volume), or because of possible adverse pharmacologic effects. The drug must then be injected slowly, i.e., as a short intravenous infusion. After the infusion is stopped, drug concentrations in the plasma decline in the usual bi-exponential manner except that the distributive phase is less prominent. The distributive phase may be very difficult to recognize under certain conditions after relatively long infusion periods. When possible, constants α and β from post-infusion data may be determined in the usual manner (by feathering), but the constants A and B cannot be determined directly. Rather, the extrapolation and feathering yield the intercepts A' and B', which may be converted to A and B in the following way (LOO and RIEGELMAN, 1970)

$$A = \frac{\alpha T}{1 - \exp(-\alpha T)} \cdot A' \tag{68}$$

and

$$B = \frac{\beta T}{1 - \exp(-\beta T)} \cdot B' \tag{69}$$

in which T is the duration of infusion.

F. Repetitive Drug Administration

Repetitive administration of fixed doses of a drug by any route at constant time intervals (τ) results eventually in the attainment of an "average" steady-state concentration ($\overline{C}_{\infty}$) (GIBALDI and WEINTRAUB, 1971):

$$\overline{C}_{\infty} = \frac{\text{Area}_{\tau}}{\tau} = \frac{D}{V\beta\tau} = \frac{D}{V_c k_{10}\,\tau} \tag{70}$$

which is similar to Eq. (46) for a one-compartment system. The bioavailability equation, Eq. (49), for a one-compartment system applies equally for two compartment systems and also for systems with more than two-compartments as long as elimination occurs only in the central compartment. The accumulation equation, Eq. (52), and the loading dose equation, Eq. (54), for one-compartment systems may be used for two-compartment systems by replacing k with β, provided that maintenance doses are administered in the post-absorptive and post-distributive concentration phase of the preceding dose.

G. The "First-Pass" Effect

If a drug is administered orally, then an appreciable fraction may be metabolized during the first pass through the liver so that less than the absorbed dose enters the systemic circulation. Consequently, the area under the C versus t curve after oral administration may be considerably smaller than the area obtained

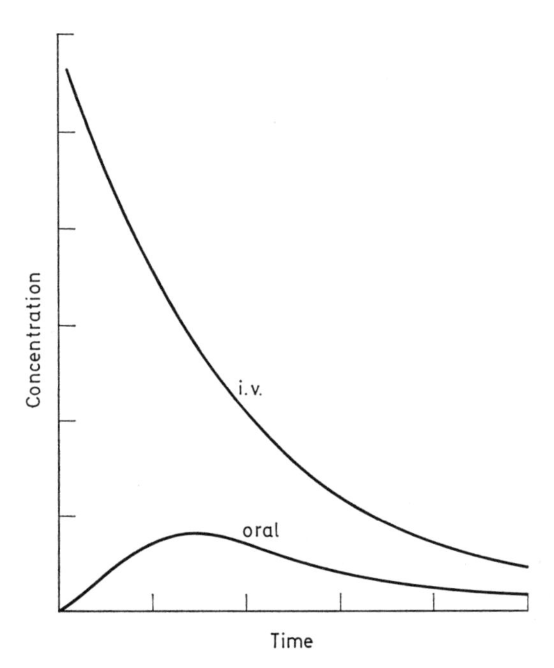

Fig. 13. First-pass effect on systemic bioavailability. Drug concentrations in the plasma after intravenous and oral administration of the same dose of a drug with a total clearance of 1.15 L/min after intravenous administration. It is assumed that the drug is eliminated only by biotransformation in the liver

after intravenous administration of the same dose (see Fig. 13), even though all of the drug was absorbed after oral administration. First-pass effects may also occur when drugs are administered rectally or intraperitoneally. The term *systemic availability* is used to indicate the extent of absorption of *unchanged* drug into the systemic circulation. The magnitude of a first-pass effect may be estimated by means of an equation that is based on a multicompartment model in which the liver is connected to the central compartment, assuming that drug elimination occurs only in the liver (GIBALDI et al., 1971 a):

$$R = 1 - \frac{\text{i.v. dose}}{\text{Area}_{\text{i.v.}}\, F} \tag{71}$$

in which R is the ratio of areas under the plasma concentration curve after oral and intravenous administration, and F is the flow rate through the liver. To be internally consistent, F should be the flow rate of plasma (approximately 0.7 L/min, in adult human males) but better agreement between calculated and experimentally determined R values has been obtained by using the blood flow rate (about 1.5 L/min) for F.

R may also be estimated without the use of $\text{Area}_{\text{i.v.}}$ data if there is evidence that the drug is completely absorbed following oral administration (GIBALDI et al., 1971 a):

$$R = \frac{F}{F + \dfrac{\text{oral dose}}{\text{Area}_{\text{oral}}}} \tag{72}$$

Equations (71) and (72) should serve primarily for approximation purposes. Implied in these equations is the assumption that the system is linear even when drug concentrations in the plasma are considerably higher than the usual therapeutic levels. Possible biotransformation of the drug in the gastrointestinal fluids or mucosa is not taken into consideration. The applicability of the equations is also affected by the rate of equilibration of drug between erythrocytes and plasma. This aspect of pharmacokinetics is relatively new and requires further study. Induction of drug metabolizing enzymes in the liver may have a pronounced effect on R and little or no effect on the biologic half-life if the rate of biotransformation of a drug in the liver is limited by certain distributional factors.

H. Other Multicompartment Models

Other open, multicompartment, linear models that have been used in pharmacokinetics include a two-compartment model with elimination from both compartments (ROWLAND et al., 1970), a three-compartment system with elimination from the central compartment (GIBALDI et al., 1972), and a three-compartment model with elimination from one of the peripheral compartments (NAGASHIMA et al., 1968). More recently developed models explicitly incorporate several physiologic parameters, such as blood flow to organs and tissues, binding of the drug to components of body tissues and fluids, and partitioning of drug between plasma and adipose tissues (BISCHOFF and DEDRICK, 1968; WAGNER, 1971).

IV. Pharmacokinetics of Non-Linear Systems

A. Elimination of Single Intravenous Doses

Drug biotransformation, secretion, and some transfer processes involve enzyme or carrier systems. These systems are more or less specific with respect to substrate and have finite capacities. The kinetics of these specialized processes can often be described by the Michaelis-Menten equation:

$$-\frac{dC}{dt} = \frac{V_m C}{K_m + C} \tag{73}$$

in which $-dC/dt$ is the rate of decline of drug concentrations at time t, C is the drug concentration at time t, V_m is the theoretical maximum rate of the process, and K_m is the Michaelis constant. K_m is equal to the drug concentration at which the rate of the process is equal to one-half its theoretical maximum rate. The Michaelis-Menten equation is not only useful for describing reaction rates in solutions of pure enzyme and substrate, but also for describing *in vivo* processes. In *in vivo* systems the constants V_m and K_m are affected by distributional and other factors and are therefore referred to as apparent *in vivo* constants. There are two limiting cases of Michaelis-Menten kinetics. If K_m is much larger than C, Eq. (73) reduces to

$$-dC/dt = \left(\frac{V_m}{K_m}\right) C \tag{74}$$

This equation has the same form as the first-order rate equation, Eq. (2). It is evident that the apparent first-order rate constant used in pharmacokinetics is often equivalent to (V_m/K_m). Consequently, if treatment with an enzyme inducer causes an increase in the amount of enzyme (and therefore of V_m), the apparent first-order rate constant of the process will also be increased. The drug concentrations in the body resulting from the usual therapeutic dosage regimens of most drugs appear to be well below the K_m of the processes involved in the disposition

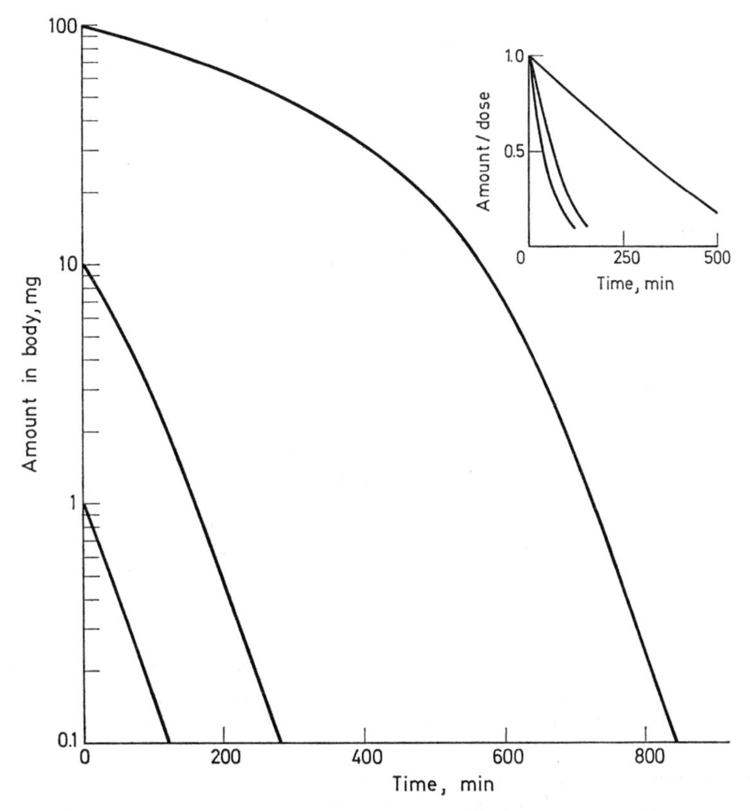

Fig. 14. Time course of drug levels in the body for a drug showing Michaelis-Menten elimination kinetics. One-compartment system, intravenous injection. Apparent *in vivo* K_M = 10 mg, V_M = 0.2 mg/min. Doses: 1.0; 10; 100 mg. Inset: the same curves, each divided by the dose to show that the principle of superposition does not apply

of these drugs. This conclusion is based on the observation that the kinetics of elimination of most drugs are apparently linear.

Another limiting case of the Michaelis-Menten equation is the one in which the drug concentrations are considerably above K_m. Equation (73) then reduces to

$$-dC/dt = V_m \tag{75}$$

Under these conditions, the rate of the process is independent of drug concentration so that the reaction proceeds at a constant rate. There are a number of drugs, notably ethanol (LUNDQUIST and WOLTHERS, 1958) and salicylate (LEVY, 1965; LEVY et al., 1972), for which the kinetics of biotransformation can approach the condition described by Eq. (75).

Figure 14 shows the time course of elimination of three different doses of a drug that is eliminated by a process with Michaelis-Menten kinetics. The lowest dose represents the case in which $K_m >> C$ and the elimination is describable by first-order kinetics. The highest dose yields initial concentrations that are considerably above K_m so that drug levels decline initially at a practically constant rate (Fig. 15). Figure 14 illustrates the important differences between elimination by apparent first-order kinetics as described in Section B of this chapter, and

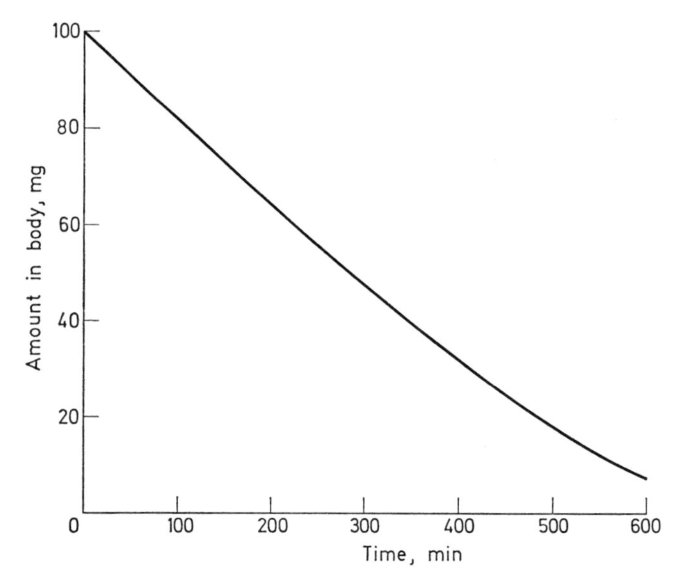

Fig. 15. The curve as in Fig. 14 for the 100 mg dose except that the data are plotted on linear coordinates. Note the almost constant rate of decline of the drug level for the first 300 minutes

elimination by apparent Michaelis-Menten kinetics. In the latter case, the principle of superposition does not apply and the time required for an initial drug concentration to decrease by 50 percent is not independent of dose but increases with increasing dose.

The plasma concentration curve in the post-distribution phase permits an estimation of apparent *in vivo* K_m and V_m values. This involves the determination of the rate of change of the plasma concentration from one sampling time to the next, $\Delta C/\Delta t$, as a function of the plasma concentration (C) at the midpoint of the sampling interval. For graphic determinations, the data are usually plotted according to one of the linearized forms of the Michaelis-Menten equation, such as the Lineweaver-Burk expression

$$\frac{1}{(\Delta C/\Delta t)} = \frac{K_m}{V_m C} + \frac{1}{V_m} \tag{76}$$

so that a plot of the reciprocal of $\Delta C/\Delta t$ versus the reciprocal of C yields a straight line with intercept, $1/V_m$, and slope, K_m/V_m. Another, often preferred, type of plot is based on the relationship

$$\frac{C}{(\Delta C/\Delta t)} = \frac{K_m}{V_m} + \frac{C}{V_m} \tag{77}$$

so that a plot of $C/(\Delta C/\Delta t)$ versus C yields a straight line with a slope of $1/V_m$ and an intercept of K_m/V_m. These plots are based on the assumption that only a single drug elimination process is operative. If there are two or more parallel Michaelis-Menten type elimination processes, these plots will also be practically linear provided that the respective V_m and K_m values are similar in magnitude (WAGNER, 1972). Although the apparent *in vivo* V_m and K_m values determined from such plots are hybrid constants, they are useful for describing and predicting the time course of drug concentrations in the body.

The time course of drug levels in the body after intravenous injection of a drug that is eliminated only by Michaelis-Menten kinetics can be calculated for a one-compartment system from the integrated form of Eq. (73):

$$t = \frac{1}{V_m}\left[C_0 - C + K_m \ln (C_0/C)\right] \tag{78}$$

in which C_0 is drug concentration at zero time. Unfortunately, it is not possible to solve this equation explicitly for C. Rather, one must determine the time (t) at which the initial concentration C_0 has decreased to C.

Drug elimination often involves Michaelis-Menten kinetics in parallel with apparent first-order kinetics. The time course of drug levels under these conditions may be determined by means of an equation that is analogous to Eq. (78) (KRÜGER-THIEMER and LEVINE, 1968; LUNDQUIST and WOLTHERS, 1958):

$$t = \frac{1}{k_* K_m + V_m}\left[K_m \ln (C_0/C) + \frac{V_m}{k} \ln \frac{(C_0 + K_m)\, k_* + V_m}{(C + K_m)\, k_* + V_m}\right] \tag{79}$$

In this equation, k_* is the rate constant for a parallel apparent first-order elimination process or the sum of rate constants for several such processes.

While the composition of excretion products is independent of dose for any given route of administration in the case of apparent first-order kinetics, this is not so for drugs that are eliminated in part by processes with limited capacities. If a drug is eliminated by parallel Michaelis-Menten and apparent first-order kinetics yielding metabolites X and Y as well as unchanged drug (X being the metabolite formed by Michaelis-Menten kinetics), then the fraction of a dose converted to X decreases with increasing dose while the fractions converted to Y and excreted unchanged increase accordingly. The relationship between dose and metabolite composition is therefore an important and useful criterion for determining the type of elimination kinetics (LEVY, 1968).

B. Various Processes Resulting in Non-Linear Elimination Kinetics

Dose-dependent elimination kinetics may be due to effects other than a limited capacity of biotransformation or excretion processes. If a drug is partly reabsorbed from renal tubules by a specialized process with limited capacity, then the elimination of large doses proceeds relatively more rapidly than the elimination of smaller doses (JUSKO and LEVY, 1970). Similarly, less binding of a drug to constituents of plasma or tissues at higher drug concentrations may result in relatively more rapid drug elimination than is observed at lower drug concentrations (KEEN, 1971; LÈVY, 1970a). The pharmacokinetic aspects of tissue binding have been discussed elsewhere in detail (WAGNER, 1971).

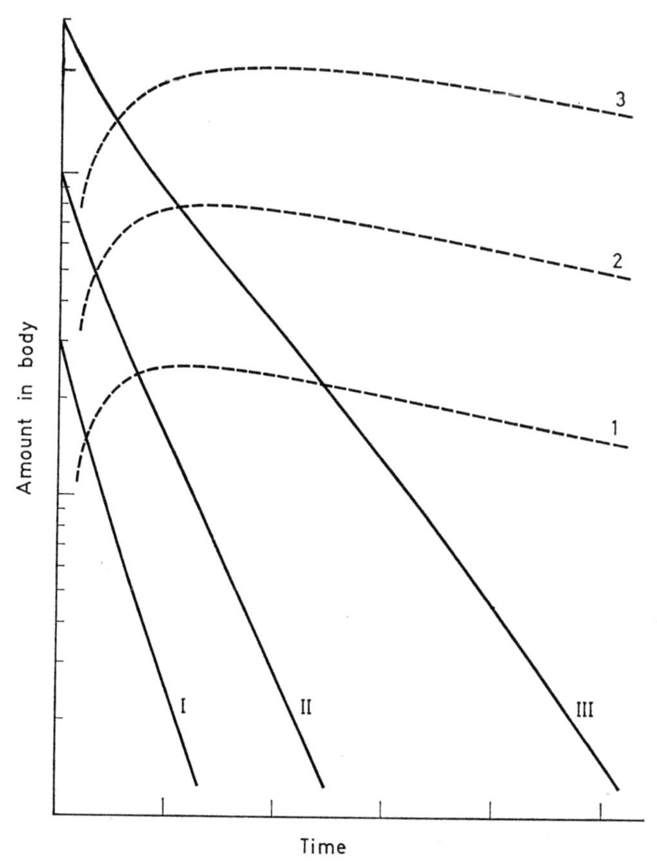

Fig. 16. Simulation of dose-dependent drug elimination due to product inhibition. Amount of drug in the body after rapid intravenous injection of doses I, II, and III, resulting in metabolite levels 1, 2 and 3, respectively

There is evidence that some drug metabolites can inhibit their own formation. This process of product inhibition can also cause dose-dependent effects, with large doses being relatively more slowly eliminated than small doses (PERRIER et al., 1973). However, while the rate of decline of drug concentrations in the post-distributive phase at any given level of drug in the body will be independent of dose in the case of simple Michaelis-Menten kinetics, this rate will tend to decrease with increasing dose in the case of product inhibition. Drug elimination may appear to be first-order but with half-lives *increasing* with increasing dose if the initial drug levels are lower than K_m and if elimination of the inhibiting metabolite is relatively slow (Fig. 16). This type of dose-dependence has been observed in man and animals (DAYTON and PEREL, 1971; O'REILLY et al., 1964). Lastly, some drugs may exert dose-dependent effects on blood circulation, urine pH, and on other physiologic processes that may affect drug disposition.

C. Non-Linear Absorption Kinetics

Some substances are absorbed from the gastrointestinal tract by specialized processes with limited capacity. This limited capacity will be reflected by a less than proportional increase with increasing dose in the urinary recovery, area under the plasma concentration-time curve, or maximum absorption rate (LEVY, 1968). Here, too, non-linear effects may also be due to factors other than capacity-limited processes, such as changes in gastric emptying rate or limited solubility of the drug in gastrointestinal fluids.

D. Repetitive Drug Administration

The plateau level of drug in the body during repeated administration of fixed doses at constant intervals is directly proportional to the size of the maintenance dose if drug elimination is apparent first-order [Eq. (46)]. In contrast, plateau levels of drugs that are eliminated by capacity-limited processes increase *more* than proportionally with increasing dose as shown in Fig. 17 (TSUCHIYA and LEVY,

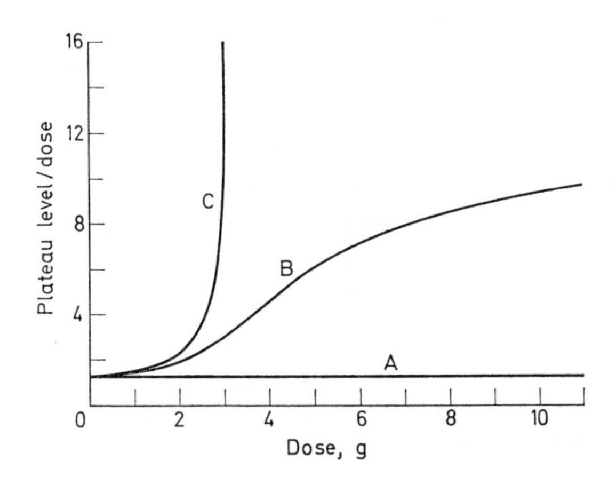

Fig. 17. Relationship between plateau levels immediately after intravenous injection and dose, given every 8 hours, if drug elimination proceeds by first-order (A), parallel first-order and Michaelis-Menten (B) and Michaelis-Menten kinetics only (C). Plotted are the ratios of the amount of drug in the body divided by the dose, as a function of dose. The following constants were used $k = 0.2\ \text{h}^{-1}$ for A; $k = 0.01\ \text{h}^{-1}$, $V_M = 0.38$ g/h, $K_M = 2.0$ g for B; and $V_M = 0.4$ g/h, $K_M = 2.0$ g/h for C

1972). Under certain conditions, doubling of the maintenance dose may cause a several-fold increase in plateau levels (LEVY and TSUCHIYA, 1972). Also, the time required to reach the plateau level will increase with increasing maintenance dose. For drugs that are eliminated by parallel Michaelis-Menten and apparent first-order kinetics, plateau levels may be calculated implicitly as a function of maintenance dose (D) and dosing interval (τ) using a redefined form of Eq. (79):

$$\tau = \frac{1}{k_* K_m + V_m}\left[K_m \ln \frac{A_{\max}}{A_{\max} - D} + \frac{V_m}{k_*} \ln \frac{(A_{\max} + K_m)\,k_* + V_m}{(A_{\max} - D + K_m)\,k_* + V_m}\right] \qquad (80)$$

in which A_{max} is the amount of drug in the body immediately after intravenous injection of a maintenance dose in the steady state. This equation applies to one-compartment systems and will be useful also for approximating plateau levels after oral administration if the drug is relatively rapidly absorbed. A_{min}, the amount of drug in the body immediately *prior* to the time of administration of the next dose at the steady state, is equal to $(A_{max} - D)$. The determination of A_{max} from Eq. (80) requires a process of successive estimations (iterations) that may be carried out manually but is most readily done by computer.

V. Kinetics of Reversible Pharmacologic Effects

A. One-Compartment Systems

If the intensity of a reversible pharmacologic effect at any time is a function of drug concentration at a receptor site, and if the drug concentration at that site is proportional to the drug concentration in the plasma, then it is possible to calculate the time course of pharmacologic effects as a function of some of the pharmacokinetic constants defined in Section II of this chapter. The relationship between drug concentration and intensity of pharmacologic effect is often approximately log-linear from about 20 to about 80 percent of the maximum intensity of effect. This linear relationship may be expressed by the equation

$$E = m \log C + e \tag{81}$$

which, upon rearrangement, yields

$$\log C = \frac{E - e}{m} \tag{82}$$

in which E is the intensity of the effect, and e and m are constants. Combining this equation with Eq. (5) for first-order elimination, and rearranging this expression, yields

$$E = E_0 - \frac{km}{2.303} t \tag{83}$$

in which E_0 is the intensity of effect immediately after intravenous injection of a drug (LEVY, 1966). This equation shows that, under the stated conditions, the intensity of effect decreases at a constant rate (R), which is a function of the apparent first-order rate constant (k) and the slope of the effect versus log C curve (m):

$$R = \frac{k\,m}{2.303} \tag{84}$$

Equation (83) is a zero-order expression and the time required for a pharmacologic effect to decrease to one-half of its initial intensity increases therefore with increasing dose.

Re-expression of Eq. (5) in terms of amount of drug in the body (A), rather than plasma concentration (C), and redefinition of t as the duration of effect (t_d) and therefore of A as the minimum amount of drug in the body (A_{min}) required to elicit the effect, yields

$$t_d = \frac{2.303}{k} \log A_0 - \frac{2.303}{k} \log A_{min} \tag{85}$$

in which A_0 is the intravenous dose. Thus a plot of duration of effect versus log A_0 is linear, with a slope of $2.303/k$. This slope is independent of the intensity of effect used as the endpoint for determining t_d, and of the relationship (linear, log-linear, etc.) between E and C. Equation (85) may be useful for determining k by pharmacologic means, i.e., without the need for chemical assays.

Since the rate of decline of a pharmacologic effect is directly proportional to k [Eq. (84)], and since the duration of effect is inversely proportional to k [Eq. (85)], Eqs. (84) and (85) may be combined to yield

$$t_d R = m (\log A_0 - \log A_{\min}) \tag{86}$$

This equation shows that $t_d R$, the product of duration and rate of decline of an effect, is independent of the elimination rate constant k. Thus, it is possible to determine the reason why an individual may show a quantitatively unusual response to a drug. An unusually long or short duration of effect may be ascribed to an unusually high or low k value if $t_d R$ is normal. If $t_d R$ is not normal, then the unusual response is due to an abnormal value of m, $A_{\min}$, m and $A_{\min}$, or m, $A_{\min}$ and k. These four possibilities can be distinguished, if necessary, by determining the relationship between E and $\log C$. Determinations of $t_d R$ have been found useful for the analysis of clinical effect data (LEVY, 1970b).

The total pharmacologic activity of a single dose of a drug has sometimes been represented as the area under the intensity of effect versus time curve. This index of total activity has shortcomings for many drugs in that it does not define the maximum intensity or duration of effect. It is useful however for quantitating such effects as diuresis, electrolyte excretion, and weight loss. The relative pharmacologic activity of a drug (i.e., the total area under the effect versus time curve divided by the dose) usually changes as a function of dose. This is due to the usually non-linear relation between the amount of drug in the body and the intensity of effect (WAGNER, 1968). Consequently, the total effect of a fixed amount of a drug per day may be affected by the dosage regimen (number of doses per day).

B. Multicompartment Systems

The time course of drug action in multicompartment systems depends upon the location of the site of action. Mathematically, the site of action may be located in the central compartment or in a peripheral compartment, or it may require representation as a separate compartment. The location of the site of action may be determined by examining the relationship between the intensity of effect and the amount or concentration of drug in each of the pharmacokinetic compartments (GIBALDI et al., 1972). The relationship between duration of action and log dose is *not* linear in multicompartment systems (GIBALDI et al., 1971b). For example, if the site of action is in the central compartment, then the duration of effect may increase considerably more than proportionally with log dose.

A peripheral compartment of a multicompartment system may be relatively poorly accessible to the central compartment. Drug moves in and out of such "deep" compartments rather slowly. If the site of drug action is in the "deep" compartment, the pharmacologic effect will be delayed and prolonged, and the relationship between drug levels in the plasma and effect may not be readily apparent. With this type of drug, repeated intravenous administration of equal doses at constant time intervals will yield the concentration versus time patterns shown in Fig. 18 for the central and "deep" peripheral compartments. This simulation, with the assumed minimum detectable drug concentration in the central compartment and minimum pharmacologically effective drug concentration in the "deep" compartment, reveals certain clinically significant characteristics. The pharmacologic effect appears only after the third dose and the intensity of this effect increases beyond the tenth dose since drug levels in the deep compartment do, in fact, accumulate. When drug administration is stopped, the effect persists well beyond the last dose. There are pronounced pharmacologic effects at a time when there is no detectable drug concentration in the plasma.

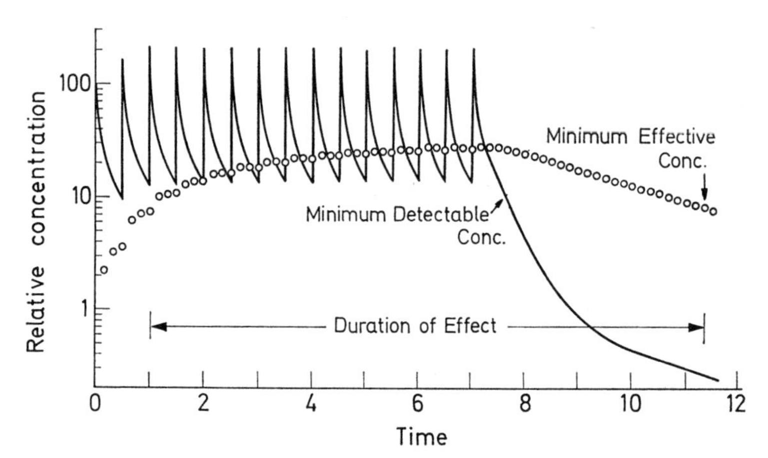

Fig. 18. Relative concentrations of a drug in the central and "deep" peripheral compartments of a three-compartment system during repetitive administration of equal doses at equal intervals. The curve and the circles represent the central and "deep" peripheral compartments, respectively. Elimination occurs from the central compartment

Some pharmacologic effects are not the result of direct effects of the drug but represent the net effect of two opposing processes of which only one may have been affected by the drug. A case in point is the hypoprothrombinemic effect of coumarin anticoagulants, which inhibit the synthesis of vitamin K-dependent clotting factors but have no effect on the degradation of these factors. Such indirect effects may also be analyzed pharmacokinetically (NAGASHIMA et al., 1969).

Recent advances in the quantification of pharmacologic effects have stimulated considerable interest in the pharmacokinetic analysis of drug action. These efforts are reviewed in a number of recent publications (LEVY and GIBALDI, 1972; LEVY, 1973).

VI. Some Useful Suggestions for the Experimentalist

The quality of a pharmacokinetic analysis can be no better than the quality of the experimental data that serve as input for such analyses. The specificity and sensitivity of analytical methods therefore require careful attention.

1. Whenever possible, determine the pharmacokinetics of elimination of a drug after rapid intravenous injection or intravenous infusion.

2. Determine plasma concentrations in the post-distributive phase over at least two or three half-lives whenever possible.

3. Include different doses in one study to determine dose-dependence.

4. Obtain at least three or four blood samples in the distributive phase if a multicompartment analysis is to be done.

5. If urinary excretion data are to be obtained, collect urine until there is no further excretion of drug or metabolites.

6. Take blood samples at midpoints of urine collection periods to facilitate determination of renal clearance.

7. Express excretion data as rates and amounts rather than as concentrations.

8. Examine the relationship between urine pH or flow rate and renal clearance.

9. When feasible, verify experimentally by multiple-dose studies the accumulation kinetics of a drug predicted from the kinetic constants obtained from single dose studies.

10. Determine bioavailability, including possible first-pass effects, following oral administration.

11. In summarizing data obtained from more than one subject, it is usually preferable to analyze individual sets of data and to average the pharmacokinetic constants thus obtained.

12. Generally average the logarithmic values of the plasma concentration or urinary excretion rate data for purposes of graphic representation. Even then it should be ascertained that the average curve is a reasonable representation of the characteristics of the individual curves.

A useful booklet entitled Guidelines for Biopharmaceutical Studies in Man is available from the A. Ph. A. Academy of Pharmaceutical Sciences, Washington, D. C. A number of digital computer programs are available for the analysis and simulation of pharmacokinetic data. Among these are SAAM (Berman and Weiss, 1967), NLIN (Marquardt, 1964), NONLIN (Metzler, 1969), and MIMIC (Control Data Corp., 1968).

References

Berman, M., Weiss, M. F.: SAAM manual, U. S. Public Health Service publication No. 1703, U. S. Government Printing Office, Washington 1967.

Bers, L.: Calculus, p. 413—416. New York: Holt, Rinehart and Winston 1969.

Bischoff, K. B., Dedrick, R. L.: Thiopental pharmacokinetics. J. pharm. Sci. **57**, 1346—1351 (1968).

Control Data Corporation: Control Data Mimic Reference Manual, Publication No. 44610400, 1968.

Dayton, P. G., Perel, J. M.: Physiological and physico-chemical bases of drug interactions in man. Ann. N. Y. Acad. Sci. **179**, 67—87 (1971).

Dettli, L.: Multiple dose elimination kinetics and drug accumulation in patients with normal and with impaired kidney function. In: Advances in the biosciences, Vol. 5 (Raspé, G., Ed.), p. 39—54. New York: Pergamon Press 1970.

Dost, F. H.: Grundlagen der Pharmakokinetic, p. 155, 169, 190. Stuttgart: Thieme 1968.

Gibaldi, M.: Effect of mode of administration on drug distribution in a two-compartment open system. J. pharm. Sci. **58**, 327—331 (1969).

Gibaldi, M., Boyes, R. N., Feldman, S.: Influence of first-pass effect on availability of drugs on oral administration. J. pharm. Sci. **60**, 1338—1340 (1971a).

Gibaldi, M., Levy, G., Hayton, W.: Kinetics of the elimination and neuromuscular blocking effect of D-tubocurarine in man. Anesthesiology **36**, 213—218 (1972).

Gibaldi, M., Levy, G., Weintraub, H.: Drug distribution and pharmacologic effects. Clin. Pharmacol. Ther. **12**, 734—742 (1971b).

Gibaldi, M., Nagashima, R., Levy, G.: Relationship between drug concentration in plasma or serum and amount of drug in the body. J. pharm. Sci. **58**, 193—197 (1969).

Gibaldi, M., Weintraub, H.: Some considerations as to the determination and significance of biologic half-life. J. pharm. Sci. **60**, 624—626 (1971).

Jusko, W. J., Levy, G.: Pharmacokinetic evidence for saturable renal tubular reabsorption of riboflavin. J. pharm. Sci. **59**, 765—772 (1970).

Keen, P.: Effect of binding to plasma proteins on the distribution, activity, and elimination of drugs. In: Handbook of Experimental Pharmacology, Vol. XXVIII (Brodie, B. B., Gillette, J. R., Eds.), Part I, Chapter 10. Berlin-Heidelberg-New York: Springer 1971.

Krüger-Thiemer, E., Levine, R. R.: The solution of pharmacological problems with computers. Part 8: Non first-order models of drug metabolism. Arzneimittel-Forsch. **18**, 1575—1579 (1968).

Levy, G.: Pharmacokinetics of salicylate elimination in man. J. pharm. Sci. **54**, 959—967 (1965).
Levy, G.: Kinetics of pharmacologic effects. Clin. Pharmacol. Ther. **7**, 362—372 (1966).
Levy, G.: Dose-dependent effects in pharmacokinetics. In: Importance of fundamental principles in drug evaluation (Tedeschi, D. H., Tedeschi, R. E., Eds.), p. 141—172. New York: Raven Press 1968.
Levy, G.: Effect of protein binding on the distribution and metabolism of a highly plasma protein bound drug (bishydroxycoumarin). In: Proc. Fourth Int. Congr. on Pharmacol., Vol. IV, p. 134—143. Basel: Schwabe 1970a.
Levy, G.: Pharmacokinetics of succinylcholine in newborns. Anesthesiology **32**, 551—552 (1970b).
Levy, G.: Correlation between drug concentration and drug response in man: pharmacokinetic considerations. In: Proc. Fifth Int. Congr. Pharmacol., Vol. 2. Basel: Karger 1973.
Levy, G., Gibaldi, M.: Pharmacokinetics of drug action. Ann. Rev. Pharmacol. **12**, 85—98 (1972).
Levy, G. Tsuchiya, T.: Salicylate accumulation kinetics in man. New Engl. J. Med. **287**, 430—432 (1972).
Levy, G., Tsuchiya, T., Amsel, L. P.: Limited capacity for salicyl phenolic glucuronide formation and its effect on the kinetics of salicylate elimination in man. Clin. Pharmacol. Ther. **13**, 258—268 (1972).
Loo, J. C. K., Riegelman, S.: New method for calculating the intrinsic absorption rate of drugs. J. pharm. Sci. **57**, 918—928 (1968).
Loo, J. C. K., Riegelman, S.: Assessment of pharmacokinetic constants from postinfusion blood curves obtained after i.v. infusion. J. pharm. Sci. **59**, 53—55 (1970).
Lundquist, F., Wolthers, H.: The kinetics of alcohol elimination in man. Acta pharmacol. (Kbh.) **14**, 265—289 (1958).
Marquardt, D. W.: DPE-NLIN, Share General Library Program No. 7—1354, 1964.
Mayersohn, M., Gibaldi, M.: Mathematical methods in pharmacokinetics. II. Solution of the two-compartment open model. Amer. J. pharm. Educ. **35**, 19—28 (1971).
Metzler, C. M.: NONLIN, technical report No. 7292/69/7293/005. Kalamazoo: The Upjohn Company 1969.
Nagashima, R., Levy, G., O'Reilly, R. A.: Comparative pharmacokinetics of coumarin anticoagulants. IV. Application of a three-compartment model to the analysis of the dose-dependent kinetics of bishydroxycoumarin elimination. J. pharm. Sci. **57**, 1888—1895 (1968).
Nagashima, R., O'Reilly, R. A., Levy, G.: Kinetics of pharmacologic effects in man: the anticoagulant action of warfarin. Clin. Pharmacol. Ther. **10**, 22—35 (1969).
Nelson, E., O'Reilly, I.: Kinetics of sulfisoxazole acetylation and excretion in humans. J. Pharmacol. exper. Ther. **129**, 368—372 (1960).
O'Reilly, R. A., Aggeler, P. M., Leong, L. S.: Studies on the coumarin anticoagulant drugs: a comparison of the pharmacodynamics of dicumarol and warfarin in man. Thrombos. Diathes. haemorrh. **11**, 1—22 (1964).
Perrier, D., Ashley, J. J., Levy, G.: Effect of product inhibition on kinetics of drug elimination. J. Pharmacokin. Biopharm. **1**, 231—242 (1973).
Rowland, M., Benet, L. Z., Riegelman, S.: Two-compartment model for a drug and its metabolite: application to acetylsalicylic acid pharmacokinetics. J. pharm. Sci. **59**, 364—367 (1970).
Tsuchiya, T., Levy, G.: Relationship between dose and plateau levels of drugs eliminated by parallel first-order and capacity-limited kinetics. J. pharm. Sci. **61**, 541—544 (1972).
Wagner, J. G.: Kinetics of pharmacologic response. I. Proposed relationships between response and drug concentration in the intact animal and man. J. theor. Biol. **20**, 173—201 (1968).
Wagner, J. G.: Biopharmaceutics and relevant pharmacokinetics, Hamilton: Drug Intelligence Publications, Chapter 40, 1971.
Wagner, J. G.: Personal communication 1972.
Wagner, J. G., Nelson, E.: Percent absorbed time plots derived from blood level and/or urinary excretion data. J. pharm. Sci. **52**, 610—611 (1963).
Wagner, J. G., Northam, J. I., Alway, C. D., Carpenter, O. S.: Blood levels of drug at the equilibrium state after multiple dosing. Nature (Lond.) **207**, 1301—1302 (1965).
Wagner, J. G., Novak, E., Leslie, L. G., Metzler, C. M.: Absorption, distribution, and elimination of spectinomycin dihydrochloride in man. Int. J. clin. Pharmacol. **1**, 261—285 (1968).

Chapter 60

Other Aspects of Pharmacokinetics

J. R. Gillette

With 7 Figures

I. Introduction

In the previous chapter (Part 3, Chapter 59), Levy and Gibaldi have considered the models that are most frequently used in pharmacokinetics and the ways that they may be used to assess the rates of absorption, distribution, metabolism and elimination of drugs, to determine the bioavailability of pharmaceutical preparations and to predict plasma levels obtained after multiple doses from data obtained with single doses. In this chapter, I shall review many of the assumptions that have been made in the derivation of the equations, discuss various processes that are reflected by the values obtained with the equations, and describe some of the situations in which the models are valid and invalid.

II. Factors that Affect the Rate of Distribution of Drugs in Tissues

The rate at which a drug enters and leaves a tissue depends on a host of interrelated factors. Among these are: 1) the properties of the capillary bed within the organ, 2) the properties of the plasma membrane of cells in the organ, 3) the lipid solubility of the drug and its degree of ionization at pH values within the blood and the extracellular and intracellular spaces, 4) the degree of binding of the drug to various macromolecules in the blood and within the extracellular and intracellular spaces, 5) relationships between the dissociation rate constant of the various drug-macromolecule complexes, the rates of diffusion of the drug across capillary walls and into tissue cells and the transit time of the blood through the capillary bed, 6) the relationships between the blood flow rate and the permeability coefficients of the capillary bed and the tissue cells, 7) relationships between the blood flow rate and the apparent volumes of distribution of the drug in the extracellular and intracellular spaces of the tissue, and 8) the rate of mixing of the drug within the extracellular and intracellular spaces.

A. Permeability of Capillaries and Cells

The endothelium of most capillary beds, such as that of skeletal muscle, have relatively large intercellular pores that allow free passage of most drugs up to the size of plasma proteins (Part 1, Chapter 1 by Hogben, 1971). Consequently, even polar drugs readily pass between the blood and the interstitial fluids. There are two noteable exceptions to this general situation, however. The capillaries of the central nervous system, with the exception of certain areas such as the area postremia and the subfornical body, do not possess intercellular pores and thus are rather impermeable to polar drugs (see Part 1, Chapter 12 by Rall, 1971).

On the other hand, the capillary bed of the liver and to a lesser extent that of the gastrointestinal tract have comparatively large pores which permit the passage of plasma proteins into the interstitial space.

In contrast to vascular endothelium, tissue cells appear to have micropores that permit the passage of only low molecular weight molecules. Usually, these micropores exclude substances having molecular weights greater than 100, although hepatocytes are readily permeated by mannitol (M. W. 180) and certain other substances of greater size (Part 1, Chapter 4 by SCHOU, 1971).

B. Lipid Solubility and Ionization of Drugs

In addition to passing through the pores of the capillaries and tissue cells, lipid soluble drugs dissolve in the plasma membranes of cells and thus may pass rapidly through the endothelial cells of the capillaries and into tissue cells. However, most drugs are either weak acids or weak bases, and thus their rates of passage through the capillary endothelial cells and into tissue cells also depend on their degree of ionization at the pH values of the blood, of the extracellular space and within tissue cells. The rate of passage of a weak acid or a weak base from the blood plasma to the extracellular space of the tissue, therefore, depends on the sum of the following clearance constants: 1) the clearance constant for the passage of the nonionized form of the drug through the capillary endothelial cells times the proportion of the total drug that exists in the nonionized form at the pH of the blood plasma; 2) the clearance for the passage of both the ionized and the nonionized forms of the drug through the intercellular pores of the capillary wall.

C. Differences in pH between Cells and Blood

Owing to the production of lactic acid and carbon dioxide during intermediary metabolism within cells, the pH within cells is slightly lower than that in the extracellular space and the blood plasma. Thus, a weak base, which is more highly ionized at lower than at higher pH values, diffuses out of cells less rapidly than it enters them. When the interior and exterior concentrations of the nonionized form are equal, therefore, the total concentration of the ionized and nonionized form of the weak base will be somewhat higher inside than outside the cells. It is also noteworthy that alterations in plasma pH can alter the tissue-to-plasma ratios of weak acids and weak bases by altering their rates of entrance into tissues. Because most tissues have pores in their capillary walls, however, most of the change would be expected between tissue cells and the extracellular fluid rather than between the extracellular fluid and the blood plasma. An exception is the change in the brain-to-plasma ratio, which would be expected to occur mainly between the extracellular fluid and the blood plasma because of the blood brain barrier.

D. Reversible Binding of Drugs to Proteins and Other Components

1. General Considerations

The binding of drugs to plasma proteins obviously decreases the concentration of both the ionized and nonionized forms of drugs in plasma and therefore decreases the rate at which the drug enters tissues. In most tissues the formation of drug-protein complexes will result in a decrease in the apparent rate constant of diffusion between the plasma and the extracellular fluid. Because the large pores in the capillary beds of the liver and the intestine permit the passage of plasma proteins into the interstitial space of these organs, however, the effect of the formation of drug-protein complexes would not affect the rate of diffusion nearly as much as it decreases the apparent rate constant of diffusion from the extracellular space into tissue cells. On the other hand, the formation of drug-protein complexes with plasma proteins would tend to hasten the passage of drugs from the tissues by maintaining a greater concentration gradient between the tissues and the plasma.

Similarly, the binding of drugs to various macromolecules and insoluble substances, such as proteins, DNA, melanin granules, phospholipids and neutral fat within tissues, would tend to hasten the passage of drugs from the plasma into tissues but to slow the release of the drugs from tissues.

Because of the wide diversity of physicochemical characteristics of drugs and the wide variety of possible macromolecules with which drugs may form complexes, the relative localization of drugs between tissues differ markedly from one drug to another. Moreover, the concentration of unbound drug can be considerably lower than either the total plasma level or the total tissue level of the drug. In fact, measurements of the relative amounts of drug in the plasma and various tissues frequently give the wrong impression of the extent of binding, because the concentrations in the tissues and plasma may be nearly equal even when the drug is highly bound in both. Thus, the rate of transfer of a drug may be slow even when the drug is very lipid soluble.

2. Dissociation Rate Constants and Transit Times

In considering possible mechanisms that limit the passage of a drug from the plasma into tissues, one might think that all of the unbound drug in the plasma could completely pass into tissues but that none of the bound form would. For this mechanism to occur, however, the half time for the dissociation of the complex would have to equal or exceed the mean transit time of the blood through the capillary bed of any given tissue. Since these transit times usually range between one and ten seconds, rate constants of dissociation for the drug protein complexes would have to be between 0.07 and 0.7 sec^{-1} or less. However, widely different equilibrium constants can result from rate constants of dissociation of this order of magnitude (Table 1). If the rate of formation of the drug-protein complex were limited solely by the diffusion of the drug molecule to the protein molecule, the rate constant for the formation of the complex would be about 2×10^9 M sec^{-1} (TAYLOR, 1972). Thus, complexes that have equilibrium association constants

Table 1. *Rate constants for the formation and dissociation of drug-protein complexes*

Drug	Protein	k_1 (M^{-1} sec^{-1})	k_{-1} (sec^{-1})	k_1/k_{-1} (M^{-1})
2,4-Dinitrolysine	γ-globulin	8×10^7	1	8×10^7
4,5-Dihydroxy-3-(p-nitrophenylazo)-2,7-naphthalene disulfonic acid	γ-globulin	1.8×10^8	7.6×10^2	2.4×10^5
4[(5,7-Disulfonic acid naphth-2-ol)-1-azo]benzene sulfonamide	Carbonic anhydrase	5.8×10^5	0.0075	7.7×10^6
1-Naphthol-2-sulfonic acid-4-[4-(4'-azobenzene-azo)] phenylarsonic acid	Serum albumin	3.6×10^5	2.5	1.4×10^5

Because the transit time of blood in liver is about 10 sec, approximately 50% of the drug could be removed if $k_{-1} = 0.07$ sec^{-1}. (Data taken from TAYLOR, 1972.)

greater than 3×10^{10} M^{-1} cannot be completely cleared from blood as it passes through the capillary beds of the tissues (GILLETTE, 1973). But the rate constants for the formation of drug-protein complexes are usually smaller than the theoretical maximum by several orders of magnitude and thus it is impossible to predict from the equilibrium association constants alone when the rate of dissociation of the drug-protein complex will limit the rate of the passage of drug into tissues. It seems probable, however, that the rates of dissociation of drug-protein complexes seldom are rate-limiting in the transfer of drugs from the blood into tissues.

3. Diffusion into Extracellular Spaces

The rate of diffusion of the unbound drug from the plasma through the extracellular space and then into tissue cells is frequently the rate-limiting step in the transfer of highly bound drugs between the plasma and tissue cells. When this occurs, the bound form of the drug in the plasma remains in equilibrium with its free form as the drug concentration in plasma declines during the passage of blood through tissue capillaries. In this regard, however, the fact that the pores in hepatic sinusoids are large enough to permit the passage of plasma proteins into the interstitial space becomes important especially when the drugs are rapidly metabolized by enzymes in the liver. Under these conditions, the plasma proteins are in intimate contact with the hepatic cells, and therefore the distance that a drug molecule has to traverse from the plasma proteins to hepatocytes is considerably shorter than the distance it has to travel to gain access to cells in most tissues.

4. Effect of Reversible Binding on the Clearance of Drugs by Liver

The formation of complexes of drugs with serum proteins may hasten the metabolism of drugs in liver by increasing their rate of removal from other tissues and carrying them to the liver. Moreover, it is also possible that drug-binding proteins in the cytoplasm of hepatocytes hasten metabolism by maintaining the concentration gradient across the membranes of the hepatocytes and by increasing the amount of drug available to those enzyme molecules located the furthest away from the plasma membrane of the cells. Indeed, ligandin proteins (FLEISCHNER et al., 1972; LITWACK et al., 1971; MOREY and LITWACK, 1969) are thought to act as carriers of sulfobromophthalein and other anions. But, whether they actually act as carriers or as tissue stores of the anions depends on a number of factors that include the association constants and the concentrations of the proteins as well as the maximum rate (V_{max}) and the K_m values of the drug-metabolizing enzymes and the relative diffusibilities of the unbound drug and drug-protein complexes. Because of the complexities of these interrelationships, it is frequently difficult to determine whether binding sites in hepatocytes serve as transport mechanisms or as storage mechanisms.

Whether plasma proteins serve as transport carriers of drugs depends not only on the activity of drug metabolizing enzymes in tissues and the effectiveness of active transport systems in kidney and liver but also on the blood flow rate through the tissues. When the activities of these systems are so high that their intrinsic capacity to clear the unbound form of drugs exceeds the blood flow rate through the tissues, then the dissociation of the drug-protein complexes become an important source of the drug for the elimination mechanisms. Under these circumstances, the plasma proteins actually serve as transport carriers for drugs. But when the intrinsic capacity of these systems to metabolize the unbound form of the drug is markedly below the blood flow rate through the tissues, the plasma proteins serve simply as storage sites.

E. Diffusion Barriers in the Extracellular Space

As a drug leaves the blood and enters the tissues, its concentration in the tissues obviously will rise until the concentrations of the freely diffusible form becomes equal in both or until its rate of transfer into the tissue equals its rate of metabolism or elimination. The rate at which the drug concentration reaches the steady-state phase, therefore, depends on a number of factors including differences in binding of the drug and differences in the pH within the tissues and the blood. But among the most important of these factors is the diffusibility of drugs through the extracellular space of the tissue and the degree of mixing that occurs therein. For example, there is evidence that polar compounds diffuse from the extracellular space of muscles at rates that are inversely proportional to their molecular weight, which suggests that the transfer of drugs in tissues is limited by tissue barriers and the mean distance between the capillaries within the tissues (SCHOU, 1971). Moreover, the finding that hyaluronidase increases the rate of diffusion of polar substances suggests that the tissue barriers for these substances are connective tissue interstitial gels containing hyaluronic acid.

Substances in blood thus equilibrate with the tissues at widely different rates. At one extreme, they may diffuse into and out of tissues so rapidly that the rate of accumulation of the drug within the tissue depends on the rate of perfusion of the tissue by the blood. At the other extreme, they may diffuse into and out of tissues so slowly that the rate of accumulation is virtually independent of the blood flow rate through the tissue.

F. Active Transport of Drugs

In some tissues, the drug may be metabolized, or actively taken up into special storage sites, such as the storage vesicles in adrenergic neurons, or be actively transported through capillary walls by special transport systems, e.g., subarachnoid villi. Usually these transport systems are relatively specific for amino acids, purines, lipids and other vitally important substances including choline and catecholamines and thus transport only those drugs that closely mimic their endogenous counterparts. Other transport systems are relatively nonspecific; for example, those systems in kidney and liver that secrete substances into urine and bile. Lipid soluble, as well as polar drugs, may sometimes be transported by these systems (see Part 1, Chapters 18 and 19 by WEINER, 1971, and SMITH, 1971), but because of their lipid solubility they pass back into the extracellular space and thence into blood and therefore their unbound forms do not accumulate in tissue cells or in the urine or bile. Indeed, the fact that they interact with the transport systems frequently may be discerned only because they inhibit the active transport of the endogenous substances that are transported or inhibit the transport of polar drugs.

G. Time-Delay Factors

There are also time-delay processes that complicate the development of pharmacokinetic models. Among the most obvious are the rate of dissolution of pharmaceutical preparations, and the passage of drugs through various parts of the gastrointestinal tract, such as the emptying time of the stomach. Moreover, prodrugs may be converted to their active form either in the gastrointestinal tract or by enzymes in various tissues (see Part 3, Chapter 61, DIGENIS and SWINTOSKY and Chapter 62, GARATTINI et al., 1974). Some substances may be actively secreted into bile and then reabsorbed from the intestine (see Part 1, Chapter 19 by SMITH, 1971; Part 3, Chapter 63 by PLAA, 1974). Other drugs may be converted in liver to metabolites that are secreted into bile, but the metabolites may then be converted back to the parent drugs in the intestine by enzymes in the intestinal mucosa or in the gut flora.

There are also more subtle time-delay factors which complicate the development of pharmacokinetic models. The clearance value for polar drugs entering the brain may be less than the rate of formation of the cerebral spinal fluid. In this situation the flow of the cerebral spinal fluid may be an important mechanism for controlling the concentration of the drugs in brain (Part 1, Chapter 12 by RALL, 1971). Thus, after the intravenous administration of compounds that are rapidly cleared from the blood by other mechanisms, the drug concentration in the cerebral spinal fluid that was formed immediately after the injection of the drug may approach the initial drug concentration in the blood plasma, but by the time the cerebral spinal fluid reaches the arachnoid granulations and passes back into blood, the drug concentration in the fluid may markedly exceed that in the blood plasma. Thus the drug in the fluid entering the blood may increase the drug concentration in the plasma. Similarly, the amount of drug that passes through the lymphatic system may occasionally affect the plasma concentration of drugs that are rapidly cleared from the blood after their intravenous administration. For example, a drug that is highly bound to plasma protein may be carried by the lymph formed in the liver to the thoracic duct. But when the concentration of the drug in blood at the time the lymph reaches the duct is markedly lower than its concentration at the time the lymph was formed, the concentration of drug in the blood will increase (DINGELL and GILLETTE, unpublished data).

Still another kind of time-delay can occur with highly lipid-soluble drugs that are administered intravenously. At the site of injection, a part of the dose of the drug may diffuse through the venous wall and become localized in the subcutaneous space (DINGELL and GILLETTE, unpublished observations). Since this portion of the dose is absorbed slowly, it persists in the tissues for a relatively long time and thus can perturb the plasma levels of compounds that are rapidly cleared by other mechanisms.

These time-delay effects of the cerebrospinal fluid flow, the lymphatic flow and the site of intravenous administration of drugs, are so rarely important in the disposition of most drugs that they are usually neglected in the development of pharmacokinetic models. When they are important, however, they can be very confusing to the investigator. They also point out the importance of withdrawing blood samples from sites other than injection sites.

III. Some Problems in the Development of Pharmacokinetic Models for Tissues and Organs

A. General Aspects

Despite the many factors that can affect the kinetics of drug distribution in tissues, the effects of these factors may be summarized by relatively simple pharmacokinetic models. For example, the rate at which a drug enters and leaves a tissue may be described by the filling and emptying of 3 compartments: 1) the apparent volume of blood within the capillaries of the tissue, 2) the apparent volume of the extracellular space, and 3) the apparent volume of the intracellular space.

B. Closed Model for a Tissue

In order to gain an insight into various aspects of this model, suppose for the moment that the drug is injected directly into the capillary bed and that it distributes instantaneously through the volume of the bed but that there is no blood flow through the capillary bed. The rate of diffusion will then depend on 1) the properties of the capillary bed and the tissue cells, 2) the degree of ionization of the drug at the pH values of blood, extracellular fluid and intracellular fluid, and 3) the degree of binding of the drug to various components in blood and the tissue. The initial rate of diffusion of the drug from plasma into the extracellular fluid thus depends on a) the total concentration of drug in plasma, b) the volume of blood in the capillaries, c) the proportion of the total drug concentration that is not bound to plasma proteins, d) the proportion of the unbound drug that is nonionized, e) the permeability coefficient for the passage of the nonionized form through the endothelial cells, and f) the diffusion constants for the passage of both nonionized and the ionized forms through the capillary pores.

As the drug diffuses into the extracellular compartment, a portion of the drug diffuses back into the blood, a portion becomes bound to extracellular components and a portion diffuses to the immediate environment of the tissue cells and then enters the tissue cells. Even if a completely valid mathematical description of the kinetics of these events were possible, it still would be virtually useless because the values of the constants in the equations would be impossible to determine accurately. On the other hand, if it is assumed that the extracellular compartment is mixed instantaneously and that the diffusion barriers in the extracellular space are a part of the capillary wall and the plasma membrane of the tissue cells, the mathematics describing the events become relatively simple.

According to Fick's Law of Diffusion, the equation for the change in the amount of drug in the capillary compartment may be written as follows (Appendix 1):

$$dQ_b/dt = C_e f_{e1} f_{e2} k_{eb} - C_b f_{b1} f_{b2} k_{be} \tag{1}$$

in which dQ_b/dt is the rate of change in the amount of drug in the capillary compartment, C_b is the total drug concentration in the capillary compartment at any given time, f_{b1} is the fraction of the drug in the capillary compartment that is not bound to protein, f_{b2} is the fraction of the unbound drug in the capillary compartment that is the nonionized form, C_e is the total concentration of drug in the extracellular compartment at any given time, f_{e1} is the fraction of the drug in the extracellular compartment that is not bound to extracellular components, and f_{e2} is the fraction of unbound drug that is in its nonionized form. The constants, k_{be} and k_{eb}, have the dimensions of volume per minute and are the clearance values for the diffusion of the drug from the capillary compartment to the extracellular compartment (k_{be}) and from the extracellular compartment to the capillary (blood) compartment (k_{eb}). Contrary to common belief, however, these clearance values are not always identical. They differ when the ionized form of the drug diffuses through the capillary pores at a significant rate and the pH values in the extracellular space and the blood are significantly different. In this situation, $k_{be} = k_{b1} + k_{b2} + f_{b3} k_{b3}$ and $k_{eb} = k_{b1} + k_{b2} + f_{e3} k_{b3}$ in which k_{b1} is the clearance of the nonionized form of the drug through the endothelial cells, k_{b2} is the clearance of the nonionized form of the drug through the intercellular pores, k_{b3} is the clearance of the ionized form of the drug through the intercellular pores, f_{b3} is the ratio of the ionized to the nonionized form in the blood and f_{e3} is the ratio of the ionized to the nonionized form of the drug in the extracellular space.

If neither the ionized nor the nonionized form of the drug enter or leave tissue cells through the micropores, the equation for the change in the amount of drug in the extracellular compartment may be written as follows:

$$\frac{dQ_e}{dt} = (C_b f_{b1} f_{b2} k_{b2} - C_e f_{e1} f_{e2} k_{eb}) + (C_i f_{i1} f_{i2} k_{ei} - C_e f_{e1} f_{e2} k_{ei}) \quad (2)$$

in which dQ_e/dt is the rate of change in the amount of drug in the extracellular compartment, and C_b, f_{b1}, $f_{b2} k_{be}$, $C_e f_{e1}$, f_{e2}, and k_{eb} are the same as defined above. C_i is the total concentration of the drug within the tissue cells at any given time, f_{i1} is the fraction of the drug in the cells that is not bound to intracellular components, f_{i2} is the fraction of the unbound drug in the cells that is not ionized and k_{ci} is the clearance value for the diffusion of the nonionized form of the drug into and out of the cells.

Similarly, the equation for the change in the concentration of the drug in the intracellular space may be written as follows:

$$dQ_i/dt = k_{ei} (C_e f_{e1} f_{e2} - C_i f_{i1} f_{i2}) - C_i f_{i1} k_{met} \quad (3)$$

in which dQ_i/dt is the rate of change in the amount of drug in the intracellular space, Q_i is the amount of drug in the tissue cells at any given time, k_{met} is the clearance value for the metabolism of the drug in the tissue and the other constants are as defined above. In this equation it is assumed that the ionized form of the drug is metabolized as rapidly as the nonionized form.

Finally, the equation for the rate of metabolism of the drug may be written as follows:

$$dQ_{met}/dt = C_i f_i k_{met}. \quad (4)$$

If the drug were not metabolized in the tissue, i.e., $k_{met} = 0$, the drug would accumulate in the extracellular and intracellular spaces until equilibrium is established. At that time there would be no net change in the concentrations of drug in the various compartments and the following relationships would be established:

$$C_b f_{b1} f_{b2} k_{be} = C_e f_{e1} f_{e2} k_{eb}$$

$$C_e f_{e1} f_{e2} k_{ei} = C_i f_{i1} f_{i2} k_{ei}.$$

On rearrangement and combination these equations become:

$$C_b \left(\frac{f_{b1} f_{b2} k_{be}}{f_{e1} f_{e2} k_{eb}} \right) = C_e = C_i \left(\frac{f_{i1} f_{i2}}{f_{e1} f_{e2}} \right).$$

However, the total amounts of drug in each compartment at equilibrium depend on the volumes of the compartments.

Thus,

$$Q_b \left(\frac{V_e}{V_b} \right) \left(\frac{f_{b1} f_{b2} k_{be}}{f_{e1} f_{e2} k_{eb}} \right) = Q_e = Q_i \left(\frac{V_e}{V_i} \right) \left(\frac{f_{i1} f_{i2}}{f_{e1} f_{e2}} \right) \quad (5)$$

$$Q_b \left(\frac{V_i}{V_b} \right) \left(\frac{f_{b1} f_{b2} k_{be}}{f_{i1} f_{i2} k_{eb}} \right) = Q_i. \quad (6)$$

The volume of distribution of the drug in the tissue at equilibrium, (Q_{tot}/C_b), is:

$$V_{d(eq)} = \frac{Q_b \left[1 + \frac{V_e}{V_b} \frac{f_{b1} f_{b2} k_{be}}{f_{e1} f_{e2} k_{eb}} + \frac{V_i}{V_b} \frac{f_{b1} f_{b2} k_{be}}{f_{i1} f_{i2} k_{eb}} \right]}{Q_b / V_b} \quad (7)$$

$$V_{d(eq)} = V_b + V_e \frac{f_{b1} f_{b2} k_{be}}{f_{e1} f_{e2} k_{eb}} + V_i \frac{f_{b1} f_{b2} k_{be}}{f_{i1} f_{i2} k_{eb}}.$$

As these equations show, the concept of the volume of distribution, even for a single tissue, represents a number of interrelated events.

In deriving equations for multicompartment systems, pharmacokineticists usually number the compartments in sequence and define the first order rate constants in the direction of the flow of the drug (Fig. 1). Thus, the capillary compartment may be designated Compartment 1; the extracellular space, Compartment 2; and the intracellular space, Compartment 3. Because the blood compartment is not in direct contact with Compartment 3, the model is called a caternary model.

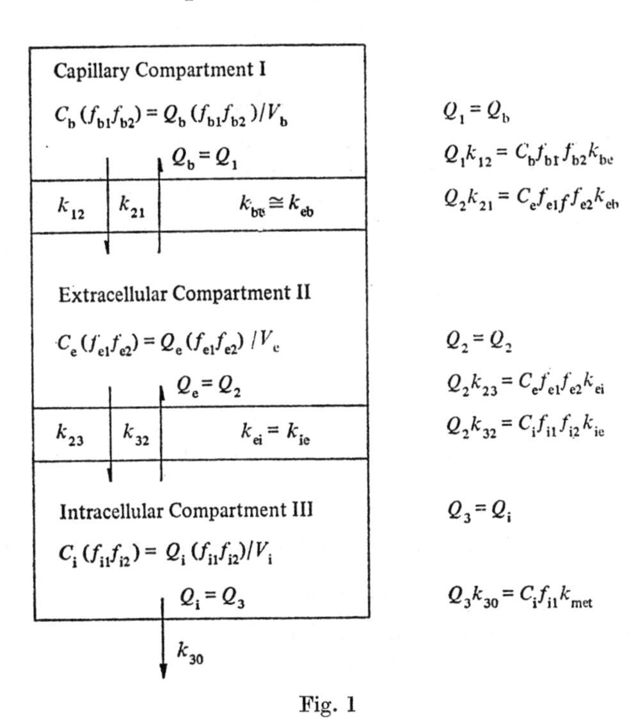

Fig. 1

After a series of transformations the equation for the rate of change of the amount of drug in the capillary compartment (Compartment 1) may be rewritten as follows: Recall Eq. (1).

$$\frac{dQ_b}{dt} = C_e f_{e1} f_{e2} k_{eb} - C_b f_{b1} f_{b2} k_{be}. \tag{1}$$

By setting $Q_1 = Q_b$, $Q_2 k_{21} = C_e f_{e1} f_{e2} k_{eb}$, and $Q_1 k_{12} = C_b f_{b1} f_{b2} k_{be}$, Eq. (1) becomes:

$$dQ_1/dt = Q_2 k_{21} - Q_1 k_{12}. \tag{8}$$

Similarly, the equation for the rate of change in the amount of drug in the extracellular compartment (Compartment 2) may also be rewritten as follows: Recall Eq. (2).

$$\frac{dQ_e}{dt} = C_b f_{b1} f_{b2} k_{be} - C_e f_{e1} f_{e2} k_{eb} - C_e f_{e1} f_{e2} k_{ei} + C_i f_{i1} f_{i2} k_{ei}. \tag{2}$$

By setting $Q_2 = Q_e$; $Q_1 k_{12} = C_b f_{b1} f_{b2} k_{be}$; $Q_2 k_{21} = C_e f_{e1} f_{e2} k_{eb}$; $Q_2 k_{23} = C_e f_{e1} f_{e2} k_{ei}$; and $Q_3 k_{32} = C_i f_{i1} f_{i2} k_{ei}$, Eq. (2) becomes:

$$dQ_2/dt = Q_1 k_{12} - Q_2 k_{21} - Q_2 k_{23} + Q_3 k_{32}. \tag{9}$$

Similarly, the equation for the rate of change in the amount of drug in the intracellular compartment (Compartment 3) is rewritten as follows:

$$\frac{dQ_i}{dt} = k_{ei}\,(C_e f_{e1} f_{e2} - C_i f_{i1} f_{i2}) - C_i f_{i1} k_{met}. \tag{3}$$

By setting $Q_3 = Q_i$; $Q_2 k_{23} = C_e f_{e1} f_{e2} k_{ei}$; $Q_3 k_{32} = C_i f_{i1} f_{i2} k_{ei}$; and $Q_3 k_{30} = C_i f_{i1} k_{met}$ Eq. (3) becomes:

$$dQ_3/dt = Q_2 k_{23} - Q_3 k_{32} - Q_3 k_{30}. \tag{10}$$

Similarly,

$$dQ_{met}/dt = Q_3 k_{30}. \tag{11}$$

Notice that k_{12}, k_{21}, k_{23}, and k_{32} have the dimensions of t^{-1} and consist of the clearance value divided by the apparent volumes of the appropriate compartments. Thus, the ratio of the rate constants describing the passive transfer of drug between two compartments usually is the ratio of the apparent volumes of the two compartments.

$$k_{21}/k_{12} = \frac{V_b/f_{b1} f_{b2}}{V_e/f_{i1} f_{i2}} = \frac{V_b'}{V_e'} \tag{12}$$

and

$$k_{32}/k_{23} = \frac{V_e/f_{e1} f_{e2}}{V_i/f_{i1} f_{i2}} = \frac{V_e'}{V_i'}. \tag{13}$$

For these reasons it is sometimes useful to substitute some of the kinetic constants by their equivalent values in order to gain a better insight into the mathematical model.

C. Perfusion Systems (Nonrecirculatory)

1. General

In order to illustrate some of the relationships between the rate of blood flow through a tissue and the filling and emptying of the tissue compartments for a given drug, it is useful to visualize a model in which a tissue is perfused with blood containing a constant concentration of drug. The model differs from most practical perfusion models in that the blood is not recirculated through the tissue. Initially, the capillary bed and the tissue are devoid of any drug. As the blood containing the drug enters the capillary bed, some of the drug will begin to pass from the blood into the extracellular space and thence into the tissue cells. The rate at which the drug leaves the blood, however, will obviously depend on the factors described in the previous model. But the relationship between the concentration of drug in the venous blood and the average drug concentration in the capillary bed will also vary with the drug. With polar drugs that slowly diffuse into the extracellular and intracellular spaces the difference between the drug concentrations in the arterial and venous blood will be small and the average concentration of drug in the capillary bed may be estimated from the average of the drug concentrations in the arterial and venous blood. But with lipid soluble drugs that rapidly diffuse into tissues, the drug concentration in the extracellular space may approach that in the venous blood, and thus the average of the drug concentrations in the arterial and venous blood may not reflect the permeability index of either the capillary bed or the plasma membrane of the tissue cells.

In the development of kinetic models describing the perfusion of tissues, different investigators have assumed either of these mechanisms depending on whether they believed that the rate of filling of the tissue was diffusion-limited or limited by the rate of perfusion by the blood. Unfortunately, however, many investigators do not provide corroborating evidence in support of either mechanism, even though the interpretation of pharmacokinetic relationships may depend on which mechanism is correct with any given drug.

2. Model 1

A general model for perfusion of tissues is shown in Fig. 2. According to Fick's principle, which is simply a statement of the conservation of matter, the

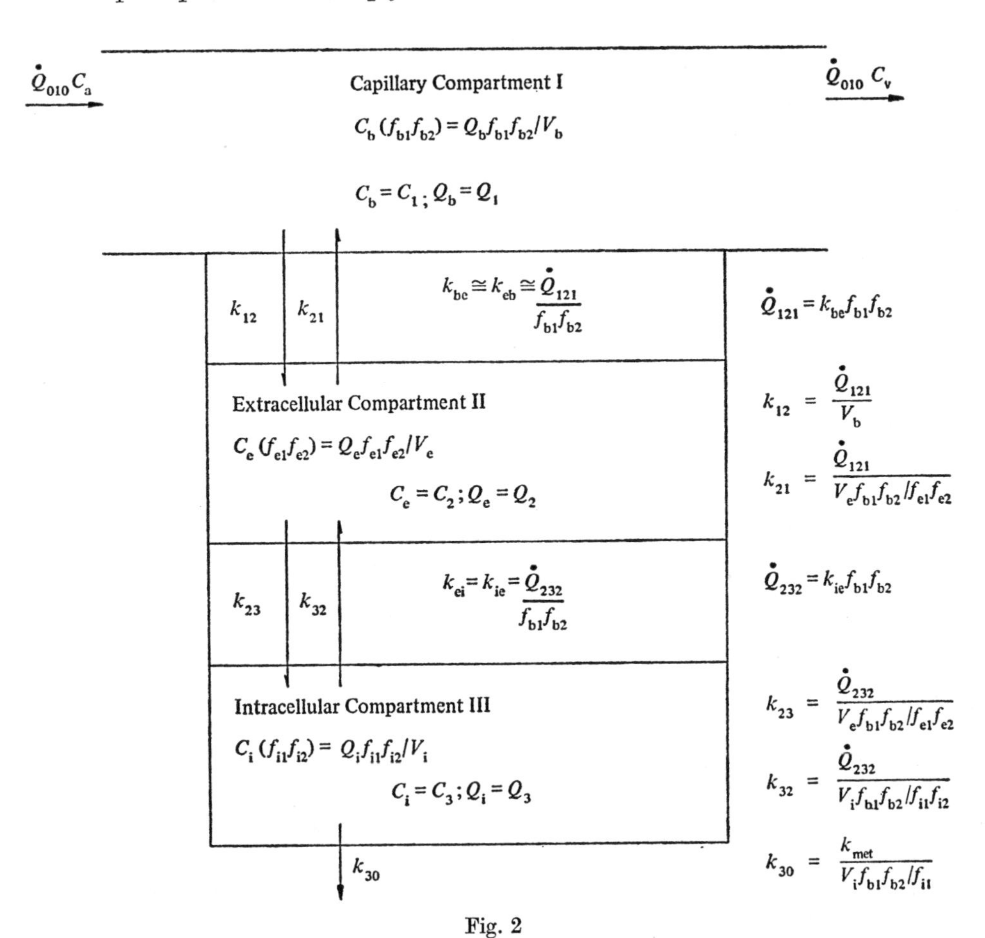

$\dot{Q}_{010} C_a$
Capillary Compartment I
$C_b (f_{b1} f_{b2}) = Q_b f_{b1} f_{b2} / V_b$
$C_b = C_1 ; Q_b = Q_1$
$\dot{Q}_{010} C_v$
k_{12}
k_{21}
$k_{be} \cong k_{eb} \cong \dfrac{\dot{Q}_{121}}{f_{b1} f_{b2}}$
$\dot{Q}_{121} = k_{be} f_{b1} f_{b2}$
Extracellular Compartment II
$C_e (f_{e1} f_{e2}) = Q_e f_{e1} f_{e2} / V_e$
$C_e = C_2 ; Q_e = Q_2$
$k_{12} = \dfrac{\dot{Q}_{121}}{V_b}$
$k_{21} = \dfrac{\dot{Q}_{121}}{V_e f_{b1} f_{b2} / f_{e1} f_{e2}}$
k_{23}
k_{32}
$k_{ei} = k_{ie} = \dfrac{\dot{Q}_{232}}{f_{b1} f_{b2}}$
$\dot{Q}_{232} = k_{ie} f_{b1} f_{b2}$
Intracellular Compartment III
$C_i (f_{i1} f_{i2}) = Q_i f_{i1} f_{i2} / V_i$
$C_i = C_3 ; Q_i = Q_3$
$k_{23} = \dfrac{\dot{Q}_{232}}{V_e f_{b1} f_{b2} / f_{e1} f_{e2}}$
$k_{32} = \dfrac{\dot{Q}_{232}}{V_i f_{b1} f_{b2} / f_{i1} f_{i2}}$
k_{30}
$k_{30} = \dfrac{k_{met}}{V_i f_{b1} f_{b2} / f_{i1}}$

Fig. 2

differential equations for the changes in the amounts of drug in each of the compartments are as follows:

$$\frac{dQ_1}{dt} = Q_0 k_{01} + Q_2 k_{21} - Q_1 (k_{12} + k_{10}), \tag{14}$$

$$\frac{dQ_2}{dt} = Q_1 k_{12} + Q_3 k_{32} - Q_2 (k_{21} + k_{23}), \tag{15}$$

$$\frac{dQ_3}{dt} = Q_2 k_{23} - Q_3 (k_{30} + k_{32}), \tag{16}$$

in which $Q_0 k_{01}$ is the rate at which the drug in arterial blood enters the capillary compartment and $Q_1 k_{10}$ is the rate at which the drug in the blood leaves the capillary compartment by way of the veins. The difference between $Q_0 k_{01}$ and $Q_1 k_{10}$ equals the blood flow rate ($\dot{Q}_{010}$) through the tissue multiplied by the difference between the arterial and venous drug concentrations.

We may transform Eqs. (14) through (16) into equations that describe the rates of passage of drug from one compartment to another, according to the apparent clearance constants of the capillary bed and the plasma membranes of the cells.

$$\dot{Q}_{010}\,(C\mathrm{a} - C\mathrm{v}) = \frac{dQ_1}{dt} + (Q_1 k_{12} - Q_2 k_{21}). \tag{17}$$

Moreover, after adding Eqs. (15) through (17) and rearranging the resultant equation, the following relationship is obtained:

$$\dot{Q}_{010}\,(C\mathrm{a} - C\mathrm{v}) = \frac{dQ_1}{dt} + \frac{dQ_2}{dt} + (Q_2 k_{23} - Q_3 k_{32}). \tag{18}$$

Similarly, Eq. (16) may be added to Eq. (18) to give another relationship:

$$\dot{Q}_{010}\,(C\mathrm{a} - C\mathrm{v}) = \frac{dQ_1}{dt} + \frac{dQ_2}{dt} + \frac{dQ_3}{dt} + Q_3 k_{30}. \tag{19}$$

From the conversion of Eqs. (1) through (4) to Eqs. (8) through (11), we note that:

$$Q_1 k_{12} = C_\mathrm{b} f_{\mathrm{b}1} f_{\mathrm{b}2} k_{\mathrm{be}};\; Q_2 k_{21} = C_\mathrm{e} f_{\mathrm{e}1} f_{\mathrm{e}2} k_{\mathrm{eb}};$$
$$Q_2 k_{23} = C_\mathrm{e} f_{\mathrm{e}1} f_{\mathrm{e}2} k_{\mathrm{ei}} \text{ and } Q_3 k_{32} = C_\mathrm{i} f_{\mathrm{i}1} f_{\mathrm{i}2} k_{\mathrm{ie}}.$$

If we let $k_{\mathrm{be}} = k_{\mathrm{eb}}$ and $k_{\mathrm{ei}} = k_{\mathrm{ie}}$, we may substitute these values into Eqs. (17) through (19) and obtain the following:

$$\dot{Q}_{010}\,(C\mathrm{a} - C\mathrm{v}) = \frac{dQ_1}{dt} + (C_\mathrm{b} f_{\mathrm{b}1} f_{\mathrm{b}2} - C_\mathrm{e} f_{\mathrm{e}1} f_{\mathrm{e}2})\, k_{\mathrm{be}}, \tag{17'}$$

$$\dot{Q}_{010}\,(C\mathrm{a} - C\mathrm{v}) = \frac{dQ_1}{dt} + \frac{dQ_2}{dt} + (C_\mathrm{e} f_{\mathrm{e}1} f_{\mathrm{e}2} - C_\mathrm{i} f_{\mathrm{i}1} f_{\mathrm{i}2})\, k_{\mathrm{ei}}, \tag{18'}$$

$$\dot{Q}_{010}\,(C\mathrm{a} - C\mathrm{v}) = \frac{dQ_1}{dt} + \frac{dQ_2}{dt} + \frac{dQ_3}{dt} + C_\mathrm{i}\,(f_{\mathrm{i}1} f_{\mathrm{i}2}/f_{\mathrm{i}2})\, k_{\mathrm{met}}. \tag{19'}$$

In turn these equations can be transformed as follows:

$$\dot{Q}_{010}\,(C\mathrm{a} - C\mathrm{v}) = \frac{dQ_1}{dt} + \left[C_\mathrm{b} - C_\mathrm{e}\left(\frac{f_{\mathrm{e}1} f_{\mathrm{e}2}}{f_{\mathrm{b}1} f_{\mathrm{b}2}}\right)\right] f_{\mathrm{b}1} f_{\mathrm{b}2} k_{\mathrm{be}}, \tag{20}$$

$$\dot{Q}_{010}\,(C\mathrm{a} - C\mathrm{v}) = \frac{dQ_1}{dt} + \frac{dQ_2}{dt} - \left[C_\mathrm{e}\left(\frac{f_{\mathrm{e}1} f_{\mathrm{e}2}}{f_{\mathrm{b}1} f_{\mathrm{b}2}}\right) - C_\mathrm{i}\left(\frac{f_{\mathrm{i}1} f_{\mathrm{i}2}}{f_{\mathrm{b}1} f_{\mathrm{b}2}}\right)\right] f_{\mathrm{b}1} f_{\mathrm{b}2} k_{\mathrm{ei}}, \tag{21}$$

$$\dot{Q}_{010}\,(C\mathrm{a} - C\mathrm{v}) = \frac{dQ_1}{dt} + \frac{dQ_2}{dt} + \frac{dQ_3}{dt} + C_\mathrm{i}\left(\frac{f_{\mathrm{i}1} f_{\mathrm{i}2}/f_{\mathrm{i}2}}{f_{\mathrm{b}1} f_{\mathrm{b}2}}\right) f_{\mathrm{b}1} f_{\mathrm{b}2} k_{\mathrm{met}}. \tag{22}$$

Let $f_{\mathrm{e}1} f_{\mathrm{e}2}/f_{\mathrm{b}1} f_{\mathrm{b}2} = x$ and $f_{\mathrm{i}1} f_{\mathrm{i}2}/f_{\mathrm{b}1} f_{\mathrm{b}2} = y$.

$f_{\mathrm{b}1} f_{\mathrm{b}2} k_{\mathrm{be}} = \dot{Q}_{121}$ and $f_{\mathrm{b}1} f_{\mathrm{b}2} k_{\mathrm{ei}} = \dot{Q}_{232}$ and substitute these values into Eqs. (20) through (22):

$$\dot{Q}_{010}\,(C\mathrm{a} - C\mathrm{v}) = \frac{dQ_1}{dt} + (C_\mathrm{b} - x C_\mathrm{e})\, \dot{Q}_{121}, \tag{23}$$

$$\dot{Q}_{010}\,(C\mathrm{a} - C\mathrm{v}) = \frac{dQ_1}{dt} + \frac{dQ_2}{dt} + (x C_\mathrm{e} - y C_\mathrm{i})\, \dot{Q}_{232}, \tag{24}$$

$$\dot{Q}_{010}\,(C\mathrm{a} - C\mathrm{v}) = \frac{dQ_1}{dt} + \frac{dQ_2}{dt} + \frac{dQ_3}{dt} + (y/f_{\mathrm{i}2})\, C_\mathrm{i} \dot{Q}_{30}. \tag{25}$$

Note that $\dot{Q}_{121}$, $\dot{Q}_{232}$, and $\dot{Q}_{30}$ have the dimensions of volume/time, and therefore may be viewed as apparent flow rates, or apparent clearance constants.

As the perfusion fluid containing the drug passes through the capillary bed, the average drug concentration in the capillary compartment will obviously increase. But the rate at which it increases will depend on the rate at which the drug passes into the extracellular space and thence into the intracellular space. If the apparent clearance constant ($\dot{Q}_{121}$) is very small, the drug concentration in the capillary compartment will rise rapidly and the venous drug concentration will soon nearly equal the arterial drug concentration. On the other hand, if the apparent clearance constants ($\dot{Q}_{121}$ and $\dot{Q}_{232}$) for the diffusion of the drug into the extracellular and the intracellular compartments are very large, the average drug concentration in the capillary compartment may increase slowly and indeed the rates of increase of the drug concentrations in the extracellular and intracellular compartments, xC_e and yC_i, may approach that in the capillary compartment. Because the capillary compartment is usually small compared with the other compartments, however, either the amount of drug in the capillary compartment will rapidly reach a steady-state value or the rate of change in the amount will soon be negligible compared with the rate of change in the extracellular and intracellular spaces. In either case, dQ_1/dt may then be omitted from the equations and the following set of relationships will then be valid:

$$\dot{Q}_{010}\,(C\text{a} - C\text{v}) = \dot{Q}_{121}\,(C_b - xC_e), \tag{26}$$

$$\dot{Q}_{010}\,C\text{a} - \dot{Q}_{010}\,C\text{v} = \dot{Q}_{121}\,C_b - \dot{Q}_{121}\,xC_e, \tag{26'}$$

$$\dot{Q}_{010}\,C\text{a} + \dot{Q}_{121}\,xC_e = \dot{Q}_{121}\,C_b + \dot{Q}_{010}\,C\text{v}, \tag{26''}$$

and

$$C\text{v} = \frac{C\text{a}\,\dot{Q}_{010} + xC_e\,\dot{Q}_{121}}{\dot{Q}_{010} + (C_b/C_v)\,\dot{Q}_{121}}. \tag{27}$$

When the passage of the drug into the extracellular space is limited mainly by the diffusion of the drug through the capillary wall, that is $\dot{Q}_{010} \gg \dot{Q}_{121}$, the venous concentration will nearly equal the arterial drug concentration after the pseudo steady-state is established, because under these conditions $\dot{Q}_{010}\,C\text{a} \gg xC_e\dot{Q}_{121}$ and $\dot{Q}_{010} \gg (C_b/C\text{v})\,\dot{Q}_{121}$.

Thus, Eq. (27) degenerates to:

$$C\text{v} = \dot{Q}_{010}\,C\text{a}/\dot{Q}_{010} = C\text{a}.$$

On the other hand, when the passage of the drug into the extracellular space is limited mainly by the rate of blood flow through the tissue, that is $\dot{Q}_{010} \ll \dot{Q}_{121}$, the drug concentration in both the extracellular compartment and the venous blood approach the average drug concentration in the capillary bed. Under these conditions, $\dot{Q}_{010}\,C\text{a} \ll xC_e\,\dot{Q}_{121}$ and $\dot{Q}_{010} \ll (C_b/C\text{v})\,\dot{Q}_{121}$ and Eq. (27) degenerates to $C\text{v} = xC_e\,\dot{Q}_{121}/(C_b/C\text{v})\,\dot{Q}_{121} = C\text{v}\,(xC_e/C_b) = C\text{v}$.

As the perfusion is continued, the change in the concentration of drug in the extracellular compartment will approach that in the capillary compartment and dQ_2/dt will decrease to a negligible value and an equation similar to Eq. (27), may be derived to describe the relationship between the blood flow rate and the clearance constant $\dot{Q}_{232}$. On further perfusion, C_i will approach C_e and thus the drug concentrations in the capillary, extracellular and intracellular spaces will be either in equilibrium or in a steady-state depending on whether the drug is metabolized in the tissue. If no metabolism occurs, the system will be in equilibrium and Ca will equal C_b, Cv, xC_e, and yC_i. Under these conditions the volume of distribution of the drug will be identical to that in the closed model [see Section III B, Eq. (7)]. However, when the drug is metabolized in the tissue, the drug concentrations in the three compartments are in steady-state and Eqs. (17) through

(19) degenerate to:

$$\dot{Q}_{010}\,(C\mathrm{a} - C\mathrm{v}) = Q_1 k_{12} - Q_2 k_{21}, \tag{28}$$

$$\dot{Q}_{010}\,(C\mathrm{a} - C\mathrm{v}) = Q_2 k_{23} - Q_3 k_{32}, \tag{29}$$

$$\dot{Q}_{010}\,(C\mathrm{a} - C\mathrm{v}) = Q_3 k_{30}. \tag{30}$$

Although Eq. (30) indicates the relationship between the rate of drug metabolism and the amount of drug within the tissue, it does not show the factors which control the relative distribution of the drug within the tissue. For this, we may solve the simultaneous equations for Q_1, Q_2, and Q_3 and substitute these values into the following equation:

$$\dot{Q}_{010}\,(C\mathrm{a} - C\mathrm{v}) = \frac{Q_\mathrm{t} Q_3 k_{30}}{Q_1 + Q_2 + Q_3}$$

and obtain the following equation:

$$\dot{Q}_{010}\,(C\mathrm{a} - C\mathrm{v}) = \frac{Q_\mathrm{t} k_{12} k_{23} k_{30}}{k_{30}\,(k_{12} + k_{23} + k_{21}) + k_{12}\,(k_{23} + k_{32}) + k_{21} k_{32}}. \tag{31}$$

From the identities [see Eqs. (17) through (19)]

$$k_{12} = C_\mathrm{b} f_{\mathrm{b}1} f_{\mathrm{b}2} k_\mathrm{be}/Q_1 = k_\mathrm{be}/V_1;$$

$$k_{21} = C_\mathrm{e} f_{\mathrm{e}1} f_{\mathrm{e}2} k_\mathrm{eb}/Q_2 = k_\mathrm{eb}/V_2;$$

$$k_{23} = C_\mathrm{e} f_{\mathrm{e}1} f_{\mathrm{e}2} k_\mathrm{ei}/Q_2 = k_\mathrm{ei}/V_2;$$

$$k_{32} = C_\mathrm{i} f_{\mathrm{i}1} f_{\mathrm{i}2} k_\mathrm{ie}/Q_3 = k_\mathrm{ie}/V_3;$$

$$k_{30} = C_\mathrm{i} f_{\mathrm{i}1} f_{\mathrm{i}2} k_\mathrm{met}/f_{\mathrm{i}2} Q_3 = k_\mathrm{met}/f_{\mathrm{i}2} V_3.$$

in which V_1, V_2, and V_3 are the apparent volumes of the compartments relative to the concentration of the nonionized, unbound form of the drug.

Setting $k_\mathrm{be} = k_\mathrm{eb}$ and $k_\mathrm{ie} = k_\mathrm{ei}$ and substituting the values into Eq. (31), we obtain:

$$\dot{Q}_{010}\,(C\mathrm{a} - C\mathrm{v}) = \frac{k_\mathrm{met}\, Q_\mathrm{t}}{k_\mathrm{met}\left[\dfrac{V_1 + V_2}{k_\mathrm{ie}} + \dfrac{V_1}{k_\mathrm{be}}\right] + f_{\mathrm{i}2}\,(V_1 + V_2 + V_3)}. \tag{32}$$

Since under steady-state conditions Eq. (25) degenerates to Eq. (25a): $\dot{Q}_{010}\,(C\mathrm{a} - C\mathrm{v}) = (y/f_{\mathrm{i}2})\, C_\mathrm{i}\, \dot{Q}_{30}$, it follows that:

$$(y/f_{\mathrm{i}2})\, C_\mathrm{i}\, \dot{Q}_{30} = \frac{k_\mathrm{met}\, Q_\mathrm{t}}{k_\mathrm{met}\left[\dfrac{V_1 + V_2}{k_\mathrm{ie}} + \dfrac{V_1}{k_\mathrm{be}}\right] + f_{\mathrm{i}2}\,(V_1 + V_2 + V_3)}.$$

Thus, the drug concentration within the tissue cells is dependent not only on the apparent volume of distribution of the drug but also on the relative clearance constants for metabolism and diffusion across the capillary wall and into cells.

It is perhaps more useful, however, to determine these relationships in terms of clearance of the drug from blood and the extraction ratio as the drug passes through the tissue. Let us set:

$$p\, C\mathrm{v}\, \dot{Q}_{30} = (y/f_{\mathrm{i}2})\, C_\mathrm{i} \dot{Q}_{30} = \frac{k_\mathrm{met}\, Q_\mathrm{t}}{k_\mathrm{met}\left[\dfrac{V_1 + V_2}{k_\mathrm{ie}} + \dfrac{V_1}{k_\mathrm{be}}\right] + f_{\mathrm{i}2}\,(V_1 + V_2 + V_3)}$$

and substitute this value into Eq. (25a).

$$\dot{Q}_{010}\,(C\mathrm{a} - C\mathrm{v}) = p\, C\mathrm{v}\, \dot{Q}_{30}.$$

On rearrangement:

$$\dot{Q}_{010}\,Ca = \dot{Q}_{010}\,Cv + p\,Cv\,\dot{Q}_{30} = Cv\,(\dot{Q}_{010} + p\,\dot{Q}_{30})$$

$$Cv = \frac{\dot{Q}_{010}\,Ca}{\dot{Q}_{010} + p\,\dot{Q}_{30}}. \tag{33}$$

The equation for the extraction ratio (ER):

$$ER = \frac{Ca - Cv}{Ca}. \tag{34}$$

On substitution of Eq. (33) into Eq. (34) and rearrangement:

$$ER = \frac{p\,\dot{Q}_{30}}{\dot{Q}_{010} + p\dot{Q}_{30}}. \tag{35}$$

Similarly, the equation for the clearance of the drug from blood by metabolism is:

$$\dot{Q}_{010}\,\frac{(Ca - Cv)}{Ca} = \frac{\dot{Q}_{010}\,p\,\dot{Q}_{30}}{\dot{Q}_{010} + p\,\dot{Q}_{30}} = \text{clearance (total)}. \tag{36}$$

The usefulness of this equation will be discussed in Part III, Chapter 69, in which $p\,\dot{Q}_{30}$ is designated C_e, the clearance of elimination. This equation was originally derived for a 2-pool recirculating perfusion system in which the concentration of unbound drug in all the compartments of the tissue allways equaled that in the venous blood and in which the drug was injected into the blood (ROWLAND et al., 1973). For this model, p is simply y/f_{i2}, that is a partition between the blood and the tissue concentrations. But as discussed above, the magnitude of "p" may also depend on the activity of the enzyme and on the diffusivity of the drug as well as on its volume of distribution. Thus, "p" is a kinetic term rather than a static term as is implied by the term "partition".

3. Model 2

When $\dot{Q}_{121} \gg \dot{Q}_{010}$, the filling of the extracellular space will occur almost instantaneously and thus the 3-pool model degenerates to a 2-pool model because the capillary and the extracellular spaces may be combined to form a single compartment. Under these conditions, the differential equations for the system would be:

$$\frac{d(Q_1 + Q_2)}{dt} = Q_0 k_{01} + Q_3 k_{32} - (Q_1 + Q_2)\,(k'_{10} + k'_{23}), \tag{37}$$

$$\frac{dQ_3}{dt} = (Q_1 + Q_2)\,k'_{23} - Q_3\,(k_{32} + k_{30}), \tag{38}$$

in which

$$k'_{10} = \frac{\dot{Q}_{010}/f_{b1}f_{b2}}{(V_b/f_{b1}f_{b2}) + (V_e/f_{e1}f_{e2})} = \frac{\dot{Q}_{010}/f_{b1}f_{b2}}{V_1 + V_2},$$

and

$$k'_{23} = \frac{k_{ei}}{(V_b/f_{b1}f_{b2}) + (V_e/f_{e1}f_{e2})} = \frac{k_{ei}}{V_1 + V_2}.$$

Moreover, Eq. (32) degenerates to:

$$\dot{Q}_{010}\,(Ca - Cv) = \frac{k_{met}\,Q_t}{k_{met}\left[\dfrac{V_1 + V_2}{k_{ie}}\right] + f_{i2}\,[V_1 + V_2 + V_3]}. \tag{39}$$

4. Model 3

On the other hand, when $\dot{Q}_{232} \gg \dot{Q}_{121}$ the filling of the extracellular and intracellular compartments occur simultaneously. Under these conditions the extracellular and intracellular compartments may be combined to form a single compartment and thus the model is a 2-pool system. Under these conditions the differential equations are:

$$\frac{dQ_1}{dt} = Q_0 k_{01} + (Q_2 + Q_3)\, k'_{21} - Q_1 (k_{10} + k_{12}), \tag{40}$$

$$\frac{d(Q_2 + Q_3)}{dt} = Q_1 k_{12} - (Q_2 + Q_3)\, (k'_{21} + k'_{30}), \tag{41}$$

in which

$$k'_{21} = \frac{k_{eb}}{(V_e/f_{e1} f_{e2}) + (V_i/f_{i1} f_{i2})} = \frac{k_{eb}}{V_2 + V_3},$$

and

$$k'_{30} = \frac{f_{i2} k_{met}}{(V_e/f_{e1} f_{e2}) + (V_i/f_{i1} f_{i2})} = \frac{f_{i2} k_{met}}{V_2 + V_3}.$$

For this model, Eq. (32) degenerates to:

$$\dot{Q}_{010}\,(Ca - Cv) = \frac{k_{met}}{k_{met}\left(\frac{V_1}{k_{be}}\right) + f_{i2}\,(V_1 + V_2 + V_3)} \tag{42}$$

5. Model 4

When both $\dot{Q}_{121}$ and $\dot{Q}_{232}$ are much larger than $\dot{Q}_{010}$, the capillary, extracellular and intracellular compartments fill and empty simultaneously. The rate at which these compartments fill and empty is thus limited by the blood flow rate through the tissue. The model is a one compartment system that is described by the following differential equation:

$$\frac{d(Q_1 + Q_2 + Q_3)}{dt} = Q_0 k_{01} - (Q_1 + Q_2 + Q_3)\, (k''_{10} + k''_{30}) \tag{43}$$

in which

$$k''_{10} = \frac{\dot{Q}_{010}/f_{b1} f_{b2}}{(V_b/f_{b1} f_{b2}) + (V_e/f_{e1} f_{e2}) + (V_i/f_{i1} f_{i2})} = \frac{\dot{Q}_{010}/f_{b1} f_{b2}}{V_1 + V_2 + V_3},$$

and

$$k''_{30} = \frac{f_{i2} k_{met}}{(V_b/f_{b1} f_{b2}) + (V_e/f_{e1} f_{e2}) + (V_i/f_{i1} f_{i2})} = \frac{f_{i2} k_{met}}{V_1 + V_2 + V_3}.$$

For this model, Eq. (32) degenerates to:

$$\dot{Q}_{010}\,(Ca - Cv) = \frac{k_{met}}{f_{i2}(V_1 + V_2 + V_3)}. \tag{44}$$

ROWLAND et al. (1973) used this model in their derivation of Eq. (36).

6. Emptying of Tissue Compartments

If the perfusion fluid is now changed to one containing no drug, the value of $\dot{Q}_{010}\,Ca$ becomes zero and the drug concentration in the tissue compartments begins to decrease. If there is no metabolism of the drug in the tissue the differential equations for a Model 1 system become:

$$\dot{Q}_{010}\,Cv = \frac{dQ_1}{dt} + (C_b - xC_e)\,\dot{Q}_{121}, \tag{45}$$

$$\dot{Q}_{010}\,Cv = \frac{dQ_1}{dt} + \frac{dQ_2}{dt} + (xC_e - yC_i)\,\dot{Q}_{232}, \tag{46}$$

$$\dot{Q}_{010}\,Cv = \frac{dQ_1}{dt} + \frac{dQ_2}{dt} + \frac{dQ_3}{dt}. \tag{47}$$

The initial event, of course, is a decrease in the average drug concentration in the capillary compartment, and thus a negative concentration gradient is established between the capillary space and the extracellular space. Thus,

$$\dot{Q}_{010}\, C\text{v} = (x C_{\text{e}} - C_{\text{b}})\, \dot{Q}_{121}. \tag{45'}$$

On rearrangement

$$\dot{Q}_{010}\, C\text{v} = x C_{\text{e}} \dot{Q}_{121} - C\text{v}\, (C_{\text{b}}/C\text{v})\, \dot{Q}_{121}$$

$$C\text{v}\, (\dot{Q}_{010} + (C_{\text{b}}/C\text{v})\, \dot{Q}_{121} = x C_{\text{e}}\, \dot{Q}_{121}$$

$$C\text{v} = \frac{x C_{\text{e}}\, \dot{Q}_{121}}{\dot{Q}_{010} + (C_{\text{b}}/C\text{v})\, \dot{Q}_{121}}. \tag{48}$$

When $\dot{Q}_{121}$ is much less than $\dot{Q}_{010}$ the venous drug concentration may decrease rapidly until a pseudo steady-state is reached. This is shown by the observation that when $(C_{\text{b}}/C\text{v})\, \dot{Q}_{121} \ll \dot{Q}_{010}$ the equation degenerates to:

$$C\text{v} = x C_{\text{e}}\, (\dot{Q}_{121}/\dot{Q}_{010}).$$

On the other hand when the drug diffuses rapidly through the capillary wall as it does in Model 2, Eq. (48) may degenerate to:

$$C\text{v} = x C_{\text{e}} \dot{Q}_{121}/(C_{\text{b}}/C\text{v})\, \dot{Q}_{121} = x C_{\text{e}} C\text{v}/C_{\text{b}} = C\text{v}$$

and C_{b} and $C\text{v}$ approximately equal $x C_{\text{e}}$. Under these conditions the rate at which the drug concentration in the extracellular space decreases will depend on the blood flow rate through the tissue.

As in Model 4, the clearance of a lipid soluble drug from the intracellular space into the extracellular space may also be very high compared with the blood flow rate through the tissue, and thus the concentration gradient between the intracellular space and the extracellular and capillary spaces may become very small. Under these conditions, Eq. (43) degenerates to:

$$\frac{d(Q_1 + Q_2 + Q_3)}{dt} = -(Q_1 + Q_2 + Q_3)\, k''_{10}, \tag{49}$$

$$-\dot{Q}_{010} C\ \text{v} = -\dot{Q}_{010} \left(\frac{Q_1 + Q_2 + Q_3}{(f_{\text{b}1}/f_{\text{b}2}\, (V_1 + V_2 + V_3)} \right), \tag{50}$$

in which k''_{10} is defined as it was in Eq. (43).

Thus, the concentrations of the drug in the capillary, extracellular and intracellular spaces may decline in parallel. The rate at which the drug leaves the tissue may depend almost entirely on the rate of blood flow through the tissue.

From this discussion, it should be evident that the kinetics for the filling and emptying of the various tissue compartments depends on a number of interrelated properties of both the drug and the tissue. In some instances the description of the kinetic events may require a 3-pool model, in others it may require a 2-pool model and in still others it may require only a single pool. In addition, the rate at which a drug enters and leaves the various compartments in some instances may be limited by diffusion across various tissue barriers, whereas in other instances the rate may be limited mainly by the blood flow rate through the tissues.

D. Pharmacokinetics of Drug Disposition and Metabolism after Intravenous Administration of Drugs

1. General

In order to gain an insight into the relationship between pharmacokinetic equations for the distribution and metabolism of an intravenously administered drug and the possible physiological events that they may describe, it is useful to review briefly the sequence of events that occur after the administration of drugs. As the drug is being injected into a vein, a bolus of drug is formed, whose size and average concentration depend on the rate of injection and the blood flow rate through the vein. The blood containing the bolus is mixed with the blood returning from other tissues to the heart, and the average drug concentration in the bolus is thereby decreased. The bolus then passes from the heart through the lung capillaries. It then passes through the heart and is distributed by way of the arteries to the capillary beds of the various organs. The boluses from each organ then return to the heart by way of the veins, but since the time required for the bolus to pass through the vascular system varies with the tissue, the various boluses will arrive at the heart at different times, and thus a series of boluses are formed for the next passage of the blood through the tissue capillary beds. Thus, the drug concentration within the capillary bed of each tissue will undergo a series of oscillations that can be quite large at first but will diminish with time until they become negligible. The drug concentration in the capillary beds of the various tissues then decline rather smoothly as the drug continues to enter and leave the various tissue compartments and is metabolized and eliminated from the body.

Owing to the complex interrelated factors discussed in Section II that determine the rate of passage of drugs into and out of a given tissue, it may seem surprising that the kinetics of drug distribution *in vivo* may fit either a 2- or 3-pool system. Clearly, the *in vivo* kinetic pattern represents the average of a number of complex physiological events occurring simultaneously in the various tissues of the body. The rate at which a drug passes through capillary walls in a tissue may differ markedly from the rate at which it passes through the plasma membranes of tissue cells. Moreover, these rates may differ markedly from one tissue to another. The apparent volumes of distribution of the drug in tissues will vary markedly from one tissue to another. These rates and volumes of distribution will also differ with the drug under investigation. It is therefore obvious that 2-compartment or even 3-compartment pharmacokinetic models describe only the major events that occur in the distribution of a drug in its major storage sites at any given time. It is equally obvious that these kinetic models may represent degenerate forms of more complex models and that the parameters of the models may represent different physiological events with different drugs. Indeed, a given tissue may be a part of the central compartment in the pharmacokinetic model for the distribution of one drug and may be a part of a peripheral compartment in the pharmacokinetic model for the distribution of another. In attempting to relate the pharmacokinetic parameters to specific physiological events, however, investigators frequently make broad generalizations about a given pharmacokinetic parameter regardless of the properties of the drug and the factors that control its rate of distribution into various tissues. Yet it is evident that many of these generalizations can be valid only with certain drugs and that many of the generalizations are contradictory. In order to gain an insight into some of the problems that arise in the interpretation of the pharmacokinetic parameters, let us consider the conditions under which various models of drug distribution in tissues might appear to be consistent with an *in vivo* system of 3 or fewer pools.

2. *In vivo* Kinetic Models

a) Model I – Caternary 3-Pool System (Fig. 3)

In Section III C 2 we discussed the kinetics of a 3-pool model for the filling of the various compartments of an organ during perfusion in which the blood was not recirculated. In the model, the drug slowly diffuses through the capillary wall into the extracellular space and slowly diffuses from the extracellular space into the intracellular space. That is $\dot{Q}_{010} \gg \dot{Q}_{121} \gg \dot{Q}_{232}$. The model would thus simulate the distribution of an ideal polar drug of rather high molecular weight that slowly enters the extracellular space through capillary pores and even more slowly passes through the plasma membranes of cells.

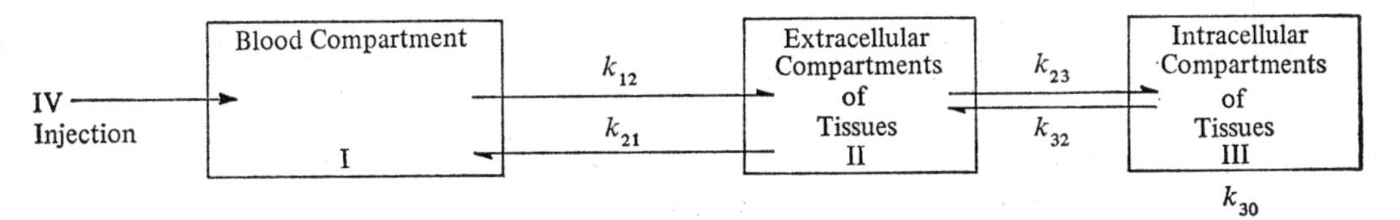

Fig. 3

The model discussed in Section III C 2, however, differs from its counterpart in the living animal. After the intravenous administration of the drug, its plasma level would continue to decline after the intracellular pool equilibrated with the extracellular pool as the drug was being metabolized and excreted. Thus, the *in vivo* model is actually a 4-pool system in which the mixing phase is the first phase. Since the capillary compartments of the various tissues would be presumably filled after the mixing phase was completed, however, the model for each tissue would have degenerated to a 3-pool system before the first blood samples are withdrawn for analysis of the drug.

If the clearance constants per gram and the apparent volumes of the tissue compartments per gram were the same for all tissues in the body, the equation that describes the changes in the plasma level of the drug would be:

$$Cp = \frac{D}{V\mathrm{a}} [A\mathrm{e}^{-\alpha t} + B\mathrm{e}^{-\beta t} + C\mathrm{e}^{-\gamma t}] \tag{51}$$

in which Cp is the concentration of drug at any given time, D is the intravenous dose of the drug and Va is the volume of the central compartment. The size of Va is estimated from the relationship, $V\mathrm{a} = D/Cpo$, in which Cpo is estimated from the equation:

$$Cpo = (DA/V\mathrm{a}) + (DB/V\mathrm{a}) + (DC/V\mathrm{a}) \tag{51a}$$

in which $DA/V\mathrm{a}$, $DB/V\mathrm{a}$, and $DC/V\mathrm{a}$ are the zero time intercept values for the α, β, and γ phases respectively. According to the model under discussion, very little drug enters the extracellular compartments of the various tissues during the initial mixing phase; Va thus nearly equals the blood volume. However, the extracellular space of tissues, such as liver, that have unusually large intercellular pores in their capillary walls, would be included in the central compartment because drugs would rapidly pass through these capillaries. Moreover, it would be expected that the conditions or treatments that result in changes in the blood volume would alter the size of the central compartment and therefore would affect the C_{po} concentration of the drug and the values of A, B, and α, but would not have a significant effect on β or on $(DB/V\mathrm{a} + DC/V\mathrm{a})$ unless the drug is rapidly metabolized. However, the rate of blood flow through the tissues would not affect the rate at which the drug entered the various tissues.

After the drug in the extracellular space equilibrates with that in blood, the equation of the 3-pool system degenerates to:

$$Cp = \frac{D}{Va'} [B e^{-\beta t} + C e^{-\gamma t}]. \tag{51'}$$

If the first blood sample were taken after the α-phase was completed and if the values for B and C were obtained by the feathering technique described in Part 3, Chapter 59, the volume of the central pool would no longer be the blood volume but would nearly equal the sum of the volumes of blood and the total extracellular spaces of the various tissues. It is also evident that after this time, changes in the volume of blood would no longer affect the distribution of the drug since any change in blood volume would be at the expense of the extracellular space. Thus, it is only during the α-phase that changes in blood volume would cause a significant effect on the kinetics of drug distribution.

After the drug in the intracellular space equilibrates with that in both the extracellular space and the blood, the 3-pool system degenerates to a 1-compartment system and thus:

$$Cp = \frac{D}{Va''} (C e^{-\gamma t}). \tag{51''}$$

If the first blood sample after both the α-phase and the β-phase had been completed, the investigator might be tempted to estimate the volume of distribution according to the equation, $Va = DC/Cpo$. But he would be unable to determine the validity of this estimate because he would not know how much of the drug was eliminated or metabolized during the α and β-phases, nor would he know whether the apparent volume of distribution was affected by the rate of drug elimination (see Section III D 4).

b) *Model Ia (a Mammillary, Degenerate Diffusion Limited System)*

In this model (Fig. 4), it is assumed that the drug in blood plasma equilibrates with some tissues faster than with others, but that the rate of passage of the drug into both kinds of tissues is limited not only by the apparent clearance constant/gm of tissue for the capillary bed but also by that of the plasma membranes of the tissue cells. If the apparent clearance constants/gm of tissue were higher in tissues that are rapidly perfused by the blood than in tissues that are poorly perfused, the system might superficially resemble a flow-limited rather than a diffusion-limited system even though mechanistically it is not flow-limited.

Because the blood compartment connects both types of tissues, the model is mammillary. But because the intracellular space of both types of tissue are not connected with the blood compartment, the model is also caternary. If we place the tissues into two groups in which the rapidly filled tissues are "G-fast" and the slowly filled tissues are "G-slow", the model would be a 5-compartment system. After the intravenous administration of the drug, changes in its blood level will be associated with 6 phases. The first phase is associated with the filling of the capillary compartments of both types of tissues. The second phase is associated with the establishment of an equilibrium between the blood and the extracellular spaces of the G-fast tissues. But the third phase could be associated with the attainment of an equilibrium between the blood and either the intracellular spaces of the G-fast tissues or the extracellular spaces of the G-slow tissues. If the investigator started to measure the plasma levels of the drug after the first three phases were completed, he would find that the kinetics for the decrease in plasma concentration would be described by the equation for a 3-compartment model:

$$Cp = \frac{D}{A}[A\,e^{-\alpha t} + B\,e^{-\beta t} + C\,e^{-\gamma t}]. \tag{52}$$

But the values and the meaning of the parameters of the equation for this model are completely different from that for Model I. In this equation, Va could exceed the blood volume by several fold, because it would include not only the blood volume, but also the apparent volume of the intracellular spaces of the

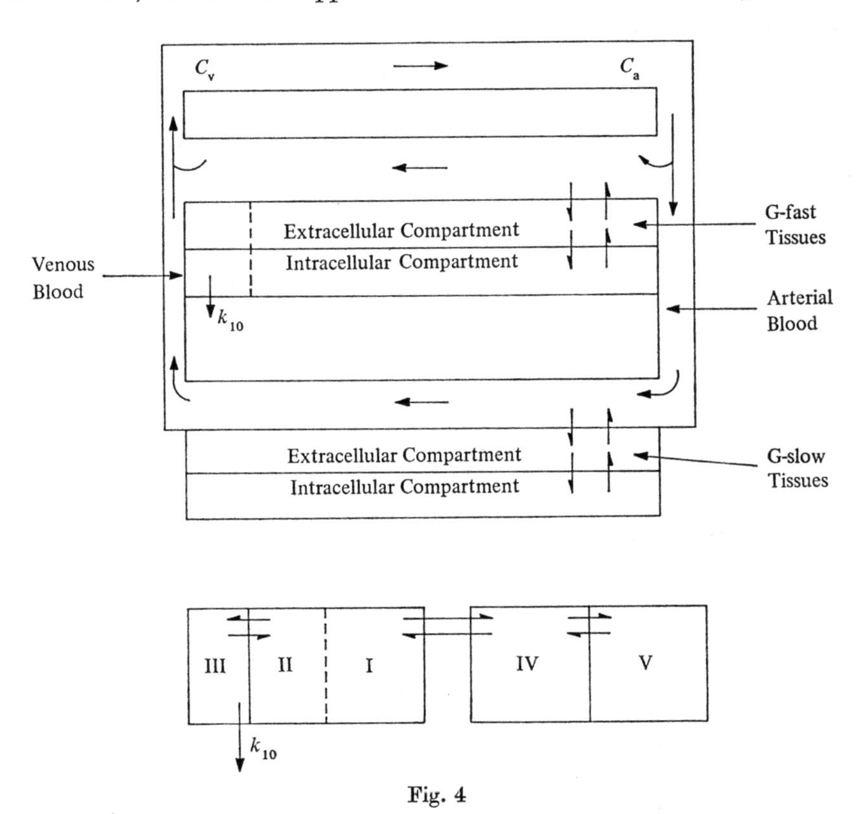

Fig. 4

G-fast tissues or the extracellular spaces of the G-slow tissues depending on which compartment became equilibrated with the blood during the third phase. Moreover, if the intracellular compartments of the G-fast tissues were not the ones equilibrated with the blood during the third phase, a part of the apparent volume of this compartment would be included in the volume of the central compartment, because the rate of diffusion of the drug into this compartment would be larger than would be predicted by the measured α-phase.

The fourth phase, i.e., the measured α-phase, would be associated with the equilibration of the blood with either the intracellular spaces of G-fast tissues or the extracellular spaces of the G-slow tissues. If the investigator started to measure the plasma levels of drug at the end of this phase, he would have found that his data could be explained by the equation for a 2-compartment system.

Thus, the volume of the central compartment would have appeared even larger than in the 3-compartment system. Moreover, he would not have known how much drug had been eliminated or metabolized during the earlier phases that could not be accounted for by the β- and γ-phases.

c) Model II – (a Mammillary, Caternary System)

In this model (Fig. 5), it is assumed that the drug instantaneously equilibrates between the blood and the extracellular spaces of the tissues, but that the drug slowly enters the cells of the various tissues. Thus, $\dot{Q}_{010} > \dot{Q}_{232}$ but $\dot{Q}_{121}$ approaches an infinitely larger value.

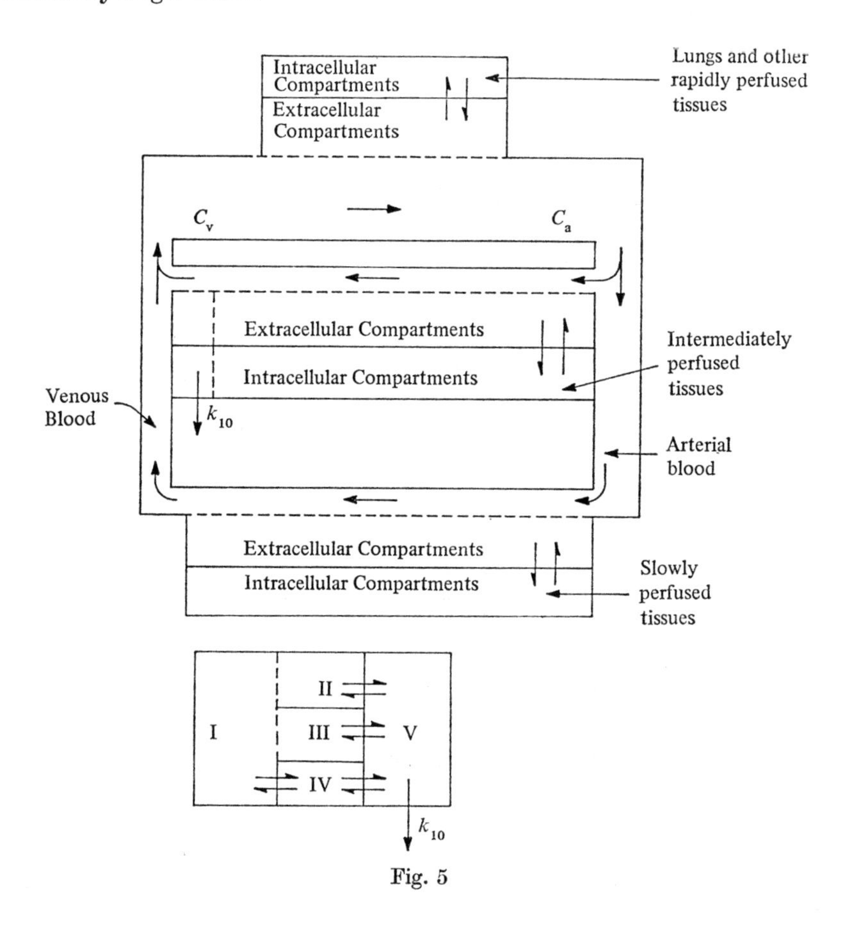

Fig. 5

This model would thus simulate the disposition of an ideal polar drug of rather small molecular weight that rapidly diffuses through the capillary pores and becomes rapidly distributed in the extracellular space but slowly enters tissue cells.

In the idealized model all of the tissues would have the same $\dot{Q}_{232}$ value per gram of tissue and the same intracellular volumes of distribution per gram of tissue. However, the rates of perfusion of the tissues by blood would differ. Immediately after intravenous administration, the drug would appear in the extracellular spaces of the various tissues as the bolus of drug passed through first the capillary bed of the lung and then the capillary beds of the other tissues. The proportion of the drug in the bolus that enters the extracellular space of any given tissue will depend on the relationship between volume of distribution of the extracellular space of the organ (V_2) and the blood flow rate through the organ ($\dot{Q}_{010}$). Before the bolus passes through a given tissue the average concentration of the drug in the bolus would be the amount of drug in the bolus divided by the volume of the bolus. But immediately after the drug passes through the capillary bed, the average concentration of the drug in the bolus would be the amount of drug in the bolus divided by the sum of the volume of the bolus (V_1) and the volume of distribution of the extracellular space of the tissue (V_2). After the bolus passes through the capillary space, some of the drug would diffuse from the extracellular space back into the blood until the next bolus of drug arrived in the capillary space. The rate of diffusion of the drug out of the extracellular space, however, would also depend on the relationship between the blood flow rate and the volume of distribution of the drug in the extracellular space of the tissue and thus each tissue would represent a single compartment. In practice, however, various organs may be grouped according to their blood flow rate per gram (Table 2). Thus, lung as the most rapidly perfused organ in the body represents one group. Group 2 comprises the adrenals, kidney and thyroid. Group 3 comprises the heart, brain, and the combination of the liver and the visera drained by the portal system. Group 4 usually comprises the skin, muscle, bone marrow, connective tissue and adipose tissue; the volume of distribution of certain drugs in the adipose tissue is so high, however, that the adipose tissue sometimes represents a separate group. The last group comprises the bones. The oscillations in the concentration of drug within the extracellular space would be greater in rapidly perfused organs, such as adrenals, thyroid and kidneys, than in intermediately perfused organs, such as the heart, liver and intestines, and in the slowly perfused

Table 2. *Blood flow rates to human tissues*

	Body weight (%)	Cardiac output (%)	Blood flow $\frac{\text{ml/g tissue}}{\text{min}}$
Group 1			
Lung	0.73	100	
Group 2			
Adrenals	0.02	1	5.5
Kidneys	0.4	24	4.5
Thyroid	0.04	2	4.0
Group 3			
Liver (hepatic)	2	5	0.2
Liver (portal)		20	0.75
Portal drained viscera	2	20	0.75
Heart	0.4	4	0.70
Brain	2	15	0.55
Group 4			
Skin	7	5	0.05
Muscle	40	15	0.03
Connective tissue	7	1	0.01
Fat	15	2	0.01

The blood flow rates per minute per gram of tissues were calculated from an assumed cardiac output of 5.6 l/min and an assumed body weight of 70 kg. In other calculations the blood volume was assumed to be 8% of the body weight and the plasma was assumed to be 4% of body weight. (The data were taken in part from Butler (1962) and Mapleson (1963).)

organs, such as connective tissue, bone marrow, muscles, fat and skin. Nevertheless, by the time the mixing phase is complete, the difference in the arterial and venous drug concentrations may be negligible in the rapidly perfused tissues but still marked in the more slowly perfused tissues. Thus, the phase represented by the filling of the extracellular spaces of the rapidly perfused organs may be complete by the time that the first blood sample is obtained, whereas the filling of the extracellular spaces of the intermediately perfused and the slowly perfused tissues may not have been complete. Therefore, the equation for the model depends to a large extent on the volumes of distribution of the drug in the extracellular space of the slowly perfused tissues and on how soon after the intravenous administration of the drug the first blood sample for the drug analysis is withdrawn. When the first blood sample is withdrawn immediately after the mixing phase is complete, the equation may represent a 4-pool system. Thus,

$$Cp = \frac{D}{Va}\left[A\,e^{-\alpha t} + B e^{-\beta t} + C e^{-\gamma t} + D e^{-\delta t}\right] \tag{53}$$

in which the α-phase is associated with the filling of the extracellular spaces of intermediately perfused organs such as the liver, heart and intestines, whereas the β-phase is associated with the filling of the poorly perfused tissues, such as skin, muscle, connective tissue, bone marrow and fat. In this equation, the volume of the central compartment is a complex term because the drug concentration will be much less in venous blood than in arterial blood. But the volume may appear to be similar to the blood volume. Changes in blood volume could alter these initial kinetic events either by changing the drug concentration in blood or by changing the relative blood flow rates to the various tissues. When the first blood sample is taken after the extracellular spaces of all of the tissues have equilibrated with the arterial blood (that is when the differences between the arterial and venous drug concentrations are small and are mainly due to the passage of drug into the cells of all of the tissues), the equation becomes:

$$Cp = \frac{D}{Va'}\left[C e^{-\gamma t} + D e^{-\delta t}\right] \tag{53'}$$

in which the apparent volume of the central pool would include the blood volume and the volumes of distribution of the drug in the extracellular spaces of all the tissues. In addition, the unbound drug concentration would be slightly higher in the cells of the rapidly perfused tissues than in those of the poorly perfused tissues and thus would contribute to the volume of the central pool. But as with the degenerate form of the equation for model I, changes in either the blood flow rate or the blood volume would have no effect after the α-phase and β-phase are completed.

d) Model III (2-Pool System)

In this model (Fig. 6), the drug diffuses slowly through the capillary walls into the extracellular space but instantaneously enters the cells of the tissues from the extracellular space. This model would simulate the kinetics of an ideal drug that diffuses slowly through the extracellular spaces of tissues but enters the cells closest to the capillaries sooner than it enters cells furthest away from the capillaries. The model would thus simulate the passage of semipolar drugs through the blood-brain barrier, as well as their passage into tissues having diffusion barriers.

As with Model I, this model requires not only that the clearances per gram of tissue through the capillary walls are similar for all of the major tissues of the body, but also that the volumes of distribution of the drug per gram of tissue are similar. If they are not similar, the equations for the distribution of the drug would have to be more complex. For the idealized model the equation for describing the changes in the drug concentration in plasma would be:

$$Cp = \frac{D}{Va}\left[A\,e^{-\alpha t} + B\,e^{-\beta t}\right] \tag{54}$$

in which the volume of the central compartment, Va, will approach the blood volume. Accordingly, changes in the blood volume should affect not only the initial con-

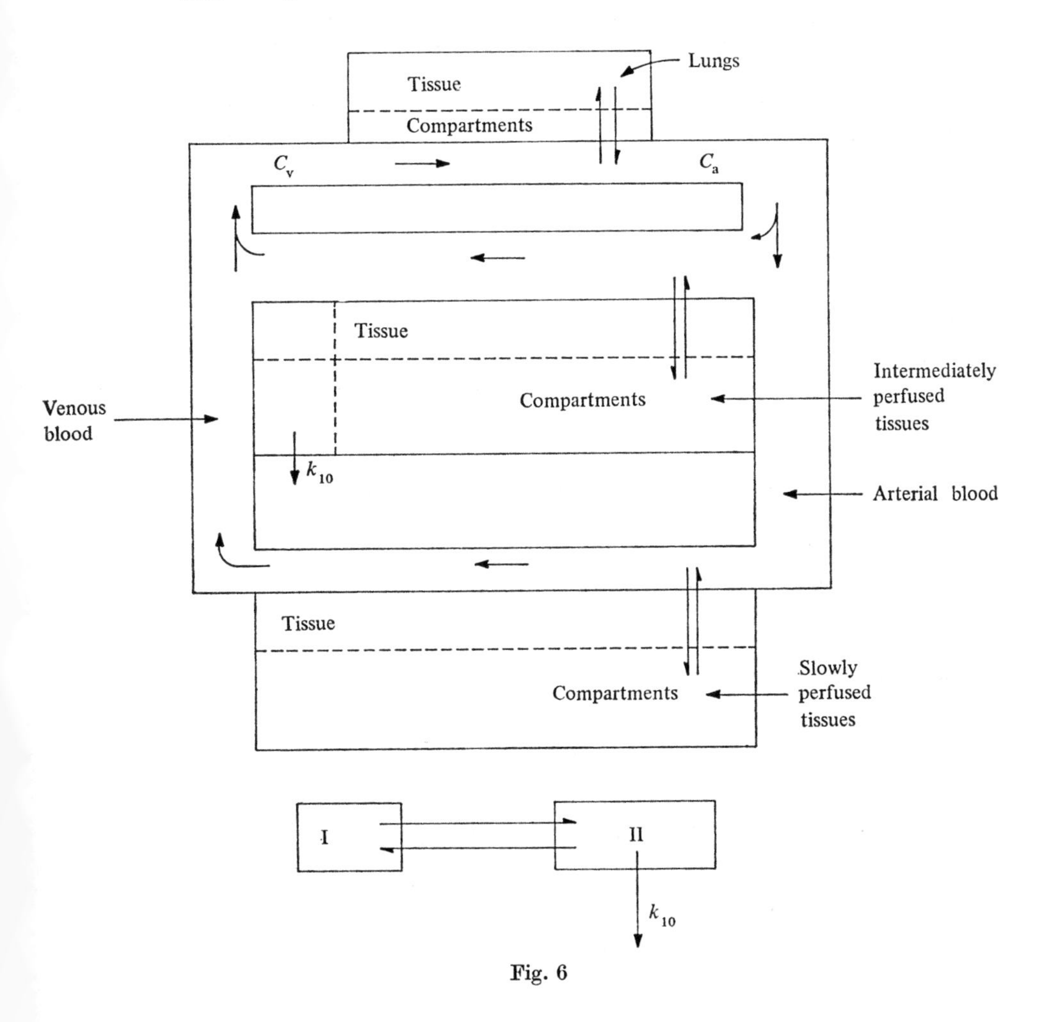

Fig. 6

centration of the drug (D/Va) but also A and α. But changes in the blood volume would not significantly affect the initial drug concentration of the β-phase, that is DB/Va, nor the value of β unless the drug is rapidly metabolized. On the other hand, the changes in the rate of blood flow through the tissues would not be expected to affect the rate at which the drug entered the tissues because the difference between the arterial and venous drug concentrations ($Ca - Cv$) would presumably be negligible for this model.

e) *Model IV (Mammillary, 3-Pool Flow-Limited System)*

In this model (Fig. 7), the drug in the blood of the capillary compartment of each tissue equilibrates instantaneously with the drug in the extracellular and intracellular compartments of the tissue. The model, therefore, implies that the concentration of the unbound nonionized drug in the tissue equals that in the

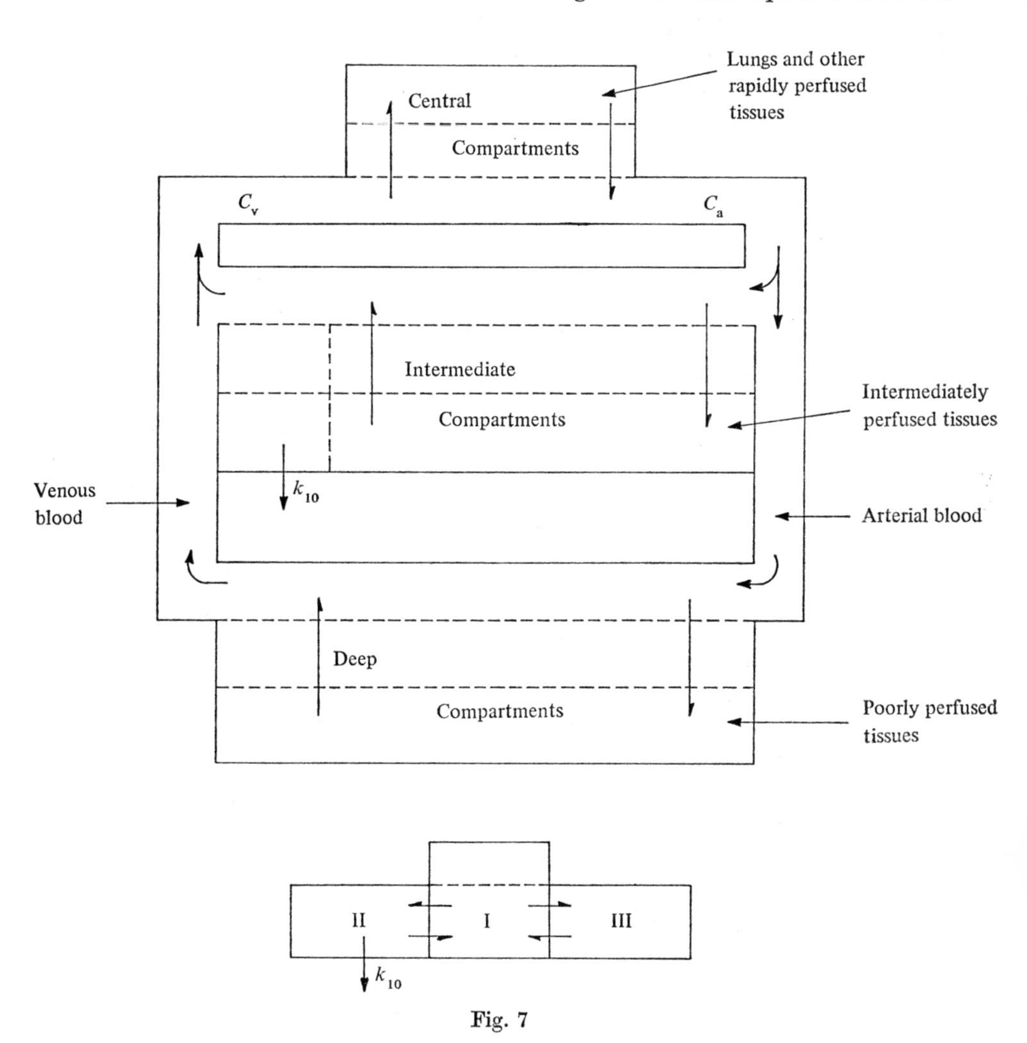

Fig. 7

capillary and venous blood draining the tissue. Thus, the blood flow rate relative to the apparent volume of distribution of the drug within the tissue limits the filling and emptying of the tissue with the drug. Since the blood flow rate per gram of tissue varies markedly from one tissue to another (Table 2), however, the rate constants, $\dot{Q}_{010}/V\mathrm{d}$ (tissue), will also vary considerably from one tissue to another. Consequently, each tissue represents a separate pharmacokinetic compartment. But as with Model II, the various organs may be grouped to form relatively few compartments.

The model presumably simulates the disposition of an ideal, highly lipid soluble drug of rather low molecular weight that readily passes through both the cells and the pores of the capillary wall, is instantaneously mixed in the extracellular compartment and readily enters the tissue cells.

If one assumes that the volume of distribution of the drug in the tissues nearly equals the tissue mass, it can be calculated that the drug in the lungs and the tissues in Group 2 will equilibrate with that in the blood during the mixing phase. Thus, if blood levels of the drug are measured immediately after the mixing phase is complete, and if the bone compartment is neglected, the equation for simulating the change in the plasma level of the drug would be as follows:

$$Cp = \frac{D}{Va}[A\,e^{-\alpha t} + B\,e^{-\beta t} + C\,e^{-\gamma t}]. \tag{55}$$

In this model the volume of the central compartment is difficult to assess. Since in the idealized model the concentration of drug in the venous blood draining any given tissue is always directly proportional to the drug concentration in the tissue, the venous blood belongs to the same compartment as the tissue, i.e., the venous blood draining a muscle belongs to the muscle compartment, whereas the venous blood from the brain belongs to the brain compartment. Therefore, the central compartment comprises only the blood in the pulmonary arteries and veins, the heart chambers and the peripheral arteries but not the blood in the peripheral veins or the capillary beds of the tissues. In addition, however, it also includes the apparent volume of the lungs and the Group 2 tissues.

According to the model, the venous blood is not part of the central pool, and thus blood samples from one of the veins in the arm presumably would reflect the filling and emptying of the various tissues such as muscle, connective tissue and skin in the lower arm, which would ordinarily be included in the deep compartment. It is, therefore, fortunate that the model seldom if ever is completely accurate in simulating the distribution of even the most highly lipid soluble compounds. For example, thiopental penetrates the blood-brain barrier so rapidly that its entrance into the brain is limited mainly by the blood flow rate to the brain (see Part 1, Chapter 12 by Rall, 1971). However, venous blood samples taken from the arm immediately after a two minute infusion of the drug indicate that the apparent volume of distribution of the drug at this time was not more than about 12 l (calculated from the data of Brodie et al., 1950). But if the deep pools were the reference compartment, the volume of distribution should have been $>$ 40 l if the drug concentration in the venous blood from the arm had been the same as that in muscle. Moreover, the plasma levels then rapidly fell at a time when the muscle and skin compartments were still being filled. Indeed, experiments in dogs revealed that the maximum drug concentration in muscle is reached between 15 to 30 min after the intravenous administration of the drug. Thus, the passage of thiopental into muscle is predominately diffusion-limited even though its passage into brain is predominately flow-limited.

Nevertheless, the model illustrates the point that the drug concentration in the venous blood may be significantly lower than that in arterial blood during the early distribution phases and higher than that in arterial blood during the terminal phases of drug disposition studies and thus, the values of the zero time intercepts for the various phases estimated with venous blood samples may be different from those estimated with arterial blood samples.

3. General Comments on the Central Compartment

It is evident from studying the various models that a variety of mechanisms of distribution may lead to 2 and 3-compartment models and that the components constituting the central compartment will differ with the dominate mechanism of distribution of the drug under study and the time at which the first blood samples are withdrawn for analysis. In some situations the volume of the central compartment may equal the blood volume. But in most instances, the pharmacokinetic study begins after the early phases of drug disposition have been completed; thus the volume of the central compartment almost invariably is larger than the blood volume. It is also evident that the 2 and 3-pool models used in pharmacokinetics are almost invariably degenerate forms of more complex kinetic systems and thus it becomes necessary to consider the effects of presampling events on the disposition of rapidly metabolized drugs (Section III D 4).

4. Concepts of Volume of Distribution

As originally defined, the volume of distribution of a drug is the total amount of substance in the body divided by its concentration at equilibrium in the reference compartment, which is usually the plasma or central compartment. However, this definition of the volume of distribution has a physiological meaning only when the drug either is not excreted or is infused into the central compartment at the same rate as it is eliminated from the central compartment. In the latter case, however, the drug must not be eliminated from the body by being metabolized in a peripheral compartment.

In pharmacokinetics, the concept of a "volume of distribution" has other definitions and interpretations. In Section III B, the terms $V_b/f_{b1}f_{b2}$, $V_e/f_{e1}f_{e2}$, and $V_i/f_{i1}f_{i2}$ may be viewed as the effective volumes of distribution of a drug in the capillary compartment, the extracellular compartment and the intracellular compartment for any given tissue, respectively. These effective volumes of distribution in tissue compartments are defined as the total amount of drug in the compartment divided by the concentration of the unbound, nonionized form of the drug. They are used in relating the apparent clearance constant of a drug for its diffusion through a membrane or compartment to the first-rate constants for the passage of the drug from one compartment to another.

Another concept of volume of distribution is useful in describing the events that occur following the intravenous injection of a drug. The "instant volume of distribution" may be defined as the total amount of drug in the body at any given time divided by the drug concentration in the central compartment at that time. Immediately after the injection of the drug the "instant volume of distribution" will increase as the drug enters the various tissues of the body. As discussed in Section III D 2, however, the drug concentration in plasma may equilibrate with some tissue compartments faster than with other compartments and with some tissues faster than with other tissues. Thus, as the drug concentration in plasma decreases, the drug may begin to leave the filled tissues at the same time as the drug is entering other tissues. However, the effective drug concentration in the capillary space must be smaller than that in the extracellular and intracellular spaces in the rapidly filled tissues in order to establish concentration gradients that permit the passage of the drug from the tissue to the blood. Under these conditions, the "instant volume of distribution" of the drug in the rapidly filled tissues would be greater than the equilibrium volume of distribution. At the same time, however, the "instant volume of distribution" for the slowly filled tissues may still be lower than the equilibrium volume of distribution.

Moreover, if initial samples for measuring the blood levels of the drug were withdrawn after the drug concentration in rapidly filled tissues had begun to decrease the value of Va, the volume of the central compartment, would reflect the "instant volume of distribution" of the drug in the rapidly filled tissues rather than the equilibrium volume of distribution in the tissues.

After the terminal phase of drug disposition is reached, the drug concentrations in all of the tissues will decrease as the drug is eliminated from the body by metabolism and excretion into urine. Thus, the "instant volumes of distribution" of all the tissue will then be at least slightly larger than their equilibrium volumes of distribution except for those tissues that metabolize or eliminate the drug. The magnitude of the difference between the "instant volume of distribution" and the equilibrium volume of distribution for each tissue, however, depends on the interrelationship between the clearance values of diffusion from the various compartments of the tissue, the effective volumes of the compartments, the blood flow rate through the tissue and the rates of metabolism and elimination of the drug from the body. Whatever these relationships are, however, the "instant volume of distribution" of the drug in the body during the terminal phase is a constant and includes the blood volume and the "instant volumes of distribution" of all the tissues.

The fact that the "instant volume of distribution" during the second phase of a 2-compartment system is somewhat larger than the volume of distribution at equilibrium has been recognized by several kineticists in the past. However, various names have been applied to it. For example, RIGGS (1963) has called it "volume distribution (remainder)", and "volume of distribution (slope)". RIEGELMAN et al. (1968) have called it "volume of distribution (area)". GIBALDI et al. (1969) have called it "volume of distribution (β)". But the concept is applicable to a multicompartment system having more than 2 pools. The names given to the volume of distribution during the last phase of drug disposition, therefore, seem either misleading or inadequately descriptive to me. For this reason, I believe that a better name would be "omega volume of distribution", or $V\text{d}\,(\omega)$.

a) Equations for $V_\text{d}(\text{eq})$, and $V_\text{d}(\omega)$ in 2 and 3 Compartment Systems

The easiest way of illustrating $V_\text{d}(\omega)$ is to write the equations for the amount of drug present in each of the compartments at any given time during the last phase and to divide the sum of these quantities by the drug concentration in the central compartment at that time.

As pointed out by LEVY and GIBALDI (Part 3, Chapter 59), the equation for the drug concentration in the central compartment of a 2-pool system in which the drug is eliminated only from the central pool is as follows:

$$Cp = \frac{D}{Va}\left[A\,e^{-\alpha t} + B\,e^{-\beta t}\right] \tag{56}$$

in which $A = (\alpha - k_{21})/(\alpha - \beta)$ (56a) and $B = (k_{21} - \beta)/(\alpha - \beta)$ (56b). In this equation,

$$\alpha = 1/2\left[k_{12} + k_{21} + k_{10} + \sqrt{(k_{12} + k_{21} + k_{10})^2 - 4\,k_{21}k_{10}}\,\right], \tag{56c}$$

$$\beta = 1/2\left[k_{12} + k_{21} + k_{10} - \sqrt{(k_{12} + k_{21} + k_{10})^2 - 4\,k_{21}k_{10}}\,\right]. \tag{56d}$$

Thus

$$\alpha + \beta = k_{12} + k_{21} + k_{10}, \tag{57}$$

$$\alpha\beta = k_{21}k_{10}. \tag{58}$$

Thus, the quantity of drug in the central compartment (Q_1) at any given time is:

$$Q_1 = Cp\,Va = D[A\,e^{-\alpha t} + B\,e^{-\beta t}]. \quad (59)$$

Moreover, the equation for the quantity of drug in the peripheral compartment (Q_2) is:

$$Q_2 = DC\,(e^{-\beta t} - e^{-\alpha t}) \quad (60)$$

in which

$$C = k_{12}/(\alpha - \beta)\,. \quad (60a)$$

But after the α-phase has been completed the equations degenerate to:

$$Cp = \frac{D}{Va}\,(B e^{-\beta t})\,, \quad (56')$$

$$Q_1 = D B e^{-\beta t}, \quad (59')$$

$$Q_2 = D C e^{-\beta t}. \quad (60')$$

On substitution of these equations in

$$V_d(\omega) = \frac{Q_1 + Q_2}{Cp} \quad (61)$$

the equation becomes:

$$V_d(\omega) = Va\,(B + C)/B. \quad (61a)$$

On substitution of the values for B and C from Eqs. (56b) and (60a) into Eq. (61a), the equations become:

$$V_d(\omega) = Va\,(k_{12} + k_{21} - \beta)/(k_{21} - \beta) \quad (62)$$

or

$$V_d(\omega) = Va\,[1 + (k_{12}/(k_{21} - \beta))]. \quad (62a)$$

By contrast the volume of distribution at equilibrium ($V_d(\text{eq})$) can be estimated by assuming that $\beta = 0$, which occurs when $k_{10} = 0$. Thus,

$$V_d(\text{eq}) = Va\,[1 + (k_{12}/k_{21})]. \quad (63)$$

In order to show why $V_d(\omega)$ is also called V_d (area) we recall from Eq. (57) that:

$$\alpha + \beta = k_{12} + k_{21} + k_{10}$$

and thus:

$$\alpha - k_{10} = k_{12} + k_{21} - \beta. \quad (57a)$$

On substitution of this equation into Eq. (62), we find that:

$$V_d(\omega) = Va\,(\alpha - k_{10})/(k_{21} - \beta).$$

If we multiply the right hand term by β/β, we obtain:

$$V_d(\omega) = Va\,\beta\,(\alpha - k_{10})/\beta\,(k_{21} - \beta)$$

$$V_d(\omega) = Va\,(\alpha\beta - k_{10}\beta)/\beta\,(k_{21} - \beta).$$

We also recall from Eq. (58) that $\alpha\beta = k_{21}k_{10}$.

Thus $V_d(\omega) = Va\,(k_{21}k_{10} - k_{10}\beta)/\beta\,(k_{21} - \beta) = Va\,k_{10}\,(k_{21} - \beta)/\beta\,(k_{21} - \beta)$.

Then $V_d(\omega) = Va\,k_{10}/\beta$ (64)

and

$$V_d(\omega)\,\beta = Va\,k_{10}. \quad (64a)$$

Moreover, after intravenous administration of the drug, the area under the curve (AUC) of the drug concentration in plasma may be calculated from the following equation:

$$AUC = \int_0^\infty C(t)\,dt = \int_0^\infty \frac{D}{Va}\,[A\,e^{-\alpha t} + B\,e^{-\beta t}]\,dt \tag{65}$$

$$AUC = \frac{D}{Va}\left[\frac{A}{\alpha} + \frac{B}{\beta}\right] = \frac{D}{Va\,k_{10}}\,.$$

On rearrangement of this equation and substitution into Eq. (64a) we find:

$$\frac{D}{AUC} = Va\,k_{10} = V_d(\omega)\,\beta. \tag{66}$$

The $V_d(\omega)$ of drug in 3-pool systems can be derived by a similar procedure. For example, the $V_d(\omega)$ for Model IV (see Section III D 2e) in which a drug passes directly from the central compartment to two peripheral compartments is given by the following equation:

$$V_d(\omega) = Va\left(1 + \frac{k_{12}}{k_{21} - \gamma} + \frac{k_{13}}{k_{31} - \gamma}\right). \tag{67}$$

By contrast, the $V_d(\omega)$ for Model I (see Section III D 2a) in which a drug passes from the central compartment to the extracellular space and then to the intracellular space is more complex.

$$V_d(\omega) = Va\left[1 + \frac{k_{12}\,(k_{32} - \gamma) + k_{12}k_{23}}{\gamma^2 - \gamma\,(k_{21} + k_{23} + k_{32}) + k_{32}k_{21}}\right]. \tag{68}$$

b) Volume of Distribution by the Extrapolation Method

In the past many investigators have assumed that the pharmacokinetics of drug distribution can be adequately explained by a 1-compartment model in which $Cp = (D/V_d)\,e^{-kt}$. They have frequently ignored the α-phase of 2-compartment models or the α and β-phases of 3-compartment models and have waited for the distribution phases to be completed before beginning to assay the drug in the plasma. They thus have assumed that the intercept value of the terminal phase is the initial drug concentration in the blood and that the volume of distribution may be estimated from the equation:

$$V_{d\,(ext)} = D/Cp_0(ext).$$

If we assume that the drug was distributed according to a 2-pool system, however, the volume of distribution measured by this method would be estimated from Eqs. (56′) and (56b) as follows:

$$V_{d\,(ext)} = D/(DB/Va) = Va\left(\frac{\alpha - \beta}{k_{21} - \beta}\right). \tag{69}$$

Recalling from Eq. (57) that:

$$\alpha + \beta = k_{12} + k_{21} + k_{10}$$

then

$$\alpha = k_{12} + k_{21} + k_{10} - \beta. \tag{69a}$$

On substituting Eq. (69a) into Eq. (69):

$$V_{d\,(ext)} = Va\left(\frac{k_{12} + k_{21} + k_{10} - 2\beta}{k_{21} - \beta}\right) \tag{70}$$

$$V_{d\,(ext)} = Va\left[1 + \frac{k_{12}}{k_{21} - \beta} + \frac{k_{10} - \beta}{k_{21} - \beta}\right]. \tag{70'}$$

Recalling from Eq. (62a) that:

$$V_{\mathrm{d}}(\omega) = V\mathrm{a}\left[1 + \frac{k_{12}}{k_{21} - \beta}\right]$$

$$\text{then } V_{\mathrm{d(ext)}} = V_{\mathrm{d}}(\omega) + V\mathrm{a}\left(\frac{k_{10} - \beta}{k_{21} - \beta}\right). \tag{70''}$$

Thus it is evident that the extrapolation method overestimates $V_{\mathrm{d}}(\omega)$ when the drug is rapidly metabolized.

c) The Partition of Areas Method

The overestimate of $V_{\mathrm{d}}(\omega)$ by the extrapolation method is due to the amount of drug metabolized during the distribution phases that is not predicted by the β-phase alone. This becomes evident from the following equations. By multiplying both sides of Eq. (69) by β we obtain the apparent clearance of the drug:

$$V_{\mathrm{d(ext)}}\,\beta = \frac{D\beta}{BD/V\mathrm{a}} = \frac{D}{(B/\beta)\,(D/V\mathrm{a})}. \tag{71}$$

Recall from Eqs. (65) and (66) that:

$$V_{\mathrm{d}}(\omega)\,\beta = \frac{D}{AUC} = \frac{D}{[(A/\alpha) + (B/\beta)]\,D/V\mathrm{a}}. \tag{72}$$

Note that the equation for $V_{\mathrm{d(ext)}}\,\beta$ does not contain the term, (A/α), present in the equation for $V_{\mathrm{d}}(\omega)\,\beta$. If we divide Eq. (72) by Eq. (71), we obtain:

$$\frac{V_{\mathrm{d}}(\omega)}{V_{\mathrm{d(ext)}}} = \frac{B/\beta}{(A/\alpha) + (B/\beta)} = X, \tag{73}$$

$$V_{\mathrm{d}}(\omega) = X\,V_{\mathrm{d(ext)}}. \tag{73'}$$

Since $(B/\beta)\,(D/V\mathrm{a})$ is the area under the concentration curve of the β-phase and $[(A/\alpha) + (B/\beta)]\,(D/V\mathrm{a})$ is the total area under the curve, the ratio, X, is the proportion of the total AUC that is associated with the β-phase. The ratio may, therefore, be viewed as the partition of areas.

If we now multiply Eq. (69) by X, we obtain

$$X\,V_{\mathrm{d(ext)}} = \frac{XD}{DB/V\mathrm{a}} \tag{74}$$

which according to Eq. (73') may be transformed to:

$$V_{\mathrm{d}}(\omega) = \frac{XD}{DB/V\mathrm{a}} \tag{75}$$

The partition of areas can be used to estimate the validity of applying the relatively simple mathematics of 1-pool systems to multicompartmental systems. The concept may also be used to assure the investigator that the elimination of the drug associated with the presampling phases of its distribution is negligible. For example, a hypothetical α-phase may be constructed by assuming that the drug is initially restricted to the plasma volume. This calculated drug concentration at zero time is plotted on semilogarithmic paper and a line is then drawn from the hypothetical concentration to the concentration of the drug found at the first assay time. The phases are resolved by the method of residuals and the area of the hypothetical α-phase is calculated. If the area of the hypothetical α-phase is negligible compared with the areas of the known phases, the investigator is assured that most of the drug is metabolized during the time that the drug was being assayed and that the $V_{\mathrm{d}}(\omega)$ obtained from the AUC of the actual data is reasonably valid. It should be emphasized, however, that the hypothetical α-

phase will overestimate any presampling α-phase, and thus the data may still be accurate even when the area of the hypothetical α-phase is significant compared with the total area of the known phases. Nevertheless, the test would suggest that the assays of the drug in blood should have been started sooner. Moreover, it should also be emphasized that the partition of areas method is valid only after the intravenous administration of the drug; it is not valid after oral, intramuscular, intraperitoneal or subcutaneous administration.

5. Area under the Curve Method Applied to Metabolites

As was pointed out by LEVY and GIBALDI (Chapter 59), the AUC method is a powerful tool in evaluating the $V_{\mathrm{d}}(\omega)$, the total body clearance of a drug and the bioavailability of drug preparations. In addition, it is also useful in evaluating various kinetic parameters of drug metabolites provided that their elimination follows first order kinetics.

Equation (29) of LEVY and GIBALDI (see Chapter 59, Section D) states that:

$$M_{\mathrm{x}} = \frac{k_{\mathrm{mx}} D}{k_{\mathrm{x}} - k} (\mathrm{e}^{-kt} - \mathrm{e}^{-k_{\mathrm{x}} t}) \tag{76}$$

in which M_{x} is the amount of metabolite present in the body at any given time, k_{mx} is the rate constant for the formation of the metabolite from the parent drug, k_{x} is the rate constant for the elimination of the metabolite from the body and k is the sum of the rate constants of all of the processes by which the parent drug is eliminated from the body including its conversion to metabolite X. We may convert the equation from a quantity of M_{x} to its concentration by dividing both sides of the equation by the volume of distribution of the metabolite.

Thus,

$$C_{\mathrm{x}} = \left(\frac{D}{V_{\mathrm{d[x]}}}\right) \left(\frac{k_{\mathrm{mx}}}{k_{\mathrm{x}} - k}\right) (\mathrm{e}^{-kt} - \mathrm{e}^{-k_{\mathrm{x}} t}) . \tag{77}$$

On the integration of the Eq. (77),

$$C_{\mathrm{x}t} = \left(\frac{D}{V_{\mathrm{d[x]}}}\right) \left(\frac{k_{\mathrm{mx}}}{k_{\mathrm{x}} - k}\right) \left(\frac{1}{k} - \frac{1}{k_{\mathrm{x}}}\right) = \left(\frac{D}{V_{\mathrm{d[x]}}}\right) \left(\frac{k_{\mathrm{mx}}}{k\, k_{\mathrm{x}}}\right) . \tag{78}$$

But k_{mx}/k is the fraction of the dose of the parent drug that is converted to the metabolite. Thus,

$$AUC = \frac{fD}{V_{\mathrm{d[x]}}\, k_{\mathrm{x}}} . \tag{78'}$$

If it is known that all of the metabolites of the drug are excreted into urine and that metabolite X is converted to relatively few metabolites, the value of "f" can be calculated by measuring the total amounts of metabolites X and its subsequent metabolites and relating them to the dose. The clearance of metabolite X may then be calculated from the equation,

$$\text{Clearance of } M_{\mathrm{x}} = V_{\mathrm{d[x]}} k_{\mathrm{x}} = \frac{fD}{AUC} . \tag{79}$$

If k_{x} can be estimated from the data, then the $V_{\mathrm{d[x]}}$ may be calculated:

$$V_{\mathrm{d[x]}} = \frac{fD}{k_{\mathrm{x}}\, AUC} . \tag{80}$$

Alternatively, if metabolite X is converted to a number of different metabolites that are difficult to assay in the urine, a dose (D_{m}) of the metabolite X may be administered and the area under the plasma curve measured.

$$AUC' = \frac{D_{\mathrm{m}}}{V_{\mathrm{d[x]'}}\, k_{\mathrm{x'}}} . \tag{81}$$

If the $V_{d[x]'} = V_{d[x]}$ and $k_{x'} = k_x$, then "f" may be calculated from the equation:

$$\frac{AUC}{AUC'} = \frac{\left(\frac{fD}{V_{d[x]}\,k_x}\right)}{\left(\frac{D_m}{V_{d[x]}\,k_x}\right)} = \frac{fD}{D_m}\,. \tag{82}$$

As pointed out by LEVY and GIBALDI (Chapter 59, Section D), when changes in the plasma level of the metabolite parallel changes in the level of the drug during the final phase, the value $k_x \gg k$. In this situation the final phase cannot be used to estimate k_x. Nevertheless, the area under the curve still provides a valid method for estimating $fD/V_{d[x]}k_x$. Therefore, when f is known from the pattern of urinary drug metabolites, the clearance of the metabolite formed from the parent drug may be compared with that found after the intravenous administration of the metabolite in order to determine whether the presence of the parent drug alters the disposition of the metabolite.

The equations illustrate the importance of using specific analytical methods for the drug when assessing the area under the plasma curve. When total radioactivity in the plasma is measured, the area under the curve of the total radioactivity is the sum of the areas under the curve for the drug and each of its metabolites present in the plasma. For example, the area under the curve for the drug and its primary metabolites would represent:

$$AUC_{(\text{total})} = D\left[\frac{1}{V_d\,k} + \frac{f_x}{V_{d[x]}\,k_x} + \frac{f_y}{V_{d[y]}\,k_y} + \frac{f_m}{V_{d[A]}\,k_m}\right]. \tag{83}$$

Thus, even a minor metabolite that persists long after the parent drug has been eliminated from the body may provide misleading results when the assay method for the drug is not specific.

6. Effects of Reversible Binding of Drugs to Macromolecules on Pharmacokinetic Parameters

a) General

As has been discussed in Section IV B, the reversible binding of drugs to plasma and tissue proteins and other macromolecules slows their rates of diffusion into and out of tissues and ultimately affects their volumes of distribution in the body. The binding of drugs in extravascular organs tends to increase their volume of distribution by decreasing their concentration in blood and thereby slows their elimination by the kidney, liver, and lungs. However, the binding of drugs to plasma proteins may either hasten or retard the elimination of drugs, depending on the mechanism of elimination (see Part 1, Chapter 10 by KEEN, 1971, and Part 3, Chapter 69 by SHAND et al., 1974). For example, the binding of drugs to plasma proteins lowers their unbound concentrations in blood and thereby decreases their rate of elimination by glomerular filtration and by inefficient transport systems in the kidney. Moreover, binding to plasma proteins would decrease the rate of elimination of lipid-soluble drugs that diffuse rapidly from glomerular filtrate back into blood, even though they are rapidly transported by active transport systems. In addition, the binding of drugs to plasma proteins would also decrease the rate of drug metabolism by relatively inactive enzyme systems in liver, especially when the Michaelis constant of the enzyme is high. However, when drugs are metabolized rapidly by liver enzymes or polar drugs are rapidly eliminated by transport systems in the kidney, their clearance by the organ may approach the blood flow rate through the organ. Under these conditions, the binding to plasma proteins and to blood cells may enhance the rate of elimination of drugs. Thus, the efficiency of the mechanisms by which a drug is eliminated determines whether the binding of drugs to plasma proteins should be considered as storage depots or as transport mechanisms (GILLETTE, 1973).

b) Effects of Reversible Binding on the Clearance of Drugs by a Single Organ

As has been discussed in Section III C 3 and in Part 3 of Chapter 69 by SHAND et al. (1974), the relationship between the blood flow rate and the clearance of drugs by active secretory transport systems or drug metabolizing enzyme systems in a given tissue may be estimated from the following equations:

$$\text{Clearance (total)} = \frac{\dot{Q}_{010}\, p\, \dot{Q}_{30}}{\dot{Q}_{010} + p\, \dot{Q}_{30}}. \tag{36}$$

From Eqs. (22), (25), and (32′), it is found that

$$\dot{Q}_{30} = f_{b1} f_{b2} k_{met}.$$

Let $p f_{b2} k_{met} = \dot{Q}_m$. On substitution into Eq. (36),

$$\text{Clearance (total)} = \frac{\dot{Q}_{010}\, \dot{Q}_m\, f_{b1}}{\dot{Q}_{010} + \dot{Q}_m\, f_{b1}} \tag{84}$$

in which:

$$f_{b1} = \frac{C_f}{C_f + \text{bound drug}} \tag{85}$$

If we assume that the drug in the blood is bound to only one kind of site on serum albumin, the equation for the concentration of bound drug is:

$$\text{bound drug} = \frac{Bta\, K_a\, C_f}{1 + K_a\, C_f} \tag{86}$$

in which Bta is the concentration of albumin in the blood plasma, C_f is the concentration of unbound drug in the plasma and K_a is the equilibrium constant of association.

The fraction of the total concentration of drug that is unbound is thus,

$$f_{b1} = \frac{C_f}{C_f + \dfrac{Bta\, K_a\, C_f}{1 + K_a\, C_f}} = \frac{1 + K_a\, C_f}{1 + K_a\, C_f + Bta\, K_a}. \tag{87}$$

Note that this equation for f_{b1} may be substituted for the f_{b1} in Eq. (84).

Thus,

$$\text{clearance (total)} = \frac{\dot{Q}_{010}\, \dot{Q}_m\, (1 + K_a\, C_f)/(1 + K_a\, C_f + Bta\, K_a)}{\dot{Q}_{010} + \dot{Q}_m\, (1 + K_a\, C_f)/(1 + K_a\, C_f + Bta\, K_a)}.$$

When $1 + K_a C_f \gg Bta\, K_a$, the concentration of the unbound form of the drug will be much greater than the bound form. The value of $Bta\, K_a$ may then be neglected and the equation degenerates to:

$$\text{clearance (total) (Phase 1)} = \frac{\dot{Q}_{010}\, \dot{Q}_m}{\dot{Q}_{010} + \dot{Q}_m}. \tag{88′}$$

The clearance of the drug will thus remain constant until the plasma level of the drug declines to a concentration at which the concentration of the bound form becomes important.

By contrast, when $K_a C_f \ll 1$, the equation degenerates to:

$$\text{clearance (total) (Phase 2)} = \frac{\dot{Q}_{010}\, \dot{Q}_m\, (1/(1 + Bta\, K_a))}{\dot{Q}_{010} + \dot{Q}_m\, (1/(1 + Bta\, K_a))}. \tag{88″}$$

The clearance will, therefore, be constant after the concentration of free drug decreases to a level at which it becomes directly proportional to the total concentration of the drug.

The clearance may also be expressed in terms of the plasma concentration of unbound drug:

$$\text{clearance (free)} = \frac{\text{clearance (total)}}{f_{b1}} . \tag{89}$$

On substitution of Eq. (89) into Eq. (88) and rearrangement:

$$\text{clearance (free)} = \frac{\dot{Q}_{010}\,\dot{Q}_{m}}{\dot{Q}_{010} + \dot{Q}_{m}\,(1 + K_{a}\,C_{f})/(1 + K_{a}\,C_{f} + Bta\,K_{a})} \tag{90}$$

When $\dot{Q}_{010} \ll p\dot{Q}_{30}$, Eq. (90) degenerates to:

$$\text{clearance (free)} = \dot{Q}_{010}/f_{b1} = \dot{Q}_{010}\,(1 + K_{a}C_{f} + Bta\,K_{a})/(1 + K_{a}C_{f}). \tag{90'}$$

In this situation the clearance increases as the unbound fraction of drug in plasma decreases; in fact clearance (free) then is larger than the blood flow rate through the organ.

By contrast, when $\dot{Q}_{010} \gg p\dot{Q}_{30}$, Eq. (90) degenerates to:

$$\text{clearance (free)} = \dot{Q}_{m}. \tag{90'}$$

Thus, in this situation the clearance (free) remains constant regardless of the binding to plasma albumin.

c) Effects of Reversible Binding on the Volume of Distribution of Drugs

When a drug is bound to plasma proteins, blood cells, and other tissues in the body, its distribution at any given time may be described by the following equation:

$$Q_{t} = C_{f}\,V_{f} + \textstyle\sum_{i=1}^{n} Bai\;V_{abi} + \sum_{i=1}^{n} Bvi\;V_{bbi} \tag{91}$$

where Q_t is the total amount of drug, $C_f\,V_f$ is the concentration times the volume of distribution of the unbound drug, $\sum_{i=1}^{n} Bai\;V_{abi}$ is the sum of the products, concentrations times the volume of distribution of drug bound to each component in blood, and $\sum_{i=1}^{n} Bbi\;V_{bbi}$ is the sum of the products, concentration times the volumes of distribution of drug bound to each component in other tissues of the body. If one assumes that the drug is bound only to a single site on plasma albumin and to a single site on a protein in muscle, the equation degenerates to:

$$Q_{t} = C_{f}\,V_{f} + Ba\,Vab + Bb\,Vbb. \tag{92}$$

The volume of distribution of unbound drug (V_{df}) as determined from the total amount of drug and its unbound concentration then becomes:

$$V_{df} = \frac{Q_t}{C_f} = V_f + \frac{Ba\;Vab}{C_f} + \frac{Bb\;Vbb}{C_f} . \tag{93}$$

But

$$Ba = \frac{Bta\;K_a\,C_f}{1 + K_a\,C_f}\,, \tag{93a}$$

and

$$B_b = \frac{Btb\;K_b\,C_f}{1 + K_b\,C_f} \tag{93b}$$

in which Bta is the concentration of serum albumin, K_a is the association equilibrium constant for the formation of the albumin-drug complex, Btb is the concentration of the binding protein in muscle and K_b is the association equilibrium constant for this binding protein-drug complex. On substitution of these values into Eq. (93) we obtain:

$$V_{\mathrm{df}} = V_{\mathrm{f}} + \frac{V\mathrm{ab}\, B\mathrm{ta}\, K_{\mathrm{a}}}{1 + K_{\mathrm{a}}\, C_{\mathrm{f}}} + \frac{V\mathrm{bb}\, B\mathrm{tb}\, K_{\mathrm{b}}}{1 + K_{\mathrm{b}}\, C_{\mathrm{f}}} \tag{94}$$

Thus, the volume of distribution of the unbound drug, V_{df}, changes with the concentration of the drug. At very high concentrations, the binding sites may be saturated and the concentration of unbound drug may exceed the concentration of the bound form; indeed the concentration of unbound form may be so high that the values of $V\mathrm{ab}B\mathrm{ta}K_{\mathrm{a}}/(1 + K_{\mathrm{a}}C_{\mathrm{f}})$ and $V\mathrm{bb}B\mathrm{tb}K_{\mathrm{b}}/(1 + K_{\mathrm{b}}C_{\mathrm{f}})$ are negligible compared with V_{f} and thus

$$V_{\mathrm{df}} \cong V_{\mathrm{f}}. \tag{94'}$$

As the drug is eliminated from the body, V_{df} increases as the concentration of the drug passes through the transition phases. When both $K_{\mathrm{a}}C_{\mathrm{f}}$ and $K_{\mathrm{b}}C_{\mathrm{f}}$ decline below about 0.1, however, the transition phases are completed and the equation becomes:

$$V_{\mathrm{df}} \cong V_{\mathrm{f}} + V\mathrm{ab}\, B\mathrm{ta}\, K_{\mathrm{a}} + V_{\mathrm{bb}} B\mathrm{tb} K_{\mathrm{b}}. \tag{94''}$$

Thus, at low drug concentrations the volume of distribution remains constant as the drug is eliminated from the body.

d) *Effects of Reversible Binding on Rate Constants of Elimination in Linear, First Order Models*

Since the reversible binding of drugs to macromolecules may alter not only the clearance of drugs, but also their apparent volumes of distribution, reversible binding will obviously affect their rates of elimination from the body. However, the effects of changes in the fraction of unbound drug in plasma (f_{b1}) depend on the activities of the drug metabolizing enzymes and blood flow rates through the organs containing the enzymes. To illustrate these effects, let us consider several theoretical models in which the drug is eliminated by a single organ in the central compartment.

(1) In Model *A* the drug is bound solely to the plasma proteins, the unbound drug is restricted to the central compartment, and the value of $K_{\mathrm{a}}C_{\mathrm{t}}$ for the drug is less than 0.1. When $K_{\mathrm{a}}C_{\mathrm{f}} < 0.1$, less than 10% of the binding sites of the protein in plasma will be occupied by the drug.

In this situation:

$$\text{clearance (free)} = \frac{\dot{Q}_{010}\, \dot{Q}_{\mathrm{m}}}{\dot{Q}_{010} + \dot{Q}_{\mathrm{m}}/(1 + B\mathrm{ta}\, K_{\mathrm{a}})}, \tag{95}$$

$$V_{\mathrm{daf}} = V_{\mathrm{af}} + V\mathrm{ab}\, B\mathrm{ta}\, K_{\mathrm{a}}, \tag{96}$$

$$k_{10} = \frac{\text{clearance (free)}}{V_{\mathrm{daf}}} = \frac{\left(\dfrac{\dot{Q}_{010}\, \dot{Q}_{\mathrm{m}}}{\dot{Q}_{010} + \dot{Q}_{\mathrm{m}}/(1 + B\mathrm{ta}\, K_{\mathrm{a}})}\right)}{V_{\mathrm{af}} + V\mathrm{ab}\, B\mathrm{ta}\, K_{\mathrm{a}}}. \tag{97}$$

Model A1:

$$\dot{Q}_{010} \ll f_{\mathrm{b1}}\dot{Q}_{\mathrm{m}}$$

$$k_{10} = \frac{\dot{Q}_{010}\,(1 + B\mathrm{ta}\, K_{\mathrm{a}})}{V_{\mathrm{af}} + V\mathrm{ab}\, B\mathrm{ta} K_{\mathrm{a}}}. \tag{97'}$$

Let $V\mathrm{ab} = f_{\mathrm{v}} V_{\mathrm{af}}$

$$k_{10} = \frac{\dot{Q}_{010}\,(1 + B\mathrm{ta}\, K_{\mathrm{a}})}{V_{\mathrm{af}} + f_{\mathrm{v}}\, V_{\mathrm{af}}\, B\mathrm{ta} K_{\mathrm{a}}} = \frac{\dot{Q}_{010}}{V_{\mathrm{af}}}\left(\frac{1 + B\mathrm{ta}\, K_{\mathrm{a}}}{1 + f_{\mathrm{v}}\, B\mathrm{ta}\, K_{\mathrm{a}}}\right). \tag{97''}$$

Since the volume of distribution of albumin is usually less than the volume of the central compartment,

$$\frac{1 + Bta\,K_a}{1 + f_v\,Bta\,K_a} > 1. \tag{97a}$$

Therefore,

$$\frac{\dot{Q}_{010}}{V_{af}}\left(\frac{1 + Bta\,K_a}{1 + f_v\,Bta\,K_a}\right) > \frac{\dot{Q}_{010}}{V_{af}}. \tag{97b}$$

Thus, reversible binding of drugs tends to increase the value of k_{10}, when $\dot{Q}_{010}$ is rate-limiting. Under these conditions, the plasma albumin may be viewed as a carrier protein.

Model A2:

$$\dot{Q}_{010} \gg f_{b1}\dot{Q}_m \tag{97''}$$

$$k_{10} = \frac{\dot{Q}_m}{V_{af} + Vab\,Bta\,K_a}.$$

But

$$\frac{\dot{Q}_m}{V_{af} + Vab\,Bta\,K_a} < \frac{\dot{Q}_m}{V_{af}}. \tag{97'''a}$$

Thus, reversible binding of drugs to plasma albumin tends to decrease the value of k_{10} when $f_{b1}\dot{Q}_m$ is rate-limiting and the albumin may be viewed as a storage protein.

(2) In Model *B*, the drug is bound solely to plasma albumin, the unbound drug distributes with body water, and the value of $K_a C_f$ for the drug is less than 0.01. In this situation:

$$\text{clearance (free)} = \frac{\dot{Q}_{010}\,\dot{Q}_m}{\dot{Q}_{010} + \dot{Q}_m/(1 + Bta\,K_a)}. \tag{98}$$

But there are at least 2 compartments in this model.

Initially,

$$V_{daf} = V_{af} + V_{ab}BtaK_a. \tag{99}$$

Ultimately

$$V_{df(eq)} = V_f + VabBtaK_a. \tag{100}$$

But from Eqs. (63) and (62a)

$$V_{df(eq)} = V_{daf}\left(1 + \frac{k_{12}}{k_{21}}\right), \tag{100'}$$

$$V_{df}(\omega) = V_{daf}\left(1 + \frac{k_{12}}{k_{21} - \beta}\right), \tag{100''}$$

in which

$$k_{12} = \frac{\dot{Q}_{121f}}{V_{af} + Vab\,Bta\,K_a}, \tag{100'a}$$

$$k_{21} = \frac{\dot{Q}_{121f}}{V_{tf}}, \tag{100'b}$$

where V_{tf} is the volume of the peripheral compartment and $\dot{Q}_{121f}$ is the clearance constant for the passage of unbound drug into and out of the peripheral compartment.

Model B1:

$$\dot{Q}_{010} \ll f_{b1}\dot{Q}_m.$$

The value of k_{10} will be identical to that in Model A1. But,

$$\beta = \frac{\text{clearance (free)}}{V_{\mathrm{df}(\omega)}} = \frac{\text{clearance (free)}}{V_{\mathrm{daf}}\left(1 + \left(\frac{k_{12}}{k_{21} - \beta}\right)\right)} = \frac{k_{10}}{1 + \left(\frac{k_{12}}{k_{21} - \beta}\right)}. \quad (101)$$

Whenever $\dot{Q}_{010} \ll f_{b1}\dot{Q}_{m}$, however, the value of $k_{21} - \beta$ will be much less than k_{21}. Therefore, $V_{\mathrm{df(eq)}}$ cannot be used to calculate β and the effects of reversible binding of drugs to serum albumin on the value of β are not obvious. Reversible binding of drugs will tend to increase k_{10} and decrease k_{12}, which will tend to increase the value of β, but as β increases, the value of $(k_{21} - \beta)$ decreases and thus the increase is not as much as might be expected from the value of $V_{\mathrm{df(eq)}}$ (see Table 3). Under these conditions plasma albumin may be viewed as a carrier protein that aids the mobilization of the drug from the tissue to the liver. However, as the bound fraction is increased, $f_{b1}\dot{Q}_{m}$ is decreased until $\dot{Q}_{010}$ is no longer rate-limiting, and thus β may be decreased when the drug is very highly bound (Table 3).

Table 3. *Effect of reversible binding of drugs to plasma albumin on various parameters of Model B1*[a]

Fraction bound	0.0	0.5	0.75	0.90	0.95	0.98	0.99
Extraction ratio (total) %	90	81.8	69.2	47.4	31.0	15.3	8.26
k_{10} (min^{-1})	0.096	0.128	0.141	0.118	0.083	0.0429	0.0236
k_{12} (min^{-1})	0.056	0.041	0.027	0.013	0.00703	0.00295	0.00150
k_{21} (min^{-1})	0.028	0.028	0.028	0.028	0.028	0.028	0.028
α (min^{-1})	0.163	0.176	0.172	0.134	0.0932	0.0497	0.0331
t-1/2 (α)	4.25	3.94	4.03	5.17	7.44	14.0	20.9
β (min^{-1})	0.0164	0.0203	0.0228	0.0246	0.0250	0.0242	0.01991
t-1/2 (β) (min)	42.2	34.2	30.4	28.2	27.8	28.7	34.8
A	0.921	0.951	0.965	0.969	0.956	0.850	0.391
B	0.079	0.049	0.035	0.031	0.044	0.150	0.609
Cp_0 (free)[b] μM	60	43.9	28.6	14.0	7.53	3.16	1.61
DB/V_{daf}[b] μM	4.72	2.17	0.987	0.439	0.334	0.474	0.979
$(B/\beta)/AUC$	0.459	0.312	0.213	0.150	0.147	0.266	0.722
V_{daf} (l)	15.0	20.5	31.5	64.5	119.5	284.5	559.5
$V_{\mathrm{df(eq)}}$ (l)	45.00	50.5	61.5	94.5	149.5	314.5	589.5
$V_{\mathrm{df}}(\omega)$ (l)	87.62	129.2	194.1	308.6	397.6	504.6	663.4

[a] The values were calculated for a 2-pool system in which $\dot{Q}_{010} = 1.6$ l/min, $\dot{Q}_{121}f = 0.84$ l/min, $\dot{Q}_{m} = 14.6$ l/min, $V_{ab} = 5.5$ l, $V_{a} = 15$ l and $V_{b} = 30$ l.
[b] The dose was assumed to be 0.9 mmoles.

Model B2:

$$\dot{Q}_{010} \gg f_{b1}\dot{Q}_{m}.$$

The value of k_{10} will be identical to that in Model A2. Moreover,

$$\beta = \frac{\text{clearance (free)}}{V_{\mathrm{df}(\omega)}} = \frac{\text{clearance (free)}}{V_{\mathrm{daf}}\left(1 + \frac{k_{12}}{k_{21} - \beta}\right)} = \frac{k_{10}}{1 + \frac{k_{12}}{k_{21} - \beta}}. \quad (102)$$

However, in this model β may be insignificant compared with k_{21} and thus

$$\beta = \frac{\dot{Q}_{m}}{V_{\mathrm{df(eq)}}} = \frac{\dot{Q}_{m}}{V_{f} + V_{ab}\,Bta\,k_{a}}. \quad (102\mathrm{a})$$

Thus, reversible binding of drugs to plasma albumin will decrease β by increasing the volume of distribution, but its effects on β will be less than its effects on k_{10} (see Table 4).

As shown in Table 4, reversible binding of a drug to plasma albumin would decrease β by only 5% when 50% of the drug in plasma is bound and by only 20% when 75% of it is bound. Indeed the decrease in β becomes important only when this kind of drug is bound to albumin by 90% or more. Thus, the general impression that marked changes in clearance (total) caused by changes either in the protein concentration of plasma or in K_a (as occurs in drug displacement) result in significant changes in β is not justified.

Table 4. *Effect of reversible binding to plasma albumin on various parameters of Model B2*[a]

Fraction bound	0	0.5	0.75	0.90	0.95	0.98	0.99
Extraction ratio (total) %	10.0	5.26	2.70	1.10	0.552	0.222	0.111
k_{10} (min^{-1}) $\times 10^3$	10.7	8.22	5.49	2.73	1.48	0.697	0.317
k_{12} (min^{-1})	0.056	0.041	0.027	0.013	0.00703	0.00330	0.00150
k_{21} (min^{-1})	0.028	0.028	0.028	0.028	0.028	0.028	0.028
α (min^{-1})	0.0914	0.0741	0.0575	0.0419	0.0353	0.0314	0.0295
t-1/2 (α) (min)	7.58	9.35	12	16.5	19.6	22.1	23.5
β (min^{-1}) $\times 10^3$	3.27	3.11	2.6	1.82	1.17	0.622	0.301
t-1/2 (β) (min)	212	223	259	381	591	1114	2302
A	0.719	0.649	0.538	0.347	0.215	0.110	0.052
B	0.281	0.351	0.462	0.653	0.785	0.890	0.948
Cp_0 (free)[b] μM	60	43.9	28.6	14.0	7.53	3.16	1.61
DB/V_{daf}[b] μM	16.9	15.4	13.2	9.11	5.91	2.82	1.52
$(B/\beta)/AUC$	0.916	0.928	0.949	0.977	0.991	0.998	0.999
V_{daf} (l)	15	20.5	31.5	64.5	119.5	284.5	559.5
$V_{df(eq)}$ (l)	45	50.5	61.5	94.5	145.5	314.5	589.5
$V_{df}(\omega)$ (l)	49.0	54.2	64.7	96.6	150.8	318.8	589.8

[a] The values were calculated for a 2-pool system in which $\dot{Q}_m = 0.1778$ l/min and the other values were those stated in Table 3.
[b] The dose was assumed to be 0.9 mmoles.

It is also noteworthy that in this model seemingly marked changes in f_{b1} do not always result in significant changes in the unbound concentration of drug (Table 4) (GILLETTE, 1973; see Part 3, Chapter 69 by SHAND et al., 1974).

(3) In Model C, the drug is bound solely to a muscle protein, the unbound drug distributes with body water and the value of $K_b C_f$ for the drug is less than 0.1. In this model,

$$\text{clearance (total)} = \text{clearance (free)} = \frac{\dot{Q}_{010}\,\dot{Q}_m}{\dot{Q}_{010} + \dot{Q}_m}\,. \tag{103}$$

V_{daf} will be affected by the reversible binding only to the extent that the muscle compartment contributes to the central compartment in degenerate multi-compartment systems (see Section III D 3). If we assume that the contribution of the drug in muscle to the central compartment is negligible, then

$$V_{da} = V_{daf} = V\text{a}, \tag{104}$$

$$V_{df(eq)} = V_f + V\text{bb}\,B\text{tb}\,K_b\,, \tag{105}$$

$$V_{df}(\omega) = V\text{a}\left(1 + \frac{k_{12}}{k_{21} - \beta}\right), \tag{106}$$

in which

$$k_{12} = \frac{\dot{Q}_{121f}}{V\text{a}}\,, \tag{106a}$$

$$k_{21} = \frac{\dot{Q}_{121f}}{V_{tf} + V\text{bb}\,B\text{ta}\,K_b}\,. \tag{106b}$$

Model C1: $\dot{Q}_{010} \ll f_{b1}\dot{Q}_m$

$$k_{10} = \frac{\text{clearance (free)}}{V_a} = \frac{\dot{Q}_{010}}{V_a}, \tag{107}$$

$$\beta = \frac{\text{clearance (free)}}{V_d(\omega)} = \frac{\dot{Q}_{010}}{V_a\left(1 + \left(\frac{k_{12}}{k_{21} - \beta}\right)\right)} = \frac{k_{10}}{1 + \left(\frac{k_{12}}{k_{21} - \beta}\right)} \tag{108}$$

As in Model B1, $(k_{21} - \beta) \ll k_{21}$ and thus $V_{df} \ll V_d(\omega)$. The value of V_{df} therefore cannot be used to estimate β directly. However, it should be noted that:

$$k_{21} = \frac{\dot{Q}_{121f}}{V_d\,(\text{tissue f})} = \frac{\dot{Q}_{121f}}{V_{tf} + V_{bb}\,B_{tb}\,K_b} \tag{108a}$$

in which $\dot{Q}_{121f}$ is the clearance value for the passage of unbound drug into and out of the peripheral compartment and V_{tf} is the volume of the compartment.

Table 5. *Effect of reversible binding to a muscle protein on various parameters of Model C1*[a]

Fraction bound	0	0.5	0.75	0.90	0.95	0.98	0.99
Extraction ratio (total) %	90	90	90	90	90	90	90
k_{10} (min^{-1})	0.096	0.096	0.096	0.096	0.096	0.096	0.096
k_{12} (min^{-1})	0.056	0.056	0.056	0.056	0.056	0.056	0.056
k_{21} (min^{-1}) $\times 10^3$	28	14	7	2.8	1.4	0.56	0.28
α (min^{-1})	0.164	0.157	0.155	0.153	0.153	0.152	0.152
t-1/2 (α) (min)	4.23	4.41	4.47	4.53	4.53	4.56	4.56
β (min^{-1}) $\times 10^3$	16.4	8.54	4.35	1.76	0.88	0.353	0.177
t-1/2 (β) (min)	42.2	81.2	160	395	787	1962	3922
A	0.921	0.963	0.982	0.993	0.997	0.999	0.999
B $\times 10^3$	78.6	36.7	17.7	6.9	3	1.4	0.68
Cp_0 (free)[b] μM	60	60	60	60	60	60	60
DB/V_{daf}[b] μM	4.72	2.20	1.06	0.41	0.18	0.084	0.041
$(B/\beta)/AUC$	0.459	0.413	0.390	0.377	0.373	0.370	0.369
V_{daf} (l)	15	15	15	15	15	15	15
$V_{df(eq)}$ (l)	45	75	135	315	615	1515	3015
$V_{df}(\omega)$ (l)	87.6	169	316	805	1619	4062	8130

[a] The values were calculated for a 2-pool system in which $\dot{Q}_{010} = 1.6$ l/min, $\dot{Q}_{121}f = 0.84$ l/min, $\dot{Q}_m = 14.6$ l/min, $V_a = 15$ l, $V_b = 30$ l, $V_{bb} = 30$ l.
[b] The dose was assumed to be 0.9 mmoles.

Thus, the binding of drug in the muscle will decrease β (Table 5). Indeed, the binding of drugs to macromolecules in the peripheral compartment may increase the $V_d(\omega)$ to an extent that the biological half-life of the drug may be several hours or days, even when its rate of metabolism is limited mainly by the blood flow rate through the liver. To illustrate the relationship between tissue binding and "first-pass" kinetics, I have calculated the half-lives of a series of hypothetical drugs that have different kinetic volumes of distribution but are completely cleared as they pass through the liver. As shown in Table 6, the minimum half-life of a drug that is eliminated solely by the liver and has a $V_d(\omega)$/kg of 20 l/kg in man would be approximately 10 h. Moreover, if 50% of the drug were cleared as it passed through the liver, the biological half-life would be about 20 h. Thus, it would be a mistake to assume that the clearance of a drug is always slow when it has a long biological half-life. In this connection it is noteworthy that the volumes of distribution and biological half-lives of a number of drugs currently on the market are consistent with the view that they are eliminated by a first-pass kinetic mechanism. For example, the clearance of nortriptyline approaches the liver blood flow in rats and is greater than 50% of the blood flow rate in many patients (ALEXANDERSON et al., 1973). Because nortriptyline is highly bound to plasma proteins, these findings suggest that plasma proteins would serve as a carrier of the drug between the tissue stores and the liver.

Table 6. *Relationships between $V_{d(k)}$, liver plasma flow, extraction ratios and t 1/2*[a]

$V_d(\omega)$ (l)	$\frac{V_d(\omega)}{kg}$ (l/kg)	Extraction ratios 100% t 1/2 (hr)	50% t 1/2 (hr)	25% t 1/2 (hr)	10% t 1/2 (hr)
5.0	0.071	0.036	0.072	0.14	0.36
15	0.214	0.11	0.22	0.44	1.1
50	0.71	0.36	0.72	1.4	3.6
70	1.0	0.5	1.0	2.0	5
140	2.0	1.0	2.0	4.0	10
280	4.0	2.0	4.0	8.0	20
700	10.0	5.0	10.0	20.0	50
1400	20.0	10.0	20.0	40.0	100
2800	40.0	20.0	40.0	80.0	200

[a] For these calculations, it was assumed that the blood flow rate was 1.6 l/min in a 70 kg man, which is equivalent to 96.0 l/hr and to 1.37 l/kg/hr.

Model C2:

$$\dot{Q}_{010} \gg f_{b1} \dot{Q}_m$$

$$k_{10} = \frac{\text{clearance (free)}}{Va} = \frac{\dot{Q}_m}{Va}, \tag{109}$$

$$\beta = \frac{\text{clearance (free)}}{V_d(\omega)} \quad \frac{\dot{Q}_m}{Va\left(1 + \left(\frac{k_{12}}{k_{21} - \beta}\right)\right)} = \frac{k_{10}}{1 + \left(\frac{k_{12}}{k_{21} - \beta}\right)}. \tag{110}$$

As in Model B2, however, β may be negligible compared with k_{21} and thus:

$$\beta = \frac{\dot{Q}_m}{V_d\,(\text{eq})} = \frac{\dot{Q}_m}{V_f + Vbb\; Btb\; k_b}. \tag{110a}$$

Thus, the reversible binding of a drug to a macromolecule in the peripheral compartment decreases the value of β (Table 7). Indeed, when the drug is bound to the same extent in plasma and tissues, the effect of the reversible binding in tissue on β would be greater than that in blood plasma because $Vab \ll Vbb$ (cf., Tables 4 and 7).

(4) Integrated equations for Models A, B, and C: Model A is a 1-compartment system in which $k_{ba}\, C_f$ is < 0.1 and thus the integrated equation is either:

$$Cp\ (\text{total}) = \frac{D}{V_{da}}\, e^{-k_{10} t} \tag{111}$$

or

$$Cp\ (\text{free}) = f_{b1}\, Cp\ (\text{total}) = \frac{f_{b1} D}{V_{da}}\, e^{-k_{10} t} \tag{112}$$

Table 7. *Effect of reversible binding of drugs to a muscle protein on various parameters of Model C2*[a]

Fraction bound	0	0.5	0.75	0.90	0.95	0.98	0.99
Extraction ratio (total) %	10	10	10	10	10	10	10
kk_{10} (min^{-1})	0.01067	0.01067	0.01067	0.01067	0.01067	0.01067	0.01067
k_{12} (min^{-1})	0.056	0.056	0.056	0.056	0.056	0.056	0.056
k_{21} (min^{-1}) $\times 10^3$	28.20	14.0	7.00	2.80	1.40	0.560	0.280
α (min^{-1})	0.0914	0.0788	0.0726	0.0690	0.0678	0.0671	0.0669
t-1/2 (α) (min)	7.58	8.80	9.54	10.04	10.22	10.32	10.36
β (min^{-1}) $\times 10^3$	3.27	1.90	1.03	0.433	0.220	0.0890	0.0446
t-1/2 (β) (min)	212	366	674	1602	3149	7791	15527
A	0.719	0.843	0.917	0.965	0.983	0.993	0.996
B $\times 10^3$	281	157	83.4	34.5	17.4	7.03	3.52
Cp_0 (free) μM	60	60	60	60	60	60	60
DB/V_{daf} μM	16.9	9.42	5.00	2.07	1.04	0.422	0.211
$(B/\beta)/AUC$	0.916	0.886	0.865	0.851	0.846	0.842	0.841
V_{daf} (l)	15	15	15	15	15	15	15
$V_{df(eq)}$ (l)	45	75	135	315	615	1515	3015
$V_d(\omega)$ (l)	49.0	84.4	156	370	727	1798	3584

[a] The values were calculated for a 2-pool system in which $\dot{Q}_m = 0.1778$ l/min and the other values were those stated in Table 5.

in which

$$k_{10} = \frac{\text{clearance (total)}}{V_{da}} = \frac{\text{clearance (free)}}{V_{daf}}, \tag{112a}$$

$$k_{10} = \frac{\text{clearance (free)}}{V_{daf}} = \left(\frac{\dfrac{\dot{Q}_{010}\,\dot{Q}_m}{\dot{Q}_{010} + \dot{Q}_m\,(f_{b1})}}{V_{af} + V_{ab}\,B\text{ta}\,K_a} \right), \tag{112b}$$

$$k_{10} = \frac{\text{clearance (free)}}{V_{daf}} = \left(\frac{\dfrac{\dot{Q}_{010}\,\dot{Q}_m}{\dot{Q}_{010} + \dot{Q}_m\,(f_{b1})}}{V_{af} + V_{ab}\left(\dfrac{1}{f_{b1}} - 1\right)} \right). \tag{112c}$$

Models B and C are 2-compartment systems in which $K_a C_f$ and $K_b C_f$ are less than 0.1 and the drug is eliminated from the central compartment.

Thus, their integrated equations are either:

$$Cp\text{ (total)} = \frac{D}{V_{da}}\,[A\,e^{-\alpha t} + B\,e^{-\beta t}] \tag{113}$$

or

$$Cp\text{ (free)} = f_{b1}\,Cp\text{ (total)} = \frac{f_{b1}\,D}{V_{da}}\,[A\,e^{-\alpha t} + B\,e^{-\beta t}] \tag{114}$$

where

$$A = (\alpha - k_{21})/(\alpha - \beta), \tag{114a}$$

$$B = (k_{21} - \beta)/(\alpha - \beta), \tag{114b}$$

$$\alpha = 1/2\,[k_{12} + k_{21} + k_{10} + \sqrt{k_{12} + k_{21} + k_{10}^2 - 4 k_{21} k_{10}}], \tag{114c}$$

$$\beta = 1/2\,[k_{12} + k_{21} + k_{10} - \sqrt{k_{12} + k_{21} + k_{10}^2 - 4 k_{21} k_{10}}], \tag{114d}$$

in which

k_{10} is the same as that in Model A.

$$k_{12} = \frac{\dot{Q}_{121}f}{V_{af} + V_{ab}\,B\text{ta}\,K_a} = \frac{\dot{Q}_{121}f}{V_{af} + V_{ab}\left(\dfrac{1}{f_{b1}} - 1\right)}. \tag{114e}$$

$$k_{21} = \frac{\dot{Q}_{121}\mathrm{f}}{V_{\mathrm{bf}} + V_{\mathrm{bb}}\,Bta\,K_{\mathrm{b}}} = \frac{\dot{Q}_{121}\,\mathrm{f}}{V_{\mathrm{bf}} + V_{\mathrm{bb}}\left(\frac{1}{ft} - 1\right)} \tag{114f}$$

where ft is the unbound fraction of the drug in the tissue.

On substitution of appropriate values into the equations for k_{10}, k_{12}, and k_{21}, the values of A, B, α, and β may be calculated in order to simulate various conditions (Tables 3 through 5 and 7).

e) Non-Linear Binding

When the value of $K_{\mathrm{a}} C_{\mathrm{f}}$ or $K_{\mathrm{b}} C_{\mathrm{f}}$ exceeds 0.1, the system no longer follows first order kinetics and thus the usual pharmacokinetic equations for the distribution of drugs cannot be used. WAGNER (1971) has derived equations for several 1 and 2-compartment models in which $K_{\mathrm{a}} C_{\mathrm{f}}$ and $K_{\mathrm{b}} C_{\mathrm{f}}$ exceed 0.1, but in these models he has assumed that the volume of distribution of the binding sites is identical to that of the unbound drug. Since the volume of the central compartment seldom equals the blood volume or the volume of distribution of serum proteins, his terms for the concentration of the binding sites should be modified by the ratio, $V_{\mathrm{ab}}/V\mathrm{a} = f_{\mathrm{v}}$. In the symbolism used in this paper, WAGNER'S equation for Model A then becomes:

$$\ln \frac{C_{\mathrm{f}}}{C_{\mathrm{f0}}} + f_{\mathrm{v}} K_{\mathrm{a}}\;Bta \quad \ln\left(\frac{C_{\mathrm{f}}\,(1 + K_{\mathrm{a}}\,C_{\mathrm{f0}})}{C_{\mathrm{f0}}\,(1 + k_{\mathrm{a}}\,C_{\mathrm{f}})}\right)$$

$$+ f_{\mathrm{v}} K_{\mathrm{a}}\;Bta\left(\frac{C_{\mathrm{f0}} - C_{\mathrm{f}}}{(1 + K_{\mathrm{a}}\,C_{\mathrm{f}})\,(1 + K_{\mathrm{a}}\,C_{\mathrm{f0}})}\right) = -\,kt. \tag{115}$$

The equation may be transformed as follows:

$$\frac{\ln C_{\mathrm{f}}/C_{\mathrm{f0}}}{1 + (K_{\mathrm{a}}\,Bta\,f_{\mathrm{v}})} + \frac{K_{\mathrm{a}}\,Bta\,f_{\mathrm{v}}}{1 + (K_{\mathrm{a}}\,Bta\,f_{\mathrm{v}})} \ln\left(\frac{C_{\mathrm{f}}\,(1 + K_{\mathrm{a}}\,C_{\mathrm{f0}})}{C_{\mathrm{f0}}\,(1 + K_{\mathrm{a}}\,C_{\mathrm{f}})}\right)$$

$$+\left(\frac{K_{\mathrm{a}}\,Bta\,f_{\mathrm{v}}}{1 + (K_{\mathrm{a}}\,Bta\,f_{\mathrm{v}})}\right)\left(\frac{C_{\mathrm{f0}} - C_{\mathrm{f}}}{(1 + K_{\mathrm{a}}\,C_{\mathrm{f}})\,(1 + K_{\mathrm{a}}\,C_{\mathrm{f0}})}\right) = \frac{-\,kt}{1 + (K_{\mathrm{a}}\,Bta\,f_{\mathrm{v}})} = -\,k_{10}t \tag{116}$$

in which C_{f} is the initial concentration of the unbound drug in plasma and $k =$ clearance (free)/Va. The equation is also valid for the degeneration forms of Models B2 and C2.

If C_{f} is plotted semilogarithmically the curve frequently may appear to be virtually identical to that of a 2-pool, first order system. However, the concentration at which the rapid rate of C_{f} decline changes to the slower rate is independent of dose in Model A, but is directly proportional to the dose in a 2-pool linear model. Moreover, when the level C_{f} drops below the concentration at which $K_{\mathrm{a}} C_{\mathrm{f}} < 0.1$, the equation degenerates to a first order system in which the rate constant is k_{10}. Thus, the extrapolation of C_{f} values associated with the apparent β-phase to zero time will give a value for the apparent β that is less dependent on the dose than is the β value of the 2-compartment linear model. In both models, however, the area under the curve of C_{f} will be directly proportional to the dose.

After the repeated administration of the drug, the transition phase of Model A will also occur at the same concentration as that of the initial dose, whereas the transition phase of the 2-pool linear model will occur at a higher concentration than initially. It should be pointed out, however, that the failure to detect an increase in the total drug concentration in a Model A system after repeated administration of the drug does not preclude the possibility that repeated drug administration results in the accumulation of the drug in a peripheral compartment.

f) Area under the Curves (AUC) of Total Drug Concentration in Plasma

In most drug disposition studies, the total drug concentration is measured without determining the proportion of the concentration that is bound to plasma proteins. Thus, the calculation of AUC will represent the sum of the areas under the curve for both the unbound and bound forms of the drug, i.e., AUC (total) = AUC (free) + AUC (bound). When the value of $K_a C_f$ is less than 0.1, changes in C_f will be proportional to changes in the concentration of the bound form (C_{ab}); thus the AUC of C_f will be proportional to the AUC of C_{ab} and therefore the AUC of Cp (total). However, when $K_a C_f > 0.1$, changes in C_f will not be proportional to changes in C_{ab} and therefore the area under the curve for the total drug concentration in plasma will not be proportional to the AUC (free). Thus, the value of D/AUC (total) will increase as the dose is increased until doses are used that result in $K_a C_f$ values greater than 10 are obtained. If larger doses are administered, and $K_a C_f > 10$, C_{ab} will remain constant until C_f decreases to $10/K_a$. Therefore, as the dose is further increased, the D/AUC (total) will approach a constant value, which equals V_a.

It is thus important to know 1) the fraction of unbound drug (f_{b1}) in drug plasma at any given time, 2) any changes in the unbound fraction as the drug is eliminated from the body, and 3) the association constant. If the fraction of unbound drug does not change, then AUC (total) values should reflect the elimination rate constant. If a drug is bound in a peripheral compartment, however, the calculated values for the apparent volume of distribution may be inappropriate when the $K_b C_f$ values for the peripheral binding exceed 0.1.

If drugs in plasma were bound only to albumin and only to one site on the albumin molecule, it would be easy to calculate when the $K_a C_f$ values would exceed 0.1. Since the concentration of albumin in plasma is about 0.6 mM, about 10% of the binding sites would be occupied when the concentration of bound drug reach about 0.06 mM. However, this simple calculation will seldom be valid because some drugs are bound only to several low affinity sites, and thus the $K_a C_f$ values may not be greater than 0.1 even when the concentration of bound drug greatly exceeds 0.06 mM Moreover, some drugs are bound not only to high affinity sites but also to several low affinity sites on albumin and other major proteins in plasma, and thus the concentration of bound drug may exceed 0.06 mM and still the $K_a C_f$ values of the high affinity sites might not exceed 0.1.

On the other hand, blood plasma contains several minor proteins that have such high affinities for certain endogenous hormones, including corticosterone, testosterone, estrogen, and thyroxine, that they are thought to act as specialized transport carriers in blood. Substances that resemble these hormones may, therefore, be bound to a certain extent by these proteins. Moreover, individual differences in the binding of drugs may occur when the drug evokes the formation of antibodies in the γ-globulin fraction. For this reason the $K_a C_f$ values of these minor components may exceed 0.1 even when the concentration of bound protein is much less than 0.06 mM. Indeed, with the development of more sensitive analytical methods, it is becoming increasingly difficult to assess the importance of the trace amount of drug in plasma that is usually found during the last phases of drug elimination. A small amount of drug bound to high affinity sites on minor components in the plasma may be neglected. But the trace amounts in plasma may reflect a storage compartment in a peripheral tissue that would lead to the accumulation of the drug after repeated administration. Thus, it is becoming evident that pharmacokinetic studies in man should be correlated with similar studies in animals in order to gain insights into the mechanisms of drug disposition and the validity of AUC calculations.

Appendix 1

Definitions of symbols used in this chapter:
Equations (1) through (50)

dQ_b/dt = The rate of change in the total amount of drug in the capillary compartment.

$Q_b = Q_1$ = The total amount of drug in the capillary compartment at any given time.

V_b = The volume of the capillary compartment.

C_b = The total drug concentration of drug in the capillary compartment of an organ.

f_{b1} = The fraction of the total amount of drug in the capillary compartment that is not bound to nondiffusible components in blood.

f_{b2} = The fraction of the unbound drug in the capillary compartment that is not ionized.

k_{be} = The initial rate of diffusion of the drug from the capillary compartment to the extracellular compartment divided by the concentration of unbound, nonionized drug in the capillary compartment. It has the dimensions of volume per minute and may be viewed as a clearance constant. It is the sum of the clearance value (k_{b1}) of the unbound, nonionized drug passing through the endothelial cells of the capillary wall, the clearance value (k_{b2}) of the unbound, nonionized drug passing through the intercellular pores of the capillary wall, and the product of the clearance value of the ionized drug passing through the intercellular pores times the ratio of the ionized to the non-ionized forms of the drug in blood. Ordinarily, it nearly equals k_{eb}.

dQ_e/dt = The rate of change in the total amount of drug in the extracellular compartment of an organ.

$Q_e = Q_2$ = The total amount of drug in the extracellular compartment at any given time.

V_e = The volume of the extracellular compartment.

C_e = The total drug concentration in the extracellular compartment.

f_{e1} = The fraction of the total amount of drug in the extracellular compartment that is not bound to nondiffusible components in the extracellular compartment.

f_{e2} = The fraction of unbound drug in the extracellular compartment that is nonionized.

k_{eb} = The effective clearance value for the diffusion of the drug from the extracellular compartment to the capillary compartment relative to the unbound, nonionized form of the drug in the extracellular compartment. It nearly equals k_{be} except when the diffusion of the ionized form of the drug through the intercellular pores is significant and the ratio of the ionized to the nonionized forms of the drug in the extracellular compartment markedly differs from the ratio in blood.

k_{ei} = The effective clearance value of the drug passing from the extracellular compartment to the intracellular compartment relative to the unbound, nonionized drug in the extracellular compartment.

dQ_i/dt = The rate of change in the total amount of drug in the intracellular compartment of an organ.

$Q_i = Q_3$ = The total amount of drug in the intracellular compartment at any given time.

V_i = The volume of the intracellular compartment.

f_{i1} = The fraction of the drug in the intracellular compartment that is not bound to nondiffusible intracellular components.

f_{i2} = The fraction of unbound drug in the intracellular compartment that is nonionized.

k_{ie} = The effective clearance value for the drug passing from the intracellular to the extracellular compartment relative to the unbound nonionized form of the drug in the intracellular compartment. It equals k_{ei}.

$k_{met} = Ek/K_m$ = The rate of metabolism (or elimination) of the drug in the intracellular compartment divided by the unbound concentration of the drug in the intracellular compartment. E is the amount of enzyme in the tissue, k is the rate of metabolism per unit of enzyme-substrate complex and K_m is the Michaelis constant.

$V_{d(eq)} = Q_{tot}/C_b$ = The volume of distribution of the drug at equilibrium, that is the total amount of drug in the tissue (or body) divided by the total concentration of drug in blood (or plasma) at equilibrium.

Q_1 $= Q_b$.

$Q_1 k_{12} = C_b f_{b1} f_{b2} k_{be} = Q_b f_{b1} f_{b2} k_{be}/V_b = Q_b k_{be}/V_1$ = The rate of passage of a drug from the capillary space to the extracellular space of a tissue at any given time.

$k_{12} = f_{b1} f_{b2} k_{be}/V_b = k_{be}/V_1$ = The first order rate constant for this passage.

Q_2 $= Q_e$.

$Q_2 k_{21} = C_e f_{e1} f_{e2} k_{eb} = Q_e f_{e1} f_{e2} k_{eb}/Ve = Q_e k_{eb}/V_2$ = The rate of passage of a drug from the extracellular space to the capillary space at any given time.

$k_{21} = k_{e1} f_{e2} k_{eb}/Ve = k_{eb}/V_2$ = The first order rate constant for this passage.

$Q_2 k_{23} = C_e f_{e1} f_{e2} k_{ei} = Q_e f_{e1} f_{e2} k_{ei}/Ve = Q_e k_{ei}/V_2$ = The rate of passage of a drug from the extracellular space to the intracellular space of a tissue at any given time.

$k_{23} = f_{e1} f_{e2} k_{ei}/V_e = k_{ei}/V_2$ = The first order rate constant for this passage.

Q_3 $= Q_i$.

$Q_3 k_{32} = C_i f_{i1} f_{i2} k_{ie} = Q_i f_{i1} f_{i2} k_{ie}/V_i = Q_i k_{ie}/V_3$ = The rate of passage of a drug from the intracellular space to the extracellular space of a tissue at any given time.

$k_{32} = f_{i1} f_{i2} k_{ie}/V_i = k_{ie}/V_3$ = The first order rate constant for this passage.

$Q_3 k_{30} = C_i f_{i1} k_{met} = Q_i f_{i1} k_{met}/V_i = Q_i k_{met}/V_3 f_{i2}$ = The rate of metabolism of a drug in the intracellular space of a tissue at any given time.

$k_{30} = f_{i1} k_{met}/V_i = Ek/K_m V_i$ = The first order rate constant for the metabolism of the drug.

$\frac{dQ_1}{dt}$ = The rate of change in the amount of drug in the capillary space of a tissue.

$\frac{dQ_2}{dt}$ = The rate of change in the amount of drug in the extracellular space of a tissue.

$\frac{dQ_3}{dt}$ = The rate of change in the amount of drug in the intracellular space of a tissue.

$Q_0 k_{01}$ = The rate at which a drug in arterial blood enters the capillary space of a tissue at any given time.

$Q_1 k_{10}$ = The rate at which a drug leaves the capillary space by way of the veins.

$\dot{Q}_{010}$ = The blood flow rate into and out of the capillary space of a tissue.

$C\mathrm{a}$ = The total concentration of drug in the arterial blood flowing into the capillary space.

C_{v} = The total concentration of drug in the venous blood flowing out of the capillary space.

$\dot{Q}_{010}(C\mathrm{a} - C\mathrm{v})$ = The rate at which a drug leaves or enters the blood flowing through the capillary space at any given time.

$x = f_{\mathrm{e}1} f_{\mathrm{e}2} / f_{\mathrm{b}1} f_{\mathrm{b}2}$ = The ratio of the fraction of the drug existing as the unbound, nonionized form in the extracellular space of the tissue to that fraction in the blood in the capillary space. Thus x is the inverse of the ratio of the total drug concentrations in the extracellular space and the capillary blood at equilibrium.

$y = f_{\mathrm{i}1} f_{\mathrm{i}2} / f_{\mathrm{b}1} f_{\mathrm{b}2}$ = The ratio of the fraction of drug existing as the unbound, nonionized form in the intracellular space to that fraction in the blood of the capillary space. Thus, y is the inverse of the ratio of the total drug concentrations in the intracellular space and the capillary blood at equilibrium.

$\dot{Q}_{121} = f_{\mathrm{b}1} f_{\mathrm{b}2} k_{\mathrm{be}}$ = The effective clearance value for the passage of a drug across the capillary wall of a tissue multiplied by the fraction of drug existing in the unbound, nonionized form in the blood. Thus, $\dot{Q}_{121}$ is k_{be} normalized to C_{b}, the total drug concentration in blood of the capillary space.

$\dot{Q}_{232} = f_{\mathrm{b}1} f_{\mathrm{b}2} k_{\mathrm{ei}}$ = The effective clearance value for the passage of a drug across the plasma membranes of tissue cells multiplied by the fraction of drug existing in the unbound, nonionized form in the blood. Thus, $\dot{Q}_{232}$ is k_{ei} normalized to C_{b}, the total drug concentration in blood of the capillary space.

Q_{t} = The total amount of drug in the tissue at any given time.

$V_1 = V_{\mathrm{b}} / f_{\mathrm{b}1} f_{\mathrm{b}2}$ = The volume of the capillary space divided by the fraction of the drug in the unbound, nonionized form in blood in the capillary space. Thus, V_1 equals the total amount of drug in the capillary space divided by the concentration of the unbound, nonionized form of the drug in the capillary space and is the effective volume of the capillary space.

$V_2 = V_{\mathrm{e}} / f_{\mathrm{e}1} f_{\mathrm{e}2}$ = The volume of the extracellular space divided by the fraction of the drug in the extracellular space that is in its unbound, nonionized form. Thus, V_2 is the effective volume of the extracellular space.

$V_3 = V_{\mathrm{i}} / f_{\mathrm{i}1} f_{\mathrm{i}2}$ = The volume of the intracellular space divided by the fraction of the drug in the intracellular space that is in its unbound, nonionized form. Thus, V_3 is the effective volume of the intracellular space.

$p = (y/f_{i2})(C_i/C_v) = (f_{i1}/f_{b1}f_{b2})(C_i/C_v)$ = The ratio of the concentration of unbound drug in the intracellular space of a tissue to the concentration of unbound, nonionized drug in venous blood under steady-state conditions.

$(Ca - C_v)/Ca$ = The extraction ratio (ER), the fraction of the total amount of drug in blood that is taken up by the tissue at any given time. It is a constant only under steady-state conditions.

$\dot{Q}_{010}(Ca - C_v)/Ca$ = The total clearance of the drug by the tissue. It too is a constant only under steady-state conditions.

Equations (51) through (118)

Cp = The concentration of drug in plasma at any given time.

D = The dose of drug administered intravenously.

Va = The apparent volume of the central compartment of a 2 or more pool system.

$V_d(\omega)$ = The total amount of drug in the body at any given time after the distribution phases are complete divided by the drug concentration in the plasma at that time. In a 2-pool system, $V_d(\omega) = V_{d(area)} = V_{d(\beta)} = V_{d(slope)} = V_{d(remainder)}$.

$V_{d(eq)}$ = The total amount of drug in the body divided by the drug concentration in plasma at equilibrium.

$V_{d(ext)}$ = The volume of distribution of a drug obtained by extrapolating a logarithmic plot of the drug concentrations in plasma at various times during the terminal phase of drug disposition back to zero time and dividing the resultant value into the amount of drug administered intravenously.

k_{12} = The first order rate constant for the passage of drug from the central compartment to the peripheral compartment of a 2-pool system.

k_{21} = The first order rate constant for the passage of drug from the peripheral compartment to the central compartment of a 2-pool system.

$k_{12}Va$ $= k_{21}(V_{d(eq)} - Va) = k_{21}V_{d(t)(eq)}$.

k_{10} = The first order rate constant for the metabolism or the excretion of a drug from the central compartment of a 2-pool system.

α = The first order rate constant of the α-phase.

β = The first order rate constant of the β-phase. In a 2-pool system the β-phase is the terminal phase.

$Q_1 = CpVa$ = The total amount of drug at any given time in the central compartment of a 2-pool system.

Q_2 = The total amount of drug at any given time in the peripheral compartment of a 2-pool system.

A $= (\alpha - k_{21})/(\alpha - \beta)$.

B $= (k_{21} - \beta)/(\alpha - \beta)$.

$\alpha + \beta$ $= k_{12} + k_{21} + k_{10}$.

$\alpha\beta$ $= k_{21}k_{10}$.

AUC = The area under a rectilinear plot of the drug concentration in plasma against time. The dimensions of AUC are (concentration) (time).

AUC_{iv} $= \frac{D}{Va}\left(\frac{A}{\alpha} + \frac{B}{\beta}\right) = \frac{D}{V}\left(\frac{\beta A + \alpha B}{\alpha\beta}\right)$: The area under the curve after the intravenous administration of a drug into the central compartment of a 2-pool system in which the drug is eliminated solely from the central compartment.

$\beta \mathrm{A} + \alpha \mathrm{B} = \frac{\beta(\alpha - k_{21}) + \alpha(k_{21} - \beta)}{(\alpha - \beta)} = \frac{k_{21}(\alpha - \beta)}{\alpha - \beta} = k_{21}$.

$\dot{Q}_{\text{m}} = p f_{\text{b}2} k_{\text{met}} = (f_{\text{il}}/f_{\text{b}1})(C_{\text{i}}/C_{\text{v}})\, k_{\text{met}} = p\dot{Q}_{30}/f_{\text{b}1}$ = The rate of metabolism of a drug in a tissue divided by the concentration of unbound drug in venous blood.

$\text{Clearance(total)} = \frac{\dot{Q}_{010}\,\dot{Q}_{\text{m}}\,f_{\text{b}1}}{\dot{Q}_{010} + \dot{Q}_{\text{m}}\,f_{\text{b}1}}$

$\text{Clearance (free)} = \frac{\dot{Q}_{010}\,\dot{Q}_{\text{m}}}{\dot{Q}_{010} + \dot{Q}_{\text{m}}\,f_{\text{b}1}}$

C_{f} = The concentration of unbound drug in the body at any given time.

Bta = Concentration of a set of binding sites on proteins in blood.

K_{a} = The equilibrium constant of association of a drug with a set of binding sites on proteins in blood.

Q_{t} = The total amount of drug in the body at any given time.

V_{f} = Volume occupied by unbound drug.

Ba = The concentration of drug bound to a set of binding sites in blood.

Vab = The volume of distribution of the set of binding sites in blood.

Bb = The concentration of drug bound to a set of binding sites in muscle.

Btb = The concentration of a set of binding sites in muscle.

K_{b} = The equilibrium constant of association of a drug with a set of binding sites in muscle.

Vbb = The volume of distribution of the set of binding sites in muscle.

$V_{\text{df(eq)}}$ = The total amount of drug in the body divided by the concentration of unbound drug in plasma at equilibrium. It is the volume of distribution (free) of the drug at equilibrium.

V_{daf} = The total amount of drug in the central compartment divided by the concentration of unbound drug in the plasma. Thus, V_{daf} is the effective volume of the central compartment.

V_{af} = The amount of unbound drug in the central compartment divided by the concentration of unbound drug in the plasma.

$\dot{Q}_{121\text{f}}$ = The rate of passage of the drug from the central compartment to a peripheral compartment divided by the concentration of unbound drug in the central compartment. It also is the rate of passage of the drug from the peripheral compartment to the central compartment divided by the concentration of unbound drug in the peripheral compartment.

V_{tf} = The volume of the peripheral compartment occupied by the unbound drug.

References

1. ALEXANDERSON, B., BORGA, O., ALVAN, G.: The availability of orally administered nortriptyline. Europ. J. clin. Pharmacol. 5, 181—185 (1973).

2. Brodie, B. B., Mark, L. C., Papper, E. M., Lief, P. A., Bernstein, E., Rovenstine, E. A.: The fate of thiopental in man and a method for its estimation in biological material. J. Pharmacol. exp. Ther. **98**, 85—96 (1950).
3. Butler, T. C.: Duration of action of drugs as affected by tissue distribution. Proc. First International Pharmacological Meeting, Vol. 6, pp. 193—205. Oxford: Pergamon Press 1962.
4. Digenis, G., Swintosky, J. V.: Drug latentiation. Handbuch der experimentellen. Pharmakologie, Vol. XXVIII. Concepts in biochemical pharmacology, Part 3, p. 86—112. Berlin-Heidelberg-New York: Springer 1974.
5. Garattini, S., Marcucci, F., Mussini, E.: Drugs probably acting through metabolites. Handbuch der experimentellen Pharmakologie, Vol. XXVIII. Concepts in biochemical pharmacology, Part 3, pp. 113—129. Berlin-Heidelberg-New York: Springer 1974.
6. Gibaldi, M., Nagashima, R., Levy, G.: Relationship between drug concentration in plasma or serum and amount of drug in the body. J. Pharm. Sci. **58**, 193—197 (1969).
7. Gillette, J. R.: Overview of drug-protein binding. N. Y. Acad. Sci. **226**, 6—17 (1973).
8. Hogben, C. A. M.: Biological membranes and their passage by drugs. Handbuch der experimentellen Pharmakologie, Vol. XXVIII. Concepts in biochemical pharmacology, Part 1, pp. 1—8. Berlin-Heidelberg-New York: Springer 1971.
9. Keen, P.: Effect of binding to plasma proteins on the distribution, activity and elimination of drugs. Handbuch der experimentellen Pharmakologie, Vol. XXVIII. Concepts in biochemical pharmacology, Part 1, pp. 213—233. Berlin-Heidelberg-New York: Springer 1971.
10. Mapleson, W. W.: An electric analogue for uptake and exchange of inert gases and other agents. J. appl. Phys. 18, 197—204 (1963).
11. Plaa, G. L.: The enterohepatic circulation. Handbuch der experimentellen Pharmakologie, Vol. XXVIII. Concepts in biochemical pharmacology, Part 3, pp. 130—149. Berlin-Heidelberg-New York: Springer 1974.
12. Rall, D. P.: Drug entry into brain and cerebrospinal fluid. Handbuch der experimentellen Pharmakologie, Vol. XXVIII. Concepts in biochemical pharmacology, Part 1, pp. 240—248. Berlin-Heidelberg-New York: Springer 1971.
13. Riegelman, S., Loo, J., Rowland, M.: Concept of a volume of distribution and possible errors in evaluation of this parameter. J. pharm. Sci. *128*, 128—133 (1968).
14. Riggs, D. S.: The mathematical approach to physiological problems, pp. 212—214. Baltimore: Williams and Wilkins 1963.
15. Rowland, M., Benet, L. Z., Graham, G. G.: Clearance concepts in pharmacokinetics. J. Pharmacokinetics Biopharmaceutics 1, 123—136 (1973).
16. Schou, J.: Subcutaneous and intramuscular injection of drugs. Handbuch der experimentellen Pharmakologie, Vol. XXVIII. Concepts in biochemical pharmacology, Part 1, pp. 47—66. Berlin-Heidelberg-New York: Springer 1971.
17. Shand, D. G., Mitchell, J. R., Oates, J. A.: Pharmacokinetic drug interactions. Handbuch der experimentellen Pharmakologie, Vol. XXVIII. Concepts in biochemical pharmacology, Part 3, pp. 272—314. Berlin-Heidelberg-New York: Springer 1974.
18. Smith, R. L.: Excretion of drugs in bile. Handbuch der experimentellen Pharmakologie Vol. XXVIII. Concepts in biochemical pharmacology, Part 1, pp. 354—389. Berlin-Heidelberg-New York: Springer 1971.
19. Taylor, P. W.: Fast reactions — flow and relaxation methods. Methods in pharmacology, Vol. 2. Physical methods, pp. 351—380. New York: Appleton-Century-Crofts 1972.
20. Wagner, J. G.: Biopharmaceutics and relevant pharmacokinetics, p. 309. Hamilton, Ill.: Drug Intelligence Publications 1971.
21. Weiner, I. M.: Excretion of drugs by the kidney. Handbuch der experimentellen Pharmakologie, Vol. XXVIII. Concepts in biochemical pharmacology, Part 1, pp. 328—353. Berlin-Heidelberg-New York: Springer 1971.

Chapter 61

Drug Latentiation

G. A. Digenis and J. V. Swintosky

I. Introduction

In 1958, Albert first introduced the term "prodrug". The term describes pharmacologic agents that undergo biotransformation producing substances which combine with receptors. Harper (1959, 1962) has used the term "latentiation" to describe the alteration of drugs to derivatives from which parent drugs are regenerated by enzymatic attack. The term "drug latentiation" will be used here in a broader sense. It will include conversion of drugs to derivatives (prodrugs) from which the parent drugs are released by either enzymatic or non-enzymatic processes *in vivo.*

In the systematic search for a prodrug, the researcher has to a) conceive structures that will undergo cleavage in the environmental conditions of the body; b) consider the intrinsic physico-chemical properties desired for the appropriate solubilities, partition coefficients, and dissolution rates; and c) select prodrug structures that have the ability to diffuse across membranes in a relatively non-toxic form and to yield the active drug in the body with sufficient rapidity to exert the desired pharmacologic or therapeutic effect. To fulfill the above requirements one must 1. chemically synthesize and characterize the new compounds (prodrugs), 2. check for proper absorption characteristics, 3. develop suitable pharmacologic, biochemical, analytical, and microbiological tests to define dose-time-action profiles, and 4. determine similarities of response between prodrugs and their parent drugs. Obviously, the systematic search for a new prodrug often requires a multidisciplinary effort.

Before discussing the various approaches to drug latentiation, some distinct attainable differences between properties of the parent drug and its prodrugs should be enumerated. These differences may be reflected by the following changes in biopharmaceutical characteristics:

1. Differences in Rates of Absorption Resulting from Conversion of a Polar to a Less Polar Molecule. The decrease in the polar character might result in increased lipid solubility; thus, a more favorable transport across lipoid barriers. Furthermore, the storage of the prodrug in body fat might be increased; thus leading, in some instances, to prolongation of pharmacologic activity.

2. Differences in Rates of Elimination and Metabolism, Both Processes being Closely Interrelated. If the drug, for example, possesses a functional group that is mainly involved in the detoxification process of the compound, the rate at which the drug is eliminated from the body may be reduced – hence, its action prolonged by chemically masking this group. In a similar situation, the conversion of an alcohol to a more polar species, which is more readily excretable (e.g., its glucuronic acid conjugate) (Williams, 1959), is not possible until the original functional group has been unmasked. Thus, elimination of the prodrug from the body may be slower than that of the parent compound.

3. Differences in Time of Onset and Duration of Action. If prodrug formation involves the modification of a functional group essential for activity, then the product may be ineffective until the original functional group has been regenerated. The period for onset of action may therefore be longer than it is for the original drug. Duration of effect may be correspondingly greater, since metabolic processes may not operate as rapidly upon the prodrug.

4. Differences in Potency. If the prodrug crosses lipid barriers more easily and is more resistant to metabolic changes, then it may concentrate at its target area more efficiently than the parent drug. If conversion to the active form occurs fairly rapidly at the site of action, potency of the derivative may be enhanced. Conversely, if transformation is slow, then decreased potency would be expected but possibly coupled with increased duration of action.

5. Differences in Irritation, Taste, Dissolution Rates, Chemical Stability, and Physical Forms. It is clear that these differences are of great practical importance in clinical pharmacology and pharmaceutics, especially as applied to drug product development.

This chapter will discuss various approaches to drug latentiation with emphasis on approaches that have yielded active compounds of therapeutic potential. Representative examples will be included during the discussions and any inadvertent omissions of references should not be taken as a lack of appreciation of the authors for other investigators' work.

II. Chemical and Structural Consideration

An investigator involved with drug product formulation or development begins with an apparently useful pharmacologic agent and tries to prepare it in a dosage form or forms that can be used in drug therapy. An effective therapeutic agent should have reproducible pharmacologic action from dose to dose, an adequate duration of action, a prolonged drug release and stability and reproducibility of action of the dosage form after prolonged storage. Oral products should be tasteless or have an acceptable taste. Drug product design of a useful pharmacologic agent provides several alternatives for attaining pharmaceutical and therapeutic objectives. Among these are: 1. appropriate formulation of the original drug into usual dosage forms, 2. alteration of the physical characteristics of the drug such as particle size, crystal form, salt form, etc., without changing the basic chemical structure of the active moiety, and 3. chemical derivatization of the drug into an inactive transport form (prodrug) which would release the active drug under the influence of biological fluids or enzymes. In this case, the investigator is obliged to consider the physical, chemical, and enzymatic factors influencing the cleavage of the prodrug and the availability of active drug to the body. Emphasis will be given to some of these matters.

For the sake of illustration, let us consider the active drug A which is chemically converted to its transport form A – C. (C here represents the main portion of the transport moiety and will be referred to as the "carrier" group throughout this chapter.)

$$A + C \longrightarrow A — C.$$

The nature of the carrier Group C is obviously expected to influence the physicochemical properties of drug A (such as solubility, rate of dissolution, distribution between hydrophilic and hydrophobic sites, absorption rates, etc.) and its rate of enzymatic and non-enzymatic cleavage. On the other hand, the rate of release of the active drug from the prodrug, A – C, will be influenced by the electronic, steric, and configurational characteristics of the carrier moiety, C. A few examples will help to illustrate this point.

The release of an alcohol from a benzoate ester under bimolecular basic hydrolysis conditions depends, among other factors, on the electronic character of the substituent at the *para* position of the phenyl group as in *1*.

$$X-C_6H_4-\overset{O}{\overset{||}{C}}-OR' + H_2O \longrightarrow X-C_6H_4-\overset{O}{\overset{||}{C}}-O^{\ominus} + R\,OH$$

1a = X = $-NH_2$
1b = X = $-NO_2$

R′OH = DRUG.

Electron releasing substituents, such as *para*-NH_2 in *1a*, retard the reaction, while electron withdrawing groups, such as *para*-NO_2 in *1b*, accelerate it (Ingold, 1969). Thus, if a benzoate ester prodrug is to be made from an alcohol drug (R′OH), the electronic character of the carrier group, C, in this case the *p*-substituted benzoic acid moiety, has to be considered. In direct agreement with the above, it was found by Levine and Clark (1955) that compound *2a* is hydrolyzed *in vitro* by human serum much more rapidly than its derivative *2b*.

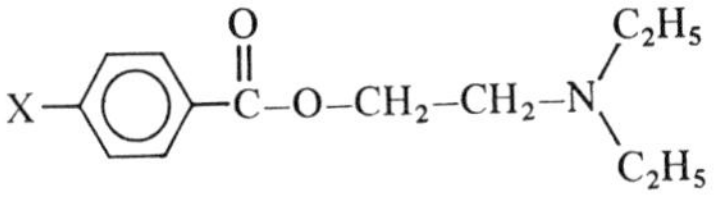

$$X-C_6H_4-\overset{O}{\overset{||}{C}}-O-CH_2-CH_2-N(C_2H_5)_2$$

2a = X = $-F$
2b = X = $-NH_2$.

It should be emphasized at this early point in the discussion that *in vivo* enzyme-mediated reactions are more complicated and at this time less predictable than those observed *in vitro*. As a consequence, the relative ease of prodrug ester cleavage, and subsequent release of the carrier group *in vivo*, may or may not be related to the rate of cleavage observed in *in vitro* tests. This is illustrated by some of the studies initiated and reported upon by Swintosky and co-workers (1966, 1968) especially in their extensive studies on the carbonate prodrug esters of some well known drugs. For example, in connection with their studies on acetaminophen (*3*), efforts were made to prepare carbonate prodrug esters (*4*) of this compound and follow their enzymatic and non-enzymatic cleavage *in vitro*. Data from these studies, as reported by Dittert et al. (1968a, b), are illustrated in Table 1.

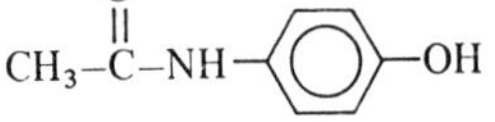

$$CH_3-\overset{O}{\overset{||}{C}}-NH-C_6H_4-OH$$

3

$$CH_3-\overset{O}{\overset{||}{C}}-NH-C_6H_4-O-\overset{O}{\overset{||}{C}}-R$$

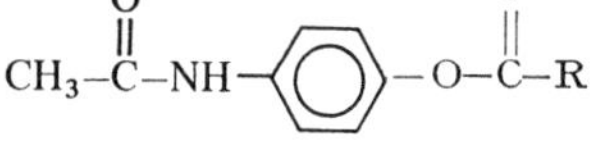

4

As can be seen from Table 1, the enzymatic cleavage of the acetaminophen carbonates by pseudocholinesterases of human plasma increases with increasing chain length up to six carbons. This agrees with the findings of Hofstee (1959) who studied the effect of acid chain lengths on the chymotrypsin-catalyzed hydrolysis of fatty acid esters. He found that seven carbons was the optimum acid chain length for this cleavage reaction. Thus, the enzyme system in human plasma responsible for ester cleavage appears to "fit" a six or seven carbon alcohol chain better than it does other chain lengths in the acetaminophen carbonate esters. In contrast, the rate of nonenzymatic cleavage of acetaminophen carbonate esters decreases with increasing chain lengths of its alcohol moiety to four carbons and thereafter remains essentially constant at a minimum value. Similar results were obtained by GORDON et al. (1948) while studying the effect of alcohol chain length on the ammonia-catalyzed cleavage of acetate esters.

Table 1. *Enzymatic and nonenzymatic cleavage behavior of acetaminophen carbonates, pH 7.4 and 37° C*

$CH_3-C(=O)-NH-C_6H_4-OCOOR$

	R	t 1/2 2% liquid human plasma (min)	Relative enzymatic cleavage rate[a]	t 1/2-buffer[b] (h)	Relative nonenzymatic cleavage rate[a]
4a	$-CH_3$ (methyl)	180	1	150	1.00
4b	$-C_2H_5$ (ethyl)	45	4	200	0.75
4c	$-C_4H_9$ (butyl)	15	12	300	0.50
4d	$-C_6H_{13}$ (hexyl)	11	16	380	0.40
4e	$-C_8H_{17}$ (octyl)	14	13	330	0.52
4f	$-CH(CH_3)_2$ (isopropyl)	55	3	1300	0.09
4g	$-CH_2-CH(CH_3)_2$ (isobutyl)	19	9.5	350	0.40
4h	chloroethyl	25	7.2	21.7	6.9
4i	phenyl	25	7.2	4.0	38
4j	4-acetaminophenyl	130	1.4	7.0	21.4

[a] Methyl ester rate is taken as 1. Relative rates higher than 1 indicate faster cleavage.
[b] Buffer — phosphate buffer (0.1 M), pH 7.4.

As mentioned before, the rate of cleavage of the carrier Group C from a prodrug (A – C) will also depend on its steric character. Table 1 shows that chain branching markedly slows both the enzymatic and non-enzymatic cleavage of the isopropyl carbonate ester of acetaminophen. This effect is probably due to steric hindrance by the methyl groups. The steric hindrance in the isobutyl case is not nearly so marked probably because the methyl groups are one carbon removed from the carbonyl group of the ester. In this case, the enzymatic and non-enzymatic cleavage reactions are slowed only slightly. Again, this data is in agreement with the ammonia-catalyzed cleavage data for straight and branched chain acetate esters (GORDON et al., 1948). It was found that isopropyl acetate was cleaved markedly slower than ethyl acetate, but isobutyl acetate was cleaved only slightly slower than *n*-butyl acetate.

Other examples illustrating the effect of steric factors on the release of the carrier Group C are found in the *in vitro* hydrolyses by human serum of compounds *5* and *6*. LEVINE and CLARK (1955) reported that the rate of hydrolysis of *5* to give the amino alcohol decreased as the steric effects of substituents placed on the α-carbon of the acid moiety of the ester were increased. Thus, *5a* hydrolyzed faster in human serum than *5b*, and *5c* was found to be the slowest.

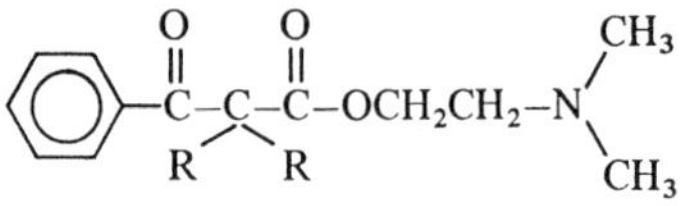

$5a = R = H$
$5b = R = CH_3$
$5c = R = CH_2\text{–}CH_2\text{–}CH_2\text{–}CH_2$ (ring)

Similarly, the rate of the release of *p*-amino benzoic acid from *6a* was found, by the same workers, to be much more rapid than that from *6b* and *6c*, respectively.

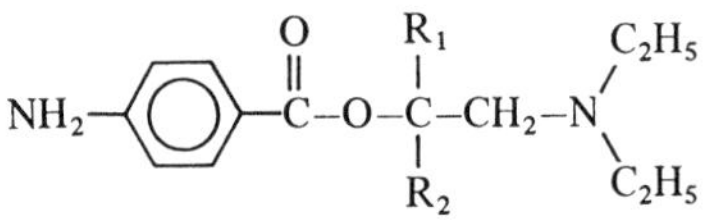

$6a = R_1 = H \quad R_2 = H$
$6b = R_1 = H \quad R_2 = CH_3$
$6c = R_1 = CH_3 \quad R_2 = CH_3$

It is common knowledge that conformational and configurational aspects influence the chemical reactivity and, hence, the biological activity of a compound. A classic example that illustrates the influence of long range conformational effects on the chemical reactivity and conversion of a molecule to a new molecular species is the acyl migration occurring in the nor-ψ-tropine bicyclic ring system. FODOR and co-workers (1953a, 1953b) reported that $N \rightarrow O$ and $O \rightarrow N$ acyl migrations occurred readily for *N*-benzoyl-nor-ψ-tropine (*7a*) and the respective *O*-benzoyl derivative (*7b*).

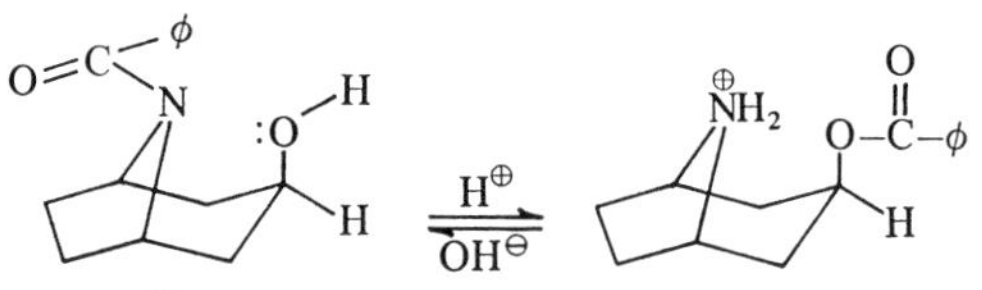

7a $\underset{OH^{\ominus}}{\overset{H^{\oplus}}{\rightleftharpoons}}$ *7b*

In contrast, the diasteriomeric *N*-benzoylnortropine (*8*) under the same conditions was recovered unchanged, and no migration was noticed when *O*-benzoylnortropine hydrochloride was dissolved in alkali:

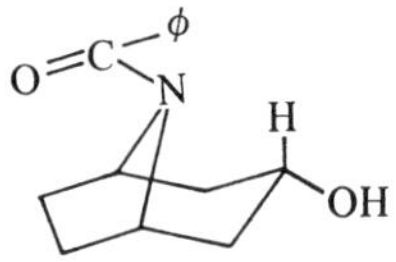

8

The occurrence of the benzoyl group migration in *7a* is evidently a result of the conformation of the molecule and the advantageous geometric orientation of the two participating sites. This intramolecular form of catalysis, particularly common in esters, is many times referred to as "neighboring group participation". In dicarboxylate esters, for example, the intramolecular effect of the carboxylate ion facilitates the hydrolysis of the half ester.

The hydrolysis of aspirin (*9*) is catalyzed by hydrogen ions below pH 4.0 and by hydroxide ion above pH 8.0. The striking feature of this hydrolytic process is that the reaction is pH-independent between pH 4 and 8 (EDWARDS, 1950;

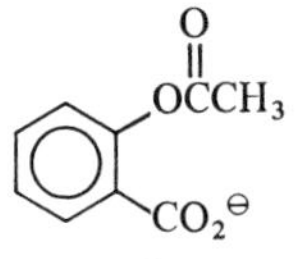

O
||
$OCCH_3$
$CO_2^{\ominus}$
9

GARRETT, 1957). The role of the carboxylate ion has been the subject of numerous studies (reviewed recently by JENCKS, 1969). Through the work of FERSHT and KIRBY (1967a and b), it is now established that the carboxylate ion group of aspirin anion (*9*) is involved in the hydrolysis of the neighboring ester group not as a nucleophile, but as a general base. However, the carboxylate ion could act as a nucleophile in cases where the salicylate moiety contains electron withdrawing groups as in 3,5-dinitro-aspirin (FERSHT and KIRBY, 1967c and d). Other similar facilitated intramolecular ester hydrolyses are known to occur with *O*-acetylsalicylamide and with *O*-carboxyaryl phosphates (JENCKS, 1969). In *O*-acetylsalicylamide (*10*) the nitrogen atom facilitates the hydrolysis of its neighboring ester through a specific base-catalyzed attack of the conjugate base of the amide (*11*) to form an imide (*12*) which undergoes subsequent hydrolysis.

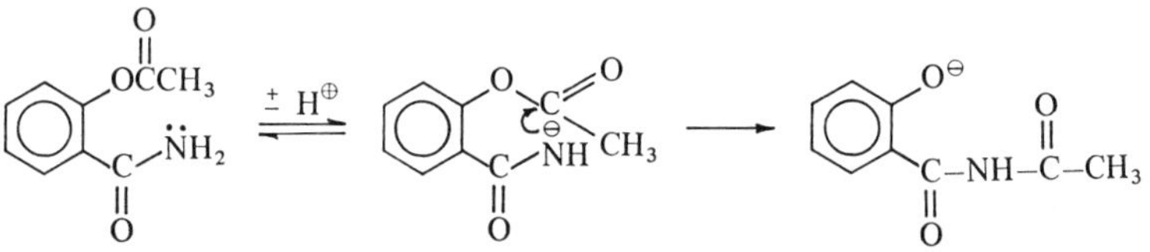

10 $\underset{}{\overset{\pm H^{\oplus}}{\rightleftharpoons}}$ *11* $\longrightarrow$ *12*

The monomolecular rate constant for imide *12* formation from *10* has been shown to be more than 60,000 times larger than the rate constant for the bimolecular reaction of an amide with phenyl acetate under the same conditions (JENCKS, 1969). In an elegant application of these findings HUSSAIN and co-workers (1972a) have shown that the ester prodrug (*13*) generates the anticonvulsant drug 5,5-diphenylhydantoin (*14*), when placed in plasma at pH 7.4 and 37° C. The half-life for this reaction in plasma was 6.8 min.

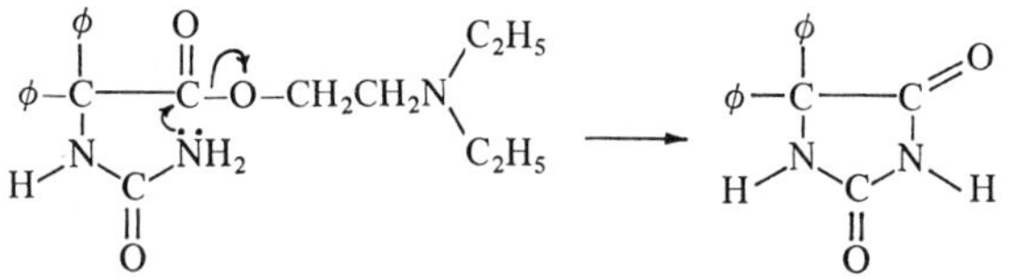

13 $\longrightarrow$ *14*

MORAWETZ et al. (1955, 1958) found that phenyl acid glutarate (*15*) and acid succinate esters (*16*) (GAETJENS and MORAWETZ, 1960) were hydrolyzed at an unusually fast rate *in vitro* in the intermediate pH region of 3 to 8.

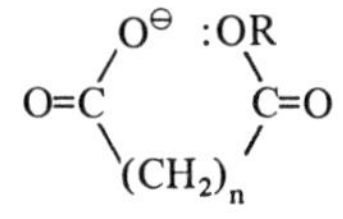

$O^{\ominus}$:OR
O=C C=O
$(CH_2)_n$

15 = n = 3
16 = n = 2

Attempts to prepare the monocholine ester of succinic acid by half-hydrolysis of succinyl choline gave the starting material and succinic acid (PHILLIPS, 1953). This illustrates the facile hydrolysis of half-esters of succinic acid (BENDER, 1960). The intramolecularly assisted hydrolysis of salicylate esters and hemiesters of glutaric and succinic acids at near neutral pH values has been exploited in the preparation of various prodrugs. Specific examples illustrating the usage of these chemical phenomena in the design of prodrugs follow below.

III. Prodrug Design

Table 2 lists a variety of types of prodrugs and potential prodrugs that have been investigated *in vitro* or *in vivo* with regard to their ability to undergo either enzymatic or nonenzymatic cleavage. The Table also lists the various carrier moieties "C" that could be attached to pharmacologically active compounds, thus reflecting some of the common chemical manipulations that can be utilized in designing prodrugs.

IV. Carboxylic Acid Drugs

The most common chemical modification of a carboxylic acid is, of course, its conversion to its ester derivative. The selection of the appropriate alcohol moiety for the derivatization of a carboxylic acid drug often can be made on the basis of *in vitro* enzymatic experiments. CHONG (1970) studied the effects of structural variations in the alkyl moiety of acetate esters with respect to their hydrolysis rates in 2% human plasma. His results showed that as the normal aliphatic chain of the alcohol moiety is lengthened, there is an increase in the rate of enzymic hydrolysis progressing from ethyl through amyl acetate. Beyond the amyl derivative, the rate decreases. Branching at the 3-position of the alkyl chain, as in isoamyl and 3,3-dimethylbutyl acetates, produces a considerable increase in the rate of hydrolysis of the esters. However, branching near the ester linkage, as in isopropyl acetate and secondary and tertiary butyl acetates, decreases the hydrolysis rate. STURGE and WHITTAKER (1950) obtained similar results with horse serum.

One of the early attempts to produce a prodrug of a carboxylic acid was the synthesis of salicylsalicylic acid (*17*) (Salysal, Persistin, Diplosal).

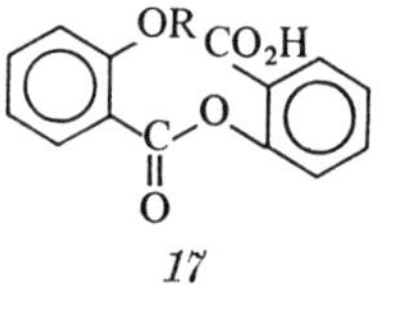

OR CO_2H
C O
O

17

A compound, such as salicylsalicylic acid, resulting from the combination of two identical drug molecules by covalent bond formation is sometimes referred to as an "identical-twin". Similarly, if two different drug molecules are united, the resulting compound is called a "nonidentical-twin" (ARIENS, 1971). The "identical-twin" salicylsalicylic acid is a tasteless, odorless, crystalline powder that is practically insoluble in water and acids but readily soluble in some less polar solvents and in free alkalies. Owing to its tastelessness, probably due to poor solubility, it is more pleasant to take than sodium salicylate. HANZLIK and PRESHO (1925) showed that within one hour the compound did not liberate free salicylic acid in acid mixtures ranging from pH 4.0 to 7.0 and in alkaline mixtures from pH 7.0 to 7.4. The highest rate of cleavage occurred at pH 8.0 and amounted to only 0.26%. *In vitro* studies by the same workers showed that gastric juice, bile, and pancreatin did not hydrolyze the drug. Full therapeutic doses ranging from 4 to 12 g in four human subjects yielded a median total salicylic acid excretion of 63%. Of the total salicylate excreted, a median of 10% consisted of the unaltered salicylsalicylate and the remainder, 90%, of ordinary salicylate. HANZLIK and PRESHO (1925) concluded that the drug escaped decomposition in the alimentary tract but exhibited good intestinal absorption and was largely converted to free salicylate in the tissues. RUBIN (1964) obtained a satisfactory and progressive increase in the mean serum salicylate level by administering a combination of 485 mg. of salicylsalicylic acid and 160 mg. of acetylsalicylic acid to humans. With such a combination, desirable therapeutic salicylate levels of 10 mg. or more per 100 ml. serum were achieved with osteoarthritic patients. Whether the liberation of salicylate from *17* is facilitated in the tissues by an intramolecular ester hydrolysis similar to the one described earlier for acetylsalicylic acid or by an enzymatic catalysis remains unknown.

A nonidentical-twin compound may be prepared by esterifying a carboxylic acid drug with the phenolic hydroxyl group of salicylic acid. The resulting ester prodrug may possess an intrinsic ability to cleave forming the molecules from which it was derived through a neighboring group participation involving the free carboxyl group of its salicylate moiety. In an attempt to design a prodrug that combined both the analgesic properties of salicylic acid and the hypnotic effects of trichloroethanol. DITTERT et al. (1968c) reported the preparation of ethyl carbonate ester of salicylic acid (*18a*).

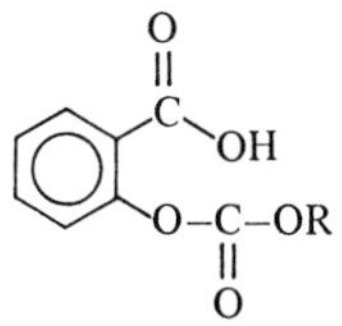

18a = R = trichloroethyl
18b = R = ethyl
18c = R = *n*-butyl
18d = R = *n*-hexyl

The nonidentical twin compound *18a*, like the aliphatic carbonate esters *18b–d*, was found to be stable in 0.1 N HCl. The prodrug *18a* was less stable toward hydrolysis in buffers at pH 7.4 and pH 12 than aspirin and the aliphatic carbonates.

In 2% human plasma in pH 7.4 buffer the aliphatic carbonates and aspirin were slowly hydrolyzed. However, *18a* hydrolyzed rapidly in 2% human plasma but remained relatively unchanged in 0.05% human pseudocholinesterase and

0.05% α-chymotrypsin from bovine pancreas. In these enzyme systems, the ethylcarbonate derivative and aspirin were hydrolyzed relatively slowly, but the butyl- and hexylcarbonates were hydrolyzed relatively rapidly. When aspirin and the butyl-, hexyl-, and trichloroethylcarbonate were administered orally to dogs and the plasma levels of total salicylate followed for 8 h, the resulting blood level curves were virtually superimposable. All the blood levels followed the same pattern and fell off at about the same rate after having reached a peak near 20 mg.-% at 2 h. These results led the authors (DITTERT et al., 1968c) to suggest that the carbonate esters of salicylic acid are as readily absorbed as aspirin despite their different aqueous and lipid solubilities and that all the drugs including aspirin were converted to free salicylate *in vivo* within 2 to 3 h after oral administration. Whether all these prodrugs hydrolyze *in vivo* via a similar mechanism has not yet been determined. These findings, however, suggest that it is possible to combine an alcohol drug and salicylic acid through a carbonate linkage, represented by structure *18*. The resulting "nonidentical-twin" would be expected to be stable in the acid environment of the stomach but would hydrolyze rapidly in human plasma near neutral pH.

Fecal blood loss studies have shown that approximately 70% of persons taking aspirin experienced gastric bleeding (STUBBE et al., 1962; PIERSON et al., 1961). The mechanism by which aspirin causes gastric hemorrhage is a matter of considerable controversy. It appears, however, that at least part of the problem is due to local irritation of gastric mucosa (DAVENPORT, 1964). LEONARDS and LEVY (1970) suggested that gastric irritation and bleeding is due to a local irritation of the mucous membranes by the very acidic (undissolved) aspirin particles. These workers suggested that certain aspirin products that contain alkaline additives cause the aspirin to dissolve more rapidly but have the disadvantages of high sodium content or limited buffer capacity.

MISHER et al. (1968) studied the pharmacology of the hexyl carbonate derivative of salicylic acid (*18d*) a compound of modest solubility in water and gastric fluids. They found that this compound was virtually free of gastric irritation in rats and dogs suggesting that if the aqueous solubility of salicylic acid derivatives is reduced sufficiently, gastric irritation can be controlled even when the carboxylic acid group is not derivatized. In an attempt to mask reversibly the acidic carboxylic acid group of aspirin, and thus reduce any possible gastric irritation induced by this function, HUSSAIN et al. (1974) prepared (1-ethoxy) ethyl-(2-acetoxy) benzoate (*19*), and its structurally related compounds *20* and *21*.

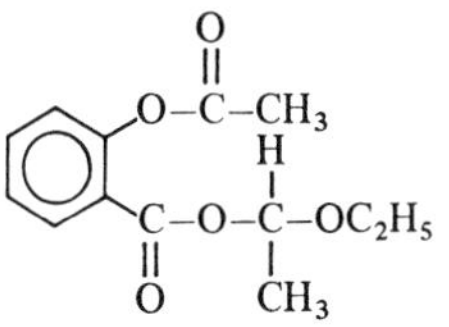

19 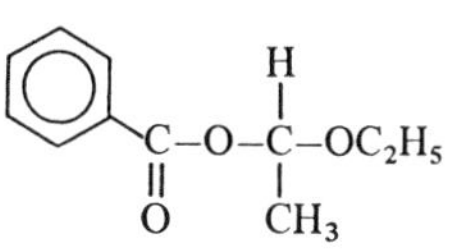 *20* 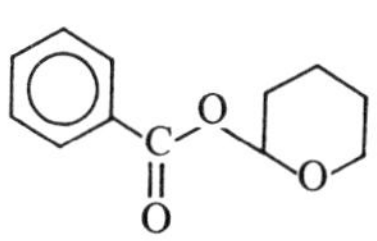 *21* 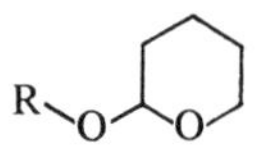 *22*

In vitro hydrolytic studies with these compounds over a wide pH range showed that they hydrolyze extremely rapidly generating the parent compound. (Their half-lives ranged from 3.1 to 16.8 sec. Furthermore, the rates of hydrolysis of arylcarboxyl tetrahydropyrans, similar to *21*, at pH's greater than 4.0 are faster by at least two orders of magnitude than the corresponding 2-alkoxy and 2-aryloxy tetrahydropyrans (FIFE et al., 1968, 1970). The latter, represented by formula *22*, are discussed later in this chapter and are referred as tetrahydropyranyl ethers (Table 2). The relatively high instability of the arylcarboxy tetrahydropyran derivatives (*21*) at pH's ranging from 2 to 12 suggests that these type derivatives are not promising as latentiated compounds for usual uses.

Table 2. *Types of prodrugs*

Drug (A)	Carrier (C)	Prodrug (A—C)	Ref.
Acid	Alcohol	Ester	[1]
	Salicylic acid	Salicylate ester *17, 18a—d*	[2]
	Ethylvinyl ether	Acylal type *19, 20*	[3]
	Dihydropyran	Alkyl or aryl carboxy tetrahydropyran *21*	[4]
	N,N-Dimethylethanolamine	Dimethylaminoethyl ester	[5]
	N,N-Diethylethanolamine	Diethylaminoethyl ester	[6]
	Chloromethylacetate	Acyloxymethyl ester *23a—d*	[7]
	Chloromethylpivalate	Substituted acyloxymethyl ester *25a*	[8]
Alcohol	Carboxylic acid	Carboxylic acid ester *26a—d, 27, 28*	[9]
	Succinic or glutaric anhydride	Hemiesters of succinic or glutaric acids *27f, 27g*	[10]
	Chloromethylmethyl ether	Methoxymethyl ethers *27i*	[11]
	Dihydropyran	Tetrahydropyranyl ethers *27m*	[12]
	Pyruvic acid	Pyruvate esters *27j*	[13]
	Chloroformate esters	Carbonate esters *4a—j, 29, 30*	[14]
	Urea	Carbamate esters *31a, 32a—i, 33a—d, 34—38*	[15]
Amine	Carboxylic acid	Amide *40*	[16]
	Aldehydes or Ketones	Schiff bases *41a* and *b*, *42a* and *b*	[17]
	Ketones with *a*-hydrogen	Enamines *44a* and *b*	[18]

References: 1. Chong (1970), Sturge and Whittaker (1950), Barnden et al. (1953), Carpenter (1948).—2. Hanzlik and Presho (1925), Rubin (1964), Dittert et al. (1968c), Ariens (1971).—3. Hussain et al. (1974).—4. Hussain et al. (1974), Fife et al. (1968, 1970).—5. Carpenter (1948).—6. Heathcote and Nassau (1951), Ungar and Muggleton (1952).—7. Jansen and Russell (1965), Agersborg et al. (1966), von Daehne et al. (1970a), Gibaldi and Schwartz (1966).—8. von Daehne et al. (1970a and b).—9. Chong (1970), Sturge and Whittaker (1950), Glick (1941), Celmer et al. (1958), Shubin et al. (1958), Celmer (1959), Smith and Soderstrom (1959), Glazko et al. (1952, 1958a), Kessler and Borman (1958).—10. Kupchan et al. (1965), Glazko et al. (1958b), Ross et al. (1958), Payne and Hackney (1958), Ceriotti et al. (1954), Rattie et al. (1970), Berger and Riley (1949).—11. Kupchan et al. (1965), Rapoport et al. (1957), Robertson (1960).—12. Kupchan et al. (1965).—13. Sudborough (1912), Kupchan et al. (1965).—14. Swintosky et al. (1966), Caldwell et al. (1967), Cram and Hammond (1964), Swintosky et al. (1968), Misher et al. (1968), Dittert et al. (1968a—d, e), Shah and Connors (1968), Rattie et al. (1970), Dittert et al. (1969), Beard and Free (1967).—15. Dresel and Slater (1952), London and Poet (1957), Huf et al. (1959), Swintosky et al. (1964), Yamamoto et al. (1962a and b).—16. Miller et al. (1939), Shapiro et al. (1958), Elslager et al. (1969a).—17. Jencks (1969), Sandler and Karo (1971), Worth et al. (1969), Elslager et al. (1969b).—18. Caldwell et al. (1971), Szmuszkovicz (1963), Heyl et al. (1953), Sandler and Karo (1971).

Esters of penicillins provide an interesting example of how esterification can profoundly alter the physico-chemical and pharmacological properties of a compound. The simple alkyl and aralkyl esters of benzyl penicillin, such as the methyl, ethyl, *n*-butyl, and benzhydryl esters, were found to hydrolyze to free penicillin in mice and rats. However, these esters attracted very little interest in the past largely because they were devoid of antibacterial activity both *in vitro* and *in vivo* when tested in monkeys, dogs, rabbits, and man (Barnden et al.,

1953; CARPENTER, 1948). Apparently these esters do not hydrolyze in man to free penicillin. CARPENTER (1948) found that the dimethylaminoethyl ester ($X = -CH_2CH_2NME_3$) of benzyl penicillin exhibited full activity in the plate

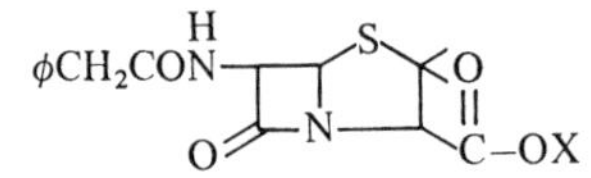

ϕCH_2CONH — penicillin nucleus — $C(=O)-OX$

bioassay due to its ready hydrolysis to free penicillin and the amino-alcohol. A greater interest was generated toward these derivatives when the dialkylamino ethyl esters, in particular the hydroiodide of the diethylamino homologue of penicillin ($X = -CH_2CH_2NEt_2$) (Estopen), were found to exhibit more pronounced selective accumulation than procaine penicillin in inflamed lung tissue of experimental animals (UNGAR and MUGGLETON, 1952). Furthermore when Estopen was tested on human subjects, it gave rise to prolonged effective levels of penicillin in their sputa (HEATHCOTE and NASSAU, 1951). More recently, JANSEN and RUSSELL (1965) achieved the synthesis of acyloxymethyl esters of penicillins (*23*). In contrast to the simple alkyl and aryl esters mentioned above, the acyloxymethyl esters of penicillins rapidly hydrolyze *in vivo*. The best studied member of this group is the acetoxy methyl ester of benzylpenicillin (Penamecillin) (*23a*). This compound is used clinically as an orally active form of penicillin G giving rise to low, but prolonged, levels of the antibiotic. After penamecillin (*23a*) is absorbed, it is assumed that it is enzymatically hydrolyzed by nonspecific esterases to the hydroxymethyl ester (*24*), which subsequently decomposes spontaneously to benzyl penicillin (*23e*) and formaldehyde (AGERSBORG et al., 1966). GIBALDI and SCHWARTZ (1966), using data obtained by AGERSBORG et al. (1966) in the dog, proposed that the prolonged benzylpenicillin serum levels observed after administration of penamecillin are chiefly a function of absorption of unchanged penamecillin through the intestinal tract.

$$R-COOCH_2OCOCH_3 \xrightarrow[\text{esterases}]{\text{nonspecific}} R-COOCH_2OH \xrightarrow{\text{spontaneous}} R-COOH + HCHO$$

23a = R = Penicillin G nucleus = *Penamecillin*
23b = R = Phenoxymethyl penicillin nucleus
23c = R = Methicillin nucleus
23d = R = Cloxacillin nucleus.

Unfortunately, with the exception of penamecillin (*23a*), esters of the type *23* appear to be absorbed very poorly, probably due to their low aqueous solubilities. More promising analogues appear to be the acyloxymethyl esters of D-α-aminobenzylpenicillin (ampicillin) (*25*). Several of these esters have been recently prepared (VON DAEHNE, 1970a). Ampicillin is relatively stable to acid, but it is absorbed rather poorly when administered orally. The acyloxymethyl esters of ampicillin possess an amino group in the side chain which renders them more acid soluble than their penicillin counterparts *23a*–*23d*. Furthermore, the acyloxymethyl esters of ampicillin were more completely absorbed than the parent compound. Of a series of these esters, the hydrochloride salt of pivaloyloxy methyl D-α-aminobenzylpenicillinate (Pivampicillin) (*25a*) rapidly hydrolyzed to ampicillin by enzymes present in the blood and tissues from various species and was better absorbed from the gastrointestinal tract of human volunteers than ampicillin. After absorption, the compound hydrolyzed rapidly to ampicillin resulting in high blood and tissue levels of the latter (VON DAEHNE et al., 1970b).

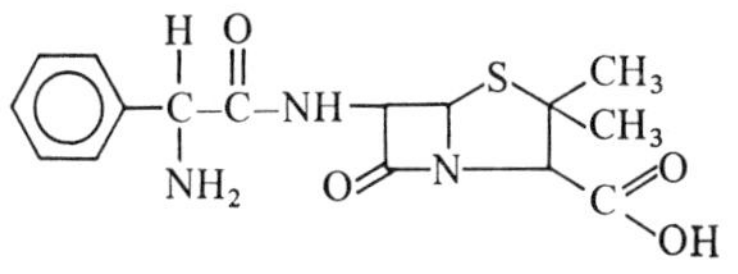

25 = Ampicillin

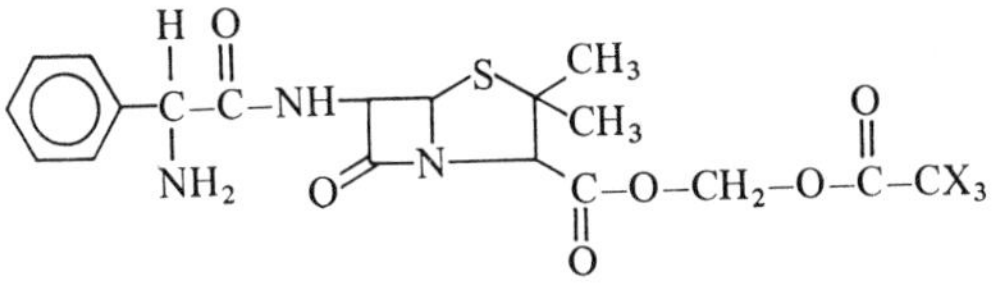

25a = X = CH_3 = pivampicillin

The chemical modification of an amphoteric penicillin, such as ampicillin (*25*), by appropriate derivatization of its carboxylic acid function would be expected to yield products that might show wider tissue distribution and higher tissue concentrations than the parent antibiotic. Ampicillin, being a very polar molecule, would tend to cross membrane barriers at a slower rate than its corresponding ester prodrugs. Furthermore, once the less polar esters are in an extravascular tissue, they are expected to cleave to the more polar ampicillin, which in turn will be less likely to cross the membrane back into the blood and will be held for a longer time in the extravascular compartment.

V. Alcohol Drugs

CHONG (1970) reported on the effects of varying the acyl chain length of ethyl and choline esters on their cleavage by human plasma esterases. His results indicate that both groups of esters hydrolyze faster as the acyl chain is lengthened up to four carbon atoms. Further lengthening or branching of the acyl chain, however, decreases the hydrolysis rates. These results are analogous to those reported by STURGE and WHITTAKER (1950) for a similar series of amyl and isoamyl esters and by GLICK (1941) for the choline esters. In contrast to the above mentioned observations, phenyl acetate could be distinguished from the other phenyl esters by its higher rate of enzymatic hydrolysis. Changing the acyl moiety of the phenyl esters from acetate to propionate resulted in a decrease in the enzymatic hydrolysis rate. Further change in structure to the butyrate produced a rate that was faster than the propionate but slower than that of the acetate. Branching of the acyl chain of phenyl esters decreased their enzymatic hydrolysis rate (CHONG, 1970).

The antibiotic oleandomycin possesses three hydroxyl groups. Marked physical, chemical, and biological changes resulted when its free hydroxyl groups were acetylated. Triacetyloleandomycin, the product of this simple esterification, exhibited increased stability, decreased water solubility, and enhancement of taste properties (CELMER et al., 1958, 1959). Administration of the triacetyl derivative to humans led to high antibiotic serum levels and recoveries in urine although it exhibited lower *in vitro* activities than oleandomycin itself (SHUBIN et al., 1958). Similar results were obtained with the homologs diacetyl and monoacetyl oleandomycin.

The substantial instability of erythromycin in acid media and gastric juice appears to result in varied and comparatively low blood levels after its oral administration to humans. The propionate ester of erythromycin, when administered to humans, produced high blood levels within half an hour after ingestion and peak concentrations approximately four times those obtained with equivalent amounts of erythromycin (SMITH and SODERSTROM, 1959). Esterification of the antibiotic appears to mask its undesirable bitter taste. Its stearate and ethyl carbonate esters have been used in the formulation of oral pediatric preparations (HARPER, 1962). In this connection, several esters of chloramphenicol have been prepared in an attempt to mask its bitter taste. The tasteless palmitate ester of chloramphenicol was found to possess no appreciable *in vitro* antimicrobial activity but was readily hydrolyzed by the enzymes of the duodenum with the release of chloramphenicol (GLAZKO, 1952). A close relationship between the rate of hydrolysis of chloramphenicol ester preparations and their rate of intestinal absorption was observed by GLAZKO et al. (1958a). When administered orally in finely divided suspensions, both the palmitate and stearate esters of the antibiotic produced good blood levels. A comparison of chloramphenicol with its palmitate and stearoylglycolate esters in human subjects indicated that the palmitate suspension produced higher blood levels than either of the other preparations, while the total urinary excretion over a longer period was about the same in all cases (GLAZKO, 1958a).

Esterification of alcohol drugs has been used in many cases as a method of increasing their durations of action. The conversion of estradiol (*26*) to its benzoate (*26a*), cyclopentylpropionate (*26b*), dipropionate (*26c*), and valerianate (*26d*)

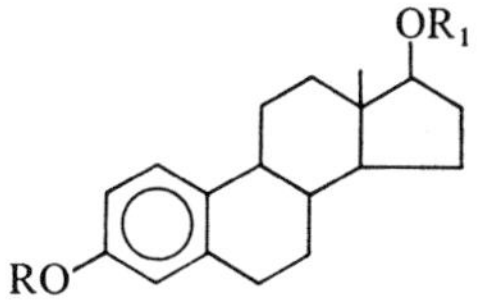

OR₁

RO

26 $= R = R_1 = H$
26a $= R = H,\ R_1 = —COC_6H_5$
26b $= R = H,\ R_1 = —COCH_2CH_2\cdot C_5H_9$
26c $= R = \quad R_1 = —COCH_2CH_3$
26d $= R = H,\ R_1 = —CO(CH_2)_3CH_3$.

ester derivatives produces estrogenic materials that are more resistant to biodegradation and that are slowly absorbed and hydrolyzed to the parent steroid (HARPER, 1962).

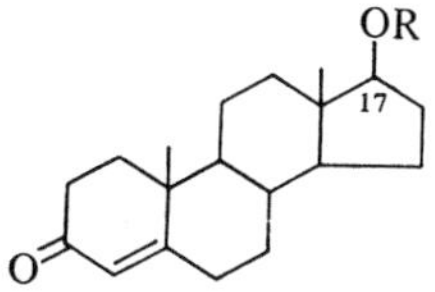

OR

17

O

27a $= R = —COCH_3$
27b $= R = —CO(CH_2)_2CH_3$
27c $= R = —CO(CH_2)_3CH_3$
27d $= R = —COCH_2OC_6H_4\cdot R$
27e $= R = —COCH_2CH_2\cdot C_6H_5$
27f $= R = —CO(CH_2)_2\cdot CO_2H$
27g $= R = —CO(CH_2)_3\cdot CO_2H$
27h $= R = —CO(CH_2)_2CONH_2$
27i $= R = —CH_2OCH_3$ (ethylene ketal)
27j $= R = —CO\cdot CO\cdot CH_3$
27k $= R = —CO(CH_2)_2COOEt$
27l $= R = —CO(CH_2)_2CON(Me)_2$

27m $= R =$

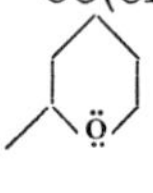

Ö

Esterification of the 17-hydroxyl group of testosterone to the acetate (*27a*), butyrate (*27b*), or valerianate (*27c*) increased the duration of testosterone effects up to two weeks and enhanced the activity of the preparation. Increased duration up to four weeks could be achieved by esterification with higher fatty acids (C_6–C_{11}), but the compounds exhibited no increased activity (Harper, 1962). Other esters of testosterone that possess enhanced and/or prolonged activity include the aryloxyacetates (*27d*), the phenylpropionates (*27e*), and others. The glutarate, adipate, and pimelate esters of testosterone were less active and less prolonged in activity than that of testosterone heptanoate. However, injection of 5 mg. of the suberate, azeloate, or sebacate exhibited activities on the 40th day after injection that were 1.1 to 1.7 times that of testosterone heptanoate (Harper, 1962).

The introduction of an α-hydroxyl group at the 17-position of progesterone leads to 17-α-hydroxy progesterone (*28a*), a product that possesses only 1/100 of the activity of its mother compound in rabbits and is ineffective in humans (Pfiffner and North, 1941; Harper, 1962).

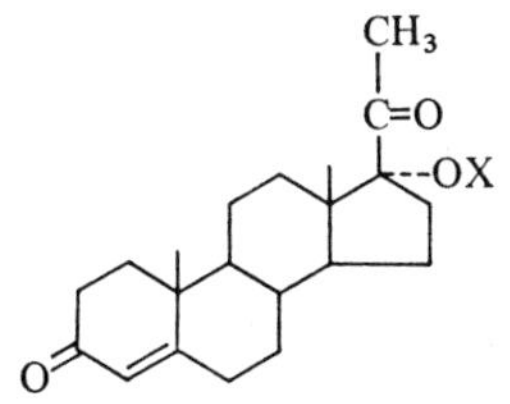

CH_3 | C=O ---OX

O

28a = X = H = progesterone
28b = X = —$CO(CH_2)_4 \cdot CH_3$.

Esterification of *28a*, however, produces marked increase in the potency and duration of activity. Thus, the 17-*n*-caproate ester (*28b*), possesses excellent depot action (Kessler and Borman, 1958).

In our previous discussions, we mentioned that the hydrolysis of hemiesters of succinic and glutaric acids at neutral pH values is assisted by their free carboxyl group. This group could, on treatment with alkali, give water soluble salts. These features of the succinic and glutaric acid hemiesters render them attractive as carriers in the design of prodrugs.

Conversion of alcohols to alkyl ethers produces derivatives that are generally stable to the action of both acids and alkalies. Furthermore, in only a few cases has the *in vivo* cleavage of an aliphatic ether been demonstrated, although aromatic methyl ethers are demethylated fairly readily (Williams, 1959). However, the methoxymethyl and tetrahydropyranyl ethers both being acetals in nature, are easily cleaved in dilute acids (Rapoport et al., 1957; Robertson, 1960).

An excellent carrier for the preparation of prodrugs of alcohols is pyruvic acid, since it is a normal physiological acid, and its liberation from its ester prodrugs should not lead to undesirable side effects. Sudborough (1912) found ethyl pyruvate to undergo substantial hydrolysis even in water.

In an attempt to utilize these chemical properties in the design of new prodrugs, Kupchan et al. (1965) prepared a variety of testosterone derivatives (*27f* to *27m*). Oral administration of the derivatives to rats revealed that testosterone acid succinate (*27f*), testosterone amido succinate (*27h*), testosterone ethylene ketal methoxymethyl ether (*27i*), testosterone pyruvate (*27j*), testosterone ethyl succinate (*27k*), testosterone *N*,*N*-dimethylamido succinate (*27l*), and testosterone tetrahydropyranyl ether (*27m*) exhibited a significant degree of androgenic activity. Only the methoxymethyl ether of testosterone ethylene ketal (*27i*) exhibited significant myotrophic activity comparable to that of methyltestosterone. Surprisingly enough, however, the glutarate ester (*27g*) failed to exhibit as significant androgenic activity as the rest of the esters. Of the seven compounds exhibiting appreciable androgenic activities, none was more potent than testosterone or methyltestosterone at a 25 mg/kg dosage level. The closeness in magnitude of the androgenic activity values for *27f* and *27h* to *27m* led the authors (KUPCHAN et al., 1965) to suggest that these compounds are latent forms of testosterone which, upon ingestion, are converted to testosterone in the gastrointestinal lumen. This conclusion is partially supported by the fact that, like testosterone, most of the compounds failed to exhibit significant myotrophic activity when administered orally to rats.

Other monoesters of dicarboxylic acids have been prepared in an effort to produce more soluble products than their corresponding parent alcohol drugs. Thus, the succinate, maleate, phthalate, and glutarate esters of erythromycin have been reported (HARPER, 1962).

The formulation of a satisfactory parenteral preparation of chloramphenicol has been handicapped by its limited solubility in water. Furthermore, the repository type of action obtained by intramuscular administration of a microcrystalline suspension of this antibiotic is unsatisfactory when high initial blood levels are required immediately after dosage (GLAZKO et al., 1958b; SCHOENBACH, 1953). In an effort to overcome this problem, concentrated solutions of chloramphenicol in various organic solvents have been employed. These preparations yield high blood levels after intramuscular administration, but they can cause irritation at the site of injection (GLAZKO et al., 1958b). To satisfy the need for a water soluble chloramphenicol preparation that could be given safely intramuscularly with a minimum of side reactions and that could be absorbed rapidly to produce bacteriostatic blood levels within a short time after administration, GLAZKO and co-workers (1958b) prepared the mono- and disuccinate esters of chloramphenicol. Both esters were found to be highly water soluble and rapidly absorbed from parenteral sites of injection. In the rat, dog, and in man, both esters could pass through the kidneys unchanged, appearing in the urine together with their hydrolysis products and with the normal metabolic products of chloramphenicol. In the rat, the monosuccinate ester appeared to be hydrolyzed in the intestinal tract and was absorbed more rapidly than the disuccinate ester, but absorption by this route was relatively poor. In human subjects, the monosuccinate ester produced good blood levels whether administered orally or parenterally. Most of the ester appeared to be hydrolyzed in the intestinal tract of man prior to absorption, but the reduced levels of the drug in urinary excretions obtained indicated incomplete intestinal absorption. However, parenteral administration of the hemiester sodium salt resulted in higher urinary through-put, indicating good absorption (GLAZKO et al., 1958b; ROSS et al., 1958). As mentioned earlier in this chapter, half-esters of succinic acid undergo a facile intramolecularly assisted hydrolysis. It is possible to hypothesize that the monosuccinate ester of chloramphenicol might undergo a rapid enzymatic or nonenzymatic hydrolysis to chloramphenicol in the intestinal tract of man prior to absorption. On the other hand, when the hemiester is administered parenterally it seems to resist hydrolysis, at least partially, perhaps due to an intermolecular association involving the free carboxyl groups with various binding sites. This suggestion is partially supported by the findings of CERIOTTI et al. (1954) who identified the presence of the monoester of chloramphenicol in human urine following parenteral administration of its disuccinate derivative, but found no evidence for the presence of the monosuccinate ester after oral administration of the disuccinate despite evidence for absorption of the compound.

Further clinical investigations with humans showed that chloramphenicol acid succinate (sodium salt) was a safe and effective form of the antibiotic causing no local or general toxicologic reactions when used by nebulization or by the intramuscular route. Early and effective bacteriostasis was achieved in a short time in the treatment of many bronchial infections (PAYNE and HACKNEY, 1958).

Carbonate diesters are derivatives of carbonic acid, the two hydrogens of which are formally replaced by alkyl or aryl groups to give the general formula ROCOOR'. The carbonate esters are relatively stable to acid hydrolysis, but readily subject to basic hydrolysis releasing the alcohol moieties. ROH and R'OH (CRAM and HAMMOND, 1964). In 1966, SWINTOSKY and co-workers announced that numerous carbonate diesters could be made of drugs containing the hydroxy group. Such derivatives exhibited different physico-chemical properties from the parent drugs, but often retained qualitatively identical pharmacologic effects and reverted to the parent drug in the body. The desirable pharmaceutical properties of the carbonate prodrugs can be illustrated with the case of *bis*-trichloroethyl carbonate, *29*:

$$CCl_3CH_2—O—CO—O—CH_2CCl_3$$

29

Trichloroethanol is a sedative, but it is a volatile liquid with an unpleasant odor and taste and therefore is not useful in therapeutics. However, its carbonate diester (*29*) is a crystalline, tasteless compound that has sedative properties and can be encapsulated or tableted (SWINTOSKY et al., 1966). Pharmacologic comparisons of trichloroethyl carbonate (*29*) (chlorethate) and trichloroethanol revealed that these compounds were about equitoxic and equipotent in depressing spontaneous motor activity in mice after oral administration (CALDWELL et al., 1967). From these observations these workers concluded that the pharmacologic activities manifested after oral administration of *29* were the result of *in vivo* regeneration of the parent drug, trichloroethanol.

In parallel double-blind studies conducted in out-patient populations given a medication for anxiety psychoneurosis, chlorethate (*29*) was more effective than the sedative phenobarbital and equivalent in efficacy to a mild tranquilizer, meprobamate (BEARD and FREE, 1967).

Acetaminophen (*3*) and trichloroethanol have unpleasant tastes, and the latter, being a liquid, is especially unsuited for convenient pharmaceutical formulation. SWINTOSKY et al. (1968) reasoned that blocking the hydroxyl groups of the drugs, through a carbonate bridge, would suppress hydrophilic properties, improve taste, and yield a chemically neutral solid with greater lipophilic character than its parent drugs. In addition, it was surmised that the resulting "nonidentical-twin" *30* would combine the complementary sedative and analgesic activities of the parent drugs. Thus, SWINTOSKY and co-workers (1968) synthesized 4-acetamido-phenyl 2,2,2-trichloroethyl carbonate (*30*) by reacting acetaminophen with 2,2,2-trichloroethyl chloroformate.

$$CH_3-\overset{O}{\overset{\|}{C}}-\overset{H}{N}-C_6H_4-O\overset{O}{\overset{\|}{C}}OCH_2CCl_3$$

30

The crystalline carbonate diester (*30*) exhibited good lipid solubility but low water solubility and was virtually tasteless. It underwent base-catalyzed hydrolysis *in vitro* and had a half-life in water at pH 7.4 of about 7 h. Compound *30* underwent slow hydrolysis at an acid pH (at pH 2 its half-life was 15 h). Its low solubility in water coupled with a slow cleavage rate in human gastric juice suggested that if stomach absorption occurs when the compound is administered orally, it would be absorbed primarily in the form of the prodrug *30*. On the other hand, hydrolysis of this compound was catalyzed by human intestinal fluid, rat intestinal mucosa, rat plasma, rat liver, and human plasma. The presence in human intestinal fluid and rat intestinal mucosa of enzymes that catalyze the cleavage of *30* suggested that absorption from the intestine of both the intact ester (*30*) and its cleavage products could occur simultaneously. The fact that esterases are abundant in the plasma and liver would suggest that in animals and humans, any carbonate that was absorbed intact would be quickly converted to acetaminophen (*3*) and trichloroethanol (SWINTOSKY et al., 1968). The prodrug *30* had the same spectrum of pharmacologic activity as a physical mixture of trichloroethanol and acetaminophen, but its analgesic and sedative properties were less intense and persisted for a longer period of time than those of the physical mixture (DITTERT et al., 1968a and b).

In a search for acetaminophen (*3*) prodrugs, DITTERT et al. (1968a) examined the physico-chemical properties of several carbonate [(*4a*) to (*4j*), Table 1] and carboxylate esters of acetaminophen. The carbonate esters (*4a*) to (*4j*) possessed lower water solubilities and higher O/W (oil:water) partition coefficients than acetaminophen itself. They were all slowly hydrolyzed in phosphate buffer at pH 7.4, but exhibited relatively fast hydrolytic rates, which were first-order in ester and in hydroxyl ion, in the alkaline pH range (RATTIE et al., 1970). Human plasma enzymes markedly accelerated their hydrolyses (DITTERT et al., 1968a). Half-lives of about 20 sec were observed when the hydrolysis of carbonate prodrug *30* was studied in a 0.03% solution of purified human pseudocholinesterase (pH 7.4), confirming that this enzyme is a very potent enzyme with respect to the hydrolysis of the carbonate diester (*30*). Proteolytic enzymes, especially the chymotrypsins and trypsin, were also efficient catalysts at pH 7.4. Other enzymes such as papain, pepsin, enterokinase, carbonic anhydrase, lipase, and penicillinase showed little or no hydrolytic activity under similar conditions (DITTERT et al., 1969). In this connection, SHAH and CONNORS (1969) have recently presented evidence, from *in vitro* kinetic experiments, suggesting that the enzymatic hydrolysis of *p*-acetamidophenyltrichloroethyl carbonate (*30*), and related carbonate diesters, involves intermediate formation of an alkyl enzyme carbonate of which the deacylation step is rate controlling.

The molar LD_{50}'s in mice, of some acetaminophen carbonates [(*4a*) to (*4c*), (*4f*), (*4g*), (*4i*), Table 1] were virtually identical. Their confidence limits overlapped each other, and those of acetaminophen, in almost all cases. Some of the compounds (*4f*, *4i*) had weak analgesic activity orally in rats. The relatively short half-lives of the carbonate esters obtained in alkaline pH ranges and their fast *in vitro* enzyme-catalyzed hydrolyses suggested that these compounds would be hydrolyzed to release acetaminophen following oral administration to animals and humans (SWINTOSKY et al., 1968; DITTERT et al., 1968a to e). This was substantiated by the finding that enzymes found in the gastrointestinal tract rapidly hydrolyzed carbonate diester *30 in vitro* (DITTERT et al., 1969). Furthermore, the toxicity and analgesic activity studies in mice and rats using various acetaminophen carbonates (*4*) (DITTERT et al., 1968a) suggested that the dissolution rates of these prodrugs, rather than their enzyme-catalyzed hydrolysis rates, probably control the availability of acetaminophen (*3*) following oral administration. The investigators (DITTERT et al., 1968d) reasoned that if the duration of action of the carbonate prodrugs is controlled by their dissolution rates in the gastrointestinal tract, then it would be expected that the particle size of their powders might have a pronounced influence on their dose-time-action profiles after oral administration. To test this hypothesis, 4-acetamidophenyl 2,2,2-trichloroethyl carbonate (*30*) powder was prepared in coarse, regular, and fine particle sizes. The powders showed significant difference in their dissolution rates in water. When suspended in a 0.5% dispersion of gum tragacanth in water and administered orally to mice, the LD_{50}'s of the carbonate powders were as follows: fine – 1796 mg/kg, regular – 2461 mg/kg, and coarse – 3340 mg/kg. All three powders produced peak plasma acetaminophen concentration at about one-hour postdrug, suggesting that in each case absorption of the drug in the mouse was essentially ended after about one hour. However, substantial differences in the plasma concentrations of acetaminophen were produced by the powders during the first two hours. Thus, the fine-particle material gave rise to a substantially higher maximum plasma concentration during the first hour than the regular material; and the latter gave rise to a substantially higher maximum plasma concentration than the coarse material (DITTERT et al., 1968d). A plot of the LD_{50}'s of the three powders versus the peak plasma concentrations of total acetaminophen, at one-hour postdrug administration, showed that there was good correlation between the toxicities of carbonate *30* powders and the peak plasma concentrations. Since higher plasma concentrations were observed for the fine-particle trichloroethyl carbonate *30* during the first hour, it was suggested that absorption was fastest from this material. It was therefore concluded that, in the mouse, there was excellent correlation between (a) particle size and acute oral toxicity, (b) particle size and peak plasma concentration, and (c) peak plasma concentration and acute oral toxicity.

Plots of total acetaminophen in plasma versus time in human subjects administered one gram of coarse, regular, or fine-particle carbonate (*30*) or 436 mg. of acetaminophen (molar equivalent of 1 mg of *30*) orally in hard gelatin capsules (DITTERT et al., 1968d). The plasma concentration curves of the regular and coarse particle carbonate *30* were nearly identical. This was attributed to the fact that in humans, where the gastrointestinal tract is much larger than in the mouse, both the large and small particles in the regular-particle carbonate *30* powder were probably completely dissolved and absorbed. In addition, the plots showed that after peak drug concentrations were reached, the rates of plasma clearance with regular and coarse particle *30* were slower (apparent half-life about 6 h) than that observed with acetaminophen (apparent half-life about 3.4 h). GIBALDI and SCHWARTZ (1966) showed that prolonged absorption of penamecillin [prodrug (*23a*)] gave a blood concentration plot of penicillin G with the appearance of prolonged excretion. In other words, the apparent half-life of the descending curve, obtained after oral administration of prodrug *23a*, was much longer than that observed following administration of penicillin G itself. Thus, the apparent half-life from the "tail" of the plot was an artifact caused by slow absorption of the prodrug during the period in which the blood concentration curve was descending. Similarly, DITTERT et al. (1968d) reasoned that the slower apparent rate of clearance of acetaminophen from the blood of humans after oral administration of regular and coarse-particle carbonate *30* was probably due to prolonged or delayed absorption of the larger particles in these powders.

The plasma concentration of acetaminophen obtained by fine-particle carbonate ester *30* reached a higher, slightly earlier peak (4 h) than those attained following administration of coarse or regular-particle powders. Furthermore, the rate of plasma clearance (half-life about 3.2 h) in the fine-particle case was almost the same as that observed for acetaminophen (half-life about 3.4 h). It was surmised that the fine-particle material *30* was completely absorbed within about five hours following administration and that prodrug *30* was completely converted to acetaminophen (*3*) by about five hours postdrug administration. Also, it was concluded that the low aqueous solubility of prodrug *30* would enable the formulator to lower the peak and extend the duration of acetaminophen, and presumably also trichloroethanol, blood levels by controlling the particle size of administered 4-acetamidophenyltrichloroethyl carbonate (*30*) powder.

In a similar study, trichloroethyl carbonate (*29*), a prodrug of trichloroethanol, was prepared in a coarse, regular, and fine-particle size powder (DITTERT et al., 1968e). When the three powders were administered orally to mice, their LD_{50}'s were as follows: fine – 1175 mg/kg, regular – 1775 mg/kg, and coarse – more than 2000 mg/kg. Although no trichloroethanol blood levels were determined in the above study, the investigators speculated (DITTERT et al., 1968e) that perhaps a similar correlation might exist between LD_{50}'s and peak trichloroethanol blood levels as that obtained with the acetaminophen prodrug *30*.

The muscle relaxant properties of mephenesin in humans (*31*) are limited by its short duration of action (and low solubility in water). In dogs, the short duration of drug action appears to be related to the loss of activity that follows the oxidation of its terminal hydroxyl group to form mephenesic acid (DRESEL and SLATER, 1952; BERGER and RILEY, 1949). Metabolism also appears to be the cause of short duration of action in humans.

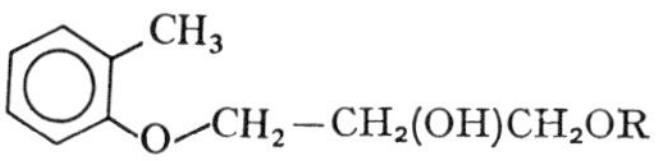

CH_3

$O-CH_2-CH_2(OH)CH_2OR$

31 = R = H
31a = R = $-CONH_2$
31b = R = $-CO(CH_2)_2CO_2H$

The acid succinate derivative of mephenesin (*31b*) had a longer duration of action than mephenesin after intraperitoneal administration (BERGER and RILEY, 1949). This difference in action was not found after oral administration (DRESEL and SLATER, 1952) of the acid succinate (*31b*).

When mephenesin (*31*) was converted to its carbamate derivative *31a*, it resisted oxidation in the dog and the appearance of its acid metabolite in the urine of dogs could not be demonstrated (DRESEL and SLATER, 1952). When mephenesin (*31*) and its carbamate ester (*31a*) were administered orally to humans and the plasma levels of the respective drugs determined at various intervals, significantly higher and more persistent plasma levels were attained by the carbamate (*31a*), as compared to those of the parent drug (*31*). No free mephenesin was found in the plasma of those subjects receiving mephenesin carbamate (*31a*) (LONDON and POET, 1957). In another study in which the human subjects received equimolar amounts of *31*, carbamate (*31a*), and methocarbamol, the carbamate of glyceryl guaiacol used in the treatment of painful muscle spasm, the latter drug produced plasma levels that exceeded significantly those obtained with *31* and *31a* (HUF et al., 1959).

In an attempt to obtain more knowledge on the enzymatic and non-enzymatic hydrolytic behavior of carbamate esters, SWINTOSKY and co-workers (1964) studied the cleavage properties of carbamate esters of phenol (Table 3) and of acetaminophen (Table 4). The data in Tables 3 and 4 show that enzyme catalysis is strongest (enzyme rate: buffer rate ratio is highest) with both phenyl and acetaminophen carbamates which are *N*-unsubstituted [(*32a*) and (*33a*)] or *N*-monoalkyl substituted [(*32d*) and (*33d*)]. Enzyme catalysis is reasonably good with monoaryl *N*-substituted acetaminophen carbamate (*33b*) but practically nil in phenyl carbamate (*32b*). When the nitrogen is disubstituted with aryl groups [(*32c*) and (*33c*)] the nonenzymatic cleavage is fast, but the presence of plasma enzymes does not appear to hasten the catalysis.

Table 3. *Enzymatic and non-enzymatic cleavage behavior of carbamate esters of phenol*

C_6H_5–OC(=O)–X

32

No.	X	Half-life in 2% liquid human plasma (min)	Half-life in phosphate buffer pH 7.4 (min)	Enzyme rate / Buffer rate
32a	—NH_2	3.5	27	8
32b	—NH—ϕ	158	285	2
32c	—N(ϕ)$_2$	2	1.3	0.7
32d	—NH-Et	42	300	7
32e	—N(Me)$_2$	no cleavage	no cleavage	—
32f	N (pyrrolidinyl)	no cleavage	no cleavage	—
32g	N (piperidinyl)	no cleavage	no cleavage	—
32h	N O (morpholinyl)	no cleavage	no cleavage	—
32i	—$NHCH_2CH_2N(Me)_2$	no cleavage	no cleavage	—

Table 4. *Enzymatic and non-enzymatic cleavage behavior of carbamate esters of acetaminophen*

CH_3–C(=O)–NH–C_6H_4–OC(=O)–X

33

No.	X	Half-life in liquid human plasma (min)	Half-life in phosphate buffer (h)	Enzyme rate / Buffer rate
33a	—NH_2	245	26	64
33b	—NH—ϕ	64	14	13
33c	—N(ϕ)$_2$	16	1	4
33d	—NH-Et	135	43	19

When the *in vitro* enzymatic and nonenzymatic cleavage behavior of the aliphatic carbamates (*34* to *37*) was studied, it was found that none were cleaved under conditions identical to those applied to *32* and *33* derivatives (SWINTOSKY et al., 1964).

$EtOCONH_2$ (*34*)

$EtOCONH(\phi)$ (*35*)

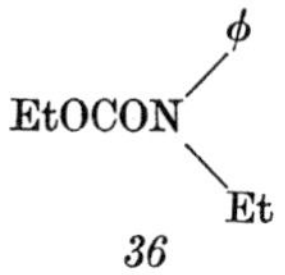

$EtOCON(\phi)(Et)$ (*36*)

$(CH_3)_2CHOCONH(\phi)$ (*37*)

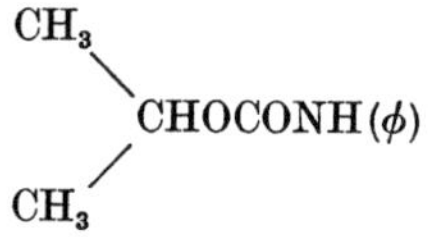

These results are interesting because the major excretion products of meprobamate (*38*), an aliphatic carbamate, in rabbits and dogs all have the carbamate

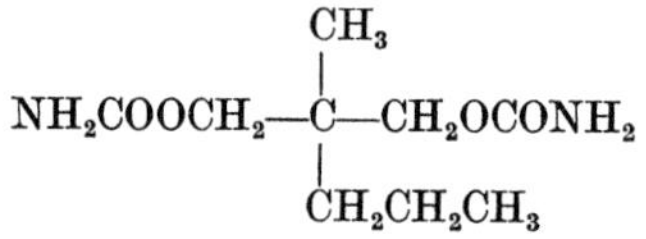

$$NH_2COOCH_2{-}\overset{\displaystyle CH_3}{\underset{\displaystyle CH_2CH_2CH_3}{C}}{-}CH_2OCONH_2$$

38

ester linkage intact (YAMAMOTO et al., 1962a and b; TSUKAMOTO et al., 1963). This indicates that at least *N*-unsubstituted carbamates of aliphatic alcohols are cleaved very slowly, if at all, *in vivo*. In this connection, the carbamate ester of mephenesin (*31a*) appears to resist hydrolytic cleavage when administered to dogs (DRESEL and SLATER, 1952) and man (LONDON and POET, 1957).

VI. Amine Drugs

There are surprisingly few types of amine prodrug derivatives. One early example of prodrug design involving amine drugs is the preparation of several *N*-acyl derivatives of sulfanilamide (MILLER et al., 1939). Some of the aliphatic acyl derivatives possessed activity as antistreptococcic agents in mice. The *n*-caproyl derivative was the most effective of the derivatives studied. Interestingly enough, the monocarboxylic acid derivatives were more effective than the dicarboxylic acid compounds. The activity appeared to increase with an increase in length of the acyl group up to and including six carbon atoms, after which it fell off rapidly. Acylation with aromatic groups, such as C_6H_5CO—, led to less active products than those obtained with saturated straight chain acyl groups. The normal acyl derivatives were also more active than the corresponding iso compounds (MILLER et al., 1939).

In the search for active structures that would not exhibit the noted cardiovascular side effects of *d*-α-methylphenethylamine, SHAPIRO and co-workers (1958) investigated the pharmacologic effects of acyl derivatives (amides) of this compound. The individual compounds, from a large series of amides, were evaluated for their excitant effect in rats. It was found that activity was substantially confined to selected structures containing an α-oxy function on the acylating group. The pharmacologic results obtained in this study were not consistent with a proposed mechanism that suggested the possibility of *in vivo* hydrolysis of the amides to free *d*-α-methylphenethylamine.

Aliphatic and aromatic aldehydes condense with aliphatic and aromatic primary amines to form *N*-substituted imines (RR″C=NR″). These compounds are also known as azomethines, anils, or more commonly, Schiff bases. A distinct chemical characteristic of Schiff bases is their susceptibility toward nucleophilic attack at their $-\overset{|}{C}=N$-carbon atom, particularly after protonation of the imine nitrogen (JENCKS, 1969; SANDLER and KARO, 1971). As a result of this property, imines can react with an amine to give another imine and a second amine. Such reactions are commonly met in biologic transaminations (KOSOWER, 1962; JENCKS, 1969).

The chemical reactivity of Schiff bases makes them attractive in the design of prodrugs. ELSLAGER et al. (1969b) and WORTH et al. (1969) have utilized the Schiff base approach to produce new derivatives of 4,4′-sulfonyldianiline (diamino diphenylsulfone, *39*) (DDS or Dapsone).

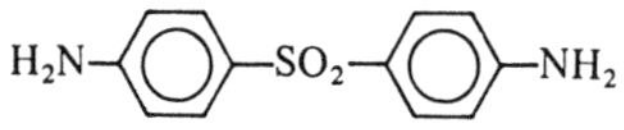

$H_2N-C_6H_4-SO_2-C_6H_4-NH_2$

39

Diamino diphenylsulfone (*39*) has occupied a preeminent position as an antileprotic agent since 1941. This, along with the observation of lower prevalence of malaria in leprosy patients under treatment with *39*, prompted further research in developing repository forms of sulfone (*39*) (ELSLAGER et al., 1969a). Several sulfanilylanilides were prepared and related compounds were investigated as potential repository antimalarial and antileprotic agents (ELSLAGER et al., 1969a). Of these 4′,4‴-sulfonylbisacetanilide (acedapsone, *40*) showed the longest

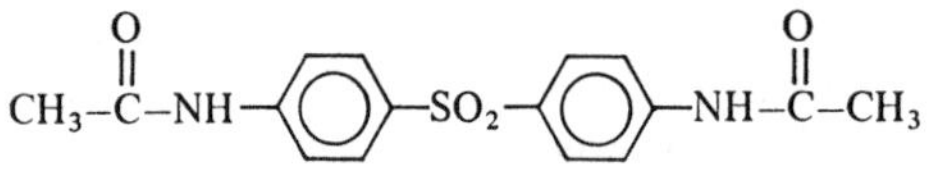

$CH_3-\overset{O}{\overset{\|}{C}}-NH-C_6H_4-SO_2-C_6H_4-NH-\overset{O}{\overset{\|}{C}}-CH_3$

40

duration of action and protected mice for 6 to 14 weeks against *Plasmodium berghei* and monkeys for 2 to 8 months against *P. cynomolgi*. Repository antimalarial effects were abolished or drastically reduced when *40* was modified by 1) replacement of the acetamide group with a formamide function, 2) replacement of both acetamide groups with amide functions containing more than two carbon atoms, 3) *N*-alkylation of one acetamide function, 4) replacement of the sulfone moiety, and 5) introduction of a chlorine atom at positions 2′ or 3′ (ELSLAGER et al., 1969a). Metabolic studies with mice revealed that sulfone (*40*) was apparently dependent upon deacylation for activity. However, the compound produced extremely low sulfone blood levels. The investigators (ELSLAGER et al., 1969b; WORTH et al., 1969) reasoned that a repository sulfone that is less dependent on enzymatic deacylation for activity and that provides higher blood sulfone levels than *40* was needed. Thus, several Schiff bases were synthesized among which *41a* was very long acting and protected mice against *P. berghei* for more

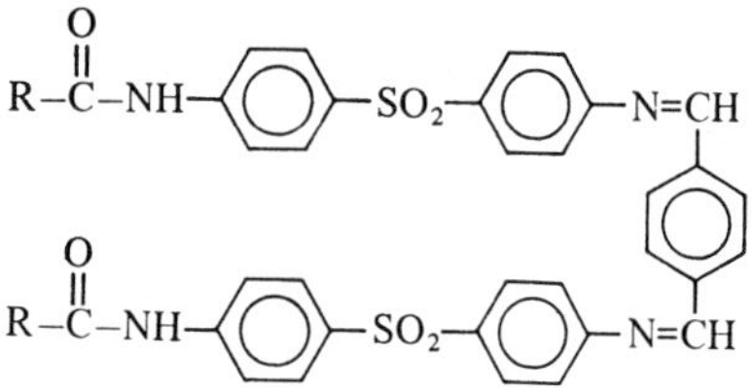

41a = R = H
41b = R = CH_3.

than 9 weeks and *41b*, for 5 to 7 weeks (ELSLAGER et al., 1969b). The intermediate periods of activity obtained by *41b*, as compared to the very long-acting *40* and short-acting *39*, suggested that *41b* provided higher sulfone blood levels than *40* as it still maintained desirable repository activity. Four other compounds [(*42a*), (*42b*), (*43a*), and (*43b*)] exhibited noteworthy repository characteristics with regard to their antimalarial and antileprotic activities.

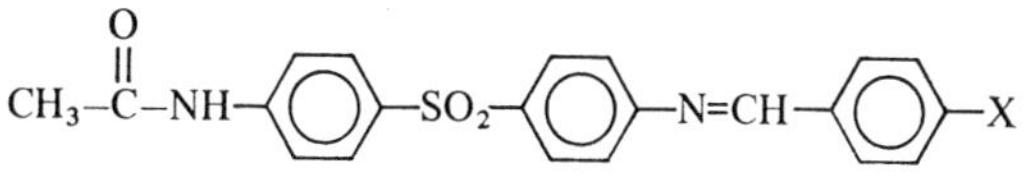

42a = X = H
42b = X = $NHCOCH_3$.

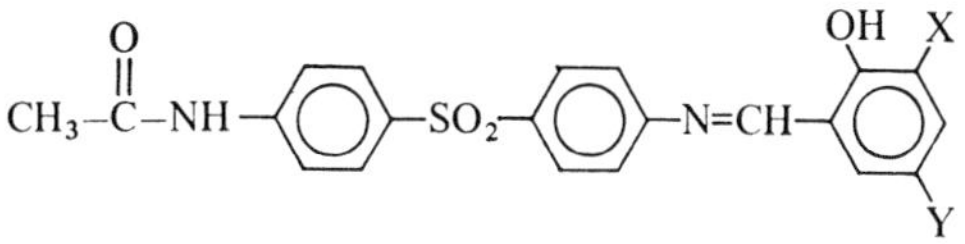

43a = X = Br, Y = Br
43b = X = Cl, Y = Cl.

The duration of protection of these compounds in mice and rats against challenge with *P. berghei* was intermediate between the short-acting *39* and the very long-acting *40*. With the exception of *43b*, the pattern of their urinary excretion in rats also was intermediate between that of *39* and *40*.

Enamines, or α,β-unsaturated amines, are used extensively as protective groups in selective organic reactions. Their structural types vary widely (SZMUSZKOVICZ, 1963) as do their hydrolysis rates (HEYL and HERR, 1953). CALDWELL et al. (1971) have recently investigated the use of this class of compounds in prodrug design. In this study, several enamines of phenylpropanolamine (*44*) were prepared in a search for derivatives that would hydrolyze at significantly different rates.

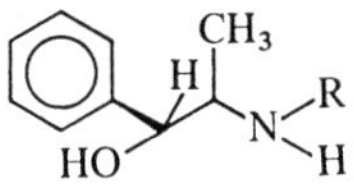

44 = R = H
44a = R = $-\underset{\displaystyle CH_3}{\underset{|}{C}}=CH=CO_2C_2H_5$
44b = R = $-CH=C(CO_2C_2H_5)_2$

In general, the compounds were found to cleave faster as the pH was lowered and exhibited different rates of hydrolysis in phosphate buffer at pH 7.4. Their hydrolysis was not accelerated in dilute plasma. The *in vitro* hydrolysis studies indicated that *44a* would hydrolyze *in vivo* to give phenylpropanolamine (*44*) but that *44b* would not hydrolyze. Acute toxicity studies in mice showed that *44a* possessed only about one-half the molar toxicity of *44*. In rats, *44a* produced pressor activity which was less than that of phenylpropanolamine (*44*), and in the antitussive test in dogs, it was about one-half as potent on a molar basis as *44*. From these data, the authors (CALDWELL et al., 1971) concluded that *44a* was a prodrug of *44*, but its *in vivo* conversion to phenylpropanolamine (*44*) was not rapid. As expected from the *in vitro* hydrolysis data, *44b* was found to be devoid of pressor activity in rats and was shown to be inactive in the antitussive test in dogs. It was thus concluded that *44b* did not cleave *in vivo* to liberate *44*.

References

AGERSBORG, H.P.K., BATCHELOR, A., CAMBRIDGE, G.W., RULE, A.W.: The pharmacology of penamecillin. Brit. J. Pharmac. Chemother. **26**, 649—655 (1966).

ALBERT, A.: Chemical aspects of selective toxicity. Nature (Lond.) **182**, 421—423 (1958).

ARIENS, E.J.: A general introduction to the field of drug design. In: ARIENS, E.J. (Ed.): Drug design, Vol. 1. New York: Academic Press 1971.

BARNDEN, R.L., EVANS, R.M., HAMLET, J.C., HEMS, B.A., JANSEN, A.B.A., TREVETT, M.E., WEBB, G.B.: Some preparative uses of benzylpenicillinic ethoxyformic anhydride. J. chem. Soc. 3733—3739 (1953).

BEARD, W.J., FREE, S.M., JR.: Two doubled-blind, controlled studies comparing chlorethate, a new sedative-tranquilizer, with meprobamate and with phenobarbital. J. clin. Pharmacol. **7**, 41—45 (1967).

BENDER, M.L.: Mechanisms of catalysis of nucleophilic reaction of carboxylic acid derivatives. Chem. Rev. **60**, 53—113 (1960).

BERGER, F.M., RILEY, R.F.: The preparation and pharmacological properties of the acid succinate of 3-(*o*-toloxy)-1,2-propanedial (myanesin). J. Pharmacol. exp. Ther. **96**, 269—275 (1949).

CALDWELL, H.C., ADAMS, H.J., RIVARD, D.E., SWINTOSKY, J.V.: Trichlorethyl carbonate. I. Synthesis, physical properties and pharmacology. J. pharm. Sci. **56**, 920—921 (1967).

CALDWELL, H.C., ADAMS, H.J., JONES, R.G., MANN, W.A., DITTERT, L.W., CHONG, C.W., SWINTOSKY, J.V.: Enamine prodrugs. J. pharm. Sci. **60**, 1810—1812 (1971).

CARPENTER, F.H.: The anhydride of benzylpenicillin. J. Amer. chem. Soc. **70**, 2964—2966 (1948).

CELMER, W.D., ELS, H., MURAI, K.: Oleandomycin derivatives. Preparation and characterization. In: Antibiotics annual 1957—1958. New York: Medical Encyclopedia, Inc. 1958.

CELMER, W.D.: Triacetyloleandomycin: Biochemical correlations. In: Antibiotics annual 1958—1959. New York: Medical Encyclopedia, Inc. 1959.

CERIOTTI, G., DE FRANCESHI, A., DE CARNERI, I., ZAMBONI, V.: Ricerche sperimentali su un derivato solubile in acqua del cloroamfenicolo: il succinato di cloroamfenicolo. Il Farmaco, Ed. Sc. **9**, 21—38 (1954).

CHONG, C.W.S.: Inhibition of human plasma esterases by 2′-dimethylaminoethyl-2,2-diphenyl valerate hydrochloride (SK & F 525A), Ph. D. thesis, Temple University, 1970.

CRAM, D.J., HAMMOND, G.S.: Organic chemistry, 2nd ed. New York: McGraw-Hill 1964.

VON DAEHNE, W., FREDERIKSEN, E., GUNDERSEN, E., LUND, F., MORCH, P., PETERSEN, H.J., ROHOLT, K., TYBRING, L., GODTFREDSEN, W.O.: Acyloxymethyl esters of ampicillin. J. med. Chem. **13**, 607—612 (1970a).

VON DAEHNE, W., GODTFREDSEN, W.O., ROHOLT, K., TYBRING, L.: Pivampicillin, a new orally active ampicillin ester. Antimicrob. Agent Chemother. 431—437 (1970b).

DAVENPORT, H.W.: Gastric mucosal injury by fatty and acetylsalicylic acids. Gastroenterology **46**, 245—253 (1964).

DITTERT, L.W., CALDWELL, H.C., ADAMS, H.J., IRWIN, G.M., SWINTOSKY, J.V.: Acetaminophen prodrugs. I. Synthesis, physico-chemical properties, and analgesic activity. J. pharm. Sci. **57**, 774—780 (1968a).

DITTERT, L.W., IRWIN, G.M., CHONG, C.W., SWINTOSKY, J.V.: Acetaminophen prodrugs. II. Effect of structure and enzyme source on enzymatic and nonenzymatic hydrolysis of carbonate esters. J. pharm. Sci. **57**, 780—783 (1968b).

Dittert, L. W., Caldwell, H. C., Ellison, T., Irwin, G. M., Rivard, D. E., Swintosky, J. V.: Carbonate ester prodrugs of salicylic acid. Synthesis, solubility characteristics, *in vitro* enzymatic hydrolysis rates, and blood levels of total salicylate following oral administration to dogs. J. pharm. Sci. **57**, 828—831 (1968c).

Dittert, L. W., Adams, H. J., Alexander, F., Chong, C. W., Ellison, T., Swintosky, J. V.: 4-Acetamidophenyl 2,2,2-trichloroethyl carbonate. Particle size studies in animals and man. J. pharm. Sci. **57**, 1146—1151 (1968d).

Dittert, L. W., Adams, H. J., Chong, C. W., Swintosky, J. V.: Trichloroethyl carbonate: Influence of particle size on oral toxicity in man. J. pharm. Sci. **57**, 1269—1270 (1968e).

Dittert, L. W., Irwin, G. M., Rattie, E. S., Chong, C. W., Swintosky, J. V.: Hydrolysis of 4-acetamidophenyl 2,2,2-trichloroethyl carbonate by esterolytic enzymes from various sources. J. pharm. Sci. **58**, 557—559 (1969).

Dresel, P. E., Slater, I. H.: Observations on the pharmacology of mephenesin carbonate. Proc. Soc. Exp. Biol. (N. Y.) **79**, 286—287 (1952).

Edwards, L. J.: The hydrolysis of aspirin. Trans. Faraday Soc. **46**, 723—735 (1950).

Elslager, E. F., Gavrilis, Z. B., Phillips, A. A., Worth, D. F.: Repository drugs. IV. 4′-4‴-Sulfonylbisacetanilide (Acedapsone, DADDS) and related sulfanilylanilides with prolonged antimalarial and antileprotic action. J. med. Chem. **12**, 357—363 (1969a).

Elslager, E. F., Phillips, A. A., Worth, D. F.: Repository drugs. V. 4′,4‴-[*p*-phenylenebis (methylidyneimino-*p*-phenylene-sulfonyl)] bisacetanilide (PSBA) and related 4′,4‴-[*Bis*(imino-*p*-phenylene-sulfonyl)] bisanilides, a novel class of long-acting antimalarial and antileprotic agents. J. med. Chem. **12**, 363—367 (1969b).

Fersht, A. R., Kirby, A. J.: Structure and mechanism in intramolecular catalysis. The hydrolysis of substituted aspirins. J. Amer. chem. Soc. **89**, 4853—4857 (1967a).

Fersht, A. R., Kirby, A. J.: Intramolecular general base catalysis of ester hydrolysis. J. Amer. chem. Soc. **89**, 4857—4863 (1967b).

Fersht, A. R., Kirby, A. J.: Intramolecular nucleophilic catalysis of ester hydrolysis by the carboxylate group. J. Amer. chem. Soc. **89**, 5960—5961 (1967c).

Fersht, A. R., Kirby, A. J.: Intramolecular general acid catalysis of ester hydrolysis by the carboxylic acid group. J. Amer. chem. Soc. **89**, 5961—5962 (1967d).

Fife, T. H., Jao, L. D.: General acid catalysis of acetal hydrolysis. The hydrolysis of 2-aryloxy-tetrahydropyrans. J. Amer. chem. Soc. **90**, 4081—4085 (1968).

Fife, T. H., Brod, L. H.: General acid catalysis and the pH-independent hydrolysis of 2-(*p*-nitrophenoxy) tetrahydropyran. J. Amer. chem. Soc. **92**, 1681—1684 (1970).

Fodor, G., Nador, K.: The stereochemistry of the tropane alkaloids. Part I. The configuration and *ψ*-tropine. J. chem. Soc. 721—723 (1953a).

Fodor, G., Kovacs, O.: The stereochemistry of the tropane alkaloids. Part II. The configurations of the cocaines. J. chem. Soc. 724—727 (1953b).

Gaetjens, E., Morawetz, H.: Intramolecular carboxylate attack on ester groups. The hydrolysis of substituted phenyl acid succinates and phenyl acid glutarates. J. Amer. chem. Soc. **82**, 5328—5335 (1960).

Garrett, E. R.: The kinetics of solvolysis of acid esters of salicylic acid. J. Amer. chem. Soc. **79**, 3401—3408 (1957).

Gibaldi, M., Schwartz, M. A.: The pharmacokinetics of penamecillin. Brit. J. Pharmac. Chemother. **28**, 360—366 (1966).

Glazko, A. J., Edgerton, W. H., Dill, W. A., Lenz, W. R.: Chloromycetin palmitate—a synthetic ester of chloromycetin. Antib. and Chemo. **2**, 234—242 (1952).

Glazko, A. J., Dill, W. A., Kazenko, A., Wolf, L. M., Carnes, H. E.: Physical factors affecting the rate of absorption of chloramphenicol esters. Antibiot. and Chemother. 8, 516—527 (1958a).

Glazko, A. J., Carnes, H. E., Kazenko, A., Wolf, L. M., Reutner, T. F.: Succinic acid esters of chloramphenicol. In: Antibiotics annual 1957—1958. New York: Medical Encyclopedia, Inc. 1958b.

Glick, D.: Some additional observations on the specificity of cholinesterase. J. biol. Chem. **137**, 357—362 (1941).

Gordon, M., Miller, J. G., Day, A. R.: Effect of structure on reactivity: 1. Ammonalysis of esters with special reference to the electron release effects of alkyl and aryl groups. J. Amer. chem. Soc. **70**, 1946—1953 (1948).

Hanzlik, P. J., Presho, N. E.: The salicylates. XV. Liberation of salicyl from excretion of salicylsalicylate. J. Pharm. exp. Ther. **26**, 61—70 (1925).

Harper, N. J.: Drug latentiation. J. Med. pharm. Chem. **1**, 467—500 (1959).

Harper, N. J.: Drug latentiation. In: Progress in drug research, Vol. 4, pp. 221—294. New York: Wiley 1962.

Heathcote, A. G. S., Nassau, E.: Concentration of penicillin in the lungs. Effects of two penicillin esters in chronic pulmonary infections. Lancet 1951 I. 1255—1257.

HEYL, F. W., HERR, M. E.: Enamine derivatives of steroidal carbonyl compounds. II. J. Amer. chem. Soc. **75**, 1918—1920 (1953).
HOFSTEE, B. H. J.: Kinetics of chymotrypsin with a homologous series of *n*-fatty acid esters as substrates. Biochem. biophys. Acta **32**, 182—188 (1959).
HUF, E. G., COLES, F. K., EUBANK, L. L.: Comparative plasma levels of mephenesin carbamate and methocarbamol. Proc. Soc. exp. Biol. (N. Y.) **102**, 276—277 (1959).
HUSSAIN, A., HIGUCHI, T., STELLA, V.: Unpublished data (1972).
HUSSAIN, A., YAMASAKI, M., TRUELOVE, J. E.: Kinetics of hydrolysis of acylals of aspirin. Hydrolysis of (1′-ethoxy) ethyl 2-acetoxybenzoate. J. pharm. Sci. **63**, 627—628 (1974).
INGOLD, C. K.: Structure and mechanism in organic chemistry, 2nd ed. Ithaca-London: Cornell University Press 1969.
JANSEN, A. B. A., RUSSELL, T. J.: Some novel penicillin derivatives. J. chem. Soc. 2127—2132 (1965).
JENCKS, W. P.: Catalysis in chemistry and enzymology. New York: McGraw-Hill, 1969.
KESSLER, W. B., BORMAN, A.: Some biological activities of certain progestogens: I. 17-α-hydroxy-progesterone 17-*n*-caproate. Ann. N. Y. Acad. Sci. **71**, 486—493 (1958).
KOSOWER, E. M.: Molecular biochemistry. New York: McGraw-Hill Inc. 1962.
KUPCHAN, S. M., CASY, A. F., SWINTOSKY, J. V.: Drug latentiation. Synthesis and preliminary evaluation of testosterone derivatives. J. pharm. Sci. **54**, 514—524 (1965).
LEONARDS, J. R., LEVY, G.: Aspirin-induced occult gastrointestinal blood loss: local versus systemic effects. J. pharm. Sci. **59**, 1511—1513 (1970).
LEVINE, R. M., CLARK, B. B.: Relationship between structure and *in vitro* metabolism of various esters and amides in human serum. J. Pharmacol. **113**, 272—282 (1955).
LONDON, I., POET, R. B.: Comparative plasma levels of mephenesin and its carbamic acid ester. Proc. Soc. exp. Biol. (N. Y.) **94**, 191 (1957).
MILLER, E., ROCK, H. J., MOORE, M. L.: Substituted sulfanilamides. I. N^4-acyl derivatives. J. Amer. chem. Soc. **61**, 1198—1200 (1939).
MISHER, A., ADAMS, H. J., FISHLER, J. J., JONES, R. G.: Pharmacology of the hexylcarbonate of salicylic acid. J. pharm. Sci. **57**, 1128—1131 (1968).
MORAWETZ, H., WESTHEAD, E. W., JR.: Acid-base catalysis in polyelectrolyte solutions. J. Polymer Sci. **16**, 273—281 (1955).
MORAWETZ, H., GAETJENS, E.: A kinetic approach to the characterization of the "microtacticity" of a polymer chain. J. Polymer Sci. **32**, 526—528 (1958).
PAYNE, H. M., HACKNEY, R. L., JR.: Studies of chloramphenicol acid succinate (sodium salt). In: Antibiotics annual 1957—1958. New York: Medical Encyclopedia Inc. 1958.
PFIFFNER, J. J., NORTH, H. B.: The isolation of 17-hydroxyprogesterone from the adrenal gland. J. biol. Chem. **139**, 855—861 (1941).
PHILLIPS, A. P.: Preparation of the monocholine ester of succinic acid and some related derivatives. J. Amer. chem. Soc. **75**, 4725—4727 (1953).
PIERSON, R. N., HOLT, P. R., WATSON, R. M., KEATING, R. P.: Aspirin and gastrointestinal bleeding. Amer. J. Med. **31**, 259—265 (1961).
RAPOPORT, H., BAKER, D. R., REIST, H. N.: Morphinone. J. Org. Chem. **22**, 1489—1492 (1957).
RATTIE, E. S., SHAMI, E. G., DITTERT, L. W., SWINTOSKY, J. V.: Acetaminophen prodrugs. III. Hydrolysis of carbonate and carboxylic acid esters in aqueous buffers. J. pharm. Sci. **59**, 1738—1741 (1970).
ROBERTSON, D. N.: Adducts of *tert*-alcohols containing an ethynyl group with dihydropyran. Potentially useful intermediates. J. Org. Chem. **25**, 931—932 (1960).
ROSS, S., PUIG, J. R., ZAREMBA, E. A.: Chloramphenicol acid succinate (sodium salt). In: Antibiotics annual 1957—1958. New York: Medical Encyclopedia Inc. 1958.
RUBIN, H. S.: Serum salicylate levels in osteoarthritis following oral administration of a preparation containing salicylsalicylic acid and acetylsalicylic acid. Amer. J. med. Sci. **248**, 31—36 (1964).
SANDLER, S. R., KARO, W.: Organic functional group preparations. New York: Academic Press 1971.
SCHOENBACH, E. B.: Pharmacology and clinical trial of a microcrystalline chloromycetin suspension. Amer. J. Med. **14**, 525—526 (1953).
SHAH, A. A., CONNORS, K. A.: Alpha-chymotrypsin-catalyzed hydrolysis of some carbonate esters. J. pharm. Sci. **57**, 282—287 (1968).
SHAPIRO, S. L., ROSE, I. M., FREEDMAN, L.: Analeptic amides of *d*-α-methylphenethylamine. J. Amer. chem. Soc. 80, 6065—6071 (1958).
SHUBIN, H., DUMAS, K., SOKMENSUER, A.: Clinical and laboratory studies on a new derivative of oleandomycin. In: Antibiotics annual 1957—1958. New York: Medical Encyclopedia Inc. 1958.
SMITH, I. M., SODERSTROM. W. H,: Clinical experience with a new erythromycin derivative, erythromycin propionate. J. Amer. med. Ass. **170**, 184—188 (1959).

STUBBE, L., PIETERSEN, J. H., VAN HEULEN, C.: Aspirin preparations and their noxious effect on the gastrointestinal tract. Brit. med. J. **1962I**, 675—680.
STURGE, L. M., WHITTAKER, V. P.: The esterases of horse blood. The specificity of horse plasma cholinesterase and ali-esterase. Biochem. J. **47**, 518—525 (1950).
SUDBOROUGH, J. J.: The formation and hydrolysis of esters of keytonic acids. J. Chem. Soc. **101**, 1227—1238 (1912).
SWINTOSKY, J. V., ADAMS, H. J., CALDWELL, H. C., DITTERT, L. W., ELLISON, T., RIVARD, D. E.: Carbonate prodrugs in formulation and therapeutics. J. pharm. Sci. **55**, 992 (1966).
SWINTOSKY, J. V., CALDWELL, H. C., CHONG, C. W., IRWIN, G. M., DITTERT, L. W.: 4-acetamidophenyl 2,2,2-trichloroethyl carbonate. Synthesis, physical properties, and *in vitro* hydrolysis. J. pharm. Sci. **57**, 752—756 (1968).
SWINTOSKY, J. V.: Unpublished data (1964).
SZMUSZKOVICZ, J.: Enamines. Advan. Org. Chem. **4**, 1—113 (1963).
TSUKAMOTO, H., YOSHIMURA, H., TATSUMI, K.: Metabolism of drugs. XXXIX. Further studies on carbamate *N*-glucuronide formation in animal body. Chem. Pharm. Bull. (Tokyo) **11**, 1134—1139 (1963).
UNGAR, J., MUGGLETON, P. W.: Accumulation of diethylaminoethanol ester of penicillin in inflammed lung tissue. Brit. med. J. **1952I**, 1211—1213.
WILLIAMS, R. T.: Detoxification mechanisms, 2nd ed. London: Chapman and Hall 1959.
WORTH, D. F., ELSLAGER, E. F., PHILLIPS, A. A.: Repository drugs. VI. 4'[*N*-(Aralkylidene-, benzylidene-, and naphthylidene) sulfanilyl] amilides, 4'-*N*-(dimethylamino) methylene sulfanilyl anilides, and related sulfanilylanilides with prolonged antimalarial and antileprotic action. J. med. Chem. **12**, 591—596 (1969).
YAMAMOTO, A., YOSHIMURA, H., TSUKAMOTO, H.: Metabolism of drugs, 28. Metabolic fate of meprobamate. (1). Isolation and characterization of metabolites. Chem. Pharm. Bull. (Tokyo) **10**, 522—528 (1962a).
YAMAMOTO, A., YOSHIMURA, H., TSUKAMOTO, H.: Metabolism of drugs. XXIX. Metabolic fate of meprobamate. 2. Further studies on the structure of the metabolites. Chem. Pharm. Bull. (Tokyo) **10**, 540—544 (1962b).

Chapter 62

Biotransformation of Drugs to Pharmacologically Active Metabolites

S. Garattini, F. Marcucci, and E. Mussini

I. Introduction

Most drugs that are administered to living organisms are extensively transformed through different metabolic pathways. The complete pattern of metabolism is known for very few drugs. Several drugs are transformed into as many as one hundred molecular species (metabolites). Depending on their chemical structures, these metabolites may be rapidly excreted, mostly through the urine and the bile. The effects of metabolites of administered drugs may be quite different from the effects of the parent compound.

This review summarizes present knowledge about metabolites that may be in fact responsible for the therapeutic effect ascribed to the administered drug. These metabolites, frequently referred to as "active metabolites", may account for the entire activity of the parent compound or only for sustaining the time or the intensity of the effects. Many factors, including animal species, strain, sex, age, circadian rhythms, route of administration, dose, previous treatments, and pathologic conditions, may markedly influence the pathway and the rate of drug metabolism.

Therefore, the designation of an "active metabolite" must be specific as to experimental conditions. Any extrapolation to other experimental conditions and especially to human therapy must be cautious and validated by suitable data.

The presence of an active metabolite should be suspected but is not proved by the following observations:

a) Unusual time lag between intravenous administration of a drug and its effect.

b) Lack of *in vitro* effect for a compound that is active *in vivo*.

c) Presence of pharmacologic activity when the drug is not present in the supposed target organ. However, "non-presence" is defined by the sensitivity of the method utilized to detect the drug.

d) Greater activity of a drug given orally than by the intravenous route, This, however, does not necessarily apply for drugs acting on the gastrointestine.

e) Alteration of pharmacologic activity by pretreatment with other compounds known to stimulate or inhibit drug metabolizing enzymes.

The definition of an active metabolite includes at least the following points:

a) The metabolite, given by any route of administration, should reproduce the pharmacologic or biochemical effects of the parent compound.

b) The minimal tissue concentration at which the metabolite is active should be comparable to the concentrations present in the same tissue after the administration of the parent compound.

c) The concentration at which the metabolite reproduces the effect of the parent drug may not necessarily refer to "whole" organ concentration, but to concentrations in given parts of the organ (for example, brain) or to subcellular concentration including those at the receptor site. This is to emphasize that in some cases an active metabolite can be formed within a cell structure, but the cell or the subcellular structure may not be permeable to that particular metabolite.

d) The supposedly active metabolite should not be significantly metabolized back to the parent compound.

Unfortunately, these criteria are seldom met as will be seen in the examples of active metabolites reported in this review.

II. Drugs Acting on the Central Nervous System

A. Hypnotics and Sedatives

a) Chloral hydrate ($Cl_3CCH(OH)_2$) one of the oldest organic sedative and hypnotic agents, is metabolized to trichloroethanol, which prolongs the duration of its pharmacologic effect (MACKAY and COOPER, 1962).

B. Anticonvulsants

a) Mephobarbital (5 phenyl-5-ethyl-1-methyl-barbituric acid) (I) undergoes *N*-demethylation to form phenobarbital (II), which retains the same therapeutic property (MARK, 1963).

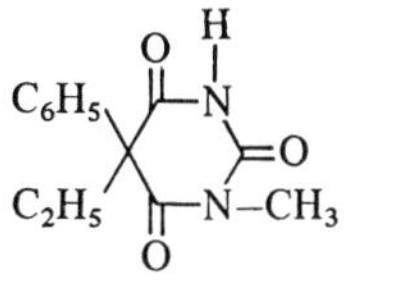

(I) 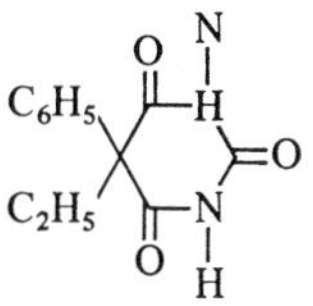 (II)

b) Primidone (2-desoxy-5-ethyl-5-phenyl barbituric acid) (III) is metabolized to phenobarbital (II). About 15% of the dose in man is oxidized to phenobarbital, which results in blood concentrations of this metabolite in the therapeutic range (MARK, 1963).

Another important metabolite of primidone is phenylethylmalondiamide (IV), which represents 50 to 70% of the administered dose and is excreted in the urine. Phenylethylmalondiamide possesses, however, only one-fortieth of the anticonvulsant activity of primidone (SPINKS and WARING, 1963).

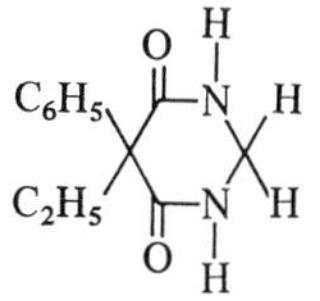

(III) 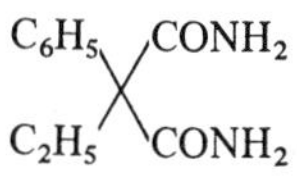 (IV)

c) Trimethadione (3,5,5-trimethyloxazolidine-2-4-dione) (V), paramethadione (3,5-dimethyl-5-ethyloxazolidine-2,4-dione) (VI) and dimidione (3-ethyl-5,5-dimethyl-oxazolidine-2,4-dione) (VII) are extensively *N*-demethylated, and it has been suggested that the *N*-demethylated metabolites are responsible for the bioactivity (SPINKS and WARING, 1963; BUTLER, 1955).

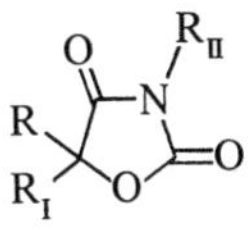

(V) $R, R_I, R_{II} = CH_3$
(VI) $R = R_{II} = CH_3; R_I = C_2H_5$
(VII) $R = R_I = CH_3; R_{II} = C_2H_5$

d) Methsuximide (*N*,α-dimethyl-α-phenylsuccinimide) (VIII), is *N*-demethylated in rats and humans to α-methyl-α-phenylsuccinimide, a pharmacologically active anticonvulsant compound (NICHOLLS and ORTON, 1971).

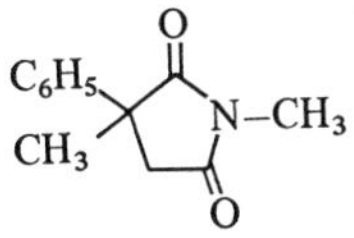

(VIII)

C. Centrally Acting Muscle Relaxant

a) Zoxazolamine (2-amino-5-chlorobenzoxazole) (IX) is metabolized in man to chloroxazone (5-chloro-2-benzoxazolinone) (X). This metabolite possesses muscle-relaxant activity (BURNS et al., 1958) and it has been subsequently introduced into clinical use.

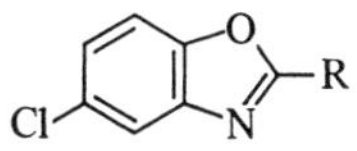

(IX) $R=NH_2$
(X) $R=O$

D. Narcotic Analgesics

a) Codeine (7,8-dehydro-4,5-epoxy-6-hydroxy-3-methoxy-*N*-methylmorphinan) (XI) in different animal species, is O-demethylated to form morphine (ADLER, 1967), a more active analgesic drug than codeine.

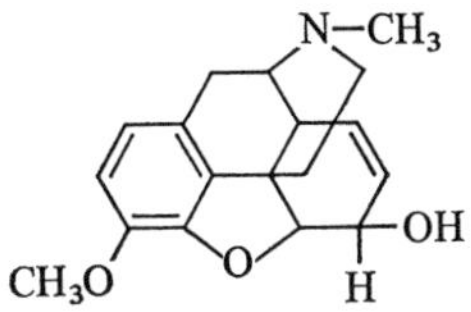

(XI)

b) Meperidine (4-carbethoxy-1-methyl-4-phenyl-piperidine) (XII), is *N*-demethylated to normeperidine (WAY, 1968) (XIII), which differs from meperidine in having greater excitant and less depressant effect.

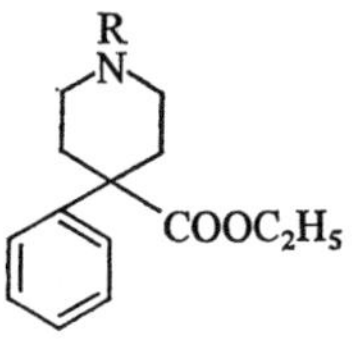

R
N
$COOC_2H_5$

(XII) = R = CH_3
(XIII)= R = H

c) Acetylmethadol (6-dimethylamino-4,4-diphenyl-3-acetoxy-heptane) (XIV) is *N*-demethylated to noracetylmethadol (XV), a very potent analgesic, in rats (McMAHON et al., 1965a).

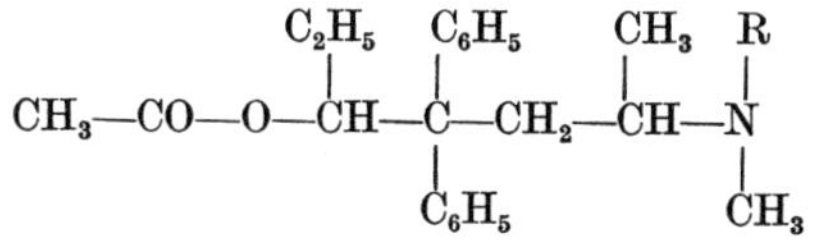

$$CH_3{-}CO{-}O{-}\underset{}{\overset{C_2H_5}{\overset{|}{CH}}}{-}\underset{\underset{C_6H_5}{|}}{\overset{C_6H_5}{\overset{|}{C}}}{-}CH_2{-}\overset{CH_3}{\overset{|}{CH}}{-}\underset{\underset{CH_3}{|}}{\overset{R}{\overset{|}{N}}}$$

(XIV) R = CH_3
(XV) R = H

d) Diphenoxylate [2,2-diphenyl-4-(4-carbethoxy-4-phenyl-1-piperidino) butyronitrile] (XVI), a well-known antidiarrheal agent is metabolized mainly by hydrolysis of the ester group to the free carboxylic acid group (difenoxine) (XVII). This metabolite is pharmacologically and clinically more active than diphenoxylate (BAROWSKY and SCHWARTZ, 1962; NIEMEGEERS et al., 1972).

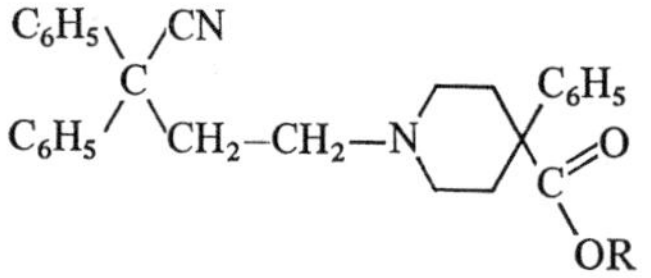

C_6H_5 CN
C
C_6H_5 $CH_2{-}CH_2{-}N$ C_6H_5
C=O
OR

(XVI) = R = C_2H_5
(XVII) = R = H

E. Analgesic-Antipyretics, Anti-Inflammatory Agents, and Inhibitors of Uric Acid Synthesis

a) Salicylates. The most important compounds in current clinical use are sodium salicylate, salicylamide, and acetyl salicylic acid. The efficacy of the different salicylates depends in part, but not entirely, on the amount of salicylic anion liberated in the body (DONE, 1960).

b) Phenacetin (*p*-Ethoxy-acetanilide) (XVIII) is rapidly biotransformed by ether cleavage to acetaminophen (4-hydroxy-acetanilide) (XIX), which is thought to be primarily responsible for the analgesic activity of phenacetin (BRODIE and AXELROD, 1949).

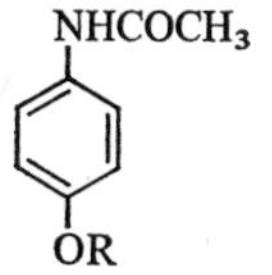

$NHCOCH_3$

OR

(XVIII) = R = C_2H_5
(XIX) = R = H

However, this view has been questioned by experiments showing that phenacetin has analgesic activity in rats even when its metabolism to acetaminophen (XIX) was inhibited (CONNEY et al., 1966).

c) Aminopyrine (4-dimethylamino-2,3-dimethyl-1-phenyl-3-pyrazoline-5-one) (XX) is demethylated to a small extent by the liver microsomal enzyme system to 4-amino antipyrine (XXI) and then acetylated to *N*-acetyl-4-amino antipyrine (XXII), which exhibits similar analgesic activity (BRODIE and AXELROD, 1950).

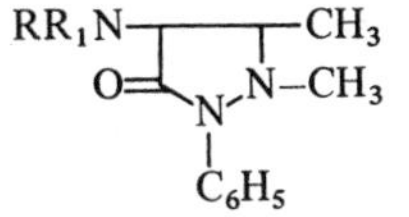

RR_1N — CH_3
O= N — N—CH_3
C_6H_5

(XX) R = R_1 = CH_3
(XXI) R = H ; R_1 = H
(XXII) R = H ; R_1 = $COCH_3$

d) Phenylbutazone (4-butyl-3,5-dioxo-1,2-diphenyl-pyrazolidine) (XXIII) is slowly metabolized in man by *p*-hydroxylation in the benzene ring to yield oxyphenylbutazone (XXIV) and by hydroxylation in the butyl side chain to yield γ-hydroxy-phenylbutazone (BURNS et al., 1960) (XXV).

The first metabolite possesses anti-inflammatory and sodium-retaining effects of the parent drug, while the second metabolite accounts for the uricosuric effect of phenylbutazone.

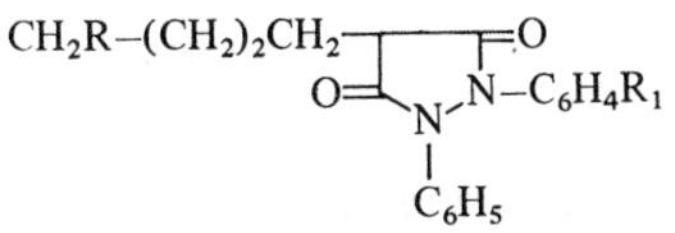

$CH_2R-(CH_2)_2CH_2$ — =O
O= N — N—$C_6H_4R_1$
C_6H_5

(XXIII) R = H ; R_1 = H
(XXIV) R = H ; R_1 = OH
(XXV) R = OH ; R_1 = H

e) Allopurinol (1-*H*-pyrazolo [3,4-*d*]pyrimidin-4-ol-) (XXVI), a drug used for the therapy of hyperuricemic patients, is metabolized to alloxanthine (XXVII), which contributes significantly to the therapeutic effect of allopurinol (ELION et al., 1966).

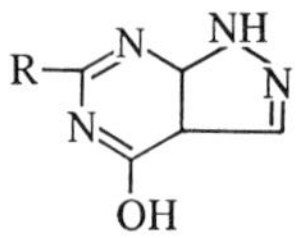

(XXVI) R = H
(XXVII) R = OH

f) (*4-phenylthioethyl*) 3,5-dioxo-1,2-diphenyl-pyrazolidine (XXVIII) is metabolized *in vivo* to its sulfoxide (XXIX), a potent uricosuric agent (BURNS et al., 1957).

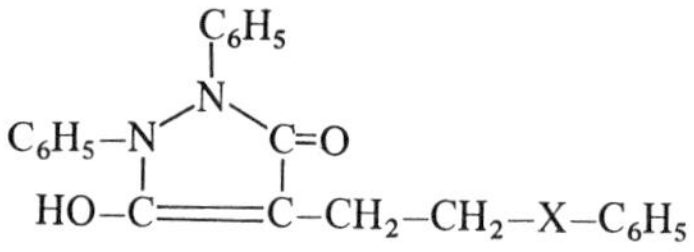

(XXVIII) X = S
(XXIX) X = S→O

F. Drugs Used in the Treatment of Psychiatric Disorders

1. Drugs for Treatment of Psychoses

a) Tetrabenazine (2-oxo-3-isobutyl-9,10-dimethoxy-1,2,3,4,6,7-hexahydro-11 *bH*-benzo [*a*] quinolizine (XXX) is biotransformed by reduction into a 2-hydroxy metabolite (RO-1-9571) (XXXI), which is also active in "protecting" norepinephrine stores from the long lasting depletion induced by reserpine (SCHWARTZ et al., 1966).

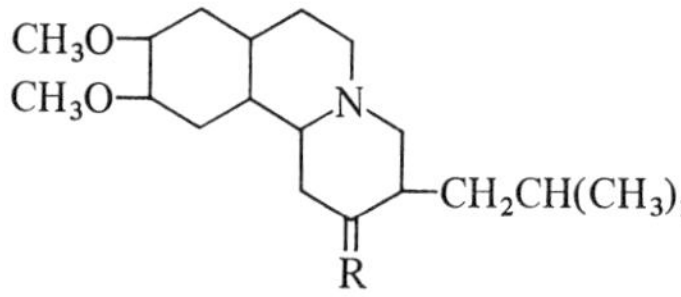

(XXX) R = O
(XXXI) R = OH

2. Drugs for Anxiety

a) Chlordiazepoxide (7-chloro-2-methylamino-5-phenyl-1,4-benzodiazepine-4-oxide) (XXXII) has been shown (KOECHLIN and D'ARCONTE, 1963; SCHWARTZ and POSTMA, 1966) to be biotransformed in dog and humans to the *N*-demethylchlordiazepoxide (Ro-5-0883/1) (XXXIII), which undergoes further deamination to form a lactame (Ro-5-2092) (XXXIV). In the dog another metabolite, oxazepam, (XXXV) has been reported to be produced by chlordiazepoxide (KIMMEL and WALKENSTEIN, 1967). These three metabolites are considered pharmacologically active (RANDALL and SCHALLEK, 1968; ZBINDEN and RANDALL, 1967).

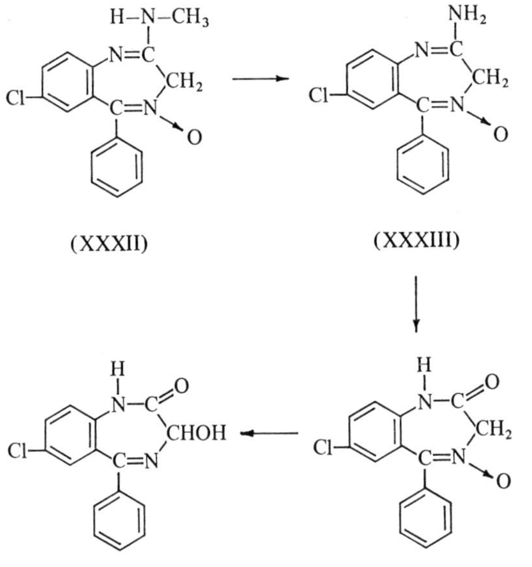

(XXXII) (XXXIII)

(XXXV) (XXXIV)

b) Diazepam (7-chloro-1,3-dihydro-1-methyl-5-phenyl-2*H*-1,4 benzodiazepine-2-one) (XXXVI) undergoes a process of N_1-demethylation to form N_1-demethyl-diazepam (XXXVII) and a process of C_3-hydroxylation to form N_1-methyl-oxazepam (XXXVIII). These two metabolites are then respectively C_3-hydroxylated and N_1-demethylated to form a common metabolite known as oxazepam (XXXV) (KVETINA et al., 1968; SCHWARTZ and POSTMA, 1968; MARCUCCI et al., 1969).

N-demethyldiazepam, *N*-methyloxazepam and oxazepam have a marked anticonvulsant effect (RANDALL et al., 1965) (see Table 1).

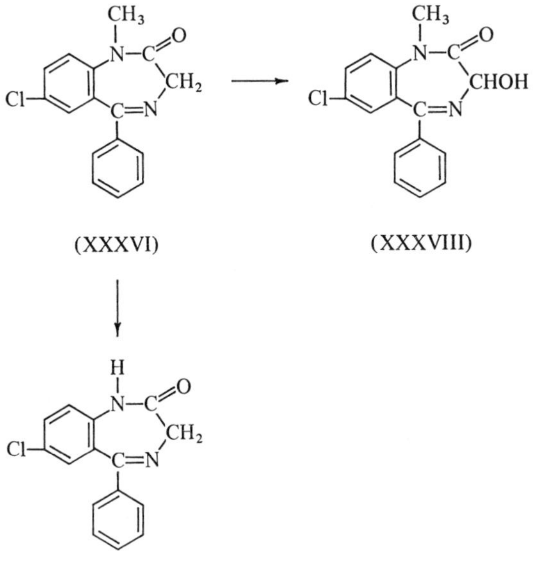

(XXXVI) (XXXVIII)

(XXXVII)

The long lasting anticonvulsant effect exerted by diazepam in mice is related to the formation and persistance in the brain of the metabolite, oxazepam. Table 2 indicates that the amounts of oxazepam found in brain 24 h after the administration of diazepam are similar to the amounts exerting a comparable anticonvulsant activity when oxazepam was administered (GARATTINI et al., 1973).

Table 1. *Anticonvulsant activity of diazepam (DZ), N-demethyl diazepam (DDZ), N-methyl-oxazepam (MOX) and oxazepam (OX) after i.v. administration to mice pretreated with pentetrazole convulsant*

Drug	Min. between drug and pentetrazole	Antipentetrazole activity (ED_{50} mg/kg i.v. and 95% fiducial limits)	LD_{50} (mg/kg i.p. and 95% fiducial limits)[a]
DZ	5	0.280 (0.233—0.336)	355 (292—382)
	180	0.680 (0.486—0.952)	
DDZ	5	0.198 (0.177—0.221)	
	180	0.640 (0.427—0.960)	290 (261—322)
MOX	5	0.155 (0.119—0.202)	
	180	0.680 (0.557—0.830)	310 (263—366)
OX	5	0.342 (0.263—0.445)	
	180	0.690 (0.460—1.035)	1500

[a] The LD_{50} was calculated 72 h after drug administration to groups of 60 mice for each drug. Pentetrazole was given i.p. at the dose of 120 mg/kg

Table 2. *Levels of brain oxazepam versus anticonvulsant activity against metrazol-induced convulsions in mice*

Drug administered (mg/kg i.v.)	Time after treatment (h)	Brain level of oxazepam (μg/g $\pm$ S.E.)	Anticonvulsant activity (% protection)[a]
Diazepam 5	20	0.17 $\pm$ 0.01	87
Oxazepam 5	18	0.15 $\pm$ 0.02	80
Oxazepam 5	24	0.11 $\pm$ 0.01	60
Oxazepam 1	15	0.06 $\pm$ 0.005	0

[a] The dose of 120 mg/kg i.p. of metrazol produced lethal convulsions in all the control mice treated only with the solvent

c) Medazepam (7-chloro-2,3-dihydro-1-methyl-5-phenyl-1-H-1,4 benzodiazepine) (XXXIX) is rapidly metabolized in rat and man (SCHWARTZ and CARBONE, 1970; DE SILVA and PUGLISI, 1970) to diazepam (XXXVI).

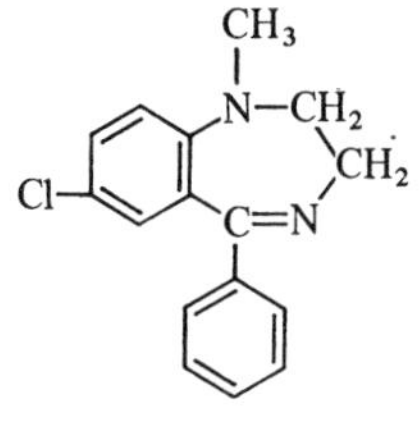

CH3 N—CH2 CH2 Cl— C=N

(XXXIX)

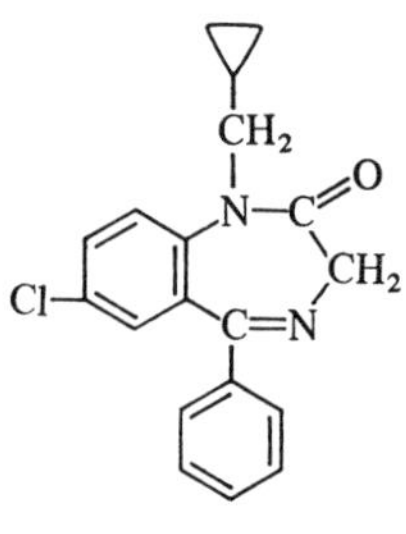

CH2 N—C=O CH2 Cl— C=N

(XL)

d) Prazepam [7-chloro-1(cyclopropylmethyl)1,3-dihydro-5-phenyl-2*H*-1,4]-benzodiazepine-2-one (XL) is metabolized in dog and man (DI CARLO and VIAU, 1970; DI CARLO et al., 1970) into at least two compounds with tranquilizing activity, namely oxazepam (XXXV) and N_1-demethyldiazepam (XXXVII) (RANDALL et al., 1965).

3. Psychotropic Drugs for Affective Disorders

a) Tricyclic Compounds Structurally Related to the Phenothiazine Class of Antipsychotic Agents. Imipramine [5-(3-dimethylaminopropyl-10,11-dihydro-5 *H*-dibenz [*b*, *f*] azepine)] [XLI) amitriptyline (10,11 dihydro-*N*,*N*-dimethyl-5*H*-dibenzo [*a*,*d*]-cycloheptene-Δ^5, γ-propylamine) (XLII), doxepine (*N*,*N*-dimethyl-dibenz-[*b*,*e*] oxepin-Δ^{11}-(6*H*)γ-propylamine) (XLIII) and prothiadene (*N*,*N*-dimethyldibenzo (*b*, *e*] thiepin-Δ^{11}-(6*H*)γ-propylamine) (XLIV) are *N*-demethylated and the metabolites (desipramine XLV, nortriptyline XLVI, demethyldoxepine XLVII and norprothiadene XLVIII) retain depressant and antireserpine activities similar to the parent drugs (GILLETTE et al., 1961; VERNIER et al., 1962; METYSOVA et al., 1965).

Recent findings have suggested that tertiary amines may be more effective than secondary amines in inhibiting the uptake of serotonin and that *N*-demethylated metabolites are usually more effective than the parent compounds as inhibitors of noradrenaline uptake (CARLSSON et al., 1966, 1969a and b; ROSS and RENYI, 1967, 1969; LIDBRINK et al., 1971; SAMANIN et al., 1972).

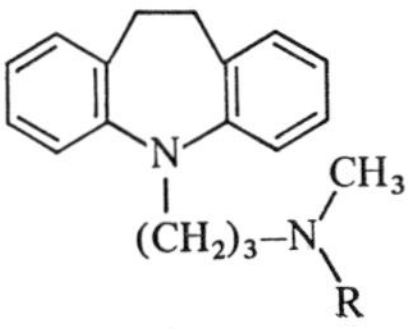

(XLI) = R = CH_3
(XLV) = R = H

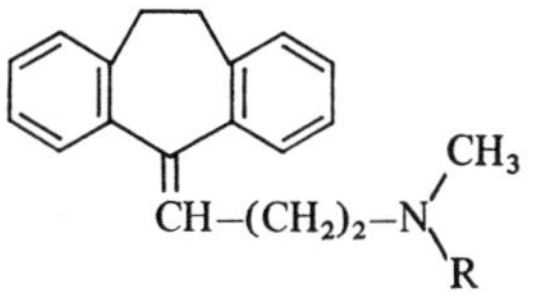

(XLII) R = CH_3
(XLVI) R = H

(XLIII) R = CH_3; X=O
(XLVII) R = H; X=O
(XLIV) R = CH_3; X=S
(XLVIII) R = H; X=S

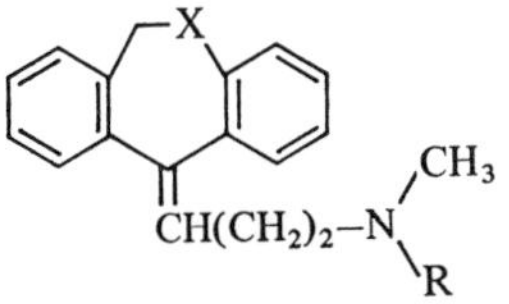

b) Monoaminoxidase (MAO) Inhibitors. Modaline sulfate (2-methyl-3-piperidino-pyrazine) (XLIX), a non-hydrazine MAO-inhibitor, requires an initial biotransformation by liver microsomal enzymes to an active unknown intermediate in order to exert its anti-MAO-like (HORITA, 1966) or imipramine-like (JORI et al., 1965) effect.

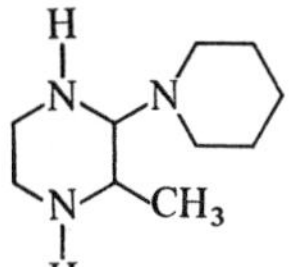

(XLIX)

III. Drugs Acting at Synaptic and Neuroeffector Functional Sites

A. Anticholinesterase Agents

a) Parathion (*O,O*-diethyl-*O*-(*p*-nitrophenyl) phosphorothionate (L) an organophosphorus insecticide, undergoes desulfuration by rat liver microsomes to paraoxon (LI), which exerts a powerful anticholinesterase activity (Davison, 1955).

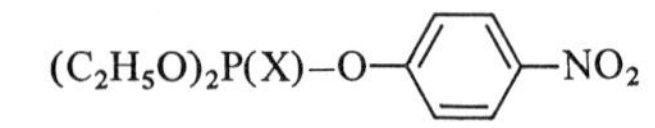

$(C_2H_5O)_2P(X){-}O{-}C_6H_4{-}NO_2$

(L) X = S
(LI) X = S

B. Drugs Acting on Postganglionic Adrenergic Nerve Endings and Structures Innervated by them (Sympathomimetic Drugs)

a) N-isopropylmethoxamine [2-(*N*-isopropylamino)-1-(2,5-dimethoxy-phenyl)-1-propanol] (LII) is dealkylated to methoxamine (LIII), which produces all the pharmacologic effects of the parent drug (Burns et al., 1967).

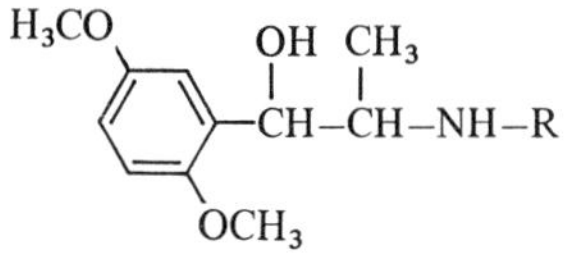

H_3CO, OCH_3 (ring substituents); side chain: $-CH(OH)-CH(CH_3)-NH-R$

(LII) R = $CH(CH_3)_2$
(LIII) R = H

b) Fenfluramine [1(3-trifluoromethylphenyl)-2-ethyl aminopropane] (LIV) a non-stimulant anorectic drug is de-ethylated in man to norfenfluramine [1(3-trifluoromethylphenyl)2-aminopropane] (LV), which possesses the same pharmacologic activity (Beckett and Brookes, 1967).

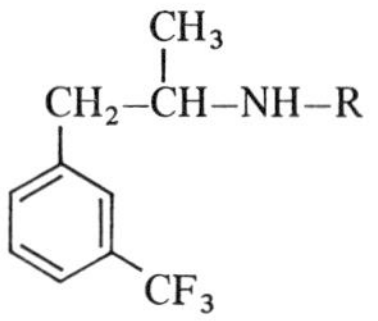

$CH_2-CH(CH_3)-NH-R$ (on ring); ring substituent: CF_3

(LIV) R = C_2H_5
(LV) R = H

c) Fenproporex (*N*-2-cyanoethylamphetamine) (LVI) is very rapidly cleaved to form amphetamine (LVII) (Tognoni et al., 1972; Beckett et al., 1972) which is responsible for the anorectic and stimulant activity of the administered compound.

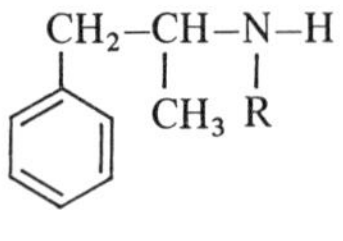

$C_6H_5-CH_2-CH(CH_3)-N(R)-H$

(LVI) R = CH_2CN
(LVII) R = H

C. Drugs Inhibiting Adrenergic Nerves and Structures Innervated by them

1. Adrenergic Neuron Blocking Agents

a) γ-methyldopa (*L*-γ-methyl-3,4-dihydroxyphenylalanine) (LVIII), a potent antihypertensive drug, acts as a substrate for the enzyme dopa-decarboxylase and leads to the formation of α-methylnorepinephrine (LIX) (KOPIN, 1968). This compound replaces norepinephrine in the storage sites.

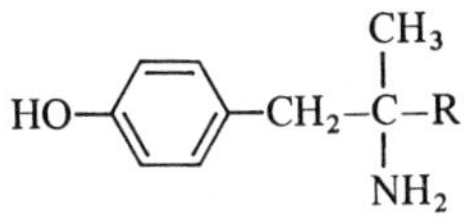

(LVIII) R = COOH
(LIX) R = H

b) γ-methyl-m-tyrosine (L-γ-methyl-3-hydroxyphenylalanine) (LX), a close analog of γ-methyldopa, is metabolized in a similar fashion via γ-methyl-*m*-tyramine, to metaraminol (LXI). This compound also can replace norepinephrine in the storage granules where it is released by nervous stimulation (BRUNNER et al., 1967).

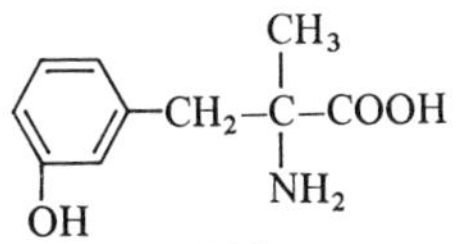

(LX) 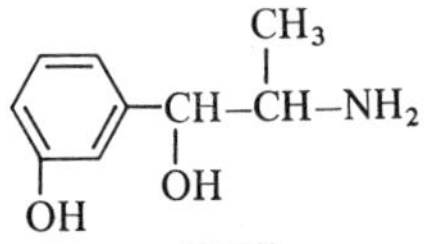 (LXI)

IV. Cardiovascular Drugs

A. Antiarrhythmic Drugs

a) Lidocaine (2-methylamino-2′,6′-acetoxylidide) (LXII) is metabolized by oxidative de-ethylation to 2-ethylamino-2′,6′-acetoxylidide (LXIII), which possesses both local anesthetic properties (EHRENBERG, 1948) and antiarrhythmic actions similar to those of lidocaine (SMITH and DUCE, 1971).

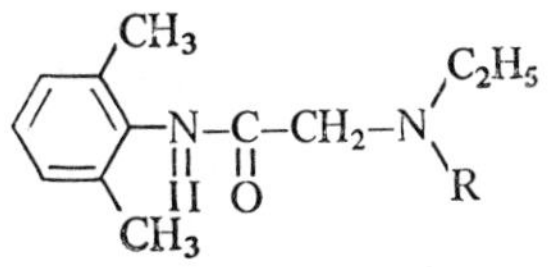

(LXII) R = C_2H_5
(LXIII) R = H

b) Propranolol (1-isopropylamino-3-(1-naphthyloxy)2-propanol-HCl) (LXIV) is biotransformed after oral administration into 4-hydroxypropranolol (LXV), which has been shown to be a β-adrenoreceptor-blocking drug with pharmacologic properties similar to those of the parent compound in man (DOLLERY et al., 1971).

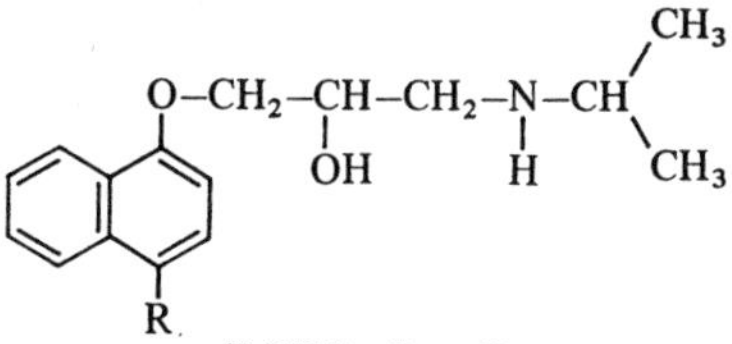

(LXIV) R = H
(LXV) R = OH

B. Vasodilator Drugs

a) Diallylmelamine [2,4-diamino-6-(diallylamino)-*s*-triazine] (LXVI) is biotransformed into a ring *N*-oxide metabolite (LXVII), which is believed to account for the hypotensive activity of the parent compound (ZINS, 1965).

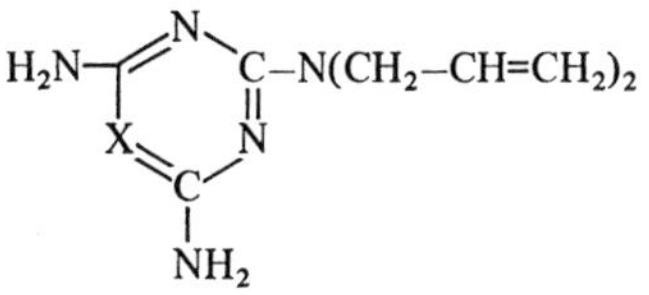

(LXVI) X = N
(LXVII) X = N→O

b) Prenylamine [*N*-(3-phenyl-2-propyl)-1,1-diphenyl-3-propylamine] (LXVIII) is biotransformed, presumably by oxidative cleavage of the propyl chain, to amphetamine (LIX), which is hydroxylated to *p*-hydroxy-amphetamine (PALM et al., 1968). These findings support the pharmacologic evidence that prenylamine acts like an indirect sympathomimetic amine, similar to amphetamine, and decreases the catecholamine content in the organs by a mechanism somewhat different from that of reserpine.

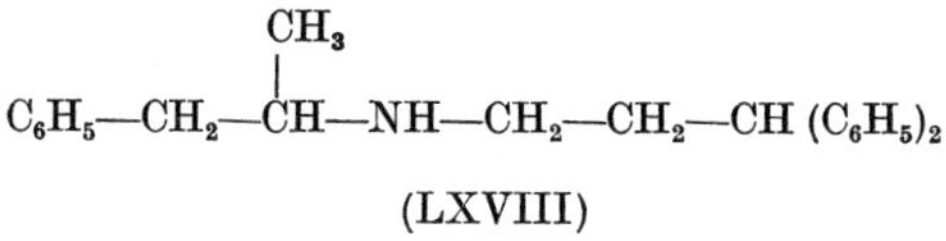

C_6H_5—CH_2—CH(CH_3)—NH—CH_2—CH_2—CH $(C_6H_5)_2$

(LXVIII)

C. Drugs Lowering Blood Lipids or Glucose

a) Nicotinic Acid (LXIX) is metabolized mainly to nicotinuric acid (LXX), which is considered to be one of the metabolites responsible for decreasing serum cholesterol levels (MILLER et al., 1960).

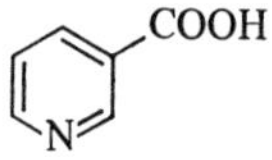

(LXIX) 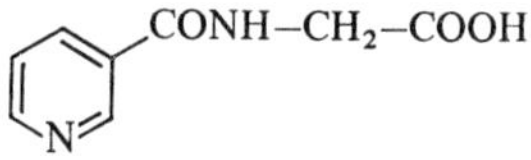 (LXX)

b) Clofibrate [ethyl-(*p*-chlorophenoxy)isobutyrate] (LXXI) is rapidly hydrolyzed by esterases to the free acid (LXXII), which can bind to albumin thereby reducing the affinity of albumin for other acids, including thyroxine and free fatty acids. This effect conceivably could lead to a decrease in triglyceride levels by altering the intrahepatic and extrahepatic balance between thyroid hormone activity and free fatty acids (THORP, 1963).

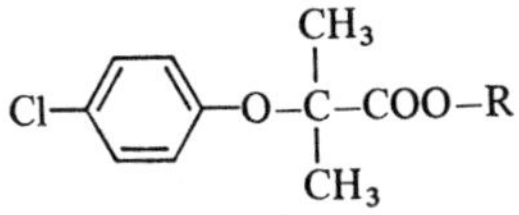

$$Cl-C_6H_4-O-C(CH_3)_2-COO-R$$

(LXXI) R = C_2H_5
(LXXII) R = H

c) 3,5-dimethylpyrazole (LXXIII) appears to be active (SMITH et al., 1965) after its metabolism to 5-methyl-pyrazole-3-carboxylic acid (LXXIV), which exerts an inhibitory effect on lipolysis.

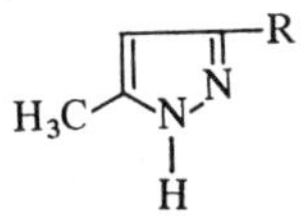

(LXXIII) R = CH_3
(LXXIV) R = COOH

d) 3,5-dimethylisoxazole (LXXV) is similarly converted to 5-carboxy-3-methyl-isoxazole (LXXVI), which shares the blockade of lipolysis and the hypoglycemic effect of the parent compound (DULIN and GERRITSEN, 1966).

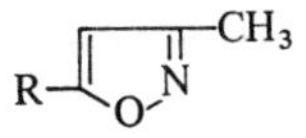

(LXXV) R = CH_3
(LXXVI) R = COOH

e) Acetohexamide (1[(L-acetylphenyl)sulfonyl]-3-cyclohexylurea) (LXXVII) in man is converted to an active metabolite, hydroxyhexamide (LXXVIII), with pharmacologic properties similar to those of the parent drug (McMAHON et al., 1965b)

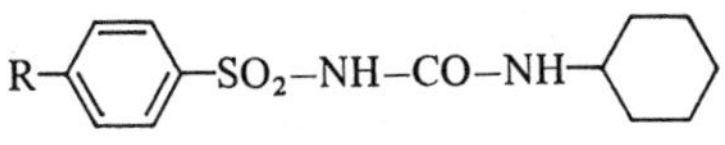

$$R-C_6H_4-SO_2-NH-CO-NH-C_6H_{11}$$

(LXXVII) R = CH_3–CO–
(LXXVIII) R = CH_3–CH(OH)–

V. Chemotherapy of Parasitic Diseases

A. Drugs Used in the Chemotherapy of Helminthiasis

a) Lucanthone (1-[(2-diethylaminoethyl)amino]-4-methylthioxanten-9-one) (LXXIX) is biotransformed in different animal species, including man (ROSI et al.,

1967) by oxidation of the methyl group to hycanthone (LXXX), which is a highly active schistomicidal agent.

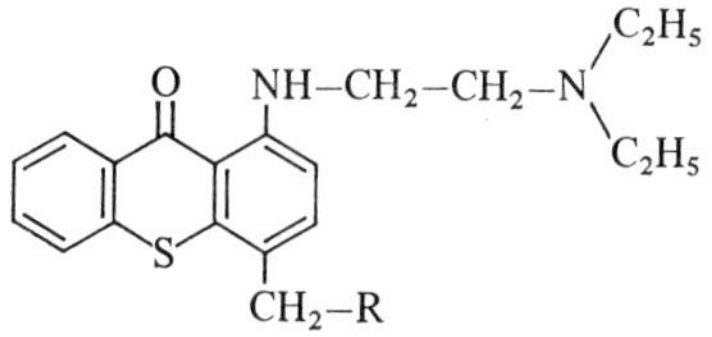

(LXXIX) R = H
(LXXX) R = OH

VI. Chemotherapy of Neoplastic Diseases

A. Alkylating Agents

a) Cyclophosphamide (2-[*bis*(2-chloroethyl)amino] tetrahydro-2*H*-1,3,2-oxazophosphorine-2-oxide) (LXXXI) shows no cytotoxic or alkylating activity unless it is converted by liver microsomal enzymes into active metabolites (BROCK and HOHORST, 1967; MELLETT, 1966). The anticancer activity of cyclophosphamide probably results from the activity of the open-ring aldehyde (LXXXII) (HILL et al., 1972).

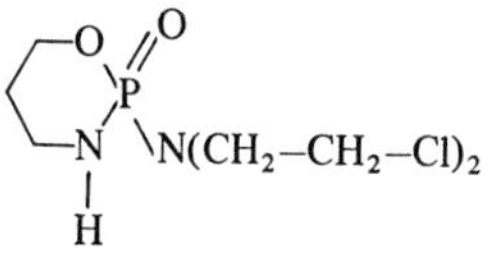

(LXXXI) 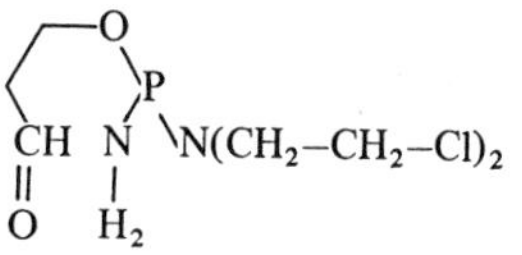 (LXXXII)

VII. Conclusions

The examples given in this chapter indicate that a large number of metabolites may exert pharmacologic effects that account for or prolong the activity of the parent compound. The study of drug metabolism is therefore important in understanding the pharmacology of several drugs. Such information is necessary for the rational use of these drugs in clinical situations in which disease or the concomitant use of other drugs may affect the rate at which active metabolites are formed or disposed of.

References

ADLER, T. K.: Studies on morphine tolerance in mice. I. *In vivo N*-demethylation of morphine and *N*- and *O*-demethylation of codeine. J. Pharmacol. exp. Ther. **156**, 585—590 (1967).

BAROWSKY, H., SCHWARTZ, S. A.: Methods for evaluating diphenoxylate hydrochloride. Comparison of its antidiarrheal effect with that of camphorated tincture of opium. J. Amer. med. Ass. **180**, 1058—1061 (1962).

BECKETT, A. H., BROOKES, L. G.: The absorption and urinary excretion in man of fenfluramine and its main metabolite. J. Pharm. Pharmacol. **19**, 42S—49S (1967).

BECKETT, A. H., SHENOY, E. V. B., SALMON, J. A.: The influence of replacement of the N-ethyl groups by the cyanoethyl group on the absorption, distribution and metabolism of ethyl amphetamine in man. J. Pharm. Pharmacol. **24**, 194—202 (1972).

BROCK, N., HOHORST, H. J.: Metabolism of cyclophosphamide. Cancer (Philad.) **20**, 900—904 (1967).

BRODIE, B. B., AXELROD, J.: The fate of acetophenetidin (phenacetin) in man and methods for the estimation of acetophenetidin and its metabolites in biological material. J. Pharmacol. exp. Ther. **97**, 58—67 (1949).

BRODIE, B. B., AXELROD, J.: The fate of antipyrine in man. J. Pharmacol. exp. Ther. **98**, 97—104 (1950).

BRUNNER, H., HEDWALL, P. R., MAITRE, L., MEIER, M.: Antihypertensive effects of alpha-methylated catecholamine analogues in the rat. Brit. J. Pharmacol. **30**, 108—122 (1967).

BURNS, J. J., SALVADOR, R. A., LEMBERGER, L.: Metabolic blockade by methoxamine and its analogs. Ann. N. Y. Acad. Sci. **139**, 833—840 (1967).

BURNS, J. J., YU, T. F., BERGER, L., GUTMAN, A. B.: Zoxazolamine physiological disposition, uricosuric properties. Amer. J. Med. **25**, 401—408 (1958).

BURNS, J. J., YU, T. F., DAYTON, P. G., GUTMAN, A. B., BRODIE, B. B.: Biochemical pharmacological considerations of phenylbutazone and its analogues. Ann. N. Y. Acad. Sci. **86**, 253—262 (1960).

BURNS, J. J., YU, T. F., RITTERBAND, A., PEREL, J. M., GUTMAN, A. B., BRODIE, B. B.: A potent new uricosuric agent, the sulfoxide metabolite of the phenylbutazone analogue, G-25671. J. Pharmacol. exp. Ther. **119**, 418—426 (1957).

BUTLER, T. C.: Metabolic demethylation of 3,5-dimethyl-5-ethyl-2,4-oxazolidinedione (paramethadione, paradione). J. Pharmacol. exp. Ther. **113**, 178—185 (1955).

CARLSSON, A., CORRODI, H., FUXE, K., HÖKFELT, T.: Effect of antidepressant drugs on the depletion of intraneuronal brain 5-hydroxytryptamine stores caused by 4-methyl-α-ethyl-meta-tyramine. Europ. J. Pharmacol. **5**, 357—366 (1969a).

CARLSSON, A., CORRODI, H., FUXE, K., HÖKFELT, T.: Effects of some antidepressant drugs on the depletion of intraneuronal brain catecholamine stores caused by 4,α-dimethyl-meta-tyramine. Europ. J. Pharmacol. **5**, 367—373 (1969b).

CARLSSON, A., FUXE, K., HAMBERGER, B., LINDQVIST, M.: Biochemical and histochemical studies on the effects of imipramine-like drugs and (+)-amphetamine on central and peripheral catecholamine neurons. Acta physiol. scand. **67**, 481—497 (1966).

CONNEY, A. H., SANSUR, M., SOROKO, F., KOSTER, R., BURNS, J. J.: Enzyme induction and inhibition in studies on the pharmacological actions of acetophenetidin. J. Pharmacol. exp. Ther. **151**, 133—138 (1966).

DAVISON, A. N.: The conversion of schradan (OMPA) and parathion into inhibitors of cholinesterase by mammalian liver. Biochem. J. **61**, 203—209 (1955).

DE SILVA, J. A. F., PUGLISI, C. V.: Determination of medazepam (Nobrium), diazepam (Valium) and their major biotransformation products in blood and urine by electron capture gas-liquid chromatography. Analyt. Chem. **42**, 1725—1736 (1970).

DICARLO, F. J., VIAU, J. P.: Prazepam metabolites in dog urine. J. pharm. Sci. **59**, 322—325 (1970).

DICARLO, F. J., VIAU, J. P., EPPS, J. E., HAYNES, L. J.: Prazepam metabolism by man. Clin. Pharmacol. Ther. **11**, 890—897 (1970).

DOLLERY, C. T., DAVIES, D. S., CONOLLY, M. E.: Differences in the metabolism of drugs depending upon their routes of administration. Ann. N. Y. Acad. Sci. **179**, 108—114 (1971).

DONE, A. K.: Salicylate intoxication. Significance of measurements of salicylate in blood in cases of acute ingestion. Pediatrics **26**, 800—807 (1960).

DULIN, W. E., GERRITSEN, G. C.: Effects of 5-carboxy-3-methylisoxazole on carbohydrate and fat metabolism. Proc. Soc. exp. Biol. (N. Y.) **121**, 777—779 (1966).

EHRENBERG, L.: The time-concentration curve of local anesthetics. Acta chem. scand. **2**, 63—81 (1948).

ELION, G. B., KOVENSKY, A., HITCHINGS, G. H., METZ, E., RUNDLES, R. W.: Metabolic studies of allopurinol, an inhibitor of xanthine oxidase. Biochem. Pharmacol. **15**, 863—880 (1966).

GARATTINI, S., MUSSINI, E., MARCUCCI, F., GUAITANI, A.: Metabolic studies on Benzodiazepines in various animal species. In: GARATTINI, S., MUSSINI, E., RANDALL, L. O. (Eds.): The Benzodiazepines. pp. 75—97. New York: Raven Press 1973.

GILLETTE, J. R., DINGELL, J. V., SULSER, F., KUNTZMAN, R., BRODIE, B. B.: Isolation from rat brain of a metabolic product, desmethylimipramine, that mediates the antidepressant activity of imipramine (Tofranil). Experientia (Basel) **17**, 417—418 (1961).

HILL, D. L., LASTER, W. R., JR., STRUCK, R. F.: Enzymatic metabolism of cyclophosphamide and nicotine and production of a toxic cyclophosphamide metabolite. Cancer Res. **32**, 658—665 (1972).

HORITA, A.: The role of microsomal enzymes in the activation and inactivation of modaline sulfate. Biochem. Pharmacol. **15**, 1309—1316 (1966).

JORI, A., CARRARA, C., PAGLIALUNGA, S., GARATTINI, S.: Pharmacological studies on modaline sulphate (W 3207). J. Pharm. Pharmacol. **17**, 703—709 (1965).

KIMMEL, H. B., WALKENSTEIN, S. S.: Oxazepam excretion by chlordiazepoxide-14C-dosed dogs. J. pharm. Sci. **56**, 538—539 (1967).

KOECHLIN, B. A., D'ARCONTE, L.: Determination of chlordiazepoxide (Librium) and of a metabolite of lactam character in plasma of humans, dogs, and rats by a specific spectrofluorometric micro method. Analyt. Biochem. **5**, 195—207 (1963).

KOPIN, I. J.: False adrenergic transmitters. Ann. Rev. Pharmacol. **8**, 377—394 (1968).

KVETINA, J., MARCUCCI, F., FANELLI, R.: Metabolism of diazepam in isolated perfused liver of rat and mouse. J. Pharm. Pharmacol. **20**, 807—808 (1968).

LIDBRINK, P., JONSSON, G., FUXE, K.: The effect of imipramine-like drugs and antihistamine drugs on uptake mechanisms in the central noradrenaline and 5-hydroxytryptamine neurons. Neuropharmacol. **10**, 521—536 (1971).

MACKAY, F. J., COOPER, J. R.: A study on the hypnotic activity of chloral hydrate. J. Pharmacol. exp. Ther. **135**, 271—274 (1962).

MARCUCCI, F., MUSSINI, E., FANELLI, R., GARATTINI, S.: The metabolism of diazepam by liver microsomal enzymes of rats and mice. Biochem. Pharmacol. **7**, 307—313 (1969).

MARK, L. C.: Metabolism of barbiturates in man. Clin. Pharmacol. Ther. **4**, 504—530 (1963).

McMAHON, R. E., CULP, H. W., MARSHALL, F. J.: The metabolism of α-*dl*-acetylmethadol in the rat: the identification of the probable active metabolite. J. Pharmacol. exp. Ther. **149**, 436—445 (1965a).

McMAHON, R. E., MARSHALL, F. J., CULP, H. W.: The nature of the metabolites of acetohexamide in the rat and in the human. J. Pharmacol. exp. Ther. **149**, 272—279 (1965b).

MELLETT, L. B.: Some aspects of the comparative pharmacology of cyclophosphamide in mice, hamsters, dogs, monkeys, and humans. In: A report to the acute leukemia task force. From the Division of Pharmacology and Toxicology, Southern Research Institute, Birmingham, Ala., Nov. 7, 1966.

METYSOVA, J., METYS, J., VOTAVA, Z.: Pharmakologische Eigenschaften der 6,11-Dihydrodibenz-(*b*, *e*)-thiepin-derivate. 2. Mitteilung. Arzneimittel-Forsch. **15**, 524—527 (1965).

MILLER, O. N., HAMILTON, J. G., GOLDSMITH, G. A.: Investigation of the mechanism of action of nicotinic acid on serum lipid levels in man. Amer. J. clin. Nutr. **8**, 480—490 (1960).

NICHOLLS, P. J., ORTON, T. C.: Absorption, distribution and excretion of methsuximide in male rats. Brit. J. Pharmacol. **43**, 459P—460P (1971).

NIEMEGEERS, C. J. E., LENAERTS, F. M., JANSSEN, P. A. J.: Difenoxine (R 15 403), the active metabolite of diphenoxylate (R 1132). Part 2: Difenoxine, a potent, orally active and safe antidiarrheal agent in rats. Arzneimittel-Forsch. **22**, 516—518 (1972).

PALM, D., GROBECKER, H., FENGLER, H.: Metabolisierung von Prenylamin (Segontin). Naunyn-Schmiedeberg's Arch. Pharmak. exp. Path. **260**, 185—186 (1968).

RANDALL, L. O., SCHALLEK, W.: Pharmacological activity of certain benzodiazepines. In: EFRON, D. H. (Ed.): Psychopharmacology. A Review of Progress, 1957—1967, pp. 153—184. Washington: Government Printing Office 1968.

RANDALL, L. O., SCHECKEL, C. L., BANZIGER, R. F.: Pharmacology of the metabolites of chlordiazepoxide and diazepam. Curr. ther. Res. **7**, 590—606 (1965).

ROSI, D., PERUZZOTTI, G., DENNIS, E. W., BERBERIAN, D. A., FREELE, H., TULLAR, B. F., ARCHER, S.: Hycanthone, a new active metabolite of lucanthone. J. med. pharm. Chem. **10**, 867—876 (1967).

ROSS, S. B., RENYI, A. L.: Inhibition of the uptake of tritiated catecholamines by antidepressant and related agents. Europ. J. Pharmacol. **2**, 181—186 (1967).

ROSS, S. B., RENYI, A. L.: Inhibition of the uptake of tritiated 5-hydroxytryptamine in brain tissue. Europ. J. Pharmacol. **7**, 270—277 (1969).

SAMANIN, R., GHEZZI, D., GARATTINI, S.: Effect of imipramine and desipramine on the metabolism of serotonin in midbrain raphe stimulated rats. Europ. J. Pharmacol. In press (1972).

SCHWARTZ, D. E., BRUDERER, H., RIEDER, J., BROSSI, A.: Metabolic studies of tetrabenazine, a psychotropic drug in animals and man. Biochem. Pharmacol. **15**, 645—655 (1966).

SCHWARTZ, M. A., CARBONE, J. J.: Metabolism of 14C-medazepam hydrochloride in dog, rat, and man. Biochem. Pharmacol. **19**, 343—361 (1970).

SCHWARTZ, M. A., POSTMA, E.: Metabolic *N*-demethylation of chlordiazepoxide. J. pharm. Sci. **55**, 1358—1362 (1966).

SCHWARTZ, M. A., POSTMA, E.: Metabolism of diazepam *in vitro*. Biochem. Pharmacol. **17**, 2443—2449 (1968).

SMITH, D. L., FORIST, A. A., DULIN, W. E.: 5-Methylpyrazole-3-carboxylic acid. The potent hypoglycemic metabolite of 3,5-dimethylpyrazole in the rat. J. med. pharm. Chem. 8, 350—353 (1965).
SMITH, E. R., DUCE, B. R.: The acute antiarrhythmic and toxic effects in mice and dogs of 2-ethylamino-2′,6′-acetoxylidine (L-86), a metabolite of lidocaine. J. Pharmacol. exp. Ther. **179**, 580—585 (1971).
SPINKS, A., WARING, W. S.: Anticonvulsant drugs. In: ELLIS, G. P., WEST, G. B. (Eds.): Progress in medicinal Chemistry. Vol. 3, pp. 261—331. Washington: Butterworths 1963.
THORP, J. M.: An experimental approach to the problem of disordered lipid metabolism. J. Atheroscler. Res. **3**, 351—360 (1963).
TOGNONI, G., MORSELLI, P. L., GARATTINI, S.: Amphetamine concentrations in rat brain and human urine after fenproporex administration. Europ. J. Pharmacol. **20**, 125—126 (1972).
VERNIER, V. G., ALLEVA, F. R., HANSON, H. M., STONE, C. A.: Pharmacological actions of amitriptyline, noramitriptyline and imipramine. Fed. Proc. **21**, 419 (1962).
WAY, E. L.: II. Distribution and metabolism of morphine and its surrogates. Ass. Res. nerv. Dis. Proc. **46**, 13—31 (1968).
ZBINDEN, G., RANDALL, L. O.: Pharmacology of benzodiazepines: laboratory and clinical correlations. In: GARATTINI, S., SHORE, P. A. (Eds.): Advances in Pharmacology, Vol. 5, pp. 213—291. New York: Academic Press 1967.
ZINS, G. R.: The *in vivo* production of a potent, long-acting hypotensive metabolite from diallylmelamine. J. Pharmacol. exp. Ther. **150**, 109—117 (1965).

Chapter 63

The Enterohepatic Circulation

G. L. PLAA

I. Introduction

Substances are said to undergo an enterohepatic circulation (EHC) when they are excreted into the bile, pass into the lumen of the intestine, are reabsorbed and then return to the liver via the circulation. Many endogenous and exogenous substances can undergo an EHC. Among the endogenous substances are the bile salts, the biliary lipids and biliary phospholipids; the degree of reabsorbability varies considerably for each of these types of substances. Other endogenous substances include estrone and estriol (SANDBERG et al., 1967), folic acid (BAKER et al., 1965; HERBERT, 1965), vitamin B_{12} (GRASBECK et al., 1958), and urobilinogen (LESTER et al., 1965). IBRAHIM and WATSON (1968) demonstrated that an EHC exists for protoporphyrin in man. The oral administration of cholestyramine, an anionic exchange resin, improves the clinical condition of patients suffering from porphyria cutanea tarda (STATHERS, 1966) or from erythropoietic protoporphyria (LISCHNER, 1966) apparently by interrupting the EHC for the porphyrins.

Biliary secretion is an important excretory pathway for a great variety of drugs and chemicals or their metabolites (STOWE and PLAA, 1968; PLAA, 1971; SMITH, 1971). BECKER and co-workers (GIBSON and BECKER, 1967; BECKER et al., 1968) reported that the lethality of ouabain, promazine, perphenazine and meprobamate is markedly increased in mice with ligated bile ducts. KLAASSEN (1973) has shown that the toxicity of colchicine, diethylstilbestrol, digoxin, ouabain, indocyanine green, rifampin and iopanoic acid was greater in mice and rats with ligated bile ducts than in normal animals.

The studies in which lethality was measured by no means established that merely the reduction in biliary excretion led to the enhanced lethality. The effect of accumulated bile salts during cholestasis on processes of drug inactivation was not evaluated. Nor were possible changes in distribution assessed. However, the data strongly suggest that alterations in biliary excretion are involved.

Several drugs exhibit an EHC. Potentially, all substances that are excreted by the biliary route could undergo such a circulation if reabsorption occurs in the intestine. KEBERLE et al. (1962) demonstrated that in the rat the presence of an EHC can have a marked effect on the persistence of glutethimide metabolites in the body. The glucuronide conjugate of chloramphenicol is excreted in the bile of rats, converted to arylamine and reabsorbed in this form; this arylamine can exert a toxic action on the thyroid (THOMPSON et al., 1954). WILLIAMS et al. (1965) point out the possibility that intestinal carcinogenic activity of aromatic amines in rats may be caused by *o*-hydroxyamines which are formed in the liver and excreted into the bile as glucuronides and finally degraded in the intestine to the free hydroxyamines, which are carcinogenic. Unfortunately, no generalizations can be made about the relative importance of the EHC in the overall response of the animal to a particular chemical agent since it depends not only upon the particular compound and the pharmacologic activity of its metabolites, but also the animal species being tested.

II. Methods for Studying the Enterohepatic Circulation

One method of studying the EHC is to compare the biologic half-lives of a substance and its metabolites in normal animals with that in animals with a bile fistula (KEBERLE et al., 1962; CHARYTAN, 1970). Another approach is to compare the amounts of substances excreted in the bile of animals with bile fistulas to the amounts excreted in the feces and urine of normal animals (WOODS, 1954; DOBBS and HALL, 1969; CALDWELL et al., 1971). Others have used "linked animals"; in this preparation, the bile cannula from one animal is inserted into the duodenum of a second animal; the substance is given to the first animal and biliary excretion of the substance in the second animal is monitored (HUCKER et al., 1966; LADOMERY et al., 1967).

Recently, two methods have been described in which repeated bile collections can be made from the uninterrupted EHC in chronic preparations. DEN-BESTEN (1971) has described a method used in dogs in which a duodenal pouch containing the intact sphincter of Oddi has been prepared. This pouch drains into the duodenum through a Gregory cannula. The duodenal section has been surgically prepared to contain a Thomas cannula through which the bile samples can be collected. During the times the bile is not collected, the cannula can be arranged to permit an uninterrupted EHC.

The second method has been devised for the collection of bile in rhesus monkeys (DOWLING et al., 1968). In this preparation, the extrahepatic biliary pathway is brought to the exterior; the bile flows through a stream-splitter connected to an electronic circuit and is collected into reservoirs. One reservoir is also connected to an electronic leveling system that controls a pump draining this reservoir and reinfuses the bile into the duodenum of the animal. When bile is needed for analysis, the stream-splitter is put in the diverted position and bile is collected directly rather than being reinfused. This particular technique obviously requires that the animal be restrained during the entire experiment.

To study the absorption of metabolic products, the metabolites can be infused into the duodenum and their subsequent biliary excretion monitored using an animal with a cannulated bile duct (FISCHER et al., 1966; ERIKSSON, 1971). The role of intestinal microorganisms in the hydrolysis of conjugates and the subsequent reabsorption of deconjugated products has been demonstrated by comparing responses obtained in animals whose intestinal lumen has been sterilized to those obtained in nonsterilized animals (DOBBS et al., 1970; CLARK et al., 1969).

III. The Enterohepatic Circulation of Bile Salts

A number of reviews deal with this subject (HOFMANN, 1965; LACK and WEINER, 1967; HOFMANN and SMALL, 1967; DIETSCHY, 1968; DOWLING, 1972). Therefore, only highlights of this information will be covered in this chapter.

The body normally conserves bile acids by reabsorption from the intestine. In man, a 3 to 5 g bile salt pool circulates through the EHC about 6 to 10 times per day (BORGSTROM et al., 1957; LINDSTEDT, 1957). Between 20 to 25% of the total bile salt pool escapes reabsorption and is excreted in the feces (BERGSTROM, 1962). In the steady state, synthesis of bile salts must equal loss and this loss is due primarily to fecal excretion. Therefore, an enhanced fecal excretion of bile salts can have a marked effect on net synthesis of these substances. NORMAN and SJOVALL

(1958) have calculated that in the rat, at any given time, about 85% of the circulating bile salts is in the intestines, about 10% is in the gut wall, and only 3 to 5% is actually in the liver.

A. Excretion of Bile Salts

Normally, the bile salts that appear in the bile are in a conjugated form, either with taurine or glycine. This conjugation occurs in the hepatic microsomes and varies according to species. Taurine conjugates predominate in the rat, while in the rabbit the glycine conjugates are more prevalent (BREMER, 1956). In man, the glycine conjugates normally predominate. However, the human liver can readily conjugate bile acids with taurine and in fact preferentially utilizes taurine, but the dietary intake of taurine is relatively small compared to that of glycine.

Bile acid synthesis in the liver is regulated by the amount of bile acids returning to the liver via the EHC. The enzyme cholesterol 7-α-hydroxylase appears to the rate-limiting enzyme in the synthesis of bile acids. This enzyme is in the endoplasmic reticulum. Its activity is increased when the EHC is interrupted but is not affected by phenobarbital induction (DOWLING, 1972). A circadian rhythm has been observed.

Since bile acids contain both hydrophobic and hydrophilic groupings, they possess surface active properties. They are able to aggregate and form micelles. In low concentrations, the bile acids exist as monomers in aqueous solution. As the concentration increases, aggregation occurs resulting in dimers, tetramers or even larger structures. For a particular bile acid, it is possible to determine the critical concentration in solution needed to promote the formation of polymers. SMALL (1967) theorized that these polymers are held together by hydrophobic bonding between the surfaces of the bile acids and that the hydrophilic portions exist on the outside surface of the resulting polymer. Larger aggregates can be formed by hydrogen bonding between the outwardly directed hydrophilic portions of the molecules. In addition to the concentration, which is different for different bile acids, aggregation is also affected by such factors as temperature, pH, and the presence of counter-ions. In general, an elevation in temperature decreases aggregation for a given bile acid, counter-ions favor aggregation, while a drop in pH increases aggregation. *In vitro* studies have established that the critical temperature for micelle formation is well above 37° C for unsubstituted cholanolic acid and for lithocholic acid, but is below body temperature for the common dihydroxy and trihydroxy bile acids. Conjugation of the acids does not seem to alter these values. The addition of phospholipid to a micellar solution of bile acids results in an expansion of the size of the micelle with a reduction of the critical concentration for micelle formation (DIETSCHY, 1968). In biologic systems, substances like cholic, deoxycholic and chenodeoxycholic acid form micelles while lithocholic does not.

B. Absorption of Bile Salts

Bile acid absorption occurs in the intestine by both active and passive transport; on a quantitative basis, the distal portion of the intestine seems to be more important (DIETSCHY, 1968). With everted gut sacs, it has been shown that only in the distal segments of the small intestine is it possible to get high serosal/mucosal ratios of bile salt after incubation. Only the ileum appears capable of moving bile acids against a concentration gradient. LACK and WEINER (1961) showed that transport across the ileum is inhibited by anoxia and by metabolic inhibitors. The process is not dependent upon the presence of glucose; phlorridizin, which blocks glucose transport, does not interfere with the uptake of bile acids (PLAYOUST and ISSELBACHER, 1964). The stereospecificity of the transport system for different bile acids has been studied in detail by LACK and WEINER (1966). Taurocholic and glycocholic acids possess the highest rate of transport among the trihydroxycholanoic acid derivatives. Conjugated dihydroxy acid derivatives and triketo derivatives exhibit slower rates. A critical structural feature is the presence of a negative charge on the bile acid side chain. Various bile acids compete with each other for transport across everted sacs. It may be concluded that these acids are served by a common transport carrier rather than by several related carriers. Active transport mechanisms are present in several species and are probably present in man (GLASSER et al., 1965).

Passive transport for bile acids, in the presence of a favorable electrochemical gradient, has also been demonstrated (DIETSCHY, 1968). Both passive ionic and nonionic diffusion have been described. Because of their low pKa values, the conjugated bile salts exist almost entirely as ions in the intestinal lumen. The ionic form of the bile salts diffuses slowly through biologic membranes. Such diffusion occurs in the small intestine and the rate of diffusion depends upon the number of hydroxyl groups in the molecules; taurocholic acid has the lowest diffusion constant and lithocholic acid the highest (DIETSCHY, 1968). With taurocholic acid, GLASSER et al. (1965) demonstrated that less than 13% of a given load of bile acids could diffuse passively through the intestine of several species. However, the experiments carried out with these various bile salts apply only to the monomers. DIETSCHY (1968) states that the micellar forms can also diffuse passively. Micellar diffusion is further complicated by the size of the aggregate and the presence of a mixed micelle; in this situation, the rate of diffusion diminishes.

Passive nonionic diffusion of unconjugated bile acids is important since these substances have a higher pKa value than the conjugated salt. Deconjugation of bile acids can occur in the intestinal lumen and in the colon. Any situation that favors deconjugation would also favor reabsorption of the nonionic species. Current data indicate that facilitated diffusion, solvent drag and exchange diffusion of bile acids do not play major roles if, indeed, they do exist (DIETSCHY, 1968).

Several investigators have studied the use of cholestyramine, an anionic exchange resin, as a means of preventing the absorption of bile salts. Oral administration of this substance increases fecal bile salt excretion in both animals and man (HUFF et al., 1963; JANSEN and ZANETTI, 1965; HASHIM and VAN ITALLIE, 1965; MOORE et al., 1968). Other investigators reported that cholestyramine has a preferential affinity for taurine-conjugated bile salts (ROE, 1962; WOOD et al., 1969). COOK et al. (1971) fed animals on diets containing high levels of taurine, and increased fecal bile salt excretion by 44 to 56% in rats and by 80% in hamsters. Taurine administration in this instance increased the amount of bile salt conjugated with taurine. This effect is particularly marked in hamsters, which have a relatively low capacity for hepatic taurine synthesis. Intestinal reabsorption of bile salts is normally very efficient (HOFMANN, 1967). COOK et al.(1972) demonstrated that cholestyramine increases fecal bile salt excretion by retarding bile salt absorption in the ileum. In these studies, they were able to demonstrate *in vivo* that cholestyramine has a greater affinity for taurine-conjugated bile salts than for the corresponding glycine conjugates.

FORMAN et al. (1968) also increased the excretion of fecal bile acids by feeding an oral hydrophilic colloid. In man, excretion increased over 300% during a 24-h period. Although the mechanism of action of this substance is not known, it is presumed that it also retards the reabsorption of bile salts in the ileum.

Neomycin also affects the EHC for bile salts. SAMUEL and STEINER (1959) demonstrated that neomycin significantly decreased serum cholesterol concentrations in man. Results with other antibiotics (kanamycin and chlortetracycline) were less consistent and less successful (SAMUEL and WAITHE, 1961; STEINER et al., 1961). Intramuscularly administered neomycin was ineffective. POWELL et al. (1962) postulated that this antibiotic acted by altering the intestinal bacterial flora. However, EYSSEN et al. (1966) showed that neomycin precipitates solutions of bile acids *in vitro* and that the cholesterol-lowering effect of neomycin occurred with a non-antibiotic derivative of neomycin, *N*-methylated neomycin. EYSSEN et al. (1966) also showed that in chicks the nonantibiotic derivative of neomycin lowered blood cholesterol and increased fecal bile acid excretion. They suggested that the effects of neomycin were due to the precipitation of bile acids and an interruption of their reabsorption from the gut. *N*-methylated neomycin also protected chicks against the hepatotoxic effects of lithocholic acid, a bile acid that is reabsorbed from the gut and can produce liver injury after absorption. This further supports the precipitation concept.

IV. Enterohepatic Circulation of Drugs

While the role of the EHC in bile salt disposition has been extensively investigated, the importance of the EHC for drugs is less well understood. The importance of the bile as a route of drug excretion is now well established. However, its role in the overall pharmacologic response to drugs cannot be described in general terms because of the diversity of the substances involved. The duration of a pharmacologic response depends upon the presence of a pharmacologically active moiety. Many drugs that are excreted into the bile are biotransformed prior to their excretion, and the metabolic products may not possess biologic activity, even if they are reabsorbed. However, deconjugations are known to occur in the intestine, and such steps can lead to the liberation of absorbable active chemical moieties; this would tend to prolong the duration of a pharmacologic effect. On the other hand, considerable variation exists between species on 1) the biotransformation products formed; 2) the quantitative importance of the biliary excretion pathway; 3) the deconjugating capacity of the intestinal flora, and 4) the reabsorptive capacity of the intestinal bed. Before one can determine whether the EHC plays an important role each of these considerations must be assessed.

Unfortunately, at times there is a tendency to equate the mere presence of biliary excretion as indicative of an EHC. This is not correct. In order to state that an EHC exists for a particular drug one must show that biliary excretion occurs and that the excreted products are reabsorbed by the intestine. An EHC consists of two aspects: biliary excretion and intestinal reabsorption. There are many studies in the literature that deal with the first phase; fewer studies deal with the second step. In the subsequent sections an attempt has been made to describe those instances in which an EHC has been established for drugs or other exogenous chemical substances.

A. Morphine

Conjugated morphine is excreted quite readily into the bile (WAY and ADLER, 1960). WOODS (1954) demonstrated that 12 h after the subcutaneous administration of morphine, 37% of the administered dose (30 mg/kg) was present in the gall bladder of dogs. In dogs with a bile fistula, 35% of the morphine was excreted into the bile within a period of 24 h. Nearly all of the morphine was excreted into the bile as a conjugate. Since the normal excretion of morphine in the feces of dogs ranges from 6 to 10%, while 35% of the dose appears in the bile, it is evident that morphine must be reabsorbed from the intestinal tract. Hydrolysis of the conjugate can occur in the intestine by the action of beta-glucuronidase.

MARCH and ELLIOTT (1954) showed that 6 h after morphine was injected into rats with bile fistula, 63% of the dose appeared in the bile and 18% in the urine. In the intact rat, 60% of the dose appeared in the urine. These results suggest that in the intact animal the morphine metabolites excreted into the intestine must be reabsorbed and eventually excreted by the kidneys.

Recently, SMITH et al. (1973) reported that in rats 63% of a dose of morphine was excreted into the bile of rats, whereas 14% was excreted in cats. These authors further studied the relative excretion rates of both morphine glucuronide (the metabolite in rats) and morphine ethereal sulfate (the metabolite in cats). After 3 h the rat excreted 81% of the morphine glucuronide, but only 17% of the morphine ethereal sulfate. The cat, after 3 h, excreted about 20% of both conjugates. The authors concluded that rates of biotransformation could not account for these differences nor could the pathway of metabolism. They proposed that morphine glucuronide and morphine ethereal sulfate are excreted into the bile by different mechanisms.

B. Methadone

The rat excretes methadone primarily via the intestines. Twenty-four hours after subcutaneous administration of ^{14}C-labeled methadone, about 70% of the label appeared in the intestines or feces; after 3 h, 17% of the label was excreted into the bile (EISENBRANDT et al., 1950). The same study revealed that when bile containing methadone and its metabolites was injected into the small intestine, absorption occurred and radioactive products appeared in the bile. This demonstrated the presence of an EHC at least for the metabolites of methadone.

Methadone is eliminated in the bile by the rat primarily as the *N*-demethylated cyclized derivative (2-ethylidine-1,5-dimethyl-3,3-diphenylpyrrolidine; EDDP), and secondarily as a highly water-soluble substance that is apparently a glucuronide conjugate but has not been identified (BASELT and CASARETT, 1972). Twenty-four hours after an intravenous dose (1 mg/kg) of ^{3}H-methadone, BASELT and CASARETT (1972) found 58% of the dose in the bile as EDDP and 27% as the highly water-soluble metabolite; they found only a trace of unchanged methadone and no methadone-*N*-oxide.

C. Etorphine

DOBBS and HALL (1969) showed that etorphine, a morphine-like substance, is excreted into the bile as a glucuronide conjugate after its intramuscular administration. Using doses ranging from 5 to 50 μg/kg, they demonstrated that 52 to 88% of the original dose was excreted as metabolites into the bile within a period of 12 h. Of the substances in the bile, 96% was characterized as etorphine glucuronide and less than 4% was etorphine. These investigators also showed that the metabolites of this substance, when administered to rats with bile fistulas, could be well absorbed from the intestine; after oral administration of 5 μg/kg, 71% to 73% of the original administered dose was excreted into the bile within a period of 24 h. The peak concentrations occurred in the bile within 5 h after the oral administration of the metabolites. Etorphine administered with bile was also well absorbed and 95% of the dose appeared in the bile within 24 h. However, the peak concentrations appeared in the bile within 1 to 2 h after its oral administration. The cumulative recoveries of etorphine and its metabolites in the bile and the feces after intramuscular administration indicated that after 8 h, 60% had been excreted into the bile whereas it took 95 h to recover 45% of the material in the feces. After 10 h, when more than 60% of the substance had been excreted via the bile, less than 5% could be found in the feces. This led to the conclusion that the major metabolite, etorphine glucuronide, was hydrolyzed by bacterial enzymes in the intestinal lumen and that the released etorphine was reabsorbed from the gastrointestinal tract. These data were certainly suggestive of an EHC for etorphine.

DOBBS et al. (1970) restudied the problem by monitoring biliary excretion of etorphine glucuronide in several animal models. They sterilized the intestinal flora by administering ampicillin, which reduced greatly the amount of beta-glucuronidase activity in the cecum of treated rats; in terms of viable bacteria, the aerobic counts were found to be reduced by a factor of 10^5 for a 24-h period. In normal animals, oral administration of radioactive etorphine glucuronide resulted in the appearance of radioactivity in the bile within a period of 4 h reaching a peak at 9 h and a diminution by 14 h. Therefore, it was clearly established that etorphine glucuronide could be reabsorbed itself or that etorphine could result from the hydrolysis of the conjugate in the gut. On the other hand, in rats treated with ampicillin, it was demonstrated that while radioactivity appeared in the bile after oral administration of labeled etorphine glucuronide, only a low undulating level was observed with no sharp maximum over a 20-h period. When the cumulative radioactivity in bile from both groups of rats was analyzed, the data showed that after 12 h normal rats had excreted 60% of the administered dose, whereas ampicillin-treated animals had excreted only 10 to 15%. After 20 h, the cumulative rate of excretion in ampicillin-treated animals only attained a level of 30% in contrast to the 60% observed in normal rats. These investigators also demonstrated that ampicillin-treated intestinal contents of rats could not hydrolyze the etorphine glucuronide as effectively as that from normal rats. These experiments clearly demonstrate how a substance excreted into the bile can be biotransformed by the intestinal flora, leading to reabsorption of an active form of a drug.

DOBBS et al. (1970) also demonstrated why species variations can occur for an EHC that depends upon biotransformation in the gut. On a relative basis, they calculated that the rat contains about 4300 units of glucuronidase activity/ml of bowel content; comparable values for the horse and cat yield values of only 107 and 392 units, respectively. Therefore, they postulated that if the limiting factor becomes hydrolysis in the gut, the rat would be much more likely to possess an EHC in contrast to other species.

D. Digitoxin

An EHC has also been demonstrated for digitoxin in several species (OKITA, 1967). After intravenous administration in man, OKITA et al. (1955) found large quantities of digitoxin and its metabolites in the gall bladder and in the small intestine. They found that digitoxin and its metabolites were reabsorbed by the intestinal mucosa. During passage through the EHC and eventually the general circulation, a large percentage of the metabolic products was removed by the kidneys. By this cycling, most of the drug and its metabolites were excreted by the kidneys. The biologic half-life of radiolabeled digitoxin was reduced from 14 to 6 h in dogs with a bile fistula; urinary excretion of the labeled material was also reduced (KATZUNG and MEYERS, 1965). KATZUNG and MEYERS (1965) proposed that water-soluble metabolites of digitoxin are converted to nonpolar lipid soluble products in the gut before intestinal reabsorption.

Recent studies in rats by GREENBERGER and THOMAS (1973) demonstrated that the biliary excretion of ^{3}H derived from ^{3}H-digitoxin is reduced when bile salt depletion occurs and is increased when bile salts are administered. Also these authors showed that pretreating rats with phenobarbital results in enhanced biliary excretion of ^{3}H associated with an increase in bile flow but no change in bile salt output. Prior treatment with ethacrynic acid diminishes bile flow, diminishes ^{3}H excretion, but has no effect on bile salt output. GREENBERGER and THOMAS (1973), on the basis of these data, have concluded that digitoxin excretion in bile is both bile salt-dependent and bile salt-independent.

CASTLE and LAGE (1973) reported that spironolactone-pretreated rats exhibit enhanced biliary excretion of ^{3}H in rats receiving ^{3}H-digitoxin; after 2 h spironolactone-treated animals excrete about 50% of the injected dose compared to 15% in the control group.

Recently, CALDWELL et al. (1971) studied the effects of cholestyramine on the EHC of digitoxin in man. In this study, the investigators gave radioactive digitoxin (1.2 mg) orally, waited 8 h to permit complete absorption, and then administered cholestyramine (4 g) 8, 12, 16 h after the digitoxin and four times daily thereafter for 5 days. The plasma disappearance of total radioactivity (digitoxin and metabolites) was calculated over a period of 7 days. The half-life was 11.5 days in those subjects not treated with cholestyramine and was reduced to 6.6 days with cholestyramine treatment. When the same type of analysis was carried out for unchanged digitoxin, the half-life was 6 days in normal individuals and 4.5 days in cholestyramine-treated individuals. Thus, cholestyramine treatment clearly resulted in enhanced clearance from the body of both digitoxin and its metabolites.

These results were also compared to those obtained using physiologic parameters of cardiac function. Ventricular ejection time and changes in electromechanical systole were studied in control and cholestyramine-treated individuals. Digitoxin increased both these values during the five-day period of study. Those individuals receiving cholestyramine exhibited shorter times for these two parameters starting on the second day after treatment. By the fourth and fifth days, these values were significantly different from those obtained in the control groups. A positive correlation between the electromechanical systole and the serum concentration of digitoxin was established. Thus, this study also showed that the decreases in plasma half-life obtained with cholestyramine treatment resulted in alterations of the pharmacologic activity of digitoxin on the heart. These results in man correlate well with those obtained in other experiments: *In vitro*, cholestyramine binds appreciable amounts of cardiac glycosides, cholestyramine treatment protects experimental animals from lethal doses of digitoxin administered subcutaneously, and cholestyramine accelerates the fecal excretion of parenterally administered digitoxin (CALDWELL and GREENBERGER, 1971).

OKITA (1967) has postulated that among the various cardiac glycosides, the nonpolar derivatives should participate to a greater degree in an EHC than should the polar glycosides. He has calculated that for digitoxin about 26% of the administered dose should be involved in an EHC, whereas for convallatoxin only 5.7% of the dose should participate. DOUGHERTY et al. (1970) has made the same calculation for digoxin in man and has arrived at a value of 6.5%. These calculations are in accord with the degree of polarity of the various derivatives. Furthermore, DOUGHERTY et al. (1970) feel that a lack of enterohepatic recycling for digoxin accounts for its short duration of action.

E. Diethylstilbestrol

Diethylstilbestrol is extensively metabolized in the rat and is excreted in the bile, mainly as a monoglucuronide conjugate; this substance undergoes an EHC (HANAHAN et al., 1953; FISCHER et al., 1966). FISCHER et al. (1966) demonstrated that 21% of the label of intraduodenally administered diethylstilbestrol monoglucuronide appeared in the bile within 4 h and that 71% appeared within 24 h. On the other hand, with diethylstilbestrol itself, 79% of the label appeared in the bile in 4 h and 95% in 24 h. Both the parent compound and the monoglucuronide have been identified in portal blood. These authors concluded that the recirculation was at least partially dependent upon hydrolysis of the monoglucuronide in the intestine. CLARK et al. (1969) demonstrated that beta-glucuronidase was implicated in the hydrolysis and that an inhibitor of this enzyme, glucaro-1,4-lactone, could decrease the rate of absorption of diethylstilbestrol monoglucuronide by 90%, whereas this inhibitor did not affect the intestinal absorption of free diethylstilbestrol. Rats pretreated with neomycin and kanamycin also had a much decreased intestinal absorption of the monoglucuronide while the absorption of diethylstilbestrol itself was not affected. The three nonglucuronide metabolites were poorly reabsorbed in the intestine.

Using everted sacs of rat intestine, FISCHER and MILLBURN (1970) demonstrated that diethylstilbestrol monoglucuronide was very poorly absorbed by these intestinal sacs, whereas free diethylstilbestrol readily entered the sacs. Further studies were carried out (FISCHER et al., 1971) in 5-day-old and 25-day-old rats. The beta-glucuronidase activity in the 5-day-old animals was very slight, whereas the large intestine of 25-day-old animals exhibited high levels of enzymic activity. Diethylstilbestrol monoglucuronide was rapidly hydrolyzed and absorbed in the intestine of 25-day-old rats. Larger amounts of the unconjugated drug were found in the intestines of 25-day-old rats than in 5-day-old animals. These studies indicate that the hydrolyzed diethylstilbestrol was being absorbed from the large bowel rather than from the small intestine. More recent evidence indicates that in 5-day-old rats diethylstilbestrol monoglucuronide can be absorbed intact from the distal portion of the intestine (FISCHER et al., 1973).

F. Steroids

ERIKSSON (1971) has also studied the absorption and EHC of neutral steroids in the rat. When pregnenolone, pregnenolone-3-sulphate, and corticosterone were administered into the duodenum, 65 to 76% of the radioactivity appeared in the bile, 8 h later. When pregnolone, pregnenolone sulphate and deoxycorticosterone sulphate were administered into the cecum, 40 to 74% of the material appeared in the bile, thus indicating significant absorption from this region. When pregnenolone sulphate was given intraduodenally, both conjugated and free steroids were found in the portal blood. Both free and conjugated steroids appeared in the bile. With pregnenolone the free-conjugate ratio was 0.85 after intragastric administration and 1.4 after intracecal administration; the conjugates consisted of glucuronides, monosulfates and diconjugates. With pregnenolone-3-sulfate the free-conjugate ratio after intracecal administration was 0.43; with deoxycorticosterone-21-sulfate the ratio was 0.8 after intracecal administration.

The excretion of corticosterone and pregnenolone was studied in "linked animals". In this particular experiment, 5 rats with bile fistulas were linked together so that the bile from one rat entered the duodenum of the next. The corticosterone or pregnolone was given into the duodenum of the first rat in the series and bile was collected from the fifth animal. Radioactivity began to appear in the bile from the last animal after 4 h, and could no longer be detected 44 h later. After measuring the fecal and urine content of the other animals, the author was able to calculate the amount of the radioactive steroid excreted in the five separate animals. About 70% of the material of the administered dose was excreted in the first animal; about 30% in the second animal; about 15% in the third animal; about 10% in the fourth animal; and about 5% in the fifth animal.

The presence of an EHC for hydrocortisone and its metabolites was demonstrated in the rat (WYNGAARDEN et al., 1955). In intact animals, 87% of the subcutaneously administered radioactivity derived from hydrocortisone was found to be excreted into the bile within 6 h, whereas only 66% appeared in the feces after 4 days; this difference indicated intestinal reabsorption. These authors then administered intragastrically the bile-containing isotope into another rat and showed that 27% of the ^{14}C was excreted into the bile of the second animal, thus showing an EHC. In guinea pigs, the results were more striking. While about 65% of the isotope was excreted into the bile, only 11 to 34% was recovered in the feces.

ARAI et al. (1962) have observed an EHC for norethynodrel in rabbits. Seven days after oral dosing, intact animals excreted 50% of the dose in the urine; however, rabbits with a bile fistula excreted only 21% in the urine. Also, the peak amounts of radioactivity present in the urine occurred after 3 to 5 days in intact animals, whereas this peak occurred by the second day in animals with bile fistulas. LAYNE et al. (1963) have reported similar observations in a very limited number of human subjects. The principal product identified in the bile was the conjugate of 17 α-ethynyl-3β-dihydroxy-5(10)-estrene.

G. Indomethacin

HUCKER et al. (1966) demonstrated that indomethacin is excreted into the bile of dogs, monkeys, and guinea pigs; after an intravenous dose (10 mg/kg) of the substance, these species excreted 56, 48, and 63% of the dose within a period of 6 h. In these species only 1.6 to 3.8% of the material appearing in bile was unchanged indomethacin. The major metabolite found in dogs was indomethacin glucuronide. The rat and guinea pig excreted indomethacin in the urine and in the feces; after 120 h, 35 to 45% of the administered dose appeared in the feces, whereas 50 to 65% appeared in the urine. The same relative distribution was found whether the drug was administered orally or intravenously. Such results indirectly suggested that the indomethacin excreted via the bile was reabsorbed in the intestine via an EHC.

These authors carried out further studies in the dog and found that 83% of an intravenous dose of indomethacin appeared in the feces and only 7.2% appeared in the urine. At first glance, indomethacin seemed to be preferentially excreted into the bile of dogs with little of the drug being reabsorbed. However, it was possible to show in the dog that a large proportion of the indomethacin excreted in the bile is reabsorbed from the gut and goes through an EHC. When "linked animals" were used, 56% of an intravenously administered dose given to the first animal appeared in the bile of the second dog. This indicated quite clearly that the excreted indomethacin in the intestine was capable of being reabsorbed. Most of the material excreted in the bile of dogs was a glucuronide conjugate of indomethacin, and this conjugate was hydrolyzed in the intestine of the recipient animal. In the dog, apparently, this cycle of conjugation in the liver, excretion into the intestine, hydrolysis in the intestine and reabsorption, continues until the compound is eliminated in the feces. Apparently, in the dog, the drug re-entering the liver after EHC is not cleared by the kidneys but is almost quantitatively returned to the bile for recirculation.

In man, it appears that an EHC for indomethacin seems to be of less importance. Only very limited data are available but about 15% of the dose is excreted in the bile in 24 h; man seems to excrete most of the administered dose in the urine (HUCKER et al., 1966).

The interspecies difference known to occur with indomethacin has also been studied by YESAIR et al. (1970). These authors were particularly interested in learning whether differences in EHC could explain the species differences; rats, dogs, and monkeys were used in their studies. The half-life of the indomethacin in plasma ranged from hours, in rats, to minutes in dogs and monkeys. Biliary excretion of indomethacin and its conjugates was extensive and rapid in dogs and monkeys, but slow in rats. In dogs, most of the intravenous dose of indomethacin appeared unchanged in the feces. Monkeys metabolized indomethacin to deschlorobenzoylindomethacin and the major metabolite found in rats was desmethylindomethacin. In rats, the plasma clearance of indomethacin by the liver was small, although 30 times greater than the clearance rate by the kidney; indomethacin was reabsorbed extensively from the intestine. The major metabolite was cleared from plasma by both the liver and kidney, but it was not reabsorbed from the intestine in rats.

In the rat, deschlorobenzoylindomethacin must be reabsorbed from the intestine since this metabolite is found only in the urine. Desmethylindomethacin, however, is cleared equally well by the kidney and the liver; since it appears in both the feces and urine in similar amounts, it is presumed that this metabolite is not readily reabsorbed from the intestine.

In the monkey, the half-life of the indomethacin was considerably shorter than that seen in rats. Indomethacin was rapidly cleared by the liver and excreted into the bile (52% of administered dose within 2 h). Ligation of the bile duct resulted in a greatly increased half-life of the drug in plasma (3 to 5 times above normal). An extensive EHC of indomethacin and its metabolites was evident in monkeys. The slow appearance of deschlorobenzoylindomethacin in the urine is attributed to this recirculation.

H. Glutethimide

KEBERLE et al. (1962) demonstrated that 90% of the dose of glutethimide administered to rats is recovered as glucuronides. When radiolabeled glutethimide was administered to rats, 68% of the radioactivity was excreted in the bile. Roughly, 95% of the radioactivity in the bile was accounted for as glucuronic acid conjugation products. When biologic half-life calculations were carried out using radioactive material and totaling all of the products excreted in the urine and feces, the half-life for the label was about 24 h. However, when this calculation was made using animals containing a bile fistula, the half-life of the label was only 6.5 h, thus indicating that much of the radioactive metabolites excreted into the bile were reabsorbed from the intestine and recirculated. KEBERLE et al. (1962) administered labeled glucuronides into the intestine and demonstrated that these substances were absorbed and excreted by the kidneys. In the same study, the half-life of the total radioactive material excreted in the urine of dogs was about 8 h compared to 24 h for rats. These half-lives included the unconjugated material plus its metabolites.

Since the data of KEBERLE et al. (1962) showed that in the rat the EHC of glutethimide and its metabolites had a marked effect on the biologic half-life of the products concerned, there has been interest in seeing whether interruption of the EHC could affect the depth or duration of glutethimide intoxication. CHARYTAN

(1970) carried out studies in dogs and man in which he monitored unmetabolized glutethimide in the bile and also the effects of biliary diversion on the plasma concentrations of this substance after oral doses were administered. He found very little unmetabolized glutethimide in the bile of dogs and of man. In both species, less than 1% of the administered dose was excreted unchanged in the bile. When biliary diversion was carried out in dogs, the blood glutethimide concentration over a period of 48 h after administration of 200 mg/kg of glutethimide was virtually identical before and after biliary diversion. Regression analysis on the blood glutethimide concentrations for the first 8 h after oral administration of this substance revealed identical regression lines for control and operated animals. From this, CHARYTAN (1970) concluded that the EHC does not have a significant effect on the disposition of unmetabolized glutethimide and that interruption of this cycle by biliary drainage was not indicated in glutethimide intoxication.

The recent work of AMBRE and FISCHER (1972) has shown that during the comatose state in man a previously undescribed hydroxylated glutethimide accumulates in larger quantities than glutethimide; there appears to be a correlation between the decline in the plasma concentration of this metabolite and recovery from coma. This metabolite is formed in only very small quantities in rats and dogs. Presently it is unknown whether this metabolite is pharmacologically active and whether it participates in an EHC. However, if it should be shown to be active the question about the utility of biliary diversion in glutethimide intoxication should be reinvestigated.

I. Amphetamine

Amphetamine and methamphetamine also participate in an EHC. CALDWELL et al. (1972) measured the biliary excretion of these substances in the rat and also the fecal excretion. Although the fecal excretion of the drugs is low, 2 to 5%, biliary excretion of the substances is appreciable. Sixteen percent of the amphetamine dose was excreted into the bile within 24 h whereas 69% appeared in the urine. Of the amount excreted into the bile 75% was 4-hydroxyamphetamine glucuronide and less than 6% was amphetamine. Eighteen percent of the methamphetamine dose was found in the bile and 54% in the urine. Of the amount excreted into the bile, 66% was 4-hydroxynorephedrine glucuronide, less than 5% was metamphetamine and 10% was amphetamine; the total output of 4-hydroxymethamphetamine and 4-hydroxyamphetamine was less than 10%. Since these glucuronides could be hydrolyzed in the intestine and the parent phenolic metabolites reabsorbed, the low fecal excretion is not surprising. Apparently, the hydrolyzed products are reabsorbed by the EHC and the substances are excreted in the urine. Although an EHC for amphetamine and metamphetamine metabolites seems apparent in the rat, CALDWELL et al. (1972) doubt that an EHC exists in man. In man the limiting molecular weight for biliary excretion is around 500, which is much higher than that for any of the metabolites of these drugs.

J. Butylated Hydroxytoluene

Butylated hydroxytoluene (BHT), an antioxidant used in food preparation, also follows an EHC in rats. LADOMERY et al. (1967) studied the biliary excretion of this substance and reported that after 4 days, 32% of the dose administered was found in the urine and 37% was excreted in the feces. However, 6 h after administration, 95% of an intravenous dose and 52% of an intraperitoneal dose appeared in the bile. These results suggested that reabsorption of the material occurred.

Reabsorption subsequently was demonstrated in a "linked animal" experiment. The bile duct of one rat was cannulated and inserted into the duodenum of the second rat. The donor rat was given an intravenous dose of BHT and the appearance of BHT derivatives in the bile of the recipient rat was monitored. Nearly 30% of the dose given to the donor rat was excreted as metabolites by the recipient rat after 10 h. This experiment clearly demonstrated the presence of an EHC. A 100-fold difference in dose-levels of BHT could still be excreted to the extent of 71% of the dose into the bile, showing that the biliary excretory system was quite efficient for BHT and its metabolites. In later experiments, HOLDER et al. (1970) indicated that the carboxylic acid derivative of BHT or its glucuronide was the recirculating compound.

K. Pentaerythritol Trinitrate

An EHC has also been postulated based on indirect experiments in which the biotransformation of pentaerythritol trinitrate in rats was studied (CREW et al., 1971). The prolonged hypotensive activity of pentaerythritol trinitrate suggested the possibility of an EHC. After the intravenous administration of this substance (10 mg/kg), about 50% of the dose was excreted 4 h later; by 24 h, 60% of the substance appeared in the bile. With a comparable dose of pentaerythritol, only 0.8% of the substance was excreted during a 24-h period. Thus, the trinitrate form is the derivative that is preferentially excreted into the bile. Examination of the bile samples indicated that the pentaerythritol trinitrate was excreted in the form of glucuronides of the mononitrate, the dinitrate, and the trinitrate derivatives. Of the three, the dinitrate was the most prominent (79%); however, 17% of the trinitrate still appeared. The 24-h urinary excretion of these substances was studied in normal rats and in animals with a bile fistula. In normal rats, roughly 31% of the administered dose was excreted in the urine within 24 h; however, in those animals containing a bile fistula, only about 13% was excreted. This difference was attributed to the intestinal reabsorption required for the EHC of the products. These authors postulated that the glucuronides of these nitrates excreted into the bile were hydrolyzed in the intestine and reabsorbed; the high aglycone content of the urine observed in intact rats in their studies support such a concept.

L. Fenamates

The major route of excretion of mefenamic acid, an anti-inflammatory agent, in the dog, is the feces (WINDER et al., 1966); in 72 h, about 65% of the dose is excreted in the feces and less than 1% in the urine. In a dog with a bile fistula, 35% appeared in the bile within 72 h. In the monkey, however, only 20% of the dose appeared in the feces and about 27% in the bile. In dogs 34% of the dose was excreted in the bile as mefenamic acid and only 1% as a metabolite. However, in monkeys 22% of the dose was excreted in the bile as metabolite and only 2% as mefenamic acid. Furthermore, the metabolite in dog bile was characterized as a glucuronide conjugate of the 3′-hydroxymethyl derivative. However, the metabolites in monkey bile were the free 3′-hydroxymethyl derivative and the free carboxyl derivative.

The presence of an EHC has been demonstrated in monkeys (WINDER et al., 1966). The metabolites of mefenamic acid in monkey bile were rapidly absorbed from the intestines of monkeys and re-excreted into the bile of a recipient animal; in 48 h, about 95% of the administered dose was recovered in the urine and bile, with about 44% of the dose in the bile. There is some indication that activated charcoal can interrupt the EHC of mefenamic acid when the charcoal is introduced into the intestinal tract.

In the discussion to the article published by WILLIAMS et al. (1965), GLAZKO pointed out a marked species difference in the biliary excretion of flufenamic acid; dogs excrete most of the radioactivity derived from labeled flufenamic acid in the feces, whereas monkeys excrete most of the radioactivity in the urine. However, both species excrete 40 to 70% of the administered dose in the bile. Apparently, an EHC for this substance exists in the monkey.

M. Phenothiazines

VAN LOON et al. (1964) studied the biliary excretion of chlorpromazine, prochlorperazine and trifluoperazine; they demonstrated the presence of an EHC for chlorpromazine and its metabolites. In dogs, 14 to 31% of an intraduodenally administered dose of ^{35}S-chlorpromazine was excreted into the bile within 10 h; 47% was excreted in 24 h. Sixty and 70% of labeled prochlorperazine and trifluoperazine, respectively, were found in the bile. Bile containing the excreted chlorpromazine and/or its metabolites was instilled in the duodenum of another dog and 30% of the radioactivity appeared in the bile of this animal after 10 h, thus showing an EHC for chlorpromazine and/or its metabolites.

N. Antibiotics

Several antibiotics are excreted into the bile (STOWE and PLAA, 1968). However, few studies have been done to determine the presence of an EHC for these substances. STEWART and HARRISON (1961) demonstrated an EHC exists for ampicillin in the rat. Normally, about 5% of the orally administered dose was excreted into the bile. However, in "linked animals", 20 to 60% of the product appearing in the bile of the donor rat was re-excreted in the bile of the recipient rat; the product excreted was unchanged ampicillin.

GLASSMAN et al. (1964) studied the biliary excretion of nafcillin in the dog. After intramuscular injection, 39% of the material was excreted into the bile within 6 h; 32% of the dose was subsequently excreted into the bile of a second dog receiving the bile from a "linked animal".

V. Enhanced Biliary Excretion of Drugs

One would think that altered biliary excretion could affect the circulation of drugs that can undergo an EHC. Although biliary excretion can be enhanced by several different mechanisms, it is yet to be demonstrated that these maneuvers alter the EHC of the substances involved.

A. Enhanced Biliary Flow

In 1967, ROBERTS and PLAA demonstrated that phenobarbital pretreatment increased the bile volume in rats and thereby increased the maximal rate of bilirubin excretion. Later studies (KLAASSEN and PLAA, 1968) demonstrated that phenobarbital pretreatment also resulted in the enhanced excretion of sulfobromophthalein (BSP) and phenol-3,6-dibromophthalein disulfonate (DBSP). Because of these data, a series of studies was initiated to determine whether microsomal induction was responsible for the enhanced biliary excretion and whether other inducers were capable of producing the same effects. KLAASSEN (1969) reported that while phenobarbital was capable of producing a significant increase in bile volume, chlordane, nikethamide, phenylbutazone and chlorcyclizine produced increases in biliary flow that were not statistically significant. Pretreatment with 3,4-benzpyrene and 3-methylcholanthrene also failed to alter biliary flow. This author concluded that induction of microsomal drug metabolizing enzymes by itself did not result in an increase in biliary flow; in these studies, only phenobarbital pretreatment increased bile flow. The effects of phenobarbital on bile flow were due to enhanced formation of the bile salt-independent fraction of canalicular bile production (BERTHELOT et al., 1970; KLAASSEN, 1971 a) and were independent of bile acid secretion in the rat (KLAASSEN, 1971 b). Recently, REDINGER and SMALL (1973) have shown that in rhesus monkeys, phenobarbital also increases bile flow. However, both bile salt-independent flow and bile salt secretion rates were increased.

The effects of phenobarbital on the biliary excretion of various drugs and chemicals were also studied in rats (KLAASSEN, 1970a). Phenobarbital treatment significantly increased the biliary excretion of amaranth and ouabain. Little or no increase in the excretion of phenol red and succinyl-sulphathiazol was observed, even though bile flow was significantly increased. Increased excretion of procaine-amide ethobromide (in the unconjugated form) was also demonstrated. These results indicated that enhanced conjugation was not necessarily involved since amaranth and ouabain were excreted unchanged; the increase in bile flow probably caused the enhanced excretion. The effects of other microsomal enzyme inducers on the excretion of BSP and DBSP were studied in rats (KLAASSEN, 1970b). Chlordane, phenylbutazone, nikethamide and chlorcyclizine did not increase biliary flow and only slightly increased excretion of these dyes into the bile. Methylcholanthrene and 3,4-benzpyrene did not affect bile flow or excretion. Again a good correlation between the ability of these agents to increase biliary flow and their ability to increase biliary excretion was found. Thus it is clearly established that drug-induced increases in bile flow can enhance biliary excretion.

B. Enhanced Formation of Metabolites

Microsomal enzyme inducers also enhance biliary excretion by increasing metabolite formation. GOLDSTEIN and TAUROG (1968) demonstrated that thyroxine glucuronide is rapidly excreted into the bile and that pretreatment with benzpyrene enhances the excretion of this substance 3 to 4-fold. This enhanced excretion occurred whether thyroxine was exogenously administered or endogenously produced. Almost all of the increase could be accounted for by enhanced excretion of thyroxine glucuronide. The authors explain the effects of benzpyrene as being due to an increase in glucuronyl transferase activity; the increase in excretion of thyroxine glucuronide occurred in animals without an increase in bile flow. These authors also studied the effects of phenobarbital pretreatment and reported a small increase in thyroxine excretion in bile that was not limited to the glucuronide alone. This slight effect was attributed to the increase in bile flow known to occur with phenobarbital treatment.

LEVINE et al. (1970) studied the effects of phenobarbital on the biliary excretion of biphenyl, diethylstilbestrol, phenolphthalein, succinylsulphathiazole and indocyanine green. Only those substances that require conjugation with glucuronide prior to excretion were enhanced by phenobarbital pretreatment; biphenyl, diethylstilbestrol and phenolphthalein were all excreted more rapidly in phenobarbital-treated rats, whereas the glucuronides of these substances were unaffected in the same rats. These studies indicated that the phenobarbital effect was due to enhanced glucuronidation. Recent work by LEVINE (1971) has demonstrated that both *N*-2-fluorenylacetamide (FAA) and *N*-hydroxy-2-fluorenylacetamide (*N*-OH-FAA) are excreted principally by way of the bile in the rat. He studied the effects of benzpyrene induction on the excretion of both substances and showed that the excretion of FAA was enhanced by benzpyrene induction, whereas the excretion of *N*-hydroxy-FAA was not enhanced. The explanation for this is that benzpyrene stimulates the metabolism of FAA to *N*-OH-FAA, which is excreted normally. Stimulation of the metabolism of FAA by benzpyrene increased the biliary excretion of the carcinogen. Conversely, pretreatment with piperonyl butoxide, an inhibitor of microsomal enzymes, depressed the metabolism of FAA and also depressed its biliary excretion. Parallel experiments carried out with *N*-OH-FAA indicated that treatments with stimulators or depressors of the microsomal enzyme systems did not affect its excretion. The results indicated that metabolism was the rate-limiting process in the excretion of FAA in the bile, but that the *N*-hydroxy derivative was not limited by alteration of its metabolism.

PAPAPETROU et al. (1972) demonstrated that intravenously administered carbimazol, methimazol and propylthiouracil are excreted rapidly in the bile of rats; after 5 h, 32, 21, and 8% respectively, were found in the bile. When the rats were pretreated with pentobarbital for 4 days prior to the administration of these substances, the excretion of methimazol and carbimazol was increased; this seemed to be due to an increased excretion of the glucuronide. The excretion of propylthiouracil was also slightly increased, but this seemed to be due to an increase in bile flow due to pentobarbital pretreatment.

C. Formation of Complexes

Chemicals that can form complexes also enhance biliary excretion. The biliary excretion of iron can be enhanced by treating animals with other substances. FIGUEROA and THOMPSON (1968) administered desferrioxamine to rats prior to the administration of an iron load and increased the biliary excretion of iron up to 32-fold. This increase in biliary iron excretion was independent of bile volume. Although the mechanism is not established, desferrioxamine apparently forms a complex with iron in tissues; the complex is then excreted via the bile.

Although cholesterol is insoluble in water, it remains in solution when excreted into the bile. Recently, HARDISON and APTER (1972) showed that micelle formation is an important mechanism in the biliary excretion of lipids. When taurocholate, a micelle-forming bile salt was infused, lipid biliary excretion increased. However, when dihydrocholate, a nonmicelle-forming salt was infused, biliary lipid excretion increased very little. They also demonstrated a close relationship between the amount of cholesterol excreted and the amount of phospholipids excreted in the bile; however, there was a very poor correlation between the amount of bile salt excreted and the amount of cholesterol excreted. It is still unclear why biliary lipid excretion is limited in spite of increasing bile salt excretion. Perhaps bile salt may require phospholipid in order to solubilize cholesterol.

References

AMBRE, J.J., FISCHER, L.J.: Glutethimide intoxication: Plasma levels of glutethimide and a metabolite in humans, dogs and rats. Res. Comm. chem. Path. Pharmacol. **4**, 307—326 (1972).

ARAI, K., GOLAB, T., LAYNE, D.S., PINCUS, G.: Metabolic fate of orally administered H^3-norethynodrel in rabbits. Endocrinology **71**, 639—648 (1962).

BAKER, S.J., KUMAR, S., SWAMINATHAN, S.P.: Excretion of folic acid in bile. Lancet **1965 I**, 685.

BASELT, R.C., CASARETT, L.J.: Biliary and urinary elimination of methadone and its metabolites in the rat. Biochem. Pharmacol. **21**, 2705—2712 (1972).

BECKER, B. A., HINDMAN, K. L., GIBSON, J. E.: Enhanced mortality of selected nervous system depressants in hypoexcretory mice. J. pharm. Sci. **57**, 1010—1012 (1968).
BERGSTROM, S.: Metabolism of bile acids. Fed. Proc. **21**, 28—32 (Suppl. No. II) (1962).
BERTHELOT, P., ERLINGER, S., DHUMEAUX, D., PREAUZ, A.-M.: Mechanism of phenobarbital-induced hypercholeresis in the rat. Amer. J. Physiol. **219**, 809—813 (1970).
BORGSTROM, G., DAHLQVIST, A., LUNDH, G., SJOVALL, J.: Studies of intestinal digestion and absorption in the human. J. clin. Invest. **36**, 1521—1536 (1957).
BREMER, J.: Species differences in the conjugation of free bile acids with taurine and glycine. Biochem. J. **63**, 507—513 (1956).
CALDWELL, J. H., BUSH, C. A., GREENBERGER, N. J.: Interruption of the enterohepatic circulation of digitoxin by cholestyramine. II. Effect on metabolic disposition of tritium-labeled digitoxin and cardiac systolic intervals in man. J. clin. Invest. **50**, 2638—2644 (1971).
CALDWELL, J. H., DRING, L. G., WILLIAMS, R. T.: Biliary excretion of amphetamine and methamphetamine in the rat. Biochem. J. **129**, 25—29 (1972).
CALDWELL, J. H., GREENBERGER, N. J.: Interruption of the enterohepatic circulation of digitoxin by cholestyramine. I. Protection against lethal digitoxin intoxication. J. clin. Invest. **50**, 2626—2637 (1971).
CASTLE, M. C., LAGE, G. L.: Enhanced excretion of digitoxin following spironolactone as it relates to the prevention of digitoxin toxicity. Res. Comm. chem. Path. Pharmacol. **5**, 99—108 (1973).
CHARYTAN, C.: The enterohepatic circulation in glutethimide intoxication. Clin. Pharmacol. Ther. **11**, 816—820 (1970).
CLARK, A. G., FISCHER, L. J., MILLBURN, P., SMITH, R. L., WILLIAMS, R. T.: The role of gut flora in the enterohepatic circulation of stilboestrol in the rat. Biochem. J. **112**, 17P (1969).
COOK, D. A., HAGERMAN, L. M., SCHNEIDER, D. L.: Effect of dietary taurine on fecal bile salt excretion in rats and hamsters fed cholestyramine. Proc. Soc. exp. Biol. (N. Y.) **138**, 830—834 (1971).
COOK, D. A., HAGERMAN, L. M., SCHNEIDER, D. L.: Preferential retention of taurine-conjugated bile salts by cholestyramine in the rat ileum. Proc. Soc. exp. Biol. (N. Y.) **139**, 70—73 (1972).
CREW, M. C., GALA, E. L., HAYNES, L. J., DICARLO, F. J.: Biliary excretion and biotransformation of pentaerythritol trinitrate in rats. Biochem. Pharmacol. **20**, 3077—3089 (1971).
DENBESTEN, L.: A technic for repeated bile collections from an intact enterohepatic circulation. Proc. Soc. exp. Biol. (N. Y.) **138**, 208—209 (1971).
DIETSCHY, J. M.: Mechanisms for the intestinal absorption of bile acids. J. Lipid Res. **9**, 297—309 (1968).
DOBBS, H. E., HALL, J. M.: Metabolism and biliary excretion of etorphine (M99-Reckitt) an extremely potent morphine-like drug. Proc. Europ. Soc. Study Drug Toxicity **10**, 77—86 (1969).
DOBBS, H. E., HALL, J. M., STEIGER, B.: Enterohepatic circulation of etorphine, a potent analgesic, in the rat. Proc. Europ. Soc. Study Drug Toxicity **11**, 73—79 (1970).
DOUGHERTY, J. E., FLANAGAN, W. J., MURPHY, M. L., BULLOCH, R. T., DALRYMPHE, G. L., BEARD, O. W., PERKINS, W. H.: Tritiated digoxin. XIV. Enterohepatic circulation, absorption, and excretion studies in human volunteers. Circulation **42**, 867—873 (1970).
DOWLING, R. H.: The enterohepatic circulation. Gastroenterology **62**, 122—140 (1972).
DOWLING, R. H., MACK, E., PICOTT, J., BERGER, J., SMALL, D. M.: Experimental model for the study of the enterohepatic circulation of bile in rhesus monkeys. J. Lab. clin. Med. **72**, 169—176 (1968).
EISENBRANDT, L. L., ADLER, T. K., ELLIOTT, H. W., ABDOU, I. A.: The role of the gastrointestinal tract in the excretion of C^{14}-labeled methadone by rats. J. Pharmacol. exp. Ther. **98**, 200—205 (1950).
ERIKSSON, H.: Absorption and enterohepatic circulation of neutral steroids in the rat. Europ. J. Biochem. **19**, 416—423 (1971).
EYSSEN, H., EVRARD, E., VANDERHAEGHE, H.: Cholesterol-lowering effects of *N*-methylated neomycin and basic antibiotics. J. Lab. clin. Med. **68**, 753—768 (1966).
FIGUEROA, W. G., THOMPSON, J. H.: Biliary iron excretion in normal and ironloaded rats after desferrioxamine and CaDTPA. Amer. J. Physiol. **215**, 807—810 (1968).
FISCHER, L. J., KENT, T. H., WEISSINGER, J. L.: Absorption of diethylstilbestrol and its glucuronide conjugate from the intestines of five- and twenty-five-day-old rats. J. Pharmacol. exp. Ther. **185**, 163—170 (1973).
FISCHER, L. J., MILLBURN, P.: Stilboestrol transport and glucuronide formation in everted sacs of rat intestine. J. Pharmacol. exp. Ther. **175**, 267—275 (1970).
FISCHER, L. J., MILLBURN, P., SMITH, R. L., WILLIAMS, R. T.: The fate of [^{14}C] stilboestrol in the rat. Biochem. J. **100**, 69P (1966).

Fischer, L. J., Weissinger, J. L., Kent, T. H.: The role of intestinal β-glucuronidase in the absorption of diethylstilbestrol glucuronide in immature rats. Toxicol. appl. Pharmacol. **19**, 396—397 (1971).

Forman, D. T., Garvin, J. E., Forestner, J. E., Taylor, C. B.: Increased excretion of fecal bile acids by an oral hydrophilic colloid. Proc. Soc. exp. Biol. (N. Y.) **127**, 1060—1063 (1968).

Gibson, J. E., Becker, B. A.: Demonstration of enhanced lethality of drugs in hypoexcretory animals. J. pharm. Sci. **56**, 1503—1505 (1967).

Glasser, J. E., Weiner, I. M., Lack, L.: Comparative physiology of intestinal taurocholate transport. Amer. J. Physiol. **208**, 359—362 (1965).

Glassman, J. M., Warren, G. H., Rosenman, S. B., Agersborg, H. P. K.: Pharmacology and distribution of WY-3277 (nafcillin): 6-(2-ethoxyl-1-naphthamido) penicillanic acid. Toxicol. appl. Pharmacol. **6**, 220—231 (1964).

Goldstein, J. A., Taurog, A.: Enhanced biliary excretion of thyroxine glucuronide in rats pretreated with benzpyrene. Biochem. Pharmacol. **17**, 1049—1065 (1968).

Grasbeck, R., Nyberg, W., Reizenstein, P.: Biliary and fecal vitamin B_{12} excretion in man. An isotope study. Proc. Soc. exp. Biol. (N. Y.) **97**, 780—784 (1958).

Greenberger, N. J., Thomas, F. B.: Biliary excretion of ^{3}H-digitoxin: Modification by bile salts and phenobarbital. J. Lab. clin. Med. **81**, 241—251 (1973).

Hanahan, D. J., Daskalakis, E. G., Edwards, T., Dauben, H. J., Jr.: The metabolic pattern of C^{14}-diethylstilbestrol. Endocrinology **53**, 163—170 (1953).

Hardison, W. G. M., Apter, J. T.: Micellar theory of biliary cholesterol excretion. Amer. J. Physiol. **222**, 61—67 (1972).

Hashim, S. A., Van Itallie, T. B.: Cholestyramine resin therapy for hypercholesteremia. J. Amer. med. Ass. **192**, 289—293 (1965).

Herbert, V.: Excretion of folic acid in bile. Lancet **1965 I**, 913.

Hofmann, A. F.: Clinical implications of physicochemical studies on bile salts. Gastroenterology **48**, 484—494 (1965).

Hofmann, A. F.: The syndrome of ileal disease and the broken enterohepatic circulation: Cholerheic enteropathy. Gastroenterology **52**, 752—757 (1967).

Hofmann, A. F., Small, D. M.: Detergent properties of bile salts: correlation with physiological function. Ann. Rev. Med. **18**, 333—376 (1967).

Holder, G. M., Ryan, A. J., Watson, T. R., Wiebe, L. I.: The biliary metabolism of butylated hydroxytoluene (3,5-di-*t*-butyl-4-hydroxytoluene) and its derivatives in the rat. J. Pharm. Pharmacol. **22**, 832—838 (1970).

Hucker, H. B., Zacchei, A. G., Cox, S. V., Brodie, D. A., Cantwell, N. H. R.: Studies on the absorption, distribution and excretion of indomethacin in various species. J. Pharmacol. exp. Ther. **153**, 237—249 (1966).

Huff, J. W., Gilfillan, J. L., Hunt, V. M.: Effect of cholestyramine, a bile acid binding polymer on plasma cholesterol and fecal bile acid excretion in the rat. Proc. Soc. exp. Biol. (N. Y.) **114**, 352—355 (1963).

Ibrahim, G. W., Watson, C. J.: Enterohepatic circulation and conversion of protoporphyrin to bile pigment in man. Proc. Soc. exp. Biol. (N. Y.) **127**, 890—895 (1968).

Jansen, G. R., Zanetti, M. E.: Cholestyramine in dogs. J. pharm. Sci. **54**, 863—867 (1965).

Katzung, B. G., Meyers, F. H.: Excretion of radioactive digitoxin by the dog. J. Pharmacol. exp. Ther. **149**, 257—262 (1965).

Keberle, H., Hoffmann, K., Bernhard, K.: The metabolism of glutethimide (Doriden). Experientia (Basel) **18**, 105—111 (1962).

Klaassen, C. D.: Biliary flow after microsomal enzyme induction. J. Pharmacol. exp. Ther. **168**, 218—223 (1969).

Klaassen, C. D.: Effects of phenobarbital on the plasma disappearance and biliary excretion of drugs in rats. J. Pharmacol. exp. Ther. **175**, 289—300 (1970a).

Klaassen, C. D.: Plasma disappearance and biliary excretion of sulfobromophthalein and phenol-3,6-dibromophthalein disulfonate after microsomal enzyme induction. Biochem. Pharmacol. **19**, 1241—1249 (1970b).

Klaassen, C. D.: Studies on the increased biliary flow produced by phenobarbital in rats. J. Pharmacol. exp. Ther. **176**, 743—751 (1971a).

Klaassen, C. D.: Does bile acid secretion determine canalicular bile production in rats? Amer. J. Physiol. **220**, 667—673 (1971b).

Klaassen, C. D.: Comparison of the toxicity of chemicals in newborn rats to bile duct-ligated and sham-operated rats and mice. Toxicol. appl. Pharmacol. **24**, 37—44 (1973).

Klaassen, C. D., Plaa, G. L.: Studies on the mechanism of phenobarbital-enhanced sulfobromophthalein disappearance. J. Pharmacol. exp. Ther. **161**, 361—366 (1968).

Lack, L., Weiner, I. M.: *In vitro* absorption of bile salts by small intestine of rats and guinea pigs. Amer. J. Physiol. **200**, 313—317 (1961).

LACK, L., WEINER, I. M.: Intestinal bile salt transport: structure-activity relationships and other properties. Amer. J. Physiol. **210**, 1142—1152 (1966).

LACK, L., WEINER, I. M.: Role of the intestine during the enterohepatic circulation of bile salts. Gastroenterology **52**, 282—287 (1967).

LADOMERY, L. G., RYAN, A. J., WRIGHT, S. E.: The excretion of [^{14}C] butylated hydroxytoluene in the rat. J. Pharm. Pharmacol. **19**, 383—387 (1967).

LAYNE, D. S., GOLAB, T., ARAI, K., PINCUS, G.: The metabolic fate of orally administered ^{3}H-norethynodrel and ^{3}H-norethindrone in humans. Biochem. Pharmacol. **12**, 905—911 (1963).

LESTER, R., SCHUMER, W., SCHMID, R.: Intestinal absorption of bile pigments. IV. Urobilinogen absorption in man. New Engl. J. Med. **272**, 939—943 (1965).

LEVINE, W. G.: Metabolism and biliary excretion of *N*-2-fluorenylacetamide and *N*-hydroxy-2-fluorenylacetamide. Life Sci. **10**, 727—735 (1971).

LEVINE, W. G., MILLBURN, P., SMITH, R. L., WILLIAMS, R. T.: The role of the hepatic endoplasmic reticulum in the biliary excretion of foreign compounds by the rat. The effect of phenobarbitone and SKF 525-A (diacetylaminoethyl diphenylpropylacetate). Biochem. Pharmacol. **19**, 235—244 (1970).

LINDSTEDT, S.: The turnover of cholic acid in man. Bile acids and steroids, 51. Acta physiol. Scand. **40**, 1—9 (1957).

LISCHNER, H. W.: Cholestyramine and porphyrin-binding. Lancet **1966 II**, 1079—1080.

MARCH, C. H., ELLIOTT, H. W.: Distribution and excretion of radioactivity after administration of morphine-*N*-methyl C^{14} to rats. Proc. Soc. exp. Biol. (N. Y.) **86**, 494—497 (1954).

MOORE, R. B., CRANE, C. A., FRANTZ, I. D., JR.: Effect of cholestyramine on the fecal excretion of intravenously administered cholesterol-4-^{14}C and its degradation products in a hypercholesterolemic patient. J. clin. Invest. **47**, 1664—1671 (1968).

NORMAN, A., SJOVALL, J.: On the transformation and enterohepatic circulation of cholic acid in the rat: Bile acids and steroids 68. J. biol. Chem. **233**, 872—885 (1958).

OKITA, G. T.: Species difference in duration of action of cardiac glycosides. Fed. Proc. **26**, 1125—1130 (1967).

OKITA, G. T., TALSO, P. J., CURRY, J. H., JR., SMITH, F. D., JR., GEILLING, E. M. K.: Metabolic fate of radioactive digitoxin in human subjects. J. Pharmacol. exp. Ther. **115**, 371—379 (1955).

PAPAPETROU, P. D., MARCHAND, B., GAVRAS, H., ALEXANDER, W. D.: Biliary excretion of ^{35}S-labelled propylthiouracil, methimazole and carbimazole in untreated and pentobarbitone pretreated rats. Biochem. Pharmacol. **21**, 363—377 (1972).

PLAA, G. L.: Biliary and other routes of excretion of drugs. In: LADU, B. N., MANDEL, H. G., WAY, E. L. (Eds.): Fundamentals of drug metabolism and drug disposition, pp. 131—145. Baltimore: Williams and Wilkins 1971.

PLAYOUST, M. R., ISSELBACHER, K. J.: Studies on the transport and metabolism of conjugated bile salts by intestinal mucosa. J. clin. Invest. **43**, 467—476 (1964).

POWELL, R. C., NUNES, W. T., HARDING, R. S., VACCA, J. B.: The influence of non-absorbable antibiotics on serum lipids and the excretion of neutral sterols and bile acids. Amer. J. clin. Nutr. **11**, 156—168 (1962).

REDINGER, R. N., SMALL, D. M.: Primate biliary physiology. VIII. The effect of phenobarbital upon bile salt synthesis and pool size, biliary lipid secretion, and bile composition. J. clin. Invest. **52**, 161—172 (1973).

ROBERTS, R. J., PLAA, G. L.: Effect of phenobarbital on the excretion of an exogenous bilirubin load. Biochem. Pharmacol. **16**, 827—835 (1967).

ROE, D. A.: The clinical and biochemical significance of taurine excretion in psoriasis. J. invest. Derm. **39**, 537—542 (1962).

SAMUEL, P., STEINER, A.: Effect of neomycin on serum cholesterol level of man. Proc. Soc. exp. Biol. (N. Y.) **100**, 193—195 (1959).

SAMUEL, P., WAITHE, W. I.: Reduction of serum cholesterol concentrations by neomycin, para-aminosalicylic acid and other antibacterial drugs in man. Circulation **24**, 578—591 (1961).

SANDBERG, A. A., KIRDANI, R. Y., BACK, N., WEYMAN, P., SLAUNWHITE, W. R., JR.: Biliary excretion and enterohepatic circulation of estrone and estriol in rodents. Amer. J. Physiol. **213**, 1138—1142 (1967).

SMALL, D. M.: Physiochemical studies of cholesterol gallstone formation. Gastroenterology **52**, 607—610 (1967).

SMITH, D. S., PETERSON, R. E., FUJIMOTO, J. M.: Species differences in the biliary excretion of morphine, morphine-3-glucuronide and morphine-3-ethereal sulfate in the cat and rat. Biochem. Pharmacol. **22**, 485—492 (1973).

SMITH, R. L.: Excretion of drugs in bile. In: BRODIE, B. B., GILLETTE, J. R., MITCHELL, J. R. (Eds.): Concepts of biochemical pharmacology, Handbook of experimental pharmacology, Vol. 28, Part 1, pp. 354—389. Berlin-Heidelberg-New York: Springer 1971.

STATHERS, G.M.: Porphyrin-binding effect of cholestyramine. Results of *in-vitro* and *in-vivo* studies. Lancet **1966 II**, 780—783.

STEINER, A., HOWARD, E., AKGUN, S.: Effect of antibiotics on the serum cholesterol concentration of patients with atherosclerosis. Circulation **24**, 729—735 (1961).

STEWART, G.T., HARRISON, P.M.: Excretion and re-excretion of a broad spectrum penicillin in bile. Brit. J. Pharmacol. **17**, 414—419 (1961).

STOWE, C.M., PLAA, G.L.: Extrarenal excretion of drugs and chemicals. Ann. Rev. Pharmacol. **8**, 337—356 (1968).

THOMPSON, R.Q., STURTEVANT, M., BIRD, O.D., GLAZKO, A.J.: The effect of metabolites of chloramphenicol (chloromycetin) on the thyroid of the rat. Endocrinology **55**, 665—681 (1954).

VAN LOON, E.J., FLANAGAN, T.L., NOVICK, W.J., MAASS, A.R.: Hepatic secretion and urinary excretion of three S^{35}-labeled phenothiazines in the dog. J. pharm. Sci. **53**, 1211—1213 (1964).

WAY, E.L., ADLER, T.K.: The pharmacologic implications of the fate of morphine and its surrogates. Pharmacol. Rev. **12**, 383—446 (1960).

WILLIAMS, R.T., MILLBURN, P., SMITH, R.L.: The influence of enterohepatic circulation on toxicity of drugs. Ann. N. Y. Acad. Sci. **123**, 110—124 (1965).

WINDER, C.V., KAUMP, D.H., GLAZKO, A.J., HOLMES, E.L.: Pharmacology of the fenamates. Ann. phys. Med. (Supplement) 7—49 (1966).

WOOD, P., SHIODA, R., ESTRICH, D., SPLITTER, S.: The influence of oral cholestyramine on human bile composition. Clin. Res. **17**, 162 (1969).

WOODS, L.A.: Distribution and fate of morphine in non-tolerant and tolerant dogs and rats. J. Pharmacol. exp. Ther. **112**, 158—175 (1954).

WYNGAARDEN, J.B., PETERSEN, R.E., WOLF, A.R.: Physiologic disposition of radiometabolites of hydrocortisone-4-C^{14} in the rat and guinea-pig. J. biol. Chem. **212**, 963—972 (1955).

YESAIR, D.W., CALLAHAN, M., REMINGTON, L., KENSLER, C.J.: Role of the enterohepatic cycle of indomethacin on its metabolism, distribution in tissues and its excretion by rats, dogs and monkeys. Biochem. Pharmacol. **19**, 1579—1590 (1970).

Chapter 64

Routes of Administration and Drug Response

C. T. Dollery and D. S. Davies

With 10 Figures

I. Introduction

The route of administration for a drug should be selected to achieve an effective concentration at the site of action as conveniently as possible. Important subsidiary objectives may be rapidity of onset and duration of drug effect and, in some instances, selectivity of action.

The possible routes of entry of drugs into the body may be divided into two classes – enteral and parenteral. In enteral administration, the drug is placed directly into a section of the gastrointestinal tract by putting it under the tongue (sublingual), by swallowing (oral, *per os, p.o.*), or by placing it in the rectum. Parenteral administration by-passes the gastrointestinal tract; this is essential to insure the effectiveness of many drugs. Of the many parenteral routes, the most frequently used are subcutaneous (s.c.), intramuscular (i.m.), and intravascular, usually by intravenous injection (i.v.). Other less frequently used parenteral routes include inhalation, either for direct action within the airways or for absorption into blood at the alveoli; injection of drugs into or near the spinal cord; insertion into the vagina; and topical application to the skin, the eyes, or the ears.

Most drugs used in therapy have plasma half-lives of less than 48 h. Thus, if the effect of a drug is to be maintained over a long period of time as, for example, in the treatment of diabetes, hypertension, rheumatoid arthritis, or depressive illness, one dose should be given at least once each day. Under these conditions, drugs that must be given by injection are troublesome. Although diabetics have been trained to give themselves insulin and, in the early 1950's, patients with severe hypertension were treated with self-injected hexamethonium, injection is unpopular and burdensome to most patients. Also, the chance of infection due to poor injection technique always exists. The parenteral route still may be practical if the duration of drug action is long, or can be rendered so by special formulation, such that infrequent injections performed by trained personnel are sufficient. For example, patients with pernicious anemia can be treated with once monthly injections of hydroxocobalamin, which is stored mainly in the liver (Herbert, 1970). Suramin, a drug used in the treatment of patients with trypanosomiasis, is eliminated slowly because of extensive binding to plasma proteins and injections of 1 g every 3 months are sufficient for prophylactic therapy. Neither hydroxocobalamin nor suramin can be taken orally because they are inadequately absorbed.

Many drugs will elicit effects of similar magnitudes when the same dose is given by a number of different routes. However, there are so many instances when this is not true that these can be important traps for the unwary pharmacologist, toxicologist, or clinician. This paper considers the main routes of drug administration and examines the reasons why the oral route cannot always be used.

Finally, we will consider several drugs with effects that vary according to the route of administration, and we will discuss the reasons for this dependence.

II. Enteral Administration of Drugs

A. The Oral Route

1. Advantages of Oral Dosing

Most drugs in general use are given by mouth simply because it is the most convenient method of drug administration available to the patient. However, oral dosage can also offer certain extra advantages over other routes, particularly parenteral administration. For example, plasma levels of rapidly eliminated drugs can be more readily sustained after oral administration than after intravenous administration because absorption from the intestine becomes the limiting process.

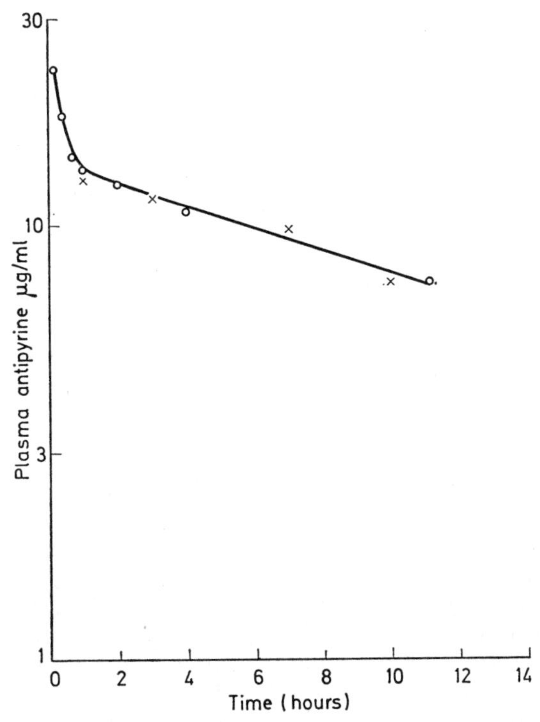

Fig. 1. Plasma levels of antipyrine following oral (×—×) and intravenous (○—○) doses of 600 mg in man (THORGEIRSSON and DAVIES, unpublished results)

This effect, in principle, can be augmented by using sustained release preparations of drugs and therefore making the rate of dissolution in the gut the limiting factor in the absorption process. Another advantage of the oral route is that it avoids the high, and often hazardous, initial drug concentrations in blood and in high blood-flow organs that occur after intravenous doses.

The oral route is suitable for drugs that readily pass from the lumen of the intestine into the systemic circulation without extensive losses. It is also assumed that the oral route is available, and that a rapid onset of action is not needed. Many lipid soluble drugs so efficiently enter the systemic circulation following oral doses that plasma concentration time curves are superimposible with those seen following i.v. administration (after the initial absorption and distribution phases). For example, plasma levels of the freely diffusable drug antipyrine are similar following intravenous and oral doses of 600 mg (Fig. 1).

2. Disadvantages of Oral Dosing

However, the oral route has many disadvantages that have necessitated the exploitation of other routes in order to make many drugs effective. The disadvantages of the oral route fall into four broad categories: (a) slow onset of action when an immediate effect of a drug is needed; (b) non-availability of the oral route due to the clinical condition of the patient or the nature of the drug; (c) poor availability of the drug in the systemic circulation due to losses in or during transfer from the lumen of the gut; and (d) systemic distribution of the drug when a selective local action is desired.

a) Slow Onset of Action

A number of factors in addition to the lipid solubility of a drug determine the time lag between dosing and the appearance of drug in the systemic circulation. These include such variable factors as the quantity of food in the gastrointestinal tract, gastrointestinal motility and gastric emptying times. This latter parameter varied from 30 to 183 min in Parkinsonian patients receiving L-dopa (BIANCHINE et al., 1971). Patients with long gastric emptying times also achieved peak serum concentrations 40% lower and up to 30 min later than those with shorter emptying times. The time lag before drugs taken orally reach their sites of absorption in the intestine can be considerable, making this route unusable in an emergency situation in which the immediate action of a drug is required.

In an emergency situation, drugs are administered parenterally, usually by intravenous injection. In a less acute situation, it may be possible to exploit the quicker onset of action reported for the two other enteral routes. In both sublingual and rectal administration, the drug is immediately placed at its site of absorption, thus eliminating any lag period. However, one cannot assume that peak levels will be achieved more rapidly through the sublingual route than through the oral route for all drugs and not even for drugs that are well absorbed. Studies of the absorption of amphetamine and nicotine (BECKETT and HOSSIE, 1971) following oral and sublingual administration showed that peak blood levels for nicotine were achieved four times faster after sublingual administration but the reverse was true for amphetamine. In studies with pentaerythritol trinitrate (DAVIDSON et al., 1971), the rate constant for absorption was about three times greater after oral than after sublingual administration. Thus the better responses to many drugs after sublingual administration than after oral administration may often be due to less metabolism during absorption rather than to a more rapid rate of absorption.

b) Nonavailability of Oral Route

Sometimes it is physically difficult to administer drugs orally. This is the case for patients who have undergone extensive gastrointestinal surgery, are unconscious, or are vomiting. Under these conditions drugs may be administered parenterally, or perhaps rectally. For example, rectal suppositories of ergotamine are available for migraine sufferers unable to take the drug orally because they are vomiting. This also applies to drugs such as theophylline, which can be given by rectal administration in order to avoid unpleasant gastric irritation when the drug is taken orally.

The oral route is not used for the antileproic drug ditophal (diethyl dithiolisophthalate, ETISUL) for a rather unusual reason. Ditophal is an ester of ethyl mercaptan to which it is readily hydrolyzed. Formation of this extremely unpleasant smelling derivative in the gastrointestinal tract would be unacceptable, so this drug was administered by rubbing into the skin. By this method more than 80% of the drug was absorbed and then hydrolyzed to ethyl mercaptan, the active form (ELLARD et al., 1963). The drug is effective when given by this route since both local leprous lesions and those remote from the site of inunction respond satisfactorily to treatment.

c) Poor Drug Availability

Occasionally the availability of drug to the systemic circulation following an oral dose is too low to be of use and other routes have to be found.

Loss of Drug in the Lumen of the Gut. Many drugs are sufficiently unstable in gut contents, either at the acid pH of stomach contents or the more alkaline pH of the intestine, that the drug is lost before it can be absorbed into the systemic circulation. Thus insulin, which is unstable in gut contents, has to be given parenterally. However, some compounds that are unstable in gut contents can still be given by an enteral route. Oxytocin is ineffective when given orally because it is inactivated by chymotrypsin. It is active by any parenteral route, but it can also be given sublingually with only a small loss of effectiveness since it is well absorbed through the buccal mucosa (BRAZEAU, 1970a).

Many penicillins are more effective when given by a parenteral route. A higher level of penicillin was found in serum and a greater proportion of drug was excreted in urine after an intramuscular dose than after equivalent oral doses (MCCARTHY and FINLAND, 1960). The relative ineffectiveness of oral penicillins has been attributed to destruction of the antibiotics in the acidic contents of the stomach. However, MCDERMOTT et al. (1946) concluded from their data on the absorption of penicillins that destruction by acid was less important than incomplete absorption in limiting the systemic availability of oral antibiotic.

Several drugs are of limited effectiveness when given orally because they are lost by adsorption or complexing with components in food. The absorption of tetracyclines is decreased by concomitant administration of aluminium hydroxide gels or calcium and magnesium salts, which form insoluble complexes with the drugs.

Inadequate Absorption. Most drugs are absorbed from the gastrointestinal tract by nonionic diffusion. Absorption, in general, takes place along the whole length of the gastrointestinal tract, but the chemical properties of each drug determine whether absorption takes place largely from the acid medium of the stomach or the more alkaline medium of the intestine. The principles governing the absorption of drugs from the gastrointestinal lumen have been reviewed in this series (SHANKER, 1971). Briefly stated, absorption is favored by a low degree of ionization and a high lipid/water partition of the nonionized form of the drug. Since the membranes of the walls of the gastrointestinal tract can be considered as lipid barriers, the lipid-soluble, nonionized form of the drug will penetrate the barrier more readily than the lipid-insoluble, ionized or charged form.

Drugs that are well absorbed when given in solution may be poorly absorbed when given in solid form because the rate of solution in gut content is limiting. A number of factors determine the rate of solution; these include particle size and crystalline form of the drug, and the rate of disintegration of the formulation in the gastrointestinal contents (LEVY, 1968). Formulation is particularly important for poorly water-soluble drugs such as corticosteroids, dicoumarol, or phenylbutazone. Indeed, the effective oral dose of corticosteroids depends very much on the particle size of the formulation.

However, even with correct formulation, the absorption of certain drugs from the gastrointestinal tract is so poor that they are largely ineffective by this route. The quaternary ammonium, ganglion-blocking drugs, hexamethonium, pentolinium and bretylium, are ionized at the pH of the intestine and are poorly absorbed (DOLLERY, 1964). Other drugs, even in the nonionized form, are not sufficiently lipid soluble to penetrate readily the lipid-like boundary of the gastrointestinal tract. Sulphonamides, such as succinyl sulphathiazole, are not readily absorbed, nor are a large number of different types of antibiotics. Indeed, the lack of absorption of sulphonamides such as succinyl sulphathiazole and antibiotics like neomycin is utilized in order to achieve selective local action with these agents in the gastrointestinal tract.

When drugs are poorly absorbed following oral administration, all enteral routes have to be abandoned since the principles governing the absorption of oral drugs apply equally to the rectal or sublingual routes. For example, the quaternary ammonium ganglion-blocking drugs and the poorly absorbed antibiotics must be given by one of the parenteral routes.

Loss of Drug during "First Pass". A more recently recognized cause of the varying effectiveness of drugs administered by different routes is the loss of drug, usually by metabolism, during the "first pass" from the site of administration into the systemic circulation. Drugs given orally are particularly susceptible to metabolism at the "first pass". Drug molecules cross the intestinal wall, a structure equipped with some drug-metabolizing enzymes, are quantitatively collected by the mesenteric veins, and pass via the hepatic portal vein directly into the liver, the organ in which most drug metabolism occurs, before reaching the systemic circulation. The other enteral routes to some extent by-pass these "first pass" metabolism barriers. Drugs given sublingually are not quantitatively conveyed to the liver. In addition, the oral mucosa may not possess, at least not to the same extent, the drug metabolizing enzymes of the intestinal wall. However, further information is needed on this point.

Rectal administration to some extent enables drug to reach the systemic circulation without passing through the liver since anatomically the rectum supplies blood to both systemic and portal circulations through the hemorrhoidal plexus. It is not known whether drugs that are conjugated in the wall of the intestine suffer a similar fate when given rectally. Information on the distribution of conjugating enzymes down the entire length of the gastrointestinal tract would be of value in deciding on suitable routes of administration of many drugs.

Loss during "first pass" through a barrier is not confined to drugs given enterally. Bronchodilator drugs inhaled into the lungs are metabolized during absorption into the systemic circulation. Extensive protein binding at the site of administration is another kind of "first pass" barrier that can diminish the effectiveness of drugs such as phenylbutazone when given intramuscularly. Diazoxide, a drug extensively bound to plasma proteins, must be given by rapid intravenous injection to exert its antihypertensive action. After rapid injection, the concentration of drug in the bolus is sufficiently high to exceed the binding capacity of the plasma protein. Thus, a higher free concentration of drug is achieved after rapid injection than after slow injection (SELLERS and KOCH-WESER, 1969). However, further studies (MROCZEK et al., 1971) have shown that rapidly injected diazoxide is more effective only in patients with accelerated hypertension, thereby casting some doubt on the explanation of SELLERS and KOCH-WESER.

The loss of drug during the "first pass" through a barrier such as the gut wall or liver limits the initial availability of drug, but it does not influence the elimination of drug when it is in the systemic circulation. A drug that is extensively metabolized during passage through the liver may still have a long half-life in plasma if it has a large volume of distribution.

For drugs extensively metabolized during "first pass", the route of administration may affect more than the availability of drug. The route also may influence the path of metabolism by determining the organ to which the highest concentration of drug is directed. This can be of great importance when metabolites possess pharmacologic activity as is illustrated with propranolol and isoproterenol in a later section.

d) Selective Local Action

The most effective means of producing selective local action is to apply the drug to the diseased tissue. Steroids, antibiotics, and antifungal agents are applied directly to skin lesions thereby avoiding high concentrations of these potentially hazardous agents in the systemic circulation. Nevertheless, steroids applied to skin are absorbed into the systemic circulation (BUTLER, 1966). In some circumstances, the amount absorbed can be sufficient to depress pituitary-adrenal function (JAMES et al., 1967).

Drugs used in the treatment of asthma are often given by inhalation for immediate local action in the airways. The catecholamine-like bronchodilator drugs are considered in a later section; here we will discuss another antiasthmatic drug that achieves a reasonable degree of local action in the airways because of its absorption characteristics.

Disodium cromoglycate (INTAL) is inhaled as a dry powder using a spin-haler. Studies with ^{14}C-labelled drug (WALKER et al., 1972) have shown that, in common with other inhalation techniques, only a small proportion reaches the airways (approximately 3 to 4%) and the remainder is swallowed. Although the drug is poorly absorbed from the gastrointestinal tract (less than 1% of the oral dose), drug deposited in the airways is rapidly absorbed. This is in accord with the findings of ENNA and SHANKER (1969) that the pulmonary epithelium is more permeable than the gastrointestinal mucosa to lipid soluble molecules and ions. Thus, for this drug by this route of administration, only that small proportion of the dose getting to the site of action enters the systemic circulation. Thus, a high degree of local action is ensured.

B. Sublingual and Rectal Administration

The advantages of sublingual and rectal administration over oral dosing have been mentioned in the previous section. These include a quicker onset of action for some drugs with a marginal advantage over the oral route and avoidance to some extent of loss by metabolism at "first pass" through the intestinal wall and liver. However, both routes are less convenient than oral dosing.

III. Parenteral Administration

A. Intravascular Injection

Drugs are most commonly introduced directly into the blood stream by injection into a vein. The obvious advantage of this route is that the drug is placed in the circulation with a minimum of delay, an important consideration in an emergency situation. We have discussed the advantages of the intravenous route for drugs that for various reasons do not reach the systemic circulation in adequate concentrations when administered enterally. In addition, since the intravenous route allows large volumes of fluid to be introduced over long periods of time, the plasma level of a drug can be held within narrow limits if necessary by means of a constant infusion pump.

However, the intravenous route has disadvantages, the most important of which is the dangerously high concentration that can be achieved in certain organs after too rapid injection. The time course of dilution of a bolus of dye injected intravenously is well understood because this method is extensively used by cardiologists to study the circulation, and especially to measure the cardiac output (FRANCH, 1966). An intravenous bolus travels around the circulation becoming progressively diluted in blood by mixing in the heart and blood vessels. If the cardiac output is high and the blood volume normal, the amount of mixing on the first transit is not very great. The peak concentration may correspond to dilution in no more than 500 to 1000 ml of blood at the time the bolus reaches the arterioles of the systemic circulation. A 10 m.Eq. intravenous bolus of potassium chloride might raise the plasma concentration of the blood perfusing the coronary arteries to 15 m.Eq/l and stop the heart. The hazards of exposing organs to such a high concentration are not usually acceptable, but this property can be made use of in some circumstances. Thiopentone is given as a bolus injection because this exposes the high blood flow organs to a very high concentration and achieves a degree of selectivity of the drug in the brain. The patient wakes up after short time due to drug redistribution to lower blood flow tissues, such as fat, long before an appreciable proportion of the dose has been metabolized (PRICE, 1960).

Normally drugs administered intravenously are rapidly redistributed into extracellular fluid and tissues. However, this redistribution into high capacity tissues, such as skeletal muscle, may be delayed under circumstances of a very low cardiac output and severe vasoconstriction (THOMSON et al., 1971). These authors studied the plasma lidocaine concentrations after intravenous administration in normal subjects and in patients in shock following myocardial infarction. The concentrations achieved in the patients in shock were twice as high as in the normal subjects. The high lidocaine concentrations in these patients may be partly due to impaired distribution, but another factor is the reduced hepatic blood flow in shock since the metabolism of lidocaine is limited by blood flow. Thus i.v. doses may result in a high concentration initially because of the bolus effect. Intravascular injection delivers a much large proportion of the total dose to high blood flow organs. If redistribution is impaired by disease, an unexpectedly high plasma concentration may result from a standard intravenous dose.

The intraarterial route is not often used. A principal use is the specialized techniques in cancer chemotherapy which require arterial infusions near to the site of action because of the reactive nature of the drugs used, such as nitrogen mustards. Several substances used in diagnostic procedures are injected intraarterially; a typical example is the injection of a radio-opaque compound into a renal or cerebral artery to visualize the circulation of the kidney or brain roentgenographically.

B. Intramuscular Injection

The advantages of the intramuscular route are similar to other parenteral routes and, in addition, unlike the intravenous route, there is no danger of achieving high concentrations of drug in high blood flow organs.

A number of factors control the rate of absorption of drugs from an intramuscular site. These include the physicochemical properties of the drug, and the vascularity of the site. Difficulties arise if a patient is in shock with vasoconstriction. Muscle blood flow may then be so low that an intramuscular dose is absorbed very slowly (HARRISON, 1971). The drug may be absorbed only when the patient has recovered sufficiently so that muscle perfusion is restored; by then, the effect of a drug such as morphine may no longer be required or even may have an adverse effect.

Frequent intramuscular injections are inconvenient and painful to the patient and, because of this, several techniques have been developed to produce depots of drug for slow release from the site of injection. For example, steroids are injected as suspensions in an oil from which they are slowly released into the systemic circulation. The effects of intramuscularly administered ADH (vasopressin) last only a few hours, but repository forms, such as vasopressin tannate in oil, are effective for 24 to 48 h after subcutaneous or intramuscular injection (BRAZEAU, 1970b).

Intramuscular administration may not produce the desired effect if the drug is highly protein bound at the site of injection. This is probably the explanation for the low initial plasma concentration of phenylbutazone that occurs after intramuscular, but not after oral, administration (Fig. 2) (BRODIE et al., 1954).

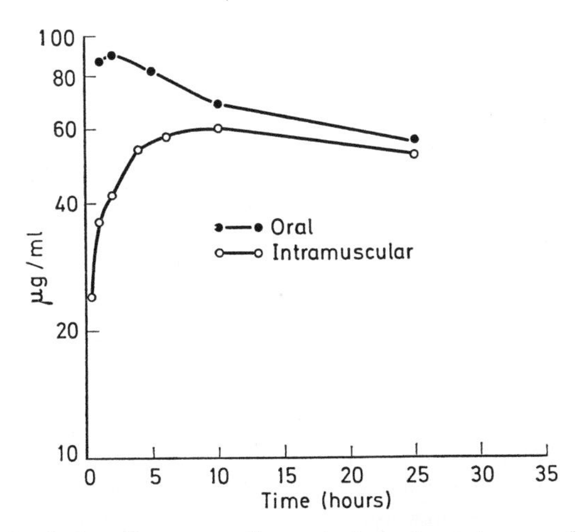

Fig. 2. Plasma levels of phenylbutazone after a single oral or intramuscular dose (800 mg) of the drug in the same subject (from BRODIE et al., 1954)

C. Subcutaneous and Percutaneous Administration

The absorption of drugs following subcutaneous administration and inunction have been reviewed in a previous volume of this series (SCHOU, 1971; KATZ and POULSON, 1971). In common with other parenteral routes, these can often be used when enteral routes are unsuitable.

D. Inhalation of Drugs

Several widely used anesthetic agents cannot be given orally because they are gases and must be inhaled. However, inhalation has important advantages when administering an anesthetic agent, primarily the fine degree of control of the depth of anesthesia. A non-volatile drug is usually eliminated from the body with a half-life of several hours. For example, the hepatic clearance of a barbiturate is only 30 to 80 ml/min, but the clearance of an anesthetic gas by the lungs may approximate the cardiac output (5000 ml/min). Especially with the less soluble anesthetic gases, such as nitrous oxide, there is a rapid equilibration between inspired gas, alveolar gas, arterial blood, and brain. Thus, both induction and termination of anesthesia are rapid. With the more soluble gases, especially ether, the problem is different. Induction is slow because the great capacity of the blood for the drug keeps the alveolar and arterial partial pressure low. The anesthetist can overcome this by increasing the inspired concentration during induction, but he cannot reverse the procedure during recovery, which is slow (PAPPER, 1964). Halothane, the most widely used volatile anesthetic, occupies an intermediate position with a blood:gas partition coefficient of 2.36 and a tissue:blood partition coefficient of 2.6 for brain and 3.5 for muscle (EGER and LARSON, 1964).

Thus, the main attraction of the inhalation route for anesthesia is the fine control that is possible by continuous adjustment of the inhaled drug concentration.

The rapid delivery of a soluble inhaled substance to high blood flow tissue is an advantage when giving anesthetics, but it may be hazardous under some circumstances. BASS (1970) reported sudden deaths in individuals who had deliberately inhaled fluorocarbon aerosol propellants. He suggested that the immediate high concentration reached by these substances in the arterial blood and heart sensitized the individuals' hearts to arrhythmias.

IV. Influence of Route of Administration on Drug Response

In this section, we will discuss several drugs with responses that depend upon the route of administration. Since it is well known that poorly absorbed drugs or drugs unstable in contents of the gastrointestinal tract exhibit different responses when given enterally or parenterally, they will not be considered here.

A. Isoproterenol

Isoproterenol (isoprenaline) is a sympathomimetic amine that has been used extensively in the treatment of patients with bronchial asthma, who usually take it by aerosol inhalation. The drug has also been used in the symptomatic treatment of heart block (Adams-Stokes syndrome), for which it has been administered sublingually (SCHUMAKER and SCHMOCK, 1954), rectally (BURCHELL, 1957), and by intravenous infusion (JONES, 1958).

The degree of β-stimulation produced by the drug (usually measured as heart rate increase) is very dependent on the route of administration. Studies in anesthetized dogs (MINATOYA et al., 1965) recorded that the drug exerted widely different potencies depending on whether it was administered intraduodenally, rectally, intravenously, or by intratracheal instillation. Dose response curves (Fig. 3) reveal that the order of potency as measured by peak increase in heart rate for intravenous, intratracheal, intestinal, or rectal administration was approximately 5000:8:1:1. Peak increase in heart rate is not a good measure of the availability of active drug in the circulation since it is dependent on the rate of absorption from the site of administration.

PORTMANN and colleagues (1965), using conscious dogs, obtained a better estimate of the availability of isoproterenol following various routes of administration by comparing the areas under the heart-rate-increase time curves. They concluded that orally administered drug had a potency of less than 1% of intravenously administered drug. Rectal administration was 2.4 times more effective than oral dosing although the rate of absorption was slower.

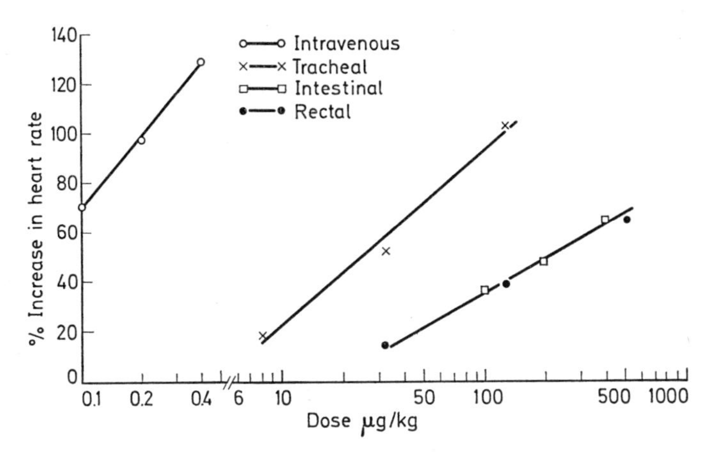

Fig. 3. Dose response curves to isoproterenol administered by various routes to anesthetized dogs

Studies in our Department (DOLLERY et al., 1971) have shown a similar discrepancy in the cardiac stimulating effects of isoproterenol administered by different routes in man. Inhalation of 400 μg of isoproterenol causes a moderate increase in heart rate in a normal man for 10 to 15 min. This can be matched by an intravenous infusion of approximately 2 μg/min during the same time period. When isoproterenol is given by mouth, a dose of more than 10 mg is required to increase heart rate measurably. An approximate ratio of potency for the chronotropic effect of isoproterenol on the human heart for the intravenous, inhaled, and oral routes is 1000:20:1.

The obvious explanation for these large differences would be that isoproterenol was incompletely absorbed from the airways and gastrointestinal tracts in dog and man. However, studies with tritiated isoproterenol in both species showed that the drug is well absorbed no matter which route of administration is used. However, the path and extent of metabolism is very dependent on the route administration (CONOLLY et al., 1972).

Isoproterenol is metabolized along two paths (Fig. 4); it may be directly conjugated, usually with sulphate, or it may be *O*-methylated by catechol-*O*-methyl transferase prior to conjugation. In man, following intravenous dosing, unchanged isoproterenol is the major component in plasma with 3-*O*-methyl isoproterenol as the only other metabolite. The urine contains unchanged isoproterenol with free and conjugated 3-*O*-methyl isoproterenol (Table 1). In contrast, following oral dosing, only the sulphate conjugate of isoproterenol can be detected in plasma and more than 80% of the total radioactivity excreted in urine is in this form with less than 10% *O*-methylated (Table 1). A similar result was obtained when the drug was inhaled as an aerosol, suggesting that a major proportion of the inhaled dose is in fact swallowed. This hypothesis received further support when it was found that drug administered intrabronchially was largely *O*-methylated (Table 1).

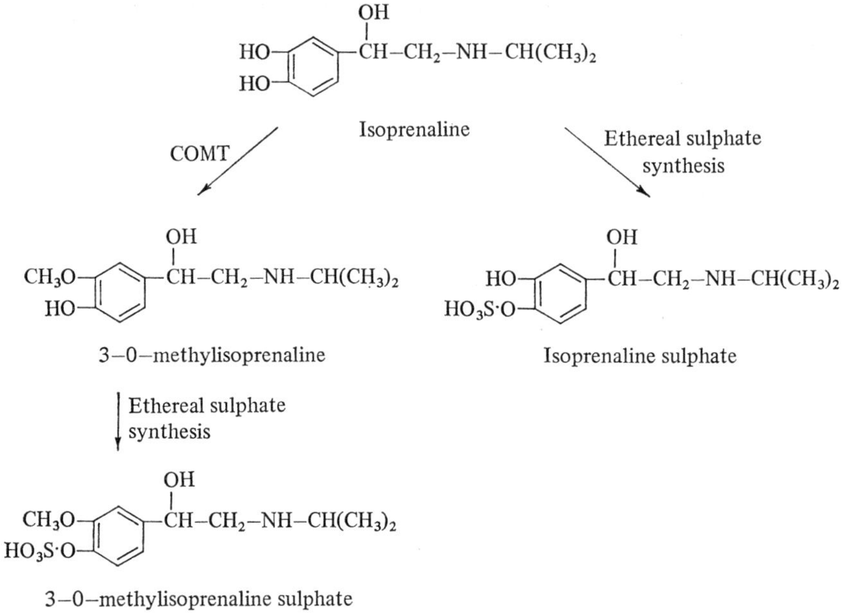

OH
HO–C₆H₃(HO)–CH–CH₂–NH–CH(CH₃)₂

Isoprenaline

COMT ↙ ↘ Ethereal sulphate synthesis

CH_3O, HO–C₆H₃–CH(OH)–CH_2–NH–$CH(CH_3)_2$

3–0–methylisoprenaline

HO, $HO_3S{\cdot}O$–C₆H₃–CH(OH)–CH_2–NH–$CH(CH_3)_2$

Isoprenaline sulphate

↓ Ethereal sulphate synthesis

CH_3O, $HO_3S{\cdot}O$–C₆H₃–CH(OH)–CH_2–NH–$CH(CH_3)_2$

3–0–methylisoprenaline sulphate

Fig. 4. Paths of metabolism of isoproterenol (isoprenaline)

These studies in man suggested that orally administered isoproterenol is inactivated largely by conjugation with sulphate either in the gut wall or in the liver during the first pass to the systemic circulation. A major proportion of a dose thought to be inhaled as an aerosol is actually swallowed and is inactivated in a similar way. In contrast, drug placed directly in the airways is well absorbed but is *O*-methylated to a greater extent than intravenously administered drug, possibly during

Table 1. *Metabolites of isoproterenol in human urine following various routes of administration*

Route	Dose (μg/kg)	Time (h)	Percentage of urine activity			
			Free I	Conj. I	Free OM-I	Conj. OM-I
Intravenous	0.06	15	65.2	0	(31.4 total)	
Oral	44	48	9.2	84.2	1.0	3.0
Inhalation[a]	5.7	48	3.5	89.6	1.2	7.2
Intrabronchial	0.03	50	8.8	6.3	6.2	59.7

[a] Pressurized aerosol

absorption from the airways. Thus, a possible explanation for the different potencies of isoproterenol given by various routes emerged. Studies with dogs confirmed this view. Intraduodenally administered isoproterenol appears in dog plasma largely as the sulphate conjugate (Fig. 5). In contrast, drug infused into the portal vein was not conjugated, which suggests that conjugation occurs mainly in the gut wall. Studies in which the total venous return from an isolated loop of dog gut was collected showed that isoproterenol is conjugated during transfer through the intestinal wall. First pass metabolism in the airways was studied using an isolated perfused dog lung. With this preparation, a large proportion of drug administered intrabronchially appeared in the circulating medium as the 3-*O*-methyl metabolite (Fig. 6). This provided further evidence of metabolism during absorption from the airways.

Isoproterenol administered as an aerosol to asthmatics produces a rapid and selective local action in the airways. Although only a small proportion (< 10%)

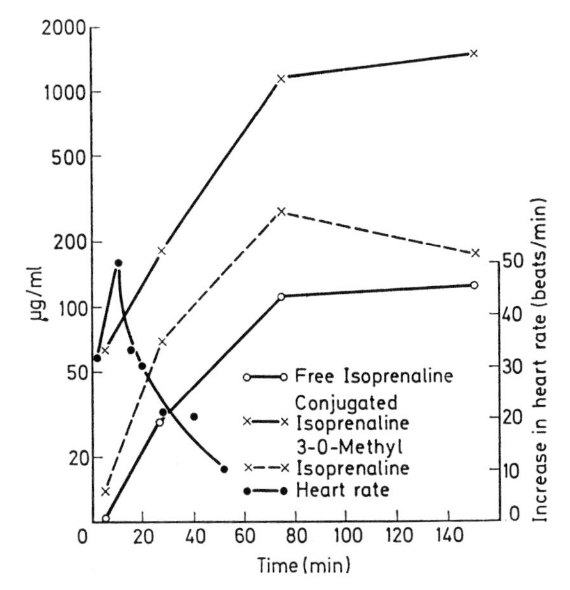

Fig. 5. Time course of changes in heart rate and plasma levels of isoproterenol and metabolites in dog following an intraduodenal dose of 1 mg/kg

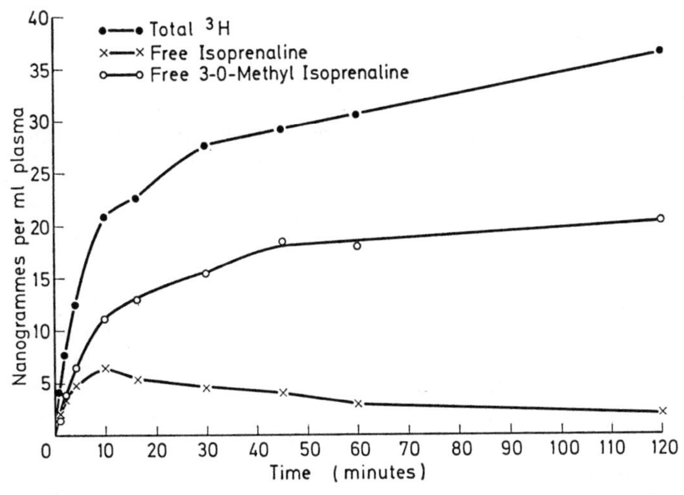

Fig. 6. Metabolism of intrabronchially administered isoproterenol in the isolated perfused dog lung (Briant and Blackwell, unpublished results)

of the drug reaches the airways, the remainder is swallowed and is largely inactivated (> 90%) in the gut wall before entering the systemic circulation. Thus the proportion of the dose entering the systemic circulation and exerting unwanted effects on the heart is very small. A proportion of the drug absorbed from the airways is also inactivated by *O*-methylation during absorption. This is important since the *O*-methyl metabolite, which appears to be formed in the lung when the drug is absorbed from the airways, has weak β-blocking properties (PATERSON et al., 1968). Since airway conductance in asthmatics can be readily reduced with β-blockers, it would be of interest to know the local concentration of this metabolite in the airways of asthmatics using excessive amounts of isoproterenol aerosols. A relationship between the rise in death rate from asthma and the use of pressurized aerosols of isoproterenol was recently reported in the United Kingdom (SPEIZER et al., 1968).

B. Chlorpromazine

Chlorpromazine was introduced into medicine about 20 years ago, but it has been realized only recently that plasma levels of the drug achieved after parenteral administration are 3 to 10 times greater (Fig. 7) than after oral doses (CURRY, 1971).

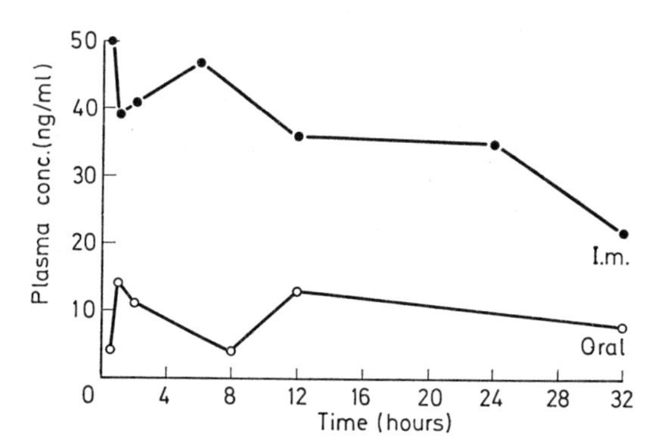

Fig. 7. Concentrations of chlorpromazine in plasma after intramuscular and oral doses of 100 mg in the same 6 subjects (from CURRY, 1971)

At first, it was thought that these differences were due to poor absorption from the gastrointestinal tract, but urinary excretion data suggested complete absorption of the drug (HOLLISTER et al., 1970). Recent studies in rats (CURRY et al., 1971) offer an explanation of these findings. In rats, as in man, plasma concentrations are higher after parenteral administration than after oral administration. Studies using perfused loops of rat intestine demonstrated that chlorpromazine was partly converted to unidentified metabolites during transfer through the intestinal wall. If metabolism during the first pass through the intestinal wall occurs *in vivo*, this would explain the discrepancy in plasma levels following oral and parenteral administration. Metabolism during first pass through the liver would also contribute to these differences, but, in rats, the intraperitoneally administered drug, which presumably reaches the systemic circulation via the liver, was completely absorbed as unchanged drug.

Extensive metabolism of chlorpromazine during transfer from the lumen of the gut has important therapeutic implications. Recent studies (CURRY, 1972) suggest that some schizophrenic patients with phenothiazine-sensitive disease failed to respond to oral chlorpromazine because they extensively metabolized the drug to inactive metabolites in the intestinal wall. These patients apparently responded to injected phenothiazines.

C. Lidocaine

The use of intravenous lidocaine (Xylocaine) for the treatment of ventricular arrhythmias has been described by several researchers (for example, JEWITT et al., 1968). Many investigators have also presented data that suggest that blood levels of 1.5 to 2.0 μg/ml are required for anti-arrhythmic activity. The treatment of cardiac arrhythmias with oral lidocaine has been considered, but major problems have been encountered.

BOYES et al. (1971) showed that plasma levels of unchanged drug are considerably lower after oral dosing than after parenteral administration of similar doses. Lidocaine is apparently well absorbed, suggesting that the discrepancy between parenteral and oral doses is again due to loss during the first pass to the systemic circulation. Studies in dogs showed that infusion of lidocaine into the portal vein (BOYES et al., 1970) results in blood levels comparable to those obtained after oral doses, suggesting that the main site of first pass metabolism is the liver. This was confirmed in studies in man. STENSON and colleagues (1971) showed that during constant infusion of lidocaine in man an average of 70% of the total body clearance of the drug could be accounted for by hepatic clearance. The total body blood clearance of lidocaine in normal subjects is high, corresponding to almost 800 ml/min in a 70 kg man (ROWLAND et al., 1971). Since liver blood flow is approximately 1380 ml/min, one would anticipate that as much as 60% of an oral dose of lidocaine would be cleared before entering the systemic circulation, thus explaining the relatively low blood levels. This agrees well with the estimates of BOYES and KEENAGHAN (1971) that an average of 35% of an oral dose of lidocaine reached the systemic circulation unchanged (Table 2).

Table 2. *Calculated percentage of an oral dose of lidocaine reaching the systemic circulation unchanged*

Subject	Dose mg	Calculated % absorbed unchanged
J.E.	250	45.0
	500	24.5
A.L.	250	21.2
	500	26.2
G.O.	250	46.2
	500	42.0
J.U.	250	43.8
	500	42.6
	Mean	36.4

With a rapidly cleared drug such as lidocaine, oral administration is a problem. Not only are large quantities of the drug lost during the first pass through the liver but high concentrations of metabolites are also obtained since large doses are needed to achieve therapeutic levels of unchanged drug. Formation of metabolites may explain the dizziness observed in patients taking oral doses of 500 mg when blood levels of unchanged drug were lower than after smaller doses administered by intravenous injection (Boyes et al., 1971).

D. Propranolol

Propranolol is a beta adrenergic receptor-blocking drug widely used to treat patients with cardiac arrhythmias and angina. It is usually given by mouth, but it may be given intravenously to patients with arrhythmias. The metabolism of propranolol is not fully understood (Paterson et al., 1970), but two metabolites that are formed in man are 4-hydroxy-propranolol, which has activity similar to the parent compound, and naphthoxylactic acid, an inactive metabolite (Fig. 8).

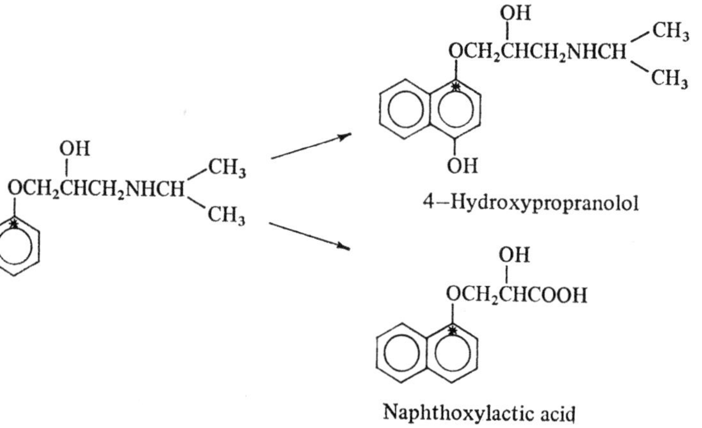

4–Hydroxypropranolol

Naphthoxylactic acid

Fig. 8. Paths of metabolism of propranolol in man

Propranolol is an example of a drug that has part of its action mediated by an active metabolite which is formed after oral but not after intravenous administration. The metabolite is probably formed in the liver (Hayes and Cooper, 1971) during the first pass of the drug through this organ. The extraction of propranolol in the liver is high (Shand et al., 1971), and after oral dosing, large amounts of 4-hydroxy-propranolol are formed in the liver and appear in the systemic circulation (Dollery et al., 1971) (Fig. 9). This does not occur following intravenous dosing (Fig. 9).

After intravenous administration the β-blockade produced by propranolol, as measured by the inhibition of the chronotropic action of infused isoproterenol on heart rate, is directly proportional to the plasma level of free propranolol and not to total radioactivity. In contrast, the β-blockade after oral administration of propranolol was 2 to 5 times greater for a given plasma level of free propranolol than after intravenous administration of the drug (Dollery et al., 1971). This is consistent with the known β-blocking activity of the metabolite, 4-hydroxy-propranolol, which is present in the plasma after oral administration of the parent drug in concentrations equal to free propranolol (Fig. 9). Similar results have been reported by Coltart and Shand (1970) who found that exercise tachycardia was inhibited more by oral administration of propranolol than by intravenous injection for a given plasma level of free drug (Fig. 10). Thus the route of administration of propranolol greatly influences the path of metabolism and the activity of the drug.

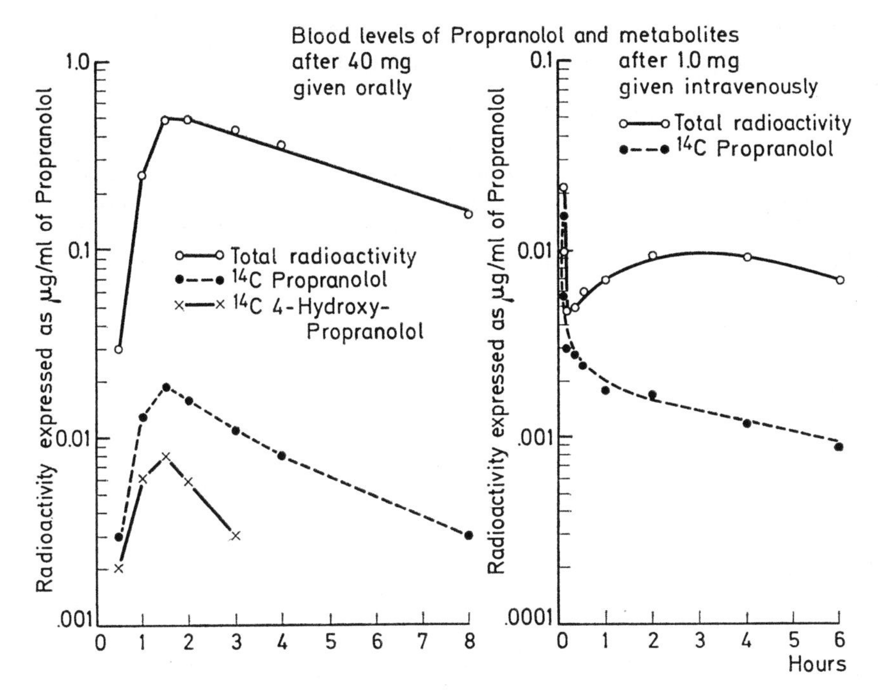

Fig. 9. Propranolol and 4-hydroxy-propranolol in plasma after oral and intravenous doses (from DOLLERY et al., 1971)

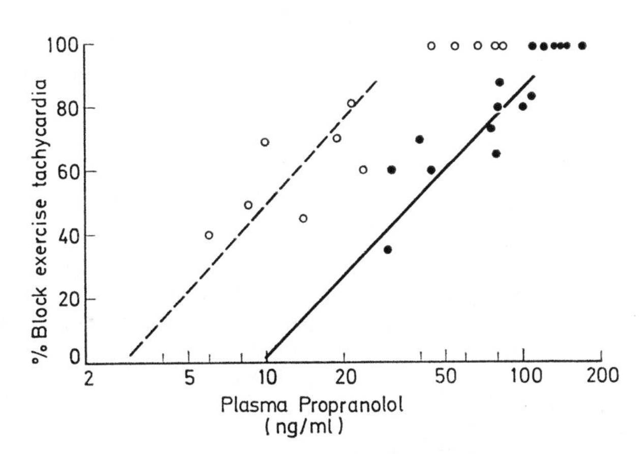

Fig. 10. Log plasma concentration/response relationship for orally administered (○) and intravenously administered (●) propranolol (from COLTART and SHAND, 1970)

V. Conclusions

Different responses are evoked by drugs given by various routes because the proportion of the dose reaching the site of action as unchanged drug is dependent on the route of administration. There are a number of reasons for this. They include poor stability at the site of administration and poor transfer to the systemic circulation due to inadequate absorption. A more recent finding is that some drugs, although adequately absorbed, fail to reach the systemic circulation in high concentration because they are metabolized during the first pass from the site of application. First pass metabolism can occur in the intestinal wall, in the liver or, for inhaled drugs, in the lung. For some drugs, first pass metabolism not only limits the availability of unchanged drug but also produces different metabolites, some of which may possess the activity of the parent drug or a different activity. The implications of "first pass" metabolism in therapeutics are great, particularly for drugs such as propranolol, lidocaine, and chlorpromazine.

References

Bass, M.: Sudden sniffing death. J. Amer. med. Ass. **212**, 2075—2079 (1970).

Beckett, A. H., Hossie, R. D.: Buccal absorption of drugs, pp. 25—45. In: Brodie, B. B., Gillette, J. R. (Eds.): Concepts in biochemical pharmacology, Part I. Berlin-Heidelberg-New York: Springer 1971.

Bianchine, J. R., Calimlim, L. R., Morgan, J. P., Dujuvne, C. A., Lasagna, L.: Metabolism and absorption of L-3, 4 dihydroxyphenylalanine in patients with Parkinson's disease. Ann. N. Y. Acad. Sci. **179**, 126—140 (1971).

Boyes, R. N., Adams, H. J., Duce, B. P.: Oral absorption and disposition kinetics of lidocaine hydrochloride in dogs. J. Pharmacol. exp. Ther. **174**, 1—8 (1970).

Boyes, R. N., Keenaghan, J. B.: Some aspects of the metabolism and distribution of lidocaine in rats, dogs and man. In: Scott, D. B., Julian, D. G. (Eds.): Lidocaine in the treatment of ventricular arrhythmias, pp. 140—151. Edinburgh: Livingston 1971.

Boyes, R. N., Scott, D. B., Jebson, P. J., Goodman, M. J., Julian, D. G.: Pharmacokinetics of lidocaine in man. Clin. Pharm. Therap. **12**, 105—116 (1971).

Brazeau, P.: Inhibitors of tubular transport of organic compounds. In: Goodman, L. S., Gilman, A. (Eds.): The pharmacological basis of therapeutics, p. 897. New York: MacMillan 1970a.

Brazeau, P.: Agents affecting the renal conservation of water. In: Goodman, S., Gilman, A. (Eds.): The pharmacological basis of therapeutics, p. 881. New York: MacMillan 1970b.

Brodie, B. B., Lowman, E. W., Burns, J. J., Lee, P. R., Cherkin, T., Goldman, A., Weiner, M., Steele, J. M.: Observations on the antirheumatic and physiologic effects of phenylbutazone (butazolidin) and some comparisons with cortisone. Amer. J. Med. **16**, 181—190 (1954).

Burchell, H. B.: Clinical problems related to surgical repair of intracardiac defects with the aid of an extracorporeal pump-oxygenator. Circulation **16**, 976—987 (1957).

Butler, J.: Percutaneous absorption of tritium-labelled Betamethazone-17-valerate. Brit. J. Derm. **78**, 665—668 (1966).

Coltart, D. J., Shand, D. G.: Plasma propranolol levels in the quantitative assessment of B-adrenergic blockade in man. Brit. med. J. **1970 III**, 731—734.

Conolly, M. E., Davies, D. S., Dollery, C. T., Morgan, C. D., Paterson, J. W., Sandler, M.: Metabolism of isoprenaline in dog and man. Brit. J. Pharmacol. In press (1972).

Curry, S. H.: Chlorpromazine: Concentrations in plasma, excretion in urine and duration of effect. Proc. roy. Soc. Med. **64**, 285—289 (1971).

Curry, S. H., D'Mello, A., Mould, G. P.: Destruction of chlorpromazine during absorption in the rat *in vivo* and *in vitro*. Brit. J. Pharmacol. **42**, 403—411 (1971).

Curry, S. H.: Action and metabolism of chlorpromazine. In: Davies, D. S., Prichard, B. N. C. (Eds.): Biological effects of drugs in relation to their plasma concentrations. London: MacMillan in press (1972).

Davidson, I. W. F., Rollins, F. O., Di Carlo, F. J., Miller, H. S., Jr.: The pharmacodynamics and biotransformation of penta-erythritol trinitrate in man. Clin. Pharmacol. **12**, 972—981 (1971).

Dollery, C. T.: Absorption, distribution and excretion of drugs used to treat hypertension. In: Binns, T. B. (Ed.): Absorption and distribution of drugs, p. 157. Edinburgh: Livingston 1964.

Dollery, C. T., Davies, D. S., Conolly, M. E.: Differences in the metabolism of drugs depending on their routes of administration. Ann. N. Y. Acad. Sci. **179**, 108—114 (1971).

EGER, E. I. II., LARSON, C. P., JR.: Anaesthetic solubility in blood and tissues: values and significance. Brit. J. Anaesth. **36**, 140—149 (1964).

ELLARD, G. A., GARROD, J. M. B., SCALES, B., SNOW, G. A.: Percutaneous absorption and routes of excretion of ditophal (Etisul). Biochem. Pharmacol. **12**, 271—281 (1963).

ENNA, S. J., SHANKER, L. S.: Drug absorption from the lung. Fed. Proc. Fed. Amer. Soc. exp. Biol. **28**, 359 (1969).

FRANCH, R. H.: Cardiac catheterization. In: HURST, J. W., LOGUE, R. B. (Eds.): The heart, pp. 164—192. New York: McGraw-Hill 1966.

HARRISON, D. C.: In: SCOTT, D. B., JULIAN, D. G. (Eds.): Lidocaine in the treatment of ventricular arrhythmias, p. 166. Edinburgh: Livingston 1971.

HAYES, A., COOPER, R. G.: Studies on the absorption, distribution and excretion of propranolol in rat, dog and monkey. J. Pharmacol. exp. Ther. **176**, 302—311 (1971).

HERBERT, V.: Drugs effective in megaloblastic anaemias. In: GOODMAN, L. S., GILMAN, A. (Eds.): The pharmacological basis of therapeutics, p. 1429, 4th ed. New York: MacMillan 1970.

HOLLISTER, L. E., CURRY, S. H., DERR, J. E., KANTER, S. L.: Studies of delayed-action medication: V. plasma levels and urinary excretion of four different dosage forms of chlorpromazine. Clin. Pharmacol. Ther. **11**, 49—59 (1970).

JAMES, V. H. T., MUNRO, D. D., FEIWEL, M.: Pituitary-adrenal function after occlusive topical therapy with Betamethasone-11-valerate. Lancet **1967 II**, 1059—1061.

JEWITT, D. E., KISHON, Y., THOMAS, M.: Lignocaine in the management of arrhythmias after acute myocardial infarction. Lancet **1968 I**, 266—270.

JONES, R. J.: Medical control of Adams-Stokes syndrome. J. Amer. med. Ass. **167**, 1840—1842 (1958).

KATZ, M., POULSEN, B. J.: Absorption of drugs through the skin. In: BRODIE, B. B., GILLETTE, J. R. (Eds.): Concepts in biochemical pharmacology, Part I, pp. 103—162. Berlin-Heidelberg-New York: Springer 1971.

LEVY, G.: Kinetics and implications of dissolution-rate-limited gastrointestinal absorption of drugs. In: ARINS, E. J. (Ed.): Physicochemical aspects of drug actions. Proceedings of the third International Pharmacological Meeting, Vol. 7, pp. 33—62. Oxford: Pergamon 1968.

MCCARTHY, C. G., FINLAND, M.: Absorption and excretion of four penicillins. New Engl. J. Med. **263**, 315—326 (1960).

MCDERMOTT, W., BUNN, P. A., BENOIT, M., DUBOIS, R., REYNOLDS, M. E.: The absorption, excretion and destruction of orally administered penicillin. J. clin. Invest. **25**, 190—210 (1946).

MINATOYA, H., LANDS, A. M., PORTMANN, G. A.: Absorption and elimination profile of isoproterenol. I. J. Pharm. Sci. **54**, 968—972 (1965).

MROCZEK, W. J., LEIBEL, B. A., DAVIDOV, M., FINNERTY, F. A.: The importance of the rapid administration of diazoxide in accelerated hypertension. New Engl. J. Med. **285**, 603—606 (1971).

PAPPER, E. M.: The pharmacokinetics of inhalation anaesthetics: clinical application. Brit. J. Anaesth. **36**, 124—128 (1964).

PATERSON, J. W., CONOLLY, M. E., DAVIES, D. S., DOLLERY, C. T.: Isoprenaline resistance and the use of pressurised aerosols in asthma. Lancet **1968 II**, 426—429.

PATERSON, J. W., CONOLLY, M. E., DOLLERY, C. T., HAYES, A., COOPER, R. G.: The pharmacodynamics and metabolism of propranolol in man. Pharmacologia Clinica **2**, 127—133 (1970).

PORTMANN, G. A., MINATOYA, H., LANDS, A. M.: Absorption and elimination profile of isoproterenol. II. J. Pharm. Sci. **54**, 973—978 (1965).

PRICE, H. L.: A dynamic concept of the distribution of thiopental in the human body. Anesthesiology **21**, 40—45 (1960).

ROWLAND, M., THOMSON, P. D., GUICHARD, A., MELMON, K. L.: Disposition kinetics of lidocaine in normal subjects. Ann. N. Y. Acad. Sci. **179**, 383—399 (1971).

SCHOU, J.: Subcutaneous and intramuscular injection of drugs. In: BRODIE, B. B., GILLETTE, J. R. (Eds.): Concepts in biochemical pharmacology, Part I, pp. 47—63. Berlin-Heidelberg-New York: Springer 1971.

SCHUMACHER, E. E. JR., SCHMOCK, C. L.: The control of certain cardiac arrhythmias with isopropylnorepinephrine. Amer. Heart J. **48**, 933—940 (1954).

SELLERS, E. M., KOCH-WESER, J.: Protein binding and vascular activity of diazoxide. New Engl. J. Med. **281**, 1141—1145 (1969).

SHAND, D.G., EVANS, G.H., NIES, A.S.: The almost complete hepatic extraction of propranolol during intravenous administration in the dog. Life Sci. Part I **10**, 1417—1421 (1971).

SHANKER, L.S.: Absorption of drugs from the gastro-intestinal tract. In: BRODIE, B.B., GILLETTE, J.R. (Eds.): Concepts in biochemical pharmacology Part I, pp. 9—23. Berlin-Heidelberg-New York: Springer 1971.

SPEIZER, F.E., DOLL, R., HEAF, P.: Investigation into use of drugs preceding death from asthma. Brit. med. J. **1**, 339—343 (1968).

STENSON, R.E., CONSTANTINO, R.T., HARRISON, D.C.: Interrelationships of hepatic blood flow, cardiac output, and blood levels of lidocaine in man. Circulation **43**, 205—211 (1971).

THOMSON, P.D., ROWLAND, M., MELMON, K.L.: The influence of heart failure, liver disease, and renal failure on the disposition of lidocaine in man. Amer. Heart J. **82**, 417—421 (1971).

WALKER, S.R., EVANS, M.E., RICHARDS, A.J., PATERSON, J.W.: The fate of ^{14}C disodium cromoglycate in man. J. Pharm. Pharmacol. In press (1972).

Chapter 65

Genetically Determined Variations in Drug Disposition and Response in Man

E. S. VESELL

With 6 Figures

I. Introduction

Physicians have long realized that patients do not react uniformly to drugs. This difference in individual response to the same dose of a drug has become a fundamental principle of pharmacology and a cornerstone of the new field of pharmacogenetics. Multiple factors have been systematically investigated and identified as contributing to individual variations in drug disposition and response. These conditions include age, sex, time of day or season of drug administration, painful stimuli, disease states affecting the heart, liver, lung or kidney, hormonal and nutritional status, and exposure to a number of agents that induce or inhibit the hepatic microsomal drug-metabolizing enzymes, including chronic administration of any one of several hundred drugs (GILLETTE, 1971; CONNEY et al., 1971; VESELL, 1968). In addition, during the last 20 years multiple genetic factors altering drug disposition and response in man have been discovered. An important question concerns the number of patients that deviate from an anticipated "average" response to drugs. One estimate of this proportion comes from WHITE (1971), who maintains that over his extensive experience as a physician at least half of his patients displayed some unusual, unanticipated response to drugs.

The pharmacogenetic entities listed in Table 1 include traditionally recognized, hereditary conditions that produce a clinically significant abnormal response to drugs. They are divided into disorders that cause abnormal handling of the drug by the body – that is, defects in the metabolism of drugs by the body – and disorders in which the drug produces an aberrant effect on the body because of altered receptor sites on which the drug acts (VESELL, 1969). This classification is convenient in considering the clinical consequences of pharmacogenetic defects, in describing approaches to their therapy, in searching for new hereditary variations in response to drugs, and in analyzing individual variations in response to new drugs.

Pharmacogenetics deals with clinically significant hereditary variations in response to drugs (VESELL, 1969; COHEN and WEBER, 1972). Its conceptual foundations were laid in 1957 by MOTULSKY; the term pharmacogenetics was coined in 1959 by VOGEL, and the first text on the subject was written by KALOW in 1962. Early in the history of pharmacogenetics it was expected that many new pharmacogenetic disorders would be discovered. Therefore, pharmacogenetics initially was limited in scope to those hereditary disorders involving adverse reactions discovered only after drug administration. This approach now seems too restrictive, and very few new pharmacogenetic conditions that conform strictly to it have been discovered recently. Furthermore, this definition excludes from consideration as pharmacogenetic the demonstration of predominantly genetic control over large differences among normal, nonmedicated individuals in drug metabolism, as well as new approaches to the therapy of other genetic disorders. Finally, it excludes several hereditary diseases that were well described before the relatively recent discovery that they are exacerbated by certain drugs: diabetes mellitus by adrenocortical steroids; acute attacks of gout by thiazide diuretics; acute intermittent porphyria by barbiturates.

II. Hereditary Conditions Affecting Drug Response Transmitted as Simple Single Factors

Traditionally classified as pharmacogenetic, the majority of the conditions enumerated in Table 1 have been summarized elsewhere (Kalow, 1962; Vesell, 1969; LaDu and Kalow, 1968; Motulsky, 1964; Vesell, 1971; Vesell, 1972) and will be described in detail below. These conditions are transmitted as single factors inherited in conformity with Mendel's laws. Whereas a condition inherited as a single factor is governed by alleles at only one genetic locus, genes at multiple loci largely control such metrical, quantitative traits as stature, intelligence quotient, blood pressure and variations in the elimination rates of certain drugs. In addition, multigenic effects have been hypothesized in certain congenital malformations such as pyloric stenosis and club foot, cleft lip and palate, and in certain common disorders such as diabetes mellitus and peptic ulcer (Roberts, 1964). This mode of inheritance is referred to as multifactorial or polygenic. In polygenically controlled traits, the multiple genes at several discrete loci each make a quantum contribution as opposed to inheritance patterns of traits transmitted as single factors. The resultant polygenically controlled trait in the individual is the sum of the contributions from each of these multiple loci. The distribution curves for polygenically controlled traits are continuous, as shown in sizable populations by the bell-shaped Gaussian distribution curves for intelligence quotients and blood pressure, whereas distribution curves of response for genetic traits inherited as single factors are discontinuous, usually being either bimodal or trimodal. Though hundreds of disease states inherited as simple single factors have been identified in man and intensively investigated (McKusick, 1971), comparatively little study has been given to genetic analysis of such prevalent disorders as cancer, arteriosclerosis, and hypertension – all of which may be partially under polygenic control. Because rare abnormal genes transmitted as single factors segregate simply in successive generations, they can be traced readily without the complexity of similar abnormal genes introduced through marriage. By contrast, genetic analysis of commonly occurring polygenically controlled disease states is technically more difficult than genetic analysis of rare conditions transmitted as simple single factors. Extensive genetic analysis of polygenically transmitted disorders is difficult in that polygenic control may be expressed only through complex interactions with etiologic factors that are environmental in nature.

Several other inborn errors of metabolism, such as phenylketonuria, galactosemia, Wilson's disease, Crigler-Najjar syndrome and Lesch-Nyhan syndrome, are associated with abnormal, toxic responses to exogenously administered compounds. No matter where the line of demarcation is drawn, the fact that many hereditary conditions cause dramatic changes in drug response has multiple implications. Undoubtedly, many additional, as yet unidentified, genetic defects result in unfavorable reactions to various drugs; and the role of heredity in controlling the expression of allergic reactions due to hypersensitivity to specific drugs has yet to be defined.

Table 1. *Twelve pharmacogenetic conditions with putative aberrant enzyme, mode of inheritance and drugs that can elicit the signs and symptoms of the disorder*

Name of condition	Aberrant enzyme and location	Mode of inheritance	Drugs that produce the abnormal response
Genetic conditions probably transmitted as single factors altering the way the body acts on drugs			
1. Acatalasia	catalase in erythrocytes	autosomal recessive	hydrogen peroxide
2. Slow inactivation of isoniazid	isoniazid acetylase in liver	autosomal recessive	isoniazid, sulfamethazine, sulfamaprine, phenelzine, dapsone, hydralazine
3. Suxamethonium sensitivity or atypical pseudocholinesterase	pseudocholinesterase in plasma	autosomal recessive	suxamethonium or succinylcholine
4. Diphenylhydantoin toxicity due to deficient parahydroxylation	? mixed function oxidase in liver microsomes that parahydroxylates diphenylhydantoin	autosomal or X-linked dominant	diphenylhydantoin
5. Bishydroxycoumarin sensitivity	? mixed function oxidase in liver microsomes that hydroxylates bishydroxycoumarin	unknown	bishydroxycoumarin
6. Acetophenetidin-induced methemoglobinemia	? mixed function oxidase in liver microsomes that deethylates acetophenetidin	autosomal recessive	acetophenetidin
Genetic conditions probably transmitted as single factors altering the way drugs act on the body			
1. Warfarin resistance	? altered receptor or enzyme in liver with increased affinity for vitamin K	autosomal dominant	warfarin

Table 1 (cont.)

2. Glucose-6-phosphate dehydrogenase deficiency, favism or drug-induced hemolytic anemia	glucose-6-phosphate dehydrogenase	X-linked incomplete codominant	a variety of analgesics [acetanilide, acetylsalicylic acid, acetophenetidin (phenacetin), antipyrine, aminopyrine (Pyramidon)], sulfonamides and sulfones [sulfanilamide, sulfapyridine, N_2-acetylsulfanilamide, sulfacetamide sulfisoxazole (Gantrisin), thiazolsulfone salicylazosulfapyridine (Azulfadine), sulfoxone, sulfamethoxypyridazine (Kynex)], antimalarials [primaquine, pamaquine, pentaquine, quinacrine (Atabrine)], non-sulfonamide antibacterial agents [furazolidone, nitrofurantoin (Furadantin), chloramphenicol, *p*-aminosalicylic acid], and miscellaneous drugs [naphthalene, vitamin K, probenecid, trinitrotoluene, methylene blue, dimercaprol, (BAL), phenylhydrazine, quinine, quinidine]
3. Drug sensitive hemoglobins a) Hemoglobin Zürich	arginine substitution for histidine at the 63rd position of the β-chain of hemoglobin	autosomal dominant	sulfonamides
b) Hemoglobin H	hemoglobin composed of 4 β-chains	autosomal recessive	same drugs as listed for G6PD deficiency
4. Inability to taste phenylthiourea or phenylthiocarbamide	unknown	autosomal recessive	drugs containing the N—C=S group such as phenylthiourea methyl and propylthiouracil
5. Glaucoma due to abnormal response of intraocular pressure to steroids	unknown	autosomal recessive	corticosteroids
6. Malignant hyperthermia with muscular rigidity	unknown	autosomal dominant	such anesthetics as halothane, succinylcholine, methoxyfluorane, ether, and cyclopropane

In an attempt to apply pharmacogenetic principles to analyzing and reducing the increasing number of harmful reactions to commonly prescribed drugs, the scope of pharmacogenetics recently has been extended considerably beyond those conditions listed in Table 1. Estimates from several medical centers that one in twenty hospital admissions results from an adverse reaction to a prescribed drug (Cluff et al., 1965) led the Task Force on Prescription Drugs to identify adverse reactions to commonly prescribed drugs as a major medical problem (PHS Pub. No. 4380, 1968). Because most conditions listed in Table 1 are rare and produce toxicity after relatively few drugs, they probably do not substantially add to the total number of serious unfavorable reactions. A notable exception is glucose-6-phosphate dehydrogenase deficiency, which afflicts one in ten male blacks in this country.

A. Genetic Conditions Transmitted as Single Factors Affecting the Manner in which the Body Acts on Drugs

1. Acatalasia

The history of acatalasia (no cases of which have been reported in the United States) illustrates how clinical observations can contribute to pharmacogenetics. In 1946, the Japanese otorhinolaryngologist Takahara discovered acatalasia in an 11-year-old Japanese girl after surgery for a friable granulating tumor in the right nasal cavity and maxillary sinus (Wyngaarden and Howell, 1966). After excising the necrotic areas, Takahara applied hydrogen peroxide to sterilize the wound but the usual bubbles of oxygen – liberated by the action of catalase on hydrogen peroxide – did not evolve and the tissue turned dark, then black, presumably through oxidative denaturation of hemoglobin by the drug. Believing that silver nitrate had been used by accident, Takahara rewashed the wound and applied hydrogen peroxide from a fresh bottle, but the same bizarre response recurred. Takahara then suspected that his patient lacked the enzyme catalase and in a series of classic studies, he demonstrated that this explanation was correct and that the defect was transmitted as an autosomal recessive trait (Takahara, 1952, 1954; Takahara et al., 1952; Takahara and Doi, 1958, 1959). His original patient lacked catalase activity in her oral mucosa and erythrocytes, as did three of her five siblings. As is oftimes observed in rare inborn errors of metabolism, consanguinity existed; the patient's parents were second cousins.

The term "acatalasemia" has been altered to "acatalasia" since the former suggests restriction of the defect to blood, whereas the enzyme is also deficient in such tissues as mucous membrane, skin, liver, muscle, and bone marrow. Because trace levels of catalase activity occur in some patients, however, for whom the term "severe hypocatalasia" seems more appropriate (Wyngaarden and Howell, 1966), "acatalasia" is also not an entirely accurate designation. Heterozygotes who usually show values of catalase activity between those of the homozygous recessives and normal individuals would be classified as having "intermediate hypocatalasia". In certain Japanese kindred, some heterozygotes do not exhibit intermediate levels of catalase activity but rather values that overlap with the normal range (Hamilton and Neel, 1963).

By 1959, intensive screening for individuals with acatalasia had gathered 38 subjects from 17 families (Takahara and Doi, 1958, 1959). These cases were scattered throughout Japan in a manner suggesting sizable geographic variation, with "pockets" containing frequencies as high as 12%. Much lower gene frequencies (approximately 0.3%) were usual in other regions of Japan.

Mild, moderate, and severe expressions of acatalasia have been described (TAKAHARA et al., 1960). The mild form of acatalasia is characterized by ulcers of the dental alveoli; in moderate types alveolar gangrene and atrophy occur, and in the severe form recession of alveolar bone develops with exposure of the necks of teeth and eventual loss of teeth (TAKAHARA et al., 1960).

Studies of catalase activity in 66 members of five affected families (NISHIMURA et al., 1959) revealed that males and females were equally affected and that the distribution of activity was trimodal, corresponding to three phenotypes designated acatalasemic, hypocatalasemic, and normal.

In 1959, YATA reported a Korean patient with acatalasia, the first non-Japanese subject to be described. Two years later, AEBI and associates published their studies in which 3 individuals with acatalasia were found in screening 73,661 blood samples from Swiss Army recruits (AEBI et al., 1961). All three were in good health and showed none of the dental defects typical of the Japanese cases. The Swiss "acatalasics", unlike the Japanese, exhibited some residual catalase activity, possibly protecting them against the hydrogen peroxide formed by certain microorganisms that are thought to be responsible for the oral lesions.

Table 2 from AEBI (1967a) shows a classification of the different types of acatalasia reported from several countries. As indicated by differences in catalase activity, gene frequency, geographical distribution and clinical manifestations, at least several distinguishable forms of acatalasia probably exist. Genetic heterogeneity is also suggested by the different properties of the catalase molecule isolated from individuals with different forms of the disease (SHIBATA et al., 1967).

In the beginning, enzymatic and antigenic properties of catalase molecules were reported to be identical in normal individuals and subjects with the Swiss type of acatalasia (MICHELI and AEBI, 1965; AEBI et al., 1964). However, subsequent studies revealed that the catalase from Swiss patients differed from the catalase of normal individuals (AEBI, 1967a and b). After further purification of the enzyme, electrophoretic differences between the catalases of normal and deficient individuals were detected. In addition, differences were established in pH and heat stabilities and it was found that normal catalase differed from the catalase of deficient subjects in sensitivity toward the inhibitors aminotriazole and azide. All these findings suggested that in Swiss families acatalasia was a structural, rather than a controller, gene mutation (AEBI, 1967b).

In human erythrocytes, catalase exists in electrophoretically distinguishable forms or isozymes (PRICE and GREENFIELD, 1954; HOLMES and MASTERS, 1965; NISHIMURA et al., 1964; THORUP et al., 1964; BAUR, 1963). All catalase isozymes exhibit molecular weights of about 250,000. The description of a minor component of catalase which, though reacting with rabbit antihuman catalase serum, lacks enzyme activity (SHIBATA et al., 1967), further complicates the research. The minor component with a molecular weight of about 60,000 (SHIBATA et al., 1967) may be a subunit or precursor of catalase. Other data suggest that bovine liver catalase is made up of three or four identical chains (SCHROEDER et al., 1964; TANFORD and LOVRIEN, 1962). The minor inactive component in acatalasics may be structurally different from that in normocatalasics. Such a structural abnormality might render the subunits unable to assemble into the polymeric form of the "apocatalase" molecule, though an alternative explanation postulates that in acatalasics the precursor subunits are entirely normal, but a defect exists in a hypothetical "coupling" enzyme that may be required to join the subunits prior to addition of the prosthetic group (SHIBATA et al., 1967).

Table 2. *Cases of acatalasia and related anomalies reported in literature until 1965*[1]

Type (year of detection)	Origin (No. of families)	Number of homozygotes (Hom) and heterozygotes (Het)	Residual catalase activity percentage (normal = 100)	Remarks
I (1947)	Japan (31) Korea (1)	Hom: 66 Het: > 100	Hom: 0— 3.2 Het: 37—56	Incomplete recessive inheritance; oral gangrene (TAKAHARA's disease) in ~ 50% of homozygotes; activity: trimodal distribution curve (no overlap)
II (1959)	Japan (1) family 13 MI	Hom: 1 (male) Het: —	Hom: 3.2 Het: ~ 100	Complete recessive inheritance (involvement of modifier or suppressor genes?)
IIIa (1962)	Japan (1) kindred 29 OHH	Hom: 3 (Het: 17)	Hom: 0 (?) Het: > 56	Overlap between heterozygous carrier and normals (dual allelic control?)
IIIb (1961)	Switzerland (3) families V.B. and G.	Hom: 11 (Het: ~ 30)	Hom: 0.1—1.3 Het: 15—85	Synthesis of two different types of catalase in heterozygotes (normal catalase + unstable variant), all homozygotes in good state of health
IV (1963)	Israel (1) Iranian born	Hom: 1 (male) Het: 15	Hom: 8 Het: 49—67	Combination with deficiency of G6PD; intolerance to fungicide
V (1963)	United States (1) Scandinavian and British extraction	Hom: 0 Het: 6	All: ~ 100	Allocatalasia: synthesis of a variant catalase; activity and stability as normal catalase

[1] From AEBI (1967a).

2. Slow Inactivation of Isoniazid

Although synthesized in 1912 by Meyer and Mally, isoniazid's bacteriostatic effect on Mycobacterium tuberculosis was not discovered until 1952 (Grunberg et al., 1952; Robitzek et al., 1952). Soon great differences were reported in the metabolism of isoniazid in man (Bönicke and Reif, 1953; Hughes, 1953; Hughes et al., 1954, 1955). Further research on the excretory products of isoniazid revealed that all the drug appeared in the urine either as acetyl isoniazid, isonicotinic acid, unchanged isoniazid, or small amounts of other derivatives (Hughes et al., 1955). Even though great differences occurred among individuals in the amount of free or acetylated urinary products (which were inversely related within every subject), each individual maintained unchanged his pattern of excretion during long-term therapy.

Isoniazid polymorphism might be expected to result in a less favorable response to treatment by rapid inactivators than by slow inactivators. Harris (1961b) reported that in 775 patients with pulmonary tuberculosis on standardized isoniazid regimens, cavity closure and sputum conversion were generally noted earlier in slow inactivators than in rapid inactivators – which would lend credence to this expectation. However, after 6 months of treatment, no clinically detectable differences between slow and rapid phenotypes were observed. If administered only once a week, isoniazid results in a diminished clinical response in rapid inactivators as compared with slow inactivator patients with tuberculosis (Price Evans, 1968). Apparently, neither the slow nor the rapid acetylase genotype is more prone to resistance to tubercle bacilli (Harris, 1961b) or reversion (Gow and Price Evans, 1964).

Polyneuritis does occur more frequently in slow inactivators and is the primary clinical problem associated with this polymorphism. During isoniazid therapy, polyneuritis developed in four of five slow inactivators, but in only two of ten rapid inactivators (Hughes et al., 1954). Devadatta et al. (1960) confirmed these conclusions with their studies in Madras. Peripheral neuritis can be prevented by the administration of pyridoxine simultaneously with isoniazid (Carlson et al., 1956). Neuritis develops after isoniazid administration because of pyridoxine deficiency, which is due to inactivation of pyridoxine and removal of the coenzyme from tissues through chemical interaction of isoniazid with pyridoxine. Isoniazid may also compete with pyridoxal phosphate for the enzyme apotryptophanase (Ross, 1958; Robson and Sullivan, 1963).

Thus, genetically controlled variations in rates of isoniazid acetylation do carry the clinical implication that on long-term therapy slow inactivators tend to develop polyneuritis more frequently than rapid inactivators. Slow acetylation is inherited as an autosomal recessive trait, and rapid acetylation of isoniazid is transmitted as an autosomal dominant trait. The half-life of isoniazid ranges from 45 to 80 min in the plasma of rapid inactivators, whereas the half-life extends from 140 to 200 min in slow inactivators (Kalow, 1962). Although slow acetylators may excrete unchanged 30% of a dose, rapid acetylators may excrete unchanged only 3% (Hughes et al., 1954; Peters, 1959, 1960a and b). These differences between slow and rapid inactivators are unrelated to intestinal absorption, protein binding, renal glomerular clearance, or renal tubular reabsorption (Jenne et al., 1961). Slow inactivators have reduced activity of acetyl transferase (Price Evans and White, 1964; Peters et al., 1965a and b), the liver supernatant enzyme mainly responsible for the metabolism of sulfamethazine and of isoniazid, as well as other monosubstituted hydrazines such as phenelzine and hydralazine. (Price Evans, 1965). Acetylation of other drugs such as *p*-aminosalicylic acid and sulfanilamide is monomorphic and accomplished probably by an acetylase different from that which acetylates isoniazid.

WHITE and PRICE EVANS (1968) found that sulfamethoxypyridazine was acetylated much less than was sulfamethazine. They reported that rapid inactivators acetylated a greater percentage of sulfamethoxypyridazine than did slow inactivators. Surprisingly, serum concentrations of free sulfamethoxypyridazine did not vary appreciably in individuals with different acetylator phenotype, presumably because of the influence of other more important factors affecting the drug's elimination from plasma.

Numerous genetic investigations of isoniazid acetylation have been reported. A twin study showed the amount of free isoniazid in 24-h samples of urine to be similar in identical twins, but much larger in fraternal twins (BÖNICKE and LISBOA, 1957). In the percentage of isoniazid dose excreted unchanged in urine, bimodal distributions were observed, implying genetically distinct methods of handling the drug (BIEHL, 1956, 1957). This concept is further supported by the large variations between Caucasian and Japanese subjects in the frequency of rapid inactivators (HARRIS et al., 1958) and by a study of 20 families that found that slow inactivation of isoniazid no doubt was recessive to rapid inactivation (KNIGHT et al., 1959). Six hours after an oral dose of 9.7 mg/kg, isoniazid concentrations in plasma of 267 members of 53 Caucasian families were bimodally distributed so that individuals could be classified as either rapid or slow inactivators (PRICE EVANS et al., 1960). The discovery that the mean concentration of isoniazid in plasma was lower in heterozygotes than in rapid inactivators homozygous for the dominant gene established a dosage effect for the trait (PRICE EVANS et al., 1960). Direct determination of the acetylation genotype can be accomplished by sensitive microbiologic and chemical assays (SUNAHARA, 1961; ELLARD et al., 1972). Diverse geographical and racial distributions of the gene have been reported. Most uncommon in Eskimos, deficiency is only slightly more frequent in Far Eastern populations (MOTULSKY, 1964). Deficiency is common in Negroes and European populations, where 70 to 80% of the individuals possess the aberrant gene either in homozygous or heterozygous state (ARMSTRONG and PEART, 1960; HARRIS, 1961a; SUNAHARA, 1961; PRICE EVANS, 1962; MITCHELL et al., 1960; DEVADATTA et al., 1960; GANGADHARAM and SELKON, 1961; SCHMIEDEL, 1961; SZEINBERG et al., 1963; MOTULSKY, 1964).

Natural selection may perpetuate polymorphisms such as the rapid and slow acetylation of isoniazid. Insight into such hypothetical advantages of certain genotypes under various environmental circumstances with respect to isoniazid acetylation must await further comprehension of the function of the acetylase *in vivo*. Little is known about naturally occurring hydrazine compounds, but PETERS and co-workers (1965b) measured dimethylaminobenzaldehyde-reacting materials in urine of drug-free individuals and reported below 0.8 mg total hydrazine equivalents per 12 hours. Such a low value suggests that the body does not encounter many naturally occurring hydrazines. Apparently, neither hexosamine nor tryptophan metabolites are natural substrates for isoniazid acetylase (WHITE and PRICE EVANS, 1967a and b).

Studies *in vitro* of the acetylase that transfers an acetyl group from acetyl coenzyme A to isoniazid have been reported by PRICE EVANS (1962), PRICE EVANS and WHITE (1964), JENNE (1965), and WEBER et al. (1968). WEBER et al. (1968) purified the enzyme 300-fold to 500-fold from the 100,000 g liver homogenate supernatant, which has all the acetyltransferase activity. They then established a ping-pong mechanism of action for the enzyme. This study suggests that the mechanism of isoniazid (INH) acetylation may be pictured as follows:

Acetyl-CoA ↓	CoA ↑	INH ↓	Acetyl-INH ↑

N-Acetyl-transferase — Acetyl-*N*-acetyltransferase — *N*-acetyl-transferase

The idea that INH acetylation includes the two step process of manufacture and decay of an acetylated enzyme is suggested by the following partial reactions: N-acetyltransferase $+$ acetyl $-$ CoA $\rightleftharpoons$ acetyl N-acetyltransferase $+$ CoA acetyl-N-acetyltransferase $+$ INH $\rightleftharpoons$ acetyl $-$ INH $+$ N-acetyltransferase.

Although purified acetylase from rapid INH inactivators exhibit approximately twice the amount of enzyme as that present in purified acetylase from slow INH inactivators, the enzymes isolated from the two acetylator phenotypes are similar with respect to pH optima, heat stability, kinetic behavior and substrate specificity (JENNE, 1965; WEBER et al., 1968; WEBER, 1971). These findings might imply that in man and in rabbit, only the amount of acetylase present in the liver differentiates the rapid and slow INH inactivators and that, qualitatively, the enzymes are similar. However, a possible basis for a structural gene defect was uncovered in a comparison of the electrophoretic properties of the rabbit enzyme. Acetylase activity exhibits two distinguishable bands in both rapid and slow INH inactivators on electrophoretic separation (LADU, 1972). One of the two components migrates faster toward the anode in the slow than in the rapid acetylator, whereas the slower migrating component has similar electrophoretic mobility in both phenotypes (LADU, 1972).

Because phenelzine is polymorphically acetylated, its side effects were observed both in rapid and slow acetylators; severe side effects, such as blurred vision and psychosis, were confined chiefly to slow acetylators. An explanation of why toxic effects of the hydrazine drug phtivazid occurred infrequently in those patients who excreted the acetylated form in high concentrations may be the reduced incidence of toxicity from a polymorphically acetylated drug in rapid as compared to slow acetylators (SMIRNOV and KOZULITZINA, 1962).

Slow acetylators are far more prone to peripheral neuropathy and a syndrome resembling systemic lupus erythematosus than are rapid acetylators (PERRY et al., 1967). Of 57 hypertensive patients receiving 200 to 3000 g of hydralazine for 1 to 15 years, all 12 who developed symptoms of hydralazine toxicity were slow acetylators (PERRY et al., 1970). Though the variance was insignificant statistically, in 153 tuberculous patients on isoniazid, antinuclear antibodies to whole nuclei, nucleoprotein, soluble nucleoprotein and isoniazid-altered soluble nucleoprotein were detected by complement fixation tests more frequently in slow than in fast acetylators (ALARCÓN-SEGOVIA et al., 1971). PERRY et al. (1971) postulate a relationship between autoimmune reactions to hydralazine and methyldopa, claiming that "Caucasian, slow acetylators also have an increased susceptibility to develop IgG antiglobulin reactions following exposure to methyldopa".

3. Succinylcholine Sensitivity or Atypical Pseudocholinesterase

Shortly after the muscle relaxant succinylcholine was introduced in 1952 and its use became widespread, occasional patients were discovered to be extraordinarily sensitive to it; several deaths associated with its use were reported (BOURNE et al., 1952; PRICE EVANS et al., 1952). This drug (also called Suxamethonium, Suxethonium, Scoline, and Anectine) was initially described in 1906 by HUNT and TAVEAU; but its properties as a muscle relaxant were not discovered until 43 years later by BOVET et al. (1949). Ordinarily employed during general anesthesia, suxamethonium is also used in treatment of tetanus and in electroconvulsive therapy. Its principal therapeutic advantage is the brevity of its action; the usual dosage of 30 to 100 mg produces muscle paralysis and apnea for approximately 2 min. However, the duration of action is 2 to 3 h in atypical patients. The brevity of action in normal cases is due to the exceedingly rapid hydrolysis of succinylcholine by plasma pseudocholinesterase (BOVET-NITTI, 1949). The choline radicals are removed by this enzyme one at a time, with formation of the relatively inactive intermediate succinylmonocholine (LEHMANN and SILK, 1953; WHITTAKER and WIJESUNDERA, 1952). Effective treatment of prolonged apnea in patients with atypical pseudocholinesterase can be achieved by transfusion of normal plasma or of a highly purified preparation of the human enzyme (GOEDDE et al., 1968).

Reduced serum pseudocholinesterase activity was noted when the initial reports of bizarre reactions were published in 1952. It became clear with the accumulation of increasing numbers of these rare cases that such abnormal individuals were otherwise healthy and, therefore, had low pseudocholinesterase not because of liver disease, poisoning by organophosphorus compounds, malnutrition, or severe anemia – all of which can diminish plasma pseudocholinesterase activity (LEHMANN and RYAN, 1956) – but rather because of an inherited defect (FORBAT et al., 1953). On the basis of a study of the families of five unrelated succinylcholine-sensitive probands, LEHMANN and RYAN (1956) proposed that the abnormality was inherited as an autosomal recessive trait. Extensive overlap at that time prevented identification of the three phenotypes simply by measurement of plasma pseudocholinesterase activity.

The nature of the enzymatic abnormality was explained by KALOW and his associates, who determined that it did not result just from smaller amounts of the normal pseudocholinesterase but rather from the presence of a structurally altered enzyme with kinetic properties decidedly different from those of the usual enzyme (KALOW and GENEST, 1957; KALOW and STARON, 1957; KALOW and DAVIES, 1959; DAVIES et al., 1960). Much lower avidity for the abnormal than for the normal enzyme is exhibited by succinylcholine and other substrates. The abnormal enzyme exerts no measurable effect on the drug at concentrations of succinylcholine present during anesthesia, whereas the normal enzyme exerts marked hydrolytic activity (DAVIES et al., 1960). The atypical enzyme was more resistant than the normal enzyme to many pseudocholinesterase inhibitors (KALOW and DAVIES, 1959). Fluoride and organophosphorus compounds were each shown to inhibit the normal and atypical enzyme differentially. At first, differential inhibitors of the normal and atypical enzyme were thought to require a positively charged nitrogen molecule (HARRIS and WHITTAKER, 1961). It was suggested that the positively charged portion of the inhibitor combined with the anionic site of the enzyme because of the importance of a positive charge on many inhibitors; only the anionic site on the atypical enzyme was considered to be defective either in accessibility or magnitude of charge (KALOW and DAVIES, 1959). Two sites previously had been described on the cholinesterase molecules: an esteratic site into which the acid portion of the substrate was positioned during hydrolysis and an anionic site that accommodated the positively charged choline radical of the substrate (WILSON, 1954). The observations of CLARK et al. (1968) support the conclusion that the anionic site of the atypical enzyme is altered. They reported that the pK of the atypical enzyme is lower than that of the usual enzyme, that choline alters the pK of the usual but not of the atypical enzyme, and that choline has a lower affinity for the atypical enzyme. The esteratic site may also be altered in the atypical enzyme. This conclusion is implied by the facts that choline stimulates the dephosphorylation step of the usual enzyme, but not of the atypical enzyme, and that differential rates exist for dephosphorylation by sodium fluoride. Since alteration of a single residue in the structure of the atypical pseudocholinesterase could modify both the anionic and esteratic sites, two distinct point mutations on the atypical pseudocholinesterase need not be postulated in order to accommodate these data. Development of tests to distinguish the three phenotypes, which could not be satisfactorily separated simply by measuring plasma pseudocholinesterase activity, was thus promoted by such detailed studies of the aberrant enzyme. KALOW and GENEST (1957) utilized dibucaine (cinchocaine), a differential inhibitor of normal and atypical pseudocholinesterase, to separate the three phenotypes.

The percentage inhibition of pseudocholinesterase activity produced by 10^{-5} M dibucaine was designated the "dibucaine number" or "DN". Whereas atypical pseudocholinesterase is inhibited only 20%, the normal enzyme is inhibited approximately 80% and heterozygotes exhibit between 52 and 69% inhibition (KALOW and GENEST, 1957). Enzyme concentration does not affect the degree of inhibition.

The discovery of additional genetic variants resulted from the use of sodium fluoride as an inhibitor (HARRIS and WHITTAKER, 1962a). Tetracaine, unlike other previously studied compounds, is hydrolyzed faster by atypical than by normal pseudocholinesterase, and an even larger separation of phenotypes apparently can be achieved with the procaine-tetracaine ratio than with the DN (FOLDES, 1968).

Family studies suggest that inheritance of various types of atypical pseudocholinesterase occurs through allelic codominant genes at a single locus (KALOW and STARON, 1957; HARRIS et al., 1960; BUSH and ROTH, 1961). Four alleles have been identified with the resulting ten genotypes:
$E_1^uE_1^u$, $E_1^uE_1^a$, $E_1^aE_1^a$, $E_1^sE_1^u$, $E_1^sE_1^s$, $E_1^sE_1^a$, $E_1^fE_1^u$, $E_1^fE_1^f$, $E_1^fE_1^a$, $E_1^fE_1^s$ where E_1 signifies the pseudocholinesterase genetic locus and u, a, s, and f indicate the "usual", "atypical", "silent", and "fluoride" sensitive alleles, respectively. Although the genes apparently vary in expression, penetrance is complete (LEHMANN and LIDDELL, 1964). One homozygous affected individual was estimated in every 2800 or 0.019 to 0.017; and approximately 3.8% of individuals in various populations are heterozygotes (KALOW and GUNN, 1959; KATTAMIS et al., 1962). Occurrence is not as rare as the initial estimates of 1 in 4000 suggested; more intensive population studies showed the number of homozygous recessive individuals for the atypical allele to be about 1 in 2500 (LADU, 1972). These results underlined the wisdom of screening all individuals for their capacity to metabolize succinylcholine before administering the drug. KALOW (1965) found that about half of the patients studied because of unusual sensitivity to succinylcholine had a normal activity and type of pseudocholinesterase; the unusual response to succinylcholine in these subjects may be due to inhibition of the enzyme by other drugs.

LEHMANN and LIDDELL (1964) discuss a series of four families in which the dibucaine values do not follow the typical pattern of inheritance. Presumably, these individuals are heterozygous for a rare, so-called silent gene. There are a few exceedingly rare individuals with complete absence of serum and liver pseudocholinesterase activity (HODGKIN et al., 1965). Apparently normal otherwise, these individuals lacked all four of the usual isozymes of serum pseudocholinesterase; the absence of antigenically cross-reacting material was revealed by immunodiffusion and immunoelectrophoretic studies (HODGKIN et al., 1965).

Heterozygotes for the silent gene exhibit two thirds of the normal serum cholinesterase activity; they widely overlap normal values (HODGKIN et al., 1965; HARRIS et al., 1963). Such silent mutations may affect the controlling element of the gene, MOTULSKY (1964) suggests, thereby completely disrupting protein production; he discusses also the possibility that a single structural mutation affects both the active site and the antigenic determinants. GOEDDE and ALTLAND (1968) studied five subjects who were homozygous recessive for the silent gene. They reported residual enzymatic activity and antigenic determinants in three of the subjects, each of whom revealed a single band on starch gel electrophoresis at the C-4 position when their sera were concentrated 6-fold. In the other two subjects, results similar to those of HODGKIN et al. (1965) were recorded.

The incidence of atypical pseudocholinesterase, unlike the polymorphism affecting isoniazid acetylation, remains comparatively constant in varying geographical areas; however, GUTSCHE et al. (1967) discovered an exceptionally high incidence of the silent mutation in a population of southern Eskimos. Further investigation of the apnea that developed in 2 Eskimo children after a single low dose of succinylcholine disclosed nineteen cases in 11 Eskimo families. Prior to this survey in Alaska (SZEINBERG et al., 1966), only 10 individuals homozygous for the silent gene had been described. The gene frequency of 0.12 in this locality, which extended from Hooper Bay to Unalakleet and centered on the lower Yukon River, suggested that 1.5% of this Alaskan population was sensitive to succinylcholine. The isolation and consequent inbreeding of these Eskimos, the authors contended, resulted in the high frequency of the rare silent gene in this, but not other, regions of Alaska. However, only 2 of these 11 Eskimo families are known to be related. Another theory is that certain characteristics of the environment may have favored the gene. According to a method adapted to analyze greater volumes of sera, 8 of the 17 deficient Eskimos had pseudocholinesterase activities of 2 to 8 units, whereas 9 of them exhibited no activity whatever. Possibly trace pseudocholinesterase activity results from a mutation other than that characterized by no detectable activity.

Similarity of gene frequencies of atypical pseudocholinesterase in most populations suggests either that little selective advantage is conferred now by the various genotypes or that the contributing environmental factors are common to widely differing countries. A potent differential cholinesterase inhibitor present in such solinaceous plants as tomatoes and potatoes (ORGELL et al., 1958) was shown by HARRIS and WHITTAKER (1962b) to be the glycoalkaloid solanine. Since atypical pseudocholinesterase is less responsive to inhibition by this naturally occurring product than is the normal enzyme, the atypical genotype might be at a selective advantage in cases of solanine poisoning. Before the pseudocholinesterase phenotypes could be readily identified, however, several outbreaks of solanine poisoning were reported (WILSON, 1959; WLLIMOTT, 1933; HARRIS and COCKBURN, 1918).

In several disease states, such as thyrotoxicosis, schizophrenia, hypertension, acute emotional disorders, and after concussion, plasma pseudocholinesterase activities may be elevated. Increases also are observed as a genetically transmitted condition without apparent clinical consequences but associated with an electrophoretically slower migrating C_4 isozyme (HARRIS et al., 1963; NEITLICH 1966). One individual was discovered (in a study of 1029 male military personnel between ages 17 and 35) whose plasma pseudocholinesterase activity of 1278 units was more than 3 times higher than the mean for all the volunteers (NEITLICH, 1966). Further investigation of the propositus's family revealed that his sister and daughter had values of 1518 and 1237 plasma pseudocholinesterase units, respectively, and his mother had 566 units. This abnormality was termed the Cynthiana variant. It was reinvestigated by YOSHIDA and MOTULSKY (1969) and, as expected, the exceptionally high pseudocholinesterase activity was associated with resistance to the pharmacological effects of succinylcholine. The Cynthiana variant may result from a defect of a regulator gene controlling pseudocholinesterase activity rather than from a structural gene abnormality, because the physiochemical and kinetic properties of the enzyme are normal. KALOW and GENEST (1957) and KALOW and STARON (1957) previously described an individual with 2.5 times the average pseudocholinesterase activity of 1556 subjects, but they did not pursue a family study. Slightly higher than normal pseudocholinesterase activity associated with a retarded electrophoretic mobility of the main isozyme was found in approximately 10% of a random sample of the British population (HARRIS et al., 1963). This slower moving band was designated C_5. The greatly elevated total plasma pseudocholinesterase activity of the American variants distinguished them from the variants described in England, although NEITLICH'S pseudocholinesterase variant also exhibited slower electrophoretic mobility than the normal C_4 isozyme. Understandably, individuals possessing markedly elevated plasma pseudocholinesterase activity are resistant to the usual doses of succinylcholine.

Recent studies on the isolation of the multiple electrophoretic forms of serum pseudocholinesterase suggest that they are interconvertible and exhibit similar kinetic behavior (LADU, 1972). The electrophoretic characteristics of the atypical pseudocholinesterase parallel those of the normal enzyme (LADU, 1972).

4. Deficient Parahydroxylation of Diphenylhydantoin

Many lipid-soluble drugs are rendered more water-soluble, and hence more excretable, through metabolism by enzyme systems in liver microsomes (GILLETTE, 1963, 1966). Most of these liver microsomal enzymes are oxidases requiring oxygen, NADPH, and cytochrome P-450; because these enzymes are exceedingly sensitive, losing most of their activity when removed from the endoplasmic reticulum, their characterization remains to be accomplished. The earliest published example of a genetic defect of mixed function oxidases in humans is deficient parahydroxylation of diphenylhydantoin (KUTT et al., 1964a). The drug is metabolized in man mainly by parahydroxylation of one of the phenyl groups to yield 5-phenyl-5'-parahydroxyphenylhydantoin, HPPH, which is conjugated with glucuronic acid and then eliminated in the urine (BUTLER, 1957; WOODBURY and ESPLIN, 1959; MAYNERT, 1960).

Since its introduction, diphenylhydantoin has become one of the most commonly used anticonvulsants. However, it can cause multiple toxic reactions including nystagmus, ataxia, dysarthria, and drowsiness. YAHR et al. (1952) report that 77% of patients develop toxicity on a daily dose of 0.6 gm, the usual amount recommended by these and other authors (YAHR and MERRITT, 1956). KUTT et al. (1964b) have shown these toxic reactions to be clearly dose related (Fig. 1). Toxic symptoms developed in the propositus, W. J., on a commonly used dosage of 4.0 mg/kg, but not on a dose of 1.4 mg/kg. Abnormally low urine levels of the metabolite HPPH occurred in combination with prolonged high blood levels of unchanged diphenylhydantoin (KUTT et al., 1964a). A study of two generations of the family of W. J. (Fig. 2) revealed 2 affected and 3 unaffected individuals, suggesting that low activity of diphenylhydantoin hydroxylase exhibits dominant transmission.

Apparently, drugs such as phenobarbital and phenylalanine are parahydroxylated by enzymes different from those that hydroxylate diphenylhydantoin, since the patient's capacity to parahydroxylate other compounds was normal. KUTT et al. (1964a) suggested that phenobarbital is less likely to accumulate than diphenylhydantoin, since unaltered phenobarbital (BUTLER, 1956) is excreted in the urine in higher amounts (reaching 30% of the daily intake) than is diphenylhydantoin (which attains only 5% of daily intake). A hydroxylation defect of phenobarbital, therefore, may be masked.

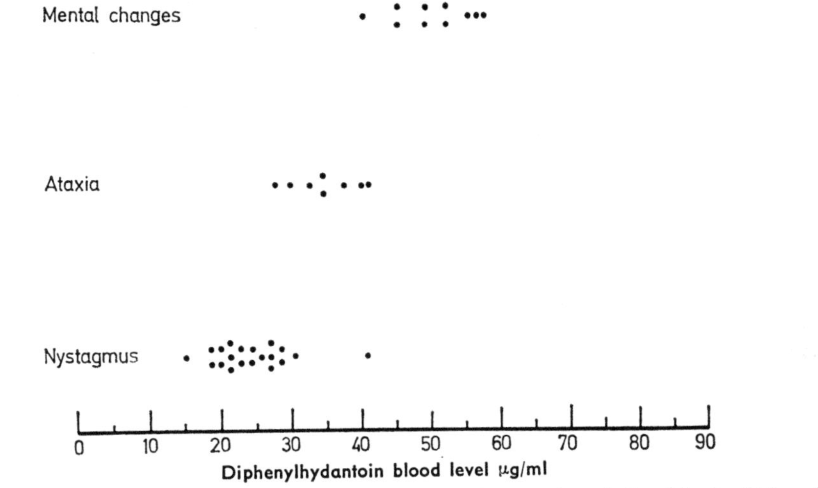

Fig. 1. The onset of nystagmus, ataxia, and mental changes in relationship to diphenylhydantoin blood concentrations, reproduced from KUTT et al., 1964b

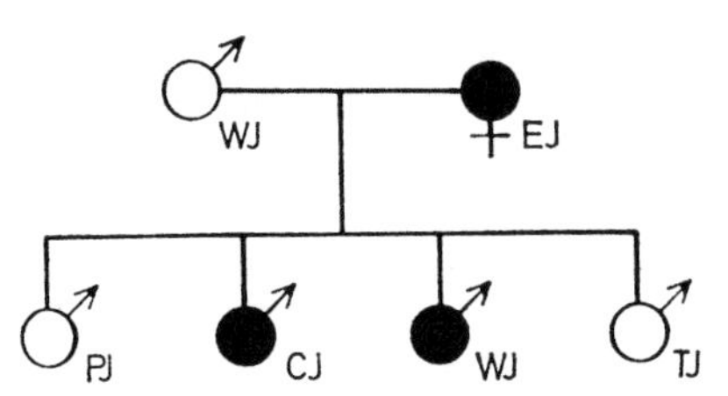

Fig. 2. Pedigree of deficient parahydroxylation of diphenylhydantoin with propositus W. J. and his affected brother C. J. and mother E. J., reproduced from KUTT et al., 1964a

Blood concentrations and urinary metabolites of diphenylhydantoin should be measured when toxic symptoms appear in patients receiving the drug, especially when low dosages are involved. If the patient is exceptionally sensitive to diphenylhydantoin and is deficient in hydroxylating capacity, the drug need not be discontinued; rather, the dosage should be adjusted to give the desired blood concentrations.

Recently, slow inactivation of isoniazid has been identified as a more important cause of diphenylhydantoin intoxication than heritable deficiency of parahydroxylase activity (BRENNAN et al., 1968). In a series of 29 individuals receiving diphenylhydantoin and isoniazid, all 5 patients who developed clinically evident diphenylhydantoin toxicity were slow isoniazid inactivators. Both isoniazid and *p*-aminosalicylic acid interfered with diphenylhydantoin parahydroxylation in rat liver microsomes (KUTT et al., 1968).

5. Bishydroxycoumarin Sensitivity

Bishydroxycoumarin sensitivity in a patient who received the drug for treatment of an acute myocardial infarction was reported by SOLOMON (1968). On a dose of 150 mg, the patient's plasma bishydroxycoumarin half-life of 82 h compared with normal values of 27 ± 5 h. The patient's mother suffered a spinal cord hematoma, causing permanent paraplegia, while she was receiving a small weekly dose of 2.5 to 5 mg of warfarin. Although family studies were not performed because of unwillingness to cooperate, this unfortunate event in the treatment of the patient's mother suggests not only the possibility of hereditary transmission of bishydroxycoumarin sensitivity, but also the desirability of measuring drug concentrations in blood during long-term therapy.

Warfarin and bishydroxycoumarin are extensively hydroxylated in the rat (IKEDA et al., 1966; CHRISTENSEN, 1966), but their metabolites in man have not been fully characterized. This patient with bishydroxycoumarin sensitivity and his mother may have a metabolic defect involving a deficiency of a hepatic microsomal hydroxylase.

Genetic factors influence responsiveness to anticoagulants in rabbits (SMITH, 1939; LINK, 1944; SOLOMON and SCHROGIE, 1966), as they do in rats, where resistance to warfarin as a rodenticide is transmitted as an autosomal dominant trait (Editorial, 1966; GREAVES and AYRES, 1967).

Increased sensitivity to coumarin anticoagulants also can result from acquired conditions, including vitamin K deficiency, increased turnover of plasma proteins, and numerous forms of liver disease that impair capacity to produce vitamin K-dependent clotting factors (O'REILLY et al., 1968).

Various drugs can increase the prothrombinopenic response to coumarin anticoagulants. Cinchophen may damage liver cells; phenothiazine may produce cholestasis, thereby diminishing absorption of vitamin K; phenylbutazone increases sensitivity by displacing warfarin from plasma albumin (AGGELER et al., 1967); and phenyramidol inhibits the hepatic microsomal enzymes responsible for metabolism of coumarin drugs (O'REILLY and AGGELER, 1965).

6. Acetophenetidin-Induced Methemoglobinemia

SHAHIDI (1967) reported severe methemoglobinemia and hemolysis in a 17 year old girl after she had taken phenacetin (acetophenetidin). Several erythrocyte components were normal, including G6PD, 6-phosphogluconate dehydrogenase, diaphorase, glutathione reductase and the concentration of reduced glutathione. Extracorpuscular compounds seemed to be the factors causing hemolysis and as much as half of the patient's hemoglobin was occasionally in the form of methemoglobin. No physiochemical abnormalities of hemoglobin were detected. After administration of phenacetin, large amounts of 2-hydroxyphenetidin and 2-hydroxyphenacetin derivatives were discovered in this patient's urine. In normal individuals, a dose of 2 g of phenacetin produced no more than 2.8% methemoglobin, and more than 70% of the dose appeared in the urine as *N*-acetyl-*p*-amino-phenol with only minute amounts of the hydroxylated products so prevalent in this patient's urine. Another sister, a brother and both parents of the patient revealed the normal response to phenacetin, but her 38 year old sister likewise responded abnormally to phenacetin.

These observations suggested an autosomal recessive inheritance of a defect in which the patient's hepatic microsomal mixed function oxidases were deficient in deethylating capacity. Phenacetin was hydroxylated in the patient and her 38 year old sister instead of being deethylated, as in normal individuals. The toxicity observed after phenacetin administration was probably produced by these hydroxylated products of phenacetin identified in their urine. If this hypothesis were correct, induction of the hepatic microsomal phenacetin hydroxylating enzymes prior to phenacetin administration should worsen the condition by increasing the concentration of toxic hydroxylated phenacetin metabolites. To test the hypothesis, the patient was pretreated with phenobarbital and then given phenacetin. Severe neurological symptoms, including bilateral positive Babinski responses, developed and profound methemoglobinemia ensued (SHAHIDI, 1967). A normal volunteer after the same pretreatment developed neither methemoglobinemia nor neurological changes.

B. Genetic Conditions Probably Transmitted as Single Factors Altering the Way Drugs Act on the Body

1. Warfarin Resistance

Genetically controlled resistance to warfarin occurs rarely in man. Such resistance in a 71-year-old patient receiving anticoagulants for a myocardial infarction was carefully documented and his family studied for the defect (O'REILLY et al., 1964). Physical and laboratory examinations of the patient showed no abnormalities other than a reproducible reduction in his one-stage prothrombin concentration to approximately 60% of normal. Anticoagulants were initially withheld because of the patient's low prothrombin time. They were administered after one month, at which time he proved to be resistant, rather than sensitive, to dicoumarol. A daily dose of 20 mg of warfarin failed to achieve any prothrombinopenic response, and a daily dose of 145 mg was required to reduce the prothrombin concentration to therapeutic levels. The mean daily dose of warfarin was 6.8 ± 2.8 mg in 105 patients on long-term anticoagulant therapy (O'REILLY et al., 1968), so that the resistant patient was 49 standard deviations above the mean.

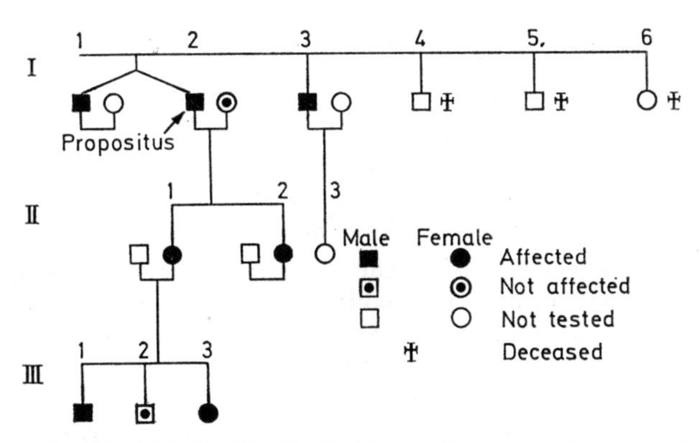

Fig. 3. Pedigree of family M indicating the incidence of resistance to coumarin anticoagulant drugs, reproduced from O'REILLY and AGGELER (1965)

Five other members of the family in three generations were also resistant to warfarin (Fig. 3). The facts that both sexes were equally affected and that representatives from all three generations were resistant to warfarin suggested autosomal dominant transmission of the trait.

Various environmental conditions lead to resistance to coumarin anticoagulant drugs. ELIAS (1965) reported decreased sensitivity to the prothrombinopenic effect of coumarin drugs in hyperthyroid patients treated with propylthiouracil, in patients with congestive heart failure relieved by operative procedures or medications, and in patients with liver disease being treated medically. Large doses of coumarin drugs may be required in pregnancy to offset increased levels of vitamin K-dependent clotting factors present during pregnancy (O'REILLY et al., 1968). Resistance to the coumarin anticoagulants can be produced by simultaneous administration of the natural antidote, vitamin K, and other therapeutic agents. Inducing agents that can reduce the blood concentration of anticoagulant drugs by stimulating their metabolism are barbiturates, glutethimide, chloral hydrate, and griseofulvin.

O'REILLY et al. (1964) performed various pharmacodynamic studies on their patient to identify the mechanism of the defective response to warfarin, before initiating long-term therapy. Blood concentrations of the anticoagulant were determined serially after the standard oral dose of 1.5 mg of warfarin sodium per kg of body weight. After an oral dose, the plasma warfarin concentrations rose rapidly, indicating that the drug was absorbed normally from the gastrointestinal tract. A normal volume of distribution and a normal rate of metabolism of the drug (O'REILLY et al., 1968) were suggested by the normal concentration of warfarin attained in plasma and its rate of elimination from plasma. Electrophoretic studies showed that warfarin was bound exclusively to albumin, as in normal subjects. Even after administration of very high doses, warfarin was not excreted unchanged in urine or stools. Amounts of a metabolite of warfarin similar to those recovered from the urine of normal subjects given equivalent amounts of drug were recovered from the patient's urine. The patient was shown to be unusually sensitive to vitamin K while on high doses of warfarin, showing elevations of prothrombin time on doses of vitamin K to which normal individuals failed to respond. Resistance to bishydroxycoumarin and the indanedione anticoagulant phenindione, but not to heparin, was also shown in the patient.

The existence of an enzyme or receptor site with altered affinity for vitamin K or for anticoagulant drugs was considered by O'REILLY et al. (1968) to be the mechanism responsible for resistance to warfarin in this patient. Because it accounted both for decreased responsiveness to anticoagulants and for increased responsiveness to vitamin K, the former mechanism was favored, since the great responsiveness of the patient to vitamin K would be difficult to explain by altered affinity for the coumarin anticoagulants.

How a genetic defect can affect the way drugs act on the body rather than the way the body transforms or metabolizes drugs is exemplified by the resistance to coumarin anticoagulants in this patient. In an enzyme system responsible for the synthesis of clotting Factors II, VII, IX, and X (O'REILLY et al., 1968), coumarin anticoagulants are antimetabolites that function by competing with the natural substrate, vitamin K, for receptor sites. Much higher concentrations of anticoagulant would be required to compete effectively with vitamin K if the receptor site is so altered that its avidity for vitamin K is enhanced.

A second large kindred involving 18 patients with warfarin resistance in 2 generations was reported in 1970 by O'REILLY. The pedigree revealed autosomal dominant transmission of the trait. The affected individuals were exquisitely sensitive to vitamin K, though the metabolism of sodium warfarin was normal. The earlier interpretation that warfarin resistance resulted from a mutation of the receptor site for vitamin K and oral anticoagulants was further substantiated by the data from this kindred (O'REILLY, 1970).

2. G6PD Deficiency, Primaquine Sensitivity or Favism

G6PD (glucose-6-phosphate dehydrogenase) deficiency, the most common hereditary enzymatic abnormality in man, is transmitted as an X-linked incomplete dominant. Actually a series of disorders, G6PD deficiency involves more than 80 physicochemically discrete variants of the molecule, each variant being associated with a slightly different clinical picture (MOTULSKY et al., 1971). Thus, G6PD deficiency illustrates the important principle of genetic heterogeneity of conditions initially considered homogeneous. Observations of new, chemically distinct mutations affecting the same protein often follow the initial discovery that a certain structural alteration in an enzyme produces a particular clinical syndrome. The new mutation may have clinical signs and symptoms quite different from those accompanying the old mutation. As with certain mutations affecting the hemoglobin molecule, these clinical differences may be so marked that they form clearly separable disease states, each requiring a different therapy and possessing a different prognosis. For example, a mild self-limited anemia is associated with the commonly encountered Negro variant of G6PD in which the drugs listed in Table 1 can continue to be given without danger, since only the susceptible older red cells are removed from the circulation by hemolysis; these are rapidly replaced by resistant younger cells. In various Mediterranean G6PD variants, however, the lesion is more severe; hemolysis affects a larger proportion of the total erythrocyte population and occurs more rapidly after administration of smaller doses of drugs. Drug administration, therefore, should not be continued after the advent of hemolysis in patients with the Mediterranean form of G6PD deficiency. Separation of the 80 chemically distinct mutations affecting the G6PD molecule according to the severity of clinical symptoms is useful for these reasons. The severity of enzyme deficiency has been associated with the type of G6PD variant by MOTULSKY et al. (1971). Several properties characterize the variants, including the total G6PD activity in erythrocytes, electrophoretic mobility of the G6PD, Kms for NADP and G6P, rate of utilization of substrate analogues, and heat stability. In addition, the specific amino acid substitution in G6PD A^+ and in G6PD Hektoen (YOSHIDA, 1970) was elucidated by microfingerprinting techniques (YOSHIDA, 1967). Tables 3 to 6 relate the severity of hemolysis to the amount of G6PD activity in erythrocytes. In the common Negro type of deficiency, only erythrocytes older than 50 days are hemolyzed. However, in the Mediterranean type, a larger fraction of the erythrocytes is destroyed due to the fact that in these variants the mutation is expressed by severe reduction of G6PD activity in relatively young cells. Sufficient G6PD activity to withstand the oxidant effect of drugs is present in the young erythrocytes of Negro variants.

The many contributions to our understanding of this important pharmacogenetic disorder need not be restated here, as several reviews of G6PD deficiency have been published (MOTULSKY, 1964; VESELL, 1969; MOTULSKY et al., 1971; MOTULSKY, 1972). G6PD deficiency has been dealt with previously as a prototype X-linked condition confirming the Lyon hypothesis and as a balanced polymorphism conferring a selective advantage in certain environments by way of protection against severe falciparum malaria.

Several important problems concerning this disorder remain unresolved despite numerous genetic studies on different G6PD variants, including the exact biochemical mechanisms by which drug administration causes hemolysis and development of a successful *in vitro* test to determine the potential of new drugs for producing hemolysis in subjects possessing different G6PD variants. The

Table 3. *Variants with very mild or no* G6PD *deficiency*

Variant	Population origin	Population frequency
Inhambane	African Bantu	Rare
Steilacoom	Negro	
A^+	Negro	Common
Levadia	Greek	
Lourenzo Marques	African Bantu	Rare
King County	Negro	Rare
Thessaly	Greek	
Karditsa	Greek	
Western	Greek	
Manjacaze	African Bantu	Rare
Baltimore-Austin	Negro	Rare
Ijebu-Ode	Negro	Rare
Minas Gerais	Brazilian	Rare
Tacoma	Negro	
Madrona	Negro	Rare
Ibadan-Austin	Negro	Rare
Ita-Bale	Negro	Rare

From MOTULSKY et al. (1971) with permission.

Table 4. *Variants with moderate to mild* G6PD *deficiency*

Variant	Population origin	Population frequency
Barbieri	Italian	Rare
Puerto Rico	Puerto Rican	Rare
A^-	Negro	Common
Constantine	Arab	Common
Taipei-Hakka	Chinese	
Kabyle	Algerian	
Chibuto	Negro-Bantu	Rare
Melissa	Greek	
Canton	South Chinese	
Columbus	Negro	Rare
Athens	Greek	Common
Washington	Negro	
Benevento	Italian	
West Bengal	Asiatic Indian	Rare
Mexico	Mexican	
Seattle	Welsh-Scottish	Rare
Kerala	Asiatic S.E. Indian	Rare
Tel Hashomer	Tunisian Jew	Rare
Capetown	Cape Colored/Norwegian	

From MOTULSKY et al. (1971) with permission.

factors resulting in drug-induced hemolytic anemia remain somewhat obscure, despite numerous studies of the dilemma. Though metabolized in a normal fashion by the body, the drugs themselves, or perhaps their hydroxylated metabolites, cause damage because of increased fragility of G6PD-deficient erythrocytes. Through glutathione reductase, which by means of NADPH, regenerates GSH from oxidized glutathione (GSSG), erythrocytes normally withstand oxidative compounds and maintain their glutathione in a reduced state. Both G6PD and 6-phosphogluconate dehydrogenase (6-PGD) form NADPH; these two enzymes

Table 5. *Variants with severe* G6PD *deficiency*

Variant	Population origin	Population frequency
Hualien-Chi	Taiwan	
San Juan	Puerto Rican	Rare
Markham	New Guinea	Common
Union	Filipino	Common
Teheran	Iran	
Hualien	Taiwan	
Indonesia	Indonesia	
Camplellpur	Pakistani	Common
Mediterranean	Greek, Sardinian, Sephardic Jew, Asiatic	Common
Corinth	Greeks, S.E. Asian	May be common
Panay	Filipino	May be common
Orchomenos	Greek	Common
Lifta	Iraqui Jew	Rare
Carswell	Irish	Rare

From MOTULSKY et al. (1971) with permission.

Table 6. *Variants with severe* G6PD *deficiency and chronic non-spherocytic hemolytic anemia*

Variant	Population origin	Population frequency
Ohio	Italian	Rare
Torrance	U.S.	
Bat-Yam	Iraqui Jew	Rare
Albuquerque	U.S. White	Rare
Bangkok	Thai	
Oklahoma	West Europe	Rare
Duarte	U.S. White	Rare
Hong Kong	Chinese	
Chicago	West Europe	Rare
Tripler	U.S. White	
Alhambra	Finnish/Swedish	
Milwaukee	Puerto Rican White	Rare
Ramat-Gan	Iraque Jew	Rare
Ashdod	North African Jew	Rare
Freiburg	German	Rare
Worcester	U.S. White	

From MOTULSKY et al. (1971) with permission.

are the first in the hexose monophosphate shunt or phosphogluconate oxidative pathway. Only 10% of the total metabolic energy of the erythrocyte is provided by this oxidative route, the major part being accounted for by the EMBDEN-MYERHOF pathway. Such factors as the pH of the suspending medium, the G6PD activity and the rate of NADPH oxidation appreciably alter the relative rates of glycolysis. The amount of NADPH available to the erythrocyte and, thereby, the concentration of GSH are reduced by deficiency of G6PD activity. Apparently, GSH is necessary to maintain sulfhydryl groups on critical proteins of the erythrocytes in a reduced state (BARRON and SINGER, 1943); GSH deficiency with normal G6PD activity causes a non-spherocytic, congenital hemolytic anemia with drug sensitivity (OORT et al., 1961; PRINS et al., 1963). The importance of GSH in the metabolic economy of the erythrocytes would thus be indicated. Conversely, by complexing most of the GSH in erythrocytes with *N*-ethylmaleimide, red cell survival, however, is not decreased (JACOB and JANDL, 1962).

The effect on mechanical fragility of exposing normal and G6PD deficient erythrocytes to various compounds and their metabolites suggested the following sequence of events in drug-induced hemolysis (FRASER and VESELL, 1968; FRASER et al., 1971a and b): first, the drug is metabolized to a product more susceptible to further oxidation. The erythrocyte converts this metabolite to an oxidant intermediate. The oxidant intermediate then damages the erythrocyte membrane (particularly in old cells), perhaps by oxidation of smaller sulfhydryl groups (WEED and REED, 1966). An index of this membrane damage is mechanical fragility. Although splenic sequestration also probably participates as an additional mechanism (WEED and REED, 1966), fragmentation of old erythrocytes ("hemolysis") in the circulation no doubt results from this membrane damage.

The metabolism of the erythrocyte is unusual in that it must function without the benefit of a nucleus. Apparently, the red cell cannot synthesize protein but does synthesize certain simpler substances such as GSH, DPN, and ATP. However, it requires energy sources for maintaining concentration gradients of sodium and potassium, and for continual reduction of methemoglobin.

This energy source is provided by the glycolytic and oxidative pathways of glucose metabolism. Certain enzymes, including G6PD, lose activity as the normal cell ages (BEUTLER, 1966). G6PD activity declines with age at a faster than normal rate in G6PD deficient cells (MARKS and GROSS, 1959). Therefore, older cells of individuals possessing mutations of their G6PD are more susceptible to hemolysis than younger cells.

The view that older cells are more vulnerable to the hemolytic action of drugs than are younger ones is corroborated by clinical studies of individuals bearing various G6PD mutations. A mild form of disorder is observed in some Negroes. Though no hemolysis occurs for two or three days when a Negro subject receives 30 mg primaquine daily, the urine may turn black. In more severe cases there may be weakness, abdominal and back pain, icterus, and black urine, but further signs may not develop. HEINZ (inclusion) bodies may appear in the erythrocytes (BEUTLER, 1966). Reticulocytosis and anemia supervene. This "acute hemolytic phase" is followed in about a week by the "recovery phase" (even in the face of continued drug administration) (BEUTLER, 1966). Though the drug is still administered, the patient feels improved and the abnormalities noted above regress. Although the mechanical fragility is increased, the Coombs' test is negative and the osmotic fragility is normal (FRASER and VESELL, 1968). A change in composition of the erythrocyte population has been held responsible for this refractory state, rather than any change in the metabolism of the drug or in the reactivity of the erythrocytes, as in immunological phenomena. With their higher G6PD activities, the younger cells resist the osmotic and oxidant effects of various drugs and their metabolites (DERN et al., 1954), while the older, more sensitive cells with their greater relative deficiency have been eliminated. A similar, but more severe, clinical course is evident in subjects with Mediterranean G6PD deficiency. The anemia is not self-limited in some patients (SALVIDIO et al., 1963); in others, it is (LARIZZA et al., 1958). The spectrum of drugs causing hemolysis in Caucasians is wider than in Negro subjects and includes fava beans and chloramphenicol (BEUTLER, 1966).

Determining *in vitro* the hemolytic potential of new drugs continues to be of prime importance. A number of tests have been suggested, including measurement of the mechanical fragility of normal and G6PD deficient erythrocytes (FRASER and VESELL, 1968; FRASER et al., 1971a and b), and measurement of

$^{14}CO_2$ evolution from glucose-1-^{14}C in erythrocytes removed from normal volunteers, both before and after primaquine ingestion (WELT et al., 1971). Ingestion of primaquine significantly increased $^{14}CO_2$ evolution. This technique might eventually prove useful, even though the degree of stimulation was variable and small, and should be investigated in G6PD deficient individuals. Any test for hemolytic potential *in vivo* founded wholly on studies *in vitro* fails to take into account the diversity of *in vivo* mechanisms that also depend on events such as splenic sequestration. Measurement of changes in the survival of G6PD-deficient-radioactively labelled cells transfused into G6PD normal subjects before and after administration of the new drug is a technically sound, but somewhat dangerous, test obviously not suitable for the routine screening of new drugs.

Hemolysis apparently may occur spontaneously or during infection in certain G6PD variants; obviously enough stress can be placed on the metabolism of G6PD deficient erythrocytes to cause hemolysis by several environmental alterations in addition to those produced by drug administration. The action of hepatic microsomal drug-metabolizing enzymes in converting alkaloids present in foods to more oxidant compounds capable of damaging erythrocyte membranes should be investigated with this in mind; during fever, perhaps the activity of these enzymes is increased and higher concentrations of oxidant metabolites develop (SONG et al., 1971).

3. Drug-Sensitive Hemoglobins

A life-threatening hemolytic anemia developed in a two year old girl and her father after administration of sulfa drugs; studies of these patients resulted in the discovery of a new hemoglobin and a new pharmacogenetic entity (HITZIG et al., 1960; FRICK et al., 1962). At age 2, after receiving sulfadimethoxine for a fever of unknown origin, the girl developed severe hemolysis. She had experienced a milder episode at 7 months of age when sulfonamides were given for an ear infection. The father had repeated but mild occurrences of jaundice and dark urine since childhood, episodes not invariably associated with drug administration. After receiving a sulfonamide for dysuria, he had a severe hemolytic crisis.

Both father and daughter registered an abnormal hemoglobin – 20 to 30% of the total pigment with electrophoretic mobility between that of hemoglobins A and S. An especially important abnormality was discovered in the β-chain, arginine substituted for the usual histidine residue at the sixty-third position, where the heme group is attached to the β-chain (MULLER and KINGMA, 1961; HUISMAN et al., 1961).

Fifteen of the 65 relatives examined showed the abnormal hemoglobin, designated hemoglobin Zürich, a defect transmitted as an autosomal dominant trait (FRICK et al., 1962). The half-life of transfused erythrocytes was calculated to be 11 days instead of the normal 120. The transfused cells disappeared rapidly after administration of either sulfonamides or primaquine (FRICK et al., 1962). In another family discovered in Maryland with the same substitution at the sixty-third position of the β chain, the severity of the hemolytic episodes was less than in the Swiss cases (RIEDER et al., 1965).

Another drug-sensitive hemoglobin designated hemoglobin H is a special form of α-thalassemia. Composed of four β chains (BAGLIONI, 1963), hemoglobin H is sensitive to the oxidant drugs described under G6PD deficiency. In certain regions of the world, such as Thailand, the frequency of hemoglobin H disease that is transmitted as an autosomal recessive is high, reaching one of 300 individuals born in Bangkok. The levels of hemoglobin H in affected subjects vary from

5 to 30%. Since hemoglobin H is almost incapable of carrying oxygen at physiological tensions, subjects with a high proportion of it are at a great disadvantage. In hemoglobin H disease, α chain synthesis is considerably slower than β chain synthesis.

4. Taste of Phenylthiourea or Phenylthiocarbamide (PTC)

The ability to taste phenylthiourea (also called phenylthiocarbamide or PTC) is transmitted as an autosomal dominant trait. Since inability to taste PTC is inherited as an autosomal recessive trait, tasters are either heterozygous or homozygous (SNYDER, 1932; BLAKESLEE, 1932; KALMUS and HUBBARD, 1960). This polymorphism was discovered in 1932 when FOX, who synthesized the compound, noted that he could not detect a bitter taste from dust of the compound arising as it was poured into a container, whereas a colleague working in the same room complained of the bitter taste (FOX, 1932).

By making 14 serial dilutions of PTC with water, HARRIS and KALMUS (1950a) developed a refined test for PTC-tasting and quantitated thresholds. Tumblers containing PTC in increasing concentrations were alternated with tumblers of water. Females detect PTC in greater dilutions than males; tasting sensitivity decreases with age; and various other compounds containing the N–C=S grouping also exhibit a bimodality in taste perception so that the N–C=S group in PTC seems to be responsible for such differences (HARRIS and KALMUS, 1950b).

In several ways, ability to taste PTC is related to thyroid disease. PTC administration can produce goiter in the rat (RICHTER and CLISBY, 1942). Compounds related to PTC by possessing the N–C= S group, such as the antithyroid drugs methyl and propylthiouracil, also show the same bimodality in taste perception exhibited by subjects to PTC. HARRIS et al. (1949) found that 41% of 134 patients with nodular goiter were nontasters, an observation confirmed by KITCHIN et al. (1959) in 447 individuals submitted to thyroidectomy for various reasons. In male patients with multiple thyroid adenomas, a marked increase in nontasting frequency was noted. Cyclic changes of thyroid involution and hyperplasia occurring with the menstrual cycle in females may be a contributing factor in thyroid disease and, in fact, may mask the actual connection of PTC-tasting and thyroid disorders (KITCHEN et al., 1959). Markedly low frequencies of nontasters occur in toxic diffuse goiter in patients of either sex (KITCHEN et al., 1959). Apparently, non-tasters are more susceptible to athyreotic cretinism (FRASER, 1961; SHEPARD and GARTLER, 1960) and also to adenomatous goiter. Tasters develop toxic diffuse goiter more frequently than nontasters.

A compound containing the grouping S–C= N, which is goitrogenic, has been isolated from turnip, cabbage, brussels sprout, kale, and rape (GREER, 1957; CLEMENTS and WISHART, 1956). The product containing this grouping (which is closely related to the N–C= S grouping of PTC) has been identified as 1-5-vinyl-2-thiooxazolidone, a compound generated by an enzyme in the plant from an inactive precursor. Though cooking reduces the concentration of the goitrogen, during the winter cattle may consume several of these plants and transmit the goitrogen in their milk (CLEMENTS and WISHART, 1956).

Compared with persons in the general population, patients with open angle glaucoma are more often PTC nontasters and those with anglo closure glaucoma are less often PTC nontasters (BECKER and MORTON, 1964).

The frequency of tasting capacity shows geographical variation: 31.5% of Europeans (SALDANHA and BECAK, 1959), 10.6% of Chinese, and 2.7% of Africans are nontasters (BARNICOT, 1950). The reasons are obscure for these geographical variations in the gene for PTC tasting. Tasters and nontasters reveal no differences in capacity to metabolize methylthiouracil and thiopentone (PRICE EVANS et al., 1962). The need for further study of this enigma is apparent. Perhaps there is relevance in the discoveries that patients with adrenal cortical insufficiency show more sensitivity than normal to taste and to olfaction, sensitivities that can be returned to normal by the administration of carbohydrate-active steroids (HENKIN et al., 1963; HENKIN and BARTTER, 1966).

5. Responses of Intraocular Pressure to Steroids: Relationship to Glaucoma

A polymorphism exists in the response of ocular pressure of normal subjects to topical steroids (ARMALY, 1968). Elevations in intraocular pressure in 80 normal individuals after local administration of 0.1% ophthalmic solution of dexamethasone 21-phosphate exhibit a trimodal distribution (ARMALY, 1968). The steroid was applied to the right eye daily for 4 weeks, ocular pressure was measured weekly, and elevations in pressure after 4 weeks were determined. The trimodal distribution of individuals in the extent of their increases in intraocular pressure over this 4-week period is shown in Table 7. The existence of three genotypes: P^LP^L for low elevations of 5 mm Hg or less, P^LP^H for intermediate increases from 6 to 15 mm Hg, and P^HP^H for high increment in pressure of 16 or more mm Hg is suggested by a genetic hypothesis and confirmed by family studies.

Table 7. *Genotype classification of dexamethasone hypertension and frequency distributions in different clinical categories*[a]

Category	No. of subjects tested	Percent frequency of genotypes Low (P^LP^L) ΔP < 6 mm Hg	Intermediate (P^LP^H) ΔP 6—15 mm Hg	High (P^HP^H) ΔP > 15 mm Hg
Limits of pressure rise (mm Hg)		5 or less	6—15	16 or more
Mean pressure rise (mm Hg)		1.96	10.0	19.5
Standard deviation (mm Hg)		± 2.00	± 2.5	[b]
Genotype		P^LP^L	P^LP^H	P^HP^H
Random sample	80	66%	29%	5%
Open-angle hypertensive glaucoma	33	6%	48%	44%
Low-tension glaucoma	15	7%	53%	40%
Normal eye in recessed-angle glaucoma	15	—	53%	47%
Normal eye in angle recession without glaucoma	4	75%	25%	—

[a] From ARMALY (1968).
[b] Range in sample 18—22 mm Hg.

In 1968, ARMALY indicated an association between certain types of response and glaucoma. In a sample of patients with open-angle hypertensive glaucoma and also with low-tension glaucoma, the distribution of responses (Table 7) differed from that in the random sample of normal subjects shown in Table 7 (ARMALY, 1968). A marked reduction in P^LP^L genotypes and a corresponding increase in P^LP^H and P^HP^H genotypes occurred (Table 7) in both conditions and surprisingly in the uninvolved eye of patients with unilateral post-traumatic glaucoma. Family studies indicated that the response of high elevations of intraocular pressure after dexamethasone administration was inherited as an autosomal recessive trait. While affliction with glaucoma can occur with genotypes other than P^HP^H and P^HP^L, ARMALY concluded that the P^H gene is closely associated with the development of the types of glaucoma listed in Table 7.

Genetic factors have been described that determine chamber angles (KELLERMAN and POSNER, 1955) and chamber depths (TÖRNQUIST, 1953) in acute angle-closure glaucoma (also called narrow-angle glaucoma). These same studies showed that environmental factors can exert appreciable effects on chamber depths. Dilatation of the pupils can precipitate acute attacks of glaucoma in individuals who have inherited narrow chambers (GRANT, 1955). In genetically susceptible individuals, several mydriatic agents can produce these acute attacks. Among these drugs are adrenergic compounds such as epinephrine, phenylephrine, ephedrine, and cocaine; in addition, drugs that block the effects of cholinergic nerves, such as atropine and scopolamine, also may cause an attack of angle-closure glaucoma (KALOW, 1962).

6. Malignant Hyperthermia with Muscular Rigidity

DENBOROUGH et al. (1962) reported that fatal hyperthermia had occurred in 10 of the 38 members of a family who had received anesthesia for various surgical procedures. This was the first indication that the rare, hitherto seemingly sporadic, malignant hyperthermia afflicting individuals exposed to various anesthetic agents might be genetically transmitted. Almost 200 cases of malignant hyperthermia have been identified and shown to be hereditary in basis (BRITT and KALOW, 1970; KALOW, 1971, 1972). The condition appears to be transmitted as an autosomal dominant and is associated with muscular rigidity (KALOW, 1972). It appears during anesthesia with nitrous oxide, methoxyflurane, halothane, ether, cyclopropane or combinations of these, and is more common in association with the use of succinylcholine as a preanesthetic agent. During anesthesia body temperature rises rapidly, occasionally reaching 112°.

The incidence of malignant hyperthermia is in the range of 1 in 20,000 cases of general anesthesia, exhibits no sex preference but occurs more in young than in older anesthetized patients (BRITT and KALOW, 1970; KALOW, 1972). Approximately two-thirds of the cases end fatally, usually with cardiac arrest. Observed consistently are tachycardia, tachypnea, hypoxia, respiratory and metabolic acidosis, hyperkalemia, and hypocalcemia (BRITT and KALOW, 1970). The degree of rigidity is variable, differing from patient to patient and sometimes being absent. This variability may indicate that the term malignant hyperthermia refers to several discrete disease states. Occasionally, rigidity is so marked that the body becomes as stiff as a board, progressing without interruption into rigor mortis. Recently, intravenous procaine or procainamide has been reported to relieve the rigidity and fever in certain cases of malignant hyperthermia (KALOW, 1972). The facts that curare does not relieve the rigidity and that a limb under tourniquet does not become rigid suggest the existence of a peripheral rather than a central lesion. The basic mechanism responsible for the lesion in malignant hyperthermia with rigidity has been related to the hypersensitivity of skeletal muscle from affected individuals to caffeine and the relief of the rigidity by procaine; apparently intracellular calcium stores in the muscles of these subjects are abnormally inhibited by those general anesthetics that produce the rigidity and fever (KALOW, 1972).

Animal models have been produced in dogs treated with halothane and dinitrophenol (GATZ et al., 1970) and in landrace pigs (KALOW, 1972). In man, there seems to be an association between the rigidity in malignant hyperthermia and the previous existence of certain muscular disorders. Muscle biopsies from patients with malignant hyperthermia exhibit enhanced response to caffeine and halothane.

III. Atypical Liver Alcohol Dehydrogenase

Atypical alcohol dehydrogenase (ADH), a variant of the enzyme that metabolizes ethanol, has been described in man (VON WARTBURG and SCHÜRCH, 1968). The atypical enzyme occurs in sufficiently high frequencies in Swiss and English populations to be designated a polymorphism. Exceptionally active, the variant occurred in 20% of 59 liver specimens from a Swiss population and in 4% of 50 livers from an English population.

In the pH rate profiles, the ratio of the activity at pH 10.8 to that at pH 8.8 is greater than 1 for the normal enzyme and less than 1 for the atypical enzyme. Whereas pyrazol inhibits the atypical more than the normal ADH, a chelator of zinc in the ADH molecule, *o*-Phenanthroline, inhibits the normal more than the atypical ADH. The substrates *N*-butanol, benzyl alcohol, and cyclohexanol are oxidized faster by the normal ADH than by the atypical ADH.

On agar gel electrophoresis or ion exchange column chromatography, three distinct peaks of both normal and atypical ADH occur (VON WARTBURG and SCHÜRCH, 1968; BLAIR and VALLEE, 1966). The ADH isozyme patterns of individual livers vary considerably in distribution of total ADH activity among the three bands; only two bands are evident in some livers. The three normal and atypical ADH isozymes have approximately the same electrophoretic mobility at pH 9.0 on agar gels (VON WARTBURG and SCHÜRCH, 1968).

Factors in the development of alcoholism attributable to differences in the atypical and normal ADH have yet to be established. The two enzymes showed wide variation in their rates of ethanol metabolism in these studies, but the mode of inheritance of the trait should be explored by family studies.

Possibly because another factor such as reoxidation of coenzyme 1 becomes rate-limiting, the atypical enzyme enhances alcohol metabolism by only 40 to 50% *in vivo*, although exhibiting 5- to 6-fold more ADH activity than an equal amount of normal ADH *in vitro* (VON WARTBURG and SCHÜRCH, 1968). Acetaldol (β-hydroxybutyraldehyde) is reduced five times faster by the atypical ADH than by the normal ADH.

After intravenous infusion of ethanol, attempts were made to correlate rates of degradation of the drug with liver ADH typed from biopsies obtained at surgery (EDWARDS and PRICE EVANS, 1967). Of 23 subjects, 2 had atypical ADH: capacity to metabolize alcohol was no different in the male subject with atypical ADH from that in males with typical ADH, whereas capacity to degrade ethanol was greater in the female subject with atypical ADH than in a small group of females who had typical ADH. The question of whether individuals with atypical ADH possess increased capacity to degrade ethanol and, possibly, to resist alcoholic cirrhosis of the liver remains unresolved by this study.

IV. Ethanol Metabolism in Various Racial Groups

For a long time it had been suspected that racial groups varied considerably in their ability to metabolize ethanol and that individuals from far eastern countries had less ethanol-metabolizing capacity than had subjects from western countries. Until now, however, such impressions had not been documented. It is also possible that genetic differences might be obscured by elevated rates of ethanol metabolism as a result of chronic ethanol ingestion. LIEBER and DE CARLI (1968, 1970) established induction of ethanol metabolism by its chronic administration. Many of the environmental factors influencing ethanol metabolism have been resolved by a recent study wherein a careful selection of subjects took into account differences in drinking habits (FENNA et al., 1971). After an overnight fast, ethanol in a final concentration of 10% was infused intravenously in normal saline solution at 8:00 a.m. When a blood ethanol concentration of approximately 125 mg-% was attained, infusions were stopped. Male volunteers – 21 Eskimo, 26 Indian and 17 white – were used for the study. Some were patients in Canadian hospitals recovering from acute infections or fractures. The rate of decrease in blood ethanol among the whites (0.370 mg-%/min) was much higher than in either the Indians (0.259 mg-%/min) or Eskimos (0.264 mg-%/min). Furthermore, the rate of ethanol metabolism was much higher in whites (0.1449 g/kg/h) than in Indians (0.1013 g/kg/h) or in Eskimos (0.1098 g/kg/h). A tendency toward adaptation to ethanol was evident in each ethnic group, but significantly so only in the Indians. The faster rate of ethanol metabolism in whites was attributed to genetic factors, since neither previous intake of ethanol nor general diet appeared to be responsible (FENNA et al., 1971).

Recently, additional evidence has been adduced to document racial differences in ethanol sensitivity. WOLFF (1972) reported that Japanese, Taiwanese, and Koreans exhibited marked facial flushing and mild to moderate symptoms of intoxication after drinking amounts of ethanol that produced no detectable effect on Caucasians. These differences in ethanol responsiveness, present since birth, were attributed to variations in autonomic reactivity (WOLFF, 1972).

V. Correlation of Certain Genetic Factors with Adverse Reactions to Various Drugs

Correlating certain genetic factors in individuals with their risk of contracting disease has resulted in the discovery of unusual AB0 blood type distributions in patients with disorders such as gastric carcinoma, duodenal ulcer, rheumatic heart disease, pernicious anemia, and diabetes mellitus, among others (VOGEL and KRUGER, 1968; VOGEL et al., 1969). To determine possible correlations between certain kinds of adverse reactions to drugs and various genetic factors for which 4500 of the patients have been typed, JICK et al. (1969) and LEWIS et al. (1971) have surveyed large populations of hospitalized patients (7000) in Boston. In young women developing venous thromboembolism while on oral contraceptives, they discovered a significant deficit of blood Group 0 individuals relative to those possessing Groups A and AB combined. A significantly lower risk of developing a thromboembolism exists for women of blood Group 0 who take antifertility agents than for women of blood Group A or AB. Perhaps their reduced level of antihemophilic globulin, in comparison to subjects with blood Group A, and their greater risk of developing bleeding with peptic ulcers, in comparison with ulcer patients with A, B or AB blood groups, have a relationship to the reduced liability to thromboembolism of individuals with blood Group 0. The JICK group (1972) reported a correlation between AB0 blood groups and the development of arrhythmias after digoxin, with a decreased risk occurring in 0 patients relative to non-0 patients.

A correlation between adverse reactions to prednisone and low serum albumin concentrations has also been established by the Boston group (LEWIS et al., 1971). The amount of prednisone that can be bound to this protein is correspondingly reduced when albumin levels are low and more of the drug is free and active, thereby increasing chances of side effects from the drug.

Several possible applications are offered by attempts to correlate genetic factors and adverse drug reactions. The availability of such correlations may furnish important clues as to the pathogenesis of these aberrant responses to drugs in addition to alerting physicians to the increased risk of drug toxicity in patients who are of certain genotypes.

VI. Reduced Drug Binding Capacity in Fetal and Newborn Blood

Studies on the binding of sulfaphenazole to fetal, neonatal and adult human plasma albumin are presented as an example of the need to resist the attribution to genetic factors of alterations in drug response produced by environmental changes (CHIGNELL et al., 1971). When compared to adult albumin prepared in a similar fashion, fetal and neonatal albumin purified by electrophoretic separation and extensive dialysis exhibited reduced sulfaphenazole binding. This unusual binding behavior suggested the existence of a structurally and hence genetically distinct fetal albumin. However, other experiments revealed that the binding properties developed not from physiochemical differences in fetal and adult albumin but because in the fetus, and for several months after birth, albumin binds very avidly to certain compounds, probably either bilirubin, fatty acids or both. After the albumin was passed over charcoal to remove these compounds, its binding properties were identical to those of adult albumin (CHIGNELL et al., 1971).

VII. Variation among Individuals in Rate of Drug Elimination

A. Genetic Control of Variations in Drug Clearance

Drug accumulation to toxic levels in certain individuals who are unable to eliminate the usual doses of a therapeutic agent as fast as normal individuals is a major cause of adverse reactions. Twenty years ago, studies disclosed large variations in the rates at which healthy, nonmedicated volunteers clear antipyrine (SOBERMAN et al., 1949), bishydroxycoumarin (WEINER et al., 1950), and phenylbutazone (BURNS et al., 1953) from their plasma. After a standard dose of the drug, plasma concentrations of chlorpromazine (CURRY and MARSHALL, 1968), propranolol (SHAND et al., 1970), warfarin (O'REILLY and AGGELER, 1970), diphenylhydantoin (KUTT et al., 1964b), or procainamide (KOCH-WESER, 1971) vary widely among normal subjects. If a given drug were administered to a population in the same dose and by the same route, it formerly was assumed that all subjects would develop similar blood and receptor site concentrations of the drug and, hence, that all subjects would respond similarly to the therapeutic agent. However, the principle of wide variations in the disposition of and responsiveness to therapeutic agents is now being proclaimed, in light of these large individual differences in rates of elimination of many drugs (SOBERMAN et al., 1949; WEINER et al., 1950; BURNS et al., 1953; SHAND et al., 1970; O'REILLY and AGGELER, 1970; KUTT et al., 1964a and b; KOCH-WESER, 1971). Practicing physicians have long respected this principle of pharmacologic individuality.

The application of an old technique recently elucidated the mechanism responsible for large individual variations in plasma half-lives of several commonly used drugs. Sir FRANCIS GALTON (1875) introduced the twin method to estimate the relative contribution of environmental and genetic factors to a trait. Based on a comparison for any given trait of the intratwin differences for identical and fraternal twins, the method assumes environmental equality. Intratwin differences should be of similar magnitude in identical and fraternal twins for traits controlled primarily by environmental factors. There should be a difference in the magnitude of intratwin differences for traits controlled primarily by genetic factors, however, since identical twins share all their genes while fraternal twins have in common, on the average, only half of their genes. Thus, intratwin differences should be less in identical twins than in fraternal twins if a trait is appreciably influenced by genetic factors. The following fraction (NEEL and SCHULL, 1954)

$$\frac{\text{variance within pairs of fraternal twins} - \text{variance within pairs of identical twins}}{\text{variance within pairs of fraternal twins}}$$

provides a range of values from 0, indicating negligible hereditary and complete environmental control, to 1, indicating virtually complete hereditary influence. On the basis of extensive, systematic investigations in experimental animals (GILLETTE, 1971; VESELL, 1968), appreciable environmental contributions to individual differences in drug metabolism were expected in man. Exposure to inducing agents, degree of health or illness, and hormonal or nutritional status are known to alter the rates at which animals metabolize certain drugs. Responsiveness in mice to a drug such as hexobarbital differs according to age, sex, litter, painful stimuli, ambient temperature, degree of crowding, time of day of drug administration, and type of bedding (VESELL, 1968). In man, such experiments would imply that environment is an important factor in variations in drug metabolism.

In healthy, nonmedicated twins, almost complete genetic and negligible environmental control of large individual variations in metabolism of phenylbutazone (VESELL and PAGE, 1968a) and antipyrine (VESELL and PAGE, 1968b) was observed (Figs. 4 and 5). For the reasons just developed these results were unexpected in human studies. In the formula given above, the value for hereditary control was 0.99 and 0.98 for phenylbutazone and antipyrine, respectively. Furthermore, predominantly genetic control over large individual differences in the metabolism of bishydroxycoumarin (VESELL and PAGE, 1968c), halothane (CASCORBI et al., 1971), ethanol (VESELL et al., 1971a), and nortriptyline (ALEXANDERSON et al., 1969) was established in twin studies. Although these investigations demonstrated that in healthy, nonmedicated subjects an appreciable genetic influence existed over large variations in drug elimination, the twin studies alone did not definitively reveal the mode of inheritance of these variations in drug elimination. Distribution curves of the twin data offer clues to the mode of inheritance (Fig. 6). Twin data suggest a polymodal distribution curve (BÖNICKE and LISBOA, 1957) for isoniazid excretion in urine, indicating single factor inheritance, later firmly established independently by family studies. Furthermore, the twins who would be most genetically informative if family studies were to be performed later are suggested by the distribution curve of urinary excretion of isoniazid. The distribution curves for antipyrine and nortriptyline are closer to the unimodal, continuous curves observed in polygenically controlled traits, in contrast to the polymodal curve for isoniazid excretion. The results were in substantial agreement with this genetic hypothesis when family studies were performed for bishydroxycoumarin (MOTULSKY, 1964), nortriptyline (ÅSBERG et al., 1971), and phenylbutazone (WHITTAKER and PRICE-EVANS, 1970). A significant

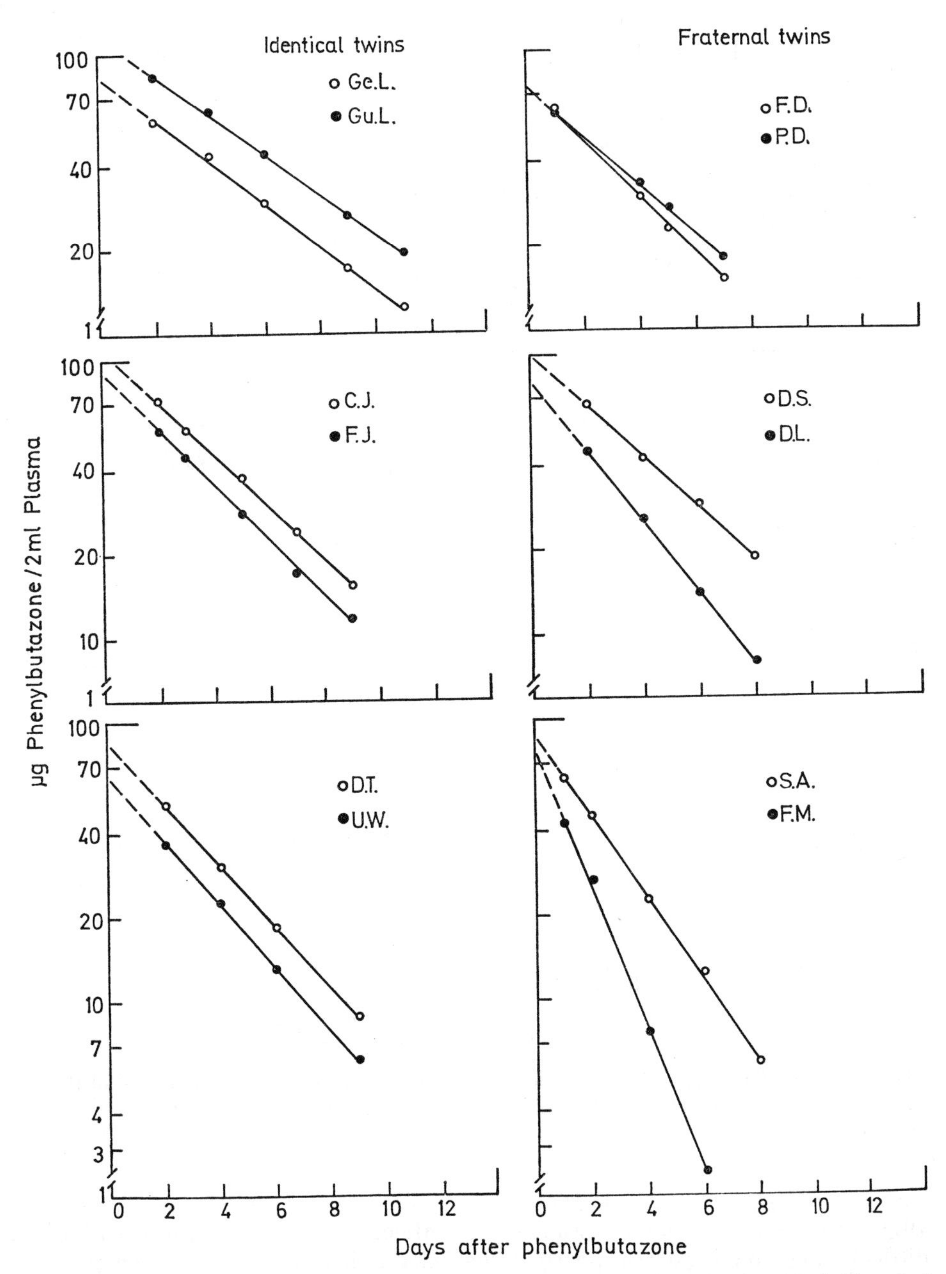

Fig. 4. Decline of phenylbutazone in the plasma of three sets of identical twins (left) and of three sets of fraternal twins (right) after a single oral dose of 6 mg/kg

regression of mean offspring value on mid-parent value was revealed by the family studies, a result consistent with polygenic control.

For several reasons, family studies of drug elimination are less satisfactory than twin studies: 1) Metabolism rates of certain therapeutic agents change

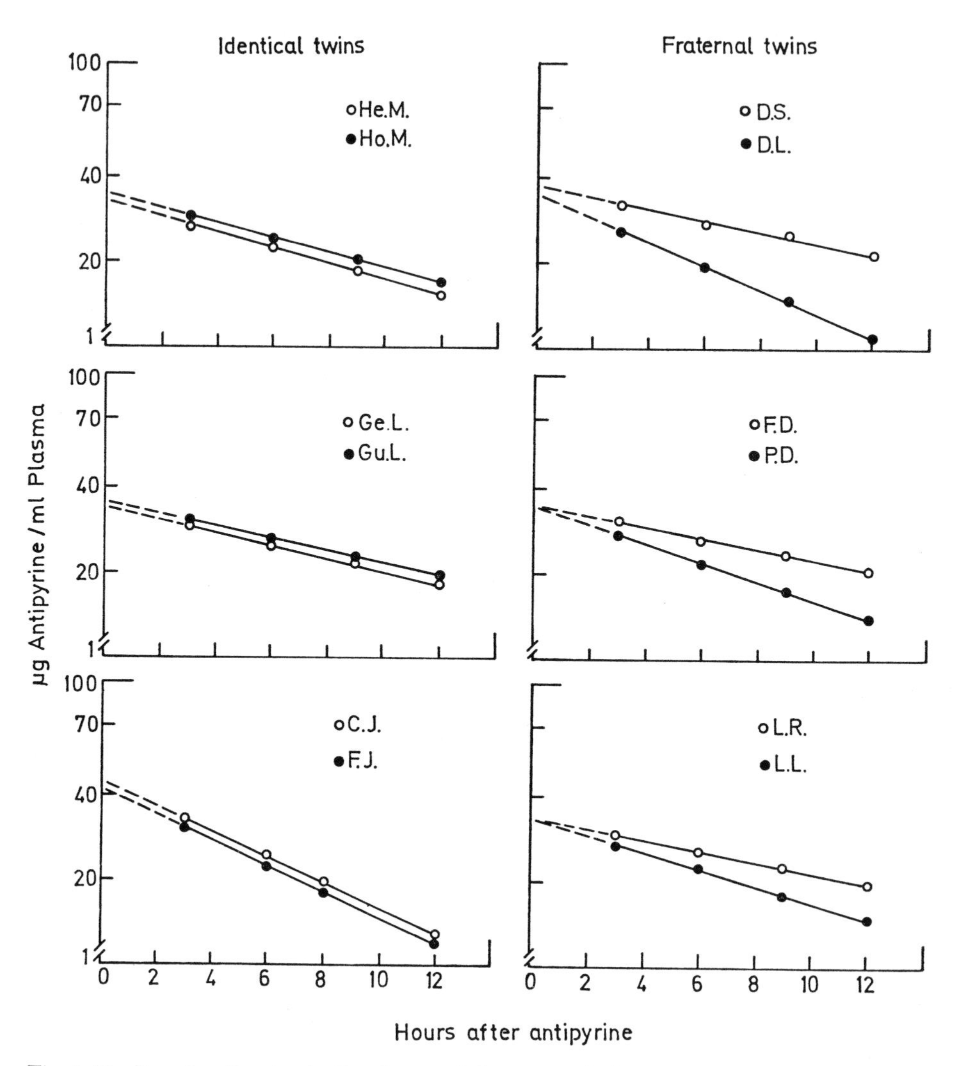

Fig. 5. Decline of antipyrine in the plasma of three sets of identical twins (left) and of three sets of fraternal twins (right) after a single oral dose of 18 mg/kg

with age (O'MALLEY et al., 1971) and vary according to sex (O'MALLEY et al., 1971). The results of family studies are difficult to interpret because of the difficulty in correcting for the poorly defined change in drug metabolizing capacity with age, whereas twin studies are by definition age corrected. 2) In experimental animals or in man, rates of drug metabolism have been altered by such environmental constituents as caffeine (MITOMA et al., 1969), nicotine (WENZEL and BROADIE, 1966), 3-methylcholanthrene (CONNEY et al., 1971), 3,4-benzpyrene (CONNEY et al., 1971) and various insecticides (CONNEY et al., 1971). Therefore, changes in drug-metabolizing capacity observed in family studies could result partially from the closer environmental similarity of children compared to parents. In one family study, the correlation that occurred between healthy, non-medicated

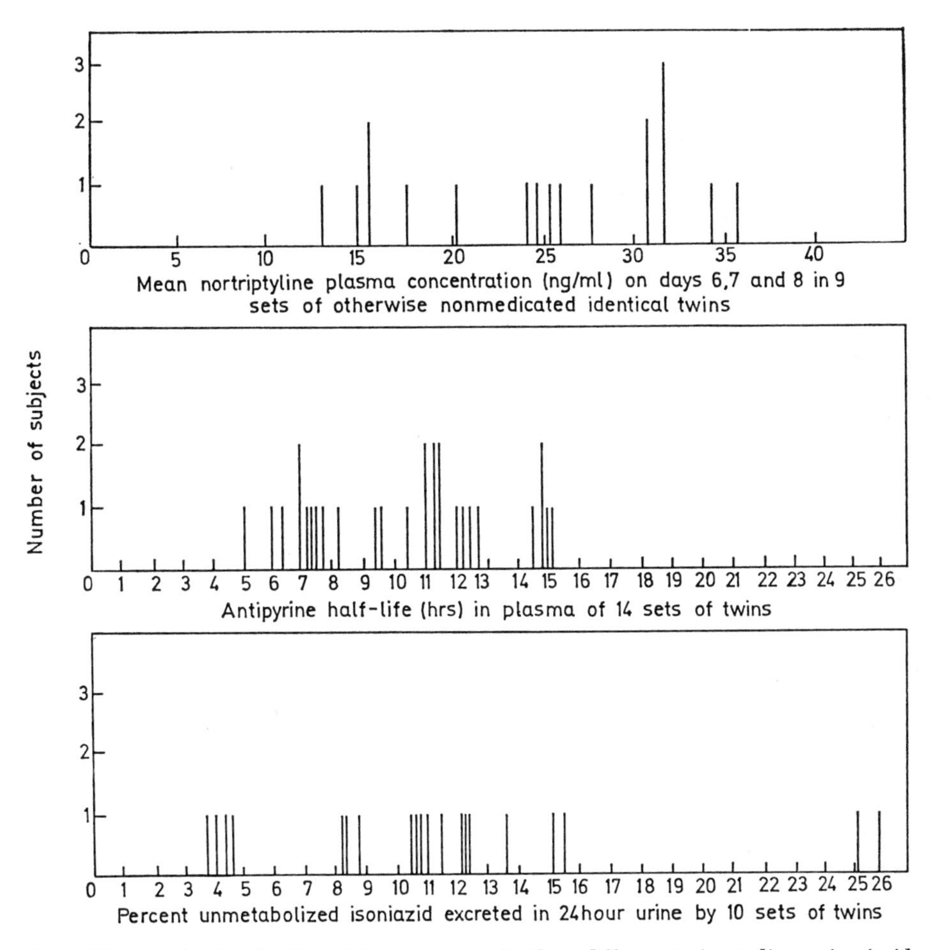

Fig. 6. Plots of the distribution of drug responses in three different twin studies on isoniazid, antipyrine, and nortriptyline

husbands and wives in plasma half-lives of phenylbutazine may be explained by the influence exerted by numerous environmental constituents on drug-metabolizing capacity (WHITTAKER and PRICE EVANS, 1970).

B. Environmental Effects on Drug Action and Genetic Control of their Expression

Several potentially useful implications are evident in the genetic control of large variations among healthy, non-medicated volunteers in rates of drug elimination. 1) Because rates of drug elimination are genetically, rather than environmentally, controlled in healthy, non-medicated subjects, they should be highly reproducible, stable values. This has been proven for several drugs. 2) As a guide to adjusting dosage according to individual requirements, determination of drug half-lives in plasma might be ascertained prior to long-term drug administration, thereby helping to reduce frequent occurrences of toxicity or undertreatment encountered when the same dose of a drug is given to all subjects.

3) Other simpler, more convenient and rapid biochemical tests for ascertaining the drug-metabolizing capacity of an individual might be developed for clinical purposes, since determination of drug half-lives in plasma is a laborious, time-consuming task requiring 4 or 5 venipunctures. Two procedures proposed for this purpose – urinary glucaric acid concentrations (HUNTER et al., 1971) and urinary ratios of cortisol to 6-8-hydroxycortisol (KUNTZMAN et al., 1968) – have certain problems requiring resolution before they can be widely utilized. 4) Correlations might exist within an individual in his capacity to metabolize various drugs. Drugs could be grouped into categories if such correlations existed. An individual's rate of handling all the drugs in a category could be readily calculated, once an individual's rate of metabolism of one drug in a category were ascertained. 5) Physicians should watch for the introduction of environmental factors that might change patients' genetically determined basal drug biotransforming capacity. For example, long-term administration of many drugs either alone or in combination; environmental exposure on a large scale to such agents as insecticides, nicotine, 3,4-benzpyrene and caffeine; and several diseases affecting cardiovascular, renal or hepatic status, can influence the duration and intensity of drug action. Sites for drug absorption, distribution, receptor interaction, biotransformation and excretion can be affected by these and other environmental factors. Furthermore, either synergistic or antagonistic effects can be produced by interactions occurring simultaneously at several of these sites. Nor is there full awareness that the nature and extent of drug interactions vary considerably according to the dose of drug, the route of administration and the length of time for which the therapeutic agents are given. Conversely, some drugs produce inhibitory effects on drug metabolism for 6 h after their administration, and enhancement (induction) 24 h later (CONNEY, 1967). Only limited clinical applicability of the genetic studies might be suggested by these multiple potential sources of environmental interference with the predominantly genetic control of variations in rates of drug elimination. Indeed, in severely ill patients, measurements of drug concentrations in serum have been advocated as a practical means of assessing the role exerted by continually shifting environments on drug metabolizing capacity (VESELL and PASSANANTI, 1971; KOCH-WESER, 1972). Several investigators have suggested that genetic factors also operate to some extent in regulating the magnitude of drug interactions and that drug metabolism within an individual may be a more uniform process, even for chemically unrelated drugs, than previously suspected. Search was made for a pattern to the rates at which a normal, nonmedicated individual cleared different drugs from his blood. Correlations between the plasma half-lives of bishydroxycoumarin and phenylbutazone were established by one study (VESELL and PAGE, 1968c); correlations between steady state blood concentrations of desmethylimipramine and nortriptyline and the plasma half-life of oxyphenylbutazone were reported in another study (HAMMER et al., 1969); a highly significant correlation between the plasma half-life of antipyrine measured after several doses and the plasma half-life of phenylbutazone was established by a third study (DAVIES and THORGEIRSSON, 1971). Therefore, for certain chemically unrelated drugs, a common rate-limiting step may exist in the several reactions required for drug oxidation within liver microsomes. The surprising correlations recently observed in rates of elimination of several chemically unrelated drugs could be the result of an enzyme, possibly cytochrome P-450 reductase, catalyzing a rate-limiting step common in the oxidation of several drugs.

The extent of individual differences in response to an inducing agent is another pharmacogenetic problem concerning drug interactions – particularly induction that has been utilized as one approach to the treatment of hyperbilirubinemia (Yaffe et al., 1966). An almost 2-fold range in the extent of induction (Vesell and Page, 1969) was revealed in a study in which phenobarbital was administered daily in an oral dose of 2 mg/kg for 14 days. As indicated by the small intratwin differences in identical twins but large differences within fraternal twinships, these large individual differences in inducibility were almost exclusively under genetic control (Vesell and Page, 1969). During the two week period of its administration, blood concentrations of phenobarbital were similar within a twinship so that differences in induction among fraternal twins could not be attributed to different blood concentrations of phenobarbital. The initial plasma half-life of antipyrine and the extent of its shortening by chronic phenobarbital administration were found to be related. The longer the initial plasma antipyrine half-life before phenobarbital administration, the greater the phenobarbital-induced shortening (Vesell and Page, 1969; Breckenridge and Orme, 1971). Once an individual's basal uninduced drug-metabolizing activity is ascertained, the degree to which induction can increase his drug-metabolizing capacity thus appears to be predictable.

Few therapeutic agents inhibit drug metabolism, although several hundred compounds may act as inducing agents (Conney, 1967). Under certain circumstances, however, recent studies show that the following drugs have prolonged antipyrine or bishydroxycoumarin metabolism in man: disulfiram (Vesell et al., 1971 b), allopurinol (Vesell et al., 1970), nortriptyline (Vesell et al., 1970), and L-dopa (Vesell et al., 1971 c). Individual differences reaching a 10-fold range were observed in the extent to which chronic administration of the same dose of inhibiting drug prolonged antipyrine half-lives in the healthy, non-medicated volunteers studied. How genetic and environmental factors compare in producing these individual differences in inhibition remain to be determined. Undoubtedly, many other drugs currently in use inhibit drug metabolism; physicians should be aware of the possibility that this form of interaction has the potential of causing toxicity through drug accumulation. The reproducibility of an individual's response to an inhibiting agent remains unknown, as do the roles of genetic and environmental factors in regulating the magnitude of inhibition. Further studies are needed to illuminate these areas.

References

Aebi, H., Heiniger, J.P., Bütler, R., Hässig, A.: Two cases of acatalasia in Switzerland. Experientia (Basel) **17**, 466 (1961).

Aebi, H.E.: Inborn errors of metabolism. Ann. Rev. Biochem. **36**, 271—306 (1967a).

Aebi, H.E.: The investigation of inherited enzyme deficiencies with special reference to acatalasia. Proc. 3rd Intern. Congr. Hum. Genet. Chicago 1966, 189—205 (1967b).

Aebi, H., Baggiolini, M., Dewald, B., Lauber, E., Suter, H., Micheli, A., Frei, J.: Observations in two Swiss families with acatalasia. II. Enzymol. Biol. Clin. (Basel) **4**, 121—151 (1964).

Aggeler, P.M., O'Reilly, R.A., Leong, L., Kowitz, P.E.: Potentiation of anticoagulant effect of warfarin by phenylbutazone. New Engl. J. Med. **276**, 496—501 (1967).

Alarcón-Segovia, D., Fishbein, E., Alcala, H.: Isoniazid acetylation rate and development of antinuclear antibodies upon isoniazid treatment. Arthr. and Rheum. **14**, 748—752 (1971).

Alexanderson, B., Price Evans, D.A., Sjöqvist, F.: Steady-state plasma levels of nortriptyline in twins: influence of genetic factors and drug therapy. Brit. med. J. **1969** IV, 764—768.

Armaly, M.F.: Genetic factors related to glaucoma. Ann. N. Y. Acad. Sci. **151**, 861—875 (1968).

Armstrong, A.R., Peart, H.E.: A comparison between the behavior of Eskimos and Non-Eskimos to the administration of isoniazid. Amer. Rev. resp. Dis. **81**, 588—594 (1960).

ÅSBERG, M., PRICE-EVANS, D. A., SJÖQVIST, F.: Genetic control of nortriptyline kinetics in man: a study of relatives of propositi with high plasma concentrations. J. Med. Genet. **8**, 129—135 (1971).
BAGLIONI, C.: In: TAYLOR, J. H. (Ed.): Molecular genetics. New York: Academic Press 1963.
BARNICOT, N. A.: Taste deficiency for phenylthiourea in African Negroes and Chinese. Ann. Eugen. **15**, 248—254 (1950).
BARRON, E. S. G., SINGER, T. P.: Enzyme systems containing active sulfhydryl groups. The role of glutathione. Science **97**, 356—358 (1943).
BAUR, E. W.: Catalase abnormality in a Caucasian family in the United States. Science **140**, 816—817 (1963).
BECKER, B., MORTON, W. R.: Phenylthiourea taste testing and glaucoma. Arch. Ophthal. **72**, 323—327 (1964).
BEUTLER, E.: In: STANBURY, J. G., WYNGAARDEN, J. B., FREDERICKSON, D. S. (Eds.): The metabolic basis of inherited disease. New York: McGraw-Hill 1966.
BIEHL, J. P.: The role of the dose and the metabolic fate of isoniazid in the emergence of isoniazid resistance. Trans. 15th Conf. Chemother. Tuberc., pp. 729—782 (1956).
BIEHL, J. P.: Emergence of drug resistance as related to the dosage and metabolism of isoniazid. Trans. 16th Conf. Chemother. Tuberc., pp. 108—113 (1957).
BLAIR, A. H., VALLEE, B. L.: Some catalytic properties of human liver alcohol dehydrogenase. Biochemistry **5**, 2026—2034 (1966).
BLAKESLEE, A. F.: Genetics of sensory thresholds: taste for phenyl thio carbamide. Proc. natl. Acad. Sci. (Wash.) **18**, 120—130 (1932).
BÖNICKE, R., REIF, W.: Enzymatische Inaktivierung von Isonicotinsäurehydrazid im menschlichen und tierischen Organismus. Naunyn Schmiedebergs Arch. exp. Path. Pharmak. **220**, 321—333 (1953).
BÖNICKE, R., LISBOA, B. P.: Über die Erbbedingtheit der intraindividuellen Konstanz der Isoniazidausscheidung beim Menschen (Untersuchungen an eineiigen und zweieiigen Zwillingen.) Naturwissenschaften **44**, 314 (1957).
BOURNE, J. G., COLLIER, H. O. J., SOMERS, G. F.: Succinylcholine (Succinoylcholine): muscle-relaxant of short action. Lancet 262, 1225—1229 (1952).
BOVET, D., BOVET-NITTI, F., GUARINO, S., LONGO, V. G., MAROTTA, M.: Proprieta farmacodinamiche di alcuni derivati della succinilcolina dotati di azione curarica. Esteri di trialchiletanolammonio di acidi bicarbossilici alifatici. Rendic. Ist. Super. Sanita **12**, 106—137 (1949).
BOVET-NITTI, F.: Degradazione di alcune sostanze curarizzanti per azione di colinesterasi. Rendic. Ist. Super. Sanita **12**, 138—157 (1949).
BRECKENRIDGE, A., ORME, M.: Clinical implications of enzyme induction. Ann. N. Y. Acad. Sci. **179**, 421—431 (1971).
BRENNAN, R. W., DEHEJIA, H., KUTT, H., MCDOWELL, F.: Diphenylhydantoin intoxication due to slow inactivation of isoniazid. Neurology (Minneap.) **18**, 283 (1968).
BRITT, B. A., KALOW, W.: Malignant hyperthermia: aetiology unknown! Canad. Anesth. Soc. J. **17**, 316—330 (1970).
BURNS, J. J., ROSE, R. K., CHENKIN, T., GOLDMAN, A., SCHULERT, A., BRODIE, B. B.: The physiological disposition of phenylbutazone (Butazolidin) in man and a method for its estimation in biological material. J. Pharm. exp. Ther. **109**, 346—357 (1953).
BUSH, G. H., ROTH, F.: Muscle pains after suxamethonium chloride in children. Brit. J. Anesth. **33**, 151—155 (1961).
BUTLER, T. C.: The metabolic hydroxylation of phenobarbital. J. Pharmacol. exp. Ther. **116**, 326—336 (1956).
BUTLER, T. C.: The metabolic conversion of 5,5-diphenyl hydantoin to 5-(*p*-hydroxyphenyl)-5-phenyl hydantoin. J. Pharmacol. exp. Ther. **119**, 1—11 (1957).
CARLSON, H. B, ANTHONY, E. M., RUSSELL, W. F., JR., MIDDLEBROOK, G.: Prophylaxis of isoniazid neuropathy with pyridoxine. New Engl. J. Med. **255**, 118—122 (1956).
CASCORBI, H. F., VESELL, E. S., BLAKE, D. A., HELRICH, M.: Halothane biotransformation in man. Ann. N. Y. Acad. Sci. **179**, 244—248 (1971).
CHIGNELL, C. F., VESELL, E. S., STARKWEATHER, D. K., BERLIN, C. M.: The binding of sulfaphenazole to fetal, neonatal and adult human plasma albumin. Clin. Pharmacol. Ther. **12**, 897—904 (1971).
CHRISTENSEN, F.: Crystallization and preliminary characterization of a dicoumarol metabolite in the faeces of dicoumarol-treated rats. Acta pharmacol. (Kbh.) **24**, 232—242 (1966).
CLARK, S. W., GLAUBIGER, G. A., LADU, B. N.: Properties of plasma cholinesterase variants. Ann. N. Y. Acad. Sci. **151**, 710—722 (1968).
CLEMENTS, F. W., WISHART, J. W.: A thyroid-blocking agent in the etiology of endemic goiter. Metabolism **5**, 623—639 (1956).

CLUFF, L. E., THORNTON, G., SEIDL, L., SMITH, J.: Epidemiological study of adverse drug reactions. Trans. Ass. Amer. Phycns. 78, 255—268 (1965).
COHEN, S. N., WEBER, W. W.: Pharmacogenetics. Pediat. Clin. N. Amer. **19**, 21—36 (1972).
CONNEY, A. H.: Pharmacological implications of microsomal enzyme induction. Pharm. Rev. **19**, 317—366 (1967).
CONNEY, A. H., WELCH, R., KUNTZMAN, R., CHANG, R., JACOBSON, M., MUNROFAURE, A. D., PECK, A. W., BYE, A., POLAND, A., POPPERS, P. J., FINSTER, M., WOLFF, J. A.: Effects of environmental chemicals on the metabolism of drugs, carcinogens, and normal body constituents in man. Ann. N. Y. Acad. Sci. **179**, 155—172 (1971).
DAVIES, D. S., THORGEIRSSON, S. S.: Mechanism of hepatic drug oxidation and its relationship to individual differences in rates of oxidation in man. Ann. N. Y. Acad. Sci. **179**, 411—420 (1971).
DAVIES, R. O., MARTON, A. V., KALOW, W.: The action of normal and atypical cholinesterase of human serum upon a series of esters of choline. Canad. J. Biochem. **38**, 545—551 (1960).
DENBOROUGH, M. A., FORSTER, J. F. A., LOVELL, R. R. H., MAPLESTONE, P. A., VILLIERS, J. D.: Anaesthetic deaths in a family. Brit. J. Anaesth. **34**, 395—396 (1962).
DERN, R. J., BEUTLER, E., ALVING, A. S.: The hemolytic effect of primaquine. II. The natural course of the hemolytic anemia and the mechanism of its self-limited character. J. Lab. clin. Med. **44**, 171—176 (1954).
DEVADATTA, S., GANGADHARAM, P. R. J., ANDREWS, R. H., FOX, W., RAMAKRISHNAN, C. V., SELKON, J. B., VELU, S.: Peripheral neuritis due to isoniazid. Bull. WHO **23**, 587—598 (1960).
Editorial: Rats become resistant to anticoagulants. WHO Chron. **20**, 29—31 (1966).
EDWARDS, J. A., PRICE EVANS, D. A.: Ethanol metabolism in subjects possessing typical and atypical liver alcohol dehydrogenase. Clin. Pharmacol. Ther. 8, 824—829 (1967).
ELIAS, R. A.: In: NICHOL, E. S. (Ed.): Anticoagulant therapy in ischemic heart disease. New York: Grune and Stratton 1965.
ELLARD, G. A., GAMMON, P. T., WALLACE, S. M.: The determination of isoniazid and its metabolites acetylisoniazid, monoacetylhydrazine, diacetylhydrazine, isonicotinic acid and isonicotinylglycine in serum and urine. Biochem. J. **126**, 449—458 (1972).
EVANS, F. T., GRAY, P. W. S., LEHMANN, H., SILK, E.: Sensitivity to succinylcholine in relation to serum-cholinesterase. Lancet **262**, 1229—1235 (1952).
FENNA, D., MIX, L., SCHAEFER, O., GILBERT, J. A. L.: Ethanol metabolism in various racial groups. Canad. med. Ass. J. **105**, 472—475 (1971).
FOLDES, F. F.: Pharmacogenetics. Ann. N. Y. Acad. Sci. **151**, 751—752 (1968).
FORBAT, A., LEHMANN, H., SILK, E.: Prolonged apnoea following injection of succinyldicholine. Lancet **265**, 1067—1068 (1953).
FOX, A. L.: The relationship between chemical constitution and taste. Proc. nat. Acad. Sci. (Wash.) **18**, 115—120 (1932).
FRASER, G. R.: Cretinism and taste sensitivity to phenylthio carbamide. Lancet **1961** I, 964—965.
FRASER, I. M., TILTON, B. E., VESELL, E. S.: Effects of some metabolites of hemolytic drugs on young and old, normal and G6PD-deficient human erythrocytes. Ann. N. Y. **179**, 644—653 (1971a).
FRASER, I. M., TILTON, B. E., VESELL, E. S.: Alterations in normal and G6PD-deficient human erythrocytes of various ages after exposure to metabolites of hemolytic drugs. Pharmacology **5**, 173—187 (1971b).
FRASER, I. M., VESELL, E. S.: Effects of drugs and drug metabolites on erythrocytes from normal and glucose-6-phosphate dehydrogenase-deficient individuals. Ann. N. Y. Acad. Sci. **151**, 777—794 (1968).
FRICK, P. G., HITZIG, W. H., BETKE, K.: Hemoglobin Zurich. I. A new hemoglobin anomaly associated with acute hemolytic episodes with inclusion bodies after sulfonamide therapy. Blood **20**, 261—271 (1962).
GALTON, F.: The history of twins as a criterion of the relative powers of nature and nurture. J. Brit. Anthropol. Inst. **5**, 391—406 (1875).
GANGADHARAM, P. R. J., SELKON, J. B.: The association between response to treatment with isoniazid and the rate of inactivation of isoniazid in South Indian patients with pulmonary tuberculosis. Proc. 16th Int. Tuberc. Conf. **2**, 556—561 (1961).
GATZ, E. E., HULL, M. J., BENNETT, W. G., JONES, J. R.: Effects of pentobarbital upon 2,4 dinitrophenol-induced hyperpyrexia during halothane anesthesia. Fed. Proc. **29**, 483 Abs. (1970).
GILLETTE, J. R.: Metabolism of drugs and other foreign compounds by enzymatic mechanisms. Recent Progr. Brain Res. **6**, 11—73 (1963).
GILLETTE, J. R.: Biochemistry of drug oxidation and reduction by enzymes in hepatic endoplasmic reticulum. Advan. Pharmacol. **4**, 219—261 (1966).

Gillette, J. R.: Factors affecting drug metabolism. Drug metabolism in man. Ann. N. Y. Acad. Sci. **179**, 43—66 (1971).
Goedde, H. W., Altland, K.: Evidence for different "silent genes" in the human serum pseudocholinesterase polymorphism. Ann. N. Y. Acad. Sci. **151**, 540—544 (1968).
Goedde, H. W., Altland, K., Schloot, W.: Therapy of prolonged apnea after suxamethonium with purified pseudocholinesterase: New data on kinetics of the hydrolysis of succinyldicholine and succinylmonocholine and further data on *N*-acetyltransferase-polymorphism. Ann. N. Y. Acad. Sci. **151**, 742—752 (1968).
Gow, J. G., Price Evans, D. A.: A study of the influence of the isoniazid inactivator phenotype on reversion in genito-urinary tuberculosis. Tubercle **45**, 136—143 (1964).
Grant, W. M.: Physiological and pharmacological influences upon intraocular pressure. Pharmacol. Rev. **7**, 143—182 (1955).
Greaves, J. H., Ayres, P.: Heritable resistance to warfarin in rats. Nature (Lond.) **215**, 877—878 (1967).
Greer, M. A.: Goitrogenic substances in food. Amer. J. clin. Nutr. **5**, 440—444 (1957).
Grunberg, E., Leiwant, B., D'Ascensio, I.-L., Schnitzer, R. J.: On the lasting protective effect of hydrazine derivatives of isonicotinic acid in the experimental tuberculosis infection of mice. Dis. Chest. **21**, 369—377 (1952).
Gutsche, B. B., Scott, E. M., Wright, R. C.: Hereditary deficiency of pseudocholinesterase in Eskimos. Nature (Lond.) **215**, 322—323 (1967).
Hamilton, H. B., Neel, J. V.: Genetic heterogeneity in human acatalasia. Amer. J. hum. Genet. **15**, 408—419 (1963).
Hammer, W., Martens, S., Sjöqvist, F.: A comparative study of the metabolism of desmethylimipramine, nortriptyline and oxyphenylbutazone in man. Clin. Pharmacol. Ther. **10**, 44—49 (1969).
Harris, F. W., Cockburn, T.: Alleged poisoning by potatoes. Analyst **43**, 133—137 (1918).
Harris, H., Hopkinson, D. A., Robson, E. B., Whittaker, M.: Genetical studies on a new variant of serum cholinesterase detected by electrophoresis. Ann. Hum. Genet. **26**, 359—382 (1963).
Harris, H., Kalmus, H.: The measurement of taste sensitivity to phenylthiourea (P. T. C.). Ann. Eugen. **15**, 24—31 (1950a).
Harris, H., Kalmus, H.: Chemical specificity in genetic differences of taste sensitivity. Ann. Eugen. **15**, 32—45 (1950b).
Harris, H., Kalmus, H., Trotter, W. H.: Taste sensitivity to phenylthiourea in goitre and diabetes. Lancet 257, 1038—1039 (1949).
Harris, H., Whittaker, M.: Differential inhibition of human serum cholinesterase with fluoride; recognition of two new phenotypes. Nature (Lond.) **191**, 496—498 (1961).
Harris, H., Whittaker, M.: The serum cholinesterase variants. A study of twenty-two families selected via the "intermediate" phenotype. Ann. Hum. Genet. **26**, 59—72 (1962a).
Harris, H., Whittaker, M.: Differential inhibition of the serum cholinesterase phenotypes by solanine and solanidine. Ann. Hum. Genet. **26**, 73—76 (1962b).
Harris, H., Whittaker, M., Lehmann, H., Silk, E.: The pseudocholinesterase variants. Esterase levels and dibucaine numbers in families selected through suxamethonium sensitive individuals. Acta Genet. (Basel) **10**, 1—16 (1960).
Harris, H. W.: Isoniazid metabolism in humans: genetic control, variation among races and influence on the chemotherapy of tuberculosis. Proc. 16th Intern. Tuberc. Conf. **2**, 503—507 (1961a).
Harris, H. W.: High-dose isoniazid compared with standard-dose isoniazid with PAS, in the treatment of previously untreated cavity pulmonary tuberculosis. Trans. 20th Conf. Chemother. Tuberc., 39—68 (1961b).
Harris, H. W., Knight, R. A., Selin, M. J.: Comparison of isoniazid concentrations in the blood of people of Japanese and European descent. Therapeutic and genetic implications. Amer. Rev. Tuberc. **78**, 944—948 (1958).
Henkin, R. I., Bartter, F. C.: Studies on olfactory thresholds in normal man and in patients with adrenal cortical insufficiency: the role of adrenal cortical steroids and of serum sodium concentration. J. clin. Invest. **45**, 1631—1639 (1966).
Henkin, R. I., Gill, J. R., Jr., Bartter, F. C.: Studies on taste thresholds in normal man and in patients with adrenal cortical insufficiency: the role of adrenal cortical steroids and of serum sodium concentration. J. clin. Invest. **42**, 727—735 (1963).
Hitzig, W. H., Frick, P. G., Betke, K., Huisman, T. H. J.: Hemoglobin Zurich: a new hemoglobin anomaly with sulfonamide-induced inclusion body anemia. Helv. paediatr. Acta **15**, 499—514 (1960).
Hodgkin, W. E., Giblett, E. R., Levine, H., Bauer, W., Motulsky, A. G.: Complete pseudocholinesterase deficiency: genetic and immunologic characterization. J. clin. Invest. **44**, 486—493 (1965).

HOLMES, R. S., MASTERS, C. J.: Catalase heterogeneity. Arch. Biochem. Biophys. **109**, 196—197 (1965).
HUGHES, H. B.: On the metabolic fate of isoniazid. J. Pharmacol. exp. Ther. **109**, 444—452 (1953).
HUGHES, H. B., BIEHL, J. P., JONES, A. P., SCHMIDT, L. H.: Metabolism of isoniazid in man as related to the occurrence of peripheral neuritis. Amer. Rev. Tuberc. **70**, 266—273 (1954).
HUGHES, H. B., SCHMIDT, L. H., BIEHL, J. P.: The metabolism of isoniazid; its implications in therapeutic use. Trans. 14th Conf. Chemother. Tuberc. pp, 217—222 (1955).
HUISMAN, T. H. J., HORTON, B., BRIDGES, M. T., BETKE, K., HITZIG, W. H.: A new abnormal human hemoglobin, Hb: Zurich. Clin. chim. Acta **6**, 347—355 (1961).
HUNT, R., TAVEAU, R. DE M.: On the physiological action of certain cholin derivatives and new methods for detecting cholin. Brit. med. J. **1906** II, 1788—1791.
HUNTER, J., MAXWELL, J. D., CARRELLA, M., STEWART, D. A., WILLIAMS, R.: Urinary *d*-glucaric acid excretion as a test for hepatic enzyme induction in man. Lancet **1971** I, 572—575.
IKEDA, M., SEZESNY, B., BARNES, M.: Enhanced metabolism and decreased toxicity of warfarin in rats pretreated with phenobarbital, DDT or chlordane. Fed. Proc. **25**, 417 (1966).
JACOB, H. S., JANDL, J. H.: Effects of sulfhydryl inhibition on red blood cells. II. Studies *in vivo*. J. clin. Invest. **41**, 1514—1523 (1962).
JENNE, J. W.: Partial purification and properties of the isoniazid transacetylase in human liver. Its relationship to the acetylation of *p*-aminosalicylic acid. J. clin. Invest. **44**, 1992—2002 (1965).
JENNE, J. W., MACDONALD, F. M., MENDOZA, E.: A study of the renal clearances, metabolic inactivation rates and serum fall-off interaction of isoniazid and para-aminosalicylic acid in man. Amer. Rev. resp. Dis. **84**, 371—378 (1961).
JICK, H., SLONE, D., WESTERHOLM, B., INMAN, W. H. W., VESSEY, M. P., SHAPIRO, S., LEWIS, G. P., WORCESTER, J.: Venous thromboembolic disease and AB0 blood type. A cooperative study. Lancet **1969** I, 539—542.
JICK, H., SLONE, D., SHAPIRO, S., HEINONEN, O. P., LAWSON, D. H., LEWIS, G. P., JUSKO, W., BALLINGALL, D. L. K., SISKIND, V., HARTZ, S., GAETANO, L. F., MACLAUGHLIN, D. S., PARKER, W. J., WIZWER, P., DINAN, B., BAXTER, C., MIETTINEN, O. S.: Relation between digoxin arrhythmias and AB0 blood groups. Circulation **45**, 352—357 (1972).
KALMUS, H., HUBBARD, S. J.: The chemical senses in health and disease. Springfield: Thomas 1960.
KALOW, W.: Pharmacogenetics: Heredity and the response to drugs. Philadelphia: Saunders 1962.
KALOW, W.: Contribution of hereditary factors to the response to drugs. Fed. Proc. **24**, 1259—1265 (1965).
KALOW, W.: Topics in pharmacogenetics. Ann. N. Y. Acad. Sci. **179**, 654—659 (1971).
KALOW, W.: Succinylcholine and malignant hyperthermia. Fed. Proc. **31**, 1270—1275 (1972).
KALOW, W., DAVIES, R. O.: The activity of various esterase inhibitors towards atypical human serum cholinesterase. Biochem. Pharmacol. **1**, 183—192 (1959).
KALOW, W., GENEST, K.: A method for the detection of atypical forms of human serum cholinesterase. Determination of dibucaine numbers. Canad. J. Biochem. **35**, 339—346 (1957).
KALOW, W., GUNN, D. R.: Some statistical data on atypical cholinesterase of human serum. Ann. Hum. Genet. **23**, 239—250 (1959).
KALOW, W., STARON, N.: On distribution and inheritance of atypical forms of human serum cholinesterase, as indicated by dibucaine numbers. Canad. J. Biochem. **35**, 1305—1320 (1957).
KATTAMIS, C., ZANNOS-MARIOLEA, L., FRANCO, A. P., LIDDELL, J., LEHMANN, H., DAVIES, D.: Frequency of atypical pseudocholinesterase in British and Mediterranean populations. Nature (Lond.) **196**, 599—600 (1962).
KELLERMAN, L., POSNER, A.: The value of heredity in the detection and study of glaucoma. Amer. J. Ophthal. **40**, 681—685 (1955).
KITCHIN, F. D., HOWEL-EVANS, W., CLARKE, C. A., MCCONNELL, R. B., SHEPPARD, P. M.: PTC taste response and thyroid disease. Brit. med. J. **1959** I, 1069—1074.
KNIGHT, R. A., SELIN, M. J., HARRIS, H. W.: Genetic factors influencing isoniazid blood levels in humans. Trans. 18th Conf. Chemother. Tuberc. pp, 52—58 (1959).
KOCH-WESER, J.: Pharmacokinetics of procainamide in man. Ann. N. Y. Acad. Sci. **179**, 370—382 (1971).
KOCH-WESER, J.: Drug therapy: serum drug concentrations as therapeutic guides. New Engl. J. Med. **287**, 5, 227—231 (1972).
KUNTZMAN, R., JACOBSON, M., LEVIN, W., CONNEY, A.H.: Stimulatory effect of *N*-phenylbarbital (phetharbital) on cortisol hydroxylation in man. Biochem. Pharmacol. **17**, 565—571 (1968).

KUTT, H., VEREBELY, K., MCDOWELL, F.: Inhibition of diphenylhydantoin metabolism in rats and in rat liver microsomes by antitubercular drugs. Neurology (Minneap.) **18**, 706—710 (1968).

KUTT, H., WOLK, M., SCHERMAN, R., MCDOWELL, F.: Insufficient parahydroxylation as a cause of diphenylhydantoin toxicity. Neurology (Minneap.) **14**, 542—548 (1964a).

KUTT, H., WINTERS, W., KOKENGE, R., MCDOWELL, F.: Diphenylhydantoin metabolism, blood levels and toxicity. Arch. Neurol. **11**, 642—648 (1964b).

LADU, B. N.: Isoniazid and pseudocholinesterase polymorphisms. Fed. Proc. **31**, 4, 1276—1285 (1972).

LADU, B. N., KALOW, W. (Eds.): Pharmacogenetics. Ann. N. Y. Acad. Sci. **151**, 691—1001 (1968).

LARIZZA, P., BRUNETTI, P., GRIGNANI, F., VENTURA, S.: I fabici sono sensibili alla primachina. Possibilità di provocare nei fabici l'insorgenza di episodi emolitici con la somministrazione di primachina. Minerva Med. **49**, 3769—3773 (1958).

LEHMANN, H., LIDDELL, J.: Genetical variants of human serum pseudocholinesterase. Progr. Med. Genet. **3**, 75—105 (1964).

LEHMANN, H., RYAN, E.: The familial incidence of low pseudocholinesterase level. Lancet **1956** II, 124.

LEHMANN, H., SILK, E.: Succinylmonocholine. Brit. med. J. **1953** I, 767—768.

LEWIS, G. P., JICK, H., SLONE, D., SHAPIRO, S.: The role of genetic factors and serum protein binding in determining drug response as revealed by comprehensive drug surveillance. Ann. N. Y. Acad. Sci. **179**, 729—738 (1971).

LIEBER, C. S., DECARLI, L. M.: Ethanol oxidation by hepatic microsomes: Adaptive increase after ethanol feeding. Science **162**, 917—918 (1968).

LIEBER, C. S., DECARLI, L. M.: Hepatic microsomal ethanol-oxidizing system: *In vitro* characteristics and adaptive properties *in vivo*. J. biol. Chem. **245**, 2505—2512 (1970).

LINK, K. P.: The anticoagulant from spoiled sweet clover hay. Harvey Lect. **39**, 162—216 (1943—1944).

MAYNERT, E. W.: The metabolic fate of diphenylhydantoin in the dog, rat and man. J. Pharmacol. exp. Ther. **130**, 275—284 (1960).

MCKUSICK, V. A.: Mendelian inheritance in man, catalogs of autosomal dominant, autosomal recessive and X-linked phenotypes. 3rd ed. Baltimore: Johns Hopkins Press 1971.

MEYER, H., MALLY, J.: Über Hydrazinderivate der Pyridincarbon-Säuren. Monatsh. Chem. **33**, 393—414 (1912).

MICHELI, A., AEBI, H.: Recherche immunochimique de la catalase érythrocytaire dans l'hémolysat acatalasique. Rev. franc. Etud. clin. biol. **10**, 431—433 (1965).

MITCHELL, R. S., BELL, J. C., RIEMENSNIDER, D. K.: Further observations with isoniazid inactivation tests. Trans. 19th Conf. Chemother. Tuberc. 62—66 (1960).

MITOMA, C., LOMBROZA, L., LEVALLEY, S. E., DEHN, F.: Nature of the effect of caffeine on the drug-metabolizing enzymes. Arch. Biochem. Biophys. **134**, 434—441 (1969).

MOTULSKY, A. G.: Drug reactions, enzymes, and biochemical genetics. J. Amer. med. Ass. **165**, 835—837 (1957).

MOTULSKY, A. G.: Pharmacogenetics. Progr. Med. Genet. **3**, 49—74 (1964).

MOTULSKY, A. G.: Hemolysis in glucose-6-phosphate deficiency. Fed. Proc. **31**, 1286—1292 (1972).

MOTULSKY, A. G., YOSHIDA, A., STAMATOYANNOPOULOS, G.: Variants of glucose-6-phosphate dehydrogenase. Ann. N. Y. Acad. Sci. **179**, 636—644 (1971).

MULLER, C. J., KINGMA, S.: Haemoglobin Zurich: $\alpha 2A\beta 2$-63Arg. Biochim. biophys. Acta (Amst.) **50**, 595 (1961).

NEEL, J. V., SCHULL, W. J.: Human heredity. Chicago: University of Chicago Press 1954.

NEITLICH, H. W.: Increased plasma cholinesterase activity and succinylcholine resistance; a genetic variant. J. clin. Invest. **45**, 380—387 (1966).

NISHIMURA, E. T., CARSON, S. N., KOBARA, T. Y.: Isozymes of human and rat catalases. Arch. Biochem. Biophys. **108**, 452—459 (1964).

NISHIMURA, E. T., HAMILTON, H. B., KOBARA, T. Y., TAKAHARA, S., OGURA, Y., DOI, K.: Carrier state in human acatalasemia. Science **130**, 333—334 (1959).

O'MALLEY, K., CROOKS, J., DUKE, E., STEVENSON, I. H.: Effect of age and sex on human drug metabolism. Brit. med. J. **1971** III, 607—609.

OORT, M., LOOS, J. A., PRINS, H. K.: Hereditary absence of reduced glutathione in the erythrocytes—a new clinical and biochemical entity? (Preliminary communication) Vox sang. **6**, 370—373 (1961).

ORGELL, W. H., VAIDYA, K. A., DAHM, P. A.: Inhibition of human plasma cholinesterase *in vitro* by extracts of solanaceous plants. Science **128**, 1136—1137 (1958).

O'REILLY, R. A.: The second reported kindred with hereditary resistance to oral anticoagulant drugs. New Engl. J. Med. **282**, 1448—1451 (1970).

O'Reilly, R. A., Aggeler, P. M.: Coumarin anticoagulant drugs: hereditary resistance in man. Fed. Proc. **24**, 1266—1273 (1965).

O'Reilly, R. A., Aggeler, P. M.: Determinants of the response to oral anticoagulant drugs in man. Pharmacol. Rev. **22**, 35—96 (1970).

O'Reilly, R. A., Aggeler, P. M., Hoag, M. S., Leong, L. S., Kropatkin, M. L.: Hereditary transmission of exceptional resistance to coumarin anticoagulant drugs. The first reported kindred. New Engl. J. Med. **271**, 809—815 (1964).

O'Reilly, R. A., Pool, J G , Aggeler, P M.: Hereditary resistance to coumarin anticoagulant drugs in man and rat. Ann. N. Y. Acad. Sci. **151**, 913—931 (1968).

Perry, H. M., Jr., Chaplin, H., Jr., Carmody, S., Haynes, C., Frei, C.: Immunologic findings in patients receiving methyldopa: a prospective study. J. Lab. clin. Med. **78**, 905—917 (1971).

Perry, H. M., Jr., Sakamoto, A., Tan, E. M.: Relationship of acetylating enzyme to hydralazine toxicity. Proc. Central Soc. Clin. Res., Fortieth Annual Meeting, Chicago **40**, 81—82 (1967).

Perry, H. M., Jr., Tan, E. M., Carmody, S., Sakamoto, A.: Relationship of acetyl transferase activity to antinuclear antibodies and toxic symptoms in hypertensive patients treated with hydralazine. J. Lab. clin. Med. **76**, 114—125 (1970).

Peters, J. H.: Relationship between plasma concentration and urinary excretion of isoniazid. Trans. 18th Conf. Chemother. Tuberc. 37—45 (1959).

Peters, J. H.: Studies on the metabolism of isoniazid. 1. Development and application of a fluorometric procedure for measuring isoniazid in blood. Amer. Rev. resp. Dis. **81**, 485—497 (1960a).

Peters, J. H.: Studies on the metabolism of isoniazid. 2. The influence of para-aminosalicylic acid on the metabolism of isoniazid by man. Amer. Rev. resp. Dis. **82**, 153—163 (1960b).

Peters, H. J., Gordon, G. R., Brown, P.: Studies on the metabolism of isoniazid in subhuman primates. Proc. Soc. exp. Biol. (N. Y.) **120**, 575—579 (1965a).

Peters, J. H., Miller, K. S., Brown, P.: The determination of isoniazid and its metabolites in human urine. Anal. Biochem. **12**, 379—394 (1965b).

Price, V. E., Greenfield, R. E.: Liver catalase. II. Catalase fractions from normal and tumor-bearing rats. J. Biol. Chem. **209**, 363—376 (1954).

Price Evans, D. A.: Pharmogenetique. Med. et Hyg. (Geneve) **20**, 905—909 (1962).

Price Evans, D. A.: Individual variations of drug metabolism as a factor in drug toxicity. Ann. N. Y. Acad. Sci. **123**, 178—187 (1965).

Price Evans, D. A.: Genetic variations in the acetylation of isoniazid and other drugs. Ann. N. Y. Acad. Sci. **151**, 723—733 (1968).

Price Evans, D. A., Kitchin, F. D., Riding, J. E.: The metabolism of methyl thiouracil and thiopentone in tasters and non-tasters of P. T. C. Ann. Hum. Genet. **26**, 126—133 (1962).

Price Evans, D. A., Manley, K. A., McKusick, V. A.: Genetic control of isoniazid metabolism in man. Brit. med. J. **1960 II**, 485—491.

Price Evans, D. A., White, T. A.: Human acetylation polymorphism. J. Lab. clin. Med. **63**, 394—403 (1964).

Prins, H. K., Oort, M., Loos, J. A., Zurcher, C., Beckers, T.: Proc. 9th Congr. Eur. Soc. Haematol. Lisbon II/1, 721 (1963).

Richter, C. P., Clisby, K. H.: Toxic effects of the bitter-tasting phenylthiocarbamide. Arch. Path. **33**, 46—57 (1942).

Rieder, R. F., Zinkham, W. H., Holtzman, N. A.: Hemoglobin Zürich: clinical, chemical, and kinetic studies. Amer. J. Med. **39**, 4—20 (1965).

Roberts, J. A. F.: Multifactorial inheritance and human disease. Progr. Med. Genet. **3**, 178—216 (1964).

Robitzek, E. H., Selikoff, I. J., Ornstein, G. G.: Chemotherapy of human tuberculosis with hydrazine derivatives of isonicotinic acid. Quart. Bull. Sea View Hosp. **13**, 27—51 (1952).

Robson, J. M., Sullivan, F. M.: Antituberculosis drugs. Pharmacol. Rev. **15**, 169—223 (1963).

Ross, R. R.: Use of pyridoxine hydrochloride to prevent isoniazid toxicity. J. Amer. med. Ass. **168**, 273—275 (1958).

Saldanha, P. H., Becak, W.: Taste thresholds for phenylthiourea among Ashkenazic Jews. Science **129**, 150—151 (1959).

SALVIDIO, E., PANNACCIULLI, I., TIZIANELLO, A.: Proc. 9th Congr. Eur. Soc. Haematol. Lisbon II/1, 707 (1963).

SCHMIEDEL, A.: The problem of rapid isoniazid inactivation. Proc. 16th Int. Tuberc. Conf. **2**, 508—512 (1961).

SCHROEDER, W. A., SHELTON, J. R., SHELTON, J. B., OLSON, B. M.: Some amino acid sequences in bovine-liver catalase. Biochim. biophys. Acta (Amst.) **89**, 47—65 (1964).

SHAHIDI, N. T.: Acetophenetidin sensitivity. Amer. J. Dis. Child. **113**, 81—82 (1967).

SHAND, D. G., NUCKOLLS, E. M., OATES, J. A.: Plasma propranolol levels in adults with observations in four children. Clin. Pharmacol. Ther. **11**, 112—120 (1970).

SHEPARD, T. H., II, GARTLER, S. M.: Increased incidence of non-tasters of phenylthiocarbamide among congenital athyreotic cretins. Science **131**, 929 (1960).

SHIBATA, Y., HIGASHI, T., HIRAI, H., HAMILTON, H. B.: Immunochemical studies on catalase. II. An anticatalase reacting component in normal, hypocatalasic, and acatalasic human erythrocytes. Arch. Biochem. Biophys. **118**, 200—209 (1967).

SMIRNOV, G. A., KOZULITZINA, T. I.: Interrelation between toxic effect and metabolic patterns of phtivazid. Vop. Med. Khim. **8**, 401—406 (1962).

SMITH, W. K.: The alleged protective action of alfalfa against the hemorrhagic sweet clover disease. J. Agr. Res. **59**, 211—215 (1939).

SNYDER, L. H.: Studies in human inheritance. IX. The inheritance of taste deficiency in man. Ohio J. Sci. **32**, 436—440 (1932).

SOBERMAN, R., BRODIE, B. B., LEVY, B. B., AXELROD, J., HOLLANDER, V., STEELE, J. M.: The use of antipyrine in the measurement of total body water in man. J. biol. Chem. **179**, 31—42 (1949).

SOLOMON, H. M.: Variations in metabolism of coumarin anticoagulant drugs. Ann. N. Y. Acad. Sci. **151**, 932—935 (1968).

SOLOMON, H. M., SCHROGIE, J. J.: The effect of phenyramidol on the metabolism of bishydroxycoumarin. J. Pharmacol. Exp. Ther. **154**, 660—666 (1966).

SONG, C. S., GELB, N. A., WOLFF, S. M.: Fever and drug metabolism: influence of pyrogen-induced fever on salicylamide metabolism in man. Clin. Res. **19**, 468 (1971).

SONG, C. S., GELB, N. A., WOLFF, S. M.: Influence of pyrogen-induced fever on salicylamide metabolism in man. J. clin. Invest. **51**, 2959—2966 (1972).

SUNAHARA, S.: Genetical, geographical, and clinical studies of isoniazid metabolism. Proc. 16th Int. Tuberc. Conf. **2**, 513—541 (1961).

SZEINBERG, A., BAR-OR, R., SHEBA, C.: In: GOLDSCHMIDT, E. (Ed.): The genetics of migrant and isolate populations. Baltimore: Williams and Wilkins 1963.

SZEINBERG, A., PIPANO, S., OSTFELD, E., EVIATOR, L.: The silent gene for serum cholinesterase. J. Med. Genet. **3**, 190—193 (1966).

TAKAHARA, S.: Progressive oral gangrene probably due to lack of catalase in the blood (acatalasemia). Report of nine cases. Lancet 263, 1101—1104 (1952).

TAKAHARA, S.: Progressive oral gangrene due to acatalasemia. Laryngoscope **64**, 685—688 (1954).

TAKAHARA, S., DOI, K.: Statistical observation on 35 cases of acatalasemia appearing in literature. Japan. J. Otol. **61**, 1727—1736 (1958).

TAKAHARA, S., DOI, K.: Statistical study of acatalasemia (a review of thirty-eight cases appearing in the literature). Acta Med. Okayama **13**, No. 1 (1959).

TAKAHARA, S., HAMILTON, H. B., NEEL, J. V., KOBARA, T. Y., OGURA, Y., NISHIMURA, E. T.: Hypocatalasemia: a new genetic carrier state. J. clin. Invest. **39**, 610—619 (1960).

TAKAHARA, S., SATO, H., DOI, M., MIHARA, S.: Acatalasemia. III. On the heredity of acatalasemia. Proc. Japan. Acad. **28**, 585—595 (1952).

TANFORD, C., LOVRIEN, R.: Dissociation of catalase into subunits. J. Amer. chem. Soc. **84**, 1892—1896 (1962).

THORUP, O. A., JR., CARPENTER, J. T., HOWARD, P.: Human erythrocyte catalase: demonstration of heterogeneity and relationship to erythrocyte ageing *in vivo*. Brit. J. Haematol. **10**, 542—550 (1964).

TÖRNQUIST, R.: Shallow anterior chamber in acute glaucoma: A clinical and genetic study. Acta ophthal. (Suppl.) (Kbh.) **39**, 1—74 (1953).

VESELL, E. S.: Factors altering the responsiveness of mice to hexobarbital. Pharm. **1**, 81—97 (1968).

VESELL, E.S.: Recent progress in pharmacogenetics. Advanc. Pharmacol. Chemother. **7**, 1—52 (1969).

VESELL, E.S. (Ed.): Drug metabolism in man. Ann. N. Y. Acad. Sci. **179**, 1—773 (1971).

VESELL, E.S.: Introduction: Genetic and environmental factors affecting drug response in man. Fed. Proc. **31**, 1253—1269 (1972).

VESELL, E.S., PAGE, J.G.: Genetic control of drug levels in man: phenylbutazone. Science **159**, 1479—1480 (1968a).

VESELL, E.S., PAGE, J.G.: Genetic control of drug levels in man: antipyrine. Science **161**, 72—73 (1968b).

VESELL, E.S., PAGE, J.G.: Genetic control of dicumarol levels in man. J. clin. Invest. **47**, 2657—2663 (1968c).

VESELL, E.S., PAGE, J.G.: Genetic control of phenobarbital-induced shortening of plasma antipyrine half-lives in man. J. clin. Invest. **48**, 2202—2209 (1969).

VESELL, E.S., PAGE, J.G., PASSANANTI, G.T.: Genetic and environmental factors affecting ethanol metabolism in man. Clin. Pharmacol. Ther. **12**, 192—201 (1971a).

VESELL, E.S., PASSANANTI, G.T., LEE, C.H.: Impairment of drug metabolism in man by disulfiram. Clin. Pharmacol. Ther. **12**, 785—792 (1971b).

VESELL, E.S., NG, L., PASSANANTI, G.T., CHASE, T.N.: Inhibition of drug metabolism by L-dopa in combination with a dopa decarboxylase inhibitor. Lancet **2**, 370 (1971c).

VESELL, E.S., PASSANANTI, G.T.: Utility of clinical chemical determination of drug concentrations in biological fluids. Clin. Chem. **17**, 851—866 (1971).

VESELL, E.S., PASSANANTI, G.T., GREENE, F.E.: Impairment of drug metabolism in man by allopurinol and nortriptyline. New Engl. J. Med. **283**, 1484—1488 (1970).

VOGEL, F.: Moderne Probleme der Humangenetik. Ergebn. Inn. Med. Kinderheilkd. **12**, 52—125 (1959).

VOGEL, F., KRUGER, J.: Statistische Beziehungen zwischen den AB0-Blutgruppen und Krankheiten mit Ausnahme der Infektionskrankheiten. Blut **16**, 351—376 (1968).

VOGEL, F., KRUGER, J., SONG, Y.K., FLATZ, G.: AB0 blood groups, leprosy, and serum proteins. Humangenetik **7**, 149—162 (1969).

VON WARTBURG, J.P., SCHÜRCH, P.M.: Atypical human liver alcohol dehydrogenase. Ann. N. Y. Acad. Sci. **151**, 936—946 (1968).

WEBER, W.W.: Acetylating, deacetylating and amino acid conjugating enzymes. In: BRODIE, B.B., GILLETTE, J.R. (Eds.): Handbook of pharmacology, pp. 564—583. Berlin, Heidelberg, New York: Springer 1971.

WEBER, W.W., COHEN, S.N., STEINBERG, M.S.: Purification and properties of *N*-acetyltransferase from mammalian liver. Ann. N. Y. Acad. Sci. **151**, 734—741 (1968).

WEED, R.I., REED, C.F.: Membrane alterations leading to red cell destruction. Amer. J. Med. **41**, 681—698 (1966).

WEINER, M., SHAPIRO, S., AXELROD, J., COOPER, J.R., BRODIE, B.B.: The physiological disposition of dicumarol in man. J. Pharm. exp. Ther. **99**, 409—420 (1950).

WELT, S.I., JACKSON, E.H., KIRKMAN, H.N., PARKER, J.C.: The effects of certain drugs on the hexose monophosphate shunt of human red cell. Ann. N. Y. Acad. Sci. **179**, 625—635 (1971).

WENZEL, D.C., BROADIE, L.L.: Stimulatory effect of nicotine on the metabolism of meprobamate. Toxicol. Appl. Pharmacol. **8**, 455—459 (1966).

WHITE, P.D.: Harvard Med. Alumni Bull. July—August 4—5 (1971).

WHITE, T.A., PRICE EVANS, D.A.: Urinary hexosamine and acetyl-hexosamine excretion in human acetylator phenotypes. Clin. chim. Acta **18**, 161—168 (1967a).

WHITE, T.A., PRICE EVANS, D.A.: The acetylation of tryptophan metabolites by rapid and slow acetylators of sulfamethazine. Experientia (Basel) **23**, 959—960 (1967b).

WHITE, T.A., PRICE EVANS, D.A.: The acetylation of sulfamethazine and sulfamethoxypyridazine by human subjects. Clin. Pharmacol. Ther. **9**, 80—88 (1968).

WHITTAKER, J.A., PRICE EVANS, D.A.: Genetic control of phenylbutazone metabolism in man. Brit. med. J. **1970 IV**, 323—328 (1970).

WHITTAKER, V.P., WIJESUNDERA, S.: The hydrolysis of succinylcholine by cholinesterase. Biochem. J. **52**, 475—479 (1952).

WILLIMOTT, S.G.: Investigation of solanine poisoning. Analyst **58**, 431—439 (1933).

WILSON, G. S.: A small outbreak of solanine poisoning. Monthly Bull. Minist. Health (Lond.) Serv. **18**, 207—210 (1959).

WILSON, I. B.: The active surface of the serum esterase. J. Biol. Chem. **208**, 123—132 (1954).

WOLFF, P. H.: Ethnic differences in alcohol sensitivity. Science **175**, 449—450 (1972).

WOODBURY, D. M., ESPLIN, D. W.: Neuropharmacology and neurochemistry of anticonvulsant drugs. Res. Publ. Ass. Res. Nerv. Ment. Dis. **37**, 24—56 (1959).

WYNGAARDEN, J. B., HOWELL, R. B.: In: STANBURY, J. B., WYNGAARDEN, J. B., FREDERICKSON, D. S. (Eds.): The metabolic basis of inherited disease. New York: McGraw-Hill 1966.

YAFFE, S. J., LEVY, G., MATSUZAWA, T., BALIAH, T.: Enhancement of glucuronide-conjugating capacity in a hyperbilirubinemic infant due to apparent enzyme induction by phenobarbital. New Engl. J. Med. **275**, 1461—1465 (1966).

YAHR, M. D., MERRITT, H. H.: Current status of the drug therapy of epileptic seizures. J. Amer. med. Ass. **161**, 333—338 (1956).

YAHR, M. D., SCIARRA, D., CARTER, S., MERRITT, H. H.: Evaluation of standard anticonvulsant therapy in three hundred nineteen patients. J. Amer. med. Ass. **150**, 663—667 (1952).

YATA, H.: A case of acatalasemia. Nihou Shika Hyoron **204**, 7—12 (1959).

YOSHIDA, A.: A single amino acid substitution (asparagine to aspartic acid) between normal (B+) and the common Negro variant (A+) of human glucose-6-phosphate dehydrogenase. Proc. Natl. Acad. Sci. (U. S. A.) **57**, 835—840 (1967).

YOSHIDA, A.: Amino acid substitution (histidine to tyrosine) in a glucose-6-phosphate dehydrogenase variant (G6PD Hektoen) associated with over-production. J. Mol. Biol. **52**, 483—490 (1970).

YOSHIDA, A., MOTULSKY, A. G.: A pseudocholinesterase variant ($E_{\text{Cynthiana}}$) associated with elevated plasma enzyme activity. Amer. J. Hum. Genet. **21**, 486—498 (1969).

Chapter 66

Aging Effects and Drugs in Man*

W. W. WEBER and S. N. COHEN

With 3 Figures

I. Introduction

The life of an individual is a continuous series of events that includes periods of growth, development, and senescence. From before birth to maturity, progress is marked primarily by growth and development, but a sharp distinction between these two processes is not possible. Growth refers mainly to increases in size resulting from cellular multiplication of intracellular substance, while development refers to maturation of structures and functions associated with progress. Taken together, they suggest a complex set of factors having a broader meaning than either term alone.

Patterns of growth and development are diverse but tend to occur in predictable periods and cycles. Most tissues tend to participate during periods of rapid growth so that increases in measurements in any one of them tend to reflect the general pattern. Notable differences in magnitude and timing occur from one person to another, and in different tissues and parts of the body. For certain tissues, the rate of change chronologically is comparatively rapid, as it is between birth and two years of age in neural tissue, or in lymphoid tissue toward the end of the first decade of life (BARNETT, 1968). The rate of change for other tissues is more gradual. For instance, the relative proportions of skeletal muscle and subcutaneous tissue change markedly toward the end of the first decade, but little alteration in the weight or height of the individual occurs during this period.

At certain stages of growth, particularly at the onset of adolescence, a profusion of physiologic changes occur with such rapidity that the age of the individual ceases to be a useful scale of reference. For example, nutritional requirements, basal metabolic rate, skeletal muscle development, and storage of calcium and nitrogen may serve as more accurate indicators of an individual's development than chronologic time.

It is evident that growth and development are accompanied by wide variation in human physiologic attributes during the several recognizable periods of childhood. Within every period, variation is so large that growth and development norms are better expressed as a range of values than as averages, and for certain attributes separate norms are needed for the sexes.

The rates of structural and functional changes abate as individuals approach maturity (HOLLINGSWORTH et al., 1965). Since the rates of aging of physiologic attributes in different individuals and in different systems of the body differ greatly, the problem of sorting normal from abnormal development in the immature person is replaced later in life by the problem of sorting involutional from pathological change (NORRIS and SHOCK, 1966). Furthermore, certain maturational

* Guttman Laboratory of Human Pharmacology and Pharmacogenetics, New York University School of Medicine. Partially supported by PHS Grant GM 17184.

changes may continue in some tissues while involutional processes have already commenced in others. For instance, while most individuals reach sexual maturity by 15 years of age, the ability for visual accommodation has begun to decline by this age. Maturation and involution even may occur simultaneously in the tissue. Coexistence of atrophy and hypertrophy in the gastric mucosa is a familiar example of this phenomenon.

Our current knowledge of the effects of aging at cellular and subcellular levels is inadequate, but is expanding rapidly. Certain types of age-related degenerative change appear to exhibit aging at a chemical level without accompanying histological change, while the opposite occurs in other situations. Loss of near vision with advancing age exemplifies the former and disseminated cerebral cortical atrophy, in which cells disappear but vital function is maintained, exemplifies the latter.

Biologists have long been intrigued by problems of size and shape in living organisms, and the limitations they impose on the physiologic attributes of individuals (MCMAHON, 1973; STAHL, 1965). The principle that many physiologic functions vary in direct proportion to the body surface area (or body weight$^{0.7}$) was recognized more than a century ago (BUTLER and RICHIE, 1960) and documented for various toxic and therapeutic agents (DREYER and WALKER, 1914). The principle has been accepted widely as a basis for adjusting drug dosage between individuals of different sizes and also of different ages (DONE, 1964). Its value in therapeutics is based on the fact that the extracellular water compartment, a part of the body of special interest in drug distribution, tends to vary in proportion to the body surface area. Therefore, drugs whose apparent volume of distribution approximates this space would reach similar concentrations in persons of different sizes if the doses were given on the basis of this criterion. The calculation of dosage for drugs which are distributed in a space that exceeds the extracellular water space cannot be made from the body area of the individual because total body water or any fraction in excess of extracellular water does not vary directly with the body area. Thus, our perspectives on physiologic changes during growth, development and senescence, and on the constitutional variability from one individual to another suggest that we should adopt a conservative attitude toward lengthy extensions of this principle, particularly with respect to extrapolations from mature individuals to infants and children.

This review is concerned primarily with the changes in the disposition of drugs that accompany normal human maturation and aging. Age-related factors affecting certain aspects of drug absorption, distribution, and elimination have been emphasized since we have considered only those drugs that are excreted from the body unchanged. We have relied on observations and experiments on individuals and human populations at different ages for this purpose, since no longitudinal pharmacologic studies have been made that cover an appreciable part of the human lifespan. Some information of this sort is available from animal studies (International Symposium on Comparative Pharmacology, 1967) but deficiencies in our knowledge of developmental and aging processes in animals are so great that it is virtually impossible to apply this information to man.

II. Absorption

Gastrointestinal morphology and function vary from the time of birth until senescence. These alterations obviously can influence the rate and pattern of drug absorption, but the variables that determine the rate and pattern are numerous and many are interrelated (WAGNER, 1968). Therefore we can tell little about the effects that fluctuations of individual factors with age may have on drug absorption.

By the later stages of fetal life the capacity to absorb substances, including intact protein and other large molecules, from the gastrointestinal tract is already well-developed. A number of abrupt changes occur at birth that can affect drug absorption. Bacteria and other exogenous materials enter the gut; certain vessels atrophy, and readjustments in splanchnic circulation take place. Twenty-four to forty-eight hours after birth the ability of the gut to absorb intact proteins diminishes markedly. The activities of enzyme and transport systems change, as does the composition of the digestive secretions (SPENCER, 1964; SMITH, 1951).

Drugs can be inactivated or transformed in the gut lumen prior to absorption by bacteria that take up residence there. These bacterial populations constitute an important part of the physiologic environment of an individual's gastrointestinal tract. They are mentioned here for this reason although the extent to which they are a factor that affects drug absorption is not established (WALSH and LEVINE, 1974). The complexity of the bacterial population is emphasized by the variety of species of microorganisms normally present and by the diverse factors involved in their development and maintenance. More than 60 species have been isolated from the enteric tract or feces of apparently healthy, adult individuals (DONALDSON, 1964; GORBACH et al., 1967). The character of the intestinal flora depends upon age. It is generally agreed that the gastrointestinal tract of the fetus is sterile and is colonized after birth by the oral route. Initially, the intestine becomes sparsely populated by enterococci and coliform organisms. The bacterial census of the intestine rises markedly with the initiation of feeding over the next one to four days. This transition period is associated in breast-fed infants with a rapid proliferation of L. *bifidus* and in bottle-fed infants with a mixed flora in which lactobacilli and nonsporulating anaerobes predominate. Ordinarily, weaning is associated with an increase in coliforms and enterococci as well as other organisms. In adults, only the large intestine contains bacteria in significant numbers. Contents of the stomach or proximal small intestine are sterile or sparsely populated with Gram-positive cocci except when contaminants are transported there from the mouth or respiratory tract. Colonic organisms may be found in the terminal ileum but even in the transition from small to large bowel, intestinal contents are sterile in about 50% of healthy persons. One of the determinants of bacterial species in the intestine is diet. In animals, diets rich in lactose increase the numbers of lactobacilli, while meat protein, gluten or casein decrease lactobacilli and increase coliforms and enterococci. Feeding a diet rich in butter fat appears to inhibit the growth of *E. coli* and *P. vulgaris*.

Absorption of many drugs and organic foreign substances from the gastrointestinal tract into the circulation is attributed to pH-dependent diffusion (SCHANKER, 1971). Numerous studies have been made of the gastric pH in full term infants from birth through the newborn period. They indicate that in most infants shortly after birth the gastric secretions have a pH within the neutral range, between 6.5 and 8. Characteristically, this changes during the first 24 h so that infants attain pH's between 1 and 3 by the end of the first day, with the volume of acid secretion closely approaching that of an average adult (MILLER, 1941). After the gastric acidity reaches its maximum, it falls under fasting conditions to a very low level, approaching achlorhydria, during the next 9 to 10 days. Premature infants tend to secrete less gastric acid and histological studies show a corresponding lack of mucosal development in the stomach. Gastric acidity reaches adult levels by 3 years of age. With advancing age, gastric acidity may decrease in both men and women, and achlorhydria may even be present in a small proportion of young adults (IVY and GROSSMAN, 1952). In one study, achlorhydria was observed in 5.3% of persons 20 to 29 years of age, 16.7% at 40 to 49 years of age, and 35.4% of persons older than 60 years (BLOOMFIELD and POLLAND, 1933).

Differences in the absorption of acid labile drugs at different ages may be attributed to differences in gastric acidity. Serum levels of penicillin G after oral administration were considerably higher in the newborn than in older infants (HUANG and HIGH, 1953). Results within the newborn period with the semi-synthetic penicillins, ampicillin, and nafcillin, follow a similar pattern (O'CONNOR et al., 1965; GROSSMAN and TICKNOR, 1965). Orally administered penicillin produces higher concentrations in premature infants than in older infants or older persons in accordance with the observation that premature infants have hypochlorhydria (HUANG and HIGH, 1953). Adults with low gastric acidity also tend to show greater absorption of penicillin G from the gastrointestinal tract (RAMMELKAMP and HELM, 1943). Higher serum levels are sustained in hypochlorhydric persons for longer periods (4 to 6 h) than in persons with normal levels of gastric acidity (2 to 4 h) (FINLAND et al., 1945).

Gastrointestinal transit time, the area available for absorption from different segments of the gut and the permeability and vascularity of different segments also vary with age (DAVIDSON, 1968; SMITH, 1951) and could affect drug absorption. Gastric emptying occurs more slowly in the newborn period than at any other time in life. Many newborn infants require eight hours for complete gastric emptying, and in some instances it may be much longer. Compared to adults, upper gastrointestinal motility is slow until about 6 months of age. In general, movements in this period are irregular and unpredictable, and are subject to a variety of influences including nutritional status, diet and feeding pattern.

Assessment of age effects on drug absorption is well illustrated in a series of studies measuring riboflavin absorption in individuals of various ages (JUSKO et al., 1970a and b, 1971). Riboflavin is absorbed by a site-specific (proximal small intestine), saturable transport process. Healthy neonates 5 to 6 days old, older infants, children, and adults to 40 years were studied using oral doses of riboflavin-5'-phosphate that were at least five times as high as needed to saturate the absorption process. Striking differences were noted in the course of absorption with time between the 5 to 6 day old neonates and the older subjects. In the 5 to 6 day old infants, absorption was slow and prolonged, yielding maximal blood levels only about one-fifth of those observed in older subjects. Since the duration of absorption was longer, the total amount of riboflavin absorbed (based on surface area) was similar to that absorbed by the older infants and adults The prolonged course of absorption was attributed to a slower transit rate through the intestinal tract due to lower intestinal motility in this age group. It was concluded that this maintained riboflavin at intestinal absorption sites for a longer period of time. Other factors that account for these findings were not excluded, such as the existence of age-related alterations in the permeability of the gastrointestinal tract or in the spatial distribution of transport processes along the gastrointestinal tract. Recent studies in newborn rats and mice make the latter hypothesis an attractive one (BATT and SCHACHTER, 1969). These showed that transport mechanisms for calcium, 3-*O*-methylglucose and L-proline were present at low levels of activity throughout the neonatal intestinal tract, including the colon. These mechanisms are known to be localized in different parts of the mature rat's small intestine. As the age of the neonatal animals increased, changes occurred in the site of localization so that the adult site was approached. These results, along with those concerning riboflavin absorption in the human, suggest that a similar developmental pattern might exist in the human neonate to account for the pattern of absorption of riboflavin shortly after birth. In older infants, children and adults, the findings with riboflavin indicate that there is a small but definite increase in absorption of the vitamin over the age range 3 months to 40 years. Analysis of the data shows that this increase is probably not accounted for by changes in maximal absorptive capacity, or by the existence of other specialized age-dependent transport processes such as that postulated in the very young neonate (JUSKO et al., 1970b). The increases in absorption of riboflavin in these older subjects is attributed to an increase in retention at intestinal absorption sites primarily because intestinal transit time is slower in older subjects than in children.

Few other studies have been carried out on the effect of age on drug absorption. One less extensive study on the absorption of xylose suggests that its absorption is similar to riboflavin (JUSKO et al., 1970a and b). A two-fold increase in D-xylose absorption occurred over an age range similar to that studied with riboflavin (LANZKOWSKY et al., 1963a and b).

In view of the lack of information describing the effect of age on drug absorption for persons beyond middle age, it is interesting to note a study carried out many years ago on the influence of age on the intestinal absorption of galactose (MEYER et al., 1943). Individuals whose ages ranged from adolescence to over 100 years, and who were judged to be in good health according to standards for their age, were tested under fasting conditions. The results clearly demonstrated that intestinal absorption of galactose was delayed in elderly persons. The rate of decline of absorption with advancing age was a linear function of age with the younger individuals (mean age 20.5 years) absorbing approximately 50% more of the sugar in 30 min than those in the older group (mean age 76.6 years). Ninety minutes after administration, galactose absorption had terminated in the younger group and was still continuing in the elderly.

The observations on galactose absorption were not correlated with variations in specific gastrointestinal functions so that one can only speculate about the factors which might have been responsible for the age-dependency observed. The rate of galactose absorption is dependent upon enzymatically catalyzed phosphorylation. If this is age-dependent, it might be the rate-limiting step in galactose absorption. Other intestinal enzymes exhibit age-dependency (SPENCER, 1964) and might limit the absorption of other compounds. A number of hemodynamic changes occur with advancing age (BENDER, 1965). Splanchnic blood flow decreases to a greater degree than blood flow to certain other regions of the body in elderly persons (BENDER, 1965). At age 65, for example, hepatic blood flow might be reduced 40 to 45% compared to the flow at age 25 and it appears to fall at a more rapid rate with advancing age than does cardiac output. If age-dependent changes in blood flow to the gastrointestinal region parallel those to the liver, as seems reasonable, absorption of drugs also might be impaired on this basis. Other factors that vary with age such as segmental bowel activity, intestinal motility, and rate of mucosal cell renewal could also affect the rate of drug absorption.

III. Distribution and Elimination

The size of body water compartments, their chemical composition, and renal excretory capacity are attributes of the individual that are of particular interest in understanding the effects of age on distribution and elimination of drugs (BUTLER, 1971). Since each of these attributes can vary in a different manner according to the time that its development is initiated and the time it reaches maturity, it is not surprising to find that changes in distribution and elimination that occur with age are complex.

Prominent gaps in the information about these physiologic characteristics exist at nearly every age. Some quantitative predictions can be made about drug distribution and elimination at certain ages of interest. Existing mathematical relationships and model systems that have been described by others have been used for this purpose (GOLDSTEIN et al., 1968a; KEEN, 1971).

A. Body Water Compartments

Total body water, extracellular water and intracellular water are functional compartments of special interest because the distribution of different classes of drugs in the body often corresponds to one of these spaces. Experimental data characterizing the size of these body water compartments during the span of life from birth to adulthood are summarized in Table 1 and Fig. 1 (FRIIS-HANSEN, 1956, 1961). At birth, the magnitude of the total body water compartment is approximately 78% of the body weight, and it decreases sharply to 60% at one year after birth, a value which closely approximates that at maturity.

Table 1. *Comparison of body water compartments in the average newborn infant, one year old child, and adult*[a]

	Average newborn infant at birth	Average 1 year old child	Average adult
Body weight (gm)	3400	10800	70000
Total body water			
(%)	78	60	58
(ml)	2650	6500	41000
Extracellular water			
(%)	45	27	17
(ml)	1530	2900	12000
Intracellular water			
(%)	34	35	40
(ml)	1160	3800	28000
Plasma water			
(%)	4 to 5	4 to 5	4 to 5
(ml)	140	430	3000

[a] Values derived from FRIIS-HANSEN, 1961.

Extracellular water also decreases with age but to an even more pronounced degree, from 45% of body weight at birth to 27% at one year. This change is due almost entirely to changes in interstitial volume, since the size of the plasma water compartment, a component part of the extracellular water, remains relatively unaltered in the range of 4 to 5% of the total body weight throughout the lifespan of the individual. The change in the size of the intracellular water compartment initially is in the opposite direction to that seen in total body and extracellular water, increasing from a value of 34% of body weight at birth to a maximum of 43% at 3 months of age and then decreasing to a value quite close to that at birth. Beyond one year of age and up to three years all three body water compartments increase. After 3 years of age, the size of each compartment gradually declines until it reaches its adult value. The size of a compartment is an individual characteristic and varies 1 to 2% at a particular age. Variations from one person to another, especially at birth, are large but do not overlap those of older children or adults.

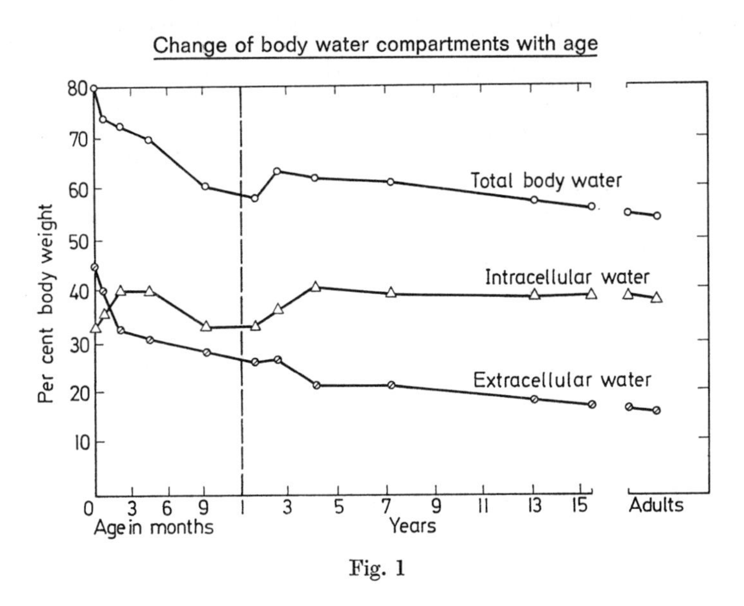

Fig. 1

Changes in the relative abundance of different tissues that are characteristic of normal growth at different ages can also affect the relative size of these body water compartments (FRIIS-HANSEN, 1961). The amount of total body water, for example, varies inversely with the amount of fat tissue present in the body and the relative amount of body fat increases greatly with age. At a fetal age of 5 months, only 0.5% of the body weight is fat. This increases to 12 to 16% in the full term newborn infant. Beyond this age few quantitative data are available, but further rapid increases in growth of fat tissue appear to occur within the first year after birth followed by a gradual decrease in rate until early adolescence.

The pattern of age-related variations in the relative sizes of these body water compartments estimated for the body as a whole may be different from the pattern within a single tissue, or a group of tissues (FRIIS-HANSEN, 1961). Growth of the liver consists mainly of cell duplication without large alterations in fat or extracellular water content. In contrast, the variations within skeletal muscle, skin and central nervous system tissue are quite different. In the newborn infant and the adult, all these tissues account for about 61 to 62% of the total body water. The extracellular water relative to body weight is about twice as large for the body as a whole in the infant at birth compared to the adult, while it is only about 50% greater within these three tissues.

B. Binding

Differences in drug distribution attributable to plasma protein binding can be due to differences in the amount of binding protein, in the strength of protein-drug interaction, or a combination of both these factors (GOLDSTEIN et al., 1968a).

1. Serum Proteins

Levels of these proteins have been studied at various ages from the fetal period into senescence (Table 2). Prior to birth, the levels of albumin and other serum proteins are dependent upon the maturity of the fetus and placenta (GITLIN, 1971). Albumin, the major drug binding component of plasma, and many other proteins are present in fetal serum by four weeks of gestation. Albumin levels increase in direct proportion to gestational age and reach an average level of 1.8 gm/100 ml of serum by 22 to 24 weeks gestation. After a full term (40 to 41 weeks) period of gestation, albumin is at a level equivalent to that found in adults (GITLIN and BOESMAN, 1966; HYVARINEN et al., 1973). Plasma albumin levels are sufficiently characteristic indicators of fetal maturity that they can be used clinically to assess the gestational age of the newborn infant at birth (HYVARINEN et al., 1973). Levels of less than 2.5 gm/100 ml of serum, for example, are taken as evidence of prematurity. Although albumin levels in childhood have not been studied in detail, they do not appear to fluctuate appreciably. In the adult, the level of serum albumin declines slowly with age. Estimates of the extent of the decrease vary somewhat from one source to another (LIBOW, 1963; SALATKA et al., 1971). One report finds that an average decrease of 20% occurs between a young adult and an elderly person (Table 2).

Table 2. *Serum albumin and* IgG *globulin values at different ages*

Age	Albumin (gm/100 ml)	IgG globulin (mg/100 ml)
Gestational age (weeks)		
4	+	—
6.6	—	+
22	—	< 200
26 to 27	1.80 ± 0.13	330 ± 6
30 to 31	2.51 ± 0.33	566 ± 143
36 to 37	3.17 ± 0.59	823 ± 135
40 to 41	3.57 ± 0.37	937 ± 175
Adult ages (years)		
20 to 30	3.34 (2.70 to 3.80)	2.93 (2.60 to 3.15)
60 to 70	2.84 (2.21 to 3.28)	3.55 (3.21 to 4.30)

Serum globulins, though of less importance than albumin in plasma drug binding, do participate in interactions with some drugs (BARAKA and GABALI, 1968; GLUECK et al., 1969; HOLGER-MADSEN, 1962; PARKE and LINDUP, 1973) and other anionic substances (GOLDSTEIN et al., 1968a). Gamma globulin (IgG), the predominant human serum immunoglobulin (GRUBB, 1970), is a participant in these interactions. Its concentration in adult serum is usually in the range of 700 to 1500 mg/100 ml serum (ALLANSMITH et al., 1968). IgG present at birth is mainly of maternal origin (GITLIN, 1971). Its serum level falls until one to three months of age and then begins to rise as synthesis by the infant increases. It reaches adult levels between seven and twelve years of age (ALLANSMITH et al., 1968; WEST et al., 1962) and continues to increase into old age. After attaining adult levels, it also varies according to sex (GRUNDBACHER and SHREFFLER, 1970).

There is very little conclusive information on whether intrinsic changes in the serum proteins occur with aging and can affect drug binding and disposition. No such changes appear to occur in the older age groups. For the younger age groups, there is one report of differences in binding characteristics for salicylate and bilirubin between umbilical cord serum and adult serum (KRASNER et al., 1973). Fetal albumin appears to bind salicylate with somewhat lower avidity than adult serum albumin although it binds bilirubin more strongly. The association constants derived from the salicylate data are 1.7 to 2.9×10^5 M^{-1} for fetal albumin and 4.0×10^5 M^{-1} for adult albumin. The corresponding association constants for the interaction with bilirubin are 5.2×10^7 M^{-1} and 2.4×10^7 M^{-1}. The possible significance of differences in affinity to drug distribution and elimination is considered below in connection with the effect of protein binding on the apparent volume of distribution of a compound and its elimination half-time.

2. Plasma Lipids

The possibility that other physiologic constituents of plasma may vary with age and have a bearing on drug distribution and elimination must be considered. The influence of free fatty acids (FFA) on drug binding deserve special comment. FFA's are a mixture of long-chain fatty acids that comprise an important group of metabolites transported by plasma albumin (SPECTOR et al., 1973). Binding studies have shown that albumin possesses three classes of sites for fatty acids and other large organic anions (GOODMAN, 1958). The primary sites have specificity for long-chain fatty acids such as oleic and linoleic acids, and in usual physiologic concentrations these are bound almost entirely to primary sites. Secondary and tertiary sites are much less specific and are shared by many organic ligands. Until recently it has been accepted that physiologic changes in the plasma FFA concentration have little or no effect on drug binding. Several reports have appeared, however, which show that physiologic changes in FFA did reduce the ability of albumin to bind a variety of compounds including tryptophan (McMENAMY and ONCLEY, 1958), thyroxine (TABACHNICK, 1964), and triiodothyronine (NATHANIELSZ, 1969). *In vitro* studies of even more recent origin on the capacity of fatty acids such as palmitic or oleic to displace the hypolipidemic drugs chlorophenoxyisobutyrate, (CPIB), and *p*-chlorophenyl-(*m*-trifluoromethylphenoxy) acetate (halofenate) from albumin provide additional evidence that FFA can influence drug binding by this protein (SPECTOR et al., 1973). The results support the general conclusion that FFA's disrupt interactions at secondary and tertiary binding sites in a manner that depends in part on the drug, and in part on the affinity of albumin for the FFA.

The extent to which these characteristics might affect drugs in individuals at different ages is not entirely clear. FFA concentrations are quite variable and rapidly responsive to changes in nutritional status, physical activity and environmental stimuli. Measurements in children show that FFA levels can fluctuate over a wide range (165 to 680 μEq/l) and that there are some differences between obese and non-obese children (COURT et al., 1971). Studies in adults have documented that FFA levels oscillate rapidly during exercise (RODAHL et al., 1964), and with the intake of coffee and caffeine (BELLETT et al., 1968), and ethanol (JONES et al., 1965). Comparative studies to evaluate the effects of these factors on FFA levels with age have not been done and would be difficult to design.

There is some information on changes of plasma lipid composition with age. Distinct age-related differences exist in the concentration of certain cholesterol fatty acid esters. Younger individuals (6 to 10 years) have a significantly smaller proportion of oleic acid esters and more linoleic acid esters than older individuals (60 to 87 years) (SWELL et al., 1960). The younger individuals also have lower levels of total cholesterol.

Cholesterol levels change beyond childhood in a complex way. They are subject to change by such factors as basal metabolism, obesity and diet in addition to age. Over the ages 17 to 30 years, cholesterol levels in men and women increase on the average by 2.2 mg total cholesterol/100 ml of serum per year from an average value of approximately 177 to 206 mg/100 ml of serum. There is a curvilinear relation between age and serum cholesterol concentration in men between 17 and 78 years. Levels reach a maximum in the sixth decade. Standard tables describing these changes with age are available (KEYS et al., 1950).

3. Other Binding Components

Reversible interactions with cell membranes and various intracellular constituents represent other types of binding with the potential for influencing drug distribution and elimination. Very little has been reported about these phenomena and their age-dependency. A relevant exception concerns the selective ability of the liver to take up and concentrate bromsulphophthalein and certain endogenous substances such as bilirubin. Two soluble proteins, Y (ligandin) and Z, have been isolated from liver and show a high affinity for these substances and for corticosteroids, azo dyes and drug anions. It has been suggested that the proteins have a role in the hepatic uptake of these substances (LITWACK et al., 1971). These proteins show similar patterns of development with maturation (LEVI et al., 1970) in newborn guinea pigs, monkeys, and man. Z protein develops very quickly during fetal life and reaches adult levels at birth. In contrast, Y protein is almost absent at birth but reaches mature levels within 2 to 3 weeks after birth. Z protein is also present in the cytoplasm of intestinal mucosa cells and appears to have a role in intestinal absorption of fatty acids (MISHKIN et al., 1972; OCKNER et al., 1972). There is evidence that Y protein is the major intracellular hepatic anion binding protein, and that it binds substances with a fairly high molecular weight which are weakly anionic, highly protein bound, and rapidly eliminated in the bile after conjugation. A number of antimicrobial substances including penicillin, chloramphenicol, and tetracycline also have been reported to bind to this protein (KUNIN et al., 1973). Steady state levels of Y protein are modulated physiologically in some manner by endogenous thyroid hormone, and are subject to induction by phenobarbital (REYES et al., 1970). The full significance of such acceptor proteins to the distribution and elimination of drugs has only begun to be explored and their importance for pharmacologic effects that vary with age is unknown at present.

C. Renal Excretion

Drugs are eliminated from the body by a number of different routes. The renal systems constitute the major pathway for elimination of most drugs, but they also may be eliminated through sweat, saliva, bile, expired air, milk and semen. Various intrinsic characteristics of the drug molecules may govern the route by which a specific agent is excreted, but in general it is the net functional capacity of the renal system that controls the rate of drug elimination.

A number of age-related factors may alter the efficiency of the kidneys as organs of drug excretion. Some of these operate upon the kidney directly, some operate only extrinsic to the kidney, and some influence renal drug elimination via a combination of direct and indirect effects. During early life, both anatomical and physiological attributes of the kidney change with age, but maturation of existing structures is the major factor responsible for age-related changes in renal function in young animals. After birth, there is a great deal of species variation with age in the developmental status of the kidneys and renal excretory mechanisms. Thus generalizations from individuals of one species about age-related changes in drug excretory ability may be misleading.

The rat kidney is incompletely formed at the end of gestation and continues to develop anatomically for two or three months after birth. Many glomeruli that were present in incomplete form at birth mature, new glomeruli appear in the renal cortex (ARATAKI, 1926; ADOLPH, 1957), and the process of differentiation of the tubular system and ascending and descending limbs of Henle's loops is completed (ADOLPH, 1957; DICKER, 1970). A similar pattern of change occurs in the kidney of the newborn rabbit (WACHSTEIN and BRADSHAW, 1965). Young mature, and senile rats have been studied to evaluate the effect of advanced age upon the microscopic anatomy of the kidney. At 300 days of age the rat kidney contains approximately 31,000 glomeruli of about 115 μ diameter. This has diminished in 500 day old animals to about 20,000 glomeruli that have a slightly larger diameter (124 μ) (ARATAKI, 1926). Glomeruli of older animals (868 to 1170 days) all demonstrated changes in the basement membrane and had greater dilatation of the glomerular capillaries than those of younger, mature rats but frankly fibrotic glomeruli were rare (ANDREW and PRUETT, 1957).

In the human, all glomeruli are present by the end of 36 weeks of gestation, and postnatal development involves only elongation and maturation of the tubules of some of the nephrons (MACDONALD and EMERY, 1959). Thus, there is apparent structural integrity in the kidney of the full-term human infant at birth, but the premature infant is born with a deficiency in glomeruli and has short, immature renal tubules. Senescent changes in kidneys from aged individuals cannot be differentiated from senile arteriosclerotic changes. Changes such as glomerular fibrosis, tubular atrophy and interstitial sclerosis were attributed to aging in some studies, while in others they were thought to be secondary to generalized vascular disease (ANDREW, 1971). Thus, the effect of senescence upon the structural integrity of the human kidney cannot be evaluated at present.

Alterations in renal function that occur as an individual develops and ages have been studied extensively in man. In the newborn infant, the glomerular filtration rate, measured as inulin or mannitol clearance, may be as low as 3 ml/min (Table 3) (GOLDSTEIN et al., 1968b; WEIL, 1955). The glomerular filtration rate in an average young adult male is 130 ml/min. Infants generally achieve an adult glomerular filtration rate by one year of age. Some infants even have "normal" adult mannitol clearances by 10 weeks of age (WEST et al., 1962). The glomerular filtration rate remains constant between about 1 year of age and 20 to 25 years of age. Afterward it declines to approximately 65 ml/min at 90 years of age according to the equation $Cl_1 = 153.2$ to $0.96 \times$ age (SHOCK, 1952).

The significance of age-related differences in glomerular filtration to drug elimination can be illustrated by antibiotic agents of the aminoglycoside group such as streptomycin, kanamycin, and gentamicin. Each of these drugs is excreted from the body unchanged, mainly in this way. The elimination half-life of these drugs early in life is prolonged in comparison to that in adults. In the case of kanamycin or gentamicin, the half-life in premature infants is longer than in full-term infants, and longer in young infants than older infants (Table 4).

The serum concentration of kanamycin should remain at or slightly above 5 μg/ml to provide a therapeutic benefit. In the healthy young adult, the administration of 7.5 mg/kg every 6 h yields therapeutic concentrations of the drug throughout the day. This dosage regimen replaces excreted drug at such a rate that no significant accumulation occurs. In the premature infant, however, this

Table 3. *Relationships between clearance and elimination half-time in the absence of protein binding in a newborn infant and adult*[a]

Clearance	Drug distributed in					
	Plasma water		Extracellular fluid		Body water	
	Newborn (140 ml)	Adult (3000 ml)	Newborn (1530 ml)	Adult (12000 ml)	Newborn (2650 ml)	Adult (41000 ml)
Tubular secretion						
PAH clearance (ml/min)	12	650	12	650	12	650
$t_{1/2}$ (min)	8.0	3.0	88.5	13	153	44
Glomerular filtration						
Inulin clearance (ml/min)	3	130	3	130	3	130
$t_{1/2}$ (min)	32	16	354	64	612	218

[a] Body weight of infant at birth taken as 3400 gm and for the adult is taken as 70000 gm.

Table 4. *Serum half-life of aminoglycoside antibiotics as a function of age*[a]

Drug	Age				
	Premature infant		Full term infant		Adult
	≦ 1 week	> 1 week	≦ 1 week	> 1 week	
Kanamycin	18.0[b]	6.0[b]	4.0[c]	3.5[c]	2.0
Streptomycin	7.0[b]				2.7
Gentamycin[c]	6.25	3.25	4.5	3.15	3.15

[a] $t_{1/2}$ in h.
[b] AXLINE and SIMON, 1964.
[c] McCRACKEN and JONES, 1970.

regimen would lead to rapid accumulation of the kanamycin in the body and toxicity. As a consequence of knowledge of the effect of age upon the excretion of this drug, it is administered to premature infants in 12 hourly 7.5 mg/kg doses. With this dose and time schedule for administration, toxicity is rare and therapeutic levels are maintained throughout therapy (AXLINE and SIMON, 1964). Similar alterations in dosage and time schedule are necessary for the other drugs of this group when they are to be administered to newborn infants (McCRACKEN and JONES, 1970; WEST et al., 1962).

There have been no studies on the requirement to alter dose and/or therapeutic schedule for the administration of these agents to aged individuals. It may be predicted that failure to adjust the dose properly because of the diminished glomerular filtration rate in such individuals would result in drug accumulation in these individuals.

The secretory activity of the renal tubule is very low in the newborn infant (WEST et al., 1962). The Tm PAH in such infants is 12 ml/min (Table 3). This measure of secretory activity rises to the young adult level by about 30 weeks after birth but falls again to about 30% of this peak by 60 to 65 years of age (MILLER et al., 1951). The rate of excretion of drugs that are eliminated from the body mainly through renal tubular secretory activity will vary with age as this physiological process matures and ages.

Penicillin G and a number of the semisynthetic penicillins are excreted unchanged by renal processes, mainly by tubular secretion. Penicillin G, methicillin, oxacillin, and ampicillin are all excreted at a very slow rate by newborn and premature infants (BARNETT et al., 1949; AXLINE et al., 1967). Their rates of excretion in the urine are directly related to post-natal age and inversely related to their half-lives of elimination. The elimination half-life and rate of urinary excretion both reach adult levels by 3 to 4 weeks of age (AXLINE et al., 1967). The dose and time of administration of the penicillins are usually altered when these drugs are administered to infants during the first few weeks of life. Whether such an alteration is indicated for elderly individuals has not been investigated. There have been no reports linking penicillin administration to newborn infants with stimulation of the tubular secretory mechanism as has been shown to occur in neonatal rabbits (HIRSCH and HOOK, 1969).

One consequence of immature tubular secretory function in the newborn infant is that renal plasma flow is difficult to estimate accurately at this age. This is a result of the poor extraction of *p*-aminohippurate by the newborn kidney; only 60% of a load of PAH is extracted in the neonate, while more than 90% is extracted in older children and adults. The use of cPAH as a measure of renal plasma flow is misleading under these circumstances and it may be underestimated in infants by as much as 50% when measured in this way (CALCAGNO and RUBIN, 1963).

D. Relations Affecting Distribution and Elimination

Extensive protein binding can affect the action of a drug by restricting distribution, decreasing penetration into different tissues and parts of the body, and by delaying renal excretion. These effects can be understood more readily from relationships between the apparent volume of distribution, renal clearance and the rate of elimination of the drug.

These relationships have been described for adults in the absence of protein for several specific values of the apparent volume of distribution (GOLDSTEIN et al., 1968b). Corresponding relationships for the newborn infant are presented for comparison with the adult values in Table 3. In general, the trends in elimination half-time are the same in newborns as in adults, but their magnitude is quite different. The time required for drug elimination by glomerular filtration is 2 to 6 times longer in the infant than in the adult. For a drug cleared in this way and distributed in tissues, values of the elimination half-time are in the range of 354 to 612 min (6 to 10 h) in the infant compared to 64 to 218 min (1 to 3.6 h) in the adult. The difference between the age groups is about the same for a drug excreted by tubular secretion. The fastest possible route of elimination for individuals of both age groups would be for a drug restricted to plasma water and secreted by the tubules. In infants this value is 8 min in contrast to 3 min for adults. Drugs distributed into tissues would be eliminated more slowly by either group of subjects.

Potential effects of plasma protein binding on the apparent volume of drug distribution, and the dynamics of drug elimination are particularly pertinent to our discussion. Binding to plasma protein results in a reduction in the free drug concentration in plasma, and as a consequence, the apparent volume of distribution increases in proportion to the extent of binding. Since drug elimination half-time increases with the volume of distribution, the time of elimination is lengthened also. The extent of these effects for drugs excreted solely by glomerular filtration and exhibiting various degrees of binding has been described quantitatively for adults in terms of a mathematical model (KEEN, 1971). This model can be used also to estimate the possible effects of growth related changes in the size of body water compartments (Table 1) and renal excretion (Table 3) on the elimination of drugs given in a single dose and excreted by glomerular filtration. Estimates for newborn infants and adults are summarized in Table 5. They show the factor (F) by which plasma protein binding might be expected to increase the elimination half-time assuming the initial free concentration of drug is the same in each case.

Table 5. *Relationships between protein-binding, volume of distribution and half-life of elimination for drugs excreted by glomerular filtration in the newborn infant and adult*[a]

Protein binding[b]	Drug distributed in								
	Plasma water (140 ml)			Extracellular water (1530 ml)			Total body water (2650 ml)		
(% bound)	V_D	$t_{1/2}$	F	V_D	$t_{1/2}$	F	V_D	$t_{1/2}$	F
Newborn infant (3400 gm)	ml	min							
0	140	32	—	1530	354	—	2650	612	—
50	280	64	2	1670	387	1.10	2790	645	1.05
80	700	160	5	2090	484	1.37	3210	740	1.21
90	1400	320	10	2790	646	1.84	3910	900	1.47
95	2800	640	20	4190	970	2.74	5310	1220	2.00
Adult (70000 gm)	(3000 ml)			(12000 ml)			(41000 ml)		
0	3000	16	—	12000	64	—	41000	218	—
50	6000	32	2	15000	80	1.25	44000	234	1.07
80	15000	80	5	24000	128	2.00	53000	281	1.29
90	30000	160	10	39000	208	3.25	68000	362	1.66
95	60000	320	20	69000	368	5.75	98000	521	2.38

[a] Calculated from Eqs. (5) and (12) of KEEN (1971). F is the factor by which plasma binding increases the length of time for which the free drug concentration exceeds any given value assuming the initial free drug concentration is the same in each case.

[b] Protein-binding could be expressed in terms of protein-drug association (or dissociation) constants instead of percent drug bound. An association constant of 10^4 M^{-1}, or less, does not affect distribution or elimination appreciably (KEEN, 1971). When the association constant is greater than 10^4 M^{-1}, binding becomes important and we can do no better than evaluate its effect on elimination half-time for each drug individually. This can be done provided information is available on the concentration of the protein, the number of binding sites it possesses for the drug, and the concentration of free drug, through the equation relating these quantities to the fraction of drug bound (GOLDSTEIN et al., 1968a).

The greatest potential effect of protein binding on elimination half-time is for a drug restricted to plasma water in both newborn infants and adults. For a highly bound drug (90% or more) there is a ten-fold or greater increase in elimination half-time. The increase with protein binding is the same in newborn infants and adults because the plasma water space represents the same proportion of body weight, about 4 to 5% (see Table 1), in both groups. When the drug is distributed into extracellular fluid, or total body water, plasma binding has much less effect on its rate of elimination. The elimination half-time is lengthened 2.74 times for a highly bound drug distributed in extracellular fluid of the newborn infant. Elimination half-time is lengthened 5.75 times in the adult under the same circumstances. The increase is dampened more effectively in the newborn infant because the extracellular fluid accounts for such a large proportion of the body weight (45%, see Table 1) in the newborn compared to that in adults (17%). A similar but smaller effect occurs with a drug distributed in total body water. Differences between newborn infants and adults for a drug distributed in this compartment can be explained by similar reasoning.

For a child one year old the potential effect of protein binding on elimination of a drug excreted by glomerular filtration and distributed in total body water, or in plasma water, should be approximately the same as for an adult. This similarity is due to the similarity in the size of these body water compartments in the one year old and adult. On the other hand, extracellular water space does not reach the adult level by one year of age, and there is still a difference in elimination half-time of a drug distributed in extracellular water as protein binding increases between a one year old and an adult.

The relatively greater lengthening of elimination half-time for drugs distributed in extracellular water compared to that in plasma water or total body water is more apparent in Fig. 2.

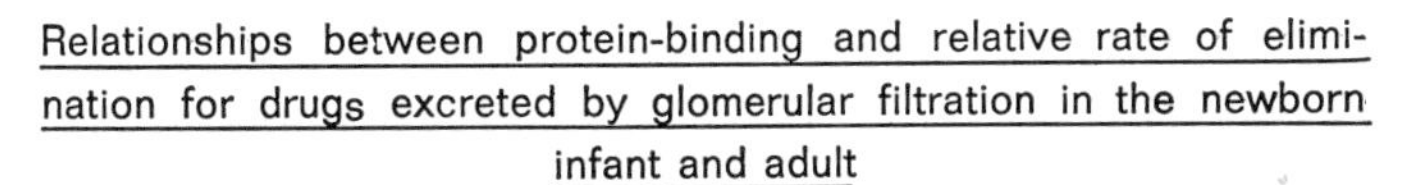
Relationships between protein-binding and relative rate of elimination for drugs excreted by glomerular filtration in the newborn infant and adult

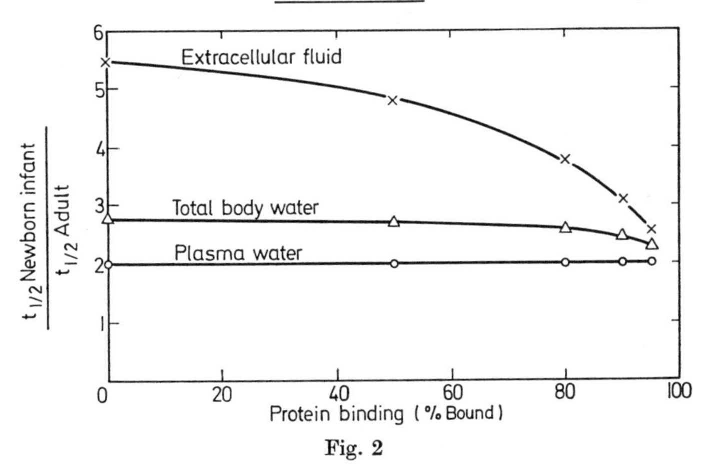

Fig. 2

The mathematical model referred to above also can be used to examine potential effects of age-related changes in body water distribution and renal function on the concentration of drug in plasma when repeated doses are administered. Table 6 shows the factor (F′) by which the peak drug level would change with protein binding in infants and adults. It also shows the maximum and minimum levels attained assuming that the initial free concentration of the drug is the same in each case. The greatest potential effects of protein binding for drugs given in multiple doses are seen when the drug is restricted to plasma water in both the newborn and adult. For a highly bound drug there is a 50% decrease in the peak blood level of the drug, and the minimum level of free drug in the blood can decrease to zero between doses. When the drug is distributed in extracellular water, or in total body water, plasma protein binding has a smaller effect on the peak and minimum levels. The effects of plasma protein binding on elimination half-life (Table 5) and on levels of drug attained with repeated doses of drug (Table 6) follow a similar pattern.

Table 6. *Relationships between protein-binding, half-life of elimination and plasma levels attained with repeated doses of drug in the newborn infant and adult*[a]

	Drug distributed in											
	Plasma water				Extracellular water				Total body water			
Protein binding (% bound)	$t_{1/2}$	Max blood level	F'	Variation	$t_{1/2}$	Max blood level	F'	Variation	$t_{1/2}$	Max blood level	F'	Variation
Newborn infant (3400 gm)												
0	32	200		200→100	304	200	1	200→100	612	200	1	200→100
50	64	133	0.67	133→ 33	387	187	0.94	187→ 87	645	193	0.97	193→ 93
80	160	103	0.52	100→ 3	484	163	0.82	163→ 63	740	176	0.88	176→ 76
90	320	100	0.50	100→ 0	646	139	0.70	139→ 39	900	156	0.78	156→ 56
95	640	100	0.50	100→ 0	970	117	0.59	117→ 17	1220	133	0.67	133→ 33
Adult (70000 gm)												
0	16	200	1	200→100	64	200	1	200→100	218	200	1	200→100
50	32	133	0.67	133→ 33	80	172	0.86	172→ 72	234	191	0.96	191→ 91
80	80	103	0.52	103→ 3	128	133	0.67	133→ 33	281	169	0.85	169→ 69
90	160	100	0.50	100→ 0	208	112	0.56	112→ 12	362	146	0.73	146→ 46
95	320	100	0.50	100→ 0	368	102	0.51	102→ 2	521	124	0.62	124→ 24

[a] Calculated assuming a single dose of drug gives a blood level of 100 units and is eliminated from the blood by a first order process. For a first order process with a rate constant k (time^{-1})

$$kt = 2.303 \log \frac{a}{a-x}$$

where a = amount of drug present at time zero; x = amount gone after time t, and $a - x$ = amount remaining after time t. It can be shown that

$$t_{1/2} = \frac{2.303 \log 2}{k}\,.$$

If we set $a = 1$, then

$$t_{1/2} = -\frac{\log (1 - x)}{0.301}$$

where $(1 - x)$ is the fraction of drug remaining. Also, since dose (here 100) = (fraction drug gone at t) (maximum drug level) then

$$\text{maximum drug level} = \frac{100}{x}$$

and the variation in blood level is

$$\frac{100}{x} - 100$$

when the drug is given at intervals of time t. F' is the factor by which plasma binding decreases the maximum blood level assuming the initial free drug concentration is the same in each case, as described in Table 5.

IV. Concluding Remarks

Maturation and aging have obvious effects on the absorption, distribution, and elimination of drugs excreted unchanged from the body. The changes in these processes tend to occur in an orderly sequence at predictable stages in the life of an individual, but their onset and extent can vary widely from one person to another. During infancy and childhood these changes are usually rapid and either periodic or cyclic. They are more gradual and progressive throughout the adult years.

Figure 3 emphasizes how patterns of maturation associated with aging differ for several physiologic attributes. The age-related mean value of the attribute relative to its adult norm is depicted from birth to old age. Serum albumin attains an adult level at birth and this is maintained with little further change into old age. Glomerular filtration and tubular secretion are both relatively low at birth and undergo maturation at markedly different rates. Beyond early adulthood both begin to decline and are diminished significantly again in the elderly.

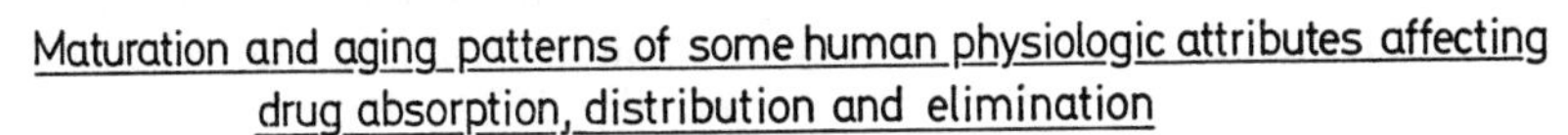

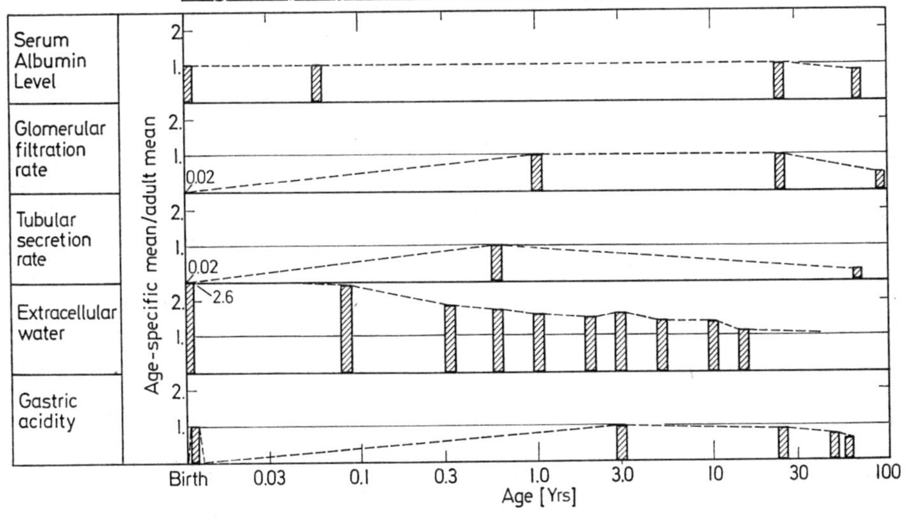

Fig. 3

The extracellular water space is much greater in comparison to body weight in the newborn infant than in a one year old child. It decreases further until it attains the adult standard at approximately 3 years of age. Changes in gastric acidity are cyclic, showing initially a low acidity which rises abruptly to adult levels on the first day after birth, and then declines to an achlorhydric state by about the tenth day after birth. Thereafter the content of gastric acid gradually rises again until it attains adult levels at approximately 3 years, and it may decline slowly later in life.

Much of the knowledge we have reviewed has been derived, of necessity, from studies of different individuals and sample populations at various ages. It is quite possible that a sample of persons studied at one age would not be representative of one at other ages, and that results related to age in this way would not typify those obtained in a longitudinal study of the same population. For example, as noted above, serum cholesterol appears to decline with age after the sixth decade. This could mean that the cholesterol level actually declines in old age. On the other hand, it is more likely that the young population that was studied consisted of several subpopulations and that the older persons are survivors of a subpopulation that had low cholesterol levels at an earlier age. Such an alternative should be considered before concluding that observations that are apparently related to age represented real biologic change with age.

There is a rather long period extending from early childhood into early senescence when the physiologic factors controlling drug absorption, distribution, and elimination remain quite constant. Very young and very old persons differ in their capacity to carry out these pharmacologic functions. It has not been possible to obtain this information directly at the extremes of life and the need for such information has led to the use of extrapolated data obtained during the period of stable function. Recently new instrumentation and analytical methods have been developed that have profoundly increased our ability to evaluate drug disposition in the very young and very old. Studies using gas-liquid chromatography-mass spectrometry (HORNING, E. et al., 1973; HORNING, M. G. et al., 1973), drugs labeled with stable (non-radioactive) isotopes (KNAPP and GAFFNEY, 1972; PROX, 1973) and immunopharmacology (COTTEN 1973), have greatly enhanced our ability to measure drugs and their metabolites in human tissues at any age after birth. Innovative studies using these approaches and others such as modern spectroscopy (CHIGNELL, 1972) and affinity chromatography (CUATRECASAS, 1972) to study the effects of drugs on tissues isolated from the body (VENTER et al., 1972) and on cells in culture (NEBERT, 1973) have led to a deeper appreciation of factors which regulate drug absorption, distribution and elimination. Also, these approaches, supplemented by new methods for monitoring and diagnosing the fetus, are beginning to provide information for evaluating these processes at various stages of fetal development (Drugs and the Unborn Child, 1973; BOREUS, 1973; MIRKIN, 1972).

References

ADOLPH, E. F.: Ontogeny of physiological regulations in the rat. Quart. Rev. Biol. **32**, 89—137 (1957).

ALLANSMITH, M., MCCLELLAN, B. H., BUTTERWORTH, M., MALONEY, J. R.: The development of immunoglobulin levels in man. J. Pediat. **72**, 276—290 (1968).

ANDREW, W.: The anatomy of aging in man and animals. New York: Grune and Stratton 1971.

ANDREW, W., PRUETT, D.: Senile changes in the kidneys of Wistar Institute rats. Amer. J. Anat. **100**, 51—80 (1957).

ARATAKI, M.: On the postnatal growth of the kidney with special reference to the number and size of glomeruli (albino rat). Amer. J. Anat. **36**, 399—436 (1926).

AXLINE, S. G., SIMON, H. J.: Clinical pharmacology of antimicrobials in premature infants. I. Kanamycin, streptomycin and neomycin. Antimicrob. Agents Chemother. 135—141, **1964.**

AXLINE, S. G., YAFFE, S. J., SIMON, H. J.: Clinical pharmacology of antimicrobials in premature infants. II. Ampicillin, methicillin, oxacillin, neomycin, and colistin. Pediatrics **39**, 97—107 (1967).

BARAKA, A., GABALI, F.: Correlation between tubocurarine requirements and plasma protein pattern. Brit. J. Anaesth. **40**, 89—93 (1968).

BARNETT, H. L.: Pediatrics, 14th edition, pp. 21—26; 238—252. New York: Appleton Century Croft 1968.

BARNETT, H. L., MCNAMARA, H., SHULTZ, S., TOMPSETT, R.: Renal clearances of sodium penicillin G, procaine penicillin G, and inulin in infants and children. Pediatrics **3**, 418—422 (1949).

BATT, E. R., SCHACHTER, D.: Developmental pattern of some intestinal transport mechanisms in newborn rats and mice. Amer. J. Physiol. **216**, 1064—1068 (1969).

BELLET, S., KERSHBAUM, A., FINCK, E. M.: Response of free fatty acids to coffee and caffeine. Metabolism **17**, 702—707 (1968).

BENDER, A. D.: The effect of increasing age on the distribution of peripheral blood flow in man. J. Amer. Geriat. Soc. **13**, 192—198 (1965).

BLOOMFIELD, A. L., POLLAND, W. S.: Gastric anacidity: its relation to disease, p. 55. New York: MacMillan 1933.

BOREUS, L. O. (Ed.): Fetal pharmacology. New York: Raven Press 1973.

BUTLER, A. M., RICHIE, R. H.: Simplification and improvement in estimating drug dosage and fluid and dietary allowances for patients of varying sizes. New Engl. J. Med. **262**, 903—908 (1960).

BUTLER, T.: The distribution of drugs. In: LADU, B. N., MANDEL, H. G., WAY, E. L. (Eds.): Fundamentals of drug metabolism and drug disposition, pp. 44—62. Baltimore: Williams and Wilkins 1971.

CALCAGNO, P. L., RUBIN, M. I.: Renal extraction of para-aminohippurate in infants and children. J. clin. Invest. **42**, 1632—1639 (1963).

CHIGNELL, C. F.: Application of physicochemical and analytic techniques to the study of drug interactions with biological systems. Critical Rev. Toxicol. **1**, 413—465 (1972).

COTTEN, M. DE V. (Ed.): Immunopharmacology. Pharmacol. Rev. **25**, 157—363 (1973).

COURT, J. M., DUNLOP, M. E., LEONARD, R. F.: High-frequency oscillation of blood free fatty acid levels in man. J. appl. Physiol. **31**, 345—347 (1971).

CUATRECASAS, P.: Affinity chromatography of macromolecules. Advan. Enzymol. **36**, 29—89 (1972).

DAVIDSON, M.: In: COOK, R. E. (Ed.): Biologic basis of pediatric practice, Chapters 79 and 80, pp. 812—824. New York: McGraw Hill 1968.

DICKER, S. E.: Mechanisms of urine concentration and dilution in mammals, Chapter 9, pp. 133—148. London: Edward Arnold Ltd. 1970.

DONALDSON, R. M., JR.: Normal bacterial populations of the intestine and their relation to intestinal function. New Engl. J. Med. **270**, 938—945 (1964).

DONE, A. K.: Developmental pharmacology. Clin. Pharmacol. Therap. **5**, 432—479 (1964).

DREYER, G., WALKER, E. W. A.: Therapeutical and pharmacological section: dosage of drugs, toxins and antitoxins. Proc. roy. Soc. Med. **7**, 51—70 (1914).

Drugs and the Unborn Child: Clin. Pharmacol. Therap. **14**, 619—770 (1973).

FINLAND, M., MEADS, M., ORY, E. M.: Oral penicillin. J. Amer. med. Ass. **129**, 315—320 (1945).

FRIIS-HANSEN, B.: Changes in body water compartments during growth. Acta paediat. Supplement **110**, 1—68 (1956).

FRIIS-HANSEN, B.: Body water compartments in children: changes during growth and related changes in body composition. Pediatrics **28**, 169—181 (1961).

GITLIN, D.: Development and metabolism of the immune globulins. In: KAGAN, B. M., STIEHM, E. R. (Eds.): Immunologic incompetance, pp. 3—13. Year Book Medical Publishers 1971.

GITLIN, D., BOESMAN, M.: Serum α-fetoprotein, albumin, and γ G-globulin in the human conceptus. J. clin. Invest. **45**, 1826—1838 (1966).

GLUECK, C. J., LEVY, R. I., GLUECK, H. I., GRALNICK, H. R., GRETEN, H., FREDRICKSON, D. S.: Acquired type I hyperlipoproteinemia with systemic lupus erythematosus, dysglobulinemia, and heparin resistance. Amer. J. Med. **47**, 318—324 (1969).

GOLDSTEIN, A., ARONOW, L., KALMAN, S. M.: Principles of drug action: the basis of pharmacology, pp. 136—145. New York: Hoeber Med. Div., Harper and Row 1968a.

GOLDSTEIN, A., ARONOW, L., KALMAN, S. M.: Principles of drug action: the basis of pharmacology, pp. 194—202. New York: Hoeber Med. Div., Harper and Row 1968b.

GOODMAN, D. S.: The interaction of human serum albumin with long-chain fatty acid anions. J. Amer. chem. Soc. **80**, 3892—3898 (1958).

GORBACH, S. L., PLAUT, A. G., NAHAS, L., WEINSTEIN, L., SPANKNEBEL, G., LEVITAN, R.: Studies of intestinal microflora. II. Microorganisms of the small intestine and their relations to oral and fecal flora. Gastroenterology **53**, 856—867 (1967).

GROSSMAN, M., TICKNOR, W.: Serum levels of ampicillin, cephalothin, cloxacillin, and nafcillin in the newborn infant. Antimicrob. Agents Chemother. 214—219 (1965).

GRUBB, R.: The genetic markers of human immunoglobulins, pp. 1—2. Berlin, Heidelberg, New York: Springer 1970.

GRUNDBACHER, F. J., SHREFFLER, D. C.: Changes in human serum immunoglobulin levels with age and sex. Z. Immunitätsforsch. **141**, 20—26 (1970).

HIRSCH, G. H., HOOK, J. B.: Maturation of renal organic acid transport: substrate stimulation by penicillin. Science **165**, 909—910 (1969).

HOLGER-MADSEN, T.: Reduction of heparin activity by plasma globulins in patients with increased heparin resistance. Acta haemat. **27**, 157—170 (1962).

HOLLINGSWORTH, J. W., HASHIZUME, A., JABLON, S.: Correlations between tests of aging in Hiroshima subjects—an attempt to define "physiologic age". Yale J. Biol. Med. **38**, 11—26 (1965).

HORNING, E. C., HORNING, M. G., CARROLL, D. I., DZIDIC, I., STILLWELL, R. N.: Chemical ionization mass spectrometry. In: COSTA, E., HOLMSTEDT, B. (Eds.): Advances in biochemical psychopharmacology, Vol. 7, pp. 15—31. New York: Raven Press 1973.

HORNING, M. G., HARVEY, D. J., NOWLIN, J., STILLWELL, W. G., HILL, R. M.: The use of gas chromatography-mass spectrometry methods in perinatal pharmacology. In: COSTA, E., HOLMSTEDT, B. (Eds.): Advances in biochemical psychopharmacology, Vol. 7, pp. 113—124. New York: Raven Press 1973.

HUANG, N. N., HIGH, R. H.: Comparison of serum levels following the administration of oral and parenteral preparations of penicillin to infants and children of various age groups. J. Pediat. **42**, 657—668 (1953).

HYVARINEN, M., ZELTZER, P., OH, W., STIEHM, E. R.: Influence of gestational age on serum levels of α-1-fetoprotein, IgG globulin and albumin in newborn infants. J. Pediat. **82**, 430—437 (1973).

International Symposium on Comparative Pharmacology: Fed. Proc. **26**, 964—1265 (1967).

IVY, A. A., GROSSMAN, M. I.: Digestive system. In: LANSING, A. I. (Ed.): Cowdry's problems of aging biological and medical aspects, 3rd ed., Chapter 20, pp. 481—526. Baltimore: Williams and Wilkins 1952.

JONES, D. P., PERMAN, E. S., LIEBER, C. S.: Free fatty acid turnover and triglyceride metabolism after ethanol ingestion in man. J. Lab. clin. Med. **66**, 804—813 (1965).

JUSKO, W. J., LEVY, G., YAFFE, S.: Effect of age on intestinal absorption of riboflavin in humans. J. Pharm. Sci. **59**, 487—490 (1970a).

JUSKO, W. J., KHANNA, N., LEVY, G., STERN, L., YAFFE, S. J.: Riboflavin absorption and excretion in the neonate. Pediatrics **45**, 945—949 (1970b).

JUSKO, W. J., LEVY, G., YAFFE, S. J., ALLEN, J. E.: Riboflavin absorption in children with biliary obstruction. Amer. J. Dis. Child. **121**, 48—52 (1971).

KEEN, P.: Effect of binding to plasma proteins on the distribution, activity, and elimination of drugs. In: BRODIE, B. B., GILLETTE, J. R. (Eds.): Handbook of pharmacology, Vol. 28, Part I, pp. 213—233. Concepts in Biochemical Pharmacology. Berlin, Heidelberg, New York: Springer 1971.

KEYS, A., MICKELSON, O., MILLER, E. v. O., HAYES, E. R., TODD, R. L.: Concentration of cholesterol in the blood serum of normal man and its relation to age. J. clin. Invest. **29**, 1347—1353 (1950).

KNAPP, D. R., GAFFNEY, T. E.: Use of stable isotopes in pharmacology-clinical pharmacology. Clin. Pharmacol. Therap. **13**, 307—316 (1972).

KRASNER, J., GIACOIA, G. P., YAFFE, S. J.: Drug-protein binding in the newborn infant. Ann. N. Y. Acad. Sci. **226**, 101—114 (1973).

KUNIN, C. M., CRAIG, W. A., KORNGUTH, M., MONSON, R.: Influence of binding on the pharmacologic activity of antibiotics. Ann. N. Y. Acad. Sci. **226**, 214—224 (1973).

LANZKOWSKY, P., LLOYD, E. A., LAHEY, M. E.: The oral D-xylose test in healthy infants and children: a micro-method of blood determination. J. Amer. med. Ass. **186**, 517—519 (1963a).

LANZKOWSKY, P., MADENLIOGLU, M., WILSON, J. F., LAHEY, M. E.: Oral D-xylose test in healthy infants and children. New Engl. J. Med. **268**, 1441—1444 (1963b).

LEVI, A. J., GATMAITAN, Z., ARIAS, I. M.: Deficiency of hepatic organic anion-binding protein, impaired organic anion uptake by liver and "physiologic" jaundice in newborn monkeys. New Engl. J. Med. **283**, 1136—1139 (1970).

LIBOW, L. S.: In: BIRREN, J. E., BUTLER, R. N., GREENHOUSE, S. W., SOKOLOFF, L. (Eds.): Human aging, a biological and behavioural study, Chapter 5, pp. 37—56. Public Health Publication 986 (1963).

LITWACK, G., KETTERER, B., ARIAS, I. M.: Ligandin: a hepatic protein which binds steroids, bilirubin carcinogens and a number of exogenous organic anions. Nature (Lond.) **234**, 466—467 (1971).

MACDONALD, M. S., EMERY, J. L.: The late intrauterine and postnatal development of human renal glomeruli. J. Anat. **93**, 331—340 (1959).

MCCRACKEN, G. H., JR.: Clinical pharmacology of gentamicin in infants 2 to 24 months of age. Amer. J. Dis. Child. **124**, 884—887 (1972).

MCCRACKEN, G. H., JR., JONES, L. G.: Gentamicin in the neonatal period. Amer. J. Dis. Child. **120**, 524—533 (1970).

MCMAHON, T.: Size and shape in biology. Science **179**, 1201—1204 (1973).

MCMENAMY, R. H., ONCLEY, J. L.: The specific binding of L-tryptophan to serum albumin. J. biol. Chem. **233**, 1436—1447 (1958).

MEYER, J., SORTER, H., OLIVER, J., NECHELES, H.: Studies in old age. VII. Intestinal absorption in old age. Gastroenterology **1**, 876—881 (1943).

MILLER, J. H., MCDONALD, R. K., SHOCK, N. W.: The renal extraction of *p*-aminohippurate in the aged individual. J. Gerontol. **6**, 213—216 (1951).

MILLER, R. A.: Observations on the gastric acidity during the first month of life. Arch. Dis. Childhood **16**, 22—30 (1941).

MIRKIN, B. L.: Symposium on developmental pharmacology. Fed. Proc. **31**, 43—80 (1972).

MISHKIN, S., STEIN, L., GATMAITAN, Z., ARIAS, I.M.: The binding of fatty acids to cytoplasmic proteins: binding to Z protein in liver and other tissues of the rat. Biochem. Biophys. Res. Commun. **47**, 997—1003 (1972),

NATHANIELSZ, P.W.: The effect of free fatty acids on resin uptake of tri-iodothyronine from human, rat, rabbit, and guinea-pig serum and human serum albumin. J. Endocrinol. **45**, 489—493 (1969).

NEBERT, D.W.: Use of fetal cell culture as an experimental system for predicting drug metabolism in the intact animal. Clin. Pharmacol. Therap. **14**, 693—699 (1973).

NORRIS, A.H., SHOCK, N.W.: Aging and variability. Ann. N.Y. Acad. Sci. **134**, 591—601 (1966).

OCKNER, R.K., MANNING, J.A., POPPENHAUSEN, R.B., HO, W.K.L.: A binding protein for fatty acids in cytosol of intestinal mucosa, liver, myocardium, and other tissues. Science **177**, 56—58 (1972).

O'CONNOR, W.J., WARREN, G.H., EDRADA, L.S., MANDALA, P.S., ROSENMAN, S.B.: Serum concentrations of sodium nafcillin in infants during the perinatal period. Antimicrob. Agents Chemother. 220—222, **1965.**

PARKE, D.V., LINDUP, W.E.: Quantitative and qualitative aspects of the plasma protein binding of carbenoxolone, an ulcer healing drug. Ann. N.Y. Acad. Sci. **226**, 200—213 (1973).

PROX, A.: Some applications of mass spectrometry in drug metabolism studies. Xenobiotica **3**, 473—492 (1973).

RAMMELKAMP, C.H., HELM, J.D., JR.: Studies on absorption of penicillin from stomach. Proc. Soc. exp. Biol. (N.Y.) **54**, 324—327 (1943).

REYES, H., LEVI, J.A., GATMAITAN, Z., ARIAS, I.M.: Studies of Y and Z two hepatic cytoplasmic organic anion-binding proteins: effects of drugs, chemicals, hormones, and cholestasis. J. clin. Invest. **30**, 2242—2252 (1970).

RODAHL, K., MILLER, H.I., ISSEKUTZ, B., JR.: Plasma free fatty acids in exercise. J. appl. Physiol. **19**, 489—492 (1964).

SALATKA, K., KRESGE, D., HARRIS, L., JR., EDELSTEIN, D., OVE, P.: Rat serum protein changes with age. Exp. Geront. **6**, 25—36 (1971).

SCHANKER, L.S.: In: LADU, B.N., MANDEL, H.G., WAY, E.L. (Eds.): Drug absorption. In: Fundamentals of drug metabolism and drug disposition. Baltimore: Williams and Wilkins 1971.

SHOCK, N.W.: Age changes in renal function. In: LANSING, A.I. (Ed.): Cowdry's problems of aging, 3rd ed., pp. 614—630. Baltimore: Williams and Wilkins 1952.

SMITH, C.A.: The physiology of the newborn infant, 2nd ed., pp. 180—198. Springfield, Ill.: Charles C. Thomas 1951.

SPECTOR, A.A.: Metabolism of free fatty acids. Progr. Biochem. Pharmacol. **6**, 130—176 (1971).

SPECTOR, A.A., SANTOS, E.C., ASHBROOK, J.D., FLETCHER, J.E.: Influence of free fatty acid concentration on drug binding to plasma albumin. Ann. N.Y. Acad. Sci. **226**, 247—258 (1973).

SPENCER, R.P.: Variation of intestinal activity with age: a review. Yale J. Biol. Med. **37**, 105—129 (1964).

STAHL, W.R.: Organ weights in primates and other mammals. Science **150**, 1039—1042 (1965).

SWELL, L., FIELD, H., JR., TREADWELL, C.R.: Relation of age and race to serum cholesterol ester fatty acid composition. Proc. Soc. exp. Biol. (N.Y.) **105**, 129—131 (1960).

TABACHNICK, M.: Thyroxine-protein interactions. III. Effect of fatty acids, 2,4-dinitrophenol and other anionic compounds on the binding of thyroxine by human serum albumin. Arch. Biochem. Biophys. **106**, 415—421 (1964).

VENTER, J.C., DIXON, J.E., MARKORS, P.R., KAPLAN, N.O.: Biologically active catecholamines covalently bound to glass beads. Proc. nat. Acad. Sci. (Wash.) **69**, 1141—1145 (1972).

WACHSTEIN, M., BRADSHAW, M.: Histochemical localization of enzyme activity in the kidneys of three mammalian species during their postnatal development. J. Histochem. Cytochem. **13**, 44—56 (1965).

WAGNER, J.: Biopharmaceutics: influence of formulation on therapeutic activity. 1. Definition, scope, and relationship to gastrointestinal physiology. Drug Intelligence **2**, 30—34 (1968).

WALSH, C.T., LEVINE, R.R.: Drug absorption in rats pretreated with antibiotics. J. Pharmacol. exp. Ther. **188**, 277—286 (1974).

WEIL, W.B., JR.: The evaluation of renal function in infancy and childhood. Amer. J. med. Sci. **229**, 678—694 (1955).

WEST, J.R., SMITH, H.W., CHASIS, H.: Glomerular filtration rate, effective renal blood flow, and maximal tubular excretory capacity in infancy. J. Pediat. **32**, 10—18 (1948).

WEST, C.D., HONG, R., HOLLAND, N.H.: Immunoglobulin levels from the newborn period to adulthood and in immunoglobulin deficiency states. J. clin. Invest. **41**, 2054—2064 (1962).

Chapter 67

Pathological and Physiological Factors Affecting Drug Absorption, Distribution, Elimination, and Response in Man

L. F. PRESCOTT

With 8 Figures

I. Introduction

Although drugs are used primarily for the treatment of disease, surprisingly little is known of the way in which disease states or physiological factors modify drug action in man. In contrast, extensive studies in laboratory animals have shown that these factors can have striking effects which if applied to man would have enormous clinical significance. The failure to apply this basic knowledge to clinical medicine amounts virtually to negligence on the part of the medical profession.

Physiological and pathological factors can modify drug response through effects on receptor sensitivity or changes in the concentration of active drug at receptors due to abnormalities in drug absorption, distribution or elimination. This review is written essentially from a clinical point of view. Emphasis therefore will be placed on adequately documented examples of clinical importance rather than data obtained from studies in laboratory animals.

II. Drug Absorption

A. Parenteral Administration

Clinicians know from long experience that direct intravenous injection is the most reliable route of drug administration and that absorption following subcutaneous or intramuscular injection can be slow and erratic in patients with peripheral circulatory failure. However, little detailed information is available concerning the kinetics of drug absorption from these sites under different conditions. The site of an intramuscular injection may be important; for example, lidocaine (lignocaine) injected into the upper arm muscles is absorbed more than twice as rapidly as the same dose injected into the thigh (MEYER and ZELECHOWSKI, 1971). The rate of absorption of local anesthetics is influenced by the effects of the specific agent on local blood vessels (e.g., vasodilation), the site of injection, total dose administered, and the presence of vasoconstrictors (COVINO, 1972). The absorption of drugs injected either subcutaneously or intramuscularly is likely to be accelerated in patients with peripheral vasodilatation caused by exercise, fever, anemia, thyrotoxicosis, Paget's disease, hypercarbia, liver disease etc., but this does not seem to have been investigated. In dermatological practice, drugs are often applied directly to the skin. Systemic absorption from intact skin may be negligible, but if drugs are applied to extensive areas of raw or inflamed skin enough may be absorbed to cause intoxication (WECHSELBERG, 1968).

B. Oral Administration

Physiological and pathological factors that can influence the absorption of drugs from the gastrointestinal tract of laboratory animals include the volume, pH, temperature, viscosity, surface tension and composition of secretions and contents, the presence or absence of food, bile salts and bacterial flora, the splanchnic blood flow, previous diet and food intake and gastrointestinal motility (LEVINE, 1970). Almost nothing is known of the effects of these variables in man in health or disease, yet striking intersubject variation in the absorption of orally administered drugs has been observed repeatedly both in hospital patients (ARMSTRONG et al., 1970; KOCH-WESER, 1971; NELSON et al., 1972) and in healthy volunteers under carefully controlled conditions (PRESCOTT et al., 1970; SCHRÖDER and CAMPBELL, 1972; BEERMANN et al., 1972). An example of intersubject variation in the absorption of phenacetin is shown in Fig. 1. We also have observed more than a 7-fold range in the amount of tetracycline absorbed by 6 fasting healthy subjects (PRESCOTT and NIMMO, 1971a).

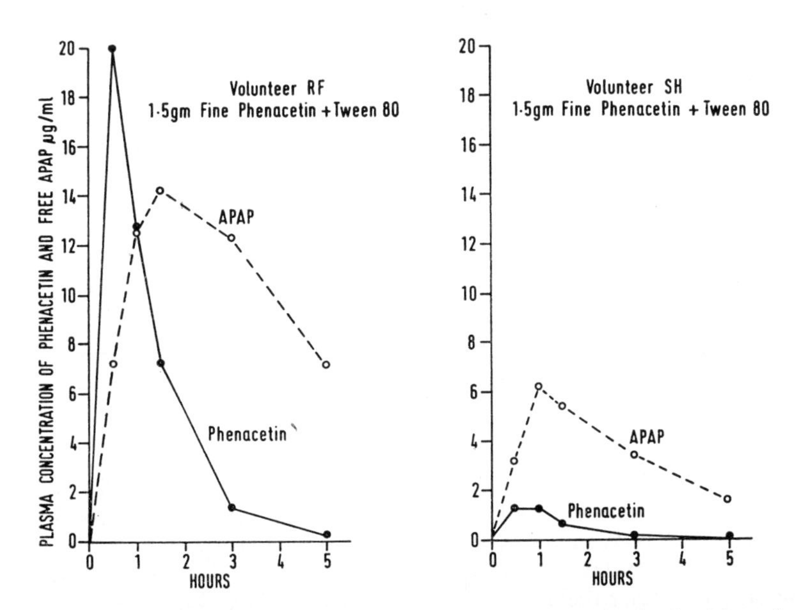

Fig. 1. Individual variation in the absorption of phenacetin in two fasting healthy volunteers receiving 1.5 g of the drug in a fine suspension. Plasma concentrations of acetaminophen (paracetamol) (APAP), the major metabolite of phenacetin, are also given. (From PRESCOTT et al., 1970)

1. Effects of Food

The effects of diet and food on drug absorption have been reviewed recently by PLACE and BENSON (1971). Food can have variable effects on drug absorption, depending on the drug and the type of meal taken. The absorption of many drugs is slowed but the total amount absorbed may not be reduced to a clinically significant extent (MACDONALD et al., 1967; GOWER and DASH, 1969; WHITE et al., 1971; JAFFE et al., 1971). On the other hand, food can significantly inhibit the absorption of some drugs (PLACE and BENSON, 1971); tetracycline absorption is grossly impaired when it is taken with dairy products (SCHEINER and ALTEMEIER, 1962). Paradoxically, the absorption of griseofulvin is greatly increased by a fatty meal (CROUNSE, 1961). Griseofulvin is poorly soluble and it has been suggested that dissolution is enhanced by the increased secretion of bile following ingestion of fats. Doctors rarely give their patients precise instructions concerning the taking of drugs in relation to food. Prolonged fasting has been shown to reduce the rate of drug absorption in animals (DOLUISIO et al., 1969) and in this context it is perhaps significant that we have observed grossly impaired absorption of acetaminophen (paracetamol) in a patient with anorexia nervosa. Diurnal variation in absorption also may occur since significantly less acetaminophen is absorbed in healthy subjects during the night than during the day (McGILVERAY and MATTOK, 1972).

2. Gastrointestinal pH

According to the pH-partition hypothesis (BRODIE, 1964), weak acids are readily absorbed from the acid gastric contents while bases are absorbed primarily from the more alkaline contents of the upper small intestine. However, in practice, acidic drugs such as warfarin, aspirin and barbiturates are absorbed much more slowly from the stomach than from the small intestine (SIURALA et al., 1969; KEKKI et al., 1971; KOJIMA et al., 1971), presumably because of the much greater relative surface area of the latter. Even ethanol is absorbed much more rapidly from the small intestine (SIEGERS et al., 1972), and the stomach does not seem to be an important site of drug absorption. As a result, changes in the pH of gastric contents have much less effect than the rate of gastric emptying on the overall rate of drug absorption. Achlorhydria occurs in patients with pernicious anemia, gastric carcinoma and atrophic gastritis and is also found in about 50% of elderly subjects without gastrointestinal disease (ANDERSSON, 1971). The absorption of salicylamide and aspirin has been studied in patients with pernicious anemia, achlorhydria and atrophic gastritis, but the results were conflicting and absorption could not be related clearly to gastric pH (HARTIALA et al., 1963; SIURALA et al., 1969).

3. Gastric Emptying Rate

The rate of gastric emptying is one of the most important factors controlling the rate of absorption of orally administered drugs (LEVINE, 1970; PRESCOTT and NIMMO, 1971b). We have shown that the rate of absorption of acetaminophen in man is directly related to the rate of gastric emptying as measured by an isotope scintiscanning technique (HEADING et al., 1973). Rapid gastric emptying was associated with rapid absorption while absorption was slow when gastric emptying was delayed. Much of the intersubject variation in drug absorption probably can be attributed to differences in the rate of gastric emptying since this is influenced by physiological factors such as posture, autonomic activity and the temperature, volume, viscosity, tonicity, and composition of gastric contents. The significance of this variation can be judged from our observation of a 100-fold range in plasma acetaminophen concentrations one hour after ingestion of a dose under controlled conditions in patients without gastrointestinal disease. Gastrointestinal motility is often abnormal in ill patients, and we have found that acetaminophen absorption is markedly impaired in patients with gastric stasis and pyloric stenosis. In such circumstances it is probably futile to give any drug orally.

Not surprisingly, drugs which influence gastric emptying can have important effects on the rate of drug absorption. Thus propantheline inhibits gastric emptying and slows down the absorption of acetaminophen while metoclopramide, a drug that stimulates gastric emptying, increases the rate of acetaminophen absorption (NIMMO et al., 1973) (Figs. 2 and 3). In other studies we have shown that the absorption of orally administered lidocaine is markedly delayed in patients premedicated with atropine prior to laparoscopy (ADJEPON-YAMOAH et

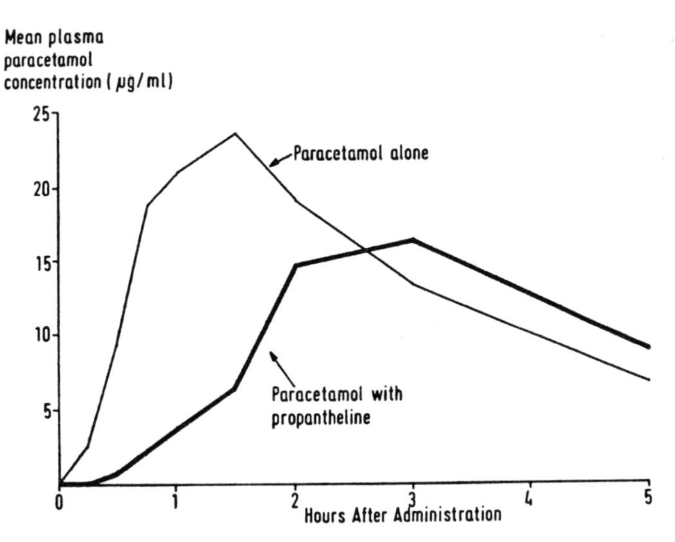

Fig. 2. Inhibitory effect of propantheline (30 mg i.v.) on the absorption of paracetamol (acetaminophen) in 6 patients given an oral dose of 1.5 g. (From NIMMO et al., 1973)

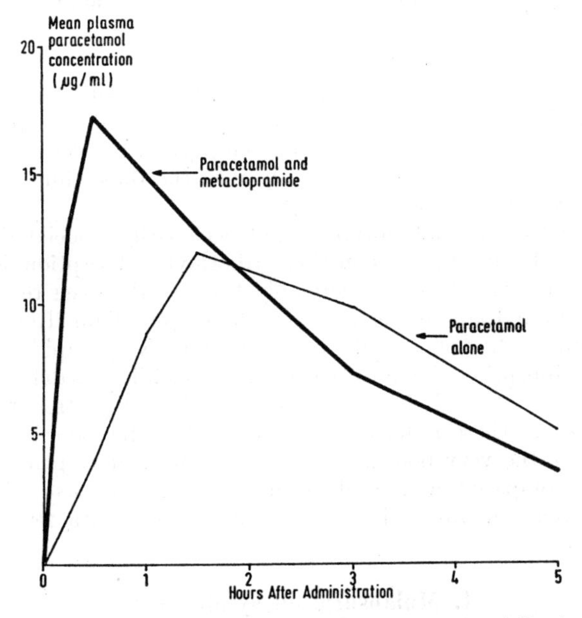

Fig. 3. Increased rate of paracetamol (acetaminophen) absorption following 10 mg of metoclopramide i.v. in 5 healthy volunteers given 1.5 g of paracetamol orally. (From NIMMO et al., 1973)

al., 1973) (Fig. 4). Poor and delayed absorption of drugs such as acetaminophen (GWILT et al., 1963) can probably be explained largely by slow gastric emptying, and this may also account in part for the effects of food on drug absorption. The absorption of basic drugs and compounds absorbed by active transport in the small intestine will obviously depend on the rate of gastric emptying. For

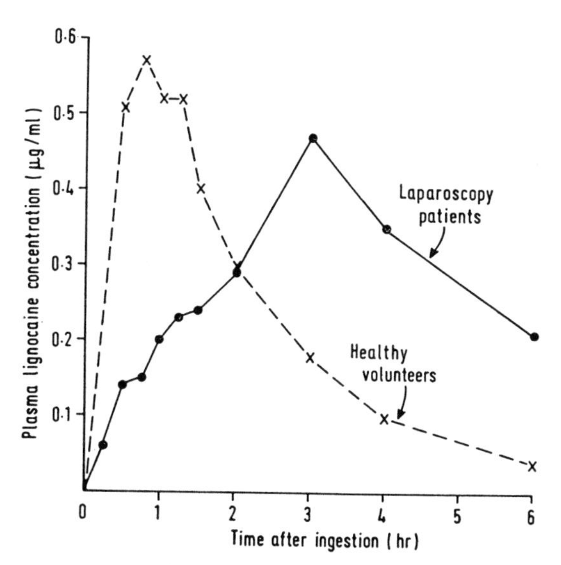

Fig. 4. Mean plasma lignocaine (lidocaine) concentrations after ingestion of 400 mg of lignocaine hydrochloride in 7 healthy volunteers and in 12 laparoscopy patients premedicated with atropine only. (From ADJEPON-YAMOAH et al., 1973)

example, not only is L-dopa absorbed in the small intestine by active transport but it is also extensively metabolized in the stomach and therapeutic failure of L-dopa has been attributed to slow gastric emptying. Conversely, absorption is rapid when the drug is introduced directly into the duodenum and in patients with gastrectomy (BIANCHINE et al., 1971).

Decreased gastrointestinal motility is not necessarily associated with reduced drug absorption. Under these conditions riboflavin absorption is actually increased (although it is delayed) presumably because the drug remains longer in contact with a limited area of actively absorbing epithelium (LEVY et al., 1972). Drug absorption might conceivably be reduced by very rapid gastrointestinal transit and the absorption of nalidixic acid and ampicillin was delayed or impaired in 30 to 50% of infants and children with acute shigella gastroenteritis (NELSON et al., 1972). Often, there is no obvious explanation for poor drug absorption. Examples include the very poor absorption of sulfonamides placed directly into the duodenum compared with oral administration (PETERSON and FINLAND, 1942) and the reduced absorption of iron in children with fever (BERESFORD et al., 1971).

4. Malabsorption Syndromes

Although the absorption of nutrients, vitamins, and minerals has been studied extensively in patients with malabsorption syndromes, the absorption of drugs has received little attention. Reduced absorption of cortisol, penicillin, and digoxin has been described (SCHEDL and CLIFTON, 1963; DAVIS and PIROLA, 1968; HEIZER et al., 1971), and our own studies have revealed essentially normal absorption of acetaminophen and methaqualone but reduced absorption of tetracycline (NIMMO et al., unpublished). Enterotoxic drugs such as neomycin, *p*-aminosalicylic acid and colchicine can cause malabsorption syndromes, and marked impairment of vitamin B_{12} absorption has been observed in patients receiving 2.6 mg of colchicine daily (WEBB et al., 1968).

5. Gastrointestinal Surgery

Conflicting results have been obtained in absorption studies in patients who have had gastrointestinal surgery. In one study *p*-aminosalicylic acid and isoniazid absorption was unaffected by gastrectomy for peptic ulcer, but in some patients there was complete failure of ethionamide absorption (MATTILA et al., 1969). In another study, the absorption of sulfisoxisole, quinidine and ethambutol was unaffected by gastrectomy unless gastric emptying had been slowed by vagotomy (VENHO et al., 1972). The colon is not normally considered an important site for drug absorption, however, failure of salazopyrin absorption has recently been observed in patients with ulcerative colitis who have had ileostomy or colonic resection (DAS and EASTWOOD, 1973). Salazopyrin is normally extensively metabolized to sulphapyridine and 5-aminosalicylic acid through azo reduction and cleavage prior to absorption, and this reaction is apparently carried out by bacteria in the large intestine (SCHRÖDER and CAMPBELL, 1972). Many other drugs are known to be extensively metabolized by the gastrointestinal mucosa or intestinal bacterial flora. Examples include salicylamide, isoprenaline, L-dopa, chlorpromazine, propranolol, propantheline, and cyclamate. We know little about the effects of physiological factors or disease on the enteric metabolism of such compounds.

6. Mesenteric Blood Flow and Biliary Tract Disease

The absorption of drugs such as the cardiac glycosides is highly dependent on portal blood flow in animals (HAASS et al., 1972), but apart from the limited observations of HEIZER et al. (1971) nothing is known of the effects of mesenteric ischemia on drug absorption in man.

The hydrolysis of propantheline to xanthene carboxylic acid is pH-dependent and is catalyzed by a heat-labile factor in duodenal bile (BEERMANN et al., 1972). The biliary excretion of propantheline is greatly stimulated by injection of cholecystokinin, and presumably the biliary excretion of other high molecular

Table 1. *Mean concentrations of cephalexin in bile and gall bladder wall in patients with biliary tract disease*

Gall bladder pathology	Cephalexin concentrations		Number of patients
	Gall bladder bile (µg/ml)	Gall bladder wall (µg/ml)	
Functioning[1]	44.2	4.5	6
Non-functioning[1]	9.4	2.5	7
Complete biliary obstruction	0	—	2

[1] Assessed radiologically. (From SALES et al., 1972.)

weight polar drugs and metabolites is influenced by physiological stimuli to bile production such as ingestion of a fatty meal. The clinical significance of this effect is unknown, but biliary excretion of drugs may be greatly reduced in patients with biliary tract disease and obstructive jaundice (SALES et al., 1972). The effect of gall bladder disease on the biliary excretion of cephalexin is shown in Table 1.

III. Drug Distribution

The kinetics of drug distribution may influence the time of onset, magnitude and duration of drug action. Distribution can be modified by physiological factors and disease states since it can depend on such factors as regional tissue blood flow, cardiac output, pH gradients, plasma protein binding, and the permeability of cell membranes. From a practical point of view, abnormalities of drug distribution are important for two main reasons: (1) the concentration of active drug at receptors may be changed and (2) the rate of drug elimination depends on the volume of distribution.

A. Regional Tissue Distribution

Antibiotics pass more readily into tissues in the presence of inflammation (e.g., meningitis) because of increased vascularity and increased permeability of the blood-cerebrospinal fluid (CSF) barrier. However, the protein content of body fluids such as CSF, exudates and effusions may determine the total concentrations of drugs that are extensively protein bound and the active unbound concentrations may not differ from those in plasma. Ampicillin, for example, is not extensively bound to plasma proteins (18%) and concentrations in synovial fluid are only slightly lower than those in plasma. In contrast, cloxacillin is highly bound (94%), and although total concentrations in synovial fluid are much lower than in plasma, the concentrations of unbound drug are comparable (HOWELL et al., 1972). Similarly, CSF concentrations of diphenylhydantoin correspond to the concentration of unbound drug in the plasma (LUND et al., 1972). The local distribution of drugs in the kidney may be greatly influenced by the state of hydration. Thus increasing concentration gradients from renal cortex to medulla and papilla have been observed with drugs such as acetaminophen, penicillin and cephalothin in hydropenic dogs (BLUEMLE and GOLDBERG, 1968; WHELTON et al., 1971). These gradients are abolished by hydration, and increased renal medullary concentrations of potentially toxic drugs and metabolites during water deprivation are probably significant factors in the etiology of analgesic nephropathy (NANRA et al., 1970).

The distribution of drugs between intracellular and extracellular water may be altered by pathological changes in acid-base balance, particularly with drugs that have pKa values close to physiological blood pH (7.4). In such circumstances, small changes in pH have a disproportionate effect on the ionization of weak organic acids and bases. HAYES (1971) suggested that the myocardial uptake and efficacy of lidocaine (pKa 7.85) is reduced by severe acidosis, and WADDELL and BUTLER (1957) described the reverse effect in dogs receiving phenobarbital (pKa 7.2).

B. Volume of Drug Distribution

The apparent volume of drug distribution is an important pharmacokinetic concept and is a measure of the overall tissue uptake. The volumes of distribution of lidocaine and procainamide are reduced in patients with cardiac failure (THOMSON et al., 1971; KOCH-WESER, 1971), presumably because of reduced cardiac output and poor perfusion of peripheral tissues. However, drug concentrations in adequately perfused tissues such as brain and myocardium are increased with a greater risk of toxicity. On the other hand, the volume of distribution may be increased in patients with a high cardiac output, vasodilatation and increased tissue perfusion. Thus the volume of distribution of amobarbital is apparently increased by exercise or inhalation of amyl nitrite (BALASUBRAMANIAM et al., 1970). NIMMO in our group has observed significantly higher plasma concentrations of acetaminophen after oral doses in convalescent hospital patients in bed than in healthy ambulant volunteers (Fig. 5). This effect could not be attributed to differences in body weight or rates of acetaminophen absorption or elimination. The most likely explanation is an increase in the volume of distribution produced by exercise performed by volunteers. It is most satisfying to be able to explain on a pharmacokinetic basis the firmly established belief that a headache is more likely to be relieved rapidly if the sufferer takes a dose of analgesic and lies down.

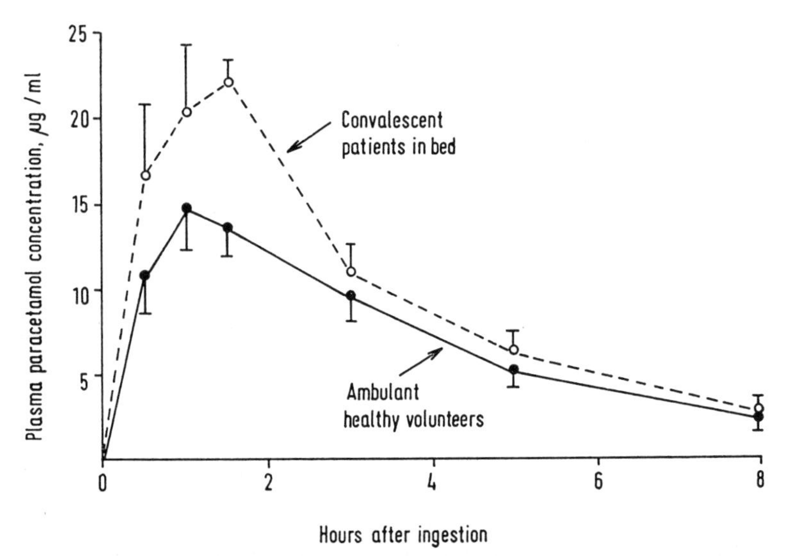

Fig. 5. Plasma concentrations of paracetamol (acetaminophen) following an oral dose of 1.5 g in 7 ambulant healthy volunteers and 12 convalescent hospital patients in bed. Values given are means ± S.E.M. (NIMMO, unpublished)

C. Plasma Protein Binding

Many illnesses are complicated by abnormalities of plasma proteins, and this may result in significant changes in the distribution of drugs that are extensively protein-bound. Theoretically, the fraction of unbound drug is increased if plasma protein concentrations are reduced and effects will be correspondingly enhanced since response appears to be related to the concentration of unbound drug. Reduced binding of sulfonamides to plasma albumin has been observed in various disease states (ANTON and COREY, 1971), in the very young and in the elderly (CHIGNELL et al., 1971; SERENI et al., 1968). The binding of amobarbital to serum albumin was reduced in patients with chronic liver disease who had serum albumin concentrations below 3.5 g/100 ml, but there was no obvious change in clinical response (MAWER et al., 1972). On the other hand, thiopental anesthesia was significantly prolonged in patients with hypoalbuminemia due to chronic liver disease (SHIDEMAN et al., 1949). This effect cannot be attributed to impaired metabolism, since only a very small fraction of the dose could possibly have been metabolized during the period of anesthesia. Another study showed that the frequency of side effects from prednisone was doubled when the serum albumin concentrations were below 2.5 g/100 ml (LEWIS et al., 1971). Reduced binding of salicylate to plasma proteins has been described in children with Down's syndrome (mongolism) but the mechanism is unknown (EBADI and KUGEL, 1970).

Table 2. *Plasma protein binding of diphenylhydantoin and desmethylimipramine in patients with poor renal function*

	Diphenylhydantoin unbound (%)	Desmethylimipramine unbound (%)
Patients with renal disease	16.6 ± 4.7	12.1 ± 1.9
Normal subjects	7.4 ± 0.7	9.5 ± 1.4

Values given are means ± S.D. (From REIDENBERG et al., 1971.)

Significantly reduced binding of acidic drugs such as diphenylhydantoin and sulfonamides has been observed in patients with impaired renal function (ANTON and COREY, 1971; REIDENBERG et al., 1971; FISCHER, 1972). The binding of diphenylhydantoin may be reduced by a factor of 2 or 3 in patients with poor renal function (Table 2) and the drug may be effective at total plasma concentrations that would normally be quite ineffective (ODAR-CEDERLÖF et al., 1970). Reduced binding was related more to the degree of azotemia than to serum albumin concentrations and was unaffected by dialysis *in vitro*. Basic drugs, such as desmethylimipramine, quinidine and dapsone, seem to bind almost normally to plasma proteins in patients with renal disease (REIDENBERG et al., 1971; REIDENBERG, 1973). Reduced binding of thiopental to plasma proteins may account for the much smaller total dose required to maintain anesthesia for prostatectomy in patients with azotemia (DUNDEE and RICHARDS, 1954). On the other hand, drug effects can be reduced if binding to plasma proteins is increased. For example, *d*-tubocurarine is normally bound to plasma globulins, and resistance to this drug in patients with hypergammaglobulinemia due to liver disease has been attributed to increased protein binding (BARAKA and GABALI, 1968).

IV. Drug Metabolism

Most drugs are inactivated by conversion to more polar metabolites with little or no pharmacological activity. The liver is the most important site of drug metabolism and liver disease therefore might be expected to have important effects on the duration of drug action. Although impairment of drug metabolism usually results in exaggerated and prolonged effects, metabolites may be active, and in some cases are responsible for toxicity (MITCHELL et al., 1973). Many physiological factors and pathological conditions have been shown to modify drug metabolism in laboratory animals (Table 3). However, little is known of the

Table 3. *Some factors which influence the activity of hepatic drug metabolizing enzymes in animals*

Physiological factors	Pathological factors
Individual variation	Dehydration
Genetic factors	Hemorrhagic shock
Age	Endotoxin shock
Sex	Adjuvant arthritis
Diet	Malnutrition
Temperature	Vitamin deficiency
Stress	Endocrine disease
Pregnancy	Alloxan diabetes
Diurnal variation	Liver damage
Reduced barometric pressure	Obstructive jaundice
Hepatic blood flow	Intestinal obstruction
Enzyme induction	Malignant disease
	Drug withdrawal

clinical importance of these factors, and this is all the more surprising since many of the disease states studied in animals have direct counterparts in clinical medicine. Apart from the effects of age, sex, genetic factors, and exposure to microsomal enzyme-inducing agents, little information is available concerning the physiological control of drug metabolism in man (O'MALLEY et al., 1971; PRICE-EVANS, 1971; CONNEY et al., 1971).

A. Liver Disease

The subject of liver disease and drug metabolism in man has recently been reviewed (PRESCOTT and STEVENSON, 1973). The position is confused. Although both acute and chronic liver lesions are associated with reduced rates of drug metabolism in laboratory animals, many patients with advanced liver disease seem able to metabolize drugs at normal rates. Occasional patients with severe liver disease exhibited impaired metabolism of tolbutamide, diphenylhydantoin, acetanilide, phenylbutazone, isoniazid, meprobamate, lidocaine, and barbiturates, but in many patients the rates of metabolism were normal (UEDA et al., 1963; KUTT et al., 1964; HAMMAR and PRELLWITZ, 1966; LEVI et al., 1968; HELD and OLDERSHAUSEN, 1969; THOMSON et al., 1971; MAWER et al., 1972). Other investigators have failed to demonstrate reductions in the rates of metabolism of barbiturates, tolbutamide, chloramphenicol, and *p*-aminosalicylic acid in patients with cirrhosis or chronic hepatitis (SESSIONS et al., 1954; NELSON, 1964; HELD and OLDERSHAUSEN, 1971). Indeed, tolbutamide was eliminated more rapidly than normal in patients with acute hepatitis, possibly because of displacement from plasma albumin by bilirubin (HELD and OLDERSHAUSEN, 1971).

According to LEVI et al. (1968) prolongation of the isoniazid half-life in patients with liver disease was usually quantitatively less important than basic differences in acetylator phenotype, while previous intake of microsomal inducing drugs counterbalanced the effects of liver disease on phenylbutazone metabolism. Depression of drug metabolism is not closely related to the clinical severity of liver disease, but correlations with serum bilirubin concentration, bromsulphthalein excretion and plasma albumin concentration have been claimed by some investigators (LEVI et al., 1968; MAWER et al., 1972). In keeping with the variable effects of liver disease on drug metabolism in man, SCHOENE et al. (1972) found normal cytochrome P-450 content and *N*-demethylase activity in liver biopsy samples from patients with mild and moderate hepatitis. Cytochrome P-450 content was reduced to 50% of normal only in patients with severe hepatitis and cirrhosis. NADPH cytochrome C reductase activity was normal even in patients with severe liver disease.

There are several possible explanations for the apparent lack of effect of chronic liver disease on drug metabolism. From a functional point of view, the effects of acute and chronic liver pathology are quite different. However, this distinction is rarely made, and almost all studies have been carried out in patients with chronic liver disease and cirrhosis. Extensive hepatocellular necrosis occurs with acute liver injury and must necessarily be associated with loss of microsomal enzymes. On the other hand, cirrhosis produces interlobular fibrosis with disruption of the normal hepatic architecture and changes in local blood flow. But at the same time there is usually marked *hepatocellular hyperplasia* so that even if the amount of enzyme per gram of liver is reduced, the total amount of enzyme in the liver may be normal or even increased. Another factor that has not always been taken into account is the almost inevitable consumption of inducing drugs by patients with chronic liver disease. Finally, the plasma drug half-life usually has been used to assess drug metabolism. This is not necessarily a reliable index since the plasma half-life depends on the plasma clearance and the volume of distribution, both of which may be abnormal in advanced liver disease.

B. Acetaminophen-Induced Acute Hepatic Necrosis

In overdosage, acetaminophen causes acute centrilobular hepatic necrosis. Drug metabolism is depressed and the plasma half-life of acetaminophen itself is significantly correlated with the clinical severity of liver damage (PRESCOTT et al., 1971). We have also observed significant impairment of metabolism of barbiturates, diphenylhydantoin and antipyrine in patients with acute hepatic necrosis following acetaminophen overdosage (PRESCOTT and STEVENSON, 1973) (Table 4).

Table 4. *Mean plasma half-life of acetaminophen, antipyrine, and barbiturates in normal subjects and in patients with acute hepatic necrosis following acetaminophen overdosage*

Drug	Plasma half-life in normal subjects (h)	Plasma half-life in patients with acetaminophen-induced hepatic necrosis (h)	Number of patients
Acetaminophen	2.0 ± 0.1[a]	7.9 ± 1.3	23
Antipyrine	12.1 ± 0.4[b]	23.9 ± 2.2	14
Barbiturates	21.1 ± 1[c]	43 ± 9[d]	3

[a] PRESCOTT et al., 1971.
[b] O'MALLEY et al., 1971.
[c] MAWER et al., 1972.
[d] 2 patients took amobarbital and one took butobarbital.

Apparently, acetaminophen metabolism was affected more than that of antipyrine and the barbiturates. However, the acetaminophen half-life was measured within 36 h of overdosage whereas studies with the other drugs were of necessity carried out subsequently at varying intervals. Combined overdosage of acetaminophen with other drugs that are metabolized by the liver is particularly dangerous.

For example, a 27 year old man who took amobarbital with acetaminophen developed severe liver damage and remained in coma for a week. In contrast, a 33 year old woman who took approximately the same dose of amobarbital alone recovered consciousness in 36 h (PRESCOTT and STEVENSON, 1973).

Acetaminophen causes hepatic necrosis by the covalent binding of active intermediate metabolites to liver cell proteins, and the severity of necrosis depends on the activity of microsomal enzymes (MITCHELL et al., 1973; JOLLOW et al., 1973; POTTER et al., 1973).

We have shown that the hepatotoxicity of acetaminophen is significantly enhanced in patients who have previously taken drugs which are likely to cause microsomal enzyme induction such as alcohol and barbiturates (WRIGHT and PRESCOTT, 1973). In contrast to the rather variable effects of chronic liver disease on drug metabolism in man, acute liver damage seems to be associated with significant depression of drug metabolism.

C. Drug Metabolism in Other Pathological Conditions

Impaired hepatic drug metabolism has also been observed in ill patients in the absence of gross liver pathology. For example, the plasma clearance of lidocaine is reduced in patients with cardiac failure and this was attributed to reduced liver blood flow (THOMSON et al., 1971). However, we have also observed markedly reduced plasma clearances of lidocaine in patients with acute myocardial infarction *without* cardiac failure or hypotension (PRESCOTT and NIMMO, 1971c). In 6 such patients the mean plasma clearance of lidocaine was 421 ± 64 ml/min, compared with clearances of 701 ± 41 ml/min in normal subjects and 358 ± 147 ml/min in patients with cardiac failure quoted by THOMSON et al. (1971). Since the apparent volume of lidocaine distribution in our patients was normal (1.87 ± 0.31 l/kg), it is unlikely that cardiac output and peripheral perfusion were greatly reduced. The combination of reduced volume of distribution and depression of drug metabolism is particularly dangerous and a progressive rise in plasma lidocaine concentrations to toxic levels in a patient with severe cardiac failure and shock is shown in Fig. 6. We have observed abnormally slow metabolism of another drug in patients with cardiac pathology. The mean plasma half-life of a new anti-arrhythmic agent (2′,6′-dimethyl-phenoxy-2-aminopropane) was 18.7 ± 1.8 h in 15 patients with ventricular arrhythmias (most of whom had acute myocardial infarction) compared with 10.2 ± 0.8 h in 6 healthy volunteers (CLARKE et al., 1973). In the absence of liver blood flow measurements it is not possible to explain these findings. Prolongation of the half-life of drugs in patients with cardiac failure is often attributed to reduced liver blood flow and hepatic congestion, but a change in liver blood flow is likely to have a significant effect only on the metabolic clearance of drugs with a high hepatic extraction ratio (GILLETTE, 1971). Warfarin has a very low hepatic extraction ratio; in this context, it is interesting to note that the anticoagulant response and plasma half-life of warfarin did not seem to be abnormal in patients with acute myocardial infarction (RISTOLA and PYÖRÄLÄ, 1972). Other unexplained examples of impaired drug metabolism include decreased conjugation of salicylate with glycine in children with Down's syndrome (EBADI and KUGEL, 1970), and decreased hydroxylation of nalidixic acid in infants and children with acute shigellosis (NELSON et al., 1972).

Significant stimulation of hepatic drug metabolism has been described in clinical conditions in which patients have taken large doses of drugs that cause induction of liver microsomal enzymes. This effect has been observed in epileptics receiving anticonvulsants, patients dependent on barbiturates or the combination of methaqualone plus diphenhydramine ("Mandrax") (BALLINGER et al., 1972), patients who have recovered from acute barbiturate overdosage (PRESCOTT et al., 1973) and chronic alcoholics (KATER et al., 1969). An increased rate of clearance of tolbutamide from the plasma of patients with chronic respiratory disease and hypercapnia was attributed to hypoxemia and multiple drug therapy (SOTANIEMI

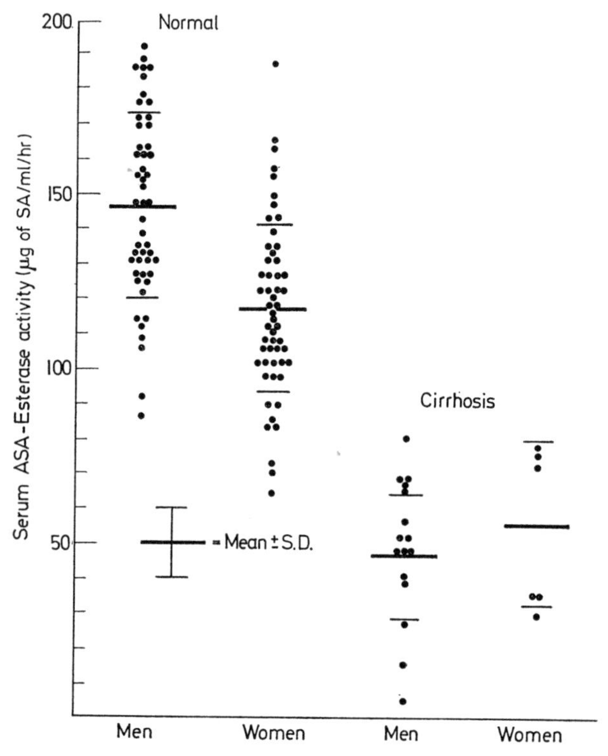

Fig. 6. Serum acetylsalicylic acid (ASA) esterase activity in normal subjects and in patients with biopsy-proven portal cirrhosis. (From MENGUY et al., 1972)

et al., 1971). In another report, resistance to digoxin in a patient with rheumatic heart disease was attributed to abnormal metabolism that resulted in the formation of metabolites with relatively little cardiac action (LUCHI and GRUBER, 1968). The mechanism is unknown, but full details of other drug therapy were not mentioned.

D. Extrahepatic Drug Metabolism

Little is known of the effects of disease on extrahepatic drug metabolism. However, pseudocholinesterase is synthesized by the liver, and ester hydrolysis may be depressed in patients with liver disease. Pseudocholinesterase activity may also be reduced in patients with poor renal function, but the mechanism is obscure. The duration of action of succinylcholine is prolonged in patients with chronic liver disease; the hydrolysis of procaine is abnormally slow in the newborn and in adults with liver disease or impaired renal function (REIDENBERG et al., 1972). Similarly, the hydrolysis of aspirin to salicylic acid is significantly impaired in patients with portal cirrhosis (MENGUY et al., 1972) (Fig. 7).

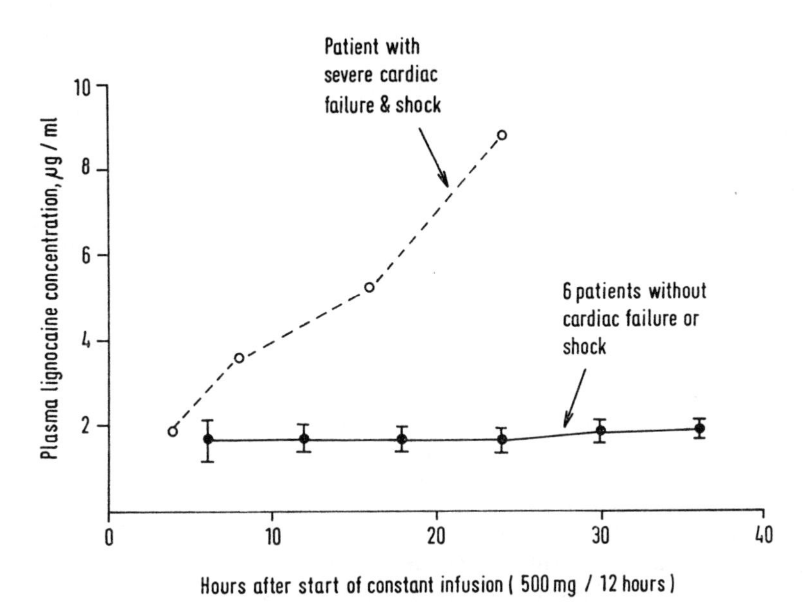

Fig. 7. Mean (± S.E.M.) plasma concentrations of lignocaine (lidocaine) during constant infusion in 6 patients without cardiac failure and in one patient with severe cardiac failure and shock. Lignocaine was given because of the occurrence of ventricular arrhythmias following acute myocardial infarction

V. Renal Excretion of Drugs

Renal disease may alter the response to drugs through several different mechanisms. Intrinsic receptor sensitivity may be modified by abnormal electrolyte or acid-base balance, plasma protein binding of acidic drugs may be reduced and the distribution of weak electrolytes may be altered by the metabolic acidosis associated with renal failure. Ester hydrolysis and acetylation of drugs may be abnormally slow and the permeability of the "blood brain barrier" may be increased (REIDENBERG, 1971). In addition, drug response may be enhanced as a result of accumulation of those drugs or their active metabolites that are normally excreted to a significant extent ($>30\%$) unchanged into the urine (Table 5).

A. Renal Failure and Drug Excretion

Fortunately, the action of most drugs is terminated by hepatic metabolism or redistribution rather than renal excretion. Nevertheless, therapeutic accidents still occur because clinicians do not know the normal mechanisms of elimination of the drugs that they use (PRESCOTT, 1972). Maintenance dosage of the drugs listed in Table 5 should normally be reduced in patients with poor renal function. Drugs with a short biological half-life can accumulate very rapidly if renal function is impaired; toxicity has been observed during the first day of therapy with procainamide in such patients (KOCH-WESER and KLEIN, 1971). Particular problems arise with drugs that have a low therapeutic ratio, such as digoxin and the

Table 5. *Drugs that are normally excreted unchanged to a significant extent in the urine*

Digoxin	Diazoxide	Methyldopa
Barbital	Acetazolamide	Quinidine
Amphetamine	Thiazide diuretics	Practolol
Procainamide	Amiloridine	Vancomycin
Tubocurarine	Chlorpropamide	Penicillins
Gallamine	Phenformin	Cephalosporins
Hexamethonium	Colistin	Lincomycin
Mecamylamine	Polymixin B	Tetracycline
Neostigmine	Aminoglycosides	Sulfonamides
Atropine	*p*-Aminosalicylic acid	Cycloserine
Methotrexate	Ethambutol	Sulfinpyrazone
	Allopurinol[a]	Acetohexamide[a]

[a] Active metabolites excreted in urine.

aminoglycoside antibiotics, and ideally plasma concentrations should be monitored in patients with renal disease. Further therapeutic difficulties arise in patients undergoing hemodialysis. Long lists have been published of drugs said to be removed by dialysis (SCHREINER, 1970), but with the exception of the antibiotics, much of the data on which these claims have been made is quite inadequate, and non-specific analytical methods for drug assay have often been used. The apparent volume of distribution is probably the most useful guide to the removal of drugs during hemodialysis. If the volume of distribution is small (i.e., less than about 0.5 l/kg) distribution is largely extracellular and appreciable loss may occur during dialysis. Conversely, weak organic bases usually have a very large volume of distribution and negligible amounts will be removed. It is often said that highly protein-bound drugs are not readily dialyzable. Nevertheless, extensively bound drugs such as salicylate can be removed rapidly by hemodialysis and the dissociation rate constant of the drug-protein complex may be more important than the degree of binding.

Prolongation of the plasma half-life of drugs in patients with renal failure is usually inversely related to the glomerular filtration rate, and dosage is often calculated according to the endogenous creatinine clearance (CUTLER and ORME, 1969; DETTLI et al., 1970; BELLET et al., 1971; MCHENRY et al., 1971). However, many drugs are secreted by the tubules in addition to filtration at the glomerulus, and the relationship between plasma half-life and creatinine clearance is neither linear nor simple.

B. Urine Flow and pH

The renal clearance of weakly acidic and basic lipid soluble drugs may be influenced by changes in urine flow and pH. However, these factors will have little or no effect on the clearance of drugs that are not reabsorbed. The clinical significance is uncertain, but forced alkaline diuresis is standard therapy for salicylate and phenobarbital intoxication. Forced acid diuresis has been used for the treatment of overdosage with weakly basic drugs such as fenfluramine. An increased plasma half-life of sulfonamides in children with gastroenteritis was attributed to decreased renal clearance in acid urine produced in response to metabolic acidosis (GLADTKE, 1971). The plasma half-life of drugs such as amphetamine and pseudoephedrine can be varied by a factor of about three by changes in urinary pH (KUNTZMAN et al., 1971). Drug response is usually not closely related to urine flow rate, but unpredictable toxic reactions to camptothecin were explained by changes in renal clearance due to variation in urine flow rate (Creaven and Allen, 1972).

C. Increased Hepatic Drug Metabolism in Renal Failure

Paradoxically, the plasma half-life of drugs that are normally metabolized by the liver may be significantly shorter than normal in patients with renal failure. This effect has been observed with amobarbital and diphenylhydantoin and may be due to reduced plasma protein binding (Mawer et al., 1972; Letteri et al., 1971). Similarly, the plasma half-life of propranolol is significantly reduced in patients with renal failure, and a direct correlation exists between the half-life and the creatinine clearance (Fig. 8) (Thompson et al., 1972). Contrary to established belief, the plasma half-life of drugs normally excreted largely unchanged in the urine may be normal in patients with poor renal function, implying a greatly increased metabolic clearance. For example, about 40% of a dose of pindolol is normally excreted unchanged in the urine, but the plasma half-life is not prolonged in patients with impaired renal function (Ohnhaus, 1973). In another study in patients undergoing regular hemodialysis, the plasma half-life of cephaloridine became shorter with increasing duration of maintenance hemodialysis (Curtis and Marshall, 1970). In these cases it is difficult to implicate reduced plasma protein binding, and compensatory metabolic routes of elimination are apparently developed.

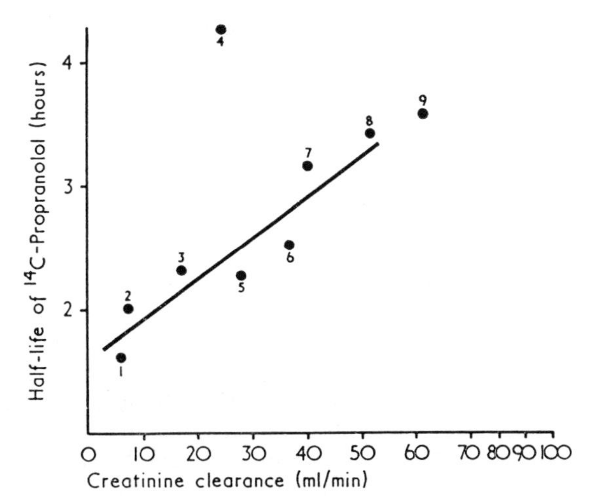

Fig. 8. The relationship between creatinine clearance and the plasma half-life of ^{14}C-propranolol following intravenous injection in patients with renal failure. (From Thompson et al., 1972)

In uremia, abnormal drug responses cannot always be explained simply by retention of the parent drug. There may be changes in receptor sensitivity or an accumulation of minor active metabolites. For example, sensitivity to the hypotensive action of methyldopa is increased, but this cannot be attributed to accumulation of unchanged drug (Stenbaek et al., 1972; Myhre et al., 1972). Other examples are the hydroxylated barbiturate metabolites, which are normally considered to be pharmacologically inactive but seem to have central nervous system depressant effects in patients with renal disease (BALASUBRAMANIAM et al., 1972).

Drug metabolites must accumulate to a remarkable degree in patients with renal failure and there is an increased risk of adverse reactions. Renal function may deteriorate further following administration of drugs such as methoxyflurane, phenylbutazone or tetracyclines. It is particularly important to avoid nephrotoxic combinations such as cephaloridine and furosemide.

VI. Receptor Sensitivity

Physiological and pathological factors can sometimes be shown to alter the sensitivity of receptors to drugs. More often, a change in response is assumed to be due to this cause in the absence of any other plausible explanation. The effects of disease on drug response in man are not well documented and often rest on anecdote or clinical impressions. Many examples could be cited, and the following have been chosen to illustrate some of the mechanisms involved.

A. Diminished Responses to Drugs

In some pathological conditions drugs cannot exert their characteristic actions because the tissues are unable to respond. Glyceryl trinitrate causes vasodilatation with increased coronary flow in healthy subjects, but in patients with ischemic heart disease the thickened atheromatous vessels cannot dilate, even when under the maximum physiological stimulus of ischemia. Glyceryl trinitrate may relieve angina, but it is thought to do so by decreasing circulatory resistance and thereby reducing cardiac work and oxygen consumption. Similarly, in severe juvenile diabetics there is loss of functioning pancreatic β-cells, and sulfonylureas are ineffective since they act largely by stimulating the release of insulin from the β-cells (BOGOCH et al., 1959).

Drugs may become less effective because of the development of tolerance or because of the natural progression of disease. True tolerance can develop rapidly following repeated administration of drugs such as nitrates, sympathomimetic amines, narcotic analgesics and some central nervous system depressants. For example, barbiturate addicts may be walking around with plasma barbiturate concentrations claimed by some to be lethal or requiring hemodialysis (PRESCOTT et al., 1973). Diabetics may become resistant to insulin because of the formation of antibodies or the development of ketoacidosis. Similarly, diuretics may lose their effect in patients with severe congestive cardiac failure, the nephrotic syndrome or cirrhosis with ascites because of electrolyte loss from previous diuretic therapy, the development of secondary hyperaldosteronism or decreasing renal function. Thiazide diuretics have little or no diuretic or blood pressure-lowering effects in patients with advanced renal failure (DREIFUS et al., 1960) and in such patients hypertension may be difficult to control because of salt and water retention.

Certain drug effects can be demonstrated only in the presence of disease. For example, phenformin has no significant effect on the blood glucose in non-diabetics and the antipyretic analgesics do not lower body temperature in the absence of fever.

B. Enhanced Response to Drugs

Both the therapeutic and toxic effects of drugs may be enhanced in certain disease states. Morphine is particularly dangerous in patients with bronchial asthma since it depresses respiration and in addition can aggravate bronchospasm through the release of histamine. Adrenergic β-blocking drugs, such as propranolol, normally have no significant effect on bronchial smooth muscle in healthy individuals, but they can cause severe bronchospasm in patients with asthma or left ventricular failure. Corticosteroids and phenylbutazone cause salt and water retention; while this effect is normally of little importance, cardiac failure may be precipitated in patients with borderline cardiac function. The coumarin anticoagulants act by competing with vitamin K in the hepatic synthesis of clotting factors. Their effects are exaggerated in patients with severe liver disease because the synthesis of vitamin K-dependent clotting factors is already depressed. In addition, patients with obstructive jaundice are particularly sensitive to coumarins because of decreased vitamin K absorption. Increased sensitivity to warfarin has also been described in patients with hyperthyroidism, but the mechanism is unknown (VAGENAKIS et al., 1972). Patients with hyperthyroidism are also particularly sensitive to sympathomimetic amines (HARRISON, 1964); their tachycardia and anxiety may be readily controlled by propranolol although the underlying hyperthyroid state is not improved.

C. Acid-Base and Electrolyte Balance

Abnormalities of acid-base and electrolyte balance are common, and can have profound effects on the actions of drugs such as cardiac glycosides, anti-arrhythmic drugs, diuretics, anti-hypertensive agents and muscle relaxants. Changes in the ratio between the intracellular and extracellular concentrations of K^+ and Ca^{++} are particularly important. Hypokalemia potentiates the effects of digitalis and antagonizes or even reverses the action of quinidine, lidocaine and diphenylhydantoin on the myocardium. Conversely, Ca^{++} potentiates the cardiotoxicity of digitalis, and succinylcholine may cause ventricular fibrillation in digitalized patients with hypercalcemia (SMITH and PETRUSCAK, 1972). In addition to the effects of abnormal electrolyte balance, the actions of digitalis are modified by thyroid function, autonomic activity, hypoxia, and myocardial pathology (MASON and BRAUNWALD, 1968).

Hypokalemia causes hyperpolarization of the muscle motor end-plate and antagonizes the actions of acetylcholine. The effects of non-depolarizing muscle relaxants such as *d*-tubocurarine are therefore enhanced. There may also be extreme sensitivity to these drugs in patients with neuromuscular diseases such as myesthenia gravis, carcinomatous myopathy and familial periodic paralysis. On the other hand, non-depolarizing relaxants are ineffective in myotonia dystrophica while the characteristic failure of muscle relaxation in this condition is aggravated by depolarizing relaxants such as succinylcholine (GILBERTSON and BOULTON, 1967). The aminoglycoside antibiotics have a weak intrinsic curare-like action that is normally of no significance, but they can cause significant weakness in patients with hypokalemia or myesthenia gravis, particularly if renal function is impaired.

Succinylcholine and decamethonium liberate K^+ from skeletal muscle and this effect is antagonized by *d*-tubocurarine. Normally, the serum K^+ rises by less than 0.5 mEq/l, but very large amounts of K^+ can be released in patients with uremia, extensive burns, severe trauma, upper motor neuron lesions, peripheral nerve injury and acute muscle degeneration. Increased sensitivity to succinylcholine may persist for as long as 6 months after neural injury, and has been attributed to denervation hypersensitivity (TOBEY et al., 1972). In such circumstances cardiac arrhythmias and arrest have occurred shortly after injection of succinylcholine, and in one patient with severe burns, the serum K^+ rose rapidly from normal to more than 8 mEq/l (TOLOMIE et al., 1967).

Malignant hyperpyrexia is a rare and often fatal complication of anesthesia and is characterized by the sudden onset of pyrexia, extreme muscular rigidity, cyanosis, tachycardia, acidosis, hyperkalemia, and occasionally myoglobinemia. It is precipitated by muscle relaxants and is an inherited abnormality associated with a mild or subclinical myopathy (KING et al., 1972). Interestingly, succinylcholine often causes myoglobinemia in children, and this response diminishes with the onset of puberty (RYAN et al., 1971).

The activity of muscle relaxants is influenced by hypothermia, depending on the type of block present. Depolarizing block produced by succinylcholine or decamethonium is potentiated while the competitive block of *d*-tubocurarine and gallamine is antagonized (BIGLAND et al., 1958).

Often, there is no obvious explanation for the enhanced effects of drugs in ill patients. Patients with severe hypothyroidism, hepatic pre-coma and respiratory failure are particularly sensitive to central nervous system depressants. Morphine and chlorpromazine have been shown to cause prolonged clinical deterioration associated with slowing of the electroencephalogram in patients with advanced liver disease, while diazepam causes only transient drowsiness with an increase in fast wave activity (LAIDLAW et al., 1961; MURRAY-LYON et al., 1971). The duration of thiopental anesthesia is increased in anemic patients (DUNDEE, 1956), and the response to central nervous system depressants and anesthetic agents is unpredictable in very ill patients.

VII. Conclusions

Physiological and pathological factors may have profound effects on the response to drugs. Unfortunately, our knowledge of the clinical importance of these factors is minimal, and the mechanisms involved are often poorly understood. Studies of drug action and disposition should be carried out in patients with the diseases for which the drug is indicated rather than in laboratory animals and healthy volunteers.

Acknowledgement

Research for this department was supported by a grant from the Scottish Hospitals Endowment Research Trust.

References

ADJEPON-YAMOAH, K. K., SCOTT, D. B., PRESCOTT, L. F.: Impaired absorption and metabolism of oral lidocaine in patients undergoing laparoscopy. Brit. J. Anesth. **45**, 143—147 (1973).

ANDERSSON, S.: Variation in gastrointestinal functions influencing the absorption of drugs. Acta pharmacol. (Kbh.) **29**, Suppl. 3, 103—108 (1971).

ANTON, A. H., COREY, W. T.: Interindividual differences in the protein binding of sulfonamides: the effect of disease and drugs. Acta pharmacol. (Kbh.) **29**, Suppl. 3, 134—151 (1971).

ARMSTRONG, B. K., UKICH, A. W., GOATCHER, P. M.: Plasma salicylate levels in rheumatoid arthritis produced by four different salicylate preparations. Med. J. Aust. **2**, 181—182 (1970).

BALASUBRAMANIAM, K., MAWER, G. E., POHL, J. E. F., SIMONS, P. J. G.: Impairment of cognitive function associated with hydroxyamylobarbitone accumulation in patients with renal insufficiency. Brit. J. Pharmacol. **45**, 360—367 (1972).

BALASUBRAMANIAM, K., MAWER, G. E., SIMONS, P. J.: The influence of dose on the distribution and elimination of amylobarbitone in healthy subjects. Brit. J. Pharmacol. **40**, 578—579 P. (1970).

BALLINGER, B., BROWNING, M., O'MALLEY, K., STEVENSON, I. H.: Drug-metabolizing capacity in states of drug dependence and withdrawal. Brit. J. Pharmacol. **45**, 638—643 (1972).
BARAKA, A., GABALI, F.: Correlation between tubocurarine requirements and plasma protein pattern. Brit. J. Anaesth. **40**, 89—93 (1968).
BEERMANN, B., HELLSTRÖM, K., ROSEN, A.: On the metabolism of propantheline in man. Clin. Pharmacol. Ther. **13**, 212—220 (1972).
BELLET, S., ROMAN, L. R., BOZA, A.: Relation between serum quinidine levels and renal function. Amer. J. Cardiol. **27**, 368—371 (1971).
BERESFORD, C. H., NEALE, R. J., BROOKS, O. G.: Iron absorption and pyrexia. Lancet **1971 I**, 568—572.
BIANCHINE, J. R., CALIMLIM, L. R., MORGAN, J. P., DUJUVNE, C. A., LASAGNA, L.: Metabolism and absorption of L-3,4-dihydroxyphenylalanine in patients with Parkinson's disease. Ann. N. Y. Acad. Sci. **179**, 126—140 (1971).
BIGLAND, B., GOETZEE, B., MACLAGAN, J., ZAIMIS, E.: The effect of lowered muscle temperature on the action of neuromuscular blocking agents. J. Physiol. (Lond.) **141**, 425—434 (1958).
BLUEMLE, L. W., GOLDBERG, M. J.: Renal accumulation of salicylate and phenacetin; possible mechanisms in the nephrotoxicity of analgesic abuse. J. clin. Invest. **47**, 2507—2514 (1968).
BOGOCH, A., DAVIS, T. W., JOW, E., WRENSHALL, G. A.: The clinical response and amount of insulin extractable from the pancreas in diabetic patients treated with oral hypoglycemic drugs. Canad. med. Ass. J. **81**, 347—356 (1959).
BRODIE, B. B.: Physicochemical factors in drug absorption. In: BINNS, T. B. (Ed.): Absorption and distribution of drugs, pp. 16—48. Baltimore: Williams and Wilkins 1964.
CHIGNELL, C. F., VESELL, E. S., STARKWEATHER, D. K., BERLIN, C. M.: The binding of sulfaphenazole to fetal, neonatal, and adult human plasma albumin. Clin. Pharmacol. Ther. **12**, 897—901 (1971).
CLARKE, R. A., JULIAN, D. G., NIMMO, J., PRESCOTT, L. F., TALBOT, R.: Clinical pharmacological studies of Kö 1173—a new antiarrhythmic agent. Brit. J. Pharmacol. Abstr. **47**, 622—623 P. (1973).
CONNEY, A. H., WELCH, R., KUNTZMAN, R., CHANG, R., JACOBSON, M., MUNRO-FAURE, A. D., PECK, A. W., BYE, A., POLAND, A., POPPERS, P. J., FINSTER, M., WOLFF, J. A.: Effects of environmental chemicals on the metabolism of drugs, carcinogens, and normal body constituents in man. Ann. N. Y. Acad. Sci. **179**, 155—172 (1971).
COVINO, B. G.: Local anesthesia. New Engl. J. Med. **286**, 1035—1042 (1972).
CREAVEN, P. J., ALLEN, L. M.: Renal clearance of camptothecin sodium in man; effect of urine volume. Fifth International Congress on Pharmacology, San Francisco, Abstr. No. 283, p. 48 (1972).
CROUNSE, R. G.: Human pharmacology of griseofulvin: The effect of fat intake on gastrointestinal absorption. J. invest. Derm. **37**, 529—533 (1961).
CURTIS, J. R., MARSHALL, M. J.: Cephaloridine serum levels in patients on maintenance hemodialysis. Brit. med. J. **1970 II**, 149—151.
CUTLER, R. E., ORME, B. M.: Correlation of serum creatinine concentration and kanamycin half-life: Therapeutic implications. J. Amer. med. Ass. **209**, 539—542 (1969).
DAS, K. M., EASTWOOD, M. A.: Personal communication.
DAVIS, A. E., PIROLA, R. C.: Absorption of phenoxymethyl penicillin in patients with steatorrhoea. Aust. Ann. Med. **17**, 63—65 (1968).
DETTLI, L., SPRING, P., HABERSANG, R.: Drug dosage in patients with impaired renal function. Postgrad. med. J. **46**, Suppl., 32—35 (1970).
DOLUISIO, J. T., TAN, G. H., BILLUPS, N. F., DIAMOND, L.: Drug absorption. II. Effect of fasting on intestinal drug absorption. J. pharm. Sci. **58**, 1200—1202 (1969).
DREIFUS, L. S., DUARTE, C., KODAMA, R., MOYER, J. H.: The effect of thiazide diuretics on the abnormal kidney. Ann. intern. Med. **53**, 1170—1179 (1960).
DUNDEE, J. W.: Thiopentone and other thiobarbiturates, p. 137. Edinburgh: Livingstone 1956.
DUNDEE, J. W., RICHARDS, R. K.: The effect of azotemia upon the action of intravenous barbiturate anesthesia. Anesthesiology **15**, 333—346 (1954).
EBADI, M. S., KUGEL, R. B.: Alteration in metabolism of acetylsalicylic acid in children with Down's syndrome: Decreased plasma binding and formation of salicyluric acid. Pediat. Res. **4**, 187—193 (1970).
FISCHER, E.: Renal excretion of sulfadimidine in normal and uraemic subjects. Lancet **1972 II**, 210—212.
GILBERTSON, A. A., BOULTON, T. B.: Anesthesia in difficult situations. The influence of disease on pre-operative preparation and choice of anesthetic. Anesthesia **22**, 607—630 (1967).
GILLETTE, J. R.: Factors affecting drug metabolism. Ann. N. Y. Acad. Sci. **179**, 43—66 (1971).
GLADTKE, E.: Beeinflussung der Arzneimitteltherapie durch pathologische Zustände. J. mond. Pharm. (La Haye) **14**, 230—250 (1971).

GOWER, P. E., DASH, C. H.: Cephalexin: human studies of absorption and excretion of a new cephalosporin antibiotic. Brit. J. Pharmacol. **37**, 738—747 (1969).

GWILT, J. R., ROBERTSON, A., GOLDMAN, L., BLANCHARD, A. W.: The absorption characteristics of paracetamol tablets in man. J. Pharm. Pharmacol. **15**, 445—453 (1963).

HAASS, A., LÜLLMANN, H., PETERS, T.: Absorption rates of some cardiac glycosides and portal blood flow. Europ. J. Pharmacol. **19**, 366—370 (1972).

HAMMAR, C.-H., PRELLWITZ, W.: Die Glucuronidbildung nach peroraler Belastung mit Acetanilid bei chronischer Hepatitis und Lebercirrhose. Klin. Wschr. **44**, 1010—1014 (1966).

HARRISON, T. S.: Adrenal medullary and thyroid relationships. Physiol. Rev. **44**, 161—185 (1964).

HARTIALA, K., KASANEN, A., RAUSSI, M.: The absorption of salicylamide in pernicious anemia, gastric achylia, and peptic ulcer. Ann. Med. exp. Fenn. **41**, 549—553 (1963).

HAYES, A. H.: Intravenous infusion of lidocaine in the control of ventricular arrhythmias. In: SCOTT, D. B., JULIAN, D. G. (Eds.): Lidocaine in the treatment of ventricular arrhythmias, pp. 189—199. Edinburgh: Livingstone 1971.

HEADING, R. C., NIMMO, J., PRESCOTT, L. F., TOTHILL, P.: The dependence of paracetamol absorption on the rate of gastric emptying. Brit. J. Pharmacol. **47**, 415—421 (1973).

HEIZER, W. D., SMITH, T. W., GOLDFINGER, S. E.: Absorption of digoxin in patients with malabsorption syndromes. New Engl. J. Med. **285**, 257—259 (1971).

HELD, H., OLDERSHAUSEN, H. F.: Zur Pharmakokinetik von Meprobamat bei chronischen Hepatopathien und Arzneimittelsucht. Klin. Wschr. **47**, 78—80 (1969).

HELD, H., OLDERSHAUSEN, H. F.: Drug metabolism in acute and chronic liver disease. Digestion **4**, 151 (Abstr.) (1971).

HOWELL, A., SUTHERLAND, R., ROBINSON, G. N.: Effect of protein binding on levels of ampicillin and cloxacillin in synovial fluid. Clin. Pharmacol. Ther. **13**, 724—732 (1972).

JAFFE, J. M., COLAIZZI, J. L., BARRY, H.: Effects of dietary components on GI absorption of acetaminophen tablets in man. J. pharm. Sci. **60**, 1646—1650 (1971).

JOLLOW, D. J., MITCHELL, J. R., POTTER, W. Z., DAVIS, D. C., GILLETTE, J. R., BRODIE, B. B.: Acetaminophen-induced hepatic necrosis. II. Role of covalent binding *in vivo*. J. Pharmacol. exp. Ther. **187**, 195—202 (1973).

KATER, R. M. H., ROGGIN, G., TOBON, F., ZIEVE, P., IBER, F. L.: Increased rate of clearance of drugs from the circulation of alcoholics. Amer. J. med. Sci. **258**, 35—39 (1969).

KEKKI, M., PYÖRÄLÄ, K., MUSTALA, O., SALMI, H., JUSSILA, J., SIURALA, M.: Multicompartment analysis of the absorption kinetics of warfarin from the stomach and small intestine. Int. J. clin. Pharmacol. **5**, 209—211 (1971).

KING, J. O., DENBOROUGH, M. A., ZAPF, P. W.: Inheritance of malignant hyperpyrexia. Lancet **1972 I**, 365—370.

KOCH-WESER, J.: Pharmacokinetics of procainamide in man. Ann. N.Y. Acad. Sci. **179**, 370—382 (1971).

KOCH-WESER, J., KLEIN, S. W.: Procainamide dosage schedules, plasma concentrations and clinical effects. J. Amer. med. Ass. **215**, 1454—1460 (1971).

KOJIMA, S., SMITH, R. B., DOLUISIO, J. T.: Drug absorption. V. Influence of food on oral absorption of phenobarbital in rats. J. pharm. Sci. **60**, 1639—1641 (1971).

KUNTZMAN, R. G., TSAI, I., BRAND, L., MARK, L. C.: The influence of urinary pH on the plasma half-life of pseudoephedrine in man and dog and a sensitive assay for its determination in human plasma. Clin. Pharmacol. Ther. **12**, 62—67 (1971).

KUTT, H., WINTERS, W., SCHERMAN, R., MCDOWELL, F.: Diphenylhydantoin and phenobarbital toxicity. Arch. Neurol. (Chic.) **11**, 649—656 (1964).

LAIDLAW, J., READ, A. E., SHERLOCK, S.: Morphine tolerance in hepatic cirrhosis. Gastroenterology **40**, 389—396 (1961).

LETTERI, J. M., MELLK, H., LOUIS, S., KUTT, H., DURANTE, P., GLAZKO, A.: Diphenylhydantoin metabolism in uremia. New. Engl. J. Med. **285**, 648—652 (1971).

LEVI, A. J., SHERLOCK, S., WALKER, D.: Phenylbutazone and isoniazid metabolism in patients with liver disease in relation to previous drug therapy. Lancet **1968 I**, 1275—1279.

LEVINE, R. R.: Factors affecting gastrointestinal absorption of drugs. Digest. Dis. **15**, 171—188 (1970).

LEVY, G., GIBALDI, M., PROCKNAL, J. A.: Effect of an anticholinergic agent on riboflavin absorption in man. J. pharm. Sci. **61**, 798—799 (1972).

LEWIS, G. P., JUSKO, W. J., BURKE, C. W., GRAVES, L.: Prednisone side-effects and serum-protein levels. Lancet **1971 II**, 778—780.

LUCHI, R. J., GRUBER, J. W.: Unusually large digitalis requirements: a study of altered digoxin metabolism. Amer. J. Med. **45**, 322—328 (1968).

LUND, L., BERLIN, A., LUNDE, P. K. M.: Plasma protein binding of diphenylhydantoin in patients with epilepsy. Clin. Pharmacol. Ther. **13**, 196—200 (1972).

MACDONALD, H., PLACE, V. A., FALK, H., DARKEN, M. A.: Effect of food on absorption of sulfonamides in man. Chemotherapia (Basel) **12**, 282—285 (1967).

MASON, D. T., BRAUNWALD, E.: Symposium on congestive heart failure. II. Digitalis: new facts about an old drug. Amer. J. Cardiol. **22**, 151—161 (1968).

MATTILA, M. J., FRIMAN, A., LARMI, T. K. I., KOSKINEN, R.: Absorption of ethionamid, isoniazid, and aminosalicylic acid from the post-resection gastrointestinal tract. Ann. Med. exp. Fenn. **47**, 209—212 (1969).

MAWER, G. E., MILLER, N. E., TURNBERG, L. A.: Metabolism of amylobarbitone in patients with chronic liver disease. Brit. J. Pharmacol. **44**, 549—560 (1972).

MCGILVERAY, I. J., MATTOK, G. L.: Some factors affecting the absorption of paracetamol. J. Pharm. Pharmacol. **24**, 615—619 (1972).

MCHENRY, M. C., GAVAN, T. L., GIFFORD, R. W., JR., GEURKINK, N. A., VAN OMMEN, R. A., TOWN, M. A., WAGNER, J. G.: Gentamicin dosages for renal insufficiency: Adjustments based on endogenous creatinine clearance and serum creatinine concentration. Ann. intern. Med. **74**, 192—197 (1971).

MENGUY, R., DESBAILLETS, L., OKABE, S., MASTERS, Y. F.: Abnormal aspirin metabolism in patients with cirrhosis and its possible relationship to bleeding in cirrhotics. Ann. Surg. **176**, 412—418 (1972).

MEYER, M. B., ZELECHOWSKI, K.: Intramuscular lidocaine in normal subjects. In: SCOTT, D. B., JULIAN, D. G. (Eds.): Lidocaine in the treatment of ventricular arrhythmias, pp. 161—168. Edinburgh: Livingstone 1971.

MITCHELL, J. R., JOLLOW, D. J., POTTER, W. Z., DAVIS, D. C., GILLETTE, J. R., BRODIE, B. B.: Acetaminophen-induced hepatic necrosis. I. Role of drug metabolism. J. Pharmacol. exp. Ther. **187**, 185—194 (1973).

MURRAY-LYON, I. M., YOUNG, J., PARKES, J. D., KNILL-JONES, R. P., WILLIAMS, R.: Clinical and electroencephalographic assessment of diazepam in liver disease. Brit. med. J. **4**, 265—266 (1971).

MYHRE, E., BRODWALL, E. K., STENBAEK, O., HANSEN, T.: Plasma turnover of methyldopa in advanced renal failure. Acta med. scand. **191**, 343—347 (1972).

NANRA, R. S., HICKS, J. D., MCNAMARA, J. H., LIE, J. T., LESLIE, D. W., JACKSON, B., KINCAID-SMITH, P.: Seasonal variation in the post-mortem incidence of renal papillary necrosis. Med. J. Aust. **1**, 293—296 (1970).

NELSON, E.: Rate of metabolism of tolbutamide in test subjects with liver disease or with impaired renal function. Amer. J. med. Sci. **248**, 657—659 (1964).

NELSON, J. D., SHELTON, S., KUSMIESZ, H. T., HALTALIN, K. C.: Absorption of ampicillin and nalidixic acid by infants and children with acute shigellosis. Clin. Pharmacol. Ther. **13**, 879—886 (1972).

NIMMO, J., HEADING, R. C., TOTHILL, P., PRESCOTT, L. F.: Pharmacological modification of gastric emptying: effects of propantheline and metoclopramide on paracetamol absorption. Brit. med. J. **1**, 587—589 (1973).

ODAR-CEDERLÖF, I., LUNDE, P., SJÖQVIST, F.: Abnormal pharmacokinetics of phenytoin in a patient with uremia. Lancet **1970 II**, 831—832.

OHNHAUS, E. E.: The pharmacokinetics of unchanged pindolol in patients with impaired renal function. Brit. J. Pharmacol. Abstr. **47**, 620—621 P. (1973).

O'MALLEY, K., CROOKS, J., DUKE, E., STEVENSON, I. H.: Effect of age and sex on human drug metabolism. Brit. med. J. **1971 III**, 607—609.

PETERSON, O. L., FINLAND, M.: The effect of food and alkali on the absorption and excretion of sulfonamide drugs after oral and duodenal administration. Amer. J. med. Sci. **204**, 581—588 (1942).

PLACE, V. A., BENSON, H.: Dietary influences on therapy with drugs. J. mond. Pharm. **14**, 261—278 (1971).

POTTER, W. Z., DAVIS, D. C., MITCHELL, J. R., JOLLOW, D. J., GILLETTE, J. R., BRODIE, B. B.: Acetaminophen-induced hepatic necrosis. III. Cytochrome P-450-mediated covalent binding *in vitro*. J. Pharmacol. exp. Ther. **187**, 203—210 (1973).

PRESCOTT, L. F.: Mechanisms of renal excretion of drugs (with special reference to drugs used by anesthetists). Brit. J. Anesth. **44**, 246—251 (1972).

PRESCOTT, L. F., NIMMO, J.: Generic inequivalence—clinical observations. Acta pharmacol. (Kbh.) **29**, Suppl. 3, 288—303 (1971a).

PRESCOTT, L. F., NIMMO, J.: Drug therapy—physiological considerations. J. mond. Pharm. (La Haye) **14**, 253—260 (1971b).

PRESCOTT, L. F., NIMMO, J.: Plasma lidocaine concentrations during and after prolonged infusions in patients with myocardial infarction. In: SCOTT, D. B., JULIAN, D. G. (Eds.): Lidocaine in the treatment of ventricular arrhythmias, pp. 168—177. Edinburgh: Livingstone 1971c.

PRESCOTT, L.F., ROSCOE, P., FORREST, J.A.H.: Plasma concentrations and drug toxicity. II. Toxicity in man. In: DAVIES, D.S., PRITCHARD, B.N.C. (Eds): Biological effects of drugs in relation to their plasma concentrations. pp. 51—81. London: Macmillan 1973.
PRESCOTT, L.F., ROSCOE, P., WRIGHT, N., BROWN, S.S.: Plasma-paracetamol half-life and hepatic necrosis in patients with paracetamol overdosage. Lancet **1971 I**, 519—522.
PRESCOTT, L.F., STEEL, R.F., FERRIER, W.R.: The effects of particle size on the absorption of phenacetin in man. Clin. Pharmacol. Ther. **11**, 496—504 (1970).
PRESCOTT, L.F., STEVENSON, I.H.: Liver disease and drug metabolism in man. In: Proceedings of the Fifth International Congress on Pharmacology. Basel: Karger **3**, 182—190, 1973.
PRICE-EVANS, D.A.: Inter-individual differences in the metabolism of drugs: the role of genetic factors. Acta pharmacol. (Kbh.) **29**, Suppl. 3, 156—163 (1971).
REIDENBERG, M.M.: Renal function and drug action, pp. 1—3, 33—37. Philadelphia: Saunders 1971.
REIDENBERG, M.M.: Effect of kidney disease on pharmacokinetics and drug response. In: Proceedings of the Fifth International Congress on Pharmacology. Basel: Karger 1973.
REIDENBERG, M.M., JAMES, M., DRING, L.G.: The rate of procaine hydrolysis in serum of normal subjects and diseased patients. Clin. Pharmacol. Ther. **13**, 279—284 (1972).
REIDENBERG, M.M., ODAR-CEDERLÖF, I., VON BAHR, C., BORGA, O., SJÖQVIST, F.: Protein binding of diphenylhydantoin and desmethylimipramine in plasma from patients with poor renal function. New Engl. J. Med. **285**, 264—267 (1971).
RISTOLA, P., PYÖRÄLÄ, K.: Determinants of the response to coumarin anticoagulants in patients with acute myocardial infarction. Acta med. scand. **192**, 183—188 (1972).
RYAN, J.F., KAGEN, L.J., HYMAN, A.I.: Myoglobinemia after a single dose of succinylcholine. New Engl. J. Med. **285**, 824—827 (1971).
SALES, J.E.L., SUTCLIFFE, M., O'GRADY, F.: Cephalexin levels in human bile in presence of biliary tract disease. Brit. med. J. **1972 III**, 441—443.
SCHEDL, H.P., CLIFTON, J.A.: Cortisol absorption in man. Gastroenterology **44**, 134—145 (1963).
SCHEINER, J., ALTEMEIER, W.A.: Experimental study of factors inhibiting absorption and effective therapeutic levels of declomycin. Surg. Gyn. Obstet. **114**, 9—14 (1962).
SCHOENE, B., FLEISCHMANN, R.A., REMMER, H.: Determination of drug metabolizing enzymes in needle biopsies of human liver. Europ. J. clin. Pharmacol. **4**, 65—73 (1972).
SCHREINER, G.E.: Dialysis of poisons and drugs—Annual review. Trans. Amer. Soc. Artif. Int. Organs **16**, 544—568 (1970).
SCHRÖDER, H., CAMPBELL, D.E.S.: Absorption, metabolism, and excretion of salicylazosulfapyridine in man. Clin. Pharmacol. Ther. **13**, 539—551 (1972).
SERENI, F., PERLETTI, L., MARUBINI, E., MARS, G.: Pharmacokinetic studies with a long-acting sulfonamide in subjects of different ages. Pediat. Res. **2**, 29—37 (1968).
SESSIONS, J.T., MINKEL, H.P., BULLARD, J.C., INGELFINGER, F.J.: The effect of barbiturates in patients with liver disease. J. clin. Invest. **33**, 1116—1127 (1954).
SHIDEMAN, F.E., KELLY, A.R., LEE, L.E., LOWELL, V.F., ADAMS, B.J.: The role of the liver in the detoxification of thiopental (pentothal) by man. Anesthesiology **10**, 421—427 (1949).
SIEGERS, C.-P., STRUBELT, O., BACK, G.: Inhibition by caffeine of ethanol absorption in rats. Europ. J. Pharmacol. **20**, 181—187 (1972).
SIURALA, M., MUSTALA, O., JUSSILA, J.: Absorption of acetylsalicylic acid by a normal and atrophic gastric mucosa. Scand. J. Gastroent. **4**, 269—273 (1969).
SOTANIEMI, E., HUHTI, E., ARVELA, P., KOIVISTO, O.: Tolbutamide clearance from the blood in hypercapnia. Scand. J. clin. Lab. Invest. **27**, Suppl. 78 (Abstr.) (1971).
SMITH, R.B., PETRUSCAK, J.: Succinylcholine, digitalis, and hypercalcemia: a case report. Anesth. Analg. Curr. Res. **51**, 202—205 (1972).
STENBAEK, Ø., MYHRE, E., BRODWALL, E.K., HANSEN, T.: Hypotensive effect of methyldopa in renal failure associated with hypertension. Acta med. scand. **191**, 333—337 (1972).
THOMPSON, F.D., JOEKES, A.M., FOULKES, D.M.: Pharmacodynamics of propranolol in renal failure. Brit. med. J. **1972 II**, 434—436.
THOMSON, P.D., ROWLAND, M., MELMON, K.L.: The influence of heart failure, liver disease, and renal failure on the disposition of lidocaine in man. Amer. Heart J. **82**, 417—421 (1971).
TOBEY, R.E., JACOBSEN, P.M., KAHLE, C.T., CLUBB, R.J., DEAN, M.A.: The serum potassium response to muscle relaxants in neural injury. Anesthesiology **37**, 332—337 (1972).
TOLOMIE, J.D., JOYCE, T.H., MITCHELL, G.D.: Succinylcholine danger in the burned patient. Anesthesiology **28**, 467—470 (1967).
UEDA, H., SAKURAI, T., OTA, M., NAKAJIMA, A., KAMII, K., MAEZAWA, H.: Disappearance rate of tolbutamide in normal subjects and in diabetes mellitus. liver cirrhosis, and renal disease, Diabetes **12**, 414—419 (1963).

VAGENAKIS, A. G., COTE, R., MILLER, M. E., BRAVERMAN, L. E., STOHLMAN, F., JR.: Enhancement of warfarin-induced hypoprothrombinemia by thyrotoxicosis. Johns Hopk. med. J. **131**, 69—75 (1972).

VENHO, V. M. K., JUSSILA, J., AUKEE, S.: Drug absorption in man after gastric suergry. Fifth International Congress on Pharmacology, San Francisco. Abstract No. 1445, 241 (1972).

WADDELL, W. J., BUTLER, T. C.: The distribution and excretion of phenobarbital. J. clin. Invest. **36**, 1217—1226 (1957).

WEBB, D. I., CHODOS, R. B., MAHAR, C. Q., FALOON, W. W.: Mechanism of vitamin B_{12} malabsorption in patients receiving colchicine. New Engl. J. Med. **279**, 845—850 (1968).

WECHSELBERG, K.: Salizylsäure-Vergiftung durch perkutane Resorption 1%iger Salizylvaseline. Pädiat. Prax. **7**, 431—433 (1968).

WHELTON, A., SAPIR, D. G., CARTER, G. G., KRAMER, J., WALKER, W. G.: Intrarenal distribution of penicillin, cephalothin, ampicillin, and oxytetracycline during varied states of hydration. J. Pharmacol. exp. Ther. **179**, 419—428 (1971).

WHITE, R. J., CHAMBERLAIN, D. A., HOWARD, M., SMITH, T. W.: Plasma concentrations of digoxin after oral administration in the fasting and postprandial state. Brit. med. J. **1971 I**, 380—381.

WRIGHT, N., PRESCOTT, L. F.: Paracetamol overdosage: potentiation of hepatotoxicity by previous drug therapy. Scot. med. J. **18**, 56—58 (1973).

Chapter 68

Absorption, Distribution, Excretion, and Response to the Drug in the Presence of Chronic Renal Failure*

M. BLACK and I. M. ARIAS

With 4 Figures

I. Introduction

Chronic renal failure (CRF) produces a complex metabolic disturbance that impairs the function of many organs involved in the absorption, distribution, excretion, and response of drugs. However, the small amount of information available in the literature regarding drugs and renal disease deals almost exclusively with the effects of reduced glomerular filtration on the renal handling of drugs. Few pharmacokinetic studies have been performed, perhaps mainly because of the traditionally poor long-term prognosis for patients with CRF. However, the increased use of extracorporeal dialysis and renal transplantation over the last decade has enabled such patients to live longer, and problems associated with long-term drug therapy have become apparent.

In the absence of comprehensive pharmacokinetic studies of patients with CRF, insight into the potential dimensions of the problem may be provided by observations made on the elimination of endogenous substrates by uremic patients or by animals with experimentally-induced uremia. The disposal of endogenous substrates is often identical with that of exogenous compounds such as drugs (CONNEY and KUNTZMAN, 1971), and abnormalities in the disposition of endogenous substrates probably reflect abnormalities in the disposition of drugs. Accordingly, this review will utilize data obtained from studies on endogenous and exogenous compounds.

II. Absorption

Little information exists concerning the influence of disease of any kind (including abnormalities of gastrointestinal function) on the absorption of drugs. HEIZER et al. (1971) demonstrated that patients with various malabsorption syndromes have difficulty in absorbing orally administered digoxin and have lower plasma levels of drug than control patients on chronic maintenance therapy. This result indicates that malabsorption of drugs can be critical in planning dosage regimens.

There is little evidence that gastrointestinal function is altered in uremia, but few observations have been made. Plasma levels of several essential and non-essential amino acids are reduced in patients with CRF (GULYASSY et al., 1968; CZERNIAK and BURZYNSKI, 1969; BURZYNSKI, 1969). GULYASSY et al. (1972)

* Research performed in the authors' laboratory was supported by a contract (No. 71-2484) from the National Institutes of Health, USPHS.

noted decreased plasma levels of tryptophan in uremic subjects when compared with normal controls after oral tryptophan loading. Malabsorption of calcium has been well-documented in patients with CRF (GOSSMANN et al., 1968; OGG, 1968) and is probably the major cause of uremic osteodystrophy (STANBURY et al., 1969). The mechanisms underlying this malabsorption are currently disputed (LUMB et al., 1971), but the demonstration of an acquired defect in the biotransformation of vitamin D (see section on Metabolism) in uremic patients has raised the possibility of a defective vitamin D effect on the active transport process by which calcium is absorbed (AVIOLI et al., 1968). LUMB et al. (1971) found normal absorption of vitamin D from the gastrointestinal tracts of uremic patients. Decreased absorption of orally administered iron occurs in patients with CRF (DUBACH et al., 1948; ESCHBACH et al., 1970), even when iron deficiency is present (LAWSON et al., 1971). LAWSON et al. (1971) showed that malabsorption of iron was present in patients who had been on a chronic dialysis program for a minimum of 6 months.

Peak plasma levels of antipyrine, given orally, were the same in uremic and normal subjects (LICHTER et al., 1973). There was no difference in the time required for the plasma concentration of drug to reach its peak. Thus, at least for this weakly basic drug (pK_a-1.4), no major problem in absorption in the presence of uremia is apparent. This may not be the case for the urinary antiseptic, nitrofurantoin, a weakly acidic drug with a pK_a of 7.0. Normally, nitrofurantoin is completely absorbed from the gastrointestinal tract and rapidly cleared from the plasma, with up to 40% appearing unchanged in the urine. Urinary recovery of the drug was reduced in patients with impaired renal function (LIPPMAN et al., 1958; SCHLEGEL et al., 1967; SACHS et al., 1968; GOFF et al., 1968), but build-up of the drug in the plasma was not observed (SCHLEGEL et al., 1967). The reason for the failure of nitrofurantoin to accumulate in the plasma of uremic patients may be rapid removal by extrarenal mechanisms as postulated by SACHS et al. (1968). However, investigations of SCHMID et al. (1967) support the possibility that malabsorption of the drug by uremic patients may also be responsible. These workers showed that less drug is recovered from the urine of uremic patients after oral dosing than after intravenous administration. Similar results were obtained in a group of patients who had previously undergone sub-total gastrectomy. Thus, malabsorption of nitrofurantoin probably occurs in the uremic patient, but further data is required to establish this fact.

In general, clinical experience with drugs in patients with CRF suggests that impaired absorption of a drug is not critical in planning dosage regimens (KUNIN, 1967b; BENNETT et al., 1970), since the reduced rate of urinary clearance of drugs tends to minimize the role of intestinal absorption in establishing steady-state levels. Yet the potential exists that malabsorption secondary to CRF occurs with drugs whose excretion is independent of renal function. This possibility should be kept in mind when considering treatment failures in patients with CRF.

III. Distribution: Protein Binding

The volume of distribution of a drug (V_d) is influenced by a variety of factors including physico-chemical properties of the drug, relative and absolute sizes of the body's fluid compartments, ease with which the drug penetrates membrane barriers, and extent of binding of the drug to serum proteins and other receptors (GOLDSTEIN et al., 1969). Though the derived values for V_d are of theoretical importance in considering dose-response relationships, sites of activity and rate of elimination, V_d is rarely calculated in clinical investigations. ORME and CUTLER (1969) found that the V_d for kanamycin, a drug that is not significantly protein-bound, was normal in patients with modest reductions in creatinine clearance. The recent observation of WHELTON et al. (1972) that the distribution of ampicillin within the kidneys of patients with chronic glomerulonephritis differed markedly from normal is an indication that, for the kidney at least, problems of drug distribution may be important in treatment failure.

Tissue concentrations of drugs that are significantly protein-bound are usually related to the plasma concentration of *unbound* drug (VERWEY and WILLIAMS, 1962a and b). Recent studies indicate that for a number of drugs, the percent of drug bound to albumin in the plasma is significantly reduced in patients with uremia. ODAR-CEDERLÖFF et al. (1970), REIDENBERG et al. (1971), and BLUM et al. (1972) found that uremic subjects have reduced binding of diphenylhydantoin in the plasma, the unbound fraction amounting to 10 to 25% of the total in comparison with the normal figure of 5 to 10%. REIDENBERG et al. (1971) showed a correlation in non-dialyzed patients between the percentage of unbound drug and the blood urea nitrogen or creatinine levels. Binding of diphenylhydantoin in plasma of uremic patients was not improved by dialysis of the plasma *in vitro*, which suggests that the abnormality did not reflect competition for binding sites by dialyzable materials present in the plasma. BLUM et al. (1972) demonstrated that the abnormality was not due to competition for binding sites by the principal metabolite of diphenylhydantoin, 5-phenyl, 5-parahydroxylphenylhydantoin. ODAR-CEDERLÖFF et al. (1970) and BLUM et al. (1972) noted the pharmacologic effect of diphenylhydantoin at significantly lower *total* plasma levels of the drug than occur in non-uremic subjects, indicating that the effective tissue level of diphenylhydantoin is preserved in uremic subjects in the presence of appreciably reduced plasma levels.

The binding of a number of other drugs to serum albumin has also been studied. ANTON and COREY (1971) showed that the binding of the sulfonamide, sulfamethazine, to albumin was reduced in 6 nephrectomized patients from a normal level of 76—79% to 53—69%. Hemodialysis did not increase the ability of the plasma to bind the drug *in vitro*, strengthening the belief that the abnormal binding of drugs in plasma of uremic patients is a manifestation of a defect in the binding properties of serum albumin. REIDENBERG (1972) extended his studies to several other drugs, including quinidine, dapsone, fluorescein, and desmethylimipramine, and concluded that most organic acids normally bound by plasma albumin show reduced binding by uremic plasma, while organic bases (e.g., desmethylimipramine, quinidine, and dapsone) show relatively normal binding. In addition to changes in binding properties of uremic plasma albumin, the plasma concentrations of albumin are often moderately reduced in uremic patients (COLES et al., 1970) and could contribute to reduced percentage binding of the drug, particularly at high plasma concentrations of drug (KEEN, 1971).

Abnormalities in binding of endogenous substances such as amino acids have also been found in uremic subjects. Tryptophan, which may be the rate-limiting essential amino acid for the synthesis of albumin by the liver (MUNRO, 1968), is normally 80% bound in the plasma. However, the amount of binding falls to 44% in uremic patients (GULYASSY et al., 1972) and is accompanied by reduction in plasma half-life from 1.9 h to 1.3—1.6 h. Plasma levels are reduced and this may be responsible for the accelerated degradation of albumin in uremia (GULYASSY et al., 1972), since tryptophan exerts a protective effect against albumin proteolysis (MARKUS, 1965).

IV. Metabolism

Since patients with CRF have difficulty in disposing of drugs normally excreted unchanged in the urine, the authors of several drug dosage guides for uremic patients (KUNIN, 1967b, 1972; BENNETT et al., 1970) have recommended the use of drugs that normally undergo significant metabolic transformation before being eliminated from the body. This preference is based upon the belief that mechanisms of drug metabolism are unimpaired in patients with uremia and that drugs which require metabolism prior to excretion are safer to use in these patients. A recent survey of drug-induced neuropsychiatric disturbance in patients with CRF (RICHET et al., 1970) showed that 11 of 18 drugs incriminated are significantly metabolized before being excreted. This experience is not unique, since increased toxicity in uremic subjects of similarly handled drugs has been documented (chloramphenicol – SUHRLAND and WEISBERGER, 1963; phenformin – MACGREGOR et al., 1972). Only SUHRLAND and WEISBERGER'S study with chloramphenicol, however, correlated toxicity with blood levels of drug; it is not possible in the other cases to exclude increased sensitivity of the uremic patient to blood levels of drug that are normally therapeutic (see section on Response to Drugs).

Although drugs are metabolized mainly in the liver, they may also be metabolized in the kidney (BOYER et al., 1960). WAN and RIEGELMAN (1972) have shown that the biotransformation of certain benzoic acid derivatives takes place primarily in the kidney. However, renal metabolism of drugs is not an important route of elimination for most drugs.

Most of the drug-metabolizing enzymes are located in the smooth endoplasmic reticulum of the liver cell. No detailed examination of the ultrastructure and organelle function of the hepatocyte in uremic patients has been reported, although we have undertaken such a study in an experimental model of chronic renal failure in the rat (BIEMPICA et al., in preparation). The preliminary results of this investigation suggest that most of the organelles of the liver cell show evidence of damage; both rough and smooth membranes of the endoplasmic reticulum are affected. These ultrastructural changes are accompanied by a prolongation of *in vivo* zoxazolamine paralysis time and ketamine narcosis time, and reductions in hepatic cytochrome P-450, and aminopyrine and benzphetamine hydroxylase activities (BLACK et al., in preparation). These observations, considered with the increasing evidence of the abnormal disposition of endogenous and exogenous substrates by the liver in uremia, suggest that uremia might be associated with an alteration in hepatic subcellular function that may significantly affect the disposal of drugs.

Vitamins A and D are converted in the liver to active metabolites (DELUCA, 1971; ROELS, 1967). In patients with CRF and the nephrotic syndrome, metabolism of vitamin A is disturbed, and increased amounts of the unchanged vitamin are found in plasma and liver (JOSEPHS, 1939; KAGAN et al., 1950; KAGAN and KAISER, 1952; SCHREINER and MALER, 1961; SMITH and GOODMAN, 1971). Similarly, when plasma samples from uremic subjects were analyzed by silicic acid column chromatography following intravenous injection of a tracer-dose of ^{3}H-labeled vitamin D_3, preferential formation of biologically inactive polar metabolites and decreased formation of the active 25-hydroxylated derivative were observed (AVIOLI et al., 1968; RITZ and JANTZEN, 1969). The possible role of this abnormality in the development of renal osteodystrophy has already been mentioned (see section on Absorption).

ENGLERT et al. (1958), using intravenous infusions of unlabeled cortisol, and BACON et al. (1970), using ^{14}C-labeled cortisol, found that the half-life of cortisol is prolonged in patients with CRF; in one patient the prolonged half-life was returned to normal following hemodialysis (BACON et al., in preparation). Feedback inhibition of cortisol production appears to take place, maintaining blood cortisol concentrations at normal levels.

The inactivation of insulin is predominantly a hepatic function (Izzo et al., 1967), and several groups have shown that the carbohydrate intolerance of uremia is associated with elevated blood levels of insulin in response to intravenously administered glucose (CERLETTY and ENGBRING, 1967; HAMPERS et al., 1968). HORTON et al. (1968) showed that the disappearance rate of administered soluble insulin was less rapid in uremic subjects as compared with control subjects, and O'BRIEN and SHARPE (1967) made a similar observation using isotopically labeled insulin. The latter authors found that the abnormality in plasma insulin clearance was corrected following kidney transplantation.

A further example of altered hepatic function in uremia concerns the chronic elevation of plasma lipids (LOSOWSKY and KENWARD, 1968; BAGDADE, 1970), which BAGDADE (1970) believes is a result of increased hepatic synthesis of lipoproteins.

Literature concerning effects of CRF upon hepatic metabolism of drugs is limited to investigations of the plasma clearance of drugs that are normally metabolized by the liver. This type of data is of limited value for differentiating altered hepatic metabolism arising from the uremic process (which are compounded by nonspecific effects such as nutrition, anemia, hypertension, etc.) from complicating factors such as changes in volume of distribution and protein binding. In addition, since many drugs are disposed of partly by renal excretion and partly by metabolic transformation, separation of renal from hepatic factors in altering plasma half-lives may be difficult. Current methodology and approaches to the study of drug metabolism in humans often permit only a superficial examination of the disposition of the drugs. For instance, the altered pattern of metabolism of vitamin D observed in uremic patients (see previous section) suggests that decreased urinary excretion of a compound may cause the liver to produce more polar metabolites that are excreted more readily by the residual kidney tissue or via the bile or gut. No evidence has been presented to test this hypothesis for compounds other than vitamin D yet the fact that drugs, such as furosemide, that normally are totally excreted unchanged in the urine do not remain indefinitely in the plasma of anuric or anephric patients indicates that alternate pathways of elimination must exist (CUTLER et al., 1974).

With respect to methodology, a recent experience with the estimation of antipyrine in the plasma of uremic subjects (LICHTER et al., 1973) provides a good example of the difficulties that may be encountered. The determination of antipyrine in plasma samples is normally a simple procedure (BRODIE et al., 1949). The sample is deproteinized, acidified and, after addition of sodium nitrite, the optical density of the nitroso-derivative of antipyrine is read spectrophotometrically at 350 nm. When this technique was used, blank values for many uremic specimens (i.e., the O.D. prior to the addition of sodium nitrite) were extremely high, a finding previously recognized by EDWARDS (1959). Furthermore, according to the procedure used by BRODIE et al. (1949), the reaction with sodium nitrite was frequently incomplete at 20 min, and in occasional samples, optical densities were still increasing 60 min after the addition of the nitrite. While a value for antipyrine could be calculated taking into account the high blank value, results were erratic and were highly inaccurate compared with results obtained using gas-liquid chromatography. Conventional recovery experiments using uremic plasma *in vitro* failed to predict difficulties with the spectrophotometric method. Figure 1 shows results of an antipyrine disappearance study in a uremic patient in which both methods for measuring antipyrine were employed. Values for the antipyrine half-life ($T\frac{1}{2}$) were different; a figure of 12 h was obtained by the spectrophotometric method and 7.6 h by the GLC method. Figure 2 compares the two methods as well as the effects of hemodialysis. While

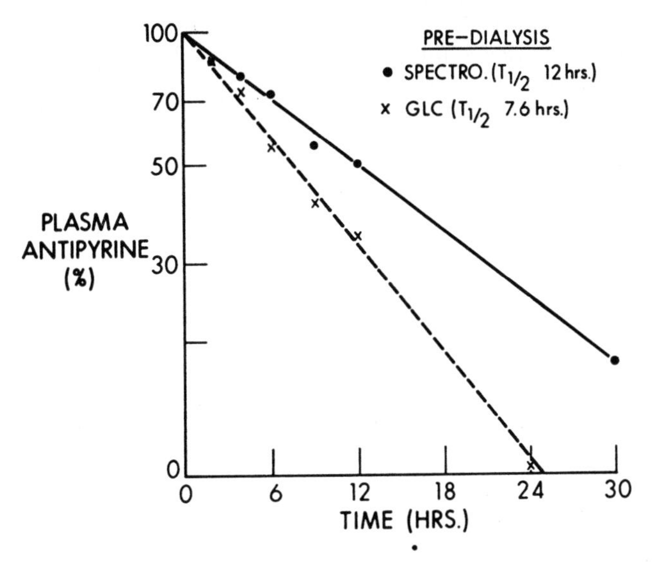

Fig. 1. Comparison of spectrophotometric and gas-liquid-chromatographic (GLC) methods for measuring plasma antipyrine in the uremic subject: effect on antipyrine half-life (T 1/2) in one patient

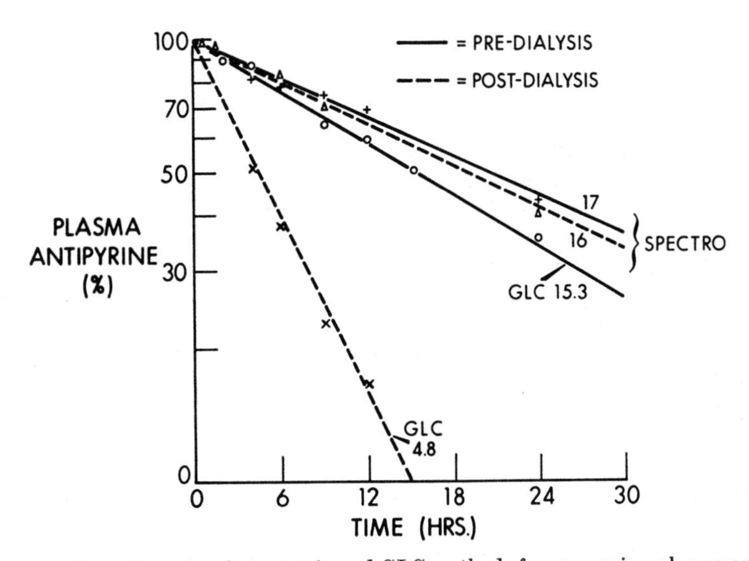

Fig. 2. Comparison of spectrophotometric and GLC methods for measuring plasma antipyrine in the uremic subject: effect of hemodialysis of the patient on antipyrine T 1/2 (numbers refer to T 1/2 in hours)

the spectrophotometric determinations failed to show an effect of the procedure, the more accurate GLC method revealed a substantial change in clearance. No consistent relationship between results obtained by the two methods was found, and the discrepancy shown in Fig. 2 between post-dialysis half-lives was the largest seen in the series.

A further example of limitations of current methodology is provided by observations on cephalothin kinetics in uremic patients. KABINS and COHEN (1964) used a microbiological assay for determination of serum levels of the drug and noted that cephalothin, which is deacetylated before being excreted, had a long half-life in uremic patients (often greater than 18 h compared with 0.5 h in normal controls), and that its clearance from the plasma was biphasic. More recently, KIRBY et al. (1971) repeated the study using a different assay system capable of measuring not only cephalothin but also its principal metabolite, desacetylcephalothin. The early phase of the "biphasic" curve reported by the previous workers was due to removal of cephalothin from the plasma (with a half-life that was slightly more prolonged than in normal subjects). However, the latter part of the curve resulted from the progressive appearance and delayed renal clearance of desacetylcephalothin, which also possesses antimicrobial activity and could not be separated from the parent drug by the assay system of KABINS and COHEN.

Although information currently available may not represent the final answer in many instances and can be variously interpreted in others, observations have been made in uremic subjects on the plasma clearance of the following drugs that are excreted predominantly in a metabolically altered form: pentothal, phenacetin, sulfisoxazole, chloramphenicol, and diphenylhydantoin.

DUNDEE and RICHARDS (1954) reported that narcosis with pentothal was prolonged in azotemic patients undergoing prostatectomy for chronic prostatic obstruction; several of these patients again underwent operations when renal function had improved and showed a more normal response to a second administration of the drug. RICHARDS et al. (1953) investigated this phenomenon in uremic rabbits, and showed that the metabolism of the barbiturate was impaired in the presence of uremia. DUBACH (1968) performed extensive studies in patients with severe renal damage resulting from excessive consumption of phenacetin. He demonstrated that the half-life of *n*-acetyl-*p*-aminophenol, the major product of phenacetin metabolism, is prolonged when compared with normal controls. REIDENBERG et al. (1969) studied the plasma clearance of sulfisoxazole, a frequently used sulfonamide that is acetylated prior to excretion. Analysis of elimination kinetics revealed an abnormality of the hepatic metabolism of the drug.

The data on chloramphenicol metabolism in uremia are conflicting. This drug is removed by glucuronidation, and both KUNIN and co-workers (1959a and b) and LINDBERG et al. (1966) found a normal serum half-life of the non-conjugated, biologically active drug. On the other hand, SUHRLAND and WEISBURGER (1963) found that blood levels of the non-metabolized drug were elevated in uremic patients receiving prolonged courses of the drug, and that this often resulted in erythropoietic depression.

Diphenylhydantoin metabolism has been investigated in uremic patients by LETTERI et al. (1971). The plasma half-life of the drug was shortened in patients with CRF, probably as a result of reduced protein-binding of the drug (REIDENBERG et al., 1971), Their data are not conclusive in this regard, however, and further observations, preferably utilizing non-colorimetric methods for drug analysis, are required to establish whether or not the plasma half-life of diphenylhydantoin is shortened in uremic patients.

Antipyrine has been used as a model compound for investigation of drug metabolism in man (VESELL and PAGE, 1968, 1969). It is well absorbed when given by mouth, not significantly protein-bound, and hydroxylated by the liver microsomal enzyme system prior to excretion. Preliminary studies in a group of patients on a chronic dialysis program (LICHTER et al., 1973) failed to show evidence of impaired antipyrine clearance but served to highlight the difficulties of performing studies of this type in uremic patients. Most of the patients were receiving several drugs on a long-term basis (antihypertensive agents, diuretics, etc.), which may influence the observed antipyrine half-life. In addition, a number of patients showed a shortening of the half-life following hemodialysis (in one patient from 15.3 h to 4.6 h, an effect that has also been observed for the plasma clearance of cortisol (BACON et al., in preparation). It is not clear whether this effect is caused by rapid changes in activity of hepatic microsomal enzymes or by other factors such as distribution of drug, hepatic blood-flow or uptake of drug by the liver cell. The observation of an effect of hemodialysis on plasma antipyrine half-life after two weeks of glutethimide therapy (which should have maximally increased hepatic microsomal enzyme activity) suggests that the mechanism may not involve liver microsomal function. Regardless of the nature of the mechanism, hemodialysis can exert a considerable influence on the clearance of compounds whose pathway of elimination involves the liver. The physiologic and pharmacologic importance of this observation merits further evaluation.

V. Renal Elimination of Drugs in Patients with CRF

The work of KUNIN and colleagues (KUNIN and FINLAND, 1959a; KUNIN, 1967a and b) and DETTLI and associates (DETTLI and SPRING, 1968; DETTLI et al.,

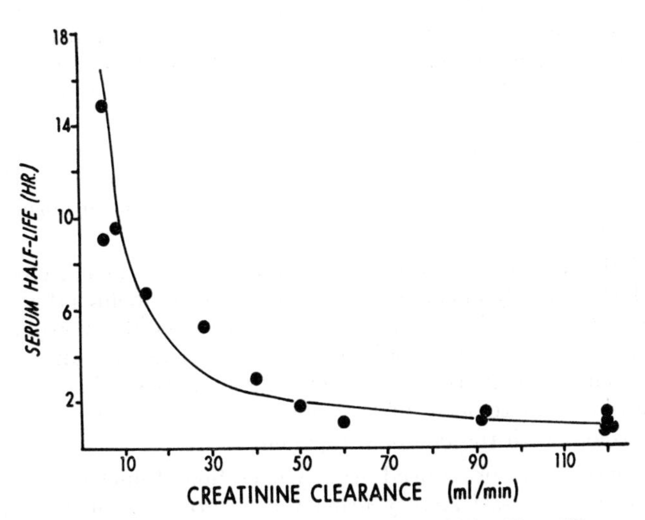

Fig. 3. Effect of impaired renal function on carbenicillin half-life. (From HOFFMAN et al., 1970, with permission of the author and Editor of Annals of Internal Medicine)

1970, 1971) established general principles concerning effects of diminished renal function on the disposition of drugs whose major pathway of elimination is by renal excretion. Figure 3 illustrates the hyperbolic relationship between the plasma half-life of such drugs (e.g., carbenicillin) and the creatinine clearance value. Thus, the plasma clearance of these drugs is grossly restricted only when creatinine clearance falls below 20 to 30 ml/min. Small further decrements of creatinine clearance then lead to a marked reduction in plasma clearance of the drugs. Studies by Dettli et al. (1971) and McHenry et al. (1971) revealed that the mean value for the plasma half-life of this type of drug in severely azotemic or anuric patients could be utilized to determine frequency of dosage in patients with CRF. This information has enabled many drugs to be used effectively and in comparative safety in patients with CRF, and several guide tables for appropriate doses of drugs and frequency of administration have been published (Kunin, 1967b, 1972; Bennett et al., 1970). However, an exception appears to be gentamicin, since uremic patients may exhibit marked individual variations in the ability to remove this antibiotic drug (Gyselynk et al., 1971; Riff and Jackson, 1971). Benner (1971) has suggested the use of individualized drug regimens derived from the rate of drug elimination by individual patients, especially for potentially toxic drugs. This sound approach, however, is limited by the necessity of having a readily available assay for measuring the plasma concentration of the drug (Sabath et al., 1971a and b).

Reference has been made to the difficulty in distinguishing hepatic and renal causes of a change in plasma half-life of a drug normally removed via both pathways. The complexities of this subject are further illustrated by the experience of Fabre et al. (1966), who determined blood levels of phenobarbital after giving standard oral doses to normal subjects and patients with various degrees of renal insufficiency. About 20 to 30% of a dose of phenobarbital was excreted unchanged in the urine of normal subjects (Butler et al., 1954), yet blood levels of the drug were similar in azotemic and non-azotemic subjects. Apparently, therefore, no general rule can be applied to the effects of chronic renal insufficiency on the disposition of most commonly used drugs. Each drug must be investigated individually in suitable patients to determine the effect of renal impairment. An important contribution has been made by Dettli who compiled a table on the clearance of 30 drugs in anuric patients (Dettli et al., 1971; Dettli, 1972). The results (Fig. 4) confirm marked variability in the effects of renal impairment on drug clearance and indicate the need for future investigation of factors influencing drug disposition in normal as well as uremic subjects.

The restriction in urinary excretion of small molecules encountered in patients with CRF affects not only unchanged drugs but also metabolites that are excreted by the kidney. For example, chloramphenicol (Kunin et al., 1959b) and cephalothin (Kirby et al., 1971) are metabolized before being excreted, and their principal metabolites accumulate in the serum for long periods of time. In such cases, biological activity and toxicity of the metabolites should be determined when planning treatment with the parent compound.

Another consequence of reduced urinary excretion of unchanged drug may be an inability to achieve satisfactory urinary concentrations of drug when this is the desired aim of therapy. Sachs et al. (1968) demonstrated a linear relationship between endogenous creatinine clearance and the amount of nitrofurantoin appearing in the urine after oral dosing with the drug. Virtually no drug was detected in the urine of patients with clearance values below 10 ml/min. Schlegel et al. (1967) reported similar results for nalidixic acid and sulfamethizole. These observations indicate that the therapeutic value of agents like urinary antiseptics may be limited in patients with poor renal function.

Drug	k_n	k_u	Q_0
Penicillin G	1.4	0.03	0.02
Ampicillin	0.6	0.11	0.18
Methicillin	1.4	0.17	0.12
Oxacillin	1.4	0.35	0.25
Carbenicillin	0.6	0.06	0.09
Cephalothin	1.4	0.06	0.04
Cephaloridine	0.4	0.03	0.08
Cephalexin	0.7	0.03	0.04
Ciba-Geigy 36278	0.7	0.03	0.04
Tetracycline	0.08	0.008	0.10
Chlortetracycline	0.12	0.08	0.67
Rolitetracycline	0.06	0.02	0.33
Doxycycline	0.03	0.03	1.0
Chloramphenicol	0.3	0.2	0.67
Streptomycin	0.27	0.01	0.04
Kanamycin	0.35	0.01	0.03
Vancomycin	0.12	0.003	0.03
Gentamycin	0.3	0.015	0.05
Colistin	0.3	0.02	0.07
Polymyxin B	0.16	0.02	0.13
Erythromycin	0.5	0.13	0.26
Lincomycin	0.15	0.06	0.40
Thiamphenicol	0.26	0.02	0.08
Sulfadiazine	0.08	0.03	0.36
Sulfisomidine	0.15	0.012	0.08
Sulfamethoxazole	0.08	0.06	0.84
Trimethoprim	0.08	0.03	0.39
Digoxin	0.019	0.006	0.31
Methyldigoxin	0.016	0.007	0.44
Digitoxin	0.0035	0.0031	0.89
Strophantin G	0.05	0.012	0.24
Peruvoside	0.01	0.01	1.0
Strophantin K	0.04	0.01	0.25
-Acetyldigoxin	0.03	0.01	0.33
5-Fluorocytosine	0.25	0.007	0.03

Fig. 4. Elimination rate constants in normal (k_n) and severely uremic (k_u) patients. Q_0 values (which are derived from a specially constructed nomograph—Dettli, 1972) indicate the extent of change of the elimination rate constant as a consequence of severe renal failure (1.0 indicates that no change has taken place, while 0.1 indicates that k has been reduced ten-fold...). (Adapted from Dettli (1972), with permission of the author and the Editor of Journal of Pharmacokinetics and Biopharmaceutics)

VI. Response to Drugs

Though toxicity due to drug accumulation in patients with CRF has been documented (Richet et al., 1966), the possibility of qualitative or quantitative differences in the response of the uremic patient to a particular drug has not been systematically investigated. Drugs whose actions augment an effect of the disease itself (e.g., emetics and sedatives) may produce pharmacologic or toxic effects at a lower plasma (or tissue) concentration of drug in uremic patients than in non-uremic subjects. Neurotoxicity with colistin occurs in azotemic patients at plasma levels of the drug that are in the normal therapeutic range (Richet et al., 1966, 1967, 1970; Wolinsky and Hines, 1962). A complicating factor in relating toxicity and plasma drug concentration is the change in protein binding of certain drugs observed in azotemic patients (see section on Volume of Distribution). This difference in protein binding may permit the tissue concentration of a drug to rise to a higher level in the azotemic patient than would be observed in non-azotemic patients with similar blood levels of drug. An example of such toxicity was documented in an azotemic patient with epilepsy who was receiving diphenylhydantoin (ODAR-CEDERLÖF et al., 1970).

More information is required to permit adequate evaluation of the increased risk of the azotemic patient to medication, particularly when these are administered chronically. When possible, both free drug and total drug concentrations in the blood should be measured in patients showing drug toxicity. The possibility of increased tissue concentrations relative to plasma concentrations of drugs can then be determined.

VII. Conclusions

The most important consideration for drug therapy of patients with CRF is the inability to excrete polar compounds normally eliminated unchanged in the urine. However, the syndrome of uremia also is associated with altered function in other organs involved in the response of the uremic subject to drugs. This review has attempted to emphasize these newer concepts without implying that these less clearly defined changes are more important than restricted renal function in the management of uremic patients. Differences in distribution, plasma protein binding, hepatic metabolism, and receptor sensitivity have been emphasized in considering drug actions in uremic patients. The need for additional data and imaginative pharmacologic studies is evident.

References

ANTON, A. H., COREY, W. T.: Plasma protein binding of sulfonamides in anephric patients. Fed. Proc. **30**, 629 (1971).

AVIOLI, L. V., BIRGE, S., WON LEE, S., SLATOPOLSKY, E.: The metabolic fate of vitamin D_3-^3H in chronic renal failure. J. clin. Invest. **47**, 2239—2252 (1968).

BACON, G. E., KENNY, F. M., MURDAUGH, H. V., RICHARDS, C.: Effect of chronic renal insufficiency on the biological half-life and production rate of cortisol in man. Clin. Res. **18**, 633 (1970).

BACON, G. E., KENNY, F. M., MURDAUGH, H. V., RICHARDS, C.: Cortisol production and metabolism in renal failure. (In preparation.)

BAGDADE, J. D.: Uremic lipemia. An unrecognized abnormality in triglyceride production and removal. Arch. intern. Med. **126**, 875—881 (1970).

BENNER, E. J.: In discussion of gentamicin and ampicillin in bile. J. infect. Dis. **124** (Suppl.), S109—S113 (1971).

BENNETT, W. M., SINGER, I., COGGINS, C. H.: A practical guide to drug usage in adult patients with impaired renal function. J. Amer. med. Ass. **214**, 1468—1475 (1970).

BIEMPICA, L., GOLDFISCHER, S., BLACK, M., BIEMPICA, S. L., GROSSMAN, S. B., ARIAS, I. M.: Cytochemical and ultrastructural appearances of the liver in experimental uremia. (Unpublished results.)

BLACK, M., GROSSMAN, S. B., ARIAS, I. M.: Biochemical changes in the livers of rats with experimental uremia, with special reference to drug metabolism. (Unpublished results.)

BLUM, M. R., RIEGELMAN, S., BECKER, C. E.: Altered protein binding of diphenylhydantoin in uremic plasma. New Engl. J. Med. **286**, 109 (1972).

BOYER, P. D., LARDY, H., MYRBÄCK, K.: The enzymes, Vol. 4, 2nd Edition. New York-London: Academic Press 1960.

BRODIE, B. B., AXELROD, J., SOBERMAN, R., LEVY, B. B.: The estimation of antipyrine in biological materials. J. biol. Chem. **179**, 25—29 (1949).

BURZYŃSKI, S.: Bound amino acids in serum of patients with chronic renal insufficiency. Clin. chim. Acta **25**, 231—237 (1969).

BUTLER, T. C., MAHAFFEE, C., WADDELL, W. J.: Phenobarbital: studies of elimination, accumulation, tolerance, and dosage schedules. J. Pharmacol. exp. Ther. **111**, 425—435 (1954).

Cerletty, J.M., Engbring, N.H.: Azotemia and glucose intolerance. Ann. Intern. Med. **66**, 1097—1108 (1967).
Coles, G.A., Peters, D.K., Jones, J.H.: Albumin metabolism in chronic renal failure. Clin. Sci. **39**, 423—435 (1970).
Conney, A.H., Kuntzman, R.: Metabolism of normal body constituents by drug-metabolising enzymes, Chapter 46. In: Brodie, B.B., Gillette, J.R. (Eds.): Handbook of experimental pharmacology, Vol. XXVIII, Concepts in biochemical pharmacology, Part 2. Berlin-Heidelberg-New York: Springer 1971.
Cutler, R.E., Forrey, A.W., Christopher, T.G., Kimpel, B.M.: Pharmacokinetics of furosemide in normal subjects and functionally anephric patients. Clin. Pharmacol. Ther. **15**, 588—596 (1974).
Czerniak, Z., Burzyński, S.: Free amino acids in serum of patients with chronic renal insufficiency. Clin. Chim. Acta **24**, 367—372 (1969).
DeLuca, H.F.: Vitamin D group, VIII. Active compounds. Chapter 7. In: Sebrell, H.W., Jr., Harris, R.S. (Eds.): The vitamins. New York-London: Academic Press 1971.
Dettli, L., Spring, P.: Factors influencing drug elimination in man. Il farmaco (Sc. ed.) **23**, 795—812 (1968).
Dettli, L., Spring, P., Habersang, R.: Drug dosage in patients with impaired kidney function. Postgrad. med. J. **46** (Suppl. Oct.), 32—35 (1970).
Dettli, L., Spring, P., Ryter, S.: Multiple dose kinetics and drug dosage in patients with kidney disease. Acta pharmacol. toxicol. scand. **29** (Suppl.), 211—224 (1971).
Dubach, R., Callender, S.T.E., Moore, C.V.: Studies in iron transportation and metabolism. VI. Absorption of radioactive iron in patients with fever and with anemias of varied etiology. Blood **3**, 526—549 (1948).
Dubach, U.C.: Absorption, Schicksal und Ausscheidung von Phenacetin und *N*-acetyl-*p*-aminophenol bei Niereninsuffizienz. Klin. Wschr. **46**, 261—264 (1968).
Dundee, J.W., Richards, R.K.: Effect of azotemia upon the action of intravenous barbiturate anesthesia. Anesthesiology **15**, 333—346 (1954).
Edwards, K.D.G.: Creatinine space as a measure of total body water in anuric subjects, estimated after single injection and haemodialysis. Clin. Sci. **18**, 455—464 (1959).
Englert, E., Jr., Brown, H., Willardson, D.G., Wallach, S., Simons, E.L.: Metabolism of free and conjugated 17-hydroxycorticosteroids in subjects with uremia. J. clin. Endocr. **18**, 36—48 (1958).
Eschbach, J.W., Cook, J.D., Finch, C.A.: Iron absorption in chronic renal disease. Clin. Sci. **38**, 191—196 (1970).
Fabre, J., de Freudenreich, J., Duckert, A., Pitton, J.S., Rudhart, M., Virieux, C.: Influence of renal insufficiency on the excretion of chloroquine, phenobarbital, phenothiazines, and methacycline. Helv. med. Acta **4**, 307—316 (1966).
Goff, J.B., Schlegel, J.U., O'Dell, R.M.: Urinary excretion of nalidixic acid, sulfamethizole and nitrofurantoin in patients with reduced renal function. J. Urol. **99**, 371—375 (1968).
Goldstein, A., Aronow, L., Kalman, S.M. (Eds.): Principles of drug action. New York-Evanston-London: Harper and Row (Hoeber Medical Division) 1969.
Gossmann, H.H., Baltzer, G., Helms, H.: Calciumstoffwechsel bei chronischer Niereninsuffizienz. Klin. Wschr. **46**, 497—503 (1968).
Gulyassy, P.F., Peters, J.H., Lin, S.C., Ryan, P.M.: Hemodialysis and plasma amino acid composition in chronic renal failure. Amer. J. clin. Nutr. **21**, 565—573 (1968).
Gulyassy, P.F., Peters, J.H., Lin, S.C., Ryan, P.M., Schoenfeld, P.: Transport and metabolism of tryptophan in uremia. Conference on "Uremia", Freiburg, W. Germany (1971). (In press.)
Gyselynck, A.M., Forrey, A., Cutler, R.: Pharmacokinetics of gentamicin: distribution and plasma and renal clearance. J. infect. Dis. **124** (Suppl.), S70—S76 (1971).
Hampers, C.L., Soeldner, J.S., Gleason, R.E., Bailey, G.L., Diamond, J.A., Merrill, J.P.: Insulin-glucose relationships in uremia. Amer. J. clin. Nutr. **21**, 414—422 (1968).
Heizer, W.D., Smith, T.W., Goldfinger, S.E.: Absorption of digoxin in patients with malabsorption syndromes. New Engl. J. Med. **285**, 257—259 (1971).
Hoffman, T.A., Cestero, R., Bullock, W.E.: Pharmacodynamics of carbenicillin in hepatic and renal failure. Ann. int. Med. **73**, 173—178 (1970).
Horton, E.S., Johnson, C., Lebovitz, H.E.: Carbohydrate metabolism in uremia. Ann. int. Med. **68**, 63—74 (1968).
Izzo, J.L., Bartlett, J.W., Roncone, A., Izzo, M.J., Bale, W.F.: Physiological processes and dynamics in the disposition of small and large doses of biologically active and inactive ^{131}I-insulin in the rat. J. biol. Chem. **242**, 2343—2355 (1967).
Josephs, H.W.: Studies in vitamin A. Relation of vitamin A and carotene to serum lipids. Bull. Johns Hopk. Hosp. **65**, 112—124 (1939).

KABINS, S. A., COHEN, S.: Cephalothin serum levels in the azotemic patient. Antimicrob. Agents and Chemother. **1964**, 207—214.

KAGAN, B. M., KAISER, E.: Vitamin A concentration in the liver in the nephrotic syndrome. J. Lab. clin. Med. **40**, 12—16 (1952).

KAGAN, B. M., THOMAS, E. M., JORDAN, D. A., ABT, A. F.: Serum vitamin A and total plasma lipid concentrations as influenced by the oral administration of vitamin A to children with the nephrotic syndrome. J. clin. Invest. **29**, 141—145 (1950).

KEEN, P.: Effect of binding to plasma proteins on the distribution, activity and elimination of drugs. Chapter 10. In: BRODIE, B. B., GILLETTE, J. R. (Eds.): Handbook of experimental pharmacology, Vol. XXVIII, Concepts in biochemical pharmacology, Part 2. Berlin-Heidelberg-New York: Springer 1971.

KIRBY, W. M. M., DE MAINE, J. B., SERRILL, W. S.: Pharmacokinetics of the cephalosporins in healthy volunteers and uremic patients. Postgrad. med. J. **47** (Suppl. Feb.), 41—46 (1971).

KUNIN, C. M.: Problems of antimicrobial therapy in renal failure. Proc. 3rd Int. Congr. Nephrol. **3**, 193—213 (1967a).

KUNIN, C. M.: A guide to use of antibiotics in patients with renal disease. A table of recommended doses and factors governing serum levels. Ann. intern. Med. **67**, 151—158 (1967b).

KUNIN, C. M.: Antibiotic usage in patients with renal impairment. Hosp. Practice **7**, 141—149 (1972).

KUNIN, C. M., FINLAND, M.: Restrictions imposed on antibiotic therapy by renal failure. Arch. intern. Med. **104**, 1030—1050 (1959a).

KUNIN, C. M., GLAZKO, A. J., FINLAND, M.: Persistence of antibiotics in blood of patients with acute renal failure. II. Chloramphenicol and its metabolic products in the blood of patients with severe renal disease or hepatic cirrhosis. J. clin. Invest. **38**, 1498—1508 (1959b).

LAWSON, D. H., BODDY, K., KING, P. C., LINTON, A. L., WILL, G.: Iron metabolism in patients with chronic renal failure on regular dialysis treatment. Clin. Sci. **41**, 345—351 (1971).

LETTERI, J. M., MELLK, H., LOUIS, S., KUTT, H., DURANTE, P., GLAZKO, A. J.: Diphenylhydantoin metabolism in uremia. New Engl. J. Med. **285**, 648—652 (1971).

LICHTER, M., BLACK, M., ARIAS, I. M.: The metabolism of antipyrine in patients with chronic renal failure. J. Pharmacol. exp. Ther. **187**, 612—619 (1973).

LIPPMAN, R. W., WROBEL, C. J., REES, R., HOYT, R.: A theory concerning recurrence of urinary infection: prolonged administration of nitrofurantoin for prevention. J. Urol. **80**, 77—81 (1958).

LOSOWSKY, M. S., KENWARD, D. H.: Lipid metabolism in acute and chronic renal failure. J. Lab. clin. Med. **71**, 736—743 (1968).

LUMB, G. A., MAWER, E. B., STANBURY, S. W.: The apparent vitamin D resistance of chronic renal failure. A study of the physiology of vitamin D in man. Amer. J. Med. **50**, 421—441 (1971).

MACGREGOR, G. A., POOLE-WILSON, P. A., JONES, N. F.: Phenformin and metabolic acidosis. Lancet **1972 I**, 69—71.

MARKUS, G.: Protein substrate conformation and proteolysis. Proc. nat. Acad. Sci. (Wash.) **54**, 253—258 (1965).

MCHENRY, M. C., GAVAN, T. L., GIFFORD, R. W., JR., GEURNIK, N. A., VAN OMMEN, R. A., TOWN, M. A., WAGNER, J. G.: Gentamicin dosages for renal insufficiency. Adjustments based on endogenous creatinine clearance and serum creatinine concentration. Ann. intern. Med. **74**, 192—197 (1971).

MUNRO, H. N.: Role of amino acid supply in regulating ribosome function. Fed. Proc. **27**, 1231—1237 (1968).

O'BRIEN, J. P., SHARPE, A. R., JR.: The influence of renal disease on the insulin I^{131} disappearance curve in man. Metabolism **16**, 76—83 (1967).

ODAR-CEDERLÖF, I., LUNDE, P., SJÖQVIST, F.: Abnormal pharmacokinetics of phenytoin in a patient with uremia. Lancet **1970 II**, 831—832.

OGG, C. S.: The intestinal absorption of ^{47}Ca by patients in chronic renal failure. Clin. Sci. **34**, 467—471 (1968).

ORME, B. M., CUTLER, R. E.: The relationship between kanamycin pharmacokinetics: distribution and renal function. Clin. Pharmacol. Ther. **10**, 543—550 (1969).

REIDENBERG, M. M.: Effect of kidney disease on pharmacokinetics and drug response. Fifth Int. Congr. on Pharmacology, San Francisco 1972. Basel-London-New York: Karger.

REIDENBERG, M. M., KOSTENBAUDER. H., ADAMS, W. P.: Rate of drug metabolism in obese volunteers before and during starvation and in azotemic patients. Metabolism **18**, 209—213 (1969).

REIDENBERG, M. M., KOSTENBAUDER, H., ADAMS, W. P., ODAR-CEDERLÖF, I., VON BAHR, C., BORGÅ, O., SJÖQVIST, F.: Protein binding of diphenylhydantoin and desmethylimipramine in plasma from patients with poor renal function. New Engl. J. Med. **285**, 264—267 (1971).

RICHARDS, R. K., TAYLOR, J. D., KUETER, K. E.: Effect of nephrectomy on duration of sleep following administration of thiopental and hexobarbital. J. Pharm. exp. Ther. **108**, 461—473 (1953).
RICHET, G., FABRE, J., DE FREUDENREICH, J., PODEVIN, R.: La tolérance médicamenteuse au cours de l'insuffisance rénale. J. Urol. Néphrol. **72**, 257—302 (1966).
RICHET, G., FABRE, J., DE FREUDENREICH, J., PODEVIN, R.: The drug tolerance of uremic patients. The need of modified dosage levels in organ insufficiency. Čas. Lék. čes. **106**, 851—854 (1967).
RICHET, G., FABRE, J., DE FREUDENREICH, J., PODEVIN, R., LOPEZ DE NOVALES, E., VERROUST, P.: Drug intoxication and neurological episodes in chronic renal failure. Brit. med. J. **1970 I**, 394—395 (1970).
RIFF, L. J., JACKSON, G. G.: Pharmacology of gentamicin in man. J. infect. Dis. **124** (Suppl.), S 98—S 105 (1971).
RITZ, E., JANTZEN, R.: Vitamin D-Aktivität im Serum urämischer Patienten. Klin. Wschr. **47**, 1112—1114 (1969).
ROELS, O. A.: Vitamins A and carotene. VII. Active compounds and vitamin A antagonists. Chapter 1. In: SEBRELL, W. H., JR., HARRIS, R. S. (Eds.): The vitamins. New York-London: Academic Press, 153—167 (1967).
SABATH, L. D., CASEY, J. I., RUCH, P. A., STUMPF, L. L., FINLAND, M.: Rapid microassay for circulating nephrotoxic antibiotics. Antimicrob. Agents and Chemother. **1970**, 83—90 (1971 a).
SABATH, L. D., CASEY, J. I., RUCH, P. A., STUMPF, L. L., FINLAND, M.: Rapid microassay of gentamicin, kanamycin, neomycin, streptomycin, and vancomycin in serum or plasma. J. Lab. clin. Med. **78**, 457—463 (1971 b).
SACHS, J., GEER, T., NOELL, P., KUNIN, C. M.: Effect of renal function on urinary recovery of orally administered nitrofurantoin. New Engl. J. Med. **278**, 1032—1035 (1968).
SCHLEGEL, J. U., GOFF, J. B., O'DELL, R. M.: Bacteriuria and chronic renal disease. Trans. Amer. Assoc. Genitourin. Surg. **59**, 32—36 (1967).
SCHMID, E., MEYTHALER, C., DVORAK, G., SCHAUDIG, H.: Clinico-pharmacologic investigations on the absorption and urinary excretion of nitrofurantoin—(Furadantin®). Verh. dtsch. Ges. int. Med. **72**, 401—403 (1967).
SCHREINER, G. E., MAHLER, J. F.: In "Uremia: Biochemistry, pathogenesis, and treatment". Springfield: Charles C Thomas 1961.
SMITH, F. R., GOODMAN, DEW. S.: The effects of diseases of the liver, thyroid, and kidneys on the transport of vitamin A in human plasma. J. clin. Invest. **50**, 2426—2436 (1971).
STANBURY, S. W., LUMB, G. A., MAWER, E. B.: Osteodystrophy developing spontaneously in the course of chronic renal failure. Arch. intern. Med. **124**, 274—281 (1969).
SUHRLAND, L. G., WEISBERGER, A. S.: Chloramphenicol toxicity in liver and renal disease. Arch. intern. Med. **112**, 747—754 (1963).
VERWEY, W. F., WILLIAMS, H. R., JR.: Relationships between the concentrations of various penicillins in plasma and peripheral lymph. Antimicrob. Agents and Chemother. **1962 a**, 476—483.
VERWEY, W. F., WILLIAMS, H. R., JR.: Binding of various penicillins by plasma and peripheral lymph obtained from dogs. Antimicrob. Agents and Chemother. **1962 b**, 484—491.
VESELL, E. S., PAGE, J. G.: Genetic control of drug levels in man: antipyrine. Science **161**, 72—73 (1968).
VESELL, E. S., PAGE, J. G.: Genetic control of the phenobarbital-induced shortening of plasma antipyrine half-lives in man. J. clin. Invest. **48**, 2202—2209 (1969).
WAN, S. H., RIEGELMAN, S.: Investigation of kidney metabolism and its contribution to overall metabolism of drugs. I. Conversion of benzoic acid to hippuric acid. J. Pharm. Sci. (In press.)
WHELTON, A., SAPIR, D. G., CARTER, G. G.: Intrarenal distribution of ampicillin in the normal and diseased human kidney. J. infect. Dis. **125**, 466—470 (1972).
WOLINSKY, E., HINES, J. D.: Neurotoxic and nephrotoxic effects of colistin in patients with renal disease. New Engl. J. Med. **266**, 759—762 (1962).

Section Seven: Drug Interactions and Adverse Drug Reactions

Chapter 69

Pharmacokinetic Drug Interactions

D. G. Shand, J. R. Mitchell, and J. A. Oates

With 4 Figures

I. Introduction

Drugs may frequently interact adversely to cause drug toxicity or render a needed therapy ineffective. The situation is compounded by the current pattern of drug usage in which approximately ten different drugs are received by the average patient in the hospital. The number of drugs employed in combination in the out-patient setting is also increasing as more chronic diseases are treated for longer periods of time. In this therapeutic milieu, surveys of drug usage have shown that more than 10% of patients were concomitantly receiving drugs considered to interact adversely with one another. This suggests a need not only for an improved awareness of drug interactions by physicians but also for careful investigative documentation of drug interactions and their clinical relevance in man in order to provide a rational education in this important aspect of therapeutics. Mere proliferation of lists containing numerous speculative and inadequately documented drug interactions is educationally counter-productive, serving only to destroy credibility and to cause many physicians to shrug off drug interactions as hobgoblins from the pharmacologist's imagination.

Interactions between drug fall into two major classes. Pharmacodynamic drug interactions include the classical examples of antagonism, synergism, and potentiation that result from alterations in the response of the target organ(s), such as the antagonism of curare by anticholinesterase drugs. The other group, pharmacokinetic drug interactions, results from alterations in the delivery of drugs to their sites of action. This chapter will focus on the latter class of interactions, especially the current investigative evidence supporting its importance in therapy. The principles underlying meaningful evaluation of these drug interactions will be reviewed. Primary attention will be given to those interactions of proven clinical significance, emphasizing the importance of understanding their mechanisms and the need for conclusive proof that they occur in man. Drug interactions that have been established in animals or *in vitro* shall be referred to as potential drug interactions in man, and data in animals will be considered only to provide background or when the information sought can only be derived in a rigorously controlled situation in experimental animals.

The necessity for proving the occurrence of a drug interaction in man and documenting its clinical relevance is dictated by a number of factors. In those drug interactions that modify the delivery of drugs to their cellular sites of action, it is misleading to infer clinical significance solely from studies in experimental animals. Metabolic pathways, protein binding and the kinetics of disposition of both the drug and its modifier usually differ widely among various species. For example, the responses of the drug metabolizing enzymes to inducing or inhibiting agents in man are not always predictable from animal studies because the dose or concentration of drug at the site of interaction in animals is often much greater than the concentrations that occur during clinical use. Moreover, it is unlikely even now that we completely comprehend all ways in which species vary. This is particularly true for the effects of disease on drug action because most pharmacologic studies are carried out in normal healthy animals.

Because a variety of factors are involved in drug disposition, many ways exist by which the delivery of a drug to its site of action can be modified (Fig. 1). From a clinical standpoint, these may be conveniently divided into those that

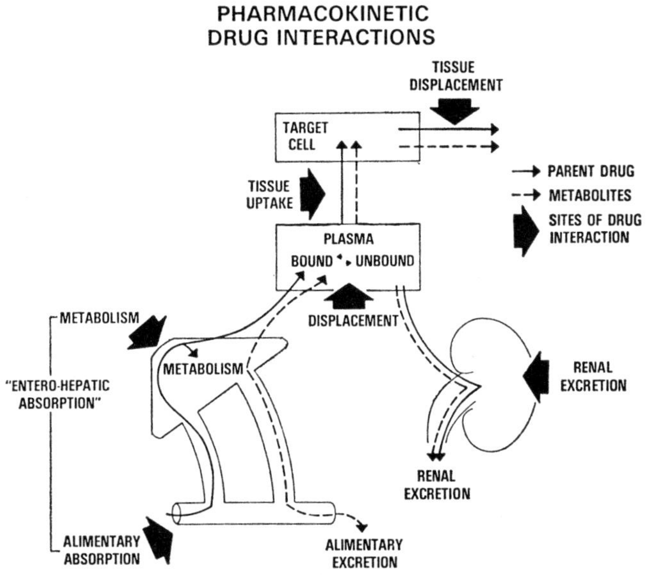

Fig. 1. Sites for potential drug interactions

increase (thereby enhancing effectiveness) or *decrease* (thereby reducing effectiveness) the delivery of free drug in the circulation to its site of action. It is necessary to consider delivery of free drug (that drug not bound to plasma proteins) in the blood because the effectiveness of a drug is, as a general rule, a function of free drug concentration at its site of action, since the equilibrium between free drug and receptor determines the pharmacologic response. Impaired delivery may occur as a result of diminished absorption from the gut or an intramuscular site; by induction of major enzyme pathways in the inactivation of a drug, thereby reducing the amount of drug reaching the systemic circulation (first-pass effect) and increasing its rate of elimination; by inhibition of drug uptake into its cellular site of action; and by displacement of the drug from its site of action. In contrast, enhanced drug delivery to the target site may occur when drug metabolism is inhibited; when induction of metabolic conversion to an active metabolite occurs; when impaired drug protein binding exists; and when the renal excretion of a drug or its active metabolite is impaired. The evaluation of the mechanism of an interaction requires consideration of all these factors, together with obtaining data that is kinetically and biologically consistent with the proposed mechanism. Furthermore, the factors that influence drug disposition do not always vary independently, and a change in one, such as altered drug binding in the blood, may affect another, such as rate of elimination. Finally, in order to gauge clinical relevance, the magnitude of the likely change is of paramount importance, and a lack of emphasis of the quantitative aspects of drug interactions has led to many misleading generalizations.

A differential diagnosis of the primary causal factors in pharmacokinetic drug interactions is listed in Table 1 according to the ways in which delivery of the drug to its site of action can be altered. Viewed from the standpoint of availability of drug at its pharmacologic target, the biological determinants of drug interactions can be readily related to the kinetic manifestations of these interactions as well as to the net clinical effect.

Table 1. *Factors altering the availability of unbound drug at its site of action*

Enhanced delivery to site of action
- (a) Increased absorption.
- (b) Decreased plasma drug binding.
- (c) Decreased elimination.
  - (i) Inhibition of drug metabolizing enzymes.
  - (ii) Decreased renal excretion.
  - (iii) Reduced hepatic blood flow.
- (d) Increased formation of active metabolites.

Decreased delivery to site of action
- (a) Decreased absorption.
- (b) Decreased uptake into (or displacement from) cellular site of action.
- (c) Enhanced elimination.
  - (i) Enzyme stimulation.
  - (ii) Increased renal elimination.
- (d) Decreased formation of active metabolites.

II. Biological Determinants of Kinetic Parameters

Pharmacokinetic drug interactions occur as a result of alterations in the biologic processes that determine drug absorption, distribution, and elimination. Changes in these processes alter free drug concentration and thereby drug action. However, in the investigation of drug interactions in man, changes in these biologic determinants can only be assessed from resultant changes in the measured kinetic parameters of plasma drug concentration. These parameters include drug half-life, clearance and volume of distribution. No single kinetic parameter, moreover, is sufficient to establish biologic mechanisms for they are all interrelated. Rather, a knowledge of all of them is required for full definition of an interaction and its differentiation from others that can produce a similar change in one parameter. For example, increased free drug concentration may be associated with both an increase in drug half-life due to inhibition of elimination and a decrease in half-life due to drug displacement from plasma binding sites. It is also important to recognize that a change in a given biological mechanism will not affect the elimination of all drugs equally: that is, the elimination of drugs may be rate-limited by different processes in different organs, and occasionally even in the same organ. The relationship of changes in these kinetic parameters to specific biologic alterations in the fate of a drug (e.g., changes in metabolism versus changes in tissue distribution) can be considered most easily in terms of the clearance concept, which relates events occurring at the organ of elimination to the resultant effects on circulating drug concentration. This approach provides an essential framework for the differential diagnosis of the several interrelated factors underlying drug interactions. Furthermore, it provides methods for the quantification of alterations in elimination processes that are essential in deciding whether a potential drug interaction will occur and, if so, what will be its likely clinical significance.

Drug clearance (Cl) by an organ of elimination can be expressed in terms of its blood flow (Q) and extraction ratio (E) as:

$$\mathrm{Cl} = Q \cdot E. \tag{1}$$

If the organ is the only one responsible for elimination, then its clearance may be equated with drug clearance from the circulation, which may be expressed as:

$$\mathrm{Cl} = \frac{0.693 \cdot Vd}{T_{1/2}} \tag{2}$$

in which $T_{1/2}$ is the drug half-life and Vd its apparent volume of distribution. Finally, drug concentration in the circulation is determined by the amount of drug reaching the circulation (FD), such that:

$$\mathrm{AUC} = \frac{\mathrm{FD}}{\mathrm{Cl}} \text{ and } \mathrm{Cl} = \frac{\mathrm{FD}}{\mathrm{AUC}} \tag{3}$$

in which AUC is the area under the concentration/time curve extrapolated to infinity, and F is the fraction of the administered dose (D) that reaches the circulation after oral administration. An equation of similar form describes the relationship between drug clearance and the average drug concentration (Cav) that is attained at steady state after chronic drug administration with a dosage interval, t:

$$\mathrm{Cl} = \frac{\mathrm{FD}}{\mathrm{Cav} \cdot t} \,. \tag{4}$$

These equations can be used to define the following series of relationships:

$$\mathrm{Cl} = Q \cdot E = \frac{0.693 \cdot Vd}{T_{1/2}} = \frac{\mathrm{FD}}{\mathrm{AUC}} = \frac{\mathrm{FD}}{\mathrm{Cav} \cdot t}$$

which may be used to describe the various factors determining drug half-life, clearance, and circulating drug concentrations.

A. Drug Half-Life

Most studies of drug interactions have emphasized the changes in drug half-life that occur. While this is an important parameter, its use as the *sole* index of drug elimination is limited because it can be influenced by factors affecting both elimination (clearance) and distribution (volume of distribution). Thus:

$$T_{1/2} = \frac{0.693 \cdot Vd}{QE} \,. \tag{5}$$

In particular, drug half-life is a poor measure of the efficiency with which a drug is eliminated by an organ. For example, a drug with a small volume of distribution may be very poorly extracted by the organ(s) of elimination, yet have the same half-life as a drug that is very efficiently removed from the circulation although it has a larger volume of distribution. At best, therefore, measurement of drug half-life alone may be used as a screening procedure; the causal factors producing the pharmacokinetic change can be determined only with the help of additional experimental and kinetic data.

B. Drug Clearance

In contrast to drug half-life, drug clearance measures only the elimination process and is unaffected by drug distribution. Drug clearance may be affected by alterations in blood flow, by the affinity of the removal process for the drug and, in some cases, by plasma drug binding. Not all drugs, however, are equally affected by such changes because their elimination may be rate-limited by different factors. When the initial extraction ratio is high, drug clearance is flow-dependent and will be relatively unaffected by alterations in, say, the activity of the drug-metabolizing enzymes. On the other hand, when extraction is low, organ clearance depends on the activity of the elimination process (e.g., drug metabolism) and is little influenced by flow. Similarly, altered plasma drug binding alters drug clearance only when elimination is limited to the free drug in the circulation. Such elimination may be termed restrictive to contrast it with non-restrictive elimination in which both bound and free forms of the drug are extractable.

For compounds that show restrictive elimination, plasma bound drug may be considered as a reservoir that tends to slow elimination and prolong drug half-life. Displacement of such a drug from its binding sites by another drug will increase free drug concentration immediately. After redistribution, this also results in accelerated elimination that compensates for the change and hence these interactions are generally transient. It should be emphasized that not all protein binding displacement interactions change free drug concentration significantly. If relatively little drug is present in the blood compared to that in the tissues (that is, the volume of distribution is large) even displacement of all the bound drug from the circulation into the tissues will not alter free drug concentration to any appreciable extent. On the other hand, total drug concentration will always be reduced after redistribution of bound drug from the circulation. In addition, total drug clearance (Cl) is related to free drug fraction (f) rather than free drug concentration. This may be understood by considering the clearance of free drug as the product of flow (Q) and the extraction ratio for free drug (e) which should be constant for restrictive elimination under first order conditions when flow is not rate limiting:

$$\text{Cl free} = Qe. \tag{6}$$

Free drug clearance may also be calculated from total drug clearance as:

$$\text{Cl free} = \frac{\text{Cl}}{f} = Qe. \tag{7}$$

A change in total drug clearance will occur at the new steady state irrespective of whether actual free drug concentration or half-life is altered. Even insignificant interactions of this type, therefore, will be accompanied by changes in *total* drug concentration, and measurement of the latter may lead to misinterpretation. Significant interactions for drugs with restrictive elimination should be accompanied by an alteration in drug half-life, which reflects changes in free drug concentration.

When drugs with non-restrictive elimination are considered, the situation is entirely different because drug bound in the circulation serves a transport function, and plasma binding therefore accelerates elimination (EVANS et al., 1973). Relevant kinetic parameters must be calculated on the basis of total drug concentration because both bound and free drug can be extracted by the eliminating organ. In this case, reduction in drug binding in the circulation cannot alter total drug clearance; rather, changes in half-life are determined by the volume of distribution of total drug. When drug is displaced from binding sites in the blood and redistributes into the tissues, the volume of distribution of total drug is increased, and the half-life is prolonged because a larger volume must be cleared at the same clearance rate. Such interactions will result in permanent changes in free drug concentration since total drug clearance, which determines total drug concentration (AUC or Cav), is unaltered and average free drug concentration is the product of total drug concentration and free drug fraction. At the present time, displacement interactions have not been described for drugs with non-restrictive elimination, although the effects of varied plasma drug binding for such drugs has been shown. The theoretical implications are worthy of discussion if only to illustrate further that the outcome of pharmacokinetic interactions depends very much on the drug in question and that the findings with one compound cannot be necessarily applied to all drugs.

C. Volume of Distribution

Biologically, the apparent volume of distribution of a drug is determined by the relative affinities of binding sites in the blood and tissues. Volume of distribution can therefore be altered by changing either tissue or plasma binding, and a change in this parameter should alert the investigator to the possibility of a redistributional interaction.

Kinetically, the apparent volume of distribution can be viewed simply as a constant required to relate the two measurable parameters of drug clearance and elimination rate constant. The methods of calculating drug clearance from area measurements (AUC) are model-independent and, consequently, the different values for calculated apparent volume of distribution depend largely on the method used to calculate elimination rate constants, that is, upon how many compartments the body is conceived of. Throughout this chapter, only a single rate constant or $T_{1/2}$ will be considered because it simplifies the mathematics enough for clinical application without recourse to computers, and the principles developed will have general validity for any chosen compartmental model or rate constant of elimination.

D. Drug Concentration

Of all kinetic parameters, the concentration of drug in the circulation is the least indicative of the mechanism of drug interactions since drug concentration is influenced by absorption, elimination, and distribution. Furthermore, changes in total drug do not always parallel those in free drug concentration when altered drug binding is involved. Drug concentration used in calculating kinetic parameters may be measured in whole blood, plasma or plasma water, the choice depending on the circumstances. If drug clearance estimates are to be used to predict organ clearance and extraction ratio, then whole blood concentrations should be used since these relate to organ blood flow (ROWLAND, 1972). The same is true of drugs with non-restrictive elimination. If plasma concentrations are used, the calculated parameters will only relate to actual organ plasma flow when drug is confined to plasma or is irreversibly bound to red cells. The use of free drug concentration is most useful when elimination is restrictive and a redistributional interaction is suspected. Thus a full kinetic evaluation requires a knowledge of the free drug fraction (f) and of the partition, uptake or binding of drug in the red cells, that is, the blood/plasma drug concentration ratio. Then all clearance and volume terms can be referenced to drug concentration in plasma, blood or plasma water, whichever is biologically the most appropriate. Finally, even when free drug concentration is measured, the complex relationship that exists between free drug and drug action may necessitate a careful assessment of changes in drug effect to determine if a drug interaction is actually of clinical importance.

E. Summary

The correct interpretation of pharmacokinetic drug interactions and their quantitative significance depends on a clear knowledge of the several factors that can influence free drug concentration at the site of drug action. The effects of altered drug disposition on the various kinetic parameters are complex because (1) any given kinetic parameter may be altered by several biological mechanisms, (2) some alterations in drug concentration may not be significant because they do not affect free drug concentration and therefore do not alter drug effect, and (3) modifying the biological processes of drug disposition will not affect all drugs equally or even in the same direction, that is, no single set of rules can be applied to all drugs. However, the clearance concept can be applied to rationalize these difficult and sometimes seemingly paradoxical effects. In particular, it can be used to classify drug elimination as high extraction, low extraction, restrictive or non-restrictive. This allows both the qualitative and quantitative effects of alterations in blood flow, drug metabolism, drug transport or drug binding to be predicted as a prelude to judging their clinical significance.

III. Mechanisms of Drug Interactions

A. Altered Absorption

The absorption of drugs from the gastrointestinal tract is a complex process influenced by several physiological and physicochemical factors (BRODIE, 1964; LEVINE, 1970; BINNS, 1971) all of which can be altered by the co-administration of other drugs. From the practical point of view, it is important to distinguish between interactions that alter the rate of drug absorption and those that change the total amount of drug absorbed. Altering the rate of absorption alone will change the drug concentration/time profile but will not affect bioavailability or average steady state drug concentration. The effects of alteration in absorption rate will therefore be of greatest importance during the administration of drugs usually given in single doses, such as analgesics. Because peak drug levels are determined by the relative magnitudes of absorption and elimination rate constants, altering the rate of absorption will have the greatest effect on drugs with short half-lives, such as procaine amide, but little effect on drugs like warfarin and those with prolonged half-lives. Conversely, altering the total fraction of absorbed drug will have an effect irrespective of the elimination rate. This is well illustrated by the interaction between phenobarbital and the antifungal agent griseofulvin. It was first shown that pretreatment with phenobarbital reduced circulating concentrations of griseofulvin (BUSFIELD et al., 1963). Because of the well-recognized effect of phenobarbital on drug metabolism, this was attributed by others to enzyme induction. However, RIEGELMAN et al. (1970) showed that the half-life of griseofulvin was unaltered after oral or i.v. administration and that the interaction was likely to have occurred on the basis of a reduction in absorption. The precise mechanism is not understood, but it is known that another barbiturate, heptabarbital, will reduce the bioavailability of dicoumarol

(AGGELER and O'REILLY, 1969). Induction of enzymes in the mucosa of the gut has not been excluded as a mechanism. These examples emphasize that drugs may alter drug disposition by more than one mechanism.

1. Alterations in Gastrointestinal pH

Drug-induced changes in the pH of gastrointestinal fluids can have complex effects on the absorption of other drugs taken concomitantly. These changes in pH can affect gut motility and tablet dissolution as well as result in unfavorable pH conditions for absorption of lipid soluble compounds. According to the classical pH-partition theory of drug absorption (BRODIE, 1964; BINNS, 1971) weak organic acids should be absorbed from the stomach, which has acid contents, whereas weak bases are absorbed best from the alkaline small intestine because drugs are primarily absorbed in the unionized, more lipid-soluble form. It appears, however, that even weakly acidic drugs, such as aspirin, warfarin, and barbiturates, or neutral compounds, such as ethanol, are absorbed much more rapidly from the small intestine than the stomach (SIURALA et al., 1969; KEKKI et al., 1971; MAGNUSSEN, 1968), presumably because of the much greater surface area of the intestine and its rich blood supply (LEVINE, 1970). Thus, the effects of pH on drug ionization and lipid solubility are often over-ridden *in vivo* by pH-dependent changes in tablet dissolution and gastric emptying time.

Increased drug absorption can occur because alkalinization may aid tablet dissolution of insoluble acidic drugs and stimulate gastric emptying. For example, aspirin is absorbed more rapidly and completely from buffered alkaline solutions than from unbuffered acidic solutions (COOKE and HUNT, 1970) because of the greater dissolution rate and aqueous solubility of aspirin in alkaline solution and the rapid gastric emptying caused by the increase in the pH of the stomach contents. Similarly, the stimulatory effect of alkali on gastric emptying probably explains the increased rate of propantheline absorption caused by sodium bicarbonate (CHAPUT DE SAINTONGE and HERXHEIMER, 1973). Magnesium hydroxide also increases the dissolution of sulfadiazine (HURWITZ, 1974).

Conversely antacids may decrease drug absorption by similar mechanisms. Sodium bicarbonate decreased the dissolution of tetracycline tablets thereby decreasing drug availability, an effect that was shown by comparison with the absorption of solutions of tetracycline (BARR et al., 1971). Magnesium hydroxide decreases the absorption of sulfadiazine sodium and pentobarbital sodium by increasing gastrointestinal pH and thereby increasing ionization of the drugs (HURWITZ, 1974), In addition to their pH-dependent effects, certain antacid drug products can alter drug absorption by changing motility and binding or chelating drugs in the gut lumen.

2. Gut Motility

The rate of gastric emptying is altered by many drugs and important drug absorption interactions are well documented. Probably the most common example is the slow absorption of many drugs when taken with food (PLACE and BENSON, 1971); this results mainly from the inhibitory effects of food on gastric emptying. Indeed, the effects of certain drugs can be markedly reduced or even abolished if gastric emptying is retarded by food (KOJIMA et al., 1971). Griseofulvin, however, is better absorbed with a fatty meal (CROUNSE, 1961). Delayed gastric emptying may be important with a drug like penicillin that is destroyed by the acidity of the stomach contents. In addition, for many drugs given orally, absorption is more rapid if given in the same dose in dilute rather than concentrated solutions, and this effect has been attributed in part to rapid gastric emptying (BOROWITZ et al., 1971). Aluminium hydroxide inhibits smooth muscle contractions and therefore gastrointestinal motility. These effects are related to the aluminium ion, which directly relaxes rat and human gastric and smooth muscle strips and partly blocks their acetylcholine-induced contractions (HURWITZ, 1974).

Extensive studies have been carried out on the drug interactions caused by drug-induced changes in the rate of gastric emptying. The rate of absorption in man of a model drug, such as acetaminophen, is directly related to the rate of gastric emptying (as measured by a radioactive isotope scintiscanning technique) although the total amount absorbed remains unchanged (HEADING et al., 1973). As expected, the mean half-life of gastric emptying, measured by the ^{113}indium technique, was prolonged several-fold by propantheline, and the rate of absorption of acetaminophen was reduced significantly. In contrast, metoclopramide significantly increased gastric emptying and also the rate of acetaminophen absorption in healthy volunteers who were known to be somewhat slow absorbers of the drug. However, these pretreatments did not alter the total urinary excretion of acetaminophen over 24 hrs nor were the total areas under the plasma concentration curves changed. Pretreatment with aluminium hydroxide affects the absorption of isoniazid (HURWITZ, 1974). The antacid lowered peak isoniazid serum levels in eight of ten tuberculous patients, and the one hour levels and, in this case, areas under the time-concentration curves were depressed. Gastric emptying of ^{51}chromium as determined by gamma camera was shown to be markedly reduced by aluminium hydroxide.

Most other studies either have been limited to experimental animals or have been inadequately carried out in man. For example, pretreatment with propantheline and metoclopramide have been reported to have completely opposite effects on the absorption of digoxin in comparison with acetaminophen (MANNINEN et al., 1973). The mean steady state serum digoxin concentration rose from 1.02 to 1.33 ng/ml after administration of propantheline for several days. These results could be explained by slow tablet dissolution and absorption of digoxin from a limited area of the intestine. In this case, the increased gastric emptying after metoclopramide would presumably reduce the effective time available for absorption while propantheline would have the reverse effect. This mechanism explains the increased total absorption of riboflavin caused by propantheline (LEVY et al., 1972). However, single blood samples for digoxin estimation were taken 24 hrs after the last dose of digoxin. For this reason, an alternative interpretation of the results is possible. The lower digoxin concentrations observed 24 hrs after metoclopramide administration could actually have been due to more rapid absorption in the beginning and lower serum levels later in the elimination phase. Conversely, the serum digoxin concentration could obviously be higher at 24 hrs when absorption was delayed by propantheline.

3. Sequestration or Metabolism in the Gut Lumen

Drugs also may interact in the gastrointestinal tract to form complexes, ion-pairs and chelates that may be absorbed more rapidly or more slowly than the parent drugs (LEVINE, 1970). The absorption of tetracycline is inhibited by the formation of insoluble chelates with aluminum hydroxide or with metals such as calcium (e.g., in milk) and iron (SCHEINER and ALTEMEIER, 1962; NEUVONEN et al., 1970). In contrast, the absorption of dicoumarol is increased by the formation of a more soluble chelate with magnesium hydroxide (AMBRE and FISCHER, 1973), and the absorption of a quaternary ammonium antiarrhythmic drug is enhanced by ion-pair formation with salicylate or trichloroacetate (GIBALDI and GRUNDHOFER, 1973). Magnesium hydroxide also decreases the absorption of drugs such as quinine, apparently by direct precipitation of the drug in the gut (HURWITZ, 1974).

The absorption of drugs such as warfarin, phenylbutazone, chlorothiazide, digitoxin, glutethimide, and thyroxine is reduced by ionic exchange resins such as cholestyramine and cholestipol (ROBINSON et al., 1970; K. J. HAHN et al., 1972; CALDWELL et al., 1971; CALDWELL and GREENBERGER, 1971; BAZZANO and BAZZANO, 1972). A similar mechanism may underlie the reduced availability of lincomycin when administered with kaolin-pectin mixtures. A potentialy useful application of this general phenomenon is the use of activated charcoal and ion exchange resins in the treatment of drug overdose. Activated charcoal has been shown to bind many drugs *in vitro* and several studies have now demonstrated this effect *in vivo*, even when the charcoal is administered some time after the drug. Therapeutically it is likely that such treatment not only prevents the absorption of unabsorbed drug but also reduces any enterohepatic cycling of drugs that are secreted in the bile, such as glutethimide and its metabolites.

Other mechanisms of altered drug absorption include an interference with micelle formation, thereby limiting the solubility of lipids. The inhibition by neomycin of the absorption of cholesterol, bile acids and vitamin A has partially been attributed to this mechanism (THOMPSON, 1970; THOMPSON et al., 1971; BARROWMAN et al., 1973). Drugs can also alter the volume and composition of gastrointestinal secretions (including bile), and changes in viscosity may modify drug absorption (LEVY and RAO, 1972).

Altered intestinal metabolism is a potential cause of drug interactions but its importance in man is not fully known (SCHELINE, 1973). Sterilization of the gut with antibiotics can potentiate the effects of coumarin anticoagulants, possibly by interfering with the synthesis of vitamin K by gut bacteria. It has been suggested that diphenylhydantoin interferes with folate absorption by inhibiting intestinal conjugase required to break down dietary folate into its absorbable, mono-folate form. However, interactions between folate and diphenylhydantoin are complex and apparently also involve mutual antagonism of both drug disposition and action.

4. Alteration in the Absorptive Process

Some drugs have a toxic effect on the intestinal mucosa and may cause a malabsorption syndrome with impaired absorption of other drugs. Perhaps the best example of this mechanism occurs after administration of high doses of neomycin. The absorption of several drugs as well as vitamin B_{12}, D-xylose and radioactive iron is affected in man (FALOON, 1970). The absorption of vitamin B_{12}, D-xylose, carotene, and possibly cholesterol also is dramatically decreased in subjects receiving colchicine (WEBB et al., 1968). Para-aminosalicylic acid (PAS) also decreases the absorption of vitamin B_{12} (TOSKES and DEREN, 1972) and drugs such as rifampin (BOMAN, 1974). SAMUELS and WAITHE (1961) noted that PAS produced an inconsistent but occasionally marked decrease in serum cholesterol, purportedly by causing a malabsorption of bile acids and cholesterol.

Drugs that are analogues of naturally occurring purines, pyrimidines, sugars and amino acids may be absorbed by active transport in the small intestine. For this reason, phenylalanine derived from the diet has been reported to reduce the absorption of *l*-dopa by competition between the two substrates for transport (BIANCHINE et al., 1971). It is also possible for one drug to inhibit enzymes involved in the active transport of another drug; such an interaction has been proposed to explain the inhibition of the absorption of *l*-dopa by chlorpromazine (RIVERA-CALIMLIM, 1972). In this context, it is interesting to note that insulin and many polypeptides greatly enhance the membrane transport of pethidine, isoniazid, salicylate and chlorpromazine drugs normally considered to cross cell membranes by passive diffusion. Not only is the intestinal uptake of isoniazid enhanced by insulin but this effect is antagonized by ouabain (DANYSZ and WISNIEWSKI, 1970). The significance of these findings for human therapy is undetermined.

5. Summary

The ability of antacids such as magnesium, calcium, and aluminium hydroxides to alter the absorption of tetracyclines and various other drugs is clearly established in humans (HURWITZ, 1974). Because the mechanisms for these interactions vary, it seems easiest simply to avoid administering most drugs simultaneously with a dose of the antacids. A similar situation exists with the decreased absorption of thyroxine, triiodothyronine, digoxin, warfarin, and probably other drugs caused by concomitant administration of cholestyramine. It is also apparent that PAS both delays and reduces the total absorption of rifampin in patients. If combined, these two drugs should be given with an interval of 8 to 12 hrs (BOMAN, 1974).

While the influence of the rate of gastric emptying on the rate of absorption of drugs has been well documented, the relevance of these findings for clinical medicine is less clear. Unfortunately, drugs that are incompletely absorbed have not been adequately studied, yet changes in total absorption of these drugs due to alterations in gastric emptying would be of much greater clinical importance than changes in absorption of drugs normally completely absorbed, particularly if they were drugs with narrow therapeutic indexes such as warfarin and digoxin.

B. Altered Elimination

The clearance of most drugs from the body is achieved either by excretion in the urine or bile or by biotransformation of the parent drug. The metabolites of drugs are usually more water soluble and, therefore, more readily excreted in the urine. All such elimination processes can be modified by the co-administration of other drugs. The clearance of volatile drugs such as anesthetics usually occurs by drug elimination in the expired air; these drugs will not be considered here.

1. Drug Metabolism

The effects of many drugs are terminated by their metabolism to inactive metabolites. A large variety of such biotransformations have been recognized and are catalyzed by an extremely versatile group of enzymes known collectively as drug metabolizing enzymes, the major component being cytochrome P-450. Because of its large size, rich blood supply and enzyme content, the liver is the major site for metabolism of both exogenously administered drugs and several endogenous substrates. Several other extrahepatic sites for drug metabolism have been identified, including the lung, intestine and kidney, but their quantitative contribution to drug elimination is generally small.

It is now clearly established that the activity of the drug-metabolizing enzymes can be either increased or decreased by the co-administration of a large number of chemical compounds, including several commonly used therapeutic agents.

Much of our knowledge of the molecular mechanisms of enzyme stimulation and inhibition has come from animal or *in vitro* experiments and has been extensively reviewed (CONNEY, 1967; KUNTZMAN, 1969; ANDERS, 1971). Investigation of such interactions in man is much less complete, but alterations in drug metabolism can, and do, occur clinically. Furthermore, the conditions required for the realization of potential interactions and their likely magnitude and clinical relevance are becoming more clearly defined.

In principle, the stimulation or inhibition of the metabolism of a drug is simple to predict when the effects are a function of circulating concentration of drug and no active drug metabolites are formed. Stimulation of drug metabolism will increase drug clearance and thereby reduce steady state drug concentration and drug action. Drug half-life and duration of action will also be decreased. Such interactions may be harmful if a drug is rendered ineffective, especially when the disease being treated is potentially fatal as in the case of a cardiac arrhythmia, severe hypertension or epilepsy. Dramatic toxicity may also occur in a patient in whom the required dose of a drug is determined in the presence of enzyme induction and the inducing agent is subsequently withdrawn.

In practice, the quantitative aspects of such interactions must be considered because drug metabolism is not always rate-limited by the activity of the drug-metabolizing enzymes. Thus, when the hepatic extraction ratio is initially very high, drug concentrations and half-life will not be greatly influenced by changes in enzyme activity after intravenous administration. On the other hand, drug concentrations after oral administration depend only on enzyme activity, so that enzyme induction and inhibition will alter drug concentrations even if drug extraction ratio is high and changes in plasma half-life are small. These seemingly surprising statements require some explanation. First, consider the effects of enzyme activity on half-life and clearance from the systemic circulation (Cl_s) after i.v. administration. ROWLAND et al. (1973) showed that the clearance of drug by an organ could be described in terms of organ blood flow (Q) and the intrinsic ability of the tissues to clear drug:

$$Cl_s = Q\left[\frac{Ce}{Q + Ce}\right] = QE. \tag{8}$$

Ce is a measure of the activity of the drug-metabolizing enzymes and was termed the intrinsic clearance by BRANCH et al. (1973b). It represents the drug clearance under first order conditions when flow is non-rate limiting. ROWLAND et al. (1973) expressed Ce as K.V.P. in which K is the first order rate constant for metabolism of free drug in the liver, V is the volume of the liver, and P is the partition ratio of drug between the tissues and effluent blood. In terms of enzyme kinetics, it can be expressed by a reduction of the Michaelis-Menten relationship to the form $V_{\max}/Km$ (WINKLER et al., 1973). The described equation quantifies the well recognized fact that the systemic elimination of highly extracted drugs is flow-dependent, whereas that of poorly extracted drugs is dependent largely on the activity of the drug-metabolizing enzymes, that is, upon intrinsic clearance. Thus, systemic drug clearance may be dependent on both flow and enzyme activity, depending on the particular drug. The effects of two-fold changes in intrinsic clearance on actual systemic clearance have been calculated for various initial extraction ratios (Table 2). As volume of distribution is assumed to be constant, the same changes will occur in elimination rate. The lower the initial extraction ratio, the more a given change in enzyme activity will alter systemic clearance and half-life. In this context, it is fortunate that most of the model drugs chosen to investigate enzyme induction and inhibition have been poorly ex-

Table 2. *The effect of doubling intrinsic clearance on the actual systemic clearance for drugs with different initial extraction ratios*

Intrinsic clearance L/min	Extraction ratio	Actual clearance[1] L/min	% Expression[0] of change
0.17	0.10	0.15	73
0.33	0.18	0.27	
0.38	0.20	0.30	54
0.75	0.32	0.50	
1.00	0.40	0.60	25
2.00	0.57	0.86	
2.25	0.60	0.90	10.2
4.50	0.75	1.13	
6.00	0.80	1.20	2.2
12.00	0.89	1.33	

[1] Assuming liver blood flow of 1.5 L/min.

[0] $\frac{\text{Calculated Change in actual clearance} \times 100}{\text{Change in intrinsic clearance}}$.

tracted, so that changes in enzyme activity gave almost proportional changes in half-life.

These changes in systemic clearance and half-life occur irrespective of the route of elimination, but similar changes in drug concentration and half-life occur only after intravenous administration. After oral administration, changes in circulating drug concentration are always proportional to changes in intrinsic clearance, irrespective of the initial extraction ratio. This occurs because drug can be eliminated by the liver during its transfer from the gut to the systemic circulation in the hepatic portal venous blood, an effect known as presystemic (or "first-pass") hepatic elimination. Presystemic elimination can be quantified by considering the relationship between drug concentration after oral administration, (AUC) oral, and systemic drug clearance, Cl_s.

$$\text{AUC oral} = \frac{\text{FD}}{\mathrm{Cl}_s}\,. \tag{9}$$

If such a drug is completely absorbed and only eliminated by the liver, then F is the fraction of the drug escaping presystemic elimination, that is,

$$F = 1 - E \tag{10}$$

in which E is the hepatic extraction ratio Cl_s/Q (GIBALDI et al., 1971; ROWLAND, 1972). On re-examination of Eq. (8),

$$E = \frac{Ce}{Q + Ce} \tag{11}$$

and rearranging,

$$Ce = \frac{QE}{(1 - E)} \tag{12}$$

that is,

$$Ce = \frac{\mathrm{Cl}_s}{F} \tag{13}$$

and from Eq. (9)

$$\text{(AUC) oral} = \frac{D}{Ce}\,. \tag{14}$$

It should be noted that the apparent clearance of an oral dose, D/AUC, is in fact equal to its intrinsic clearance, *Ce*, assuming complete absorption, but is an overestimate of actual systemic clearance, Cl_s. Furthermore, unlike systemic clearance, neither apparent clearance nor intrinsic clearance is influenced by organ blood flow. Thus, average drug concentrations after oral administration depend only upon enzyme activity and will change with altered drug metabolism irrespective of the magnitude of the change in systemic clearance or half-life. As ROWLAND (1972) points out, quite dramatic changes in circulating concentrations of highly extracted drugs should still occur although half-life changes may be very small.

These theoretical considerations may provide an explanation of why phenobarbital, for instance, lowers steady state concentrations of both low and high extraction drugs. Unfortunately, there are few data on the half-lives of highly extracted drugs under these conditions. Despite this, the theoretical arguments are compelling, and suggest, at the very least, that large changes in the steady state concentrations of high clearance drugs (or AUC after a single dose) can occur together with minimal alteration in half-life.

In reviewing the literature, acceptance of altered drug metabolism as a mechanism, therefore, requires that drug concentration, clearance and half-life are all altered appropriately. When changes in these parameters are observed, changes in *Ce* (drug metabolism) must be differentiated from displacement of drug from its binding sites in blood or tissues. This can be achieved by calculations of volume of distribution, direct measurement of free drug concentration or accurate therapeutic data on drug action. Measurement of drug action is most important because, for example, both enzyme induction and plasma binding displacement will shorten drug half-life but may have opposite effects on drug action or toxicity.

a) Stimulation of Drug Metabolism

The reduced action of oral anticoagulants during the co-administration of barbiturates was the first such interaction to be described clinically (AVELLANEDA, 1955). Many lines of evidence have confirmed that the decreased effectiveness of the oral anticoagulants warfarin and dicoumarol following phenobarbital administration is due to stimulation of drug metabolism. Both the therapeutic effect and the plasma concentrations of warfarin and dicoumarol are reduced (CUCINELL et al., 1965; CORN and ROCKETT, 1965; GOSS and DICKHAUS, 1965; MACDONALD et al., 1969) and their half-lives shortened (CORN, 1966; MACDONALD et al., 1969). Although not all criteria have been met in full for all combinations, this interaction appears to extend to most oral anticoagulants and barbiturates (DAYTON et al., 1961; BRECKENRIDGE and ORME, 1971; KOCH-WESER and SELLERS, 1971). Several studies have demonstrated that these interactions occur commonly in clinical practice (MACDONALD and ROBINSON, 1968; ROBINSON and MACDONALD, 1966).

The elimination of antipyrine also is enhanced by barbiturates with a resultant shortening of plasma half-life (VESELL and PAGE, 1969). Although altered pharmacologic action of antipyrine has not been demonstrated, this compound is not bound to plasma proteins and the concentration measured is, therefore, entirely in the free form. The steady state concentrations of diphenylhydantoin also have been shown to be reduced by phenobarbital administration (CUCINELL et al., 1965; KUTT et al., 1969) and the half-life of the drug is shortened (KRISTENSEN et al., 1969). With this interaction, altered therapeutic effect in the treatment of epilepsy cannot be demonstrated because phenobarbital is also an effective antiepileptic drug. Carbamazepine also has been shown to decrease diphenylhydantoin half-life and steady state plasma concentrations (HANSEN et al., 1971). Anticonvulsant therapy, in general, has been associated with hypocalcemia and folic acid deficiency. Enzyme induction has been implicated in the production of both these serious side effects. The interaction between folate and diphenylhydantoin is complicated and the literature somewhat conflicting as reviewed by REYNOLDS (1972). Early suggestions that folate absorption was altered by either inhibition of intestinal conjugase or alkalinization of the gut contents by diphenylhydantoin have not been confirmed. On the other hand, the reduction in serum folate by several anticonvulsants seemed to be related to the enhanced excretion of endogenously formed *d*-glucaric acid (MAXWELL et al., 1972; LATHAM et al., 1973), which has been suggested as a marker of hepatic drug metabolizing activity (HUNTER et al., 1971). Folate/diphenylhydantoin interactions are further complicated by the fact that some evidence exists that folate supplementation may enhance diphenylhydantoin metabolism, thereby lowering serum drug concentrations to the extent that seizures occurred (BAYLIS et al., 1971). It should be mentioned that a pharmacological action at the neuronal level has not been excluded. It has been proposed that the hypocalcemia results from an induction of the metabolism of vitamin D because this effect correlated with *d*-glucaric acid excretion in epileptics receiving several anticonvulsants (LATHAM et al., 1973). Furthermore, the half-life of vitamin D_3 in humans is shortened by phenobarbital (T. J. HAHN et al., 1972), and plasma levels of one of its active metabolites, 25-hydroxy D, are reduced by chronic anticonvulsant therapy (T. J. HAHN, 1973).

Phenobarbital also has been shown to reduce the steady state plasma concentrations of digitoxin and to shorten its half-life in a small number of subjects (SOLOMON and ABRAMS, 1972). Some evidence has accumulated to show that phenobarbital can stimulate one of the pathways of the metabolism of cortisol, as the urinary excretion of one of its metabolites 6-β-OH cortisol is increased (BURSTEIN and KLAIBER, 1965). But the importance of this effect on circulating concentrations of endogenous cortisol is small because it is compensated for in the presence of an intact pituitary – adrenal axis (CHOI et al., 1971; BOGDANSKI et al., 1971). There is some evidence, however, that pharmacological doses of prednisone are less effective in asthmatics receiving phenobarbital and, in addition, the half-life of another steroid, dexamethasone was decreased (BROOKS et al., 1972). Phenobarbital may also produce a small, but inconsistent effect on the elimination of thyroxine (CAVALIERI et al., 1973).

Barbiturates can also affect the disposition of some highly extracted drugs, such as the tricyclic antidepressants and chlorpromazine, when the latter are administered orally. Thus steady state concentrations of nortriptyline are lower in a group of patients receiving barbiturates compared to those not receiving these compounds (ALEXANDERSON et al., 1969). The plasma concentrations of chlorpromazine were likewise reduced by phenobarbital treatment in a single patient (CURRY et al., 1970) and in a small series the rate of urinary excretion of chlorpromazine was reduced and its conjugated metabolite enhanced, indicating an increased metabolism (FORREST et al., 1970). Although these data are suggestive of stimulation of metabolism, drug half-lives were not measured in these studies and therapeutic effects could not be judged with any degree of accuracy. It is interesting that the tricyclic antidepressants have themselves been implicated as enzyme inducers (O'MALLEY et al., 1973) although the reported shortening of antipyrine half-life was small (18%) and of doubtful biologic significance.

Much interest has been generated in the ability of enzyme inducers, particularly phenobarbital, to lower serum bilirubin concentrations in certain types of jaundice.

The potential importance of phenobarbital therapy for preventing neonatal kernicterus has been debated (for an extensive review, see WILSON, 1972). Early reports that phenobarbital can lower bilirubin concentrations in single cases of congenital nonhemolytic hyperbilirubinemia (YAFFE et al., 1966; CRIGLER and GOLD, 1966) have been confirmed. The half-life of radiolabelled bilirubin was shown to be decreased (CRIGLER and GOLD, 1969) and the effect was attributed to induction of hepatic glucuronyl transferase activity because the patients' abilities to conjugate other compounds with glucuronide were enhanced. Subsequently ARIAS et al. (1969) divided such patients into two groups, the first being more severely jaundiced and resistant to phenobarbital, while the second had milder jaundice and responded to phenobarbital treatment. It may be that only those patients with some initial enzyme activity are capable of a favorable response. Patients with Gilbert's syndrome responded to phenobarbital (BLACK and SHERLOCK, 1970) and in three patients, hepatic uridine diphosphate glucuronyl transferase activity was increased.

Several studies have confirmed the initial observation of TROLLE (1968a) that children born of mothers treated with phenobarbital showed a lower incidence of neonatal hyperbilirubinemia. It also became apparent that treatment of the infants after birth was less successful and that at least three days of treatment with phenobarbital were required for an obvious effect to occur (TROLLE, 1968b; RAMBOER et al., 1969). Furthermore, treatment of infants with severe jaundice was relatively ineffective (McMULLIN, 1968; CUNNINGHAM et al., 1969). While there is little doubt that the treatment of the mother in the last trimester is the most effective therapy, treatment of the infants themselves can often produce a favorable response and a lesser need for exchange transfusion in ABO (YEUNG and FIELD, 1969) and Rhesus incompatibilities (McMULLIN et al., 1970) if begun early enough.

Because of the availability of other forms of treatment and a natural reluctance to treat mothers and their offspring with drugs, enzyme induction has not found a place in the routine treatment of neonatal jaundice (WILSON, 1972).

In addition to the barbiturates, several other compounds can stimulate drug metabolism in general. The ability of glutethimide to reduce the effectiveness and shorten the half-life of warfarin is well established (CORN, 1966; MACDONALD et al., 1969). Diphenylhydantoin can stimulate the metabolism of the steroids, cortisol and dexamethasone. Just as with phenobarbital, the enhanced secretion of 6-β-OH cortisol during diphenylhydantoin therapy may be of limited significance as the endogenous production of cortisol is enhanced in normal subjects, thereby compensating for its increased metabolic clearance such that circulating cortisol concentrations are little affected (WERK et al., 1964; 1971, CHOI et al., 1971). These data on cortisol metabolism serve to illustrate the fact that enzyme induction is unlikely to produce dramatic effects on endogenous hormone production in patients with normal homeostasis (LIPSETT, 1971)

Plasma concentrations of metyrapone are also markedly diminished in the presence of diphenylhydantoin, so that the metyrapone test of pituitary-adrenal function is rendered falsely negative (MEIKLE et al., 1969). In the same study the kinetics of intravenously administered metyrapone were not significantly altered. Subsequent work showed that intestinal absorption in rats was unaltered and it was suggested that the reduced metyrapone levels after oral administration were due to its presystemic metabolism (JUBIZ et al., 1970). In view of the very high plasma clearance of metyrapone calculated from the data of MEIKLE et al. (1969), a presystemic hepatic effect is quantitatively consistent, and provides an explanation for the altered metyrapone kinetics after oral, but not intravenous, administration.

Phenylbutazone pretreatment causes a shortening of the half-life of antipyrine (CHEN et al., 1962) and in one patient reduced the steady state concentration of digitoxin (SOLOMON et al., 1971). The latter workers showed a similar effect in a patient receiving diphenylhydantoin. However, the possible stimulation of digitoxin metabolism has been investigated in only a few patients, and then only incompletely. Effects on absorption and enterohepatic recycling have not been excluded.

Certain environmental agents have been shown to stimulate drug elimination. Workers exposed to lindane and DDT have shorter antipyrine half-lives than normal (KOLMODIN et al., 1969; KOLMODIN-HEDMAN, 1973). The concentrations of insecticides in the blood and fat may in turn be reduced by other inducing agents, such as phenobarbital and diphenylhydantoin (DAVIES et al., 1969; McQUEEN et al., 1972). Although placental benzpyrene hydroxylase is stimulated in women with a history of smoking (WELCH et al., 1969b), there is no evidence that smoking habits or coffee drinking significantly affect the half-lives of dicumarol, antipyrine and phenylbutazone (VESELL and PAGE, 1969). Ethanol ingestion, on the other hand, may well affect drug metabolism either directly or because of the long-term liver damage that may be associated with alcoholism. Drug interactions with ethanol illustrate a well recognized finding in animal experiments that an enzyme inducer may cause inhibition of metabolism when it is present in the body, and that stimulation of drug metabolism may only become apparent when the modifying drug has been eliminated. VESELL et al. (1971) showed that antipyrine half-life was shortened following ethanol ingestion for several days in normal volunteers. Similarly, KATER et al. (1969) reported a large series of alcoholics who had been taking ethanol in large quantities for at least 3 months before admission, but who had been withdrawn from ethanol ingestion for at least 7 days before hospital admission. In these patients, the clearance of tolbutamide, diphenylhydantoin and warfarin was enhanced and drug half-life shortened in comparison with control subjects. It should be mentioned that none of the alcoholics had severe liver disease. The data on tolbutamide were confirmed by CARULLI et al. (1971) who showed, in addition, that acute ethanol administration was associated with a prolonged half-life of tolbutamide, rather than a shortening. Similarly, plasma warfarin concentration increased following alcohol ingestion in a single patient studied by BRECKENRIDGE and ORME (1971). These effects of ethanol must be distinguished from those resulting from severe alcoholic liver damage in which drug metabolism may be impaired (LEVI et al., 1968; BRANCH et al., 1973a).

b) Inhibition of Metabolism

The phenomenon of inhibition of drug metabolism is also well established in man. As might be predicted from the likely competitive nature of the inhibition, a compound whose metabolism can be inhibited may itself also act as an inhibitor. For example, dicoumarol can inhibit the metabolism of tolbutamide as judged by an increase in steady state levels, drug half-life and therapeutic effect (KRISTENSEN and HANSEN, 1968; SOLOMON and SCHROGIE, 1967). Dicoumarol also elevates serum chlorpropamide concentrations (KRISTENSEN and HANSEN, 1968) and increases the steady state concentrations and half-life of diphenylhydantoin (HANSEN et al., 1966). Phenyramidol (an analgesic now withdrawn from clinical use) can, in turn, prolong the half-life of dicoumarol and increase its anticoagulant effect (SOLOMON and SCHROGIE, 1966).

Tricyclic antidepressents can be involved similarly. Thus perphenazine administration will inhibit the metabolism of nortriptyline, increase the steady state plasma concentration and decrease the urinary excretion of its metabolites (GRAM and OVERØ, 1972). Nortriptyline itself can prolong the half-lives of both antipyrine and dicoumarol (VESELL et al., 1970).

Several other inhibitors of drug metabolism have been recognized. Chloramphenicol was shown to prolong the half-lives of tolbutamide, diphenylhydantoin and dicoumarol by CHRISTENSEN and SKOVSTED (1969). Steady state levels of both tolbutamide and diphenylhydantoin were increased as a result of this interaction. Sulphaphenazole has been shown to increase steady state plasma concentrations, therapeutic action and half-life of tolbutamide (CHRISTENSEN et al., 1963). As will be discussed later, the *in vitro* displacement of tolbutamide from its plasma binding sites by sulphaphenazole does not contribute to the interaction *in vivo*. KUTT et al. (1966) showed that patients receiving diphenylhydantoin and antituberculous drugs, had elevated plasma levels of diphenylhydantoin with associated toxicity and a reduction in the hydroxylated metabolite of diphenylhydantoin. Further investigation incriminated isoniazid as the offending agent (KUTT et al., 1970) and showed that the interaction was more likely to occur in patients who were slow isoniazid acetylators (BRENNAN et al., 1970), presumably because concentrations of the inhibitory drug would be higher. Sulthiamine can also increase steady state concentrations and prolong the half-life of diphenylhydantoin (HANSEN et al., 1968). Disulfiram can inhibit the metabolism of antipyrine and of ethanol (VESELL et al., 1971). Some reports have suggested that the metabolism of diphenylhydantoin also is inhibited, its plasma levels increased and toxicity evidenced during disulfiram therapy (KIØRBOE, 1966; OLESEN, 1967). A case report suggests a similar interaction with warfarin (ROTHSTEIN, 1968) and a preliminary study has shown that disulfiram increases warfarin concentration and prothrombin time (O'REILLY, 1971). The xanthine oxidase inhibitor, allopurinol, was developed with the idea of prolonging the effects of 6-mercaptopurine and indeed it does. It is, therefore, not surprising that it might influence the effects of azathioprine, which is converted in the body to 6-mercaptopurine as the active moiety (NIES and OATES, 1971). Allopurinol has also been shown to prolong the half-life of dicoumarol and antipyrine (VESELL et al., 1970).

Mention should be made of the possible inhibition of drug metabolism by methylphenidate. While an early report suggested that methylphenidate could inhibit diphenylhydantoin metabolism in a child (GARRETTSON et al., 1969), further studies involving greater numbers of patients failed to confirm this (KUPFERBERG et al., 1972). Similarly, the suggestion that ethyl biscoumacetate metabolism could be inhibited by methylphenidate was not confirmed by HAGUE et al. (1971).

There has been some suggestion that the metabolism of phenobarbital may be inhibited by diphenylhydantoin because its plasma concentration was increased (MORSELLI et al., 1971). As with the stimulation of diphenylhydantoin elimination by phenobarbital, the therapeutic significance and reproducibility of this interaction are open to question.

The interaction between chloral hydrate and ethyl alcohol is complex and its study represents one of the few in which concentrations of both the interacting drugs and their metabolites have been measured (SELLERS et al., 1972a). On the one hand, ethyl alcohol ingestion increased the concentrations of trichlorethanol, the hypnotic metabolite of chloral hydrate, presumably by stimulating NADH production and thereby accelerating the reduction of chloral hydrate. On the other hand, chloral hydrate ingestion increased the circulating concentrations of ethanol and reduced those of its metabolite, acetaldehyde. This latter effect appeared to be due to trichlorethanol rather than to the parent compound. Motor performance was worsened with the mixture compared to either drug alone (SELLERS et al., 1972b). In addition, the contribution of effects of the combination of two central depressants is difficult to assess, emphasizing the complexity of clinical situations that may arise. Indeed, one wonders how much more complex some apparently simple interactions might be if the kinetics of both drugs had been measured simultaneously.

c) Altered Formation of Active Drug Metabolites

When phenobarbital or other chemicals induce the production of an active metabolite from a second drug, a paradoxical situation may result in which accelerated drug metabolism causes an enhanced pharmacologic or toxic effect. Although definitive information on such interactions in man is generally lacking, the investigative approach to them requires special consideration. Whether more active metabolite is formed as a result of enzyme induction depends on whether the *proportion* of drug converted to active metabolite (in comparison to renal elimination and to other metabolic pathways) is increased, and whether or not disposition of the active metabolite is accelerated as well. For example, the pharmacologic effect of cyclophosphamide is mediated by its metabolism in hepatic microsomes to an active alkylating agent, and this pathway is induced by phenobarbital. Although it has not been determined definitively whether phenobarbital enhances the toxicity of cyclophosphamide in man, the report of excessive cyclophosphamide toxicity in a patient on phenobarbital and diphenylhydantoin (DRUMMOND et al., 1968) poses an urgency to the solution of this problem. Phenobarbital does indeed increase the *rate* at which total metabolites are formed from cyclophosphamide in man (JAO et al., 1972), but because only a negligible fraction of cyclophosphamide is excreted unchanged, this does not result in an increase in the *total* amount of metabolites formed from a given dose. The key question then is whether the proportion of active to inactive metabolites is increased. Although there is *in vitro* data suggesting formation of an alkylating agent via the 4-hydroxycyclophosphamide pathway (CONNORS et al., 1974), it is not known whether it is possible to obtain evidence in man for an enhanced formation of the active metabolite relative to alternate pathways.

Because marked inter-species variations frequently occur in the pathways of drug metabolism, the clinical relevance of accelerated formation of active metabolites particularly requires that definitive data on such interactions be obtained in man. These data should include measurement of an enhanced pharmacologic or toxic effect by a decrease in the plasma level of the parent drug or evidence for conversion of a greater proportion of parent drug to the active metabolite. The effects of both inducers and inhibitors of drug metabolism on the formation of active drug metabolites are considered at length in two subsequent chapters in this volume (GILLETTE and MITCHELL, 1974; MITCHELL et al., 1974).

2. Renal Excretion

The two basic mechanisms for renal excretion of drugs are glomerular filtration and tubular secretion. Following both these processes, however, back diffusion through the renal tubules can occur, thereby reducing their actual clearance. Drug interactions involving renal elimination can occur either by affecting tubular secretion or by altering the fraction of drug reabsorbed from the tubular fluid. Because glomerular filtrate contains no protein, alterations in plasma binding can also affect renal elimination, but this will be considered separately below. Kinetic interactions might also occur by altered glomerular filtration rate or renal blood flow, but at the moment remain only a theoretical possibility. (See Section III B3, Hemodynamic Interactions.)

a) Altered Tubular Reabsorption

The passive reabsorption of weak acids and bases can be altered by changes in urinary pH, provided that the pK_a is such that the fraction of ionized drug can be changed over the attainable range of urinary pH. This would include weak acids with a pK_a in the range of 3.0 to 7.5 and bases with a pK_a of 7.5 to 10.5. The effect of urinary pH on renal drug elimination can occur whether the drug is filtered or secreted and has been called "ion trapping" because the ionized form of the drug is not lipid soluble and cannot diffuse back through the renal tubules. Weak acids are more ionized at alkaline pH and urinary clearance can therefore be increased by alkalinization and vice versa. The opposite is true of weak bases whose elimination may be increased by urinary acidification. Quantitatively, the relationship between ionized and non-ionized species is described by the Henderson-Hasselbach equation. The effects that pH would have on the urine/plasma concentration ratio, U/P, is described by

$$\frac{U}{P} = \frac{1 + 10^{pH_u - pK}}{1 + 10^{pH_p - pK}} \quad \text{for acids} \tag{15}$$

and

$$\frac{U}{P} = \frac{1 + 10^{pK - pH_u}}{1 + 10^{pK - pH_p}} \quad \text{for bases} \tag{16}$$

in which pH_p and pH_u are the pH of plasma and urine, respectively. It can be seen that significant alterations in U/P can occur even when the pK is some 3 units different from pH_u. For example, with an acid of pK of 3.4, U/P will be one when urine pH is 7.4 (equal to plasma). If urine pH falls to 5.4, then U/P will fall one hundred fold. A similar fall in drug clearance would occur if urine volume were constant. It should be mentioned that pH changes in the plasma can also effect tissue uptake because intracellular contents are generally acid (pH 6.9). However, only minor changes in plasma pH are tolerated, so that the magnitude of this effect is smaller than that produced by changes in urine.

The effects of urinary pH changes on the renal elimination of several commonly used therapeutic agents is well recognized. In humans, weak acids such as aspirin, salicylate and phenobarbital are excreted more rapidly in alkaline urine, whereas weak bases such as quinidine and amphetamine are excreted more quickly in acid urine (Hollister and Levy, 1965; Gerhardt et al., 1969; Davis et al., 1971). An inadequate response to standard doses of these drugs or problems of toxicity from overdosage will result whenever the diets of these patients include sufficient amounts of sodium bicarbonate, acid ash or cranberry juice. A new threat comes from a current health fad – self administration of large quantities of vitamin C, a potent urinary acidifying agent. It should be emphasized that such interactions will only be significant when the bulk of drug elimination occurs by the renal route. Thus, Borga et al. (1969) point out that although urinary pH changes can alter the urinary excretion of desmethylimipramine, this is of no clinical significance because only a minute fraction of the elimination of the drug occurs by excretion into the urine.

b) Tubular Secretion

Many drugs are eliminated by active secretion into the urine. Separate transport systems are present for acidic and basic drugs. In general, administration of an acid drug secreted by the renal tubules will slow the tubular secretion of other concurrently given organic acids and administration of secreted basic drugs will slow the tubular secretion of other bases. These alterations assume clinical significance when the renal excretion of the drug is a major pathway for elimination. For example, salicylate and other anions can slow the rate of methotrexate excretion (LIEGLER et al., 1969). Likewise, phenylbutazone potentiates the hypoglycemic action of acetohexamide by interfering with the tubular secretion of an active hypoglycemic metabolite of acetohexamide (FIELD et al., 1967). Therapeutic advantage is taken of this type of interaction when probenecid is given with penicillin to prolong effective plasma concentrations of the latter.

3. Hemodynamic Drug Interactions

In view of the importance of the circulation in transporting drugs through the body, it is not surprising that recent studies have shown that hemodynamic alterations produced by one drug can alter the elimination of another. Such interactions have been described as hemodynamic drug interactions, and, although they have as yet been described only in animal studies, there is every reason to believe that they can occur in man. Taking the previously described equation for organ drug clearance:

$$\mathrm{Cl} = Q\left[\frac{Ce}{Q + Ce}\right] = QE \quad (8)$$

it can be seen that alterations in blood flow alone can influence systemic drug clearance, and again that the effects are dependent on the initial extraction ratio. The relationship between flow and actual clearance is curvilinear. When flow is low compared to intrinsic clearance, Ce, extraction ratio approaches unity and clearance is proportional to flow. As flow increases relative to Ce, extraction ratio falls and clearance changes become smaller until at very high flows clearance becomes flow-independent and equal to Ce. The effects of altering flow on the clearance of a series of drugs with varying initial extraction ratios is shown in Fig. 2. When extraction ratio is less than 0.2 to 0.3, alterations in flow produce

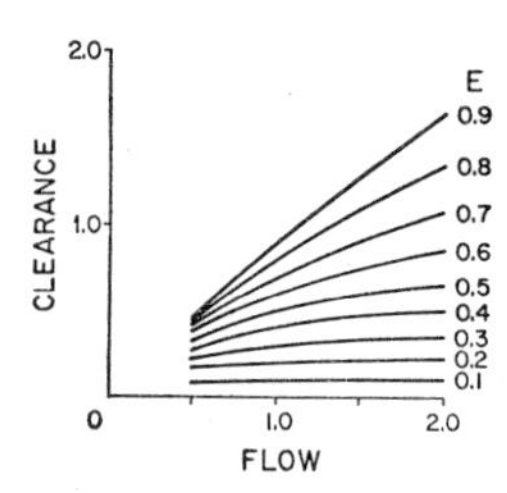

Fig. 2. Effects of altered blood flow on the systemic clearance of drugs with various extraction ratios at normal flow (i.e., when flow = 1.0)

little change in systemic clearance and elimination is essentially independent of flow. However, with highly extracted drugs, clearance becomes essentially proportional to flow, and blood flow alterations can alter drug clearance significantly. This analysis quantifies the fact that the hepatic elimination of highly extracted drugs, like lidocaine and propranolol, is flow-dependent, whereas the hepatic elimination of drugs with lower extraction ratios, like oxyphenbutazone, is less influenced by flow changes.

There are now several examples in animal studies showing that changes in hepatic blood flow produced by one drug can alter the elimination of another. The clearance of lidocaine in monkeys was reduced by the administration of propranolol, which reduces liver blood flow as a result of beta-adrenergic blockade (NIES et al., 1973). That this was due to flow changes alone (as opposed to enzyme inhibition) is supported by the fact that the extraction ratio was unaltered and that *d*-propranolol, which is devoid of beta-blocking activity, was without effect although it too is metabolized by the liver. Propranolol is an interesting compound in that the activity of the *l*-isomer reduces the clearance of the *d*-isomer when given as the racemate, the two isomers being eliminated at the same rate under conditions of constant flow (NIES et al., 1973; BRANCH et al., 1973a). This may explain the fact that the half-lives of both *l*- and *dl*-propranolol are longer than that of the *d*-isomer in man (GEORGE et al., 1972). Norepinephrine, which also reduces hepatic blood flow, has also been shown to reduce lidocaine clearance (BENOWITZ et al., 1973). In contrast, the effects of propranolol on oxyphenbutazone clearance were minor as predicted from its low initial extraction ratio. Increased drug clearance as a result of increasing hepatic blood flow by glucagon (BRANCH et al., 1973b) and isoproterenol (BENOWITZ et al., 1973) has been demonstrated for propranolol and lidocaine.

The magnitude of the changes in drug concentration resulting from hemodynamic interactions therefore depends on the change in flow and the extraction ratio of the drug in question. Poorly extracted drugs, such as antipyrine, diphenylhydantoin and oxyphenbutazone, will be little affected. The effects on high clearance drugs, like lidocaine, propranolol, chlorpromazine and tricyclic antidepressants, depend on the likely range of flow changes produced. At this time, changes in excess of 2 to 3 fold seem unlikely, and such changes could be significant only for drugs with a narrow therapeutic range of concentration. It is also possible that therapeutic manipulation of hepatic blood flow would be beneficial in the treatment of drug overdose with high clearance drugs like tricyclic antidepressants and propoxyphene.

As mentioned before, the route of administration will influence the effect that flow alterations have on steady state concentrations. Theoretically at least, flow changes should only influence concentration after intravenous administration because average concentration after oral administration is influenced only by enzyme activity. In this case, decreases in drug clearance and half-life are compensated by the increase in extraction ratio which reduces the amount of drug reaching the circulation by the presystemic effect, and vice versa. Again, although average concentration (or AUC) will be unaltered, differences in peak and trough concentrations will occur as a result of altered systemic clearance.

The effects of blood flow changes on the clearance of drugs by other eliminating organs, such as the kidney, has received little attention although it is clear that alterations in glomerular filtration rate can result in profound changes in the clearance of drugs that are eliminated by filtration alone. The fact that glomerular filtration rate can be maintained in the face of changes in renal blood flow suggests that hemodynamic drug interactions would be less significant for this group of compounds. On the other hand, the elimination of drugs that are actively secreted by the kidney could potentially be influenced by flow changes. The principles outlined above for the effects of liver flow may well also apply to the effects of renal flow on the renal secretion of drugs. Obviously, the clearance of PAH by the kidney is flow-dependent since the compound is used to measure renal blood flow (its extraction ratio being close to unity). Whether the previously outlined effects of flow on clearance of drugs such as penicillin apply to the kidney awaits further experimental data.

C. Redistribution

The binding of drugs to both blood and tissue elements can greatly influence free drug concentrations at their sites of action. Tissue uptake involving active processes can be important when the tissue represents the site of drug action. Various binding and uptake processes can be involved in drug interactions of likely therapeutic significance, but of all pharmacokinetic interactions these have been the least rigorously established. This stems from the fact that few published studies in man have measured free drug concentration in the circulation following *in vivo* drug administration. Changes in drug binding usually have been shown *in vitro* and measurements *in vivo* consisted only of total drug concentration in the blood or plasma. Changes in free drug have been inferred from alterations in drug action. While such estimates do have clinical relevance, they are not precise enough for an accurate quantification of drug redistribution within the body. Particular care must be taken in the case of interactions that lower total drug concentration because these may be due to enzyme induction, which reduces drug action, or to displacement of plasma bound drug, which can have the opposite effect of increasing drug action.

1. Drug Binding in the Blood

It is now evident that two groups of drugs can be distinguished on the basis of whether or not their elimination is dependent on the free drug concentration in the circulation (see KEEN, 1971; GILLETTE, 1971). This distinction is important because drug interactions involving alterations in plasma drug binding will have very different outcomes in the two groups.

a) Restrictive Elimination

By definition restrictive elimination occurs when the extraction ratio of the organ of elimination is less than the fraction of free drug in the blood. Such elimination applies to drugs that are excreted by glomerular filtration alone and for drugs such as diphenylhydantoin and warfarin whose hepatic metabolism is a function of free drug concentration in the circulation. Displacement of such a drug from its binding sites by another can increase free drug concentration and therefore enhance drug action (or toxicity). However, this effect is only transient because the increased free drug concentration also increases elimination rate. Thus once a new steady state is achieved, free drug concentration returns to its initial value, at least in theory (Fig. 3).

Some confusion can arise if a clear distinction is not made between free and total drug concentration, and it has recently been emphasized that only insignificant changes in free drug concentration may result from binding displacement even though total drug concentration may change dramatically (GILLETTE, 1973; WARDELL, 1974). This results because free drug concentration is inversely proportional to the volume of distribution of free drug present in plasma water, which is determined by tissue binding as well as that in blood. These effects may be quantified using the approach of GILLETTE (1971) in which the volume of distribution of free drug (Vf) is calculated in terms of the volumes of distribution of body water (V), albumin (V_a), and tissues (Vt) and the bound/free drug con-

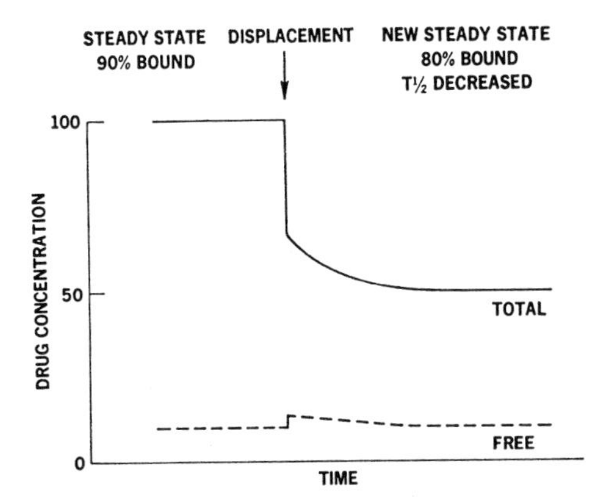

Fig. 3. The effects of altered plasma drug binding on total and free plasma concentrations of a drug with restricted elimination. The drug is assumed to be distributed in total body water and bound to albumin, but not the tissues. Immediately after displacement, a two-fold decrease in the % free drug causes only a small, 37% increase in actual free drug concentration, and total drug concentration falls to 67% of its initial value. As elimination rate is proportional to actual free drug concentration (restrictive elimination), a new steady state is established such that the actual free drug concentration returns to control values, though total drug concentration is reduced by half

centration ratios for albumin (B_a/f) and tissues (B_t/f) as:

$$Vf = V + V_a \cdot \frac{B_a}{f} + Vt \cdot \frac{Bt}{f} \tag{17}$$

in which $V = V_a + V_t$. Thus, assuming total body water of 40.1. and a volume of distribution of albumin of 5.5.1. in normal man,

$$Vf = 40 + 5.5 \frac{Ba}{f} + 34.5 \frac{Bt}{f}.$$

The volume of distribution of albumin (amount in body divided by concentration in plasma) was used instead of actual plasma volume because albumin is present outside the circulation and is assumed to be involved in displacement. In the absence of tissue binding,

$$Vf = 40 + 5.5 \frac{Ba}{f}. \tag{18}$$

It can be seen that Vf decreases as albumin binding is reduced and free drug concentration tends to rise, but never to the same extent as the % free drug. Using Eq. (18), it can be shown that the change in Vf and therefore drug concentration for any given change in f is greater the greater the initial degree of binding and that binding displacement will only be significant for compounds bound at least 90% initially. Even then its complete displacement (which increases free fraction 10-fold) will only increase free drug concentration two-fold. Such changes will be even less should there be any degree of tissue binding. For example, if a drug were 50% tissue bound ($Bt/f = 1$) then complete displacement of a drug that was 90% bound would only change free drug concentration by one-third. Thus we see that binding displacement interactions will only be significant for drugs with small volumes of distribution. On the other hand, the concentrations of total drug may change quite dramatically even though free drug concentration does not because the volume of distribution of total drug, Vd, can be expressed as:

$$V_d = Vf \times f \qquad (19)$$

in which f is the fraction of free drug $(f/B + f)$. In contrast to Vf, Vd increases as binding decreases, and there is a positive linear relationship between Vd and f with an intercept corresponding to the volume of albumin such that when all drug is free, Vd = total body water. Using the previous example of a 90% bound drug that is completely displaced, the 2-fold increase in free drug concentration reduces total drug concentration to 1/5. If, in addition, the drug is tissue bound, the slope of the relationship between Vd and f increases, and the change in Vd becomes more nearly one of proportionality. When tissue binding is 50% then total displacement, which raised free drug concentration only by one-third, would reduce total drug concentration even further to one-sixth of its initial value. Thus, the change in total drug concentration is much greater and more sensitive to alterations in plasma binding than free concentration, and can therefore be misleading. Half-life measurements are then essential because half-life is inversely proportional to free drug concentration for drugs with restrictive elimination.

Although full quantitation is generally lacking in human or even animal studies, several examples of clinically significant, plasma-binding-displacement interactions are known. Chloral hydrate in large doses can potentiate the effects of warfarin during its continuous administration (SELLERS and KOCH-WESER, 1970) because it is converted to trichloroacetic acid, which displaces the anti-coagulant from plasma binding sites, thereby increasing anticoagulation and shortening drug half-life. Epidemiological data from the Boston Collaborative Drug Surveillance Program (1972) suggests that this interaction may be transient since the warfarin dosage requirement was reduced only for the first four days of chloral hydrate administration. Thereafter warfarin dosage was the same as that used under control conditions. The likely temporary nature of this interaction may explain the failure of other clinical studies to consistently detect it (GRINER et al., 1971; UDALL, 1971). Another example comes from the studies of AGGELER et al. (1967) and O'REILLY and AGGELER (1968) who showed that phenylbutazone could displace warfarin from binding sites on human albumin *in vitro* and that phenylbutazone administration in man increased the effectiveness of a single dose of the anticoagulant and shortened its half-life. In this case, however, the change in effectiveness of warfarin is apparently not transient but continued for some 15 days (UDALL, 1971), suggesting that inhibition of metabolism might also contribute. This is supported by a quantitative examination of these interactions. From the literature, warfarin is about 97% bound (f = 0.03) with an apparent volume of distribution of total drug of about 7.5 l, giving a volume of distribution of free drug of 250 l ($Vf = Vd/f$). Using Eq. (17)

$$250 = 40 + 5.5 \cdot \frac{97}{3} + \frac{V_t B_t}{f}$$

the term $V_t B_t/f$ then equals 30 l. Thus, in order to produce a two-fold increase in free drug concentration by reducing Vf to 125 l, plasma drug binding must be reduced to 86%, which represents an increase in free drug fraction of some 4.7-fold. The two-fold increase in free drug concentration should reduce half-life to 50%, corresponding approximately to the changes seen for the interaction of warfarin with chloral hydrate (SELLERS and KOCH-WESER, 1970) and with phenylbutazone (AGGELER et al., 1967). SELLERS and KOCH-WESER (1970) suggest that this would give a relatively minor change in prothrombin time in normal subjects and is consistent with their data. The data of O'REILLY and

AGGELER (1968) then become difficult to explain solely on the basis of a change in binding since a much more dramatic effect on prothrombin time following phenylbutazone was seen with only a two-fold change in half-life. This supports our interpretation that phenylbutazone may also have been inhibiting warfarin metabolism, thereby counteracting to some extent the effects of a greater change in free drug or half-life. Certainly therapeutic plasma concentrations of phenylbutazone can decrease binding of warfarin to 60% when examined *in vitro* (SOLOMON et al., 1968). Obviously, only direct measurements of free warfarin concentration *in vivo* can resolve this problem, but the analysis shows the importance of full quantification, especially when more than one biologic process can be influenced by co-administration of another drug.

This is also well illustrated by the study of CHRISTENSEN et al. (1963) who found that sulfaphenazole could potentiate the hypoglycemic effect of tolbutamide and that plasma binding displacement took place *in vitro*. Although quoted as an example of a redistributional interaction, the authors themselves were able to show that the interaction occurred as the result of inhibition of tolbutamide metabolism because (1) tolbutamide half-life was markedly prolonged and steady state total plasma concentrations were elevated, and (2) other sulfonamides that displaced tolbutamide equally well *in vitro*, that is, they increased free drug fraction two-fold, had no effect *in vivo*. The increased effectiveness of coumarins in the presence of clofibrate also has been thought to result from drug binding displacement (SOLOMON et al., 1968). However, inhibition of metabolism has been shown in dogs (HUNNINGHAKE and AZARNOFF, 1968), and total drug concentration and half-life of warfarin are little altered in man (SCHROGIE and SOLOMON, 1967; O'REILLY et al., 1972).

Another well-established effect of binding displacement is the increased incidence of fatal kernicterus seen in premature infants after administration of sulfazoxazole/penicillin combination (SILVERMAN et al., 1956). Because the sulfonamide displaced bilirubin from albumin binding sites *in vitro*, ODELL (1959) suggested that binding displacement increased free bilirubin concentration with consequent increased brain levels and kernicterus. Animal data are supportive of this view (JOHNSON et al., 1959). Moreover, total bilirubin concentrations in serum were less in a treated group compared to a control group of children, and there was enhanced staining with bilirubin in the brain, despite the lower total serum concentration (HARRIS et al., 1958). Though free bilirubin concentrations have not been measured and quantification is therefore poor, this interaction is one of the better examples of its type. Perhaps the best example, however, is provided by the study of KUNIN (1966) on the effects of sulfonamides and aspirin on the kinetics of a number of penicillins. Not only was displacement of penicillin demonstrated *in vitro*, but an increase in free drug concentration was shown *in vivo*. In addition, total urinary excretion was unaffected. Unfortunately, the effects of redistribution on drug half-life were not measured. It should be mentioned that the renal elimination of penicillins in general is probably non-restrictive in that they are actively secreted by the renal tubules. The efficiency of their hepatic elimination is unclear, however, and it is therefore impossible to decide at this time whether their overall elimination rate would be enhanced or decreased by altered drug binding in the plasma.

b) *Non-Restrictive Elimination*

Non-restrictive elimination has been described only recently for drugs, such as propranolol, that are eliminated by hepatic metabolism (EVANS et al., 1973).

Animal data on chlorpromazine (CURRY, 1972) are also consistent with this type of elimination. The data of ALEXANDERSON (1972) and ALEXANDERSON and BORGA (1972) suggest that nortriptyline would also be eliminated under non-restrictive conditions. Drug displacement interactions have not yet been described for these types of drugs, but in theory should not affect total drug clearance or average concentration because, by definition, extraction ratio is greater than the fraction of free drug. However, the volume of distribution of total drug will increase and therefore half-life will be prolonged. In this way free drug concentration should be permanently elevated because no compensatory increase in elimination rate can occur (Fig. 4). Although plasma binding displacement has

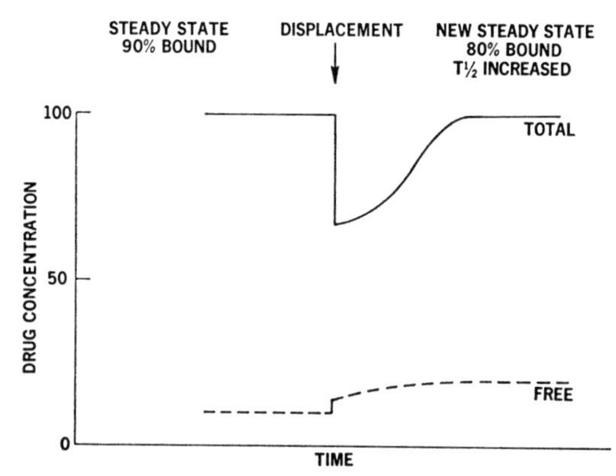

Fig. 4. Effects of displacement from albumin binding sites on the disposition of a drug with non-restrictive elimination. Conditions immediately after displacement are the same as those in Fig. 3, but because elimination rate is reduced by displacement when elimination is non-restrictive, total and free drug concentrations rise during the attainment of the new steady state

been shown *in vitro* for tricyclic antidepressants by diphenylhydantoin (BORGA et al., 1969), there are reasons to suggest that plasma binding displacement interactions may not be as clinically relevant as those involving drugs with restrictive elimination. First, those compounds known to show non-restrictive elimination are highly extracted by the liver and have widely variable plasma concentrations following oral drug administration because of the presystemic hepatic effect. Only large effects would therefore show against this background of individual variation. Second, such compounds (tricyclics, propranolol and chlorpromazine) are not bound to single siteson plasma albumin and the degree of binding usually falls to about 90% with high concentrations. This suggests a binding site with a very large capacity, but a low affinity, in which case little drug displacement would be expected, thus limiting the extent to which free drug fraction could be increased.

The interactions between sulfonamides and penicillins have been mentioned previously and undoubtedly result in redistribution. It is unclear at this time whether overall penicillin elimination is non-restrictive, though it seems likely that the renal component would be. When precise quantitation of these interactions involving drugs with non-restrictive elimination is obtained in humans, their therapeutic importance will more clearly be defined. Moreover, it should be recognized that their potential occurrence precludes the routine use of *total* drug concentrations in plasma when monitoring for kinetic drug interactions.

2. Tissue Binding or Uptake

The effects of altered tissue binding or uptake on drug action depends (1) on whether the tissue represents the site of drug action, and (2) on the effects that drug redistribution may have on elimination. Theoretically, altered tissue uptake is likely to have dramatic effects on drug disposition (GILLETTE, 1971). Displacement of drug from a tissue binding site will decrease the apparent volume of distribution. This should not affect drug clearance; drug half-life will be shortened [see Eq. (2)].

In 1948, ZUBROD et al. demonstrated an interaction between the antimalarials pamaquine and quinacrine, both of which are extensively bound in the liver. When the drugs were administered together, plasma pamaquine concentrations were markedly elevated and drug half-life prolonged. Clearly, the metabolism of pamaquine was inhibited, but the fact that drug concentration changes were greater than those in half-life suggested that perhaps displacement from binding sites, particularly in the liver, might be involved. However, such a mechanism would be difficult to prove, especially since the major binding site was present in the likely organ of metabolism.

A more clearcut example of the importance of inhibited tissue uptake of a drug to its site of action is shown with interactions involving the guanidinium adrenergic neuron blocking drugs quanethidine, bethanidine and debrisoquin. The highly selective effect of these ring-substituted bases on adrenergic neurons results from the fact that they are all substrates for the norepinephrine pump (MITCHELL and OATES, 1970). This energy-requiring membrane transport system uniquely concentrates them at their site of action. Inhibition of the norepinephrine pump will prevent the uptake of these guanidiniums into the adrenergic neuron (BRODIE et al., 1965; MITCHELL and OATES, 1970), and thereby prevent their pharmacologic effects (LEISHMAN et al., 1963; MITCHELL et al., 1970). The tricyclic antidepressants are competitive inhibitors of the norepinephrine pump, and it has been demonstrated in controlled studies that the antihypertensive effects of guanethidine, bethanidine and debrisoquin in man are almost totally abolished by concomitant administration of tricyclic antidepressants such as desipramine and protriptyline (MITCHELL et al., 1967, 1970). The antagonism of the antihypertensive effect of bethanidine is most dramatic, occurring with only two 25 mg doses of desipramine. Antagonism of the action of the guanidinium antihypertensives is produced also by imipramine (LEISHMAN et al., 1963), amitriptyline (MEYER et al., 1970) and even the least potent of the tricyclics, doxepin, when given in doses of more than 100 mg daily (FANN et al., 1971a). Chlorpromazine is less potent as an inhibitor of the norepinephrine pump, but in doses employed in psychiatric therapy, it too will antagonize the antihypertensive effects of guanethidine (FANN et al., 1972b). As chlorpromazine is a less potent competitive inhibitor, its antagonism can be partially overcome by increasing the dose of guanethidine. In patients with severe hypertension, the loss of control of blood pressure resulting from these drug interactions can lead to serious clinical complications such as stroke and malignant hypertension.

Drugs identified in animal studies as inhibitors of the norepinephrine pump must be evaluated in man to determine whether their potency and bioavailability combine to produce significant inhibition of the norepinephrine pump in clinically relevant doses in man. Inhibition of this transport mechanism in man may be evaluated in several ways. Blockade of the norepinephrine pump will antagonize the pressor response to indirectly acting amines such as tyramine, which must be transported into the adrenergic neuron in order to release norepinephrine. Thus, a rightward shift of partial dose response curves to tyramine results from administration of desipramine, protriptyline, amitriptyline (MITCHELL et al., 1970) and doxepin to humans (FANN et al., 1971a). The pressor response to norepinephrine itself must be measured concomitantly in order to exclude receptor blockade as a cause for reduced pressor sensitivity to tyramine; the pressor effects of norepinephrine are potentiated by the antidepressants due to prevention of norepinephrine inactivation by re-uptake via the neuronal pump. Inhibition of the norepinephrine pump can also be assessed by measuring the effect of drugs on the late phase of elimination of guanethidine, which is felt to represent primarily the equilibrium state determined by storage of the drug in adrenergic neurons (OATES et al., 1971). Thus, pretreatment with protriptyline will produce more rapid initial elimination of guanethidine from the body, with much less guanethidine being eliminated during the late phase of its disposition. The effect of plasma from patients receiving potential inhibitors of the norepinephrine pump on amine uptake into the rat iris, brain slices, or other tissues rich in adrenergic neurons can be used as an *in vitro* assessment (BORGA et al., 1970; SJÖQVIST et al., 1969). Many of the drugs that inhibit the norepinephrine pump will also block amine uptake into platelets and cause a reduced level of platelet serotonin after treatment for a week or more (FANN et al., 1971a). Whether the order of potency as inhibitors of the platelet amine pump corresponds with that for the norepinephrine pump in adrenergic neurons remains to be determined. For evaluating directly the antagonism of the guanidinium antihypertensives, bethanidine is an optimal test drug; because of its more rapid turnover in the human adrenergic neuron, antagonism of its antihypertensive effect is seen more abruptly following blockade of the norepinephrine pump.

In contrast to the interactions with tricyclic antidepressants, a displacement of guanethidine from its site of action appears to be the mechanism responsible for the antagonism of guanethidine by amphetamine (LAURENCE and ROSENHEIM, 1960). In this case, the administration of amphetamine results in a dramatic increase in the rate of excretion of guanethidine from the adrenergic neuronal pool (OATES et al., 1971). This suggests that amphetamine displaces guanethidine from its site of action within the adrenergic neuron.

D. Multiple Mechanisms

The general rule that a drug seldom demonstrates absolute specificity extends into the area of drug interactions, and a modifier of the disposition of another drug may act by more than one mechanism. Although phenobarbital is generally considered as an enzyme inducing agent, animal experiments have shown that it can also increase hepatic blood flow (OHNHAUS et al., 1971; BRANCH et al., 1974). While both enzyme activity and blood flow are capable of increasing drug clearance, their relative contributions will vary according to the initial extraction ratio of the drug in question. BRANCH et al. (1974) investigated these factors in the unanesthetized monkey and showed that with antipyrine (a very low extraction drug) over 90% of the change in clearance following phenobarbital was likely due to enzyme induction. On the other hand, with propranolol, which has an hepatic extraction ratio of about 0.5 in this species, the enhanced clearance could be attributed as much to the 30% increase in hepatic blood flow as to enzyme induction. These differences are entirely predictable from the described perfusion-limited model of ROWLAND et al. (1973), and suggest that increased clearance of highly extracted compounds such as indocyanine green after treatment with phenobarbital may not be due entirely to induction of ligandin transport protein, as suggested by REYES et al. (1971). Certainly indocyanine green elimination in man is largely flow-dependent (CHERRICK et al., 1960; CAESAR et al., 1961).

In addition to their effect in stimulating drug metabolism, some barbiturates can also impair drug absorption. For example, phenobarbital can reduce the absorption of griseofulvin (RIEGELMAN et al., 1970). Heptabarbital stimulates the metabolism of dicoumarol and causes a shortening of drug half-life after i.v. administration. However, it also reduces drug absorption, because the area under the plasma curve was markedly reduced after oral dosage (AGGELER and O'REILLY, 1969). In this case presystemic elimination is an unlikely cause of the discrepancy because dicoumarol has a low hepatic extraction ratio.

Multiple mechanisms also appear to be involved in the most striking drug interaction of all, the hypertensive crisis produced when sympathomimetic amines (e.g., tyramine) are taken by patients receiving monoamine oxidase (MAO) inhibitors (e.g., pargyline, phenelzine, furazolidone). Initial studies of pressor responses in humans demonstrated that patients not only were supersensitive to tyramine but also to the pressor effect of amphetamine, an indirectly-acting sympathomimetic drug that is not a substrate for MAO (PETTINGER and OATES, 1968). These findings were thought to result from a greater availability of norepinephrine for release from the adrenergic neuron by tyramine and amphetamine. Subsequent work in animals, however, has demonstrated that most inhibitors of MAO also block the metabolism of amphetamine and tyramine by hepatic enzymes when the inhibitors are given in large doses (TRINKER and RAND, 1970). Moreover, extirpation of the liver from animals resulted in the same supersensitivity to intravenous injections of tyramine and amphetamine as had pretreatment of normal animals with MAO inhibitors (RAND and TRINKER, 1968). The administration of MAO inhibitors to animals whose livers had been removed failed to increase further the pressor response to the amine drugs. Thus, it was suggested that the sensitizing effect of the MAO inhibitors was due entirely to an inhibition of MAO and of hepatic microsomal enzymes and not to an increased availability of norepinephrine for release. However, the administered doses of MAO inhibitors in these studies were huge respective to human doses, and it is unknown whether therapeutic doses of the inhibitors would block hepatic microsomal metabolism of amines in man. Moreover, SMITH (1966) observed enhanced inotropic effects of tyramine and amphetamine in isolated cardiac atria from animals treated with pargyline. Thus, it seems probable that MAO inhibitors sensitize to pressor amines in humans by making more norepinephrine available for release, by inhibiting MAO destruction of substrates such as tyramine, and possibly by blocking hepatic microsomal metabolism of the pressor amines.

Animal experiments have suggested that phenylbutazone and tolbutamide, which initially displace dicoumarol from plasma binding sites when administered concomitantly, may subsequently induce dicoumarol metabolism, since dicoumarol was much less effective when the new steady state drug concentration in plasma was achieved after redistribution (WELCH et al., 1969a). These experiments have no proven counterpart in man. An effect of tolbutamide on the metabolism of the dicoumarol-related drug warfarin has not been demonstrated clinically (POUCHER and VECCHIO, 1966). Although phenylbutazine can stimulate the metabolism of another drug, antipyrine (CHEN et al., 1962), other studies suggest that it may inhibit the metabolism of warfarin in humans (O'REILLY and AGGELER, 1968), as previously discussed (see p. 297). Thus, the relevance of the animal studies for the clinical use of these drugs is uncertain.

There are also animal data suggesting that enzyme inducers may actually inhibit drug metabolism while the inducer drug is present in the body. The contrast between the acute inhibitory effects and the long term stimulatory effects of ethanol on tolbutamide clearance indicates that this phenomenon may occur in man (CARULLI et al., 1971).

Complex interactions also are likely to occur when both components of the drug combination affect the disposition of the other. Thus, there is some evidence that the combination of diphenylhydantoin and phenobarbital results in lower plasma concentrations of diphenylhydantoin and higher concentrations of phenobarbital (MORSELLI et al., 1971). The interaction between chloral hydrate and ethanol is another example. Stimulation of chloral hydrate metabolism by ethanol leads to the production of more trichlorethanol which, in turn, inhibits ethanol elimination (SELLERS et al., 1972a and b). As previously mentioned, phenylbutazone and sulfafenazole inhibit tolbutamide metabolism and displace it from plasma binding sites *in vitro*, although this latter effect is probably unimportant in man (CHRISTENSEN et al., 1963). Few studies of drug interactions have included determinations of the circulating concentration of both components. This is unfortunate because mutual interactions may well be commoner than supposed in view of the fact that several of the mechanisms of kinetic interactions ultimately involve drug binding to some protein moiety.

IV. Investigation of Drug Interactions

Several approaches have been used to investigate kinetic drug interactions. One approach has been to evaluate the effects of modifiers of drug disposition on certain model drugs. This placed emphasis on the mechanisms of such interactions in the hope that the findings would have general applicability. It soon became evident, however, that the outcome of a potential interaction depends critically on the disposition of the drug in question. As the disposition of different drugs is rate-limited by different processes, it is clear that no single drug can provide an appropriate model to test all interactions. If we are to use model drugs at all, therefore, they should be tailored to the proposed interaction. For example, antipyrine remains an excellent drug to test altered hepatic metabolism because it is well absorbed, non-protein bound and eliminated entirely by metabolism. Furthermore, because it is so poorly extracted by the liver, its elimination is rate-limited by the activity of the drug-metabolizing enzymes. Since its presystemic elimination is minimal following the oral route of administration, changes in half-life reflect alterations in enzyme activity. Certain endogenous urinary metabolites, such as 6-β-OH cortisol and *d*-glucaric acid, have been suggested as markers of drug metabolizing activity. Because of high variability and lack of predictive precision, they can only be used as a semi-quantitative screen. Furthermore, should an effect on hepatic blood flow be suspected, then a highly extracted compound like indocyanine green, bromsulphthalein, para-amino hippurate, propranolol or even lidocaine must be used to evaluate the suspected interaction.

Investigations of this type employing selected model drugs have established potential mechanisms of drug interactions, but a pressing need exists for a more quantitative approach when examining specific interactions that might occur with clinically used drug combinations. Such testing is more difficult because several interactions are possible and multiple interactions can occur simultaneously. In the initial evaluation or screen for suspected interactions in humans, the oral route of drug administration is preferred because (1) it is the only way to detect interactions involving absorption, (2) it is the usual clinical route of drug administration, and (3) the magnitude of the interaction may be much greater than after intravenous administration for drugs eliminated rapidly by the liver. In terms of the mechanism of the interaction, additional studies with intravenous administration of the drug are essential because (1) they will confirm altered absorption, (2) they provide more accurate estimates of systemic drug clearance and half-life, and (3) they can be used to assess the extraction ratio of the organ of elimination, thereby gaining insight into possible rate limiting factors. Timed urinary collections can be used to assess the contribution of the renal route to overall elimination and to measure renal clearance. *In vitro* studies of plasma drug binding and possible displacement may suggest a redistributional interaction, provided the *in vivo* data are consistent. If this is the case, then *in vivo* measurement of free (nonbound) drug concentration should be compared against pharmacologic response during these studies because altered drug action is the final arbiter of a significant interaction.

It may be helpful to discuss the possible results of such studies, the differential diagnosis of their mechanism, and the logical progression of their full investigation. Let us assume that circulating concentrations following a single oral dose of the test drug have been measured before and after the potential modifier drug. It is advisable to measure both plasma and blood concentrations (or at least determine the blood/plasma drug concentration ratio) and to continue measurements long enough (at least 3 half-lives if possible) to calculate accurately the area under the concentration/time curve extrapolated to infinity (AUC). Such a study will aid in the detection of an interaction but will not define its mechanism. Although a change in half-life may provide a clue, such estimates are less accurate than those made following intravenous administration. Thus having detected the interaction, a study of the effects of the modifier on intravenously administered drug should be carried out. Should the interaction be due to an effect on absorption, the kinetics of intravenously administered drug should be unaltered. Appropriate changes in the amount of drug or its metabolites in the urine may provide confirmatory evidence. In addition, altered rates of absorption, as opposed to the fraction absorbed, can be detected using the kinetic analyses outlined in Chapter 59. If the drug is eliminated by metabolism it is worthwhile calculating systemic drug clearance after intravenous administration using whole blood concentration. This may allow a distinction to be made between altered absorption and changes in oral availability due to changes in presystemic hepatic elimination except in the case of highly extracted drugs. As mentioned, a reduction in AUC to one-half may occur as a result of changing hepatic extraction ratio from 80 to 90%, which would be reflected by only about a 10% reduction in half-life that might well escape detection. On the other hand, if the hepatic extraction ratio were only 0.2, then a halving of the AUC would be associated with a shortening of half-life by 45%. The Eqs. (11) to (14) can also be used to decide whether a given change in hepatic clearance is compatible with the change in oral bioavailability, thus aiding in the detection of interactions with a dual mechanism on metabolism and absorption.

Assuming that a change is found in the AUC after intravenous administration, that is, systemic drug clearance is altered, then several possibilities exist. Enhanced clearance associated with a shortening of half-life may be due to increased elimination by the liver or kidneys, or as a result of a redistributional interaction. Alteration in renal clearance can be measured directly using timed samples and calculating mean drug concentrations in blood and plasma (though steady state estimates are more accurate). Effects of pH and/or altered tubular secretion may be responsible. If the drug is not excreted in the urine, then altered metabolism may be responsible. Three mechanisms exist for enhanced hepatic clearance: increased hepatic blood flow, increased activity of hepatic enzymes or active transport processes, and decreased binding to blood proteins of drugs with restrictive elimination. The first two mechanisms decrease drug action and the last increases it. Again a knowledge of the extraction efficiency of the organ of elimination (liver) is helpful. If extraction efficiency is low, then altered enzyme activity or transport process is more likely, whereas highly extracted drugs should be more sensitive to altered blood flow. If a decrease in pharmacologic response occurs during the drug interaction, then the first two mechanisms can be separated from the third. If these data are unavailable or not conclusive, then altered drug binding must be excluded. An *in vitro* binding study should show if the test drug is (1) highly bound, and (2) can be displaced by the modifier drug. In addition, a knowledge of the percentage of free drug in plasma and of the organ extraction ratio for the drug should indicate whether elimination is likely to be restrictive and redistribution associated with accelerated elimination. Estimates of apparent volume of distribution can also help. As previously discussed, a change in this volume calculated from total drug concentration may be very large without being associated with a significant change in actual free drug concentration in the blood and therefore without change in pharmacologic action.

Finally, once an interaction has been detected and its mechanism elucidated, there is much to be said for conducting studies during chronic oral administration during which steady state kinetics exists: (1) this best mimicks the clinical use of many drugs, (2) the patient can be used as his own control by use of pretreatment and posttreatment control periods, (3) the elimination of some drugs may change quantitatively during chronic administration, (4) transient interactions can be detected and the time-course of events seen, and (5) this is the only feasible way to study patients who may require continuous administration of the test drug to control their disease. Moreover, it may be important to investigate a drug interaction in the patient for whom a given therapy is intended, because this is the only way in which the effects of the disease state on such drug interactions can be assessed. Chronic studies also may be necessary to define the clinical significance of drug binding displacement interactions for drugs with restrictive elimination. As previously mentioned, such interactions may be transient in nature and clinically insignificant after new steady state drug concentrations in plasma are reached. Studies conducted during chronic therapy must be designed to obviate variations in drug intake due to problems with patient compliance.

V. Clinical Relevance

The importance of a given drug interaction depends on the magnitude of the change in drug action that is produced compared to that which can be safely tolerated for the drug in question. The range of circulating concentrations over which a drug is effective without producing toxicity can be described as the therapeutic window and is analogous to the therapeutic ratio. The narrower the therapeutic window, the more significant kinetic drug interactions become and the more frequently they occur. Both the severity and frequency of these interactions is reduced as the therapeutic window widens, because the magnitude of the change that can occur in circulating concentration becomes smaller compared to that tolerated, not only because these drugs are safer, but because they often show marked interindividual variation in plasma levels under normal conditions. Clinical relevance is also determined by the severity of the disease being treated. In general, it may be said that the significance of drug interactions that reduce drug action is proportional to the severity of the underlying disease, whereas the importance of interactions that increase drug action is determined largely by the therapeutic ratio of the drug and the seriousness of drug toxicity.

A. Factors Determining the Clinical Significance of Pharmacokinetic Drug Interactions

1. Inherent Properties of the Drug or Disease State

Interactions involving drugs that have a narrow therapeutic window, that result in severe toxicity in overdosage, or that are used to control life-threatening diseases are always potentially significant. A classic example would be the coumarin anticoagulants used in the treatment of recurrent pulmonary or arterial embolism. Other examples include hypoglycemic agents (e.g., tolbutamide), antiarrhythmic agents (e.g., quinidine, procaineamide and lidocaine), cardiac glycosides (digitoxin) and antihypertensives (e.g., guanethidine, bethanidine). Interactions involving anticonvulsants, such as diphenylhydantoin, are especially important when drug toxicity is induced because the side effects can be debilitating and can confuse diagnosis by mimicking the underlying disease.

Optimally, avoidance of these interactions is achieved through a knowledge of those drugs known to cause them. In addition, for possible drug interactions when clinical information is incomplete, careful observation of therapeutic and adverse effects should be made whenever other drugs are added to or removed from the regimen. If clinical warning signals for either drug toxicity or inadequate drug response are not reliable, routine monitoring of plasma drug concentrations may provide useful clinical guidance.

2. Pharmacokinetic Factors

Several pharmacokinetic factors determine the clinical significance of other drug interactions. Many of these factors have already been discussed, including organ extraction, route of administration and volume of distribution of the test drug. The large interindividual variation in the disposition of many drugs is especially important. For example, 20-fold or greater variations in the plasma concentrations of tricyclic antidepressants and phenothiazines have been described. It is important to recognize this large background of variation, first, because a given change in plasma concentration will only be significant for those patients at the extremes of the normal range, and second, because these drugs generally have a fairly wide therapeutic window. In addition, their pharmacologic effects often are difficult to measure with precision. The relevance of interactions involving these compounds will then depend on the magnitude of the change in the pharmacologic action that can be produced. Though few absolute rules can be formulated, some generalizations can be made. Those factors that lower drug concentration at the receptor site by enhancing drug elimination (enzyme induction, increased organ blood flow, plasma binding displacement) are unlikely to produce more than a 2 to 3 fold change. On the other hand, inhibition of drug elimination or absorption may be much more dramatic.

The dose of the potential modifier of drug disposition can also be important for producing a drug interaction in man. BRECKENRIDGE and colleagues (1973) have recently shown that stimulation of warfarin metabolism by quinalbarbital is dose-dependent in the rat. Furthermore, they found in two patients that as quinalbarbital concentrations in plasma increased so did the effects on warfarin metabolism, until a maximum was reached. This suggests that enzyme induction may be a function of the concentration of the inducing agent reaching the liver.

However, the change in steady state warfarin levels in a group of patients was poorly correlated with quinalbarbital concentrations, and antipyrine half-lives failed to correlate well with plasma phenobarbital concentrations. These data suggest that individual variation within the population may have obscured the concentration/effect relationship that existed for each individual. The relation of induction to levels of the inducer is supported by data in the monkey showing that the change in antipyrine clearance by phenobarbital followed the plasma levels of phenobarbital (BRANCH et al., 1974).

Although dose-response relationships for inhibitors of drug metabolism have not been investigated in man, most of these are likely to be competitive inhibitors, in which case their effects should also be dose-dependent. With this background, it is easy to see why some drugs that inhibit drug metabolism in animal studies fail to result in an alteration in drug metabolism when they are administered in therapeutic doses in man since they may not be present in significant concentrations to have an effect. Concerning redistributional interactions, the circulating concentration of both drugs is an important determinant of the change in free drug concentration, as their interaction with one binding site is likely to be competitive (SOLOMON et al., 1968). Thus the dose and individual variation in the disposition of both the drug and its displacing agent can influence the final outcome. This would seem especially true of acidic, highly albumin-bound drugs, such as coumarin anticoagulants and tolbutamide.

The kinetics of the modifier drug may also be important to the time relationships of these interactions. The time taken for the full effect of changes in drug metabolizing enzymes to occur or to wear off is a function of the half-life of both the modifier drug and the drugs that are being affected because it takes about 4 half-lives for the achievement of a new steady state. Enzyme inhibition is generally assumed to be rapid and dose-dependent, therefore, the time taken to reach a new steady state will depend on whether the inhibitor or drug being inhibited has the longer half-life. For enzyme induction, it is unclear how long the synthesis of new enzyme protein takes in man. However, assuming that this was not rate-limiting, it takes about 3 weeks for phenobarbital to accumulate to its steady state ($T_{1}/_{2}$ 5 days). This may account for the fact that a new steady state for warfarin following phenobarbital administration may take 3 to 4 weeks, even though the half-life of warfarin is only one day. These kinetic considerations demonstrate that the changes resulting from alterations in drug metabolism may be quite insidious. Whether more rapid induction could be achieved by loading and maintenance doses of the inducer will be unclear until further information is available concerning the relationship between the concentration of the inducing agent and its effect, but the prospect of achieving rapid enzyme induction in the treatment of certain drug toxicities is an attractive one.

3. Pharmacogenetic Factors

In a study of the effects of phenobarbital administration on antipyrine half-life, VESELL and PAGE (1969) have shown that the greatest changes in antipyrine elimination occurred in those subjects that initially had the longest half-life, that is, the lowest drug clearance. Because antipyrine half-life was itself largely genetically controlled, it was suggested that the magnitude of enzyme induction might also be genetically determined. The interaction between diphenylhydantoin and isoniazid is another example of a genetically determined interaction (KUTT et al., 1969). The data of CORN (1966) also showed the importance of individual variation by comparing the effects of phenobarbital and glutethimide on warfarin half-life in a within-subject study. As much as a 20% change in warfarin half-life was seen without concomitant administration of the other drugs. Although mean half-life was significantly shorter following administration of the sedatives, the variation in change was even larger and 40% of the patients did not respond. These data suggest that the change in warfarin concentrations due to enzyme induction may be small enough on occasion to be obscured by daily variation even though the interaction can be clearly demonstrated in most patients.

4. Disease-Induced Pharmacokinetic Factors

All processes involved in drug disposition can be affected by disease, sometimes quite dramatically. Thus the background against which drug interactions occur may quite different and importantly so. For example, the half-lives of phenylbutazone and antipyrine may be greatly prolonged in liver disease, yet simultaneous administration of other drugs may reduce half-lives to within the normal range (LEVI et al., 1968; BRANCH et al., 1973). Thus the effects of the disease are more important than interactions involving enhanced elimination in these patients. On the other hand, inhibition of metabolism would be even more significant in patients with cirrhosis. Another example of the effects of disease on potential drug interactions is the reduction in plasma drug binding in the presence of renal or hepatic disease (REIDENBERG and AFFRIME, 1973), in which case binding displacement interactions would be of lesser importance. Also, the elimination of lidocaine is reduced in heart failure presumably because of a reduction in hepatic blood flow (STENSON et al., 1971; THOMSON et al., 1971). Such patients will be more liable to toxicity and require smaller doses. Thus, both disease and its treatment may be important to the clinical relevance of drug interactions.

B. Summary

Early enthusiasm for ascribing clinical significance to all pharmacokinetic drug interactions that could be demonstrated in experimental animals or *in vitro* has become tempered with a greater knowledge of their mechanisms and their more accurate quantitation in man. Nonetheless, these interactions can and do occur with significant frequency and often with disastrous effect. They can only be avoided by awareness and a knowledge of mechanism and the pharmacokinetics of the drugs we use. Patient treatment should be individualized, if possible, by objective measurement of pharmacologic action. Routine plasma level monitoring may be helpful when drug action is difficult to quantify or cannot be used as an index of drug concentration (e.g., therapy of seizure disorders or of cardiac arrhythmias). There is, however, no substitute for continuous review of the patient's clinical status, especially when medications are being altered. Whenever a drug interaction is suspected, it should be remembered that a pharmacokinetic mechanism must be excluded before interactions at the receptor level can be established.

References

AGGELER, P.M., O'REILLY, R.A., LEONG, L., KOWITZ, P.E.: Potentiation of anticoagulant effect of warfarin by phenylbutazone. New Engl. J. Med. **276**, 496—501 (1967).

AGGELER, P.M., O'REILLY, R.A.: Effect of heptabarbital on the response to bishydroxycoumarin in man. J. Lab. clin. Med. **74**, 229—238 (1969).

ALEXANDERSON, B.: Pharmacokinetics of nortriptyline in man after single and multiple oral doses: The predictability of steady-state plasma concentrations from single-dose plasma-level data. Europ. J. Clin. Pharmacol. **4**, 82—91 (1972).

Alexanderson, B., Borga, O.: Interindividual differences in plasma protein binding of nortriptyline in man — a twin study. Europ. J. Clin. Pharmacol. **4**, 196—200 (1972).

Alexanderson, B., Price Evans, D. A., Sjöqvist, F.: Steady-state plasma levels of nortriptyline in twins: Influence of genetic factors and drug therapy. Brit. med. J. **1969 IV**, 764—768.

Ambre, J. J., Fischer, L. J.: Effect of coadministration of aluminium and magnesium hydroxides on absorption of anticoagulants in man. Clin. Pharmacol. Ther. **14**, 231—237 (1973).

Anders, M. W.: Enhancement and inhibition of drug metabolism. Ann. Rev. Pharmacol. **11**, 37—56 (1971).

Arias, I. M., Gartner, L. M., Cohen, M., Ben Ezzer, J., Levi, A. J.: Chronic nonhemolytic unconjugated hyperbilirubinemia with glucuronyl transferase deficiency. Clinical, biochemical, pharmacologic and genetic evidence for heterogeneity. Amer. J. Med. **47**, 395—409 (1969).

Avellaneda, M.: Interferencia de los barbituricos en la accion del tromexan. Medicina (Buenes Aires) **15**, 109—115 (1955).

Barr. W. H., Adir, J., Garrettson, L.: Decrease of tetracycline absorption in man by sodium bicarbonate. Clin. Pharm. Ther. **12**, 779—784 (1971).

Barrowman, J. A., D'Mello, A., Herxheimer, A.: A single dose of neomycin impairs absorption of vitamin A (Retinol) in man. Europ. J. Clin. Pharmacol. **5**, 199—202 (1973).

Baylis, E. M., Crowley, J. M., Preece, J. M., Sylvester, P. E., Marks, V.: Influence of folic acid on blood-phenytoin levels. Lancet **1971 I**, 62—64.

Bazzano, G., Bazzano, G. S.: Digitalis intoxication. Treatment with a new steroid-binding resin. J. Amer. med. Ass. **220**, 828—830 (1972).

Benowitz, N., Rowland, M., Forsyth, R., Melmon, K. L.: Circulatory influences on lidocaine disposition. Clin. Res. **21**, 467 (1973).

Bianchine, J., Calimlim, L. R., Morgan, J. P., Dujuvne, C. A., Lasagna, L.: Metabolism and absorption of *L*-3,4-dihydroxyphenylalanine in patients with Parkinson's disease. Ann. N. Y. Acad. Sci. **179**, 126—140 (1971).

Binns, T. B.: The absorption of drugs from the alimentary tract, lungs and skin. Brit. J. Hosp. Med. **6**, 133—142 (1971).

Black, M., Sherlock, S.: Treatment of Gilbert's syndrome with phenobarbitone. Lancet **1970 I**, 1359—1361.

Bogdanski, D. F., Blaszkowski, T. P., Brodie, B. B.: Effects of enzyme induction on synthesis and catabolism of corticosterone in rats. Analysis by mean of steady-state kinetics. J. Pharmacol. exp. Ther. **179**, 372—379 (1971).

Boman, G.: In: Morselli, P. L., Garattini, S., Cohen, S. N. (Eds.): Drug interactions. New York: Raven Press (in press).

Borgá, O., Azarnoff, D. L., Forshell, G. P., Sjöqvist, F.: Plasma protein binding of tricyclic antidepressants in man. Biochem. Pharmacol. **18**, 2135—2143 (1969).

Borowitz, J. L., Moore, P. F., Yim, G. K. W., Miya, T. S.: Mechanism of enhanced drug effects produced by dilution of the oral dose. Toxicol. appl. Pharmacol. **19**, 164—168 (1971).

Boston Collaborative Drug Surveillance Program: Interaction between chloral hydrate and warfarin. New Engl. J. Med. **286**, 53—55 (1972).

Branch, R. A., Herbert, C. M., Read, A. E.: Determinants of serum antipyrine half-lives in patients with liver disease. Gut **14**, 569—573 (1973a).

Branch, R. A., Nies, A. S., Shand, D. G.: The disposition of propranolol. VIII. General implications of the effects of liver blood flow on elimination from the perfused rat liver. Drug Metab. Dispos. **1**, 687—690 (1973b).

Branch, R. A., Shand, D. G., Nies, A. S.: Increase in hepatic blood flow and *d*-propranolol clearance by glucagon in the monkey. J. Pharmacol. exp. Ther. **187**, 581—587 (1973c).

Branch, R. A., Shand, D. G., Wilkinson, G. R., Nies, A. S.: Increased clearance of antipyrine and *d*-propranolol after phenobarbital treatment in the monkey. J. clin. Invest. **53**, 1101—1107 (1974).

Breckenridge, A., Orme, M.: Clinical implications of enzyme induction. Ann. N. Y. Acad. Sci. **179**, 421—431 (1971).

Breckenridge, A., Orme, M., L. E. Davies, L., Thorgeirsson, S. S., Davies, D. S.: Dose-dependent enzyme induction. Clin. Pharmacol. Ther. **14**, 514—520 (1973).

Brennan, R. W., Dehejia, H., Kutt, H., Verebely, K., McDowell, F.: Diphenylhydantoin intoxication attendant to slow inactivation of isoniazid. Neurology **20**, 687—693 (1970).

Brodie, B. B.: Physico-chemical factors in drug absorption. In: Binns, T. B. (Ed.): Absorption and distribution of drugs, pp. 16—48. Baltimore: Williams and Wilkins 1964.

Brodie, B. B., Chang, C. C., Costa, E.: On the mechanism of action of guanethidine and bretylium. Brit. J. Pharmacol. **25**, 171—178 (1965).

BROOKS, S.M., WERK, E.E., ACKERMAN, S.J., SULLIVAN, I., THRASHER, K.: Adverse effects of phenobarbital on corticosteroid metabolism in patients with bronchial asthma. New Engl. J. Med. **286**, 1125—1128 (1972).

BURSTEIN, S., KLAIBER, E.L.: Phenobarbital-induced increase in 6 *B*-hydroxycortisol excretion: clue to its significance in human urine. J. clin. Endocr. **25**, 293—296 (1965).

BUSFIELD, D., CHILD, K.J., ATKINSON, R.M., TOMICH, E.G.: An effect of phenobarbitone on blood-levels of griseofulvin in man. Lancet **1963 II**, 1042—1043.

CAESAR, J., SHALDON, S., CHIANDUSSI, L., GUEVARA, L., SHERLOCK, S.: The use of indocyanine green in the measurement of hepatic blood flow and as a test of hepatic function. Clin. Sci. **21**, 43—57 (1961).

CALDWELL, J.H., BUSH, C.A., GREENBERGER, N.J.: Interruption of the enterohepatil circulation of digitoxin by cholestyramine. II. Effect on metabolic disposition of tritium-labeled digitoxin and cardiac systolic intervals in man. J. clin. Invest. **50**, 2638—2644 (1971).

CALDWELL, J.H., GREENBERGER, N.J.: Interruption of the enterohepatic circulation of digitoxin by cholestyramine. I. Protection against lethal digitoxin intoxication. J. clin. Invest. **50**, 2626—2637 (1971).

CARULLI, N., MANENTI, F., GALLO, M., SALVIOLI, G.F.: Alcohol-drug interactions in man: alcohol and tolbutamide. Europ. J. clin. Invest. **1**, 421—424 (1971).

CAVALIERI, R.R., SUNG, L.C., BECKER, C.E.: Effects of phenobarbital on thyroxine and triiodothyronine kinetics in Graves' disease. J. clin. Endocr. **37**, 308—316 (1973).

CHAPUT DE SAINTONGE, D.M., HERXHEIMER, A.: Sodium bicarbonate enhances the absorption of propantheline in man. Eur. J. Clin. Pharmacol. **5**, 239—242 (1973).

CHEN, W., VRINDTEN, P.A., DAYTON, P.G., BURNS, J.J.: Accelerated aminopyrine metabolism in human subjects pretreated with phenylbutazone. Life Sci. **2**, 35—42 (1962).

CHERRICK, G.R., STEIN, S.W., LEEVY, C.M., DAVIDSON, C.S.: Indocyanine green: Observations on its physical properties, plasma decay, and hepatic extraction. J. clin. Invest. **39**, 592—600 (1960).

CHOI, Y., THRASHER, K., WERK, E.E., JR., SHOLITON, L.J., OLINGER, C.: Effect of diphenylhydantoin on cortisol kinetics in humans. J. Pharmacol. exp. Ther. **176**, 27—34 (1971).

CHRISTENSEN, L.K., HANSEN, J.M., KRISTENSEN, M.: Sulphaphenazole-induced hypoglycaemic attacks in tolbutamide-treated diabetics. Lancet **1963 II**, 1298—1310.

CHRISTENSEN, L.K., SKOVSTED, L.: Inhibition of drug metabolism by chloramphenicol. Lancet **1969 II**, 1397—1399.

CONNEY, A.H.: Pharmacological implications of microsomal enzyme inductions. Pharmacol. Rev. **19**, 317—366 (1967).

CONNORS, T.A., COX, P.J., FARMER, P.D., FOSTER, A.D., JARMAN, M.: Some studies on the active intermediate formed in the microsomal metabolism of cyclophosphamide and isophosphamide. Biochem. Pharmacol. **23**, 115—129 (1974).

COOKE, A.R., HUNT, J.N.: Absorption of acetylsalicylic acid from unbuffered and buffered gastric contents. Amer. J. dig. Dis. **15**, 95—102 (1970).

CORN, M.: Effect of phenobarbital and glutethimide on biological half-life of warfarin. Thrombos. Diathes. haemorrh. (Stuttg.) **16**, 606—612 (1966).

CORN, M., ROCKETT, J.F.: Inhibition of bishydroxycoumarin activity by phenobarbital. Med. Ann. D. C. **34**, 578—579 (1965).

CRIGLER, J.F., JR., GOLD, N.I.: Sodium phenobarbital-induced decrease in serum bilirubin in an infant with congenital nonhemolytic jaundice and kernicterus. J. clin. Invest. **45**, 998—999 (1966).

CRIGLER, J.F., JR., GOLD, N.I.: Effect of sodium phenobarbital on bilirubin metabolism in an infant with congential, nonhemolytic, unconjugated hyperbilirubinemia, and kernicterus. J. clin. Invest. **48**, 42—55 (1969).

CROUNSE, R.G.: Human pharmacology of griseofulvin. The effect of fat intake on gastrointestinal absorption. J. invest. Dermatol. **37**, 529—533 (1961).

CUCINELL, S.A., CONNEY, A.H., SANSUR, M., BURNS, J.J.: Drug interactions in man. I. Lowering effect of phenobarbital on plasma levels of bishydroxycoumarin (dicumarol) and diphenylhydantoin (dilantin). Clin. Pharm. Ther. **6**, 420—429 (1965).

CUNNINGHAM, M.D., MACE, J.W., PETERS, E.R.: Clinical experience with phenobarbitone in icterus neonatorium. Lancet **1969 I**, 550—551.

CURRY, S.H.: Relation between binding to plasma protein, apparent volume of distribution, and rate constants of disposition and elimination for chlorpromazine in three species. J. Pharm. Pharmacol. **24**, 818—819 (1972).

CURRY, S.H., DAVIS, J.M., JANOWSKY, D.S., MARSHALL, J.H.L.: Factors affecting chlorpromazine plasma levels in psychiatric patients. Arch. gen. Psychiat. **22**, 209—215 (1970).

DANYSZ, A., WISNIEWSKI, K.: Control of drug transport through cell membranes. Materia Medica Polona **2**, 35—44 (1970).

DAVIES, J. E., EDMUNDSON, W. F., CARTER, C. H., BARQUET, A.: Effect of anticonvulsant drugs on dicophane (D. D. T.) residues in man. Lancet **1969 II**, 2—9.
DAVIS, J. M., KOPIN, I. J., LEMBERGER, L., AXELROD, J.: Effects of urinary pH on amphetamine metabolism. Ann. N. Y. Acad. Sci. **179**, 493—501 (1971).
DAYTON, P. G., TARCAN, Y., CHENKIN, T., WEINER, M.: The influence of barbiturates on coumarin plasma levels and prothrombin response. J. clin. Invest. **40**, 1797—1802 (1961).
DRUMMOND, K. H., HELLMAN, D. A., MARCHESSAULT, J. H. V., FELDMAN, W.: Cyclophosphamide in the nephrotic syndrome of childhood. Canad. Med. Ass. J. **98**, 524—531 (1968).
EVANS, G. H., NIES, A. S., SHAND, D. G.: The disposition of propranolol. III. Decreased half-life and volume of distribution as a result of plasma binding in man, monkey, dog and rat. J. Pharmacol. exp. Ther. **186**, 114—122 (1973).
FALOON, W. W.: Drug production of intestinal malabsorption. N. Y. State J. Med. **70**, 2189—2192 (1970).
FANN, W. E., KAUFMANN, J. S., GRIFFITH, J. D., DAVIS, J. M., JANOWSKY, D. S., OATES, J. A.: Doxepin: Effects on transport of biogenic amines. Psychopharmacologica (Berl.) **22**, 111—125 (1971a).
FANN, W. E., DAVIS, J. M., JANOWSKY, D. S., OATES, J. A.: Chlorpromazine reversal of the antihypertensive action of guanethidine. Lancet **1971II**b, 436—437.
FIELD, J. B., OHTA, M., BOYLE, C., REMER, A.: Potentiation of acetohexamide hypoglycemia by phenylbutazone. New Engl. J. Med. **277**, 889—894 (1967)
FORREST, F. M., FORREST, I. S., SERRA, M. T.: Modification of chlorpromazine metabolism by some other drugs frequently administered to psychiatric patients. Biol. Psych. **2**, 53—58 (1970).
GARRETTSON, L. K., PEREL, J. M., DAYTON, P. G.: Methylphenidate interaction with both anticonvulsants and ethyl biscoumacetate. J. Amer. med. Ass. **207**, 2053—2056 (1969).
GEORGE, C. F., FENYVESI, T., CONOLLY, M. E., DOLLERY, C. T.: Pharmacokinetics of dextro-, laevo-, and racemic propranolol in man. Europ. J. clin. Pharmacol. **4**, 74—76 (1972).
GERHARDT, R. E., KNOUSS, R. P., THYRUM, P. T., LUCHI, R. J., MORRIS, J. J.: Quinidine excretion in aciduria and alkaluria. Ann. intern. Med. **71**, 927—933 (1969).
GIBALDI, M., GRUNDHOFER, B.: Enhancement of intestinal absorption of a quaternary ammonium compound by salicylate and trichloroacetate. J. pharm. Sci. **62**, 343—344 (1973).
GIBALDI, M., BOYES, R. N., FELDMAN, S.: Influence of first-pass effect on availability of drugs on oral administration. J. Pharm. Sci. **60**, 1338—1340 (1971).
GILLETTE, J. R.: Factors affecting drug metabolism. Ann. N. Y. Acad. Sci. **179**, 43—66 (1971).
GILLETTE, J. R.: Overview of drug-protein binding. Ann. N. Y. Acad. Sci. **226**, 6—17 (1973).
GILLETTE, J. R., MITCHELL, J. R.: Drug actions and interactions: theoretical considerations (this volume).
GOLDSTEIN, A., ARONOW, L., KALMAN, S. M.: Principles of drug action: The basis of pharmacology. New York: Hoeber Medical Division, Harper and Row 1968.
GOSS, J. E., DICKHAUS, D. W.: Increased bishydroxycoumarin requirements in patients receiving phenobarbital. New Engl. J. Med. **273**, 1094—1095 (1965).
GRAM, L. F., OVERO, K. F.: Drug interaction: Inhibitory effect of neuroleptics on metabolism of tricyclic antidepressants in man. Brit. med. J. **1972 I**, 463—465 (1972).
GRINER, P. F., RAISZ, L. G., RICKLES, F. R., WIESNER, P. J., ODOROFF, C. L.: Chloral hydrate and warfarin interaction: clinical significance. Ann. intern. Med. **74**, 540—543 (1971).
HAGUE, D. E., SMITH, M. E., RYAN, J. R., MCMAHON, F. G.: The effect of methylphenidate and prolintane on the metabolism of ethyl biscoumacetate. Clin. Pharmacol. Ther. **12**, 259—262 (1971).
HAHN, K.-J., EIDEN, W., SCHETTLE, M., HAHN, M., WALTER, E., WEBER, E.: Effect of cholestyramine on the gastrointestinal absorption of phenprocoumon and acetylosalicylic acid in man. Europ. J. Clin. Pharmacol. **4**, 142—145 (1972).
HAHN, T. J., BIRGE, S. J., SCHARP, C. R., AVIOLI, L. V.: Phenobarbital-induced alterations in vitamin D metabolism. J. Clin. Invest. **51**, 741—748 (1972).
HAHN, T. J.: Anticonvulsant therapy and vitamin D. Ann. intern. Med. **78**, 308—309 (1973).
HANSEN, J. M., KRISTENSEN, M., SKOVSTED, L., CHRISTENSEN, L. K.: Dicoumarol induced diphenylhydantoin intoxication. Lancet **1966 II**, 265—266.
HANSEN, J. M., KRISTENSEN, M., SKOVSTED, L.: Sulthiame (Ospolot®) as inhibitor of diphenylhydantoin metabolism. Epilepsia **9**, 17—22 (1968).
HANSEN, J. M., SIERSBAEK-NIELSEN, K., SKOVSTED, L.: Carbamazepine-induced acceleration of diphenylhydantoin and warfarin metabolism in man. Clin. Pharmacol. Ther **12**, 539—543 (1971).
HARRIS, R. C., LUCEY, J. F., MACLEAN, J. R.: Kernicterus in premature infants associated with low concentrations of bilirubin in the plasma. Pediatrics **21**, 875—884 (1958).
HEADING, R. C., NIMMO, J., PRESCOTT, L. F., TOTHILL, P.: The dependence of paracetamol absorption on the rate of gartric emptying. Brit. J. Pharmacol. **47**, 415—421 (1973).

HOLLISTER, L., LEVY, G.: Some aspects of salicylate distribution and metabolism in man. J. Pharm. Sci. **54**, 1126—1129 (1965).

HUNNINGHAKE, D. B., AZARNOFF, D. L.: Drug interactions with warfarin. Arch. int. Med. **121**, 349—352 (1968).

HUNTER, J., MAXWELL, J. D., CARRELLA, M., STEWART, D. A., WILLIAMS, R.: Urinary D-glucaric-acid excretion as a test for hepatic enzyme induction in man. Lancet **1971 I** 572—575.

HURWITZ, A.: In: MORSELLI, P. L., GARATTINI, S., COHEN, S. N. (Eds.): Drug interactions. Raven Press (in press).

JAO, J. W., JUSKO, W. J., COHEN, J. L.: Phenobarbital effects on cyclophosphamide pharmacokinetics in man. Cancer Res. **32**, 2761—2764 (1972).

JOHNSON, L., SARMIENTO, F., BLANC, W. A., DAY, R.: Kernicterus in rats with an inherited deficiency of glucuronyl transferase. J. Dis. Child. **79**, 591—608 (1959).

JUBIZ, W., LEVINSON, R. A., MEIKLE, A. W., WEST, C. D., TYLER, F. H.: Absorption and conjugation of metyrapone during diphenylhydantoin therapy: Mechanism of the abnormal response to oral metyrapone. Endocrinology **86**, 328—331 (1970).

KATER, R. M., ROGGIN, G., TOBON, F., ZIEVE, P., IBER, F. L.: Increased rate of clearance of drugs from the circulation of alcoholics. Amer. J. med. Sci. **258**, 35—39 (1969).

KEEN, P.: Effect of binding to plasma proteins on the distribution, activity, and elimination of drugs. In: BRODIE, B. B., GILLETTE, J. R. (Eds.): Handbook of experimental pharmacology, Vol. XXVIII, Part I, Chapter 10. Berlin-Heidelberg-New York: Springer 1971.

KEKKI, M., PYÖRÄLÄ, K., MUSTALA, O., SALMI, H., JUSSILA, J., SIURALA, M.: Multi-compartment analysis of the absorption kinetics of warfarin from the stomach and small intestine. Int. J. clin. Pharmacol. **2**, 209—211 (1971).

KIØRBOE, E.: Phenytoin intoxication during treatment with Antabuse (disulfiram). Epilepsia **7**, 246—249 (1966).

KOCH-WESER, J., SELLERS, E. M.: Drug interactions with coumarin anticoagulants. New Engl. J. Med. **285**, 478—498 (1971).

KOJIMA, S., SMITH, R. B., DOLUISIO, J. T.: Drug absorption. V. Influence of food on oral absorption of phenobarbital in rats. J. pharm. Sci. **60**, 1639—1641 (1971).

KOLMODIN, B., AZARNOFF, D. L., SJOQUIST, F.: Effect of environmental factors on drug metabolism: Decreased plasma half-life of antipyrine in workers exposed to chlorinated hydrocarbon insecticides. Clin. Pharmacol. Ther. **10**, 638—642 (1969).

KOLMODIN-HEDMAN, B.: Decreased plasma half-life of phenylbutazone in workers exposed to chlorinated pesticides. Europ. J. Clin. Pharmacol. **5**, 195—198 (1973).

KRISTENSEN, J., HANSEN, J. M.: Accumulation of chlorpropamide caused by dicoumarol. Acta med. scand. **183**, 83—86 (1968).

KRISTENSEN, M., HANSEN, J. M., SKOVSTED, L.: The influence of phenobarbital on the half-life of diphenylhydantoin in man. Acta med. scand. **185**, 347—350 (1969).

KUNIN, C. M.: Clinical pharmacology of the new penicillins. II. Effect of drug which interfere with binding to serum proteins. Clin. Pharmacol. Ther. **7**, 180—188 (1966).

KUNTZMAN, R.: Drugs and enzyme induction. Ann. Rev. Pharmacol. **9**, 21—36 (1969).

KUPFERBERG, H. J., JEFFERY, W., HUNNINGHAKE, D. B.: Effect of methylphenidate on plasma anticonvulsant levels. Clin. Pharmacol. Ther. **13**, 201—204 (1972).

KUTT, H., HAYNES, J., VEREBELY, K., McDOWELL, F.: The effect of phenobarbital on plasma diphenylhydantoin level and metabolism in man and in rat liver microsomes. Neurology **19**, 611—616 (1969).

KUTT, H., WINTERS, W., McDOWELL, F. H.: Depression of parahydroxylation of diphenylhydantoin by antituberculosis chemotherapy. Neurology **16**, 594—602 (1966).

KUTT, H., BRENNAN, R., DEHEJIA, H., VEREBELY, K.: Diphenylhydantoin intoxication. A complication of isoniazid therapy. Amer. Rev. resp. Dis. **101**, 377—384 (1970).

LATHAM, A. M., MILLBANK, L., RICHENS, A., ROWE, D. J. F.: Liver enzyme induction by anticonvulsant drugs, and its relationship to disturbed calcium and folic acid metabolism. J. clin. Pharmacol. **13**, 337—342 (1973).

LAURENCE, D. R., ROSENHEIM, M. L.: Clinical effects of drugs which prevent the release of adrenergic transmitter. In: VANE, J. R., WOLSTENHOLME, G. E. W., O'CONNOR, M. (Eds.): Adrenergic mechanisms, pp. 201—208. Boston: Little, Brown 1960.

LEISHMAN, A. W. D., MATTHEWS, H. L., SMITH, A. J.: Antagonism of guanethidine by imipramine. Lancet **1963 I**, 112.

LEVI, A., SHERLOCK, S., WALKER, D.: Phenylbutazone and isoniazid metabolism in patients with liver disease in relation to previous drug therapy. Lancet **1968 I**, 1275—1279.

LEVINE, R. R.: Factors affecting gastrointestinal absorption of drugs. Digest. Dis. **15**, 171—188 (1970).

LEVY, G., RAO, B. K.: Enhanced intestinal absorption of riboflavin from sodium alginate solution in man. J. Pharm. Sci. **61**, 279—280 (1972).

LEVY, G., O'REILLY, R. A., AGGELER, P. M., KEECH, G. M.: Pharmacokinetic analysis of the anticoagulant action of warfarin in man. Clin. Pharmacol. Ther. **11**, 372—377 (1970).
LEVY, G., GIBALDI, M., PROCKNAL, J. A.: Effect of an anticholinergic agent on riboflavin absorption in man. J. pharm. Sci. **61**, 798—799 (1972).
LIEGLER, D. G., HENDERSON, E. S., HAHN, M. A., OLIVERIO, V. T.: The effect of organic acids on renal clearance of methotrexate in man. Clin. Pharmacol. Ther. **10**, 849—857 (1969).
LIPSETT, M. B.: Factors influencing the rate of metabolism of steroid hormones in man. Ann. N. Y. Acad. Sci. **179**, 442—449 (1971).
MACDONALD, M. G., ROBINSON, D. S.: Clinical observations of possible barbiturate interference with anticoagulation. J. Amer. med. An. **204**, 97—100 (1968).
MACDONALD, M. G., ROBINSON, D. S., SYLWESTER, D., JAFFE, J. J.: The effects of phenobarbital, chloral betaine, and glutethimide administration on warfarin plasma levels and hypoprothrombinemic responses in man. Clin. Pharmacol. Ther. **10**, 80—84 (1969).
MAGNUSSEN, M. P.: The effect of ethanol on the gastrointestinal absorption of drugs in the rat. Acta pharmacol. (Kbh.) **26**, 130—144 (1968).
MANNINEN, V., APAJALAHTI, A., MELIN, J., KARESOJA, M.: Altered absorption of digoxin in patients given propantheline and meteclopramide. Lancet **1973 I**, 398—401.
MAXWELL, J. D., HUNTER, J., STEWART, D. A., ARDEMAN, S., WILLIAMS, R.: Folate deficiency after anticonvulsant drugs: An effect of hepatic enzyme induction? Brit. med. J. **1972 I**, 297—299.
MCMULLIN, G. P.: Phenobarbitone and neonatal jaundice. Lancet **1968 II**, 978—979.
MCMULLIN, G. P., HAYES, M. F., ARORA, S. C.: Phenobarbitone in rhesus haemolytic disease. Lancet **1970 II**, 949—952.
MCQUEEN, E. G., OWEN, D., FERRY, D. G.: Effect of phenytoin and other drugs in reducing serum DDT levels. N. Z. med. J. **75**, 208—211 (1972).
MEIKLE, A. W., JUBIZ, W., MATSUKURA, S., WEST, C. D., TYLER, F. H.: Effect of diphenylhydantoin on the metabolism of metyrapone and release of ACTH in man. J. clin. Endocr. **29**, 1553—1558 (1969).
MEYER, J. F., MCALLISTER, C. K., GOLDBERG, L. I.: Insidious and prolonged antagonism of guanethidine by amitriptyline. J. Amer. med. Ass. **213**, 1478—1488 (1970).
MITCHELL, J. R., ARIAS, L., OATES, J. A.: Antagonism of the antihypertensive action of guanethidine sulfate by desipramine hydrochloride. J. Amer. med. Ass. **202**. 973—976 (1967).
MITCHELL, J. R., CAVANAUGH, J. H., ARIAS, L., OATES, J. A.: Guanethidine and related agents. III. Antagonism by drugs which inhibit the norepinephrine pump in man. J. clin. Invest. **49**, 1596—1604 (1970).
MITCHELL, J. R., OATES, J. A.: Guanethidine and related agents. I. Mechanism of the selective blockade of adrenergic neurons and its antagonism by drugs. J. Pharmacol. exp. Ther. **172**, 100—107 (1970).
MITCHELL, J. R., POTTER, W. Z., HINSON, J. A., SNODGRASS, W. R., TIMBRELL, J. A., GILLETTE, J. R.: Toxic drug reactions (this volume).
MORSELLI, P. L., RIZZO, M., GARATTINI, S.: Interaction between phenobarbital and diphenylhydantoin in animals and in epileptic patients. Ann. N. Y. Acad. Sci. **179**, 88—107 (1971).
NEUVONEN, P. J., GOTHONI, G., HACKMAN, R., BJORKSTEN, K.: Interference of iron with the absorption of tetracyclines in man. Brit. med. J. **1970 IV**, 532—534.
NIES, A. S., EVANS, G. H., SHAND, D. G.: The hemodynamic effects of *beta* adrenergic blockade on the flow-dependent hepatic clearance of propranolol. J. Pharm. exp. Ther. **184**, 716—720 (1973).
NIES, A. S., OATES, J. A.: Clinicopathologic conference: Hypertension and the lupus syndrome-revisited. Amer. J. Med. **51**, 812—814 (1971).
OATES, J. A., MITCHELL, J. R., FEAGIN, O. T., KAUFMANN, J. S., SHAND, D. G.: Distribution of guanidinium antihypertensives-mechanism of their selective action. Ann. N. Y. Acad. Sci. **179**, 302—308 (1971).
ODELL, G. B.: Studies in kernicterus. I. The protein binding of bilirubin. J. clin. Invest. **38**, 823—833 (1959).
OHNHAUS, E. E., THORGEIRSSON, S. S., DAVIES, D. S., BRECKENRIDGE, A.: Changes in liver blood flow during enzyme induction. Biochem. Pharmacol. **20**. 2561—2570 (1971).
OLESEN, O. V.: The influence of disulfiram and calcium carbimide on the serum diphenylhydantoin. Arch. Neurol. **16**, 642—644 (1967).
O'MALLEY, K., BROWNING, M., STEVENSON, I., TURNBULL, M. J.: Stimulation of drug metabolism in man by tricyclic antidepressants. Europ. J. clin. Pharmacol. **6**, 102—106 (1973).
O'REILLY, R. A.: Potentiation of anticoagulant effect by disulfiram (antabuse). Clin. Res. **19**, 180 (1971).
O'REILLY, R. A., AGGELER, P. M.: Phenylbutazone potentiation of anticoagulant effect. Fluorometric assay of warfarin. Proc. exp. Biol. (N. Y.) **128**, 1080—1083 (1968).

O'REILLY, R. A., SAHUD, M. A., ROBINSON, A. J.: Studies on the interaction of warfarin and clofibrate in man. Thrombos. Diathes. haemorrh. (Stuttg.) **27**, 309—318 (1972).
PETTINGER, W. A., OATES, J. A.: Supersensitivity to tyramine during monoamine oxidase inhibition in man. Clin. Pharmacol. Ther. **9**, 341—344 (1968).
PLACE, V. A., BENSON, H.: Dietary influences on therapy with drugs. J. Mond. Pharm. **14**, 261—278 (1971).
POUCHER, R. L., VECCHIO, T. J.: Absence of tolbutamide effect on anticoagulant therapy. J. Amer. med. Ass. **197**, 1069—1070 (1966).
RAMBOER, C., THOMPSON, R. P. H., WILLIAMS, R.: Controlled trials of phenobarbitone therapy in neonatal jaundice. Lancet **1969 I**, 966—968.
RAND, M. J., TRINKER, F. R.: The mechanism of the augmentation of responses to indirectly acting sympathomimetic amines by monoamine oxidase inhibitors. Brit. J. Pharmacol. **33**, 287—303 (1968).
REIDENBERG, M. M., AFFRIME, M.: Influence of disease on binding of drugs to plasma proteins. Ann. N. Y. Acad. Sci. **226**, 115—126 (1973).
REYES, H., LEVI, J. A., LEVINE, R., GATMAITAN, Z., ARIAS, I. M.: Bilirubin: a model for drug metabolism studies in man. Ann. N. Y. Acad. Sci. **179**, 520—528 (1971).
REYNOLDS, E. H.: Diphenylhydantoin: Hematologic aspects of toxicity. In: WOODBURY, D. M., PENRY, J. K., SCHMIDT, R. P. (Eds.): Antiepileptic drugs, Chapter 23, pp. 247—262. New York: Raven Press 1972.
RIEGELMAN, S., ROWLAND, M., EPSTEIN, W. L.: Griseofulvin-phenobarbital interaction in man. J. Amer. med. Ass. **213**, 426—431 (1970).
RIVERA-CALIMLIM, L.: Effect of chronic drug treatment on intestinal membrane transport of 14-C-*l*-dopa. Brit. J. Pharmacol. **46**, 708—713 (1972).
ROBINSON, D. S., BENJAMIN, D. M., MCCORMACK, J. J.: Interaction of warfarin and non-systemic gastrointestinal drugs. Clin. Pharmacol. Ther. **12**, 491—495 (1970).
ROBINSON, D. S., MACDONALD, M. G.: The effect of phenobarbital administration on the control of coagulation achieved during warfarin therapy in man. J. Pharmacol. exp. Ther. **153**, 250—253 (1966).
ROTHSTEIN, E.: Warfarin effect enhanced by disulfiram. J. Amer. med. Ass. **206**, 1574—1575 (1968).
ROWLAND, M.: Influence of route of administration on drug availability. J. pharm. Sci. **61**, 70—74 (1972).
ROWLAND, M., BENET, L. Z., GRAHAM, G. G.: Clearance concepts in pharmacokinetics. J. Pharmacokin. Biopharm. **1**, 123—136 (1973).
SAMUEL, P., WAITHE, W. I.: Reduction of serum cholesterol concentrations by neomycin, paraaminosalicylic acid, and other antibacterial drugs in man. Circulation **24**, 578—591 (1961).
SCHEINER, J., ALTEMEIER, W. A.: Experimental study of factors inhibiting absorption and effective therapeutic levels of declomycin. Surg. Gynec. Obstet. **114**, 9—14 (1962).
SCHELINE, R. R.: Metabolism of foreign compounds by gastrointestinal microorganisms. Pharmacol. Rev. **25**, 451—532 (1973).
SCHROGIE, J. J., SOLOMON, H. M.: The anticoagulant response to bishydroxycoumarin. II. The effect of D-thyroxine, clofibrate, and norethandrolone. Clin. Pharm. Ther. **8**, 70—77 (1967).
SELLERS, E. M., LANG, M., KOCH-WESER, J., LEBLANC, E., KALANT, H.: Interaction of chloral hydrate and ethanol in man. I. Metabolism. Clin. Pharmacol. Ther. **13**, 37—49 (1972a).
SELLERS, E. M., CARR, G., BERNSTEIN, J. G., SELLERS, S., KOCH-WESER, J.: Interaction of chloral hydrate and ethanol in man. II. Hemodynamics and performance. Clin. Pharmacol. Ther. **13**, 50—58 (1972b).
SELLERS, E. M., KOCH-WESER, J.: Potentiation of warfarin-induced hypoprothrombinemia by chloral hydrate. New Engl. J. Med. **283**, 827—831 (1970).
SILVERMAN, W. A., ANDERSEN, D. H., BLANC, W. A., CROZIER, D. N.: A difference in mortality rate and incidence of kernicterus among premature infants allotted to two prophylactic antibacterial regimens. Pediatrics **18**, 614—625 (1956).
SIURALA, M., MUSTALA, O., JUSSILA, J.: Absorption of acetylsalicylic acid by a normal and an atrophic gastric mucosa. Scand. J. Gastroenterol. **4**, 269—273 (1969).
SJÖQVIST, F., BERGLUND, F., BORGA, O., HAMMER, W., ANDERSSON, S., THORSTRAND, C.: The pH-dependent excretion of monomethylated tricyclic antidepressants. Clin Pharmacol. Ther. **10**, 826—833 (1969).
SMITH, C. B.: The role of monoamine oxidase in the intraneuronal metabolism of noradrenalin released by indirectly acting sympathomimetic amines or by adrenergic nerve stimulation J. Pharmacol. exp. Ther. **151**, 207—220 (1966).
SOLOMON, H. M., ABRAMS, W. B.: Interactions between digitoxin and other drugs in man. Amer. Heart J. **83**, 277—280 (1972).

SOLOMON, H. M., SCHROGIE, J. J.: The effect of phenyramidol on the metabolism of bishydroxycoumarin. J. Pharmacol. exp. Ther. **154**, 660—666 (1966).

SOLOMON, H. M., SCHROGIE, J. J.: Effect of phenyramidol and bishydroxycoumarin on the metabolism of tolbutamide in human subjects. Metabolism **16**, 1029—1033 (1967).

SOLOMON, H. M., SCHROGIE, J. J., WILLIAMS, D.: The displacement of phenylbutazone-^{14}C and warfarin-^{14}C from human albumin by various drugs and fatty acids. Biochem. Pharmacol. **17**, 143—151 (1968).

SOLOMON, H. M., REICH, S., SPIRT, N., ABRAMS, W. B.: Interactions between digitoxin and other drugs *in vitro* and *in vivo*. Ann. N. Y. Acad. Sci. **179**, 362—369 (1971).

STENSON, R. E., CONSTANTINO, R. T., HARRISON, D. C.: Interrelationships of hepatic blood flow, cardiac output, and blood levels of lidocaine in man. Circulation **43**, 205—211 (1971).

THOMPSON, G. R.: Inhibitory effect of neomycin on cholesterol absorption in germ-free pigs. Gut **11**, 1063 (1970).

THOMPSON, G. R., BARROWMAN, J., GUTIERREZ, L.: Action of neomycin on the intralumen phase of lipid absorption. J. clin. Invest. **50**, 319—323 (1971).

THOMSON, P. D., ROWLAND, M., MELMON, K. L.: The influence of heart failure, liver disease, and renal failure on the disposition of lidocaine in man. Amer. Heart J. **82**, 417—421 (1971).

TOSKES, P. P., DEREN, J. J.: Selective inhibition of vitamin B_{12} absorption by para-aminosalicylic acid. Gastroenterology **62**, 1232—1237 (1972).

TRINKER, F. R., RAND, M. J.: The effect of nialamide, pargyline, and tranylcypromine on the removal of amphetamine by the perfused liver. J. Pharm. Pharmacol. **22**, 496—499 (1970).

TROLLE, D.: Phenobarbitone and neonatal icterus. Lancet **1**, 251—252 (1968a).

TROLLE, D.: Decrease of total serum-bilirubin concentration in newborn infants after phenobarbitone treatment. Lancet **1968 II**, 705—708.

UDALL, J. A.: Letter to editor regarding Chloral hydrate and warfarin therapy. Ann. intern. Med. **75**, 141—142 (1971).

VESELL, E. S., PAGE, J. G.: Genetic control of the phenobarbital-induced shortening of plasma antipyrine half-lives in man. J. clin. Invest. **48**, 2202—2209 (1969).

VESELL, E. S., PAGE, J. G., PASSANANTI, G. T.: Genetic and environmental factors affecting ethanol metabolism in man. Clin. Pharmacol. Ther. **12**, 192—201 (1971).

VESELL, E. S., PASSANANTI, G. T., GREENE, F. E.: Impairment of drug metabolism in man by allopurinol and nortriptyline. New Engl. J. Med. **283**, 1484—1488 (1970).

WARDELL, W.: In: MORSELLI, P. L., GARATTINI, S., COHEN, S. N. (Eds.): Drug interactions. New York: Raven Press (in press).

WEBB, D. I., CHODOS, R. B., MAHAR, C. Q., FALOON, W. W.: Mechanism of vitamin B_{12} malabsorption in patients receiving colchicine. New Engl. J. Med. **279**, 845—850 (1968).

WELCH, R. M., HARRISON, Y. E., CONNEY, A. H., BURNS, J. J.: An experimental model in dogs for studying interactions of drugs with bishydroxycoumarin. Clin. Pharmacol. Therap. **10**, 817—825 (1969).

WELCH, R. M., HARRISON, Y. E., GOMMI, B. W., POPPERS, P. J., FINSTER, M., CONNEY, A. H.: Stimulatory effect of cigarette smoking on the hydroxylation of 3,4-benzpyrene and the *N*-demethylation of 3-methyl-4-monomethylaminoazobenzene by enzymes in human placenta. Clin. Pharmacol. Ther. **10**, 100—109 (1969).

WERK, E. E., JR., MACGEE, J., SHOLITON, L. J.: Effect of diphenylhydantoin on cortisol metabolism in man. J. clin. Invest. **43**, 1824—1835 (1964).

WERK, E. E., JR., THRASHER, K., SHOLITON, L. J., OLINGER, C., CHOI, Y.: Cortisol production in epileptic patients treated with diphenylhydantoin. Clin. Pharmacol. Ther. **12**, 698—703 (1971).

WILSON, J. T.: Developmental pharmacology: a review of its application to clinical and basic science. Ann. Rev. Pharmacol. **12**, 423—450 (1972).

WINKLER, K., KEIDING, S., TYGSTRUP, N.: Clearance as a guantitative measure of liver function. In: BAUMGARTNER, G., PREISIG, R. (Eds.): The liver: quantitative aspects of structure and function, pp. 144—155. Basel: S. Karger 1973.

YAFFE, S. J., LEVY, G., MATSUZAWA, T., BALIAH, T.: Enhancement of glucuronide conjugating capacity in a hyperbilirubinemic infant due to apparent enzyme induction by phenobarbital. New Engl. J. Med. **275**, 1461—1465 (1966).

YEUNG, C. Y., FIELD, C. E.: Phenobarbitone therapy in neonatal hyperbilirubinaemia. Lancet **1969 II**, 135—139.

ZUBROD, C. G., KENNEDY, T. J., SHANNON, J. A.: Studies on the chemotherapy of the human malarias. VIII. The physiological disposition of pamaquine. J. Clin. Invest. 27: Symposium on Malaria, pp. 114—120 (1948).

Chapter 70

Interactions of Cardiovascular Drugs at the Receptor Level*

L. I. GOLDBERG

I. Introduction

Because of the relative ease of measurement of response, cardiovascular drugs are excellent tools for ascertaining interactions at the receptor. The precise nature of the receptors in the cardiovascular system and the events leading to contraction of muscle cells in the heart or in the blood vessels must be included among the most important unanswered questions of molecular biology. At present, most of our information about receptors and drug-receptor interactions is obtained from mathematical interpretations of dose response curves (ARIENS et al., 1964). The present survey is based on such indirect information and is primarily concerned with potentially significant interactions that could occur when two or more cardiovascular drugs are administered simultaneously.

II. Drugs that Increase Myocardial Contractility

Despite obvious needs, physicians have remarkably few drugs that they can use to increase myocardial contractility. Although digitalis was introduced into therapeutics more than 100 years ago, it is still the primary drug used for the treatment of congestive heart failure. The only other myocardial stimulants available for clinical use are sympathomimetic amines acting on *beta*-adrenergic receptors, methylxanthine derivatives and, more recently, glucagon. Why more cardiac stimulating drugs have not been developed is an interesting question in itself. One answer may be that until quite recently digitalis was thought to exert actions that could not be duplicated by any other drug, and tremendous effort was expended in extracting and testing chemically different cardiac glycosides from plant and animal sources (HOCH, 1961). Unfortunately, all cardiac glycosides, both natural and chemically modified, appear to possess similar pharmacologic and toxic actions (WALTON and GAZES, 1971).

Cardiac glycosides were considered unique because studies had suggested that these drugs could increase myocardial contractility without increasing utilization of oxygen. More recent investigations of isolated myocardial tissues have demonstrated that this is not the case (GOUSIOS et al., 1967). The early work was based on evidence obtained in the intact failed heart, which becomes smaller under the influence of digitalis and thus uses less oxygen (BING et al., 1950). It is now apparent that there is no difference among the positive inotropic drugs with respect to the amount of oxygen required per unit of increased contractility, and that differences in myocardial oxygen utilization reported to occur with the various agents were the results of effects on heart rate and peripheral resistance

* Supported in part by U. S. Public Health Service grants GM-14270, HE-06491, and GM-1543.

(GOLDBERG, 1968). After considering the influence of non-cardiac actions, it is necessary to determine the relationship between the amount of a positive inotropic agent required to increase myocardial contractility and the amount required to produce arrhythmias. Once this relationship has been established for a specific drug, it should then be possible to determine the influence of other agents acting by different mechanisms. Possible interactions include addition, antagonism or synergism of positive inotropic actions, alterations of the contractile force/arrhythmic dose ratio, and influence of non-cardiac effects.

A. Cardiac Glycosides

Despite extensive investigations, the precise mechanisms responsible for the positive inotropic actions and the arrhythmic actions of cardiac glycosides are not known (LEE and KLAUS, 1971). ATPase, an enzyme dependent on sodium and potassium, has an extremely high affinity for cardiac glycosides. Some investigators speculate that ATPase is the receptor for the positive inotropic action of cardiac glycosides (BESCH et al., 1970). On the other hand, alternative hypotheses suggest that this enzyme is not involved with positive inotropic effects, but with production of arrhythmias (LEE and KLAUS, 1971). The preponderance of evidence today suggests that the final result of digitalis action is a change in calcium flux, and extensive research is underway to localize the site of calcium movement in the myocardium.

1. Influence of Cations

a) Potassium

Reduction of serum potassium levels lowers the threshold of digitalis-induced arrhythmias. Thus, potassium-wasting diuretics, such as thiazides, mercurials, and carbonic anhydrase inhibitors, increase the toxicity of cardiac glycosides. Conversely, the potassium-retaining diuretics, spironolactone and triamterene, may decrease toxicity. Administration of potassium salts is one of the treatments of choice for digitalis arrhythmias, but such therapy does not reverse the conduction defects produced by digitalis (FISCH et al., 1960). However, potassium does not completely reverse digitalis toxicity, which indicates that cardiac glycosides may act by more than one mechanism.

b) Calcium

Calcium increases the toxicity of digitalis, but there is controversy concerning whether calcium and digitalis have synergistic or even additive positive inotropic actions (FARAH and WITT, 1963). Calcium and digitalis are not often administered to the same patient, but occasionally the two are used in attempts to resuscitate the myocardium after cardiopulmonary bypass surgery. Chelation of calcium by use of EDTA has been employed to treat digitalis-induced arrhythmias (COHEN et al., 1959).

c) Magnesium

Magnesium ions produce effects opposite to those of calcium and antagonize the arrhythmias produced by digitalis (STANBURY and FARAH, 1950). Conversely, digitalis intoxication may occur in alcoholism, after diuresis, and in other conditions with low serum magnesium levels (SELLER et al., 1970).

2. Specific Digitalis Blocking Agents

Investigators have been intensely interested in developing a drug that will specifically antagonize the arrhythmic effects of digitalis. Two potentially important advances have been reported recently. SCHMIDT and BUTLER (1971) developed a digoxin antibody that specifically antagonizes all effects of this cardiac glycoside in animals. YEH et al. (1972) reported that a spironolactone analog, potassium canrenoate, selectively antagonized digitalis-induced arrhythmias.

3. Relationship of Digitalis Blood Levels to Cardiac Activity

The development of specific radioimmune assays for measurements of serum levels of cardiac glycosides has made it possible to correlate the arrhythmias produced by these drugs with serum levels. Although there is a general relationship between high levels of cardiac glycosides and arrhythmias, considerable overlap has been found in levels detected in toxic and non-toxic patients (SMITH et al., 1969; FOGELMAN et al., 1971).

The correlation between positive inotropic effect and blood levels is even more difficult because of inherent problems in measuring myocardial contractility in intact humans. Studies carried out with a non-invasive technique for measuring systolic time intervals suggest that the half-time for the positive inotropic effect of large doses of digitalis is similar to the half-life of the glycoside in the serum (DOHERTY, 1968). As with the toxic doses, there appears to be a relationship between the degree of shortening of the systolic time interval and blood levels, but marked inconsistencies occur in individual patients (SHAPIRO et al., 1970; CARLINER et al., 1974b). Marked variations in pathology, myocardial electrolyte content, and effects of other drugs probably prevent prediction of the magnitude of digitalis response solely on the basis of serum levels.

B. Sympathomimetic Amines

The positive inotropic action of sympathomimetic amines is much more clearly related to action on a specific receptor. Although other investigators had suggested specific receptors for sympathomimetic amines, AHLQUIST (1948) clearly classified the receptors on the basis of response. In the cardiovascular system, *alpha* receptors mediate vasoconstriction and *beta* receptors produce vasodilation and cardiac stimulation. More recently, LANDS et al. (1967) suggested that *beta* receptors should be divided into *beta*$_1$ (cardiac stimulation) and *beta*$_2$ (vasodilation). The appropriateness of this subdivision has been confirmed by discovery of selective *beta*$_1$ and *beta*$_2$ agonists and antagonists (LEVY, 1966; DUNLOP and SHANKS, 1968).

Progress has also been made at the molecular level. Sympathomimetic amines activate cyclic AMP in some tissues, and ROBISON et al. (1971) have speculated that adenyl cyclase represents the beta-adrenergic receptor. More recent evidence opposes this hypothesis and suggests that activation of cyclic AMP is only one of many steps that follows the interaction of a sympathomimetic amine with the beta-adrenergic receptor (MAYER, 1970). LEFKOWITZ and HABER (1971) reported the isolation of the beta receptor in cardiac muscle as a complex protein with an extremely high affinity for catecholamines. Proof of this identification requires demonstration of a characteristic beta-adrenergic response resulting from interaction of a sympathomimetic amine with this protein.

The widespread use of *beta*-adrenergic blocking agents for the treatment of arrhythmias, angina pectoris, and hypertension has created a group of patients in whom the usual dose of sympathomimetic amines will not exert typical responses. All of the available *beta*-adrenergic blocking agents, however, are competitive inhibitors, and thus it should be possible to overcome the effect of the blocking agents in the event sympathomimetic therapy is required.

DOLLERY et al. (1971) demonstrated the metabolic transformation of isoproterenol to a weak *beta*-adrenergic blocking agent, 3-methoxy-isoproterenol. Although this finding does not appear to have clinical significance, the endogenous production of more potent *beta*-adrenergic blocking agents from other drugs is possible. In this regard, SHAND et al. (1970) demonstrated that an active metabolite is formed after oral, but not after intravenous, administration of propranolol.

C. Methylxanthines

A definitive receptor for methylxanthines has not been described, but a common mechanism beyond the receptor has been postulated to explain the positive inotropic actions of methylxanthines and sympathomimetic amines. Methylxanthines inhibit phosphodiesterase and thus increase levels of cyclic AMP. As with sympathomimetic amines, however, it is uncertain whether elevation of cyclic AMP levels is definitively responsible for the positive inotropic action (MAYER, 1970).

D. Glucagon

The pancreatic polypeptide, glucagon, also exerts a positive inotropic effect that is not blocked by *beta*-adrenergic blocking agents (GLICK et al., 1968; LUCCHESI, 1968). Because cyclic AMP is also increased by glucagon, a relationship between activation of this nucleotide and positive inotropic actions is again suggested (GLICK, 1972). Glucagon increases cardiac output of some patients in shock and produces beneficial effects in some patients with chronic congestive heart failure. There is considerable controversy, however, concerning the clinical utility of this agent (SIMANIS and GOLDBERG, 1971). Interestingly, animal experimentation and investigation has suggested that the failed myocardium does not respond adequately to glucagon, possibly due to inability of the heart to respond by increasing cyclic AMP (LEVEY et al., 1970). More recently, glucagon administration to animals has resulted in deposition of a material with glucagon-like characteristics in the lung, and clinical studies have been terminated until significance of these findings have been determined (GLICK, 1972). Nevertheless, the finding of a positive inotropic action of glucagon may lead to the development of more appropriate polypeptides for clinical use.

E. Interaction: Therapeutic and Adverse

From the therapeutic viewpoint, there appears to be little reason at the myocardial level to administer more than one drug from a single classification. The sympathomimetic amines offer the greatest diversity of possible peripheral effects, and combinations of drugs such as isoproterenol, norepinephrine and dopamine have been utilized for this reason (GOLDBERG and TALLEY, 1971). Combined utilization of drugs of the different classes appears more rational. A few examples follow.

Aminophylline has been used for many years in combination with cardiac glycosides in treatment of patients with heart failure unresponsive to usual regimens. Reported beneficial effects, however, may be due to more than combined myocardial actions of the two drugs, since aminophylline also exerts bronchial dilating and diuretic responses.

Sympathomimetic amines and cardiac glycosides exert diverse cardiac effects and it is sometimes beneficial to administer these drugs concurrently. Digitalis prolongs transmission from the atrium to the ventricles, whereas sympathomimetic amines decrease this interval. Thus, in the patient with atrial fibrillation, cardiac glycosides may be used to prevent the excessive ventricular rate, which may be precipitated by sympathomimetic amines. The combined use of digitalis and sympathomimetic amines for the treatment of chronic congestive heart failure has not been adequately explored because of unavailability of an orally active sympathomimetic amine with beneficial peripheral actions. Dopamine increases cardiac output and renal blood flow and can cause a marked sodium diuresis in patients with severe congestive heart failure, but is restricted to intravenous use (GOLDBERG, 1972). L-dopa will increase sodium excretion when taken orally (FINLAY et al., 1971), but there is a tolerance to the effect, possibly because of decreased production of dopamine (TJANDRAMAGA et al., 1973).

Finally, the myocardium of patients treated with *beta*-adrenergic blocking agents could be stimulated by glucagon, digitalis, or methylxanthines rather than by very large doses of sympathomimetic amines.

The most dangerous adverse effect of cardiac stimulants is production of myocardial ischemia and arrhythmias, especially in patients with coronary insufficiency. It is not always possible to predict the result of combined administration of two myocardial stimulants. For example, although both isoproterenol and cardiac glycosides lower the threshold of excitability of the myocardium to ventricular arrhythmias, administration of isoproterenol could convert ventricular tachycardia produced by digitalis to a normal sinus rhythm. In this case, isoproterenol stimulation of the SA node increases the sinus rate to such an extent that sinus rhythm may predominate over the slower ventricular pacemaker. On the other hand, administration of norepinephrine in a dose that elevates blood pressure and increases vagal activity could induce arrhythmias.

III. Drugs Acting on Blood Vessels

Most chemical substances, when administered in a sufficiently large dose, will produce an effect on blood vessels. As with the myocardial stimulants, only limited information is available concerning the molecular bases of drug-receptor interactions in the blood vessel. The ultimate mechanism is considered to be the result of ionic changes (SOMLYO and SOMLYO, 1970).

A. Vasoconstricting Agents

1. Sympathomimetic Amines

Sympathomimetic amines cause vasoconstriction in both arterial and venous vascular beds by acting on *alpha*-adrenergic receptors. These effects are specifically antagonized by phenoxybenzamine, phentolamine and other *alpha*-adrenergic blocking agents. Unlike *beta*-adrenergic blocking agents, *alpha*-adrenergic blocking agents do not bear close structural similarities to sympathomimetic amines. Phenoxybenzamine is a non-competitive blocking agent and exerts its effect for a prolonged period of time. Phentolamine is non-competitive and has a shorter duration of action. A number of drugs acting on the central nervous system exert *alpha*-adrenergic blocking activity, the most prominent of which are phenothiazines and butyrophenones. The combined use of sympathomimetic amines with these drugs has both desirable and undesirable consequences. For example, norepinephrine exerts *beta*-adrenergic effects on the heart and *alpha*-adrenergic effects on the peripheral vessels, and thus combined use of *alpha*-adrenergic blocking agents could reduce its effectiveness as a vasopressor agent. On the other hand, if more cardiac action is desired, unwanted vasoconstriction could be reduced by administration of *alpha*-adrenergic blocking agents. Epinephrine exerts both cardiac and peripheral *beta* effects, and administration of an *alpha*-adrenergic blocking agent could unmask *beta*-adrenergic vasodilation and result in hypotension. Thus, epinephrine should be cautiously used in patients treated with phenothiazines.

2. Angiotensin

The polypeptide, angiotensin, is one of the most potent vasoconstrictor agents known. Its role in the initiation of renal vascular hypertension and other disease states including essential hypertension has been extensively investigated. The recent discovery of a polypeptide that specifically blocks the vasoconstrictor actions of angiotensin will undoubtedly contribute to this research. This discovery has removed angiotensin from the classification of non-specific vasoconstrictor to that of a drug with a known receptor (PALS et al., 1971). Results from human studies with angiotensin blocking agents have not yet been reported.

3. Ergot Derivatives

Gangrene caused by rye contaminated with a parasitic fungus, Claviceps Purpurea, has been reported for thousands of years. This mysterious ailment was eventually found to be the result of intense vasoconstriction produced by ergot alkaloids contaminating the grain. Ergot alkaloids are also *alpha*-adrenergic blocking agents, but this effect does not influence the intense vasoconstriction seen with large overdoses. Ergot poisoning is occasionally seen today from ingestion of contaminated grain or as a result of suicidal or accidental ingestion of more purified alkaloids.

The danger of addition of other vasoconstricting agents to the regimen of patients receiving an ergot alkaloid must always be considered. This admonition is pertinent because many patients with migraine are treated with relatively large doses of ergot alkaloids over an extended period of time (CRANLEY et al., 1963). There is no specific ergot blocking agent, but patients with ergotism can be treated with sodium nitroprusside and other vasodilators acting directly on smooth muscle (CARLINER et al., 1974a).

B. Vasodilating Agents

The utilization of vasodilating agents for the treatment of hypertension and as adjunctive therapy for the treatment of shock has greatly increased in recent years. A number of very potent "non-specific" vasodilating agents are currently being investigated for the treatment of hypertension (GILMORE et al., 1970; CLARK and GOLDBERG, 1972). In addition, several prostaglandins have been synthesized, many of which have potentially useful vasodilating properties (WEEKS, 1972).

A promising new area is the development of drugs to improve blood flow to selective vascular beds. In attaining this goal, it may be advantageous to use two or more vasodilators acting by different mechanisms, and even to combine vasoconstrictor, vasodilator and positive inotropic agents. The advantages of combined therapy for the treatment of shock has recently been discussed (GOLDBERG and TALLEY, 1971). On the other hand, combined use of two or more vasodilating agents may produce pronounced hypotension that could lead to cerebral or myocardial ischemia.

1. Sympathomimetic Amines

Sympathomimetic amines cause vasodilation by action on *beta*-adrenergic receptors and by a less defined mechanism that must be termed non-specific at the present time. The differential vasodilation caused by dopamine will be discussed in a separate section.

In general, *beta*-adrenergic vasodilation is most prominent in the skeletal muscle and splanchnic vascular beds and is relatively weak in skin and kidney. Thus, when drugs such as isoproterenol are administered intravenously, cardiac output is augmented and blood flow increases to the muscles and splanchnic vascular beds. Blood flow to the kidneys usually does not increase (ROSENBLUM et al., 1968).

The action of *beta*-adrenergic agonists on veins is not completely understood. Studies of isolated veins have shown a vasodilating action (HUGHES and VANE, 1967), but venoconstriction usually occurs in man probably because of reflex vasoconstriction initiated by the reduction in arterial pressure (KAISER et al., 1964). The vasodilating actions of *beta*-adrenergic agonists are completely blocked by *beta*-adrenergic blocking agents, and as stated above, this effect can be more selectively antagonized by specific *beta*$_2$-antagonists.

A nonspecific vasodilating action has been reported with large doses of several sympathomimetic amines (GOLDBERG et al., 1968). Insufficient data is available to derive structure-activity relationships. A detailed investigation of one of these amines, mephentermine, in the dog demonstrated similar vasodilating actions in the renal, mesenteric and femoral vascular beds. The vasodilation was not blocked by *beta*-adrenergic blocking agents or other classical antagonists (CALDWELL and GOLDBERG, 1970). In usual clinical doses, this vasodilation is masked by *alpha*-adrenergic vasoconstriction; however, a further fall in blood pressure may occur in patients in shock, and in those receiving *alpha*-adrenergic blocking agents.

2. Dopamine

Studies of HOLTZ and CREDNER (1942) suggested that dopamine exerted different vascular effects than other sympathomimetic amines. These investigators found that dopamine was a depressor in the guinea pig and rabbit but a vasopressor in the rat and cat. It is now known that dopamine is a vasodepressor in all species, including man, if administered after an *alpha*-adrenergic blocking agent (MCNAY et al., 1966).

Extensive investigations have shown that dopamine causes renal and mesenteric vasodilation by action on specific receptors (MCNAY and GOLDBERG, 1966; EBLE, 1964). Vasodilation has also been observed in the coronary vascular bed, but only after administration of large doses of *alpha*- and *beta*-adrenergic blocking agents (SCHUELKE et al., 1971). The selective renal vasodilating actions of dopamine have resulted in clinical trials for the treatment of shock, congestive heart failure, hypertension, and drug intoxication (GOLDBERG, 1972). The increased renal blood flow produced by dopamine is limited by *alpha*-adrenergic vasoconstriction, which occurs when large doses of the amine are administered. Accordingly, phenoxybenzamine has been used with dopamine to produce even greater increases in renal blood flow.

The vasodilating actions of dopamine are specifically antagonized by butyrophenones (YEH et al., 1969), phenothiazines (TSENG and WALACZEK, 1970), and bulbocapnine (Goldberg and Musgrave, 1971). Apomorphine, which is structurally similar to dopamine, appears to be both a dopamine agonist and antagonist (Goldberg et al., 1968; Goldberg and Musgrave, 1971). A study of more than 100 dopamine analogs has demonstrated that the structural requirements for activation of dopamine receptors are extraordinarily specific and thus far only the *n*-methyl derivative has been shown to exert typical dopamine-like activity (Goldberg et al., 1968).

3. Cholinergic Drugs

Acetylcholine and other drugs acting on the cholinergic receptor causes vasodilation in all vascular beds. Hypotension due to vasodilation, however, is only rarely encountered (Rongey and Weisman, 1972), because of reflex vasoconstriction and because bradycardia limits the use of large doses of acetylcholine. The vasodilating actions of cholinergic agents are completely blocked by atropine.

4. Histamine

Histamine is an arteriolar dilating agent in man and most animals, but is a vasoconstrictor in rodents (Roth and Tabachnick, 1971). Cerebral vessels are extremely sensitive to histamine and severe headaches may occur in man. Histamine is a potent capillary dilator and characteristic skin reactions may be produced. Antihistamines antagonize the arteriolar vasodilation produced by histamine, but they are not as effective against the capillary vasodilation. Epinephrine, although not a specific antagonist of histamine, is extremely valuable in treating the bronchial constriction and vasodilation seen in anaphylactic shock.

IV. Summary

This survey describes the mechanisms of action of a number of cardiovascular drugs and indicates some of the potentially significant interactions that could occur when these drugs are administered together. Most of the interactions described can be explained by the known pharmacology of the individual drugs. Investigations of unexpected reactions could lead to new information of both clinical and basic significance.

References

Ahlquist, R. P.: A study of adrenotropic receptors. Amer. J. Physiol. **153**, 586—600 (1948).
Ariens, E. J., Simanis, A. M., van Rossum, J. M.: Drug receptor interaction: interaction of one or more drugs with the receptor system. In: Molecular pharmacology. New York: Academic Press 1964.
Besch, H. R., Jr., Allen, J. C., Glick, G., Schwartz, A.: Correlation between the inotropic action of ouabain and its effects on subcellular enzyme systems from canine myocardium. J. Pharmacol. exp. Ther. **171**, 1—12 (1970).
Bing, R. J., Maraist, F. M., Dammann, J. F., Jr., Draper, A., Jr., Heimbecker, R., Daley, R., Gerard, R., Calazel, P.: Effect of strophanthus on coronary blood flow and cardiac oxygen consumption of normal and failing hearts. Circulation **2**, 513—516 (1950).
Caldwell, R. W., Goldberg, L. I.: An evaluation of the vasodilation produced by mephentermine and certain other sympathomimetic amines. J. Pharmacol. exp. Ther. **172**, 297—309 (1970).
Carliner, N. A., Denune, D. P., Finch, C. S., Jr., Goldberg, L. I.: Sodium nitroprusside treatment of ergotamine-induced peripheral ischemia. J. Amer. Med. Ass. **227** (3): 308—309 (1974a).
Carliner, N. H., Gilbert, C. A., Pruitt, A. W., Goldberg, L. I.: Effects of maintenance digoxin therapy on systolic time intervals and serum digoxin concentrations. Circulation **50**: 94—98 (1974b).

CLARK, D. W., GOLDBERG, L. I.: Guancydine: A new antihypertensive agent: use with quinethazone and guanethidine or propranolol. Ann. intern. Med. **76**, 579—585 (1972).
COHEN, B. D., SPRITZ, N., LUBASH, G. D., RUBIN, A. L.: Use of a calcium chelating agent (NaEDTA) in cardiac arrhythmias. Circulation **19**, 918—927 (1959).
CRANLEY, J. J., KRAUSE, R. J., STRASSER, E. S., HAFNER, C. D.: Impending gangrene of four extremities secondary to ergotism. New Engl. J. Med. **269**, 727—729 (1963).
DOHERTY, J. E.: The clinical pharmacology of digitalis glycosides: A review. Amer. J. med. Sci. **255**, 382—414 (1968).
DOLLERY, C. T., DAVIES, D. S., CONOLLY, M. E.: Differences in the metabolism of drugs depending upon their routes of administration. Ann. N.Y. Acad. Sci. **179**, 108—112 (1971).
DUNLOP, D., SHANKS, R. G.: Selective blockade of adrenoceptive beta receptors in the heart. Brit. J. Pharmacol. **32**, 201—218 (1968).
EBLE, J. N.: A proposed mechanism for the depressor effect of dopamine in the anesthetized dog. J. Pharmacol. exp. Ther. **145**, 64—70 (1964).
FARAH, A., WITT, P. N.: Cardiac glycosides and calcium. In: WILBRANDT, W., LINDGREN, P. (Eds.): New aspects of cardiac glycosides, Vol. 3 of Proc. 1st Int. Pharmacol. Meeting, pp. 137—171. New York: MacMillan 1963.
FINLAY, G. D., WHITSETT, T. L., CUCINELL, E. A., GOLDBERG, L. I.: Augmentation of sodium and potassium excretion, glomerular filtration rate and renal plasma flow by levodopa. New Engl. J. Med. **284**, 865—870 (1971).
FISCH, C., MARTZ, B. L., PRIEBE, F. H.: Enhancement of potassium-induced atrioventricular block by toxic doses of digitalis drugs. J. clin. Invest. **39**, 1885—1893 (1960).
FOGELMAN, A. M., LAMONT, J. T., FINKELSTEIN, S., RADO, E., PEARCE, M. L.: Fallibility of plasma-digoxin in differentiating toxic from non-toxic patients. Lancet **1971 II**, 727—729.
GILMORE, E., WEIL, J., CHIDSEY, C.: Treatment of essential hypertension with a new vasodilator in combination with beta-adrenergic blockade. New Engl. J. Med. **282**, 521—527 (1970).
GLICK, G.: Glucagon: A perspective. Circulation **45**, 513—515 (1972).
GLICK, G., PARMLEY, W. W., WECHSLER, A. S., SONNENBLICK, E. H.: Glucagon: its enhancement of cardiac performance in the cat and dog and persistence of its inotropic action despite beta-receptor blockade with propranolol. Circulation Res. **22**, 789—799 (1968).
GOLDBERG, L. I.: Use of sympathomimetic amines in heart failure. Amer. J. Cardiol. **22**, 177—182 (1968).
GOLDBERG, L. I.: Cardiovascular and renal actions of dopamine: Potential clinical applications. Pharmacol. Rev. **24**, 1—29 (1972).
GOLDBERG, L. I., MUSGRAVE, G. E.: Selective attenuation of dopamine-induced renal vasodilation by bulbocapnine and apomorphine. Pharmacologist **13**, 227 (1971).
GOLDBERG, L. I., SONNEVILLE, P. F., MCNAY, J. L.: An investigation of the structural requirements for dopamine-like renal vasodilation: Phenylethylamines and apomorphine. J. Pharmacol. exp. Ther. **163**, 188—197 (1968).
GOLDBERG, L. I., TALLEY, R. C.: Current therapy of shock. Advan. int. Med. **17**, 363—378 (1971).
GOUSIOS, A. G., FELTS, J. M., HAVEL, R. J.: Effects of ouabain on force of contraction, oxygen consumption, and metabolism of free fatty acids in the perfused rabbit heart. Circulation Res. **21**, 445—448 (1967).
HOCH, J. H.: A survey of cardiac glycosides and genins. Columbia, S. C.: University of South Carolina Press 1961.
HOLTZ, P., CREDNER, K.: Die enzymatische Entstehung von Oxytyramin im Organismus und die physiologische Bedeutung der Dopadecarboxylase. Naunyn-Schmiedeberg's Arch. exp. Path. Pharmak. **200**, 356—388 (1942).
HUGHES, J., VANE, J. R.: An analysis of the responses of the isolated portal vein of the rabbit to electrical stimulation and to drugs. Brit. J. Pharmacol. **30**, 46—66 (1967).
KAISER, G. A., ROSS, J., JR., BRAUNWALD, E.: *Alpha* and *beta* adrenergic receptor mechanisms in the systemic venous bed. J. Pharmacol. exp. Ther. **144**, 156—162 (1964).
LANDS, A. M., ARNOLD, A., MCAULIFF, J. P., LUDUENA, F. P., BROWN, T. G., JR.: Differentiation of receptor systems activated by sympathomimetic amines. Nature (Lond.) **214**, 597—598 (1967).
LEE, K. S., KLAUS, W.: The subcellular basis for the mechanism of inotropic action of cardiac glycosides. Pharmacol. Rev. **23**, 193—261 (1971).
LEFKOWITZ, R. J., HABER, E.: A faction of the ventribular myocardium that has the specificity of the cardiac beta-adrenergic receptor. Proc. nat. Acad. Sci. (Wash.) **68**, 1773—1777 (1971).
LEVEY, G. S., PRINDLE, K. H., JR., EPSTEIN, S. E.: Effects of glucagon on adenyl cyclase activity in left and right ventricles and liver in experimentally-produced isolated right ventricular failure. J. molec. Cell Cardiol. **1**, 403—410 (1970).

LEVY, B.: Dimethyl isopropylmethoxamine: a selective beta-receptor blocking agent. Brit. J. Pharmacol. **27**, 277—285 (1966).
LUCCHESI, B. R.: Cardiac actions of glucagon. Circulation Res. **22**, 777—787 (1968).
MAYER, S. E.: Adenyl cyclase as a component of the adrenergic receptor. In: PORTER, R., O'CONNOR, M. (Eds.): Symposium on molecular properties of drug receptors. London: J. A. Churchill 1970.
MCNAY, J. L., BOGAERT, M., GOLDBERG, L. I.: Hypotensive effects of dopamine after phenoxybenzamine in experimental animals and man. Abst. III. Int. Cong. Pharmacol. Sao Paolo, Brazil, p. 130 (1966).
MCNAY, J. L., GOLDBERG, L. I.: Comparison of the effects of dopamine, isoproterenol, norepinephrine and bradykinin on canine renal and femoral blood flow. J. Pharmacol. exp. Ther. **151**, 23—31 (1966).
PALS, D. T., MASUCCI, F. D., SIPOS, F., DENNING, G. S., JR.: A specific competitive antagonist of the vascular action of angiotensin. II. Circulation Res. **29**, 664—672 (1971).
ROBISON, G. A., BUTCHER, R. W., SUTHERLAND, E. W.: Cyclic AMP. New York: Academic Press 1971.
RONGEY, K. A., WEISMAN, H.: Hypotension following intraocular acetylcholine. Anesthesiology **36**, 412 (1972).
ROSENBLUM, R., BERKOWITZ, W. D., LAWSON, D.: Effect of acute intravenous administration of isoproterenol on cardiorenal hemodynamics in man. Circulation **38**, 158—168 (1968).
ROTH, F. E., TABACHNICK, I. A.: Histamine and antihistamines. In: Drill's pharmacology in medicine, 4th ed. New York: McGraw-Hill 1971.
SCHMIDT, D. H., BUTLER, V. P., JR.: Reversal of digoxin toxicity with specific antibodies. J. clin. Invest. **50**, 1738—1744 (1971).
SCHUELKE, D. M., MARK, A. L., SCHMID, P. G., ECKSTEIN, J. W.: Coronary vasodilatation produced by dopamine after adrenergic blockade. J. Pharmacol. exp. Ther. **176**, 320—327 (1971).
SELLER, R. H., CANGIANO, J., KIM, K. E., MENDELSSOHN, S., BREST, A. N., SWARTZ, C.: Digitalis toxicity and hypomagnesemia. Amer. Heart J. **79**, 57—68 (1970).
SHAND, D. G., NUCKOLLS, E. M., OATES, J. A.: Plasma propranolol levels in adults with observation in four children. Clin. Pharmacol. Ther. **11**, 112—120 (1970).
SHAPIRO, W., NARAHARA, K., TAUBERT, K.: Relationship of plasma digitoxin and digoxin to cardiac response following intravenous digitalization in man. Circulation **42**, 1065—1072 (1970).
SIMANIS, J., GOLDBERG, L. I.: The effects of glucagon on sodium, potassium and urine excretion in patients in congestive heart failure. Amer. Heart J. **81**, 202—210 (1971).
SMITH, T. W., BUTLER, V. P., JR., HABER, E.: Determination of therapeutic and toxic serum digoxin concentrations by radioimmunoassay. New Engl. J. Med. **281**, 1212—1216 (1969).
SOMLYO, A. P., SOMLYO, A. V.: Vascular smooth muscle. II. Pharmacology of normal and hypertensive vessels. Pharmacol. Rev. **22**, 249—353 (1970).
STANBURY, J. B., FARAH, A.: Effects of the magnesium ion on the heart and on its response to digoxin. J. Pharmacol. exp. Ther. **100**, 445—453 (1950).
TJANDRAMAGA, T. B., ANTON, A. H., GOLDBERG, L. I.: Preliminary investigations of levodopa as adjunctive therapy for hypertension. In: Hypertension. Recent Advances, The Third Hahnemann Symposium on Hypertensive Disease. New York: Grune & Stratton 1973.
TSENG, L. T., WALACZEK, E. J.: Blockade of the dopamine depressor response by bulbocapnine. Fed. Proc. **29**, 741 (1970).
WALTON, R. P., GAZES, P. C.: Cardiac glycosides. II. Pharmacology and clinical use. In: Drill's pharmacology in medicine, 4th ed. New York: McGraw-Hill 1971.
WEEKS, J. R.: Prostaglandins. Ann. Rev. Pharmacol. **12**, 317—336 (1972).
YEH, B. K., DE GUZMAN, N. T., PINAKATT, T.: Potassium canrenoate: a new agent in the treatment of digitalis-induced arrhythmia. Clin. Res. **20**, 405 (1972).
YEH, B. K., MCNAY, J. L., GOLDBERG, L. I.: Attenuation of dopamine renal and mesenteric vasodilation by haloperidol: evidence for a specific dopamine receptor. J. Pharmacol. exp. Ther. **168**, 303—309 (1969).

Chapter 71

Drug Interactions in Cancer Chemotherapy

B. A. CHABNER and V. T. OLIVERIO

With 3 Figures

I. Introduction

As a result of clinical research over the last ten years, the simultaneous use of multiple pharmacologic agents in the therapy of malignant disease has become the accepted form of treatment for acute leukemia (HOLLAND, 1968), advanced Hodgkin's disease (DEVITA et al., 1970), lymphosarcoma (BAGLEY et al., 1972), and other tumors (COOPER, 1969; JAMES et al., 1965). These therapeutic advances were largely the outgrowth of earlier experiences with single agents, which despite obvious antitumor activity, elicited incomplete responses or complete remissions of only brief duration. The limitations of single drug therapy were two-fold: (1) the toxicity of the drug limited the amount and duration of drug exposure tolerated by the host, and thus restricted cell kill, and (2) adaptive mechanisms allowed survival and proliferation of a fraction of resistant neoplastic cells despite a metabolic block lethal to the bulk of the tumor. Noting the effectiveness of combined chemotherapy in infectious diseases such as malaria (BLOUNT, 1967), tuberculosis (BARRY, 1964), and enterococcal bacterial endocarditis (LERNER and WEINSTEIN, 1966), oncologists reasoned that the use of several active agents in combination might produce greater cell kill, while delaying the appearance of resistance to the individual drugs. The success of such combinations has established a permanent and expanding role for chemotherapy in the treatment of human malignancy.

Prompted by the demonstrated superiority of drug combinations in the treatment of certain neoplastic diseases, pharmacologists sought to develop a conceptual and experimental framework for identifying synergistic combinations. The contributions of these studies to the problem of selection of synergistic combinations in clinical chemotherapy will be discussed in this chapter. The reader is referred to several excellent reviews for additional information regarding experimental combination chemotherapy (GOLDIN and MANTEL, 1957; SARTORELLI, 1965, 1969). In the present paper we will also discuss combinations in clinical use and general principles that have evolved for the selection of clinically effective combinations. A second type of drug interaction, the inadvertent antagonism of one antineoplastic agent by another or by non-antineoplastic drugs, will also be considered as an important factor in the planning and evaluation of chemotherapeutic regimens.

II. Use of Drug Combinations for Enhanced Antitumor Effect

A. Definition of Terms

Drug interaction may be described as additive, synergistic, or antagonistic in terms of the therapeutic response elicited. However, considerable disagreement exists as to the appropriate use of the terms. Sartorelli (1965) defined synergism as an interaction resulting in an effect greater than the sum of the effects produced by the same agents used independently. Venditti et al. (1956) proposed a broader definition of synergism, emphasizing the need to consider the overall therapeutic benefit to the host. Thus, agents that exerted a greater antitumor effect at the expense of safety were not considered synergistic, while combinations that resulted in improved therapy due to increased antitumor activity or decreased toxicity were termed synergistic. While Sartorelli's definition applies well to experimental drug interactions, a quantitative determination of synergism is difficult to make clinically and the definition proposed by Venditti states the problem in more practical terms.

B. Experimental Assessment of Drug Interaction

Various experimental designs have been employed to analyze the action of drug combinations. An isobolic diagram, in which are plotted the various pairs of drug doses required to produce a specific effect, has been used to interpret the nature of interaction of two agents, as shown in Fig. 1 (Loewe, 1953). This type of design is of limited usefulness in experimental evaluation of antineoplastic drugs. In these cases, drug effect is usually judged not in terms of a fixed end point, but rather as the percent increase in life span of a tumor-bearing animal as

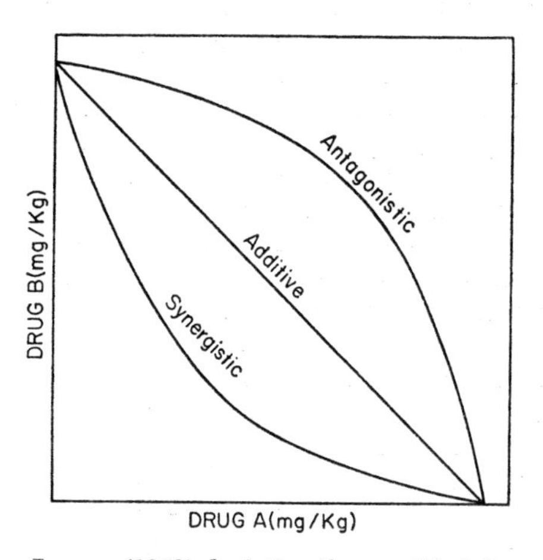

Fig. 1. Isobologram from Loewe (1953) depicting the possible interaction of two pharmacologic agents. The isobolic curves are defined by a series of points that represent the dosages of Drugs A and B required to produce a specific pharmacologic endpoint

compared to untreated controls. In interpreting this type of data, a protocol such as that outlined by Clarke (1958) is more readily employed. In this protocol, A and B are considered to be the maximum tolerated doses of two drugs. Animals with tumors were treated according to one of the schedules given in Table 1, Method 1. Synergism was demonstrated when the effect of $1/4\,A + 1/4\,B$ was equal to or greater than the independent action of either A or B alone. When the action of $1/2\,A + 1/2\,B$ was equal to the independent action of either A or B, the drugs were termed additive in their effect. Method 2 used fractional doses of one drug with full doses of the other to avoid the possibility that one of the agents might have a high threshold of activity below which no antitumor effect could be observed.

Table 1. *Protocol for determination of interaction of two antineoplastic agents in experimental animal tumor systems*

Treatment	
Method 1 Control	Method 2 Control
A	A
B	B
A + B	A + B
1/2 A	1/2 A + B
1/2 B	A + 1/2 B
1/2 A + 1/2 B	1/4 A + B
1/4 A	A + 1/4 B
1/4 B	
1/4 A + 1/4 B	

A or B in each case represents the maximum tolerated dose of Drug A or Drug B.

Such protocols for defining drug interaction provide limited information about optimal dose levels of drug combinations but no information about optimal schedules. MANTEL (1958) presented a more complete scheme for evaluating interactions. He analyzed the effects of combinations of two drugs on the survival of mice with a specified inoculum of L 1210 mouse leukemia. This method provided a very reproducible duration of survival in control animals. Agents A and B were used in various proportions and the optimal total dose of both drugs (A + B) was determined for each proportion (Fig. 2). In the example depicted, A + B are 6-mercaptopurine (6-MP) and azaserine (AZA). By inspection of the data, the optimal proportion and total dose of the combination of drugs would be identified,

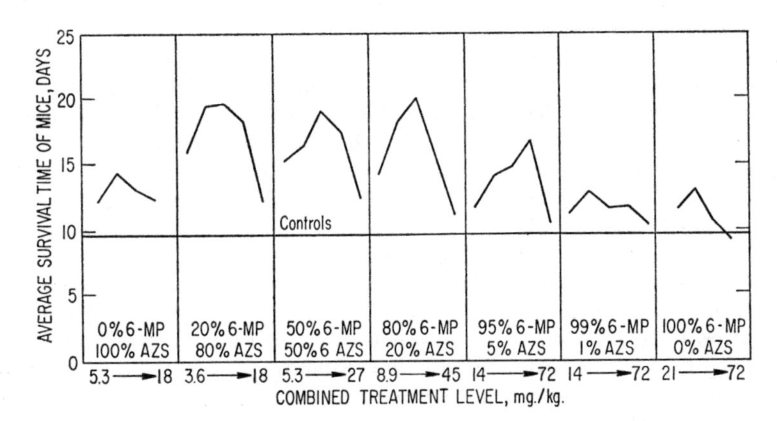

Fig. 2. A series of experiments designed to define the optimum combination of doses of 6-mercaptopurine (6-MP) and azaserine (AZA). Mice were inoculated with 10^6 leukemia L 1210 cells on Day 0; 6-MP and AZA, in various proportions and dosages, were given intraperitoneally every other day beginning on Day 1. From MANTEL (1958) with permission

as shown for azaserine and 6-MP in Fig. 3. Additional experiments would be required to define the optimal schedule of administration of the agents, including such permutations as sequential or intermittent administration. KLINE et al. (1966) illustrated that this experimental approach may yield unexpected results;

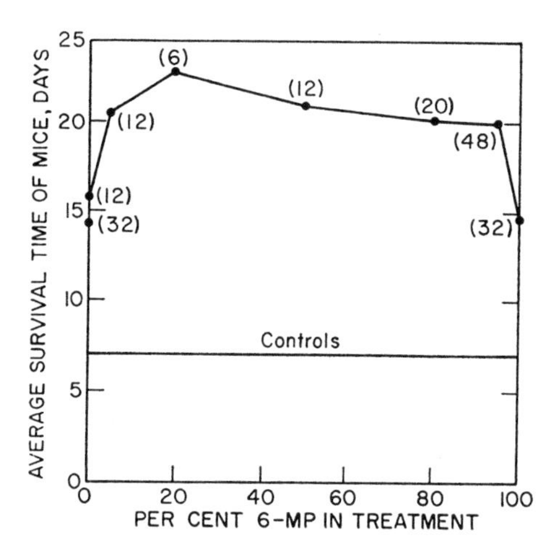

Fig. 3. A plot of the maximal survival time at the optimal total dosage for each proportion of 6-MP to AZA. The drugs in combination are superior to the separate drugs over a wide range of relative proportions. From MANTEL (1958) with permission. Numbers in ( ) indicate the total dosage of the two drugs in mg/kg that gave maximal survival at each proportion

the optimal combination of 5-fluorouracil (5-FU) and methotrexate (MTX) used in treating L 1210 leukemia included a single, low dose of 5-FU with a daily dose of MTX quite near the optimal level for treatment with MTX alone.

C. Biochemical Rationale for Combination Chemotherapy

The selection of drugs to be tested clinically for synergistic activity is often empirical. However, consideration of the biochemical mechanisms of action of antineoplastic agents can be useful in developing new combinations and as a means of explaining the success of proved combinations (Table 2). Two concepts, sequential and concurrent blockade, have been important in selecting combinations studied experimentally, but neither has been of significant value in clinical chemotherapy. POTTER (1951) defined sequential blockade as the simultaneous inhibition of sequential enzymes in a single biochemical pathway. The effectiveness of sequential blockade has been amply demonstrated in antimicrobial chemotherapy by such combinations as pyrimethamine and sulfamethazole, both inhibitors of microbial folate biosynthesis (HITCHINGS and BURCHALL, 1965). However, no such clear cut example exists in antineoplastic pharmacology. Investigators have inferred that the synergism of 5-FU and MTX might be due to blockade of dihydrofolate reductase and thymidylate synthetase (KLINE et al., 1966), sequential enzymes in the pathway of thymidylate synthesis. However, interpretation of these experiments is open to question since both agents have multiple sites of action (COHEN et al., 1958).

Concurrent blockade was first defined by ELION and co-workers (1963) as the simultaneous inhibition of parallel enzymatic pathways leading to the same end product. It has been the basis for studies of several antineoplastic combinations in experimental chemotherapy. In particular, the synergistic actions of azaserine, a glutamine antagonist that inhibits de novo purine biosynthesis, and 6-thioguanine (6-TG), which inhibits reutilization of preformed purines (SARTORELLI and LEPAGE, 1958), appear to be examples of this type of blockade.

Table 2. *Mechanism of action of selected antitumor agents*

Agent	Major action	Ref.
Actinomycin-D	Inhibits DNA-directed RNA polymerase	PERRY, 1962
Arabinosyl-cytosine	Inhibits DNA polymerase,	KIMBALL and WILSON, 1968
	incorporated into RNA	CHU and FISCHER, 1968
L-Asparaginase	Depletes extracellular L-asparagine, thus depressing protein synthesis	BROOME, 1968
1,3-*bis*-(2-chloroethyl)-1-nitrosourea	Alkylates DNA; carbamoylation	MONTGOMERY, 1970
Daunomycin	Inhibits DNA-directed RNA polymerase	DI MARCO et al., 1965
5-Fluorouracil	As nucleotide, inhibits thymidylate synthetase	HEIDELBERGER, 1963
Hydroxyurea	Inhibits ribonucleotide reductase	FRENKEL et al., 1964
6-Mercaptopurine	As nucleotide blocks PRPP amido transferase	BENNETT et al., 1963
Methotrexate	Inhibits dihydrofolate reductase	BERTINO, 1963
Nitrogen mustard	Alkylates DNA	ROBERTS et al., 1968
Prednisone	Lymphocytolytic	DOUGHERTY, 1952
Procarbazine	H_2O_2 formation, methylation of nucleic acids	BERNEIS et al., 1965 KREIS and YEN, 1965
6-Thioguanine	Replaces guanine in DNA	LEPAGE, 1960
Vinblastine and vincristine	Block mitosis in metaphase	CREASEY, 1968

SARTORELLI (1969) suggested a more generally applicable concept, complementary inhibition, to describe the use of combinations of agents that deplete the intracellular supply of a polymeric molecule, such as DNA, RNA, or protein, through independent mechanisms of action. Drugs were categorized as inhibitors of the synthesis of polymer precursors (e.g., antimetabolites), or agents that directly attack the formed molecule (alkylating agents, Actinomycin D, Mitomycin C, ribonuclease). A few drugs interfere with both the synthesis and function of the macromolecule; for example, 6-TG blocks both precursor synthesis and the replication of DNA (Table 2). Synergism was predicted for a drug combination that impedes both the synthesis and function of a specific macromolecule at multiple sites. The combination of an antimetabolite and alkylating agent offers not only the advantage of attack on both the synthesis and integrity of a given polymer, but the antimetabolite might also impede the repair of DNA cross-linked by an alkylating agent. Clinical and experimental studies have borne out the effectiveness of combinations of alkylating agents and antimetabolites (COOPER, 1969), but again the biochemical interpretation of these studies is clouded by the multiplicity of biochemical effects of the antimetabolites.

BRUCE and co-workers (1966) proposed an empirical classification of antineoplastic agents that has some predictive value in selecting synergistic combinations. The classification is based on an analysis of a drug's effect on host hematopoetic stem cells and tumor cells. Mice inoculated with AK lymphoma were sacrificed 24 h after receiving an antineoplastic agent; spleen and bone marrow cells were removed from the donor and infused into recipient mice for assay of residual viable cells. Agents fell into three categories:

1. Drugs that kill lymphoma and marrow cells to an equal extent. The log of the cell kill is related linearly to dose. (Nitrogen mustard and 1,3-*bis*-(2-chloroethyl)-1-nitrosourea (BCNU).

2. Drugs that kill lymphoma cells more than normal marrow. The cell kill is exponentially related to dose only for lower doses of drug (TdR-^{3}H, vinblastine, MTX, azaserine).

3. Drugs that kill both marrow and tumor cells exponentially over a wide range of doses and display a 10,000-fold greater effect on tumor than on marrow stem cells. (Cyclophosphamide, Actinomycin D, 5-FU.)

BRUCE proposed that the classification reflected the relative specificity of the drugs for killing cells during periods of DNA synthesis. Thus agents in category 1 were non-cell cycle dependent, while those in the second group exerted their lethal effects only during DNA synthesis and would selectively kill rapidly proliferating tissue. Drugs in Group 3 killed rapidly proliferating cells preferentially, but in all phases of the cell cycle. The classification is not consistent in all instances with current concepts of the biochemical mechanism of action of the drugs. For example, both MTX (BERTINO, 1963) and 5-FU (HEIDELBERGER, 1963) seem to inhibit the conversion of deoxyuridine to thymidine, and thus should not differ appreciably in their cell cycle specificity. Likewise, cyclophosphamide and nitrogen mustard are both alkylating agents that demonstrate cross resistance clinically and experimentally (KARNOFSKY, 1958), and yet were categorized separately. Finally, the usefulness of the system is hampered by its inability to classify prednisone and 6-MP, two clinically important agents. Undoubtedly, the particular sensitivities of the AK lymphoma are reflected in the tumor's response to the various agents.

However, the potential value of this classification was indicated by the observation that combinations of two drugs from the same class of agents did not show enhanced antitumor activity, while certain combinations of agents from different classes were synergistic. For example, cyclophosphamide and BCNU were synergistic in the treatment of AK lymphoma, while 5-FU and cyclophosphamide were not. The full potential of this work will not be realized until other useful agents, including ara-C, 6-TG, and daunomycin, have been classified and the accuracy of the system's predictions verified.

In summary, attempts to define a rational biochemical basis for combining antineoplastic agents have been hampered by uncertainty as to the mechanism of drug action, and the difficulties of predicting clinical effectiveness on the basis of activity in animal tumor systems. In addition, both theoretical and experimental approaches fail to consider pharmacologic determinants of man's response to antineoplastic agents, such as drug absorption, metabolism, excretion, and unusual toxicities, which may play a primary role in the clinical response to a given agent. Thus, a number of promising combinations, such as BCNU and ara-C (VAN EDEN et al., 1970; VOGLER, 1971), azaserine and 6-TG (HAYES et al., 1967), and glutamine antagonists with L-asparaginase (TARNOWSKI et al., 1970), all of which were supported by impressive experimental results and compelling theoretical considerations, have proved ineffective in clinical use.

III. Clinical Experience with Combination Chemotherapy

The selection of agents for clinical trial in combination is guided mainly by previous experience with the individual agents in the disease under study. The major criterion for inclusion in combinations is that each agent by itself elicit a significant effect on the specific disease. Second, the dose-limiting toxicity of the agents should not be additive, thus allowing the use of each agent in full doses. Most antineoplastic drugs produce bone marrow suppression; therefore, drugs that have little effect on the marrow, such as vincristine and prednisone, offer a particular advantage in combination with myelotoxic drugs. Finally, the use of an agent in combination may be proscribed because of its unusual pharmacologic properties. Thus, it would be disadvantageous to use BCNU, which suppresses bone marrow function 4 to 6 weeks after administration, in combination with alkylating agents or antimetabolites that produce the usual myelosuppression 10 to 14 days after administration because of the prolonged period of leukopenia and thrombocytopenia that might ensue.

The major criteria for judging the effectiveness of combinations of agents are two-fold, the percentage of complete remissions and the duration of unmaintained remission. The initial goal of chemotherapy is the eradication of all evidence of disease, or complete remission. Verification of complete remission status requires repetition of many of the original diagnostic studies. These studies often include lymphangiography, liver biopsy, or bone marrow biopsy in Hodgkin's disease, and peritoneoscopy or "second look" laparatomy in ovarian cancer. Despite careful evaluation, apparent clinical remissions may vary greatly in the extent of tumor eradication; for example, while remission status in acute leukemia signals a reduction in neoplastic cell population to 10^8 cells (SKIPPER and PERRY, 1970) or fewer, there may be an 8 log difference in the number of remaining cells. If the rate of replication of cells is the same regardless of the absolute number of cells remaining, then the time required for reappearance of the tumor, otherwise known as the duration of unmaintained remission, provides an important yardstick in judging the magnitude of tumor cell kill achieved by a given treatment. These two factors, the percentage of complete remissions and the median duration of unmaintained remission, will be focal points for the following discussion of clinical combination chemotherapy.

A. Acute Leukemia

Much of our current interest in antineoplastic drug combinations derives from the landmark clinical studies of childhood leukemia, which have been summarized in other reviews (HOLLAND. 1968; FREI and FREIREICH, 1965). Initial efforts to achieve remission in childhood acute lymphocytic leukemia resulted in the discovery of a number of agents, including prednisone, vincristine, 6-MP, and cyclophosphamide, that could produce a significant percentage of remissions (22 to 64%) in previously untreated patients. The combination of two of these agents increased the percentage of remissions in most instances (Table 3, from HENDERSON and SAMAHA, 1969); the percentages were greatest (90 to 100%) for vincristine and prednisone (HARDISTY et al., 1969).

Unmaintained remission averaged only 60 days after treatment with vincristine and prednisone (HOLLAND, 1968). The brevity of this period indicated a large residual leukemic cell population and led to the institution of a period of intensive multidrug consolidation therapy, usually with a combination of myelotoxic agents such as MTX, 6-MP, or cyclophosphamide. These agents were well tolerated at this point because of the resurgence of normal marrow during the induction of remission. The addition of consolidation therapy with MTX, 6-MP, cyclophosphamide, and BCNU following the induction of remission with vincristine and prednisone extended the duration of subsequent unmaintained remission to 160 days, a confirmation of the magnitude of additional tumor cell kill during consolidation therapy (HOLLAND, 1968). The further addition of multidrug continuous maintenance therapy, as described by PINKEL et al. (1971) (Table 4), has yielded a median duration of maintained hematologic remission of 33 months. The 5-year leukemia-free survival rate is 17%, and an increase to 50% is expected for patients receiving prophylactic craniospinal irradiation to prevent central nervous system relapse (PINKEL et al., 1971).

Table 3. *Remission induction in childhood leukemia* (HENDERSON and SAMAHA, 1969)

	P	VCR	MTX	6-MP	Cytoxan	D
Prednisone (P)	63	100	80	83	76	65
Vincristine (VCR)	100	57			80	
Methotrexate (MTX)	80		22	45		
6-Mercaptopurine (6-MP)	83		45	27		
Cyclophosphamide (Cytoxan)	76	80			40	
Daunomycin (D)	65					38

Table 4. *The St. Jude's hospital chemotherapy protocol for treatment of childhood leukemia* (PINKEL et al., 1971)

Phase	Duration	Drug schedule[a]
Remission induction	4 to 6 weeks	VCR, 1.5 mg/m²/week i.v. Prednisone, 40 mg/m²/day p.o.
Consolidation	1 week	MTX, 10 mg/m²/day × 3 i.v. 6-MP, 1 g/m²/day × 3 i.v. cyclophosphamide, 600 mg/m²/day × 1 i.v.
Maintenance	3 years or to relapse	6-MP, 50 mg/m²/day p.o. MTX, 20 mg/m²/week i.v. cyclophosphamide, 200 mg/m²/week i.v. VCR, 1 mg/m²/week i.v.

[a] Abbreviations are those used in Table 3.

In most instances, studies of remission maintenance therapy demonstrated the superiority of combination therapy over single drug therapy (HOLLAND, 1968; CHAVELIER and GLIDEWELL, 1967), and a similar superiority for high-dose over low-dose combinations (PINKEL et al., 1971). Thus, combinations of antineoplastic agents are more effective than single agents in each of the phases of treatment of acute leukemia: induction, consolidation, and maintenance. The primary theoretical benefits of combination therapy, augmentation of cell kill and the suppression of resistant cell lines, have been realized in the therapy of childhood acute lymphocytic leukemia (ALL). These benefits are reflected in the longer survival of children treated with drug combinations, and the failure of resistant cell lines to appear in a significant number of patients after up to 36 months of continuous maintenance therapy.

Progress in the treatment of acute myeloblastic leukemia (AML) has lagged far behind that achieved in childhood ALL: present efforts are still directed at the first hurdle of curative chemotherapy, achieving a high percentage of complete remissions. Two unfavorable factors influence the poorer results with AML: (1) the older average age of patients with AML, and their decreased tolerance for the infectious and hemorrhagic complications of leukemia and chemotherapy, and (2) the ineffectiveness of non-myelotoxic regimens, such as vincristine and prednisone, in inducing remission, necessitating the use of the considerably more toxic myelosuppressive drugs. Nonetheless, the availability of a number of agents with remission induction potential has led to the successful trial of various combinations. Table 5 illustrates that the most effective of these combinations to date have employed ara-C with either 6-TG (GEE et al., 1969), daunomycin (CAREY, 1970), or with vincristine, prednisone, and cyclophosphamide (FREIREICH et al., 1970).

Table 5. *Remission induction in acute myelocytic leukemia*

Drug regimen	No. patients	Complete remissions (%)	Ref.
1. Ara-C 3 mg/kg/day i.v. in p.m. 6-TG 2.5 mg/kg/day p.o. in a.m. Both given for 8 to 32 days	38	21 (55)	GEE et al., 1969
2. Ara-C + Cytoxan 150 mg/m²/day × 4d i.v. q · 2 weeks Prednisone 200 mg/day p.o. × 4 Vincristine 2 mg i.v. × 1 Regimen is repeated every 2 weeks	19	10 (53)	FREIREICH et al., 1970
3. Ara-C 100 mg/m²/day i.v., Days 1 to 5 6-TG 70 mg/m²/day, Days 8 to 12 Repeat cycle every 21 days	6	5 (83)	GUYER et al. 1971
4. Ara-C 100 mg/m²/day i.v. Daunomycin 45 mg/m²/day, Days 1 + 2 Duration of therapy not stated	not stated	(36)	CAREY, 1970
5. Ara-C 100 mg/m²/day i.v. 6-MP 2.5 mg/kg/day p.o. Duration of therapy not stated	not stated	(32.2)	CAREY, 1970

Experimental studies indicated that ara-C could be used effectively in combination therapy. Ara-C was active against murine leukemias resistant to the other commonly used antileukemic agents, including 6-MP, MTX, the vinca alkaloids, and cyclophosphamide (EVANS et al., 1964, 1969; WODINSKY and KENSLER, 1965). Further work by BURCHENAL and DOLLINGER (1967) demonstrated synergism for combinations of ara-C and 6-MP in L 1210 leukemia, and suggested the trial of ara-C and a thiopurine in human leukemia (Table 5). Considerations of cell cycle kinetics and the mechanisms of drug action have led to promising modifications in the use of 6-TG and ara-C in AML. GEE and colleagues (1969) have recently employed ara-C and 6-TG in a schedule of repeated doses given daily (Table 5) for 8 to 32 days in an effort to cover a period of time greater than the generation time of the leukemic cell, which has been estimated to be 40 to 98 h (CLARKSON et al., 1967; DEFENDI and MANSON, 1963), thus at some point during the treatment period exposing all proliferating cells to the *S*-phase specific drug during their period of DNA synthesis. GUYER and colleagues (1971), among others, have used a continuous 5-day infusion of ara-C followed by a rest interval of 2 days and then a 5-day course of 6-TG. This latter schedule is based on the supposition that the simultaneous use of ara-C, an inhibitor of DNA synthesis, and 6-TG, a drug that depends on incorporation into DNA for its activity, would diminish the effectiveness of the latter drug. This argument has been substantiated by the experimental work of LE PAGE and WHITE (1972). Additional agents probably will be added to the current combination of ara-C and 6-TG in order to improve the percentage of remission induction. In view of the experimental demonstrations of synergism for combinations of 6-TG with inhibitors of de novo purine synthesis, such as 6-MP (HENDERSON and JUNGA, 1960), the addition of such an agent to the combination merits consideration.

Treatment of the blastic phase of chronic granulocytic leukemia has been the least successful of all therapeutic efforts in leukemia research. CANELLOS et al. (1971) described the induction of remission by vincristine and prednisone in 20% of patients in blast crisis. The study demonstrated the increased sensitivity of hypodiploid blastic cell lines to the therapeutic combination.

B. Lymphomas

The success of combination therapy in childhood ALL prompted investigation of a similar approach in human solid tumors. The most notable improvement over single drug therapy has occurred in the treatment of Hodgkin's disease. Prior to the trial of combinations, several individual agents were known to produce a high incidence of partial responses and a lower percentage of usually brief complete remissions (10 to 30%) (CARBONE and SPURR, 1968). The most effective of these were the alkylating agents, nitrogen mustard and cyclophosphamide, the vinca alkaloids, and more recently, procarbazine, and BCNU. In 1965, LACHER and DURANT reported complete remissions in 40% of patients with advanced Hodgkin's disease treated with chlorambucil and vinblastine. At the same time, DEVITA et al. (1970) used a four-drug combination of cyclophosphamide, vinblastine, methotrexate, and prednisone to obtain complete remissions in 80% of patients with Stage III to IV disease. This study has become the basis for current therapy of advanced lymphoma. The regimen was subsequently modified to include vincristine instead of vinblastine, procarbazine instead of methotrexate, and nitrogen mustard in place of cyclophosphamide. The resulting combination, termed MOPP (Table 6), has produced a complete remission rate of 81% and a median duration of unmaintained remission of 36 months. No patient from the original MOPP study in remission for longer than 42 months has relapsed, which indicates the strong possibility of "cure" in long term responses.

Table 6. *The* MOPP *protocol for advanced Hodgkin's disease* (DEVITA et al., 1970)

Agent	Schedule[a]
Nitrogen mustard	6 mg/m², Days 1 + 8
Vincristine (Oncovin)	2.4 mg/m², Days 1 + 8
Procarbazine	100 mg/m², Days 1 to 10
Prednisone	60 mg/m², Days 1 to 14, Cycles 1 + 4 only

[a] Cycles are repeated at 28 day intervals. A total of 6 cycles given.

The success of the MOPP combination appears related to several factors. The non-additive toxicities of all but procarbazine and nitrogen mustard (both of which are myelosuppressive) have allowed use of the drugs in intensive, full-dose cycles of therapy. Second, the long duration of unmaintained remission following MOPP testifies to the magnitude of cell kill achieved by remission induction using agents with different mechanisms of action. Finally, patients who do relapse respond very well to repeat treatments with MOPP, which indicates that drug resistance has been averted in many patients.

Several questions remain to be answered concerning the chemotherapy of advanced Hodgkin's disease. Principal among these is whether maintenance therapy following MOPP will decrease the incidence of relapse and extend the median duration of remission; although the studies are still preliminary, maintenance with MOPP or a variety of single agents has thus far failed to show a conclusive improvement in results (DEVITA and CANELLOS, 1972). A second important question is whether the sequential use of MOPP and radical radiotherapy will improve the results of treatment of advanced Hodgkin's disease.

Progress in pathologic classification, staging, and treatment of Hodgkin's disease, and a greater understanding of the natural history of this disorder have spurred interest in related non-Hodgkin's lymphomas. A number of disorders have been defined within the general category of lymphosarcoma, including Burkitt's lymphoma, childhood lymphosarcoma, reticulum cell sarcoma (or histiocytic lymphoma), and the lymphocytic lymphomas. A separate therapeutic approach has evolved for each of these, based partly on empirical success, such as the use of cyclophosphamide in Burkitt's lymphoma (BURKITT, 1966), and partly on similarities to other disorders. For example, vincristine and prednisone are used in childhood lymphosarcoma because it clinically resembles childhood leukemia. In patients with advanced lymphosarcoma, drug combinations have produced results clearly superior to the traditional single agent therapy with alkylating agents (DEVITA and CANELLOS, 1972). The combination of cyclophosphamide with vincristine and prednisone in several different schedules has resulted in 35 to 57% complete remissions (BAGLEY et al., 1972; HOOGSTRATEN et al., 1969), as against a maximum of 26% achieved by the best previous single drug therapy, high-dose cyclophosphamide (MENDELSON et al., 1970). The most promising results have been obtained with an intensive one week course of cyclophosphamide ($400\ mg/m^2/d \times 5$)[1], vincristine ($1.4\ mg/m^2 \times 1$), and prednisone ($100\ mg/m^2 \times 5$), repeated at three week intervals until complete remission is achieved; with this regimen, 57% of patients with lymphocytic lymphoma have achieved remission status (BAGLEY et al., 1972).

Less favorable results have been reported in advanced reticulum cell sarcoma using combination therapy, the incidence of complete remission falling between 30 and 38% (BAGLEY et al., 1972; LOWENBRAUN et al., 1970). However, one recent study reported a complete remission rate of 60% in patients treated with a combination of cyclophosphamide and vincristine, followed sequentially by ara-C and MTX leucovorin (LEVITT et al., 1972).

C. Therapy of Solid Tumors

While definite progress has been made in the chemotherapy of leukemias and lymphomas, the solid tumors have responded less consistently. A few tumors, such as choriocarcinoma (HERTZ et al., 1961), Wilm's tumor (BURGERT and GLIDEWELL, 1967), Burkitt's lymphoma (BURKITT, 1966), and retinoblastoma

[1] 400 mg per square meter of body surface area per day for 5 days.

(HYMAN et al., 1968), appear to be curable with single agents, in some instances used in conjunction with surgery and radiotherapy.

Drug combinations show promise in other malignancies, particularly those commonly seen in children. Vincristine, actinomycin D, and cyclophosphamide have been used in combination with radiotherapy and local tumor excision to produce an impressive complete remission rate of 76% in children with metastatic rhabdomyosarcoma (PRATT et al., 1971). The duration of complete remission and the relapse rate have not been calculated because of the short length of time of the study at this point, but the initial results have been confirmed in at least two medical centers (WILBUR et al., 1971). A similar combination of vincristine and cyclophosphamide has achieved a complete remission rate of 25 to 56% in generalized neuroblastoma, but the mean survival time for both partial and complete responders was only approximately one year, as opposed to a mean of 4 months in untreated patients (JAMES et al., 1965; EVANS et al., 1969).

A critical examination of the successful combination therapies in current clinical use reveals a central role for the vinca alkaloids in most of these programs. This effectiveness has been neither explained nor predicted by studies on animal tumor systems. For example, the combination of vincristine and prednisone, a particularly effective combination in human leukemia and lymphoma, has shown a synergistic response in only one animal tumor, the P 1798 (VADLAMUDI et al., 1971). The most complete study of vinca alkaloids in experimental combinations has been with actinomycin D (CREASEY, 1966). This study clearly demonstrated synergistic activity of the combination of vinblastine and actinomycin D in several murine tumor systems, including the Ehrlich ascites carcinoma, L 1210 lymphoma, and sarcoma 180. In all three tumors, vinblastine was superior to vincristine in combination with actinomycin D. When given in regimens employing a single dose of each drug, optimal results were obtained by sequential use of the two agents (vinblastine first, and actinomycin D 24 h later) rather than simultaneous administration. However, simultaneous multiple daily doses of the two agents in combination produced maximal survival times. Biochemical studies of Ehrlich ascites tumor cells indicated a possible biochemical explanation for the observed interaction: the secondary increase in RNA synthesis 18 to 24 h after vinblastine was thought to be instrumental in recovery from the drug. However, this increase is inhibited by actinomycin D, an inhibitor of DNA-dependent RNA polymerase. These studies demonstrate the value of experimental pharmacology in exploring ways of modifying successful combinations used in human disease. They also raise important questions about whether the present regimens employing vincristine, rather than vinblastine, and the specific schedules of administration of vinca alkaloids with other drugs are actually the most effective.

Several new areas of investigation show promising results in the treatment of solid neoplasms in adults. A five drug combination (5-FU, cyclophosphamide, methotrexate, prednisone, and vincristine), originally studied by COOPER (1969), has produced a therapeutic response in 70% of women with metastatic breast carcinoma, a marked improvement over the 15 to 30% response to single chemotherapeutic agents. Similarly, a recent study indicated that metastatic ovarian carcinoma has an enhanced response to combined therapy with cyclophosphamide plus 5-FU or methotrexate (LI and HSU, 1970). Testicular neoplasms also respond consistently, though usually briefly, to chemotherapeutic agents. The initial report by LI et al. (1960) showed that 9 of 23 patients, or 30%, achieved complete remission with a combination of actinomycin D, an alkylating agent (usually chlorambucil), and an antimetabolite (usually methotrexate). A larger study by

MACKENZIE (1966) failed to demonstrate a difference in the response rate to actinomycin D alone versus the above combination.

From these studies, therefore, combination chemotherapy appears to be particularly valuable in the treatment of malignancies of the lymphoreticular system, leukemia, and certain solid tumors of childhood, mainly because of the non-additive toxicities and therapeutic effectiveness of vincristine and prednisone in combination with myelosuppressive inhibitors of DNA or RNA synthesis. The common solid tumors of adults, however, have more prolonged cell generation times and less differential sensitivity to inhibitors of DNA synthesis. These tumors have shown little response to the chemotherapeutic combinations and undoubtedly will require a new drug armamentarium with an entirely new focus for biochemical attack. Recent developments in immunology and viral oncology may provide such vulnerable sites for new therapy.

IV. Antagonistic Drug Interactions

The increased use of chemotherapeutic agents in combination has resulted in antagonistic drug interactions in several unexpected situations. These inadvertent interactions illustrate basic mechanisms of drug interaction discussed in preceding chapters. Several of these interactions concern the folic acid antagonist, methotrexate (MTX).

Acetylsalicylic acid has been shown to displace MTX, also a weak heterocyclic organic acid, from its binding to serum albumin (LIEGLER et al., 1969), thus increasing the concentration of free drug for a given dose. Intracellular transport of MTX proceeds by way of an energy dependent carrier system that is inhibited by several agents, including corticosteroids, cephalothin, and thiobarbituric acid. However, vincristine increases MTX uptake by 38% (ZAGER et al., 1972). Finally, the nephrotoxic antibiotics commonly used for the treatment of gram negative or fungal infections in clinical oncology, such as gentamicin (BERMAN and KATZ, 1958), kanamycin (FALCO et al., 1969), colistin (WOLINSKY and HINES, 1962), or amphotercin B (BELL et al., 1962), may disrupt renal tubular function and retard the renal excretion of agents, such as MTX, that are eliminated mainly through the kidneys. This leads to unexpected toxicity.

The high frequency of hyperuricemia in patients with leukemia has often led to the simultaneous use of allopurinol, a hypouricemic agent, with 6-mercaptopurine (6-MP). However, allopurinol, an inhibitor of xanthine oxidase, also blocks degradation of the thiopurine, necessitating a 3/4 reduction in 6-MP dosage in patients also receiving allopurinol (ELION et al., 1963).

Drug interactions mediated by effects on hepatic drug-metabolizing enzyme systems have received increasing attention in recent years, particularly with regard to the effects of various agents on cyclophosphamide metabolism. Cyclophosphamide is converted by hepatic microsomal enzymes into the active metabolite, aldophosphamide, which is further converted to carboxyphosphamide by a soluble enzyme thought to be aldehyde oxidase (HILL et al., 1972; GRAUL et al., 1967; RAUEN and KRAMER, 1964). The activation reaction is inhibited by nicotine, ephedrine, phenobarbital, and cocaine *in vitro* (HILL et al., 1972). However, pretreatment of rodents with phenobarbital *in vivo* markedly increased peak serum alkylating activity due to cyclophosphamide metabolites. Despite the increase in peak plasma alkylating activity after phenobarbital pretreatment, the barbiturate did not influence the antitumor effect of a given dose of cyclophosphamide in the treatment of the Walker 256 carcinosarcoma (SLADEK, 1972). Likewise, the effect of corticosteroids on plasma alkylating activity following cyclophosphamide has been studied in rats with contradictory results; HAYAKAWA and co-workers (1969) found decreased alkylating activity, while HANASONO and FISCHER (1972) found no effects.

The possibility of interactions of antineoplastic agents with the metabolism of other drugs has been suggested by the recent findings of DONELLI and GARATTINI (1971) and TARDIFF and DUBOIS (1969). They showed that a variety of antitumor drugs, including the alkylating agents, 5-FU, MTX, and daunomycin, may acutely depress hepatic microsomal *o*-demethylase and *p*-hydroxylase activity in rats. DONELLI and GARATTINI (1971) observed a rebound increase in these enzyme activities nine days after cessation of treatment, indicating that the effect of an agent on microsomal enzymes may be biphasic and the interaction of two agents may be dependent on the time interval between doses.

Although these findings concerning the metabolism of antineoplastic agents have not been extended to man, it is apparent that several possible interactions exist with respect to extensively metabolized agents such as cyclophosphamide, procarbazine, and nitrogen mustard (ADAMSON, 1971). These interactions could be of major significance in determining the efficacy of drug combinations currently in clinical use.

Finally, specific drug combinations are antagonistic when used simultaneously. Thus, 6-TG is probably more effective when given sequentially with ara-C, rather than simultaneously. Similarly, L-asparaginase, which limits protein synthesis through depletion of L-asparagine, inhibits the antitumor effect of methotrexate, an inhibitor of thymidylate synthesis, presumably by preventing unbalanced growth and thymine-less death of affected cells (CAPIZZI et al., 1970). The same drugs administered in sequence may have additive or synergistic antitumor activity (CAPIZZI et al., 1972; CONNORS and JONES, 1970). Undoubtedly, more instances of drug interaction will become apparent with increased knowledge of drug mechanisms of action and the clinical effects of combinations.

References

ADAMSON, R. H.: Metabolism of anticancer agents in man. Ann. N.Y. Acad. Sci. **179**, 432—441 (1971).

BAGLEY, C. M., JR., DEVITA, V. T., BERARD, C. W., CANELLOS, G. P.: Advanced lymphosarcoma: intensive cyclical combination chemotherapy with cyclophosphamide, vincristine, and prednisone. Ann. intern. Med. **76**, 227—234 (1972).

BARRY, V. C.: Chemotherapy of tuberculosis. London: Butterworths 1964.

BELL, N. H., ANDRIOLI, V. T., SABESIN, S. M., UTZ, J. P.: On the nephrotoxicity of amphotericin B in man. Amer. J. Med. **33**, 64—69 (1962).

BENNETT, L. L., JR., SIMPSON, L., GOLDEN, J., BARKER, T. L.: The primary site of inhibition by 6-mercaptopurine on the purine biosynthetic pathway in some tumors *in vivo*. Cancer Res. **23**, 1574—1580 (1963).

BERMAN, L. B., KATZ, S.: Kanamycin nephrotoxicity. Ann. N.Y. Acad. Sci. **76**, 149—156 (1958).

BERNEIS, K., KOFLER, M., BOLLAG, W., KAISER, A., LANGEMANN, A.: The degradation of deoxyribonucleic acid by new tumor inhibiting compounds: the intermediate formation of hydrogen peroxide. Experientia (Basel) **19**, 132—133 (1963).

BERTINO, J. R.: The mechanism of action of the folate antagonists in man. Cancer Res. **23**, 1286—1306 (1963).

BLOUNT, R. E.: Management of chloroquine-resistant falciparum malaria. Arch. intern. Med. **119**, 557—560 (1967).

BROOME, J. D.: Studies on the mechanism of tumor inhibition by L-asparaginase. Effects of the enzyme on asparagine levels in the blood, normal tissues, and 6C3HED lymphomas of mice: differences in asparagine formation and utilization in asparaginase-sensitive and -resistant lymphoma cells. J. exp. Med. **127**, 1055—1072 (1968).

BRUCE, W. R., MEEKER, B. E., VALERIOTE, F. A.: Comparison of the sensitivity of normal hematopoietic and transplanted lymphoma colony-forming cells to chemotherapeutic agents administered *in vivo*. J. nat. Cancer Inst. **37**, 233—245 (1966).

Burchenal, J. H., Dollinger, M. R.: Cytosine arabinoside (NSC-63878) in combination with 6-mercaptopurine (NSC-755), methotrexate (NSC-740) or 5-fluorouracil (NSC-19893) in L 1210 mouse leukemia. Cancer Chemother. Rep. **51**, 435—438 (1967).

Burgert, E. O., Jr., Glidewell, O.: Dactinomycin in Wilm's tumor. J. Amer. med. Ass. **199**, 464—468 (1967).

Burkitt, D.: African lymphoma. Observations on response to vincristine sulphate therapy. Cancer (Philad.) **19**, 1131—1137 (1966).

Canellos, G. P., Devita, V. T., Whang-Peng, J., Carbone, P. P.: Hematologic and cytogenetic remission of blastic transformation in chronic granulocytic leukemia. Blood **38**, 671—679 (1971).

Capizzi, R. L., Nichols, R., Mullins, J.: Long-term survival of leukemic mice by therapeutic synergism between asparaginase (A'ASE) and methotrexate (MTX). Fed. Proc. **31**, 553 Abs (1972).

Capizzi, R. L., Summers, W. P., Bertino, J. R.: Antagonism of the antineoplastic effect of methotrexate (MTX) by L-asparaginase (ASN'ASE) or L-asparagine (ASN) deprivation. Proc. Amer. Ass. Cancer Res. **11**, 14 (1970).

Carbone, P. P., Spurr, C.: Management of patients with malignant lymphoma: a comparative study with cyclophosphamide and vinca alkaloids. Cancer Res. **28**, 811—822 (1968).

Carey, R. W.: Comparative study of cytosine arabinoside (CA) therapy alone and combined with thioguanine (TG), mercaptopurine (MP), or daunomycin (DN) in acute myelocytic leukemia (AML). Proc. Amer. Ass. Cancer Res. **11**, 15 (1970).

Chavelier, L., Glidewell, O.: Schedule of 6-mercaptopurine and effect of inducer drugs in prolongation of remission maintenance in acute leukemia. Proc. Amer. Ass. Cancer Res. **8**, 10 (1967).

Chu, M.-Y., Fischer, G. A.: The incorporation of ^{3}H-cytosine arabinoside and its effect on murine leukemic cells (L 5178 Y). Biochem. Pharmacol. **17**, 753—767 (1968).

Clarke, D. A.: In discussion of Mantel, N.: An experimental design in combination chemotherapy. Ann. N. Y. Acad. Sci. **76**, 909—931 (1958).

Clarkson, B., Ohkita, T., Ota, K., Fried, J.: Studies of cellular proliferation in human leukemia. I. Estimation of growth rates of leukemic and normal hematopoietic cells in two adults with acute leukemia given single injections of tritiated thymidine. J. clin. Invest. **46**, 506—529 (1967).

Cohen, S. S., Flaks, J. G., Barner, H. D., Loeb, M. R., Lichtenstein, J.: The mode of action of 5-fluorouracil and its derivatives. Proc. nat. Acad. Sci. (Wash.) **44**, 1004—1012 (1958).

Connors, T. A., Jones, M.: Antagonism of the anti-tumor effects of asparaginase by methotrexate. Biochem. Pharmacol. **19**, 2927—2929 (1970).

Cooper, R. G.: Combination chemotherapy in hormone resistant breast cancer. Proc. Amer. Ass. Cancer Res. **10**, 15 (1969).

Creasey, W. A.: Tumor-inhibitory effects of combinations of the vinca alkaloids with actinomycin D. Biochem. Pharmacol. **15**, 367—375 (1966).

Creasey, W. A.: Modifications in biochemical pathways produced by the vinca alkaloids. Cancer Chemother. Rep. **52**, 501—507 (1968).

Defendi, V., Manson, L. A.: Analysis of the life-cycle in mammalian cells. Nature (Lond.) **198**, 359—361 (1963).

Devita, V. T., Jr., Canellos, G. P.: Treatment of the lymphomas. Sem. Hemat. **9**, 193—209 (1972).

Devita, V. T., Jr., Serpick, A. A., Carbone, P. P.: Combination chemotherapy in the treatment of advanced Hodgkin's disease. Ann. intern. Med. **73**, 881—895 (1970).

di Marco, A., Silvestrini, R., de Marco, S., Dasdia, T.: Inhibiting effect of the new cytotoxic antibiotic daunomycin on nucleic acids and mitotic activity of HeLa cells. J. Cell Biol. **27**, 545—550 (1965).

Donelli, M. G., Garattini, S.: Drug metabolism after repeated treatments with cytotoxic agents. Europ. J. Cancer **7**, 361—364 (1971).

Dougherty, T. F.: Effect of hormones on lymphatic tissue. Physiol. Rev. **32**, 379—401 (1952).

Elion, G. B., Callahan, S., Nathan, H., Bieber, S., Rundles, R. W., Hitchings, G. H.: Potentiation by inhibition of drug degradation: 6-substituted purines and xanthine oxidase. Biochem. Pharmacol. **12**, 85—93 (1963).

Elion, G. B., Singer, S., Hitchings, G. H.: Antagonists of nucleic acid derivatives. VIII. Synergism in combinations of biochemically related antimetabolites. J. biol. Chem. **208**, 477—488 (1954).

Evans, A. E., Heyn, R. M., Newton, W. A., Jr., Leikin, S. L.: Vincristine sulfate and cyclophosphamide for children with metastatic neuroblastoma. J. Amer. med. Ass. **207**, 1325—1327 (1969).

Evans, J. S., Musser, E. A., Bostwick, L., Mengel, G. D.: The effect of 1-β-D-arabinofuranosylcytosine hydrochloride on murine neoplasms. Cancer Res. **24**, 1285—1293 (1964).

FALCO, F. G., SMITH, H. M., ARCIERI, G. M.: Nephrotoxicity of aminoglycosides and gentamycin. J. infect. Dis. **119**, 406—409 (1969).

FREI III, E., FREIREICH, E. J.: Progress and perspectives in the chemotherapy of acute leukemia. Advanc. Chemother. **2**, 269—298 (1965).

FREIREICH, E. J., BODEY, G. P., HART, S., RODRIGUEZ, V., WHITECAR, J. P., FREI III, E.: Remission induction in adults with acute myelogenous leukemia. Rec. Res. Cancer Res. **30**, 85—91 (1970).

FRENKEL, E. P., SKINNER, W. N., SMILEY, J. D.: Studies on a metabolic defect induced by hydroxyurea (NSC-32065). Cancer Chemother. Rep. **40**, 19—22 (1964).

GEE, T. S., TU, K.-P., CLARKSON, B. D.: Treatment of adult acute leukemia with arabinosylcytosine and thioguanine. Cancer (Philad.) **23**, 1019—1032 (1969).

GOLDIN, A., MANTEL, N.: The employment of combinations of drugs in the chemotherapy of neoplasia: a review. Cancer Res. **17**, 635—654 (1957).

GRAUL, E. H., SCHAUMLÖFFEL, E., HUNDESHAGEN, H., WILMANNS, H., SIMON, G.: Metabolism, of radioactive cyclophosphamide. Animal tests and clinical studies. Cancer (Philad.) **20** 896—899 (1967).

GUYER, R. J., WINFIELD, D. A., SHAHANI, R. T., BLACKBURN, E. K.: Combination chemotherapy in acute myeloblastic leukemia. Brit. med. J. **1971 I**, 231—232.

HANASONO, G. K., FISCHER, L. J.: Plasma levels and urinary excretion of [^{14}C] cyclophosphamide and its radioactive metabolites in rats pretreated with prednisolone. Biochem. Pharmacol. **21**, 272—276 (1972).

HARDISTY, R. M., MC ELWAIN, T. J., DARBY, C. W.: Vincristine and prednisone for the induction of remissions in acute childhood leukemia. Brit. med. J. **1969 II**, 662—665.

HAYAKAWA, T., KANAI, N., YAMADA, R., KURODA, R., HIGASHI, H., MOGAMI, H., JINNAI, D.: Effect of steroid hormone on activation of endoxan (cyclophosphamide). Biochem. Pharmacol. **18**, 129—135 (1969).

HAYES, D. M., COSTA, J., MOON, J. H., HOOGSTRATEN, B., HARLEY, J. B.: Combination therapy with thioguanine (NSC-752) and azaserine (NSC-742) for multiple myeloma. Cancer Chemother. Rep. **51**, 235—238 (1967).

HEIDELBERGER, C.: Biochemical mechanisms of action of fluorinated pyrimidines. Exp. Cell Res. Suppl. **9**, 462—471 (1963).

HENDERSON, E. S., SAMAHA, R. J.: Evidence that drugs in multiple combinations have materially advanced the treatment of human malignancies. Cancer Res. **29**, 2272—2280 (1969).

HENDERSON, J. F., JUNGA, I. G.: Potentiation of carcinostasis by combinations of thioguanine and 6-mercaptopurine. Biochem. Pharmacol. **5**, 167—168 (1960).

HERTZ, R., LEWIS, J., JR., LIPSETT, M. B.: Five years' experience with the chemotherapy of metastatic choriocarcinoma and related trophoblastic tumors in women. Amer. J. Obstet. Gynec. **82**, 631—640 (1961).

HILL, D. L., LASTER, W. R., JR., STRUCK, R. F.: Enzymatic metabolism of cyclophosphamide and nicotine and production of a toxic cyclophosphamide metabolite. Cancer Res. **32**, 658—665 (1972).

HITCHINGS, G. H., BURCHALL, J. J.: Inhibition of folate biosynthesis and function as a basis for chemotherapy. Advan. Enzymol. **27**, 417—468 (1965).

HOLLAND, J. F.: Progress in the treatment of acute leukemia. In: DAMESHEK, W., DUTCHER, R. M. (Eds.): Perspectives in leukemia, pp. 217—240. New York-London: Grune and Stratton 1968.

HOOGSTRATEN, B., OWENS, A. H., LENHARD, R. E., GLIDEWELL. O, J., LEONE, L. A., OLSON, K. B., HARLEY, J. B., TOWNSEND, S. R., MILLER, S. P., SPURR, C. L.: Combination chemotherapy in lymphosarcoma and reticulum cell sarcoma. Blood **33**, 370—378 (1969).

HYMAN, G. A., ELLSWORTH, R. M., FEIND, C. R., TRETTER, P.: Combination therapy in retinoblastoma. A 15-year summary of methods and results. Arch. Ophthal. **80**, 744—746 (1968).

JAMES, D. H., JR., HUSTU, O., WRENN, E. L., JR., PINKEL, D.: Combination chemotherapy of childhood neuroblastoma. J. Amer. med. Ass. **194**, 123—126 (1965).

KARNOFSKY, D. (Ed.): Comparative clinical and biological effects of alkylating agents. Ann. N.Y. Acad. Sci. **68**, 657—1266 (1958).

KIMBALL, A. P., WILSON, M. J.: Inhibition of DNA polymerase by β-D-arabinosylcytosine and reversal of inhibition by deoxycytidine-5'-triphosphate. Proc. Soc. exp. Biol. (N.Y.) **127**, 429—432 (1968).

KLINE, I., VENDITTI, J. M., MEAD, J. A. R., TYRER, D. D., GOLDIN, A.: The antileukemic effectiveness of 5-fluorouracil and methotrexate in the combination chemotherapy of advanced leukemia. Cancer Res. **26**, 848—852 (1966).

KREIS, W., YEN, W.: An antineoplastic C^{14}-labeled methylhydrazine derivative in P 815 mouse leukemia. A metabolic study. Experientia (Basel) **21**, 284—286 (1965).

LACHER, M. J., DURANT, J. R.: Combined vinblastine and chlorambucil therapy of Hodgkin's disease. Ann. intern. Med. **62**, 468—476 (1965).
LEPAGE, G. A.: Incorporation of 6-thioguanine into nucleic acids. Cancer Res. **20**, 403—408 (1960).
LEPAGE, G. A., WHITE, S. C.: Scheduling of arabinosylcytosine (ara-C) and 6-thioguanine (6-TG) therapy. Proc. Amer. Ass. Cancer Res. **13**, 11 (1972).
LERNER, P. I., WEINSTEIN, L.: Infective endocarditis in the antibiotic era (to be concluded). New Engl. J. Med. **274**, 323—331 (1966).
LEVITT, M., MARSH, J. C., DECONTI, R. C., MITCHELL, M. S., SKEEL, R. T., FARBER, L. R., BERTINO, J. R.: Combination sequential chemotherapy in advanced reticulum cell sarcoma. Cancer (Philad.) **29**, 630—636 (1972).
LI, M. C., HSU, K.-P.: Combined drug therapy for ovarian carcinoma. Clin. Obstet. Gynec. **13**, 928—944 (1970).
LI, M. C., WHITMORE, W. F., JR., GOLBEY, R., GRABSTALD, H.: Effects of combined drug therapy in metastatic cancer of the testis. J. Amer. med. Ass. **174**, 1291—1299 (1960).
LIEGLER, D. G., HENDERSON, E. S., HAHN, M. A., OLIVERIO, V. T.: The effect of organic acids on renal clearance of methotrexate in man. Clin. Pharmacol. Ther. **10**, 849—857 (1969).
LOEWE, S.: The problem of synergism and antagonism of combined drugs. Arzneimittel-Forsch. **3**, 285—290 (1953).
LOWENBRAUN, S., DEVITA, V. T., SERPICK, A. A.: Combination chemotherapy with nitrogen mustard, vincristine, procarbazine, and prednisone in lymphosarcoma and reticulum cell sarcoma. Cancer (Philad.) **25**, 1018—1025 (1970).
MACKENZIE, A. R.: Chemotherapy of metastatic testis cancer. Cancer (Philad.) **19**, 1369—1376 (1966).
MANTEL, N.: An experimental design in combination chemotherapy. Ann. N.Y. Acad. Sci. **76**, 909—931 (1958).
MENDELSON, D., BLOCK, J. B., SERPICK, A. A.: Effect of large intermittent intravenous doses of cyclophosphamide in lymphoma. Cancer (Philad.) **25**, 715—720 (1970).
MONTGOMERY, J.: In: BURGER, A. (Ed.): Medicinal chemistry, 3rd ed., pp. 703—707. New York: Wiley-Interscience 1970.
PERRY, R. P.: The cellular sites of synthesis of ribosomal and 4 S RNA. Proc. nat. Acad. Sci. (Wash.) **48**, 2179—2186 (1962).
PINKEL, D., HERNANDEZ, K., BORELLA, L., HOLTON, C., AUR, R., SAMOY, G., PRATT, C.: Drug dosage and remission duration in childhood lymphocytic leukemia. Cancer (Philad.) **27**, 247—256 (1971).
POTTER, V. R.: Sequential blocking of metabolic pathways *in vivo*. Proc. Soc. exp. Biol. (N.Y.) **76**, 41—46 (1951).
PRATT, C. B., FLEMING, I. D., HUSTER, H. O.: Multimodal therapy of childhood rhabdomyosarcoma. Proc. Amer. Ass. Cancer Res. **12**, 26 (1971).
RAUEN, H. M., KRAMER, K.-P.: Der Gesamtalkylantien-Blutspiegel von Ratten nach Verabreichung von Cyclophosphamid und Acyclophosphamid. Arzneimittel-Forsch. **14**, 1066—1067 (1964).
ROBERTS, J. J., BRENT, T. P., CRATHORN, A. R.: The mechanism of alkylating agents on the mammalian cells. Inactivation of the DNA template and its repair. In: CAMPBELL, P. N. (Ed.): A symposium on the interactions of drugs and subcellular components in animal cells. Boston: Little, Brown 1968.
SARTORELLI, A. C.: Approaches to the combination chemotherapy of transplantable neoplasms. Prog. exp. Tumor Res. (Basel) **6**, 228—288 (1965).
SARTORELLI, A. C.: Some approaches to the therapeutic exploitation of metabolic sites of vulnerability of neoplastic cells. Cancer Res. **29**, 2292—2299 (1969).
SARTORELLI, A. C., LEPAGE, G. A.: Inhibition of ascites cell growth by combinations of 6-thioguanine and azaserine. Cancer Res. **18**, 938—942 (1958).
SKIPPER, H. E., PERRY, S.: Kinetics of normal and leukemic leukocyte populations and relevance to chemotherapy. Cancer Res. **30**, 1883—1897 (1970).
SLADEK, N. E.: Therapeutic efficacy of cyclophosphamide as a function of its metabolism. Cancer Res. **32**, 535—542 (1972).
TARDIFF, R. G., DUBOIS, K. P.: Inhibition of hepatic microsomal enzymes by alkylating agents. Arch. int. Pharmacodyn. **177**, 445—456 (1969).
TARNOWSKI, G. S., MOUNTAIN, I. M., STOCK, C. C.: Combination therapy of animal tumors with L-asparaginase and antagonists of glutamine or glutamic acid. Cancer Res. **30**, 1118—1122 (1970).
VADLAMUDI, S., PADARATHSINGH, M., WARAVDEKAR, V. S., GOLDIN, A.: Leukemia P 1798 as a possible experimental model for remission induction and remission maintenance for acute leukemia. Proc. Amer. Ass. Cancer Res. **12**, 16 (1971).

van Eden, E. B., Falkson, H. C., Falkson, G.: 1,3-*bis*-(2-chloroethyl)-1-nitrosourea (BCNU; NSC-409962) given concomitantly with cytosine arabinoside (NSC-63878) in the treatment of cancer. Cancer Chemother. Rep. **54**, 347—359 (1970).

Venditti, J. M., Humphreys, S. R., Mantel, N., Goldin, A.: Combined treatment of advanced leukemia (L 1210) in mice with amethopterin and 6-mercaptopurine. J. nat. Cancer Inst. **17**, 631—638 (1956).

Vogler, W. R.: Clinical trials of 1-β-D-arabinofuranosyl cytosine and 1,3-*bis*-(2-chloroethyl)-1-nitrosourea combination in metastatic cancer and acute leukemia. Cancer (Philad.) **27**, 1081—1088 (1971).

Wilbur, J. R., Sutow, W. W., Sullivan, M. P., Costio, J. R., Taylor, H. G.: Successful treatment of rhabdomyosarcoma with combination chemotherapy and radiotherapy. Presented at the American Society of Clinical Oncology, Chicago, April 7, 1971.

Wodinski, I., Kensler, C. J.: Activity of selected compounds in subline of leukemia L 1210 resistant to cytosine arabinoside (NSC-63878). Cancer Chemother. Rep. **43**, 1—3 (1964).

Wodinsky, I., Kensler, C. J.: Activity of cytosine arabinoside (NSC-63878) in a spectrum of rodent tumors. Cancer Chemother. Rep. **47**, 65—68 (1965).

Wolinsky, E., Hines, J. D.: Neurotoxic and nephrotoxic effects of colistin in patients with renal disease. New Engl. J. Med. **266**, 759—762 (1962).

Zager, R. F., Frisby, S. A., Oliverio, V. T.: Cellular transport and antitumor activity of methotrexate (MTX) in combination with clinically useful drugs. Proc. Amer. Ass. Cancer Res. **13**, 33 (1972).

Chapter 72

Combined Actions of Antimicrobial Drugs*

E. JAWETZ

With 5 Figures

I. Introduction

Physicians generally believe that antimicrobial drugs are widely effective and fairly harmless. Most physicians also have the vague feeling that if one antimicrobial drug is good, two should be better, and three should cure almost everybody of everything. This attitude may be responsible for the widespread use of antimicrobial drugs in combinations. At Johns Hopkins Hospital, 11% of 7094 patients admitted during a three-month period in 1963 received from two to five antimicrobial drugs (CLUFF et al., 1964). Since that time several scientific and medical organizations have strongly opposed the widespread use of fixed combinations of antimicrobial drugs (A. M. A. Council on Drugs, 1970; Natl. Res. Council, 1969). In fact, fixed combinations have been removed from the market in the United States by action of the Food and Drug Administration. No evidence exists, however, that the prescribing of several antimicrobials simultaneously has diminished. It is therefore appropriate to reexamine the experimental basis of combined antimicrobial drug action and the clinical situations in which such combined antimicrobial action might be necessary or harmful. This review draws heavily on experimental work of our own group in the Department of Microbiology, University of California, San Francisco, and on previous summaries prepared by us (JAWETZ and GUNNISON, 1953; JAWETZ, 1967; JAWETZ, 1968). The literature in this field is vast and a complete survey is impractical. Therefore, only representative individual papers are cited.

II. Possible Clinical Indications for Combined Antimicrobial Drug Treatment

Although the indiscriminate prescription of antimicrobial combinations is deplored, certain clinical situations permit, or even compel, the use of such combinations. Some of the most important of these clinical situations are the following:

a) Overwhelming, Life-Threatening Infections

When treating some acutely and seriously ill patients suspected of having bacteremia, the physician may be forced to administer two or more antimicrobials simultaneously in the hope of arresting the growth of an unknown etiologic agent. Prior to such drug treatment, all steps must be taken to make a prompt and precise laboratory diagnosis. The drugs are chosen by "aiming" at the types of organisms most likely to cause the clinical symptoms in a given patient, and are administered only until laboratory identification of the etiologic organism permits specific

---

* This chapter was completed in January 1972.

treatment. At present, sepsis caused by gram negative bacteria, particularly in hospitalized, immunodeficient patients, is the leading problem in this category. The empiric administration of drug mixtures in suspected gram negative bacteremia has been widely advocated, e.g., carbenicillin plus gentamicin (Schimpff et al., 1971), polymyxin plus kanamycin (Hodgin and Sanford, 1965; Bryant et al., 1971), or kanamycin plus ampicillin (Gotoff and Behrman, 1970) among many others (Jawetz, 1968). Naturally, preferences vary from one hospital to another and from one year to another, depending on the drug susceptibility of organisms commonly encountered in a given setting, and on the advent of new drugs.

b) Rapid Emergence of Drug-Resistant Microbial Mutants

When treating patients with certain chronic infections the rapid emergence of mutants resistant to one drug may impair the chances for cure. The addition of a second drug sometimes delays the emergence of resistant strains (Klein and Kimmelman, 1947). This effect has been unequivocally demonstrated in chronic mycobacterial infections (Cohn et al., 1959; Heaton et al., 1959; Mandel et al., 1959; Russell et al., 1959). This effect is the basis for simultaneous administration of non-cross-reacting drugs such as streptomycin plus aminosalicylic acid or isoniazid plus ethambutol when treating patients with tuberculosis. Concern for the prevention of rapidly emerging drug resistance may lead to the use of drug combinations also in acute infections, e.g., tetracycline plus streptomycin in plague, or gentamicin plus carbenicillin in pseudomonas sepsis (Smith et al., 1969).

c) Microbial Infections that May Require Drug Synergism for Eradication of an Infectious Process

This topic is discussed in detail below.

d) Mixed Infections

The rare occasions when two or more distinct microorganisms produce a systemic infection may warrant the use of two separate drugs, each aimed at one infecting organism. More commonly, a mixture of microorganisms participates in infections of surfaces, e.g., skin abrasions, wounds, mucous membrane lesions. The topical application of poorly absorbed antimicrobials (e.g., neomycin plus bacitracin plus polymyxin) to infected superficial lesions or conjunctivitis is widely practiced and probably reasonable.

e) Possible Reduction in Toxicity

Occasionally a toxic antimicrobial may be given in much smaller doses when combined with a second drug than when given alone. This may substantially reduce the risk of adverse effects. For example, the addition of ampicillin to a kanamycin regimen for the treatment of a chronic urinary tract infection caused by enterobacter may permit the use of lower doses of kanamycin.

In summary, drug combinations may be clinically effective because of the simultaneous operation of several of the above, and perhaps other features. At the same time, administration of combinations of drugs has distinct disadvantages when compared to the use of a single properly selected antimicrobial.

III. Clinical Disadvantages of Combined Antimicrobial Drug Treatment

Simultaneous use of two or more antimicrobials implies a promise of "broad spectrum" and "wide coverage". This engenders a false sense of security in the physician, and discourages strenuous and repeated efforts toward a specific etiologic diagnosis.

The more drugs are given simultaneously, the greater the likelihood of adverse effects. These include direct toxic effects, sensitization, and hypersensitivity reactions to each drug in a combination. The variety of reactions confuses the physician and harms the patient.

Simultaneous use of two or more drugs instead of a single one imposes an unnecessarily high cost. If employed in a closed environment, e.g., a hospital, it also favors the emergence of microorganisms resistant to several drugs in that environment and thus contributes to the danger of hospital-borne infections (LEPPER et al., 1956).

It is incorrect and misleading to speak of a "synergistic" drug combination without specifying a microbial strain. No fixed drug combination regularly results in such a desirable effect. This and many other considerations have led to a regulatory agency ban on fixed antimicrobial combinations in the United States (Natl. Res. Council, 1969).

Antagonism between antimicrobial drugs is a theoretical argument against the indiscriminate use of drug combinations, although such an undesirable effect is an infrequent result of clinical treatment, as described below.

IV. Problems in Defining and Measuring Effects of Antimicrobial Drugs in Combination

A central problem in the laboratory study of antimicrobial combinations has been the diversity of methods that can be used to assess antimicrobial activity, and the difficulty of relating unequivocally any event *in vitro* to a therapeutic result in human diseases (JAWETZ, 1967). Even when feasible, chemical estimates of the plasma concentration of antimicrobial drugs may not reflect their antimicrobial activity, since large amounts of the drug may be highly bound to plasma proteins and therefore inactive. For this reason, direct evaluation of antimicrobial activity often is easier and more meaningful. Several methods are used to measure antimicrobial activity, and each type of measurement determines the definition that can be applied to antimicrobial drug interactions and to their relevance in clinical therapy. The most common methods are the following: (1) Estimation of bacteriostatic effect. Endpoints are expressed as the minimum amount of drug necessary to suppress visible growth of a microorganism under a given set of laboratory conditions for a given time. (2) Estimation of bactericidal effect. Results are expressed as the bactericidal rate, i.e., the slope of the plot of viable survivors at various time intervals, under a given set of laboratory conditions. Alternatively, bactericidal effects can be expressed as the smallest concentration of drug resulting in a fixed number (or proportion) of viable survivors at a fixed time, under a given set of conditions. (3) Estimation of therapeutic effect in experimental infections (prevention of death or lesions, prolongation of life). (4) Evaluation of therapeutic effect in natural infections (clinical cure, suppression or eradication of infection).

The results of any one method of evaluation may not coincide with those of another method because each may measure different events in the test system. Even within one test system, results may depend on drug concentration, time of drug-microbe interaction, inoculum size, other laboratory variables and on the meaning attached to a given event. This is illustrated in Fig. 1. At Time I, fewer bacteria survive exposure to Drug A than to the combination of A plus B. Thus, in terms of early bactericidal effect, A plus B is less effective than A alone – an example of antimicrobial antagonism. On the other hand, at Time II, fewer bacteria survive exposure to Drugs A plus B than to Drug A alone – perhaps an example of positive drug summation. The interpretation of such test results obviously depends on whether importance is attached to early bactericidal action or to completeness of bactericidal effect.

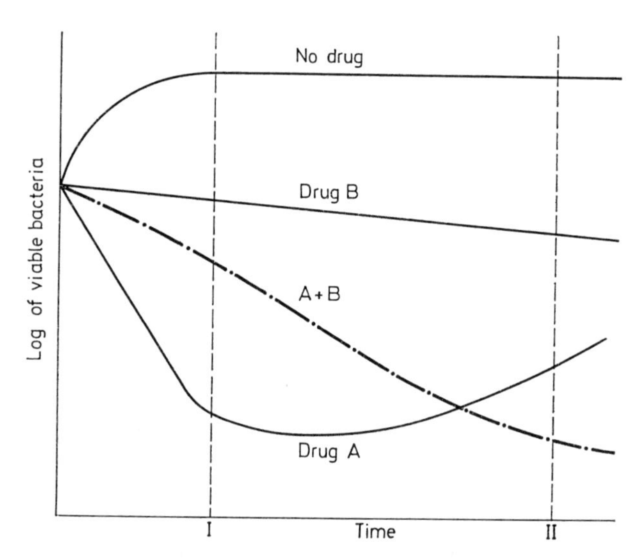

Fig. 1. Schematic representation of numbers of bacteria remaining viable at two times of incubation, with exposure to Drug A, Drug B, or Drugs A + B

The fallacy must also be considered of comparing single drugs with combinations on the basis of dose (mg/kg). Experimentally, one can determine the amount of one minimal effective dose (MED) of drug in bacteriostatic, bactericidal, or curative terms. However, the relationship between dose and effect of a drug cannot be estimated experimentally below one MED. This relationship might be X or Y in Fig. 2 rather than linear. Consequently, the effect of one-half MED by weight of Drug A plus one-half MED by weight of Drug B might be greater than, equal to, or less than the effect of one MED by weight of either drug alone. In the past, this point has been the basis for claims of "synergism", i.e., greater than additive effects, a claim that is not warranted (JAWETZ and GUNNISON, 1953).

The selection of drug pairs for clinical therapy often is a complex task. Several laboratory methods have been proposed for this purpose, but all of them are cumbersome and expensive. Most methods employ arbitrary concentrations of drugs (resembling those reached in body fluids) in various combinations, and a two-step procedure. In the first step, bacteria are exposed to single or combined drugs and, after incubation, inhibition of growth is noted. In the second step, bacteria are removed from the influence of drugs and placed in a nutrient that will permit growth if the organisms have survived the initial contact with drug. However, many variables, such as inoculum size, incubation time, and medium, can influence the result (JAWETZ et al., 1955; GARROD and WATERWORTH, 1962).

Equally important is whether the patient receiving drug combinations exhibits the desired activity in his body fluids. This is most commonly determined by assaying serum or other fluid obtained from the patient during therapy, using as a bioassay the organism isolated from the patient prior to the beginning of treatment (SCHLICHTER and MACLEAN, 1947). If the body fluid of the treated patient diluted 1:5 or more rapidly kills the patient's original microorganism, it is assumed that he is receiving adequate therapy (JAWETZ, 1962; BLOUNT, 1965).

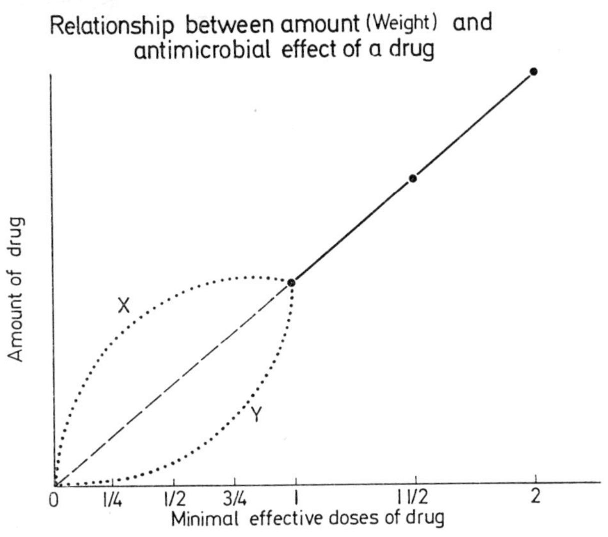

Fig. 2. Hypothetical relationship between the weight of a drug and its antimicrobial effect

V. Dynamics of Combined Antimicrobial Action

Several of the potential advantages of antimicrobial combinations in clinical therapy listed above are based on the presence of heterogeneous populations of microbes. When two antimicrobial drugs act simultaneously on a homogeneous microbial population, the resulting effect can be one of three types: indifference, synergism, or antagonism (Fig. 3). By far the most common result is indifference; in other words, the effect of two drugs is equal to that of the single more active drug, or is equal to the arithmetic sum of the effects of the two individual drugs. The same total effect can be obtained with a single drug used in a dose equivalent to that of the mixture. The vast majority of antimicrobial combinations used in clinical practice give end results in the category of indifference. The original observations on antagonism and synergism *in vitro* are shown in Fig. 4 (JAWETZ et al., 1950).

A. Antagonism

Antagonism may be defined as a combined drug effect that is smaller than the sum of the effects of the single drugs present in a mixture. Since this is difficult to measure, the term antagonism should be restricted to those instances in which a combination of antimicrobials results in a total effect smaller than that produced by the single action of the more effective member of the combination.

Antagonism can be demonstrated *in vitro* by a decrease either in the bacteriostatic activity or in the bactericidal rate (Fig. 4) of a drug combination below that of its components. To establish the phenomenon, it is essential to rule out physical

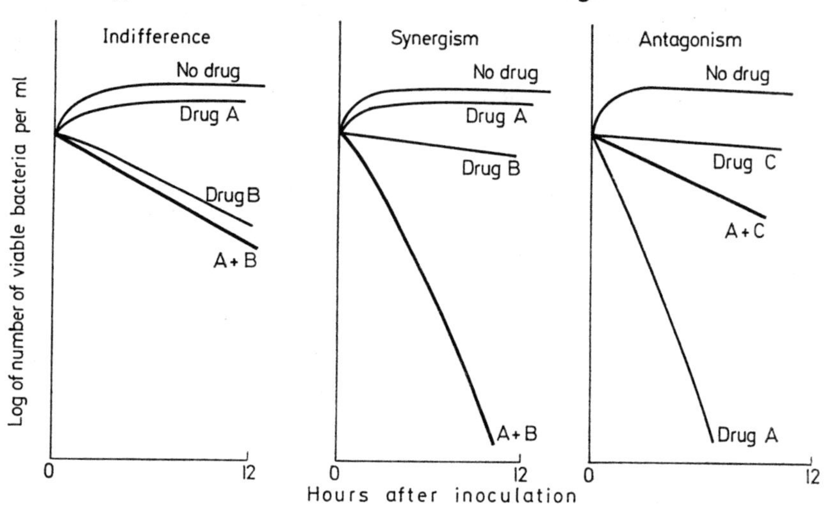

Fig. 3. Types of combined action of two antimicrobial drugs on a homogeneous microbial population. Schematic representation of bactericidal action *in vitro* showing the possible types of results seen when one drug, or two drugs, act on a homogeneous population of bacteria under conditions permitting growth

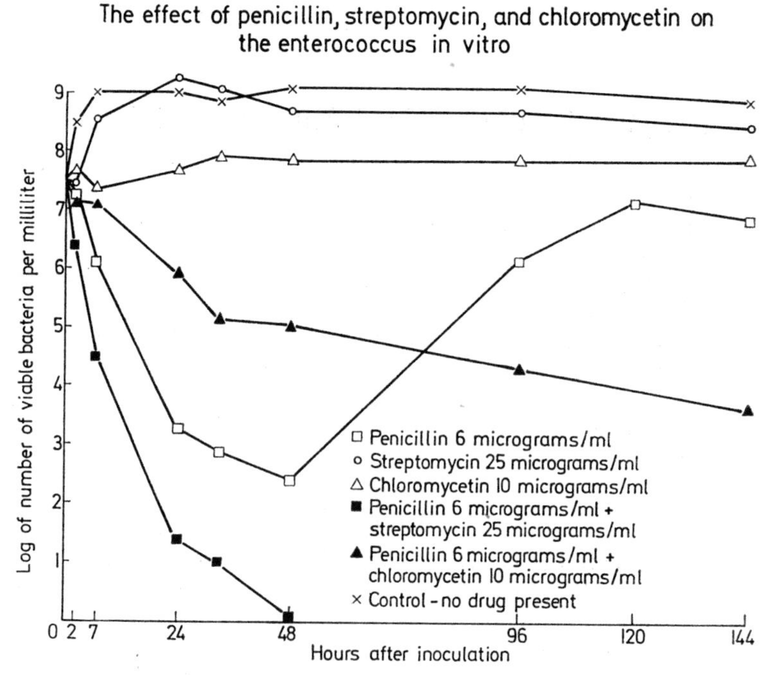

Fig. 4. The effects of penicillin (6 μg/ml), streptomycin (25 μg/ml), and chloramphenicol (10 μg/ml) alone, and in combination, on a strain of enterococcus (*Strep. fecalis*) *in vitro*. Penicillin and streptomycin are synergistic; penicillin and chloramphenicol are antagonistic (JAWETZ et al., 1950)

interaction, e.g., precipitation, of one drug with the other in the mixture. For optimal demonstration of bactericidal antagonism, a minimal bacteriostatic quantity of the antagonizing drug must be added to a bactericidal quantity of the effective drug, as illustrated in Fig. 4. A large excess of the effective drug can usually overcome antagonism; likewise, an amount of the antagonizing drug greatly in excess of a bacteriostatic quantity often obscures antagonism. Antagonism is limited not only by dose but also by time relationships; the antagonizing drug must act on microorganisms either before or at the same time as the bactericidal drug, but not after (WALLACE et al., 1967). Antagonism can be demonstrated only with organisms that are capable of multiplication under the conditions of the test, in terms of temperature, nutrients, etc. (GUNNISON et al., 1955).

Within the limits described above, antagonism is easily demonstrated *in vitro*. Antagonism also may occur readily when experimental infections are treated with single doses of antimicrobial drugs. Figure 5 illustrates the death rates in mice with streptococcal infections treated either with penicillin G alone, or with penicillin plus chloramphenicol. The death rate increases progressively the more chloramphenicol is added to the minimal 95% curative dose of penicillin (18 μg/ mouse). However, a five-fold larger dose of penicillin overcomes this antagonism.

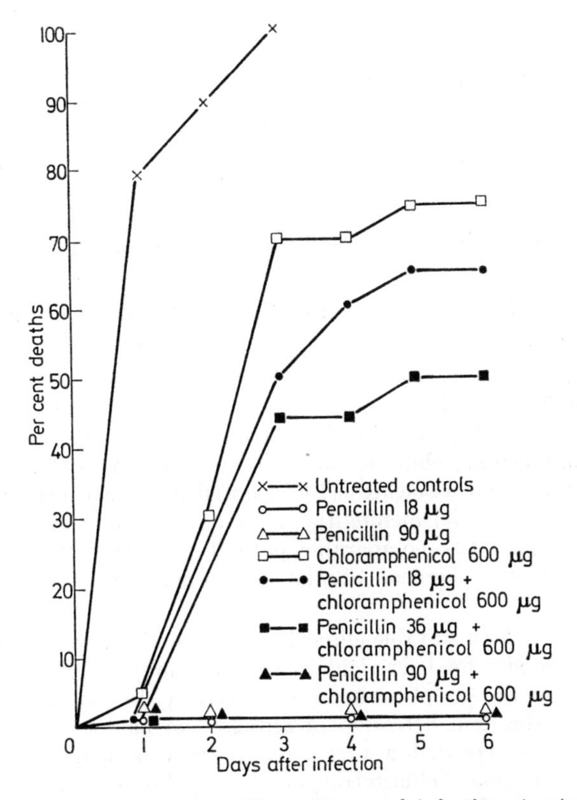

Fig. 5. Death rates observed in mice with streptococcal infections treated with penicillin alone (18 or 90 μg/mouse), chloramphenicol alone (600 μg/mouse), or penicillin in various doses administered at the same time as chloramphenicol, but injected separately. Although chloramphenicol antagonizes the action of penicillin, an excess of penicillin can overcome this antagonism

In multiple-dose treatment of experimental infections, the everchanging blood and tissue levels of the two drugs reduce the opportunity for observing antagonism as the outcome of treatment (JAWETZ and GUNNISON, 1953). In experimental pneumococcal meningitis, antagonism can be demonstrated only if the antagonizing chloramphenicol is administered prior to penicillin (WALLACE et al., 1967).

1. Mechanism of Antimicrobial Antagonism

When antimicrobial drugs are combined, physical or chemical interactions may occur *in vitro* that result in loss of activity (NOONE and PATTISON, 1971). However, such incompatibilities can be excluded as a mechanism in the bactericidal antagonism discussed here (WINTERS et al., 1971). The most frequently antagonizing drugs are tetracyclines, chloramphenicol, and macrolides, all inhibitors of protein synthesis in bacteria. These drugs can antagonize penicillins (inhibitors of cell wall synthesis), and aminoglycosides (inhibitors of ribosomal protein synthesis) (JAWETZ et al., 1951). On the other hand, drugs that act on cell membranes (e.g., polymyxins), or other drugs that inhibit cell wall synthesis (e.g., vancomycin, bacitracin), have not been shown to antagonize penicillins or aminoglycosides. It has therefore been postulated that inhibition of protein synthesis is a prerequisite for antagonism (CHANG and WEINSTEIN, 1966). However, sulfonamides can antagonize penicillins (GUNNISON et al., 1961), and these drugs do not inhibit bacterial protein synthesis. Thus, perhaps antagonism can occur whenever bacteriostatic drugs slow microbial metabolism in situations that require fully active metabolism for a maximum bactericidal effect. The modes of action of penicillins and aminoglycosides are so different that it is difficult to imagine a single specific biochemical event to interfere with their actions. Likewise, the modes of action of several antagonizing drugs (sulfonamides, tetracyclines, chloramphenicol) are so divergent as to make it improbable that a single biochemical feature could be responsible for their antagonizing capacity.

2. Antimicrobial Antagonism in the Treatment of Clinical Disease

The most dramatic examples of antimicrobial drug antagonism in clinical therapy come from the treatment of patients with bacterial meningitis. LEPPER and DOWLING (1951) noted a mortality rate of 30% in patients with pneumococcal meningitis treated with penicillin alone, compared to a mortality rate of 79% among patients receiving chloratetracycline in addition to the same regimen of penicillin. MATHIES and associates (1967) noted a fatality rate of 4.3% among 140 children with bacterial meningitis receiving ampicillin alone, and a fatality rate of 10.5% among 124 comparable patients receiving the same regimen of ampicillin plus chloramphenicol. Similar results have been recorded by OLSSON et al. (1961).

Meningitis provides a suitable clinical setting for the demonstration of antagonism in man: necessity for bactericidal action but only marginally effective drug levels attainable at the site of antimicrobial activity. In bacterial infections other than meningitis, the demonstration of antagonism has been less dramatic and more difficult. When penicillin was used to treat patients with streptococcal pharyngitis, the addition of chlortetracycline or erythromycin interfered somewhat with the eradication of streptococci (STRÖM, 1961). However, the clinical course was similar in all treated patients. In bacteriuria associated with chronic pyelonephritis, the organism was eliminated in 88% of patients if a drug combination exhibiting synergism *in vitro* was employed, but in only 14% of patients if the drug combination exhibited antagonism (McCabe, 1967). These experiences lead to the conclusion that antimicrobial antagonism can occur in man, but that with the exception of meningitis, it is rarely of sufficient magnitude to interfere with a successful outcome of treatment.

B. Synergism

There is no generally accepted definition of antimicrobial synergism that would separate it from the effects of simple summation. Goodman and Gilman's text (1965) suggests that "Synergism should be reserved to describe combined effects of heterergic drugs that are greater than those of the active component alone". In the discussion of antibiotic combinations in that text the term synergism is not used at all – probably wisely.

Tests for the estimation of positive summation of antimicrobial action *in vitro* ("synergism") may indicate: (1) enhanced bacteriostatic effect, (2) selective suppression by each drug of a separate portion of the bacterial population, (3) delay by one drug of the emergence of mutants resistant to the other drug, (4) increased bactericidal action. One or more of these actions may play a role in the clinical application of antimicrobial combinations as described in an earlier section.

1. Mechanism of Antimicrobial Synergism

The mechanism is fairly well understood in at least three types of antimicrobial synergism. Each of them is described briefly, without reference to its relative clinical importance.

a) Blocking Successive Steps in a Metabolic Sequence

Sulfonamides are antibacterial because they compete with para-aminobenzoic acid that is required by some bacteria for the synthesis of dihydrofolate. Folate antagonists, like trimethoprim or pyrimethamine, inhibit the activity of the enzyme that reduces dihydrofolate to tetrahydrofolate (dihydrofolic acid reductase). These folate antagonists inhibit the enzyme of bacteria and protozoa thousands of times more efficiently than the same enzyme of mammalian cells. The simultaneous presence of a sulfonamide and a folate antagonist results in the simultaneous block of sequential steps in the pathway leading to the synthesis of purines and nucleic acid. Such a simultaneous block can result in much more complete inhibition of growth than either component of the mixture alone. Optimal proportions of the component of the mixture have been determined *in vitro* (Hitchings, 1970) but may be difficult to maintain *in vivo*. This type of combined antimicrobial action is being applied to urinary tract infections, enteric fevers, malaria and toxoplasmosis.

b) Inhibition by one Drug of an Enzyme Capable of Destroying the Second Drug

Penicillins susceptible to hydrolysis by beta-lactamases are usually ineffective against organisms that produce such lactamases. However, if the beta-lactamase of an organism could be firmly bound, or its production inhibited, then penicillins might have a chemotherapeutic effect on this organism. Long ago a marked additive effect of penicillins and tetracyclines, when acting on some beta-lactamase-producing staphylococci, had been attributed to inhibition of adaptive enzyme formation by tetracycline (Hahn and Wisseman, 1951). Some lactamase-resistant penicillins and cephalosporins have a much higher affinity for certain beta-lactamases than do hydrolyzable penicillins (Sutherland and Batchelor,

1964; SABATH and ABRAHAM, 1964). Beta-lactamases from gram negative bacteria, e.g., pseudomonas, are bound very efficiently by methicillin, cloxacillin, or cephalosporins. These drugs are fairly resistant to the enzyme action and can protect hydrolyzable penicillins from destruction. This type of combined effect has been employed a few times in clinical therapy (SABATH et al., 1967). However, sufficient concentration of the binding drug can be achieved readily only in urine, not in the systemic circulation. In addition, the effect is observed best with low bacterial concentrations (FRAHER and JAWETZ, 1967), a fact that limits the application of this approach to combined drug action.

c) *Antimicrobial Synergism Manifested by Marked Enhancement of Early Bactericidal Rate*

This type of combined antimicrobial action has been studied most extensively, and has been reviewed repeatedly (JAWETZ and GUNNISON, 1953; JAWETZ, 1968). A few characteristics of this type of synergism are listed here. The effect extends over a fairly wide range of concentrations of each member of the drug pair and is not critically influenced by the proportions in the mixture. Only one member of the pair need exhibit antibacterial activity alone in the concentration employed in the mixture; the other may be ineffective in that concentration although active against the particular organism in a 10-fold to 1000-fold larger amount (JAWETZ and GUNNISON, 1953; MOELLERING et al., 1971). The simultaneous presence of the two drugs is essential for synergism. If one drug is removed before the second is added, synergism does not occur (MILES et al., 1951).

This type of synergism has been demonstrated with gram negative rods, staphylococci, and other organisms, but it has been studied in greatest detail with enterococci (*Strep. fecalis*). Enterococci are usually inhibited, but not killed, by penicillins (JAWETZ et al., 1950). The addition of an aminoglycoside (streptomycin, kanamycin, or gentamicin) in amounts achievable in the systemic circulation results in an immediate striking increase in the rate of bactericidal action (MILES et al., 1951) (Fig. 4). The concentration of aminoglycoside, which is synergistic with the penicillin, usually has no discernible action alone, but inhibits the organisms in a concentration 10-fold to 1000-fold higher. If the organism is genetically resistant to a high concentration of a particular aminoglycoside (e.g., streptomycin more than 6 mg/ml or kanamycin more than 3.7 mg/ml), then synergism of penicillin with that particular aminoglycoside does not occur (STANDIFORD et al., 1970; ZIMMERMANN et al., 1970; MOELLERING et al., 1971). This rule seems generally valid although some exceptions have been observed (WATANAKUNAKORN, 1971).

This type of synergism probably depends on the alteration of microbial permeability by one drug so that penetration by the second drug is enhanced. PLOTZ and DAVIS (1962) showed that brief treatment of growing *E. coli* with penicillin hastened subsequent killing of these cells by streptomycin. They suggested that penicillin damaged the cell wall, which promoted further damage by streptomycin and increased its access to intracellular sites. Subsequently, MOELLERING et al. (1971) demonstrated that enterococci exhibit a relative permeability barrier to aminoglycosides, and that this barrier can be breached by agents that inhibit cell wall synthesis or alter cell wall structure. They showed that uptake of radioactive streptomycin was enhanced in enterococci growing in the presence of penicillin or other substances that impair bacterial cell wall synthesis or structure (vancomycin, bacitracin, EDTA). However, streptomycin uptake was not affected by drugs that are known to affect the cell membrane, or by drugs that inhibit protein synthesis (MOELLERING and WEINBERG, 1971). Conversely, streptomycin does not enhance the uptake of radioactive penicillin (ROWLEY et al., 1950).

HEWITT and his associates (1966) have suggested that synergism of penicillin with an aminoglycoside might involve the production by penicillin of cell-wall deficient forms with higher susceptibility to the aminoglycoside than the parent bacteria. Arguments against this proposal were presented by MOELLERING and WEINBERG (1971). They conclude that the aminoglycoside is essential in producing the bactericidal effect on enterococci, and therefore organisms that are genetically resistant to a given aminoglycoside also are resistant to the penicillin plus aminoglycoside synergism (ZIMMERMAN et al., 1970).

Many observations indicate that cell wall inhibitors other than penicillins may be synergistic with aminoglycosides. Thus vancomycin can enhance the action of streptomycin on enterococci (MANDELL et al., 1970; WILKOWSKE et al., 1970), and cephalosporins can enhance the action of kanamycin on methicillin-resistant staphylococci (BULGER, 1967). Cephalosporins are less satisfactory than penicillins in combinations with aminoglycosides acting on enterococci (RAHAL et al., 1968). Before the advent of methicillin, JAWETZ and BRAINERD (1962) showed that lactamase-producing staphylococci were sometimes susceptible to synergistic effects of bacitracin with penicillin and other drugs.

Apparent synergism within the definitions given here has also been encountered with many other drug combinations acting on gram negative organisms. MCCABE and JACKSON (1965) demonstrated apparent bactericidal synergism of aminoglycosides with tetracycline or chloramphenicol acting on coliform organisms from urinary tract infections. Other gram negative bacteria (proteus, enterobacter, klebsiella) at times also were susceptible to ampicillin plus aminoglycoside synergism (BULGER and KIRBY, 1963; BULGER and ROOSEN-RUNGE, 1969; KAYE et al., 1961). Carbenicillin can be synergistic with gentamicin for some strains of pseudomonas (EICKHOFF, 1969; SMITH et al., 1969; SONNE and JAWETZ, 1969). Polymyxins can be synergistic with sulfonamides against some strains of serratia (GREENFIELD and FEINGOLD, 1970). The mechanism outlined above probably applies in most of these situations. Amphotericin B and 5-fluorocytosine may be synergistic against certain yeast-like organisms, but the mechanism is obscure (MEDOFF et al., 1971).

2. Antimicrobial Synergism in Clinical Disease

Combinations of sulfonamides and folic antagonists have been used widely for several years (HITCHINGS, 1971). While the experimental support for synergism between sulfadiazine and pyrimethamine in mouse toxoplasmosis is good (EYLES and COLEMAN, 1955), little evidence exists that such a conclusion applies in human toxoplasmosis (FELDMAN, 1968a and b). Nevertheless, most patients with clinically diagnosed toxoplasmosis receive such combined treatment.

For human therapy, a fixed combination of 1 part trimethoprim to 5 parts of a sulfonamide (mostly sulfamethoxazole in Britain) is usually given. CATTELL et al. (1971), among others, reported on the use of this combination in urinary tract infections; DONNO et al. (1969) reported its use in malaria; and KAMAT (1970) reported trials in the treatment of brucellosis or enteric fevers. Undoubtedly, this combination is often clinically effective and sometimes preferable to other drugs. However, I am not aware of any controlled clinical studies that clearly establish that the action of the trimethoprim-sulfonamide combination presents a positive summation manifestly beyond the range of simple additive effects of each of the components of the mixture.

The clinical application of beta-lactamase binding by certain enzyme-resistant penicillins is difficult since large amounts of penicillin are required. Treatment of urinary tract infection is an exception (SABATH at al., 1967).

The type of antimicrobial synergism that results in a large increase in the rate of bactericidal action has been applied in clinical medicine chiefly in disease states which require bactericidal action for cure. This includes bacterial endocarditis and infections by organisms, such as enterococci, that can be suppressed but not eradicated by a single drug. Many cases of endocarditis caused by viridans streptococci (the commonest etiologic agent) can be cured by penicillin. However, the simultaneous use of an aminoglycoside can greatly increase the rate of bactericidal action, making possible a shorter course of treatment. Reports of several different groups of patients in the last ten years have illustrated this point (DOWLING, 1965), most recently TAN et al. (1971).

Enterococci (*Strep. fecalis*) are the etiologic organisms in 5 to 15% of cases of bacterial endocarditis. Penicillin inhibits most strains of enterococci *in vitro*, but the bactericidal effect of penicillin is pronounced in only a few strains. Treatment with penicillin alone fails to cure a majority of patients with enterococcal endocarditis (KOENIG and KAYE, 1961; JAWETZ and SONNE, 1966). Ampicillin may be somewhat more effective than penicillin G as a single drug, but in many cases it, too, cannot be relied upon to eradicate the enterococci (BEATY et al., 1967; JAWETZ and SONNE, 1966). HUNTER and PATERSON (1956) and ROBBINS and TOMPSETT (1951) empirically discovered that patients with enterococcal endocarditis could be cured by a combination of penicillin and streptomycin. This drug combination was strikingly bactericidal for many strains of *Streptococcus fecalis* (JAWETZ et al., 1950) (Fig. 4). Subsequent clinical experience confirmed the efficacy of penicillin and streptomycin combinations in curing patients with enterococcal endocarditis, and it was often stated that enterococci were uniformly susceptible to this combined drug effect. This belief was manifestly incorrect. Occasional enterococcal strains were not killed by penicillin combined with streptomycin (HEWITT et al., 1966; JAWETZ and SONNE, 1966). Patients infected with such strains died of endocarditis unless a different bactericidal drug combination, e.g., penicillin plus kanamycin (HEWITT et al., 1966), could be given. Systematic surveys of enterococcal isolates (STANDIFORD et al., 1970; MOELLERING et al., 1971) suggest that strains with very high resistance to a given aminoglycoside are not subject to synergistic action of that aminoglycoside with penicillin. Simple determination of insusceptibility of a strain to these high concentrations of aminoglycoside (e.g., streptomycin more than 6 mg/ml, kanamycin more than 3.7 mg/ml) may provide the information that a given combination should not be used clinically (STANDIFORD et al., 1970). Most isolates of enterococci have been susceptible to small concentrations of gentamicin thus far (WATANAKUNAKORN, 1971), and this drug has exhibited synergistic action with penicillin against enterococci *in vitro* quite uniformly.

Enterococci are often resistant to cephalosporins employed alone. Although *in vitro* cephalosporin plus aminoglycoside synergism has been observed (BEATY et al., 1967), the administration of cephalosporins with streptomycin has repeatedly failed to cure patients with enterococcal endocarditis (RAHAL et al., 1968).

Bactericidal synergism has also been employed in the treatment of bacterial infections caused by other organisms. Before the advent of beta-lactamase-resistant penicillins, staphylococcal sepsis and endocarditis were frequently cured by specifically selected synergistic drug combinations (JAWETZ and BRAINERD, 1962). When treating patients with chronic urinary tract infections caused by various gram negative bacteria, MCCABE and JACKSON (1965) selected specific drug combinations that exhibited synergistic action against the offending microorganism. Treatment with such synergistic drug combinations eliminated the organism in 88% of patients, whereas treatment with drug combinations that exhibited antagonism was successful in only 14%. The significant difference between these two groups supports the faith in clinical efficacy of specifically selected synergistic antimicrobial combinations. Dozens, or perhaps hundreds, of other claims for the clinical efficacy of "synergistic" antimicrobial combinations have been made, but the validity of most of these is doubtful.

VI. Conclusion

Perhaps because of its large financial impact, the study of combined antimicrobial drug action has been filled with intense emotions, violent claims and counterclaims. An effort was made here to describe the simplest common denominators that determine combined antimicrobial activity *in vitro*, and their application to experimental and clinical treatment. A famous physician is said to have written 800 years ago: "If one can manage well with one individual drug, one should not use a compound one, ... one should use medications compounded of multiple ingredients only when compelled to do so" (BEN-MAIMON, 1958). Hopefully, this brief outline may help physicians to feel compelled less often to use combinations of antimicrobial drugs.

Acknowledgements

The investigative work performed in the author's laboratory was supported, in part, by donations from the Research Committee of the University of California and from the Burroughs Wellcome Fund.

References

A.M.A. Council on Drugs: Fixed-dose combinations of drugs. J. Amer. med. Ass. **213**, 1172—1175 (1970).

BEATY, H.N., TURCK, M., PETERSDORF, R.G.: Activity of broad spectrum antibiotics against enterococci and their efficacy in enterococcal endocarditis. Ann. N.Y. Acad. Sci. **145**, 464—471 (1967).

BEN-MAIMON, RABBI MOSES: The preservation of youth. New York: Philosophical Library 1958.

BLOUNT, J.G.: Bacterial endocarditis. Amer. J. Med. **38**, 909—922 (1965).

BRYANT, R.E., HOOD, A.F., HOOD, C.E., KOENIG, M.G.: Factors affecting mortality of gram negative rod bacteremia. Arch. intern. Med. **127**, 120—128 (1971).

BULGER, R.J.: *In-vitro* activity of cephalothin/kanamycin and methicillin/kanamycin combinations against methicillin-resistant *Staphylococcus aureus*. Lancet **1967 I**, 17—19.

BULGER, R.J., KIRBY, W.M.M.: Gentamicin and ampicillin: synergism with other antibiotics. Amer. J. med. Sci. **246**, 717—726 (1963).

BULGER, R.J., ROOSEN-RUNGE, U.: Bactericidal activity of the ampicillin/kanamycin combination against *Escherichia coli*, enterobacter-klebsiella, and proteus. Amer. J. med. Sci. **258**, 7—13 (1969).

CATTELL, W.R., CHAMBERLAIN, D.A., FRY, I.K., MCSHERRY, M.A., BROUGHTON, C., O'GRADY, F.: Long-term control of bacteriuria with trimethoprim-sulfonamide. Brit. med. J. **1971 I**, 377—379.

CHANG, T.W., WEINSTEIN, L.: Inhibitory effects of other antibiotics on bacterial morphologic changes induced by penicillin G. Nature (Lond.) **211**, 763—765 (1966).

CLUFF, L.E., THORNTON, G.F., SEIDL, L.G.: Studies on the epidemiology of adverse drug reactions. I. Methods of surveillance. J. Amer. med. Ass. **188**, 976—983 (1964).

COHN, M.I., MIDDLEBROOK, G., RUSSELL, W.R., JR.: Combined drug treatment of tuberculosis. I. Prevention of emergence of mutant populations of tubercle bacilli resistant to both streptomycin and isoniazid *in vitro*. J. clin. Invest. **38**, 1349—1355 (1959).

DONNO, L., SANGUINETI, V., RICCIARDI, M. L., SOLDATI, M.: Antimalarial activity of Kelfizina-trimethoprim and Kelfizina-pyrimethamine versus chloroquine in field trials in Nigeria. Amer. J. trop. Med. Hyg. **18**, 182—187 (1969).

DOWLING, H. F.: Present status of therapy with combinations of antibiotics. Amer. J. Med. **39**, 796—803 (1965).

EICKHOFF, T. C.: *In vitro* effects of carbenicillin combined with gentamicin or polymyxin B against *Pseudomonas aeruginosa*. Appl. Microbiol. **18**, 469—473 (1969).

EYLES, D. E., COLEMAN, N.: An evaluation of the curative effects of pyrimethamine and sulfadiazine, alone and in combination, on experimental mouse toxoplasmosis. Antibiot. and Chemother. **5**, 529—539 (1955).

FELDMAN, H. A.: Toxoplasmosis. New Engl. J. Med. **279**, 1370—1375 (1968a).

FELDMAN, H. A.: Toxoplasmosis (concluded). New Engl. J. Med. **279**, 1431—1437 (1968b).

FRAHER, M. A., JAWETZ, E.: Combined action of β-lactamase-resistant and β-lactamase-susceptible penicillins on 20 strains of *Pseudomonas aeruginosa*. Antimicrob. Agents Chemother. (Annual) 711—715 (1967).

GARROD, L. P., WATERWORTH, P. M.: Methods of testing combined antibiotic bactericidal action and the significance of the results. J. clin. Path. **15**, 328—338 (1962).

GOODMAN, L. S., GILMAN, A. (Eds.): The pharmacological basis of therapeutics, 3rd. ed. New York: Macmillan 1965.

GOTOFF, S. P., BEHRMAN, R. E.: Neonatal septicemia. J. Pediat. **76**, 142—153 (1970).

GREENFIELD, S., FEINGOLD, D. S.: The synergistic action of the sulfonamides and the polymyxins against *Serratia marcescens*. J. infect. Dis. **121**, 555—558 (1970).

GUNNISON, J. B., KUNISHIGE, E., COLEMAN, V. R., JAWETZ, E.: The mode of action of antibiotic synergism and antagonism: the effect *in vitro* on bacteria not actively multiplying. J. gen. Microbiol. **13**, 509—518 (1955).

GUNNISON, J. B., SPECK, R. S., JAWETZ, E., BRUFF, J. A.: Studies on antibiotic synergism and antagonism: the effect of sulfadiazine on the action of penicillin *in vitro* and *in vivo*. Antibiot. and Chemother. **1**, 259—266 (1951).

HAHN, F. E., WISSEMAN, C. L., JR.: Inhibition of adaptive enzyme formation by antimicrobial agents. Proc. Soc. exp. Biol. (N.Y.) **76**, 533—535 (1951).

HEATON, A. D., RUSSELL, W. F., JR., DENST, J., MIDDLEBROOK, G.: Combined drug treatment of tuberculosis. IV. Bacteriologic studies on the sputum and resected pulmonary lesions of tuberculosis patients. J. clin. Invest. **38**, 1376—1383 (1959).

HEWITT, W. L., SELIGMAN, S. J., DEIGH, R. A.: Kinetics of the synergism of penicillin-streptomycin and penicillin-kanamycin for enterococci and its relationship to L-phase variants. J. Lab. clin. Med. **67**, 792—807 (1966).

HITCHINGS, G. H.: Folate antagonists as antibacterial and antiprotozoal agents. Ann. N.Y. Acad. Sci. (Monogr.) **186**, 444—451 (1971).

HODGIN, U. G., SANFORD, J. P.: Gram-negative rod bacteremia. An analysis of 100 patients. Amer. J. Med. **39**, 952—960 (1965).

HUNTER, T. H., PATERSON, P. Y.: Bacterial endocarditis. D. M. 1—48 (1956).

JAWETZ, E.: Assay of antibacterial activity in serum. Amer. J. Dis. Child. **103**, 81—84 (1962).

JAWETZ, E.: Combined antibiotic action: Some definitions and correlations between laboratory and clinical results. Antimicrob. Agents Chemother. (Annual) 203—209 (1967).

JAWETZ, E.: The use of combinations of antimicrobial drugs. Ann. Rev. Pharmacol. **8**, 151—170 (1968).

JAWETZ, E., BRAINERD, H. D.: Staphylococcal endocarditis: results of combined antibiotic therapy in fourteen consecutive cases (1956—1959). Amer. J. Med. **32**, 17—24 (1962).

JAWETZ, E., GUNNISON, J. B.: Antibiotic synergism and antagonism: an assessment of the problem. Pharmacol. Rev. **5**, 175—192 (1953).

JAWETZ, E., GUNNISON, J. B., COLEMAN, V. R.: The combined action of penicillin with streptomycin or chloromycetin on enterococci *in vitro*. Science **111**, 254—256 (1950).

JAWETZ, E., GUNNISON, J. B., COLEMAN, V. R., KEMPE, H. C.: A laboratory test for bacterial sensitivity to combinations of antibiotics. Amer. J. clin. Path. **25**, 1016—1031 (1955).

JAWETZ, E., GUNNISON, J. B., SPECK, R. S.: Studies on antibiotic synergism and antagonism: the interference of aureomycin, chloramphenicol, and terramycin with the action of streptomycin. Amer. J. med. Sci. **222**, 404—412 (1951).

JAWETZ, E., SONNE, M.: Penicillin-streptomycin treatment of enterococcal endocarditis. A re-evaluation. New Engl. J. Med. **274**, 710—715 (1966).

KAMAT, S. A.: Evaluation of therapeutic efficacy of trimethoprim-sulfamethoxazole and chloramphenicol in enteric fever. Brit. med. J. **1970 III**, 320—322.

KAYE, D., KOENIG, M. G., HOOK, E. W.: The action of certain antibiotics and antibiotic combinations against *Proteus mirabilis* with demonstration of *in vitro* synergism and antagonism. Amer. J. med. Sci. **242**, 320—330 (1961).

KLEIN, M., KIMMELMAN, L. J.: The correlation between the inhibition of drug resistance and synergism in streptomycin and penicillin. J. Bact. **54**, 363—370 (1947).
KOENIG, M. G., KAYE, D.: Enterococcal endocarditis. Report of nineteen cases with long-term follow-up data. New Engl. J. Med. **264**, 257—264 (1961).
LEPPER, M. H., DOWLING, H. F.: Treatment of pneumococcic meningitis with penicillin compared with penicillin plus aureomycin: studies including observations on an apparent antagonism between penicillin and aureomycin. Arch. intern. Med. **88**, 489—494 (1951).
LEPPER, M. H., DOWLING, H. F., JACKSON, G. G., SPIES, H. W., MELLODY, M.: A comparison of the effect of two antibiotics (novobiocin and spiramycin) used in combination with that of spiramycin alone on the antibiotic sensitivity of *Micrococcus pyogenes*. J. Lab. clin. Med. **48**, 920—921 (1956).
MANDEL, W., HEATON, A. D., RUSSELL, W. F., JR., MIDDLEBROOK, G.: Combined drug treatment of tuberculosis. II. Studies of antimicrobially-active isoniazid and streptomycin serum levels in adult tuberculosis patients. J. clin. Invest. **38**, 1356—1365 (1959).
MANDELL, G. L., LINDSEY, E., HOOK, E. W.: Synergism of vancomycin and streptomycin for enterococci. Amer. J. med. Sci. **259**, 346—349 (1970).
MATHIES, A. W., JR., LEEDOM, J. M., IVLER, D., WEHRLE, P. F., PORTNOY, B.: Antibiotic antagonism in bacterial meningitis. Antimicrob. Agents Chemother. (Annual) 218—224 (1967).
MCCABE, W. R.: Clinical use of combinations of antimicrobial agents. Antimicrob. Agents Chemother. (Annual) 225—233 (1967).
MCCABE, W. R., JACKSON, G. G.: Treatment of pyelonephritis. Bacterial, drug, and host factors in success or failure among 252 patients. New Engl. J. Med. **272**, 1037—1044 (1965).
MEDOFF, G., COMFORT, M., KOBAYASHI, G. S.: Synergistic action of amphotericin B and 5-fluorocytosine against yeast-like organisms. Proc. Soc. exp. Biol. (N.Y.) **138**, 571—574 (1971).
MILES, C. P., COLEMAN, V. R., GUNNISON, J. B., JAWETZ, E.: Antibiotic synergism requires the simultaneous presence of both members of a synergistic drug pair. Proc. Soc. exp. Biol. (N.Y.) **78**, 738—741 (1951).
MOELLERING, R. C., JR., WEINBERG, A. N.: Studies on antibiotic synergism against enterococci. II. Effect of various antibiotics on the uptake of ^{14}C-labeled streptomycin by enterococci. J. clin. Invest. **50**, 2580—2584 (1971).
MOELLERING, R. C., JR., WENNERSTEN, C., WEINBERG, A. N.: Studies on antibiotic synergism against enterococci. I. Bacteriologic studies. J. Lab. clin. Med. **77**, 821—828 (1971).
National Research Council Division of Medical Sciences Drug Efficacy Study (National Academy of Sciences): Fixed combinations of antimicrobial agents. New Engl. J. Med. **280**, 1149—1154 (1969).
NOONE, P., PATTISON, J. R.: Therapeutic implications of interaction of gentamicin and penicillins. Lancet **1971 II**, 575—578.
OLSSON, R. A., KIRBY, J. C., ROMANSKY, M. J.: Pneumococcal meningitis in the adult. Clinical, therapeutic, and prognostic aspects in forty-three patients. Ann. intern. Med. **55**, 545—549 (1961).
PLOTZ, P. H., DAVIS, B. D.: Synergism between streptomycin and penicillin: a proposed mechanism. Science **135**, 1067—1068 (1962).
RAHAL, J. J., JR., MEYERS, B. R., WEINSTEIN, L.: Treatment of bacterial endocarditis with cephalothin. New Engl. J. Med. **279**, 1305—1309 (1968).
ROBBINS, W. C., TOMPSETT, R.: Treatment of enterococcal endocarditis and bacteremia. Amer. J. Med. **10**, 278—299 (1951).
ROWLEY, D., COOPER, P. D., ROBERTS, P. W., SMITH, E. L.: The site of action of penicillin. I. Uptake of penicillin on bacteria. Biochem. J. **46**, 157—161 (1950).
RUSSELL, W. F., JR., KASS, I., HEATON, A. D., DRESSLER, S. H., MIDDLEBROOK, G.: Combined drug treatment of tuberculosis. III. Clinical application of the principles of appropriate and adequate chemotherapy to the treatment of pulmonary tuberculosis. J. clin. Invest. **38**, 1366—1375 (1959).
SABATH, L. D., ABRAHAM, E. P.: Synergistic action of penicillins and cephalosporins against *Pseudomonas pyocyanea*. Nature (Lond.) **204**, 1066—1069 (1964).
SABATH, L. D., ELDER, H. A., MCCALL, C. E., FINLAND, M.: Synergistic combinations of penicillins in the treatment of bacteriuria. New Engl. J. Med. **277**, 232—238 (1967).
SCHIMPFF, S., SATTERLEE, W., YOUNG, V. M., SERPICK, A.: Empiric therapy with carbenicillin and gentamicin for febrile patients with cancer and granulocytopenia. New Engl. J. Med. **284**, 1061—1065 (1971).
SCHLICHTER, J. G., MACLEAN, H.: A method of determining the effective therapeutic level in the treatment of subacute bacterial endocarditis with penicillin. Amer. Heart J. **34**, 209—211 (1947).
SMITH, C. B., DANS, P. E., WILFERT, J. N., FINLAND, M.: Use of gentamicin in combinations with other antibiotics. J. infect. Dis. **119**, 370—377 (1969).

Sonne, M., Jawetz, E.: Combined action of carbenicillin and gentamicin on *Pseudomonas aeruginosa in vitro*. Appl. Microbiol. **17**, 893—896 (1969).
Standiford, H. D., de Maine, J. B., Kirby, W. M. M.: Antibiotic synergism of enterococci. Arch. intern. Med. **126**, 255—259 (1970).
Ström, J.: Penicillin and erythromycin singly and in combination in scarlatina therapy and the interference between them. Antibiot. and Chemother. **11**, 694—697 (1961).
Sutherland, R., Batchelor, F. R.: Synergistic activity of penicillins against penicillinase-producing gram-negative bacilli. Nature (Lond.) **201**, 868—869 (1964).
Tan, J. S., Terhune, C. A., Jr., Kaplan, S., Hamburger, M.: Successful two-week treatment schedule for penicillin-susceptible *Streptococcus viridans* endocarditis. Lancet **1971 II**, 1340—1343.
Wallace, J. F., Smith, R. H., Garcia, M., Petersdorf, R. G.: Studies on the pathogenesis of meningitis. VI. Antagonism between penicillin and chloramphenicol in experimental pneumococcal meningitis. J. Lab. clin. Med. **70**, 408—418 (1967).
Watanakunakorn, C.: Penicillin combined with gentamicin or streptomycin: synergism against enterococci. J. infect. Dis. **124**, 581—586 (1971).
Wilkowske, C. J., Facklam, R. R., Washington II, J. A., Geraci, J. E.: Antibiotic synergism: enhanced susceptibility of Group D streptococci to certain antibiotic combinations. Antimicrob. Agents Chemother. (Annual) 195—200 (1970).
Winters, R. E., Chow, A. W., Hecht, R. H., Hewitt, W. L.: Combined use of gentamicin and carbenicillin. Ann. intern. Med. **75**, 925—927 (1971).
Zimmermann, R. A., Moellering, R. C., Jr., Weinberg, A. N.: Enterococcal resistance to antibiotic synergism. Antimicrob. Agents Chemother. (Annual) 517—521 (1970).

Section Eight: Perspectives on the Importance of Drug Disposition in Drug Therapy and Toxicology

Chapter 73

Drug Actions and Interactions: Theoretical Considerations

J. R. GILLETTE and J. R. MITCHELL

With 6 Figures

During the past several years, it has become increasingly evident that the pharmacologic and toxicologic actions of drugs may be frequently altered by treatments that change the rates at which drugs are metabolized and eliminated. Such changes in the rates of drug metabolism can be important in the clinic because they can markedly alter the steady-state plasma levels of drugs in patients and thereby alter the intensity of the therapeutic or toxicologic responses of drugs. In addition, however, pharmacologists have utilized changes in drug metabolism to determine whether the pharmacologic responses evoked by a drug are mediated by the drug itself or by its metabolites. In these experiments, it has been frequently assumed that when the intensity of the duration or the response is increased by inhibitors of cytochrome P-450 enzymes in liver or shortened by inducers of these enzymes, the response must be mediated solely by the parent compound. Conversely, when the intensity or the duration of the response is decreased by inhibitors or increased by inducers of these enzymes, the response is assumed to be mediated by an active metabolite. Studies during the past several years have revealed, however, that changes in the activity of the enzymes involved in drug metabolism do not always result in alterations of the biologic half-lives or of the pharmacologic actions of drugs to the extent predicted by *in vitro* experiments. Indeed there are a number of physiologic and pharmacokinetic factors which must be kept in mind when attempting to extrapolate the results obtained *in vitro* to the living animal and to patients. Moreover, inducers and inhibitors of cytochrome P-450 enzymes may cause a number of other effects in addition to changing the activity of cytochrome P-450 enzymes.

In the present paper, we shall discuss some of the situations in which inducers and inhibitors of drug metabolizing enzymes would be expected to cause marked changes in the pharmacologic and toxicologic action of drugs, some situations in which the inducers and inhibitors would be expected to have little or no effect, and some situations in which the inducers and inhibitors would be expected to have an effect opposite to that which superficially might have been anticipated. In order to clarify these situations, several hypothetical conditions will be considered in which the inducer or inhibitor is assumed to evoke its action on only one enzyme system.

I. Pharmacologic and Toxicologic Effects are Mediated Solely by the Parent Drug

Let us first consider the effect of inducers and inhibitors of drug metabolizing enzymes when the pharmacologic action is caused by the parent drug and not by its metabolites. If the intensity of the response is related to the concentration of the unchanged drug in the target tissues and indirectly related to its concentration in blood plasma, the duration of response should depend on the rate at which the drug is eliminated but not on the pattern of its metabolites. On the assumption that the body may be viewed as a single compartment and that the rate of elimination of the drug can be expressed as a first order reaction, the decrease in the drug concentration can be estimated from the first order equation shown in Fig. 1. It should be pointed out, however, that the drug may be eliminated from the body by several different mechanisms and that the sum of the clearances by these mechanisms represents the total body clearance.

Pharmacokinetics of drugs whose actions are mediated solely by the parent drug

$$C = (D/Vd)\ e^{-kt}$$

Where:

k Vd = Clearance by the kidney
+ clearance by the lungs
+ clearance by metabolic pathway 1
+ clearance by metabolic pathway 2
+ clearance by metabolic pathway ....

Proportion of total clearance attributed to any given pathway, X

$$Px = \frac{kx\ Vd}{k\ \ Vd}$$

Clearance by an organ = BFR (C arterial − C venous)/C arterial
Clearance by an enzyme ≅
[Vm/(Effective Km + C venous)] (C venous/C arterial)
At low C venous
Clearance by an enzyme ≅ (Vm/Km) (C venous/C arterial)

Fig. 1. Kinetics of elimination of parent drugs

It should be obvious, however, that inducers and inhibitors of a drug metabolizing enzyme will be important in therapy only when they cause clinically significant changes in the total body clearance of the drug. Some of the questions that might be considered in evaluating the clinical importance of induction and inhibition of drug metabolizing enzyme systems are thus: 1. Is the pathway that is altered by the inducer or inhibitor a major pathway of elimination of the drug either before or after the administration of the inducer or inhibitor? For example, suppose that an inducer doubled the clearance of a drug by a particular pathway or an inhibitor halved it. Either one of these would have a negligible effect on the total body clearance if initially the clearance of the drug by the pathway were only 20% of the total body clearance. But either one would have a marked effect on the total body clearance if initially the clearance by the pathway was 90% of the total body clearance. 2. Is the pathway of elimination limited by the activity of the enzyme or is it limited by some other factor, such as the blood flow rate? For example, if 90% of a drug were cleared from the blood as it passed through the liver, doubling the activity of the drug metabolizing enzyme in liver would have little effect on the total body clearance of drug. 3. Does the treatment alter the order of the metabolic reaction? For example, after depletion of the body of a cosubstrate, the rate of metabolism by the pathway depends on the rate of formation of the cosubstrate and thus, the reaction becomes zero order. 4. Are changes in the total body clearance of the drug sufficiently large to be clinically significant? Other questions related to these are discussed below and summavized in Table 1.

Table 1. *Pharmacologic and toxicologic effects are mediated solely by the parent drugs*

A.	What proportion of the drug is excreted unchanged?
B.	Do high doses of the drug limit the elimination of the drug by saturating transport systems in the kidney, by saturating drug metabolizing enzymes in the body, or by depletion of cosubstrates used in conjugation reactions?
C.	What is the relative importance of tissues that contain drug metabolizing enzymes?
D.	Is the rate of drug metabolism limited mainly by the blood flow rate through the tissues?
E.	What is the relative importance of the pathways of drug metabolism in the tissues?
F.	Do substances cause depletion of cosubstrates used in conjugation reactions and thereby limit the rate of conjugation of the drug?
G.	What is the therapeutic index of the drug? Are changes in the plasma level of the drug therapeutically or toxicologically significant?

A. What Proportion of the Drug is Excreted Unchanged?

In the extrapolation of results from *in vitro* experiments to the living animal, it is important to determine whether the drugs are excreted by the kidneys or the lungs largely unchanged or whether the drugs are extensively metabolized before they are excreted. Obviously, inducers and inhibitors of drug metabolizing enzymes will have little effect on the biological half-life of drugs that are excreted largely unchanged by kidneys or the lungs. For example, changes in drug metabolizing enzymes usually do not result in changes in the duration of action of anesthetic gases or of long lasting barbiturates, such as barbital. Indeed, the biological half-life of a drug that is excreted largely unchanged by the kidney would be altered to a greater extent by substances that affect the kidney than it would by inducers or inhibitors of drug metabolizing enzyme systems. Thus, the biological half-lives of polar drugs that are excreted largely by kidney transport systems are prolonged by inhibitors of these transport systems. For example, probenecid prolongs the action of penicillin (Beyer et al., 1951). Moreover, the biological half-lives of lipid-soluble drugs that rapidly diffuse back into the blood stream after they are excreted by the glomerulus or the kidney transport systems may be altered by substances that change the urinary pH or the rate of urinary flow. For example, the clearance of phenobarbital is increased by diuresis or by substances that increase the urinary pH (Waddell and Butler, 1957).

On the other hand, the biological half-life of a drug that is extensively metabolized before being excreted into the urine would be altered to a greater extent by inducers and inhibitors of drug metabolizing enzymes than it would be by inhibitors of kidney transport systems or by substances that change the urinary pH or the rate of urinary flow.

B. Does Increasing the Dose of a Drug Affect its Total Body Clearance?

As the dose of a drug is increased, the relative proportion of a drug that is excreted unchanged can be altered either by decreasing the apparent clearance by the kidney or by decreasing the apparent clearance by one or more of the metabolic transformations. It is noteworthy that when this occurs, the elimination of the drug no longer follows first order kinetics but follows a composite of zero order and first order kinetics.

At high doses the proportion of the dose that is excreted unchanged would decrease when the transport systems in the kidney become saturated or in some instances when the binding sites of plasma proteins become saturated. For example, as shown in the following chapter (MITCHELL et al., 1974) a greater fraction of furosemide is metabolized at high doses than at low doses. In this situation inducers and inhibitors of the drug metabolizing enzymes might have a greater influence on the duration of action at high doses of the drug than they would at low doses.

On the other hand, the urinary clearance of the drug might increase as the dose is increased because the drug metabolizing enzyme systems become saturated. For example, LEVY (1968) has pointed out that the time required to eliminate 50% of a dose of salicylate was increased from 3 h to 20 h as the dose of salicylate was increased from 0.25 g to 20 g because the glycine conjugase becomes saturated at the higher doses. In this situation inducers and inhibitors of the conjugase could still affect the clearance of the drug from the body but the effect may not be as great at high doses as it would be at low doses.

The urinary clearance of unchanged drug might also increase because the body becomes depleted of a cosubstrate required in the conjugation of the drug. For example, BÜCH et al. (1967) found that the proportion of the dose of acetaminophen excreted as the sulfate conjugate in rats decreases as the dose is increased from 300 mg/kg to 600 mg/kg, but the decrease could be prevented by the coadministration of sulfate. In this situation, inducers and inhibitors of drug metabolizing enzymes that catalyze other routes of elimination would affect the time that the drug persists in the body but inducers and inhibitors of the conjugation system would have little effect. In fact, pretreatment of rats with phenobarbital does not increase the rate of excretion of the acetaminophen sulfate conjugate after high doses of acetaminophen, unless sodium sulfate is also administered to the animals (BÜCH et al., 1965).

C. What is the Relative Importance of Drug Metabolizing Enzymes in Different Tissues?

Another aspect that should be kept in mind is that a drug metabolite may be formed by various enzymes in a number of tissues of the body. It is, therefore, important to know the contribution each tissue makes toward the total body clearance of the drug. Although studies *in vitro* may show that a foreign compound, such as 3,4-benzpyrene, may be metabolized not only by liver but also by various extrahepatic tissues, including lung, kidney, skin and the gastrointestinal tract (WATTENBERG and LEONG, 1971), it cannot be automatically assumed that the activity of the enzymes in extrahepatic tissues contribute significantly to the total clearance of the drug from the body. Knowing the relative clearance of each tissue is especially important because inducers and inhibitors can cause different effects in different tissues. For example, pretreatment of animals with phenobarbital increases the metabolism of certain drugs by the liver (CONNEY,

1967) but has little or no effect on the metabolism of these drugs in lung or kidney (FEUER et al., 1971; GILMAN and CONNEY, 1963). On the other hand, 7,8-benzoflavone inhibits the metabolism of 3,4-benzpyrene in extrahepatic tissues but has little effect on its metabolism in liver of untreated animals (WIEBEL et al., 1971). Thus, the biological half-lives of compounds that are metabolized mainly in the liver should be shortened by pretreatment of the animals with phenobarbital but should not be affected to any significant extent by the prior administration of substances that inhibit only the enzyme in the extrahepatic tissues. By contrast, pretreatment of animals with 3-methylcholanthrene, which increases the activity of cytochrome P-450 enzymes in many extrahepatic tissues including the gastrointestinal tract, may alter not only the half-lives of certain drugs in the body but also their apparent bioavailability (PANTUCK et al., 1974).

D. Is the Rate of Drug Metabolism Limited Mainly by the Blood Flow Rate through the Tissues?

Some drugs are metabolized so rapidly by a tissue that their clearances approach the blood flow rate through the tissue. For example, it has been calculated that the metabolism of nortriptyline in certain patients is limited by hepatic blood flow (ALEXANDERSON et al., 1973). Similarly, the metabolism of propranolol is also limited by the rate of blood flow through the liver (STENSON et al., 1971; BRANCH et al., 1973). In these situations, one would not ordinarily expect that inducers of drug metabolizing enzymes would affect the biological half-life of these drugs as much as they would affect that of a slowly metabolized drug. However, phenobarbital pretreatment also causes an increase in hepatic blood flow and thus the increase in the rate of metabolism of drugs is due not only to an increase in the cytochrome P-450 enzymes in liver microsomes but also to an increase in blood flow rate (OHNHAUS et al., 1971; BRANCH et al., 1974).

E. What is the Relative Importance of the Pathways of Drug Metabolism in the Tissues?

Some drugs are metabolized along several different pathways involving reactions that are catalyzed by different enzymes. For example, morphine is not only *N*-demethylated but is also converted to a glucuronide conjugate. In fact, its conversion to the glucuronide is so rapid that any treatment that would specifically change the rate of *N*-demethylation by the cytochrome P-450 enzymes in liver microsomes would not be expected to cause a significant alteration of the half-life of the drug. Since phenobarbital increases glucuronidation (PANTUCK et al., 1974), it presumably would shorten the half-life of morphine in the body mainly by increasing its rate of glucuronidation rather than increasing its *N*-demethylation.

F. Do Substances that Deplete the Body of Cosubstrates for Conjugation Limit the Rate of Metabolism of the Drug?

Occasionally, the depletion of cosubstrates for conjugation reactions caused by large doses of certain substances can decrease the rate of conjugation of drugs and thereby prolong their duration of action. This kind of an effect, however, would occur only when the conjugate represents the major pathway of elimination of the drug. For example, the administration of acetaminophen decreases the amount of salicylamide excreted as salicylamide sulfate (LEVY and YAMADA, 1971). But acetaminophen does not appreciably affect the total body clearence of salicylamide because most of the salicylamide is excreted as its glucuronide (LEVY and YAMADA, 1971). It is noteworthy, however, that according to this mechanism, substances can lead to a decrease in the metabolism of a drug even though the substance is not an inhibitor in the usual sense of the word. Indeed, when the conjugates are formed by different transferases, this kind of drug interaction would be difficult to predict from *in vitro* experiments alone.

G. Are Changes in the Plasma Level of a Drug Therapeutically or Toxicologically Significant?

It is important to realize that the plasma levels of drugs in patients receiving drugs therapeutically may vary markedly from one patient to another. A part of the variability is due to dietary effects on drug absorption and to differences in the compliance of patients taking the drug. Moreover, clinical trials that are designed to evaluate the therapeutic effects of a drug are seldom designed to determine the lowest plasma level at which the drug evokes its maximal therapeutic and minimal toxicologic effects. Indeed, with most drugs, the physician doesn't concern himself with the plasma levels of drugs in his patients as long as they are above the therapeutic level and below the toxic level. It is not surprising, therefore, that increases and decreases in the metabolism of drugs caused by drug interactions are seldom important in the clinic when the range between the therapeutic levels and the toxic levels of the drugs in patients is large. But the finding that phenobarbital increases the metabolism of dicoumarol and diphenylhydantoin (CUCINELL et al., 1965), whose safety range is small, can be of great clinical importance. In fact, the failure to understand the mechanisms of induction and inhibition of drugs having narrow safety ranges has led to serious toxicities and in some cases death.

II. Pharmacologic and Toxicologic Effects are Mediated Solely by Reversibly Acting Metabolites of the Drug

In some instances, pharmacologically inactive substances are converted in the body to pharmacologically active metabolites. In these instances, however, the effects of inducers and inhibitors of drug metabolizing enzymes on the plasma levels of active metabolites are difficult to predict.

The pharmacokinetic equation for the formation and inactivation of an active metabolite is shown in Fig. 2. Again, it has been assumed that the body can be viewed as a single compartment and that the metabolism of the parent drug and its active metabolite follows first order kinetics. The total body clearance of the parent drug, which equals its rate constant of elimination (k_e) times its volume of distribution (V_a), represents the sum of the clearances of all the excretory and metabolic mechanisms by which the parent drug is eliminated from the body. The clearance term which represents the conversion of the parent drug to the active metabolite is obtained by multiplying the rate constant (k_a) for the formation of the active metabolite by the volume of distribution of the parent drug (V_a). The total body clearance of the active metabolite, which equals its rate constant of elimination (k) times its volume of distribution (V_b), also represents the sum of the clearances of all of the excretory and metabolic mechanisms by which the active metabolite is eliminated from the body. The time at which the plasma level of the active metabolite reaches a maximum is given by the equation: $t_{\max} = \ln (k_e/k)/(k_e\text{-}k)$. And the maximum plasma level is given by the equation $C_{b\max} = \mathrm{Ao}\ (k_a/k_e) \exp - k\, t_{\max}$, in which $\mathrm{Ao} = D_a/V_b$.

Pharmacokinetics of drugs whose actions are mediated solely by a reversibly acting metabolite of the drug

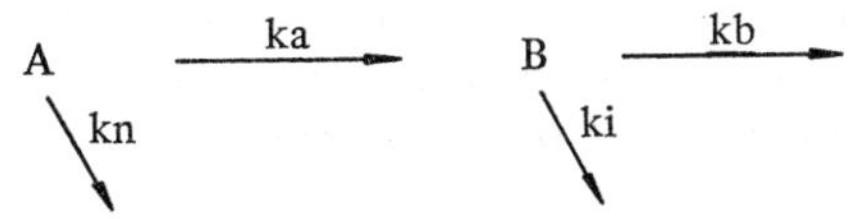

$$A \xrightarrow{ka} B \xrightarrow{kb}$$

A $\searrow$ kn; B $\searrow$ ki

$ka + kn = ke$; $kb + ki = k$

$$Cb = \frac{Da}{Va}\left(\frac{ka\,Va}{ke\,Va\,(Vb/Va) - k\,Vb}\right)\left(e^{-kt} - e^{-ket}\right)$$

Where:
ka Va = Part of the total clearance of A which leads to B.
ke Va = Total body clearance of A.
kVb = Total body clearance of B.

Fig. 2. Kinetics of elimination of active metabolites

From these equations, it becomes obvious that the time at which an active metabolite reaches its maximum plasma level depends solely on the total body clearances of both the parent compound and the active metabolite. The magnitude of the maximum plasma level of the active metabolite depends not only on these clearances but also on the proportion of the dose of the parent drug that is converted to the active metabolite.

Some of the questions that might be asked in evaluating the importance of induction and inhibition of drug metabolizing enzyme systems in changing the kinetics of reversibly acting metabolites are shown in Table 2.

Table 2. *Pharmacologic and toxicologic effects are mediated solely by reversibly acting metabolites of the Drug*

A. Do inducers or inhibitors alter the proportion of the dose that is converted to the active metabolite ?
B. Do inducers or inhibitors alter the rate of formation of the active metabolite ?
C. Do inducers or inhibitors alter the rate of elimination of the active metabolite ?
D. Do substances change the tissue levels of cosubstrates used in the conjugation reactions of the active metabolite ?
E. Do high doses of the drug alter the pattern of metabolites of the active metabolite by depleting the cosubstrate ?
F. Do inducers or inhibitors alter the rates of both the formation and inactivation of the active metabolite ?
G. Is the clearance of the active metabolite greater than the clearance of the parent compound ?

A. Do Inducers or Inhibitors Alter the Proportion of the Dose that is Converted to the Active Metabolite?

When only a small proportion of a drug is converted to its active metabolite, specific inducers or inhibitors of the enzyme that catalyzes the formation of the active metabolite should affect the magnitude of the maximum plasma level of the metabolite, but should not affect the time at which the maximum level occurs. That is inducers should increase the maximum level and inhibitors should decrease it. On the other hand, treatments that do not alter the activity of the enzyme that catalyzes the formation of the active metabolite, but markedly change the other pathways of elimination of the parent drug, should alter not only the magnitude of the maximum level but also the time at which the maximum level occurs. Thus, treatments that increase the rate of elimination along nonactivating pathways should decrease both the maximum level and shorten the time at which the maximum level occurs, whereas treatments that decrease the rate of elimination along these pathways should increase the maximum level and lengthen the time at which it occurs.

B. Do Inducers or Inhibitors Alter the Rate of Formation of the Active Metabolite without Significantly Changing the Proportion of the Dose that is Converted to the Active Metabolite?

When nearly all of the drug is converted to the active metabolite, inducers and inhibitors of the enzyme that forms the active metabolite may also affect both the maximum plasma level of the active metabolite and the time at which it appears. But the changes in the maximum concentration of the active metabolite would not be very dramatic when the metabolite is slowly eliminated from the body provided that the rate of its elimination follows first order kinetics after induction. For example, it can be calculated that when all of a drug is converted to an active metabolite and when the clearance of the parent compound is four times the clearance of the active metabolite in the untreated animal, a 20-fold increase in the clearance of the parent compound would shorten the t_{max} to about 12% of the initial t_{max} but would increase the maximum plasma level by only 50%. Moreover, if the initial relative clearances of the parent compound and the active metabolite were higher than 4-fold, then a 20-fold increase in the metabolism of the parent compound would increase the maximum plasma level even less dramatically. Thus, the principle effect of induction of the drug metabolizing enzyme in this situation would be on the time of onset of the response rather than on the intensity of the response.

A similar situation would also occur when a relatively small proportion of a drug is converted to its active metabolite and inducers increase the rate of elimination by all of the pathways to about the same extent.

C. Do Inducers or Inhibitors Alter the Rate of Elimination of the Active Metabolite?

The active metabolite may be metabolized by a number of pathways or excreted by the kidneys. Thus, the effectiveness of inducers or inhibitors that act solely on enzymes which catalyze the conversion of the active metabolite to an inactive metabolite depends on the relative importance of the mechanisms of elimination of the active metabolite. Obviously, if the active metabolite were excreted largely unchanged, induction or inhibition of an enzyme that catalyzes its inactivation would not appreciably affect either the intensity or the duration of the response. But when the active metabolite is eliminated almost solely by a single pathway of metabolism, induction or inhibition of the enzyme that catalyzes the inactivation of the metabolite by that pathway may affect both the maximum plasma level of the metabolite and the time at which it occurred. However, when the clearance of the parent compound is markedly greater than the clearance of the active metabolite in the untreated animal, inducers and inhibitors of the enzyme that inactivates the metabolite would not change the maximum plasma level nor the time at which the maximum level of the metabolite appeared as much as they would change the time at which the plasma concentrations of the metabolite decreased to ineffective levels. Thus, they would not change the intensity of the response as much as they would the duration of the response.

D. Do Substances Change the Tissue Levels of Cosubstrates Used in the Conjugation Reactions of the Active Metabolite?

As was pointed out previously (see Section I-F), large doses of some substances can deplete the tissues of a conjugation cosubstrate, such as sulfate, glycine, or glutathione, and thereby slow the conjugation of other drugs with the cosubstrate. Indeed, the rate of elimination of the drug by the conjugation reaction would no longer be first order because it would now be limited by the rate of formation of the cosubstrate from precursors in the body. When either the parent drug or the active metabolite is inactivated by one of the conjugation reactions, it is possible that large doses of the depleting substance may enhance and prolong the response even though the depleting substance may not inhibit the particular transferase that acts on the parent drug or its active metabolite.

E. Do High Doses of the Drug Alter the Pattern of Metabolites Derived from the Active Metabolite by Depleting the Cosubstrate?

In some situations the rate of conversion of high doses of a drug to its active metabolite may be rapid enough to deplete the endogenous cosubstrate consumed in the conjugation of the active metabolite. In these instances, inducers of the enzyme that catalyzes the formation of the active metabolite may also increase the maximum plasma level by a greater percentage than they would at low doses. At these high doses of the drug, however, the administration of the depleted substance or one of its precursors may decrease the maximum plasma level of the metabolite and shorten the duration of its response.

The effects of inhibitors of the enzyme that catalyzes the formation of the active metabolite would decrease the intensity of the response but their effects on the duration of action would be difficult to predict. If the cosubstrate were rapidly synthesized, the inhibitors would increase the time required to produce the maximum plasma level of the metabolite, but they could also increase the apparent clearance of the active metabolite by the conjugation reaction.

F. Do Inducers or Inhibitors Alter the Rates of Both the Formation and Inactivation of the Active Metabolite?

Inducers and inhibitors used in studies of drug metabolism are seldom specific and thus may affect both the rate of formation and the rate of inactivation of the active metabolite. The effect of the inducers and inhibitors thus depends on their relative effects on the two processes. Ordinarily, however, inducers shorten both the onset and the duration of action, whereas inhibitors delay the onset of action, but prolong the action after it appears. For example, desipramine delays the appearance of the syndrome caused by tremorine, but prolongs the syndrome caused by oxotremorine, the active metabolite of tremorine (SJÖQVIST et al., 1968).

G. Is the Clearance of the Active Metabolite Greater than the Clearance of the Parent Drug?

In the previous sections it has been assumed that the active metabolite is cleared more slowly than the parent compound. Indeed, in most situations in which it is important to know whether an active metabolite is formed, the clearance of the metabolite is relatively slow. Occasionally, however, the active metabolite is cleared much more rapidly than is the parent drug. In this situation, the pharmacokinetic equation for the concentration of the active metabolite, shown in Fig. 2, degenerates to $Cb = Ao\,(ka/k)\exp - ket)$. Thus, the apparent half-life of the active metabolite under these conditions is virtually identical to that of the parent drug, but the concentrations of the active metabolite would still be affected by treatments which change the relative rates of formation and inactivation of the active metabolite as well as those which change the total body clearance of the parent drug.

III. Pharmacologic and Toxicologic Effects are Mediated Solely by Chemically Reactive Metabolites

Many chemically inert substances can be transformed in the body to reactive intermediates that combine covalently with tissue macromolecules. In recent years, it has become increasingly evident that covalent binding to tissue macromolecules may mediate various toxicities including carcinogenesis, mutagenesis, hypersensitivity reactions and tissue necrosis.

In some instances, factors that accelerate or decelerate drug metabolism may affect the magnitude of the covalent binding in seemingly unpredictable ways. The reason for many of these results becomes clear after perusing the kinetics of covalent binding *in vivo* (Fig. 3). If the conjugates between the macromolecules and the reactive metabolites are not rapidly metabolized, then the amount of

Pharmacokinetics of covalent binding

A —ka→ B; A ↓kn; B ↓ki

B —Xkx→ C:X

B —Yky→ C:Y

B —f(G,B)→ C:G

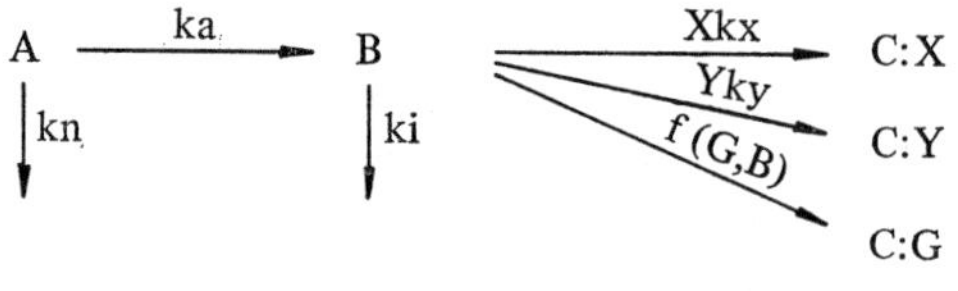

$$ka + kn = ke$$

$$Xkx + Yky + f(G,B) + ki = k$$

$$C{:}X + C{:}Y = Da\left(\frac{ka}{ke}\right)\left(\frac{Xkx + Yky}{k}\right)\left(1 + R\right)$$

$$C{:}X = Da\left(\frac{ka}{ke}\right)\left(\frac{Xkx + Yky}{k}\right)\left(\frac{Xkx}{Xkx + Yky}\right)\left(1 + R\right)$$

$$R = \frac{(ke)\,e^{-kt} - (k)\,e^{-ket}}{k - ke}$$

At $t = 0$, $R =$ minus 1.0; at $t = \infty$, $R = 0$

Fig. 3. Kinetics of covalent binding of reactive metabolites

covalently bound metabolites will accumulate until all of the drug is metabolized The amount of drug that eventually becomes covalently bound thus depends on the ratio, $Da\ (k_a/k_e)\ (Xk_x + Yk_y)/k$, after both of the exponential terms reach zero, that is when "t" equals infinity and $R = 0$ (Fig. 3). This ratio predicts that the proportion of the dose that becomes covalently bound is not always dependent on the rate at which the drug is converted to the reactive metabolite. For example, if all of the parent drug were converted to the reactive metabolite and all of the chemically reactive metabolite became covalently bound to tissue macromolecules, then all of the parent drug would become covalently bound regardless of the rate at which the drug is converted to its chemically reactive metabolite or the rate at which the chemically reactive metabolite combines with the tissue macromolecules. Thus, inducers and inhibitors of drug metabolizing enzymes do not result in changes in covalent binding simply because they change the rate of drug metabolism. Instead they alter the amount of covalent binding because these changes in rates result in changes in the proportion of the dose that becomes converted to the chemically reactive metabolite or in the fraction of the reactive metabolite that becomes covalently bound or both. Thus, it is not the rate of drug metabolism *per se*, but the pattern of metabolism that determines the magnitude of covalent binding.

The total amount of a foreign compound that becomes covalently bound is thus an indirect measure of the proportion of the drug that becomes converted to its chemically reactive metabolite and the relative rates at which the chemically reactive metabolite reacts with tissue macromolecules and is converted to chemically inactive metabolites.

Whether a given amount of covalently bound metabolite results in toxicity depends on a host of factors. If the chemically reactive metabolite reacted only with vitally important macromolecules, then there might be a direct relationship between the amount of covalently bound metabolite and the severity of the lesion regardless of the tissue in which the covalent binding occurred. But it seems likely that the covalent binding to macromolecules would be rather indiscriminate and that only a portion of the total covalently bound metabolite would be bound to vitally important macromolecules. Indeed, in some instances the covalent binding to the vitally important components of the cell may not even be a part of the material being assayed. For example, CCl_4 is thought to exert its toxic effects by an interaction between its reactive metabolite and lipids, but this interaction would not be measured by the methods used to measure the covalent binding of the reactive metabolite with tissue proteins. According to this view, therefore, the amount of covalently bound metabolite required to cause a specific kind of tissue damage, such as necrosis, will vary with the foreign compound and the tissue. Thus, the fraction of the covalently bound metabolite that is bound to vitally important macromolecules may be represented by the ratio $Xk_x/(Xk_x + Yk_y)$, shown in Fig. 3.

There are probably a number of other factors in addition to the amount of drug that becomes covalently bound to vitally important tissue macromolecules that also affect the development of tissue damage. Since many of these factors, including the rate of tissue repair, are incompletely understood, it seems unlikely that we will ever be able to predict with certainty whether a given amount of covalent binding of a drug to tissue macromolecules will result in tissue damage. Nevertheless, with any given compound that is known to cause damage in a given tissue, changes in the severity of the damage should parallel changes in the amount of covalently bound metabolite.

Some of the questions that might be asked in attempting to predict when inhibitors and inducers of drug metabolizing enzymes would change either of the ratios that determine the amount of drug that becomes covalently bound to tissue macromolecules are discussed below (Table 3).

Table 3. *Pharmacologic and toxicologic effects are mediated solely by chemically reactive metabolites*

A.	Do inducers or inhibitors alter the relative proportion of the dose that is converted to the active metabolite?
B.	Do inducers or inhibitors alter the relative proportion of the reactive metabolite that becomes covalently bound?
C.	Do substances change the tissue levels of cosubstrates used in conjugation reactions?
D.	Do high doses of the drug lead to depletion of cosubstrates for conjugation reactions?
E.	Are chemically reactive metabolites formed in different tissues?
F.	Do chemically reactive metabolites leave the tissues in which they are formed?

A. Do Inducers or Inhibitors Alter the Relative Proportion of the Dose that is Converted to the Reactive Metabolite?

When virtually all of the foreign compound is converted to the chemically reactive metabolite, the proportion of the dose that is converted to the chemically reactive metabolite would not be changed by either inhibitors or inducers of the activating enzyme unless they also markedly changed the pattern of metabolism of the parent compound by markedly increasing its conversion to nontoxic metabolites.

Even when only a portion of the dose is converted to the chemically reactive metabolite, inducers and inhibitors would not change the proportion of the dose that becomes covalently bound when they change the rate of metabolism of the parent compound by a nontoxic pathway to about the same extent as they change the rate of formation of the chemically reactive metabolite. Thus, inducers and inhibitors may not necessarily change the *amount* of covalently bound metabolites even though they markedly alter the biological half-life of the parent drug and the *rate* of covalent binding to tissue macromolecules.

On the other hand, when the parent foreign compound is excreted largely unchanged by the kidney or lung or when it is metabolized mainly along pathways that do not lead to the formation of its chemically reactive metabolite, inducers and inhibitors of the enzyme system that catalyzes the formation of the reactive metabolite would be expected to change the proportion of the dose that is converted to the chemically reactive metabolite and thereby change the amount of the metabolite that becomes covalently bound to tissue macromolecules. Indeed, inducers of the activating enzyme can increase the amount of covalently bound metabolite and inhibitors can decrease it even though they do not significantly alter the biologic half-life of the parent compound.

Treatments that lead to a decrease in the urinary excretion of the unchanged drug or inhibitors that block only those pathways of metabolism of the parent compound that do not lead to the formation of the chemically reactive metabolite would also increase the proportion of the dose that is converted to the chemically reactive metabolite. Since these treatments and inhibitors would increase the half-life of the parent drug, one might be led to the erroneous conclusion that the toxic reaction is caused by the parent drug and not by the reactive metabolite.

B. Do Inducers or Inhibitors Alter the Relative Proportion of the Reactive Metabolite that Becomes Covalently Bound?

Because the reactions between the chemically reactive metabolite and the tissue macromolecules ordinarily are nonenzymatic, inducers and inhibitors would not usually alter the vate constant of covalent binding. But inducers and inhibitors could affect the amount of covalently bound metabolites if the toxic metabolite was further metabolized by nontoxic pathways. Thus, inducers of the enzymes that catalyze the conversion of the reactive metabolite to inactive metabolites would decrease the covalent binding to macromolecules, whereas inhibitors of these enzymes would increase it.

C. Do Substances Change the Tissue Levels of Cosubstrates Used in Conjugation Reactions?

As has been pointed out previously (see Sections I-F and II-D), some substances can deplete the body of the cosubstrates required in conjugation reactions including the formation of glycine conjugates, ethereal sulfates and mercapturic acids.

In these situations, the rates of conjugation of these cosubstrates with foreign compounds as well as that of the depleting substance no longer follow first order kinetics but follow zero order kinetics because the rates of conjugation are limited by the rates at which the cosubstrates are synthesized or mobilized from tissue stores in the body.

When the products of the conjugation reactions are not chemically reactive, the depleting substances would tend to increase the covalent binding of reactive metabolites to macromolecules whether the parent compound or the reactive metabolite becomes conjugated with the cosubstrate. But the magnitude of the increase would depend on the importance of the conjugation reaction in the absence of the depleting substance. For example, when the parent drug is inactivated mainly by conjugation with the cosubstrate, the administration of the depleting substance would greatly increase the proportion of the dose of parent compound that is converted to the reactive metabolic and thereby would increase the covalent binding of the reactive metabolite to macromolecules. Similarly when the reactive metabolite is eliminated mainly by its conjugation with the cosubstrate, administration of the depleting substance would also increase the covalent binding. But when the conjugation reaction represents a relatively minor pathway of elimination for the parent compound or its reactive metabolite, the depleting substance would cause a modest increase in the steady-state level of the reactive metabolite and thereby would only slightly increase the proportion of the reactive metabolite that becomes covalently bound.

If only the parent drug were inactivated mainly by the conjugation reaction, the effects of inducers and inhibitors of drug metabolizing enzymes would be relatively simple to predict. Inducers and inhibitors that affected the enzyme that catalyzed the formation of the reactive metabolite would alter the covalent binding to a greater extent in the presence of the depleting substance than in its absence. Moreover, inducers and inhibitors of the enzymes that metabolize the drugs along nonactivating pathways but not the conjugation reaction would also change the amount of covalent binding to a greater extent in the presence of the depleting substance than in its absence. But inducers and inhibitors of the enzyme that catalyzes the conjugation reaction would usually cause little or no change in the amount of covalent binding in the presence of the depleting substances.

However, if in the absence of the depleting substance only the active metabolite were inactivated mainly by the conjugation reaction, the effects of inducers and inhibitors would depend on a number of interrelated factors and therefore would be difficult to predict. For example, suppose that the depleting substance was administered in a dose just large enough to deplete the animal of the cosubstrate and that the drug was administered just after the cosubstrate was completely consumed. In this situation, the kinetics for the formation and the elimination of the active metabolite would depend on the total elimination rate constant of the parent compound, the rate constant for the formation of the reactive metabolite, the rate constant for the inactivation of the metabolite by pathways other than that by the conjugation reaction and the rate of synthesis of the cosubstrate. Let us now suppose the rate of synthesis of the cosubstrate was negligible during the time that the parent drug and the reactive metabolite remained in the body. In this hypothetical situation, an inducer of the enzyme that catalyzes the formation of the reactive metabolite would increase the covalent binding and an inhibitor would decrease it when either of them altered the proportion of the parent compound that is converted to the reactive metabolite. But, when virtually all of the parent compound is converted to the reactive metabolite, the inducer and the inhibitor of the enzyme that catalyzes the formation of the reactive metabolite should not alter the covalent binding of the reactive metabolite to tissue macromolecules. It is possible, however, that an inducer could increase the rate of formation of the reactive metabolite and thereby increase its concentration to such an extent that the other mechanisms of elimination of the reactive metabolite become saturated.

By contrast, an inducer or an inhibitor of the enzymes that catalyze the inactivation of the reactive metabolite by nonconjugation mechanisms would markedly change the amount of covalently bound metabolite. In fact, these inducers and inhibitors would change the degree of covalent binding to a greater extent when the cosubstrate of the conjugation reaction is depleted than when it is not.

But inducers or inhibitors of the conjugation enzyme would obviously have no affect on covalent binding when the rate of synthesis of the cosubstrate was negligible.

In our series of hypothetical experiments let us now suppose that after the depleting substance had consumed all of the cosubstrate, the rate of synthesis of the cosubstrate is not negligible. Instead, let us suppose that its synthesis was rapid enough to combine with about half of the reactive metabolite formed from the parent drug during the time that the parent drug and the reactive metabolite are in the body. Again, inducers or inhibitors of the enzyme that catalyzes the formation of the reactive metabolite will alter the covalent binding when these inducers and inhibitors alter the proportion of the parent compound that is converted to the reactive metabolite. But in contrast to the situation in which the rate of synthesis of the cosubstrate was negligible, inducers and inhibitors of the activating enzyme will now alter the amount of covalent binding even when they do not alter the proportion of the parent drug that is converted to the reactive metabolite provided that the rate constant for the inactivation of the reactive metabolite is much greater than the rate constant for its formation. This occurs because the rate of synthesis of the cosubstrate is presumably constant (that is zero order) and therefore the amount of the reactive metabolite that can be inactivated by combining with the cosubstrate depends on the amount of time that the reactive metabolite is present in the body. Inhibitors of the activating enzyme would prolong the time that the reactive metabolite remains in the body and therefore they would increase the apparent clearance of the reactive metabolite. But inducers would shorten the time that the reactive metabolite remains in the body and therefore would decrease the apparent clearance of the reactive metabolite.

Inducers and inhibitors of the nonconjugation mechanisms for the elimination of the reactive metabolite would still tend to change the amount of covalent binding but their effect would not be as great as they would be when the rate of synthesis of the cosubstrate is negligible. Moreover, inhibitors of the conjugation enzyme may tend to increase the amount of covalent binding by decreasing the proportion of the reactive metabolite that is inactivated by conjugation with the cosubstrate, but their effect would be about the same as it would be when the depleting substance is not administered. By contrast, inducers of the enzyme would have little effect on covalent binding.

In our series of hypothetical experiments. let us now suppose that the dose of the depleting substance was much greater than that required to consume all of the cosubstrate initially present in the body and that the substance was present during the time that the drug was in the body. Indeed this usually occurs when the effects of depleting substances are studied. In contrast to the previous situations, however, the effects of inducers and inhibitors of the activating enzyme would now depend on the relative rates at which the cosubstrate was consumed by conjugation with the depleting substance and the reactive metabolite of the drug. If the cosubstrate were consumed much more rapidly by the depleting substance than by the reactive metabolite of the drug, the steady-state concentration of the cosubstrate may be kept at negligible levels even when the cosubstrate is rapidly synthesized. Moreover, if the formation of the conjugates of the depleting substance and the reactive metabolite were catalyzed by the same conjugation enzyme, the depleting substance would tend to inhibit the formation of the conjugate of the reactive metabolite. In these situations, the effects of inhibitors and inducers of the activating enzyme may have little or no effect on the covalent binding of the reactive metabolite when most of the parent drug is converted to the reactive metabolite.

When the effects of depleting substances are considered, it is important to realize that in some instances the reactive metabolite may be conjugated with a cosubstrate to form a metabolite that has a chemical reactivity even greater than that of the initial metabolite. For example, there is now considerable evidence that *N*-2-fluorenylacetamide (*N*-2-FAA) is first converted to *N*-hydroxy-*N*-2-fluorenylacetamide and then to *N*-0-sulfate-*N*-2-fluorenylacetamide before it becomes covalently bound to tissue macromolecules including protein and nucleic acids (WEISBURGER and WEISBURGER ,1973). In this situation, the coadministration of *N*-2-FAA and acetanilide or acetaminophen, which deplete the body of sulfate, results in a decrease in the incidence of tumor formation caused by *N*-2-FAA presumably because the sulfate conjugate can no longer be formed. In accord with this view the administration of sodium sulfate in addition to acetanilide and *N*-2-FAA, partially reverses the protective effects of acetanilide.

D. Do High Doses of the Drug Lead to Depletion of Cosubstrates Used in Conjugation Reactions?

High doses of some foreign compounds that are transformed to chemically reactive metabolites can also deplete the tissues of cosubstrates of conjugation reactions. In most instances, the proportion of dose that becomes covalently bound is increased as the size of the dose is increased. Indeed, covalent binding should be increased with high doses of the drug when either the parent drug or its reactive metabolite is conjugated with the cosubstrate. If the parent drug were conjugated, the proportion of the dose that is transformed to the reactive metabolite is increased after depletion of the cosubstrate. If the reactive metabolite is conjugated to form an inactive metabolite, its clearance is decreased after depletion of the cosubstrate.

After the cosubstrate is depleted by high doses of the drug, the situation is analogous to the hypothetical situation described previously in which a depleting substance is administered in a dose just large enough to completely consume the cosubstrate (see Section III-C). As in that situations, the effects of inducers and inhibitors on covalent binding depend on the proportion of the parent drug that is converted to the reactive metabolite and on the rate of synthesis of the cosubstrate. If only a small proportion of the parent drug were converted to the reactive metabolite or if the synthesis of the cosubstrate were not negligible, inducers of the activating enzyme should increase the covalent binding of the reactive metabolite to the tissue macromolecules. But if virtually all of the parent compound were converted to the reactive metabolite and the synthesis of the cosubstrate were negligible, induction of the activating enzyme would not change the amount of covalent binding.

On the other hand, induction of the enzymes that catalyze the inactivation of either the parent compound or the reactive metabolite by nonconjugation reactions should decrease the covalent binding. When the parent drug is partially eliminated by the conjugation reaction, induction of the enzyme that catalyzes its inactivation by the nonconjugation reaction should be more effective at high doses than at low doses. Similarly, when the reactive metabolite is partially inactivated by the conjugation reaction, induction of the enzyme that catalyzes the inactivation of the reactive metabolite by nonconjugation reactions would also be more effective at high doses than at low doses of the parent compound.

By contrast, induction of the enzyme that catalyzes the inactivation of the reactive metabolite by the conjugation reactions would decrease covalent binding at low doses of the drug but would have little or no effect at high doses.

E. Are Chemically Reactive Metabolites Formed in Different Tissues?

If a reactive metabolite combines with tissue macromolecules so rapidly that it does not leave the tissue in which it is formed, the covalently bound metabolite found in a given tissue would depend on the activity of the enzyme that catalyzes the formation of the reactive metabolite and on the activities of the enzymes that catalyze its inactivation in that tissue. In this situation, covalent binding could occur only in those tissues that contained the activating enzyme. The effects of inducers and inhibitors, on the covalent binding in any given tissue, however, would be complex. For example, an inducer may increase the activity of the activating enzyme in one tissue but not another. In this situation, the covalent binding would be increased in the tissue in which the activity of the activating enzyme is increased. But when the conversion of the parent drug to the reactive metabolite in that tissue is a major mechanism by which the drug is eliminated from the body, the inducer would decrease the covalent binding in other tissues even though it did not change the activity of the enzymes in these tissues. Thus, an inducer could increase the covalent binding in liver but decrease it in lung or kidney. On the other hand, an inhibitor of the activating enzyme in liver could decrease the covalent binding in liver but increase it in lung or kidney.

In contrast, inducers and inhibitors of enzymes that catalyze the inactivation of the reactive metabolite would affect the covalent binding only in those tissues in which they are effective.

F. Do Reactive Metabolites Leave the Tissues in which they are Formed?

If the reactive metabolite reacts slowly with macromolecules and diffuses out of the cells in which it is formed, the kinetics of covalent binding would depend on the activity of the activating enzyme and the activities of the enzymes that inactivate the reactive metabolite in all of the tissues of the body. In this situation, covalent binding could occur in tissues that do not contain enzymes that catalyze the formation of the reactive metabolite. Moreover, the effects of inducers and inhibitors on covalent binding would depend on the activities of the enzymes in tissues in which they exerted their effects. For example, if most of the reactive metabolites were formed in the liver, then inducers and inhibitors that change the activity of the activating enzyme in liver would have a greater effect than those which change only the activities of the enzyme in other tissues. Similarly, if most of the reactive metabolite were inactivated in the liver then inducers and inhibitors of the inactivating enzymes in liver would have a greater effect than those which acted solely on inactivating enzymes in other tissues.

Of course the kinetics of covalent binding could be more complex than those illustrated by either of these extreme situations. Suppose that about half of the reactive metabolite diffuses out of the tissues in which it is formed. In this situation, a portion of the reactive metabolite might be carried by the blood to a tissue that also contained the activating enzyme and the covalent binding in that tissue would depend not only on the activity of its activating enzyme but also on the activities of the activating enzyme in the other tissues. Thus, covalent binding in the tissue would be influenced mainly by activators and inhibitors of the activating enzyme in the tissue but could also be affected by inducers and inhibitors of the enzyme in other tissues. In this kind of situation, changes in the amount of covalent binding in any given tissue caused by inhibitors and inducers would be very difficult to predict with accuracy.

G. Examples of the Effects of Inducers, Inhibitors and Depleting Substances on Covalent Binding and Drug Toxicity

During the past few years, our laboratory has been evaluating the possibility that tissue necrosis caused by high doses of various drugs and other foreign compounds is mediated by the formation of chemically reactive metabolites that become covalently bound to tissue macromolecules. Our general approach has been to determine whether radiolabeled drugs at high doses are covalently bound to macromolecules in tissues that become necrotic, and to determine whether changes in the activities of drug metabolizing enzymes caused by inducers and inhibitors result in parallel changes in the amount of covalent binding and the severity of the necrosis. These studies have led to many of the concepts described in this paper.

Studies of the toxic effects of acetaminophen have been particularly revealing (for review, see following chapter by Mitchell et al.). At the usual therapeutic doses, this drug is among the safest of all minor analgesics. But in large doses, it can produce fatal centrilobular liver necrosis in man and in experimental animals. In mice, acetaminophen does not cause centrilobular necrosis in liver unless the dose is greater than about 300 mg/kg. Studies on the covalent binding of radiolabeled acetaminophen to liver protein revealed that little covalent binding occurred until the dose reached the toxic level. Moreover, the level of glutathione in the liver decreased as the dose of the drug was increased, suggesting the possibility that glutathione inactivated a chemically reactive metabolite of acetaminophen and thereby prevented the covalent binding and the liver necrosis. In accord with this view, little covalent binding and no liver necrosis occurred at doses of acetaminophen that deplete liver glutathione less than 85% (Table 4). Moreover, a time course study revealed that after the administration of a toxic dose of acetaminophen, there was a delay in the covalent binding; very little covalent binding occurred until about 85% of the glutathione in liver had been depleted.

Table 4. *Relationship between glutathione levels in liver, covalent binding of acetaminophen metabolites, and liver necrosis in mice*

Dose of Acetaminophen (mg/kg)	Hepatic Glutathione % of initial level	Covalent binding nmoles bound (mg protein)	Toxicity
100	76	0.04	none
200	41	0.08	none
375	19	0.71	minimal
750	17	1.89	extensive

Data taken from MITCHELL et al. (1973a and b).

The major routes of acetaminophen metabolism in man and experimental animals are through the formation of its glucuronide and sulfate conjugates. The finding that acetaminophen depletes liver of glutathione therefore raised the possibility that a small proportion of the dose of acetaminophen was converted to a reactive metabolite which reacted with glutathione to form a conjugate that would be ultimately excreted as a mercapturic acid. Subsequent studies confirmed this view, but showed that even in the hamster, the animal species which is most sensitive to the necrotic effects of acetaminophen, the proportion of the acetaminophen that was eliminated as the mercapturic acid was less than 15% (Table 5) (JOLLOW et al., 1975). As the dose of acetaminophen was increased in hamsters the proportion of the dose that was excreted as the mercapturic acid decreased presumably because the glutathione in liver became depleted. In addition,

Table 5. *Acetaminophen metabolites in hamsters*

Dose of Acetaminophen	Glucuronide	Sulfate	Mercapturic acid	Unchanged
	% of total urinary metabolite			
25 mg/kg	42	40	13	5
400 mg/kg	63	18	4	14

Data from JOLLOW et al. (1975).

however, the proportion of the dose excreted as the sulfate conjugate was also decreased either because the sulfate transferase became saturated or because the body became depleted of sulfate at the higher doses of acetaminophen. Accordingly, the proportion of the dose of acetaminophen excreted unchanged in the urine was increased as the size of the dose was increased.

The finding that the covalent binding of radiolabeled acetaminophen was inversely related to the liver levels of glutathione suggested the possibility that the magnitude of the covalent binding of acetaminophen metabolites and the severity of the necrosis could be enhanced by the administration of another substrate that depletes liver of glutathione. In accord with this view, the prior treatment of mice with diethyl maleate, which depletes liver glutathione without causing necrosis, increases both the covalent binding of acetaminophen metabolites to liver protein and the severity of acetaminophen-induced liver necrosis. But the prior administration of cysteine (Table 6), which leads to the synthesis of glutathione, or of cysteamine or dimercaprol, which presumably react chemically with the reactive metabolite, decreases both the covalent binding of the acetaminophen metabolite and the severity of liver necrosis. Since cysteine and cysteamine are also converted to sulfate, it is also possible that a part of the decrease in covalent binding of acetaminophen caused by these substances could have been due to an increase in the clearance of acetaminophen by sulfate conjugation, that is an increase in sulfate could have decreased the ratio, k_a/k_e. But the contribution of this mechanism of protection would be small because the total body clearance could have been decreased by only 20% as estimated by the decrease in the urinary sulfate conjugate.

Table 6. *Effect of diethyl maleate or cysteine on in vivo covalent binding of acetaminophen* (375 mg/kg) *to mouse liver protein*

Treatment[a]	Severity of liver necrosis (after pHAA)	Protein bound acetaminophen (nmol/mg protein) Liver
None	minimal	0.88 ± 0.09
Diethyl Maleate[a]	extensive	1.57 ± 0.11
Cysteine[a]	none	0.38 ± 0.08

[a] Diethyl maleate 0.3 ml/kg, i. p., 30 min before acetaminophen; cysteine 150 mg/kg, i. p., 5 min before and 20 min after acetaminophen.
Data taken from MITCHELL et al. (1973b).

Since only a small fraction of the dose of acetaminophen is converted to the reactive metabolite, it might be expected that selective changes in the activity of the enzyme system which catalyzes the formation of the reactive metabolite would alter the proportion of the acetaminophen that is converted to the reactive metabolite (k_a/k_e) and thereby alter the amount of covalently bound acetaminophen. From *in vitro* studies, it was found that the activating enzyme in liver was a cytochrome P-450 oxidase in liver microsomes (POTTER et al., 1973). Accordingly, pretreatment of mice with phenobarbital, which induces the cytochrome P-450 in liver microsomes, increases both the covalent binding of radiolabeled acetaminophen and the severity of liver necrosis, whereas treatment of the animals with cobaltous chloride or piperonyl butoxide, which decreases the activity of these enzymes, decreases both the covalent binding and the severity of the necrosis.

Since the formation of the glucuronide and the sulfate conjugates are not catalyzed by cytochrome P-450, one would not necessarily expect that inducers and inhibitors of cytochrome P-450 enzymes would alter the biological half-life of drugs which are cleared mainly by conjugation reactions. In accord with this view, pretreatment of mice with either phenobarbital or cobaltous chloride did not appreciably alter the biological half-life of acetaminophen (MITCHELL et al., 1973a). But piperonyl butoxide increased the half-life of the drug because it blocks not only cytochrome P-450 enzymes but also glucuronide conjugation (Fig. 4).

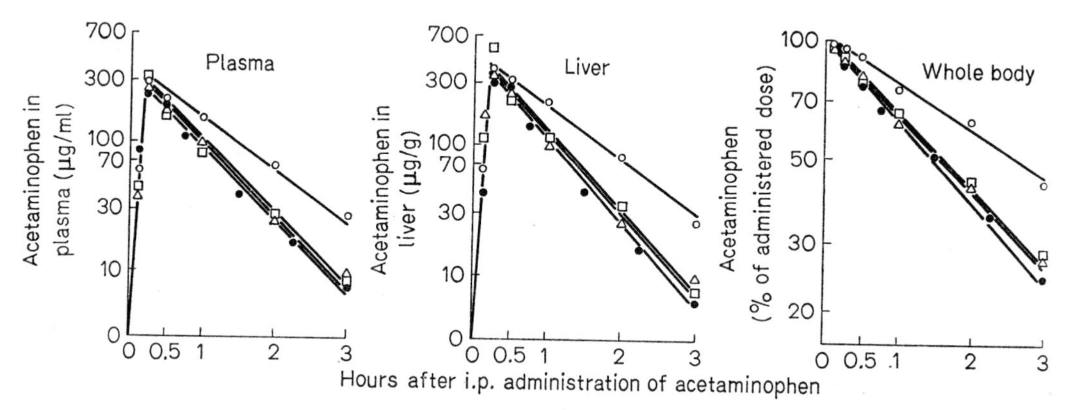

Fig. 4. The effect of treatments on the elimination of acetaminophen in mice Normal-△, phenobarbital-●, cobaltous chloride-□, piperonyl butoxide-○. (Data from MITCHELL et al., 1973a).

These findings thus illustrate that treatments can markedly change the toxicity of drugs even though they do not markedly change the biological half-life of the drug. The findings also illustrate the problem of finding an animal species that mimics humans in the metabolism and toxicity of drugs when the toxicity is mediated by metabolites formed along minor pathways.

Other concepts of the factors that affect covalent binding of drugs to macromolecules and tissue necrosis can be illustrated by our studies with bromobenzene. A cytochrome P-450 enzyme in liver microsomes catalyzes the conversion of bromobenzene to a chemically reactive epoxide which reacts with protein and other macromolecules in liver. However, liver cells possess a number of mechanisms that inactivate bromobenzene epoxide (Fig. 5): A part of the epoxide rearranges nonenzymatically to form *p*-bromophenol; an enzyme present in microsomes catalyzes the addition of water to form a dihydrodiol which in turn can be oxidized to a catechol; another enzyme in the soluble fraction of liver catalyzes the formation of a glutathione conjugate that ultimately is excreted in urine as a mercapturic acid (AZOUZ et al., 1953; KNIGHT and YOUNG, 1958; DALY et al., 1972; ZAMPAGLIONE et al., 1973; JOLLOW et al., 1974). The steady-state concentration of the epoxide in liver thus depends on the rate of formation of the epoxide and on the rate of inactivation of the epoxide by the various enzymatic and nonenzymatic reactions.

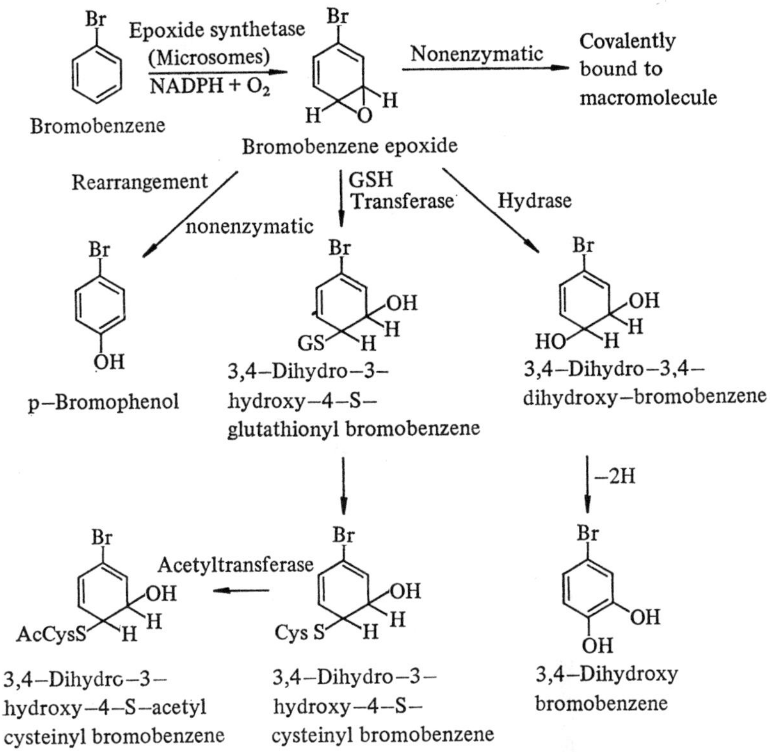

Br
Epoxide synthetase
(Microsomes)
NADPH + O_2
Bromobenzene
Br
H
O
H
Bromobenzene epoxide
Nonenzymatic
Covalently
bound to
macromolecule
Rearrangement
nonenzymatic
GSH
Transferase
Hydrase
Br
OH
p-Bromophenol
Br
OH
H
GS
H
3,4-Dihydro-3-
hydroxy-4-S-
glutathionyl bromobenzene
Br
OH
H
HO
H
3,4-Dihydro-3,4-
dihydroxy-bromobenzene
-2H
Br
OH
H
AcCysS
H
Acetyltransferase
3,4-Dihydro-3-
hydroxy-4-S-acetyl
cysteinyl bromobenzene
Br
OH
H
Cys S
H
3,4-Dihydro-3-
hydroxy-4-S-
cysteinyl bromobenzene
Br
OH
OH
3,4-Dihydroxy
bromobenzene

Fig. 5. Pathways of bromobenzene metabolism

As with acetaminophen, little covalent binding of radiolabeled bromobenzene and no necrosis occur until the liver levels of glutathione are markedly decreased (Fig. 6) (JOLLOW et al., 1974). Moreover, the severity of the necrosis and the magnitude of the covalent binding can be markedly increased by the prior administration with diethyl maleate and can be decreased by the administration of cysteine (REID and KRISHNA, 1973). Thus, glutathione protects the liver against the toxic effects of bromobenzene epoxide by combining with it to form the glutathione conjugate.

Unlike acetaminophen, however, virtually all of the bromobenzene is converted to its epoxide before it is eliminated from the body. Thus, pretreatment of rats with phenobarbital, an inducer of the cytochrome P-450 enzyme that catalyzes the formation of the bromobenzene epoxide, cannot increase the proportion of the dose that is converted to the epoxide and therefore should not ordinarily increase the covalent binding of bromobenzene epoxide to liver protein even though the pretreatment markedly increases the rate of bromobenzene metabolism. Indeed, pretreatment with phenobarbital should cause a small decrease in covalent binding because it increases the activity of the epoxide hydrase (DALY et al., 1972; JOLLOW et al., 1974). In accord with this view, phenobarbital decreases the covalent binding of bromobenzene (Table 7) provided that low, nontoxic doses of the toxicant are administered (REID and KRISHNA, 1973; GILLETTE, 1973). On the other hand, when toxic doses are administered, the treatment with phenobarbital markedly increases the covalent binding of bromobenzene metabolites. The reason for the marked increase becomes clear when it is realized that glutathione is rapidly synthesized in the body. Thus, a considerable

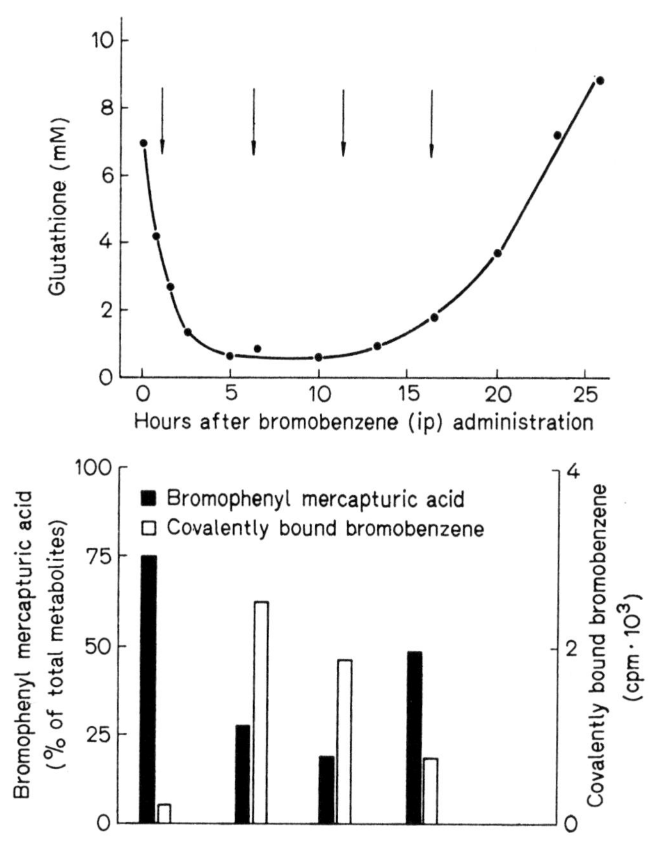

Fig. 6. Relationship *in vivo* between the concentration of glutathione in the liver, the formation of a bromobenzene-glutathione conjugate (measured in urine as bromophenylmercapturic acid) and covalent binding of a bromobenzene metabolite to liver proteins. Nonradioactive bromobenzene (10 mmoles/kg) was administered i. p. to rats and the depletion of glutathione from the liver was determined. ^{14}C-Bromobenzene (50 μCi, 11.8 mCi/mmole) was administered i. v. to additional bromobenzene-treated rats at the time intervals marked with arrows. Urinary metabolites were collected for 12 h. The rats were then sacrificed and the extent of ^{14}C-bromobenzene bound to liver protein was determined. (Data from Jollow et al., 1974)

Table 7. *Effect of phenobarbital on the covalent binding of* ^{14}*C-bromobenzene metabolites in mouse liver*[a]

Dose (mmol/kg)	Untreated (nmol/mg protein)	Treated[b] (nmol/mg protein)
0.13	0.074	0.027
1.15	0.79	0.99
4.85	0.44	9.83

[a] Data from Reid and Krishna (1973).
[b] Mice were given 80 mg/kg of phenobarbital for 3 days.

portion of the bromobenzene epoxide is still inactivated by its conjugation with glutathione even after the liver levels of glutathione have been decreased to very low levels (ZAMPAGLIONE et al., 1973; GILLETTE, 1973; JOLLOW et al., 1974). However, the rate of synthesis of the glutathione conjugate is now limited by the rate of synthesis of glutathione and thus the proportion of the bromobenzene epoxide that is inactivated by formation of the conjugate can be decreased by accelerating the rate of formation of the epoxide.

It also seems likely that after the liver is depleted of glutathione the steady-state level of the bromobenzene epoxide increases to a concentration at which it can leave the liver and be carried by the blood to other tissues. In accord with this view, the pretreatment of mice with phenobarbital increases the covalent binding of bromobenzene epoxide not only to liver proteins but also to lung proteins even though phenobarbital does not induce the enzyme that catalyzes the formation of bromobenzene epoxide in lung (REID et al., 1973).

Acknowledgement

This paper was presented at the Third International Freiburg Symposium on Liver Diseases: Drugs and the Liver, Freiburg, Germany, October 12, 1973.

References

ALEXANDERSON, B., BORGA, O., ALVAN, G.: The availability of orally administered nortriptyline. Europ. J. Clin. Pharmacol. **5**, 181—185 (1973).

AZOUZ, W.M., PARKE, D.V., WILLIAMS, R.T.: Studies in detoxification. 51. The determination of catechols in urine, and the formation of catechols in rabbits receiving halogenobenzene and other compounds. Dihydroxylation *in vivo*. Biochem. J. **55**, 146—1515 (1953).

BEYER, K.H., RUSSO, H.F., TILLSON, E.K., MILLER, A.K., VERWEY, W.F., GASS, S.R.: "Benemid." *p*-(di-*nr*-Propylsulfamyl)-benzoic acid. Its renal affinity and its elimination. Amer. J. Physiol. **166**, 625—640 (1951).

BRANCH, R.A., SHAND, D.G., WILKERSON, G.R., NIES, A.S.: The reduction of lidocaine clearance by *dl*-propranolol: An example of hemodynamic drug interaction. J. Pharmacol. exp. Ther. **184**, 515—519 (1973).

BRANCH, R.A., SHAND, D.G., WILKERSON, G.R., NIES, A.S.: Increased clearance of antipyrene and *d*-propranolol after phenobarbital treatment in the monky. Relative contributions of enzyme induction and increased hepatic blood flow. J. clin. Invest. **53**, 1101—1107 (1974).

BÜCH, H., KARACHRISTIANIDIS, G., RUDIGER, W.: Beeinflussung der metabolischen Umwandlung von Phenacetin und *N*-acetyl-*p*-aminophenol (NAPAP) durch Barbiturate. Naunyn-Schmiedebergs Arch. exp. Path. Pharmak. **251**, 107—108 (1965).

BÜCH, H., PFLEGER, K., RUDIGER, W.: Nachweis und Bestimmung von Phenacetin, *N*-acetyl-*p*-aminophenol sowie ihren Hauptumwandlungsprodukten in Harn und Serum. Z. klin. Chem. **5**, 110—114 (1967).

CONNEY, A.H.: Pharmacological implications of microsomal enzyme induction. Pharmacol. Rev. **19**, 317—366 (1967).

CUCINELL, S.A., CONNEY, A.H., SANOUR, M., BURNS, J.J.: Drug interactions in man. I. Lowering effect of phenobarbital on plasma levels of bishydroxycoumarin (Dicumarol) and diphenylhydantoin (Dilantin). Clin. Pharmacol. Ther. **6**, 420—429 (1965).

DALY, J.W., JERINA, D.M., WITKOP, B.: Arene oxides and the NIH shift: The metabolism, toxicity and carcinogenicity of aromatic compounds. Experientia (Basel) **28**, 1129—1149 (1972).

FEUER, G., SOSA-LUCERO, J.C., LUMB, G., MODDEL, G.: Failure of various drugs to induce drug-metabolizing enzymes in extrahepatic tissues of the rat. Toxicol. appl. Pharmacol. **19**, 579—589 (1971).

GILLETTE, J.R.: Factors that effect the covalent binding and toxicity of drugs. In: LOOMIS, T.A. (Ed.): Pharmacology and the future of man, Vol. 2, pp. 187—202. San Francisco 1972. Basel: S. Karger 1973.

GILMAN, A.G., CONNEY, A.H.: The induction of aminoazo dye *N*-demethylase in nonhepatic tissues by 3-methylcholanthrene. Biochem. Pharmacol. **12**, 591—593 (1963).

JOLLOW, D.J., MITCHELL, J.R., ZAMPAGLIONE, N., GILLETTE, J.R.: Bromobenzene-induced liver necrosis. Protective role of glutathione and evidence for 3,4-bromobenzene oxide as the hepatotoxic metabolite. Pharmacology **11**, 151—169 (1974).

JOLLOW, D. J., THORGEIRSSON, S. S., POTTER, W. Z., HASHIMOTO, M., MITCHELL, J. R.: Acetaminopheninduced hepatic necrosis. VI. Metabolic disposition of toxic and nontoxic doses of acetaminophen. Pharmacology 1975 (in press).

KNIGHT, R. N., YOUNG, L.: Biochemical studies of toxic agents. II. The occurrence of premercapturic acids. Biochem. J. **70**, 111—119 (1958).

LEVY, G.: Dose dependent effects in pharmacokinetics. In: TEDESCHI, D. H., TEDESCHI, R. E. (Eds.): Importance of fundamental principles in drug evaluation, pp. 141—171. New York: Raven Press N. Y. 1968.

LEVY, G., YAMADA, H.: Drug biotransformation interactions in man. 3. Acetaminophen and salicylamide. J. Pharm. Sci. **60**, 215—221 (1971).

MITCHELL, J. R., JOLLOW, D. J., POTTER, W. Z., DAVIS, D. C., GILLETTE, J. R., BRODIE, B. B.: Acetaminophen-induced hepatic necrosis. I. Role of drug metabolism. J. Pharmacol. exp. Ther. **187**, 185—194 (1973a).

MITCHELL, J. R., JOLLOW, D. J., POTTER, W. Z., GILLETTE, J. R., BRODIE, B. B.: Acetaminophen-induced hepatic necrosis. IV. Protective role of glutathione. J. Pharmacol. exp. Ther. **187**, 211—217 (1973b).

MITCHELL, J. R., POTTER, W. Z., HINSON, J. A., SNODGRASS, W. R., TIMBRELL, J. A., GILLETTE, J. R.: Toxic drug reactions. In: GILLETTE, J. R., MITCHELL, J. R. (Eds.): Handbuch der experimentellen Pharmakologie, this volume. Berlin-Heidelberg-New York: Springer.

OHNHAUS, E. E., THORGEIRSSON, S. S., DAVIES, D. S., BRECKENRIDGE, A.: Changes in liver blood flow during enzyme induction. Biochem. Pharmacol. **20**, 2561—2570 (1971).

PANTUCK, E. J., HSIAO, K. C., MAGGIO, A., NAKAMURA, K., KUNTZMAN, R., CONNEY, A. H.: Cigarette smoking substantially accelerates rate of phenacetin metabolism in man. Clin. Pharmacol. Ther. **15**, 9—17 (1974).

POTTER, W. Z., DAVIS, D. C., MITCHELL, J. R., JOLLOW, D. J., GILLETTE, J. R., BRODIE, B. B.: Acetaminophen-induced hepatic necrosis. III. Cytochrome P-450 mediated covalent binding *in vitro*. J. Pharmacol. exp. Ther. **187**, 203—210 (1973).

REID, W. D., KRISHNA, G.: Centrolobular hepatic necrosis related to covalent binding of metabolites of halogenated aromatic hydrocarbons. Exp. molec. Pathol. **18**, 80—99 (1973).

REID, W. D., ILETT, K. F., GLICK, J. M., KRISHNA, G.: Metabolism and binding of aromatic hydrocarbons in the lung. Amer. Rev. resp. Dis. **107**, 539—551 (1973).

SJÖQVIST, F., HAMMER, W., SCHUMACHER, H., GILLETTE, J. R.: The effect of desmethyl imipramine and other "anti-tremorine" drugs on the metabolism of tremorine and oxotremorine in rats and mice. Biochem. Pharmacol. **17**, 915—934 (1968).

STENSON, R. E., CONSTANTINO, R. T., HARRISON, D. C.: Interrelationships of hepatic blood flow, cardiac output, and blood levels of lidocaine in man. Circulation **43**, 205—211 (1971).

WADDELL, W. J., BUTLER, T. C.: The distribution and excretion of phenobarbital. J. clin. Invest. **36**, 1217—1226 (1957).

WATTENBERG, L. W., LEONG, J. L.: Tissue distribution studies of polycyclic hydrocarbon, hydroxylase activity. In: BRODIE, B. B., GILLETTE, J. R. (Eds.): Handbuch der experimentellen Pharmakologie, Part 2, pp. 422—430. Berlin-Heidelberg-New York: Springer 1971.

WEISBURGER, J. H., WEISBURGER, E. K.: Biochemical formation and pharmacological, toxicological, and pathological properties of hydroxylamines and hydroxamic acids. Pharmacol. Rev. **25**, 1—66 (1973).

WIEBEL, F. J., LEUTZ, J. C., DIAMOND, L., GELBOIN, H. V.: Aryl hydrocarbon (benzo(a)pyrene) hydroxylase in microsomes from rat tissues: Differential inhibition and stimulation by benzflavones and organic solvents. Arch. Biochem. Biophys. **144**, 78—86 (1971).

ZAMPAGLIONE, N., JOLLOW, D. J., MITCHELL, J. R., STRIPP, B., HAMRICK, M., GILLETTE, J. R.: Role of detoxifying enzymes in bromobenzene-induced liver necrosis. J. Pharmacol. exp. Ther. **187**, 218—227 (1973).

Chapter 74

Toxic Drug Reactions

J. R. Mitchell, W. Z. Potter, J. A. Hinson, W. R. Snodgrass, J. A. Timbrell, and J. R. Gillette

With 13 Figures

The terminology used to describe unwanted reactions to drugs is confusing. There is little consistency in the terms used by pharmacologists and physicians. It is not always clear what is meant when clinicians use expressions such as idiosyncrasy, intolerance, adverse effect, drug hazard, drug allergy, susceptibility, hypersensitivity, side effect, drug-induced disease and toxic effect. This is unfortunate because it would seem best not to label a drug reaction until the general mechanism is certain. It is often counter-productive to categorize phenomena too soon, since operational terms are more stimulating to research than labels are. Once a drug reaction is given a label, scientists from other disciplines who might be interested in studying the problem assume that it is already solved. An excellent example of this premature labeling was the classification of primaquine-induced hemolytic anemia as a hypersensitivity reaction. Once this definition appeared in the literature, no biochemist would look at the problem (basic scientists seem afraid of the word "hypersensitivity") and it was left for a very astute physician to disclose that the hemolytic anemia was caused by a genetic deficiency of glucose-6-phosphate dehydrogenase in erythrocytes (Marks and Banks, 1965).

For these reasons we prefer to think of adverse drug reactions according to pathogenesis: 1) pharmacologic reactions due to an exaggerated therapeutic effect or to an unwanted secondary drug action, 2) toxic drug reactions, 3) drug reactions of unknown etiology.

I. Exaggerated or Unwanted Drug Actions

These are the most frequent drug reactions and occur commonly in clinical practice. These may be subdivided into pharmacokinetic and pharmacodynamic reactions.

1. Pharmacokinetic Drug Reactions result from an overabundance of drug at drug receptor sites in target tissues. In this situation, drug toxicity is due either to an exaggeration of the desired therapeutic effect or to an unwanted secondary effect on another tissue. Thus, orthostatic syncope caused by a ganglionic blocking drug would be an example of the desired blockade on sympathetic ganglia displayed to excess, whereas paralytic ileus would be an example of an unwanted blockade on parasympathetic ganglia. The excessive amount of drug at the receptor site may occur from too large of a drug dose or from alterations in the pharmacokinetic processes that determine drug concentration (absorption, tissue binding and distribution, metabolism, excretion). The principles underlying pharmacokinetic drug reactions are well reviewed in several preceding chapters in this volume.

2. *Pharmacodynamic Drug Reactions* result from changes either in the responsiveness of the drug receptor in the target tissue or in the responsiveness of receptors mediating an unwanted secondary effect. The changes in receptor responsiveness occur because of physiological or pathological alterations in the condition of the patient or because of pharmacodynamic drug interactions. The increased sensitivity of patients with chronic obstructive pulmonary disease to sedatives and of patients with diuretic-induced low plasma potassium to digitalis are examples of this group. Pharmacodynamic drug reactions are discussed at length in many of the preceding chapters in this volume.

II. Toxic Drug Reactions

These rare but real drug reactions occur much less frequently than the previously described ones. However, their toxic effects are of greater clinical concern to the physician because they are often so severe that they lead to irreversible failure of the liver, kidneys, bone marrow or other organs and subsequent death of the patient. The high incidence of serious hepatitis in subjects receiving isoniazid for tuberculosis prophylaxis and the frequently lethal occurrence of massive hepatic necrosis in humans taking overdoses of acetaminophen are two important examples of this group.

The investigation of such reactions has been handicapped by the tendency to view them as isolated instances of idiosyncrasy or hypersensitivity. Recent studies in our laboratory, however, have demonstrated that many of these toxic drug reactions result from a common initiating event: the metabolic activation of chemically stable drugs to potent alkylating or arylating agents in the body (MITCHELL et al., 1973a; GILLETTE et al., 1974).

A. Cell Necrosis

1. Acetaminophen (Paracetamol)

This commonly used drug is one of the safest of all minor analgesics when taken in normal therapeutic doses, but large overdoses produce fatal hepatic necrosis in humans (PRESCOTT et al., 1971), rats (BOYD and BERECZKY, 1966; MITCHELL et al., 1973b) mice (MITCHELL et al., 1973b) and hamsters (DAVIS et al., 1974). Prior treatment of animals with inducers of drug metabolism, such as phenobarbital or 3-methylcholanthrene, markedly potentiated the severity of the necrosis (MITCHELL et al., 1973b; POTTER et al., 1974). In contrast, pretreatment with inhibitors of drug metabolism, such as piperonyl butoxide, cobaltous chloride or α-naphthylisothiocyanate (ANIT), protected against the necrosis (MITCHELL et al., 1973b; MITCHELL and JOLLOW, 1974). The lack of correlation between acetaminophen tissue levels and acetaminophen-induced liver necrosis demonstrated that a toxic metabolite rather than acetaminophen itself caused the hepatocellular damage. The hepatotoxicity was not related to drug-induced methemoglobinemia, because unlike other aniline derivatives acetaminophen does not cause significant methemoglobinemia.

Acetaminophen radiolabeled with tritium or with carbon-^{14} was given to normal mice and to mice pretreated with compounds that altered acetaminophen-induced liver necrosis. The animals were killed at various times and the livers examined for covalently bound metabolites of acetaminophen. Autoradiograms showed covalent binding of acetaminophen preferentially in the necrotic centrilobular area of the liver (Fig. 1) (JOLLOW et al., 1973). Pretreatment with an inducer of metabolism, phenobarbital, greatly increased binding, whereas pretreatment with different inhibitors of metabolism markedly decreased binding. Thus, the effect of treatments on covalent binding correlated directly with treatment effects on hepatic necrosis. Similar results were obtained when covalent binding was measured quantitatively by precipitation of tissue proteins followed

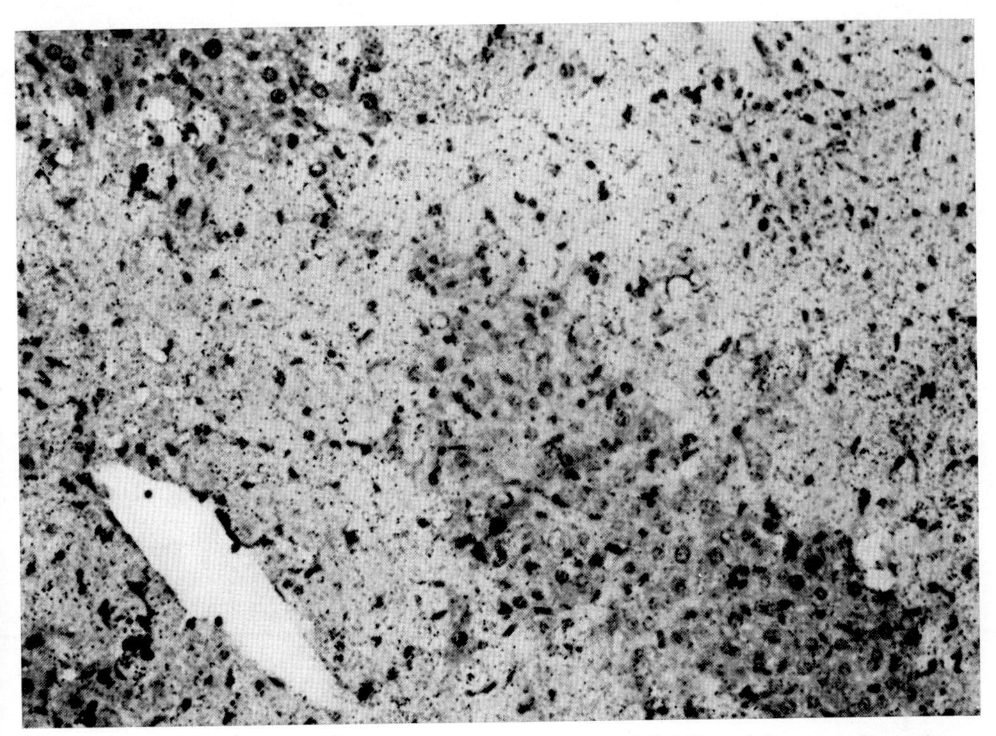

Fig. 1. Autoradiogram of a paraffin section (periodic acid-Schiff and hematoxylin stain, × 250) from liver of a 25 g mouse 24 hr after administration of ^{3}H-acetaminophen (0.5 mc; 750 mg/kg). Radioactivity is localized primarily in the necrotic centrilobular zones

by repeated extraction with various organic solvents (Table 1). Evidence of the covalent nature of the binding was obtained by protease digestion of the solvent-extracted protein precipitate and isolation of the radiolabel bound to single amino acids and peptide fragments. Since acetaminophen is a chemically stable compound, the finding of a covalent linkage with macromolecules of its target tissue, the liver, confirms that the drug is converted in the body to a reactive arylating agent.

In view of this striking correlation between the severity of hepatotoxicity and the extent of covalent binding by the arylating metabolite of acetaminophen, it was surprising that significant covalent binding did not occur until over 60% of the drug had been eliminated from the liver (Fig. 2). Glutathione was depleted from the liver of animals receiving acetaminophen because it combined with a minor metabolite of the drug and formed a readily excreted mercapturic acid (Mitchell et al., 1973a and c; Potter et al., 1974; Jollow et al., 1974a). Thus, the possibility arose that the arylating metabolite of acetaminophen initially is

Table 1. *Effect of treatments on in vivo covalent binding of acetaminophen (pHAA) to mouse tissue protein*

Treatment	Dose of pHAA	Severity of liver necrosis after pHAA[a]	Covalently bound ^{3}H-acetaminophen[b]	
			Liver	Muscle
			nmol/mg protein	
None	375	1—2+	1.02 ± 0.17	0.02 ± 0.02
Piperonyl butoxide	375	0	0.33 ± 0.05	0.01 ± 0.01
Cobaltous chloride	375	0	0.39 ± 0.11	0.01 ± 0.03
ANIT	375	0	0.11 ± 0.06	0.01 ± 0.02
Phenobarbital	375	2—4+	1.60 ± 0.10	0.02 ± 0.02
None	750	3—4+	1.89 ± 0.15	0.04 ± 0.03
Piperonyl butoxide	750	0 or 1+	0.78 ± 0.16	0.04 ± 0.02
Cobaltous chloride	750	0 or 1+	0.85 ± 0.15	0.03 ± 0.03
Phenobarbital	750	4+	2.08 ± 0.12	0.01 ± 0.01

[a] Severity of liver necrosis (1+ to 4+) taken from MITCHELL et al., 1973b.
[b] Covalently bound ^{3}H-acetaminophen determined 2 hr after ^{3}H-acetaminophen, i.p. Values are means ± S.E. of at least 4 experimental or 8 control observations. (Data from JOLLOW et al., 1973; MITCHELL and JOLLOW, 1974.)

detoxified by reacting preferentially with glutathione (Fig. 3). After the liver is depleted of glutathione, however, the metabolite then combines with liver macromolecules essential to the life of hepatocytes.

In support of this view, covalent binding and liver necrosis occurred only after doses of acetaminophen sufficiently large to exceed the availability of glutathione for detoxification (Fig. 4). Similarly, when glutathione concentrations in the liver were compared with the extent of covalent binding at various times after the administration of acetaminophen, significant binding occurred only after glutathione was markedly depleted (MITCHELL et al., 1973c; POTTER et al., 1974).

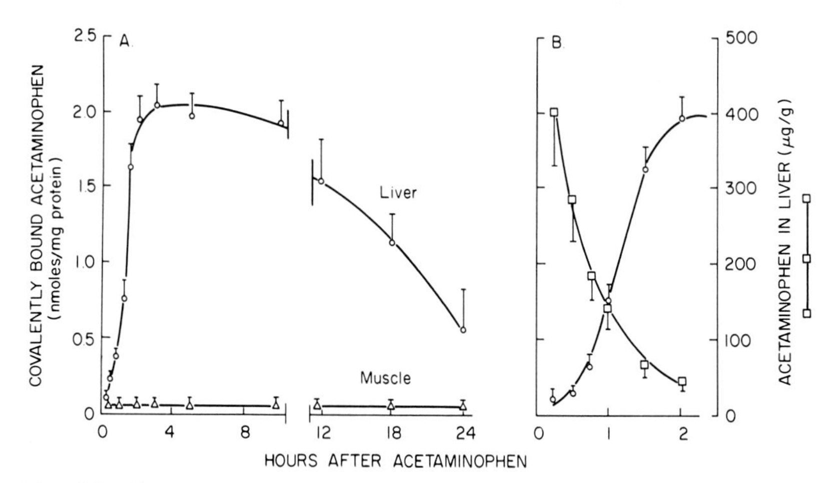

Fig. 2A and B. Time course of covalent binding of ^{3}H-acetaminophen to mouse tissues. Points represent means of four mice ± S.E. A. Covalently bound ^{3}H-acetaminophen in liver (○——○) and muscle (△——△) determined at various intervals after ^{3}H-acetaminophen (750 mg/kg, 0.5 mc, i.p.). B. Expanded time scale of liver binding from (A) with concentration of acetaminophen in the liver (□——□). (Data from JOLLOW et al., 1973)

Moreover, the effect of various pretreatments that alter the availability of glutathione also altered hepatic necrosis and covalent binding of acetaminophen. Depletion of glutathione by diethyl maleate pretreatment dramatically potentiated acetaminophen-induced hepatic necrosis and covalent binding (MITCHELL et al.,

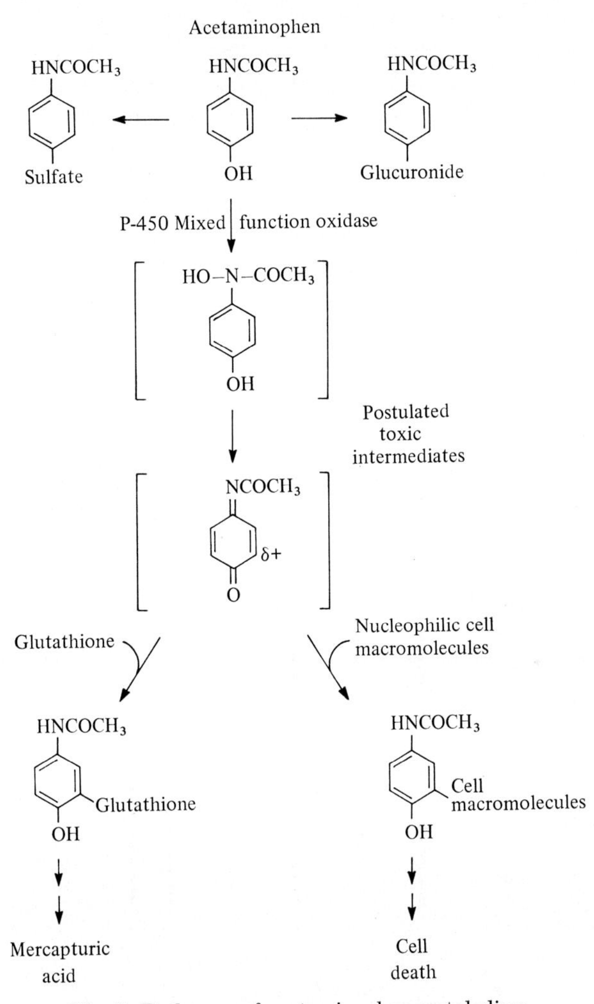

Fig. 3. Pathways of acetaminophen metabolism

1973c; POTTER et al., 1974). Conversely, treatment with cysteine, a precursor of glutathione (BOYLAND and CHASSEAUD, 1967), decreased both the hepatic necrosis and covalent binding (MITCHELL et al., 1973c). Since neither diethyl maleate nor cysteine inhibited the metabolism of acetaminophen, they apparently altered the hepatic damage via their effect on glutathione availability. Thus, glutathione was essential for the protection of thiol and other nucleophilic groups in animal tissues from the arylating metabolite of acetaminophen.

The importance of glutathione in protecting humans from acetaminophen and other hepatotoxins, however, was uncertain. Conjugation of glutathione with drugs was known to occur but to a lesser extent than in other animal species (GROVER and SIMS, 1964). However, the identification of a mercapturic acid of acetaminophen in human urine demonstrated that acetaminophen is converted to an electrophilic reactant in humans, just as in other animal species (JAGENBURG and TOCZKO, 1964; MITCHELL et al., 1974a; JOLLOW et al., 1974a).

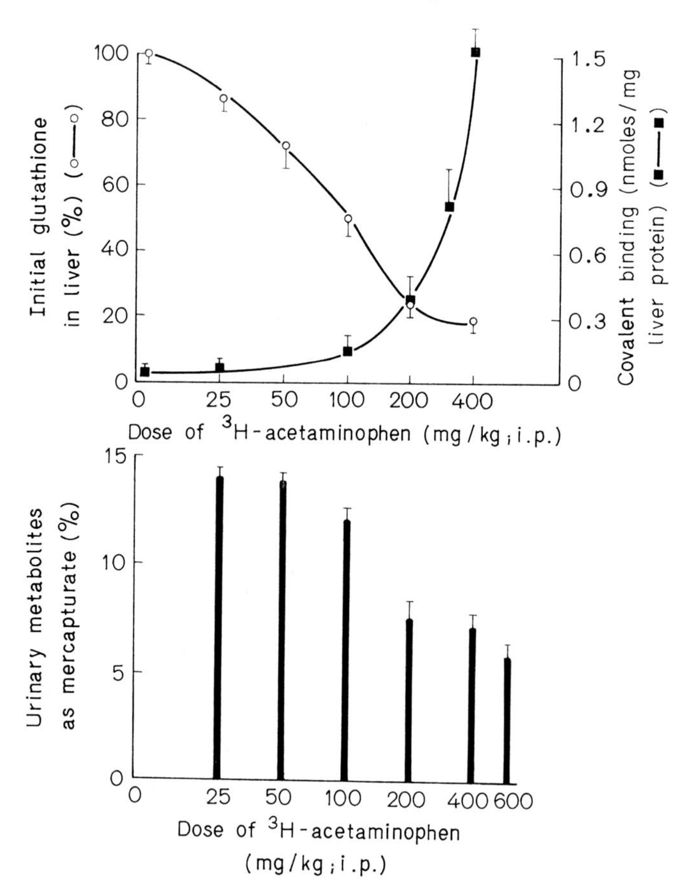

Fig. 4. Relationship *in vivo* between the concentration of glutathione in the liver, the formation of an acetaminophen-glutathione conjugate (measured in urine as acetaminophen-mercapturic acid) and covalent binding of an acetaminophen metabolite to liver proteins. Several doses of ^{3}H-acetaminophen were administered i.p. to hamsters and glutathione concentrations and covalent binding of radiolabeled material were determined 3 hr later. An additional group of hamsters were treated similarly and urinary metabolites were collected for 24 hr. (Data from POTTER et al., 1974; JOLLOW et al., 1974a)

Since the toxic metabolite of acetaminophen combines preferentially with glutathione, formation of the toxic metabolite can be estimated by measuring the acetaminophen-glutathione conjugate excreted in urine as the mercapturic acid. Phenobarbital pretreatment of human subjects markedly increased the amount of acetaminophen excreted as the mercapturate from about 4% of the dose to about 7%, indicating an increased formation of the toxic metabolite after phenobarbital induction (Fig. 5) (MITCHELL et al., 1974a). These data suggest that hepatic injury after acetaminophen overdosage might be potentiated in patients receiving inducers of liver cytochrome P-450 enzymes such as phenobarbital. A retrospective study of patients suffering from acetaminophen-induced hepatic injury supports this view (WRIGHT and PRESCOTT, 1973).

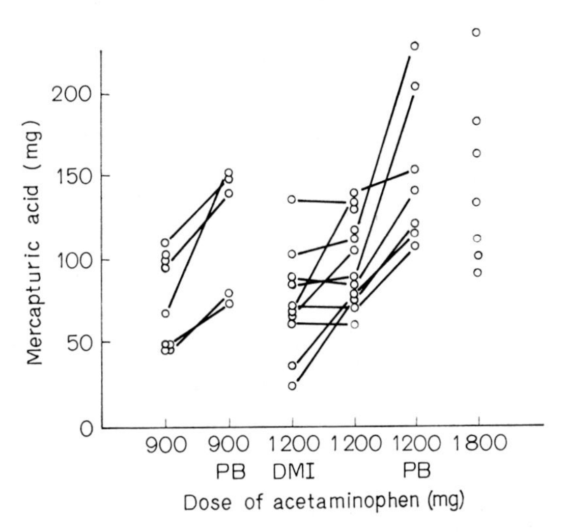

Fig. 5. Effects of dose and treatments on the excretion of acetaminophen-mercapturic acid in human urine. Acetaminophen was administered orally and urine collected over a 24-hr period. Phenobarbital (PB) and desipramine (DMI) administered where indicated. (Data from MITCHELL et al., 1974a)

The normal concentration of glutathione in liver of various animals is about 4 mM (BOYLAND and CHASSEAUD, 1969). Liver necrosis occurs in animals after doses of acetaminophen that deplete more than 70% of the hepatic glutathione (Fig. 4) (MITCHELL et al., 1973c; POTTER et al., 1974). Assuming a similar glutathione level in humans, one would expect that at least 4 mmoles of toxic metabolite is necessary to cause liver injury in man (70% $\times$ 1.5 kg liver $\times$ 4 mM glutathione). Thus normal individuals would be susceptible to acetaminophen doses over 15 g (4% $\times$ 15 g dose/molecular weight = 4 mmoles of metabolite), and patients with induced drug-metabolizing enzymes would be susceptible to doses as low as 10 g (Fig. 5). These experimentally derived predictions agree quite well with clinical observations of overdosed patients (PRESCOTT et al., 1971; CLARK et al., 1973). Thus glutathione apparently served a protective role in man as well as in other animal species.

At present there is no specific therapy for acetaminophen poisoning (PRESCOTT et al., 1971; CLARK et al., 1973). In the hope of finding an inhibitor of the formation of the arylating metabolite from acetaminophen, the effect of desipramine on acetaminophen metabolism in humans was examined, since this antidepressant agent was known to block several drug-metabolizing pathways (MITCHELL et al.,

1970; VESELL et al., 1972). Concomitant treatment with desipramine significantly decreased the excretion of acetaminophen-mercapturic acid in only 4 of 10 humans (Fig. 5). Desipramine did not decrease the rate or total absorption of acetaminophen. Since the drug did not retard clearance of acetaminophen from the body, it might be considered an ideal inhibitor to use in treatment of overdosed patients because only the toxic pathway is apparently inhibited. However, desipramine failed to protect against acetaminophen-induced liver necrosis in mice, although this failure may have resulted from the extremely rapid metabolism of desipramine in this species (DINGELL et al., 1964). For these reasons, the potential use of desipramine in humans for therapy of acetaminophen overdosage remains in doubt.

On the other hand, acetaminophen-induced hepatic damage and covalent binding in animals is prevented, but not reversed by the administration of a variety of glutathione-like nucleophiles that can penetrate hepatocytes (in contrast to exogenously administered glutathione), such as cysteine, cysteamine and dimercaprol (MITCHELL et al., 1973a and c, 1974a). Administration of these substances, therefore, provides an alternative for treatment of overdosed patients seen early after poisoning. This work has led to the successful use of cysteamine in the therapy of acetaminophen-overdosed patients (PRESCOTT et al., 1974) and emphasizes the importance of understanding biochemical mechanisms of toxicity before rational approaches to treatment can be made.

Although the identity of the arylating metabolite of acetaminophen (4-hydroxyacetanilide) is uncertain, a *N*-hydroxy derivative seems the most likely candidate. The involvement of an epoxide intermediate is unlikely since studies on the hepatotoxicity of structural analogues of acetaminophen, including acetanilide, phenacetin, 2-hydroxyacetanilide, *N*-methyl-4-hydroxyacetanilide, and 4-chloroacetanilide, generally show little or no hepatic necrosis (POTTER and MITCHELL, 1974). An examination of the metabolites of acetaminophen also failed to reveal a dihydrodiol or other derivative indicative of the intermediate formation of an epoxide (JOLLOW et al., 1974a). Indeed, the electron donating ability of the amido and hydroxyl group substituents on the phenyl ring of acetaminophen makes it improbable that an epoxide metabolite could be the reactive hepatotoxin (DALY et al., 1972). The relative lack of hepatotoxicity by 2-hydroxyacetanilide also suggests that direct oxidation (dehydrogenation) to the chemically reactive imidoquinone is not involved in the hepatotoxicity. In contrast, extensive indirect evidence suggests that the toxic metabolite is formed via *N*-hydroxylation (Fig. 3). Masking of the nitrogen group as in *N*-methyl-4-hydroxyacetanilide blocks hepatotoxicity. The *N*-hydroxylation of 4-chloroacetanilide and 2-acetylaminofluorene and the conversion of acetaminophen to its reactive metabolite are catalyzed by a cytochrome P-450-dependent mixed-function oxidase located in the microsomal fraction of the liver and are markedly enhanced by the presence of sodium fluoride in the reaction mixture (HINSON et al., 1974a and b; POTTER et al., 1973a; THORGEIRSSON et al., 1973a). In addition, pretreatment of hamsters with 3-methylcholanthrene, which increases the hepatotoxicity of acetaminophen (MITCHELL et al., 1973b; POTTER et al., 1974), correspondingly increases the *in vitro* *N*-hydroxylation of 4-chloroacetanilide and 2-acetylaminofluorene and the *in vitro* covalent binding of acetaminophen. By contrast, phenobarbital pretreatment of hamsters neither altered the rate of *N*-hydroxylation of 4-chloroacetanilide nor the covalent binding of acetaminophen *in vitro* (HINSON et al., 1974a; POTTER et al., 1974). It seems likely that the *N*-hydroxylated metabolite subsequently undergoes dehydration to a chemically reactive imidoquinone before arylating tissue macromolecules (Fig. 3).

2. Phenacetin

Since the first report by SPUHLER and ZOLLINGER in 1953, there has been considerable interest in the possible association between the regular intake of analgesics and the occurrence of a special type of interstitial nephritis called "analgesic nephropathy". Abuse of analgesic mixtures containing phenacetin, aspirin or antipyrine and caffeine became increasingly widespread after World War II. The incidence of abuse varies considerably with regard to the population group. In Denmark 29.4% of the patients admitted to a department of medicine abused analgesics (LARSEN and MØLLER, 1959). In Sweden, KASANEN et al. (1962) reported that 10% of the employees in a large industrial establishment and 20% of a series of hospital patients abused analgesics. In Australia the incidence of analgesic abuse in hospital patients was 19.1% (LAVAN et al., 1966). In Swiss watch factories 7% of the males and 17% of the female workers abused analgesics (DUBACH et al., 1968). Indeed, GAULT et al. (1968) have estimated that the *per capita* yearly consumption of phenacetin alone is as high as 40 g in Australia, 25 g in Denmark, 23 g in Switzerland and 12 g in Scotland.

Is it possible that analgesic nephropathy could be due to true drug toxicity from phenacetin? This has long been a controversial question (SHELLEY, 1967) and was recently revisited (ABEL, 1971; MICHIELSON et al., 1972). In this regard it is noteworthy that large doses of phenacetin cause liver necrosis in several animal species. Although phenacetin does not cause acute hepatic necrosis in rats (BOYD and BERECZKY, 1966), it produces centrilobular necrosis in about 20% of mice given near lethal doses (MITCHELL et al., 1973b). By contrast, the hamster, a species that rapidly converts polycyclic and monocyclic acetanilide derivatives to their *N*-hydroxylated metabolites (MILLER et al., 1964; WEISBURGER et al., 1964; THORGEIRSSON et al., 1973a; HINSON et al., 1974a and b) and is unusually susceptible to acetaminophen-induced hepatic necrosis (DAVIS et al., 1974; POTTER et al., 1974) is sensitive to the hepatotoxic effects of phenacetin (SNODGRASS and MITCHELL, 1974). As with acetaminophen (POTTER et al., 1974), phenacetin-induced necrosis in hamsters was markedly potentiated by pretreatment with 3-methylcholanthrene but not by phenobarbital. For example, phenacetin doses of 400 mg/kg produced massive centrilobular necrosis in 3-methylcholanthrene-treated animals (SNODGRASS and MITCHELL, 1974). Moreover, the severity of necrosis paralleled the magnitude of the covalent binding of radiolabeled phenacetin to hepatic proteins and the depletion of hepatic glutathione (SNODGRASS and MITCHELL, 1974). For example, little covalent binding and no hepatic necrosis occurred at doses that depleted hepatic glutathione less than 80%. However, considerable binding and necrosis occurred at doses that depleted hepatic glutathione more than 80%. Moreover, pretreatment of hamsters with 3-methylcholanthrene increased the depletion of hepatic glutathione, the covalent binding and the severity of necrosis after phenacetin, whereas pretreatment with cobaltous chloride or piperonyl butoxide decreased them. These findings indicate that glutathione in the liver prevents covalent binding and necrosis by reacting with the arylating metabolite of phenacetin.

Although the reactive metabolite of phenacetin has not been identified, it may be important that *N*-hydroxy-4-ethoxyacetanilide is formed by hamster liver microsomes (HINSON and MITCHELL, 1974). Pretreatment with 3-methylcholanthrene markedly increased the rate of formation of the metabolite, whereas pretreatment with cobaltous chloride decreased it. Thus, formation of the *N*-hydroxy metabolite correlated directly with the hepatotoxicity of phenacetin after various pretreatments. It also seems possible that the metabolite may lose the ethyl group nonenzymatically and convert to the reactive imidoquinone (*N*-acetyl-4-benzoimidoquinone), which will undergo nucleophilic addition at the 3-position of the phenyl ring (CALDER et al., 1974). This may explain the finding that the glutathione conjugate of phenacetin (4-ethoxyacetanilide) appears in the urine only as the 3-acetylcysteine, 4-hydroxy derivative (SHAHIDI, 1968).

It is unlikely that the arylating metabolite of phenacetin arose directly from acetaminophen after an initial de-ethylation of phenacetin. Phenobarbital pretreatment, which greatly accelerates the de-ethylation (BÜCH et al., 1967), failed to potentiate phenacetin-induced hepatic injury or glutathione depletion in rats (SNODGRASS and MITCHELL, 1974). In contrast, phenobarbital pretreatment potentiates acetaminophen-induced hepatic necrosis and glutathione depletion in rats (MITCHELL et al., 1973b; POTTER et al., 1974).

The reactive imidoquinone may mediate not only the hepatotoxicity, but also the nephrotoxicity produced by phenacetin. NERY (1971) has suggested the *N*-hydroxylated derivative of phenacetin as an intermediate for such minor urinary metabolites as hydroquinone and acetamide. Moreover, CALDER et al., (1973) subsequently examined the nephrotoxicity of *N*-hydroxy-phenacetin, hydroquinone and 4-benzoquinone after intravenous administration and found acute renal tubular necrosis, which was especially marked after the latter two compounds. This led them to conclude that 4-benzoquinone or hydroquinone might be the ultimate nephrotoxic metabolite.

It still seems plausible, however, that the toxic metabolite is either *N*-hydroxy-phenacetin or its reactive derivative, *N*-acetyl-4-benzoimidoquinone. We have measured covalent binding of phenacetin in the kidney after administration of both phenyl-radiolabeled and acetyl-radiolabeled drug. Although binding was low (0.2 nmole/mg protein) relative to that in the liver, the same amount of binding occurred after both isotopes, demonstrating that the reactive arylating metabolite retained the acetyl moiety of phenacetin (SNODGRASS and MITCHELL, 1974). Thus, the arylating metabolite cannot be 4-benzoquinone. We also demonstrated that phenacetin depleted renal glutathione about 40% after these doses, suggesting that glutathione may protect the kidney as well as the liver against tissue damage caused by phenacetin. It is interesting in this regard that 4-aminophenol, which produces massive renal tubular necrosis (CALDER et al., 1973), markedly depleted renal glutathione more than 80% (SNODGRASS and MITCHELL, 1974) and the necrosis was blocked by pretreatment of rats with an inhibitor of drug metabolism, cobaltous chloride (POTTER and MITCHELL, 1974). Thus, the possibility arises that phenacetin fails to produce acute renal necrosis because it is not metabolized by the kidney cells at a rate sufficient to exceed the glutathione capacity for metabolite conjugation. Supporting this possibility is the demonstration that glutathione clearly serves a protective function against renal necrosis produced by arylating metabolites of certain furan and thiophene analogues (see below).

It should also be considered that phenacetin might be *N*-hydroxylated in the liver and then be transported to the kidney where, under acid conditions (CALDER et al., 1974), it would rearrange to the reactive *N*-acetyl-4-benzoimidoquinone. For example, the *N*-hydroxy metabolites of 2-acetylaminofluorene (WEISBURGER and WEISBURGER, 1973) and of 4-chloroacetanilide (HINSON et al., 1974b) are both sufficiently stable that they are excreted as *N*-0-glucuronides in the urine. It is likely that *N*-hydroxy-phenacetin, but not *N*-hydroxyacetaminophen, would also be stable enough to be concentrated in the kidney and excreted in the urine. This might explain the ability of phenacetin to cause papillary necrosis and renal pelvic carcinoma (see below), since it is unlikely that these tissues, in contrast to the proximal tubules of the kidney, contain quantities of drug-metabolizing enzymes sufficient to form significant amounts of reactive metabolite.

3. Furosemide (Frusemide)

This frequently used diuretic drug is contraindicated in pregnancy because of its recognized teratogenic potential (Physicians Desk Reference, 1973). The drug also has been reported to potentiate renal injury when used in combination with cephaloridine (DODDS and FOORD, 1970; FOORD, 1969). Recently, the drug was noted to produce massive hepatic necrosis in mice (MITCHELL et al., 1973d, 1974b). The liver damage is probably caused by a metabolite of furosemide rather than the parent drug, because necrosis was prevented when the metabolism of furosemide was inhibited by pretreatment of mice with three different types of cytochrome P-450 enzyme inhibitors: piperonyl butoxide, cobaltous chloride and α-naphthylisothiocyanate (ANIT).

Since furosemide is a chemically stable drug, the finding of a covalent linkage with macromolecules of its target tissue, the liver, also indicated that it was converted in the body to a chemically reactive, arylating hepatotoxin (POTTER et al., 1973b; WEIHE et al., 1974). Covalent binding was measured quantitatively by extraction of tissue proteins with organic solvents or by hydrolysis of the protein and isolation of the radiolabeled material bound to single amino acids. Pretreatment of mice with piperonyl butoxide, cobaltous chloride or ANIT almost completely abolished both the *in vivo* covalent binding of furosemide and furos-

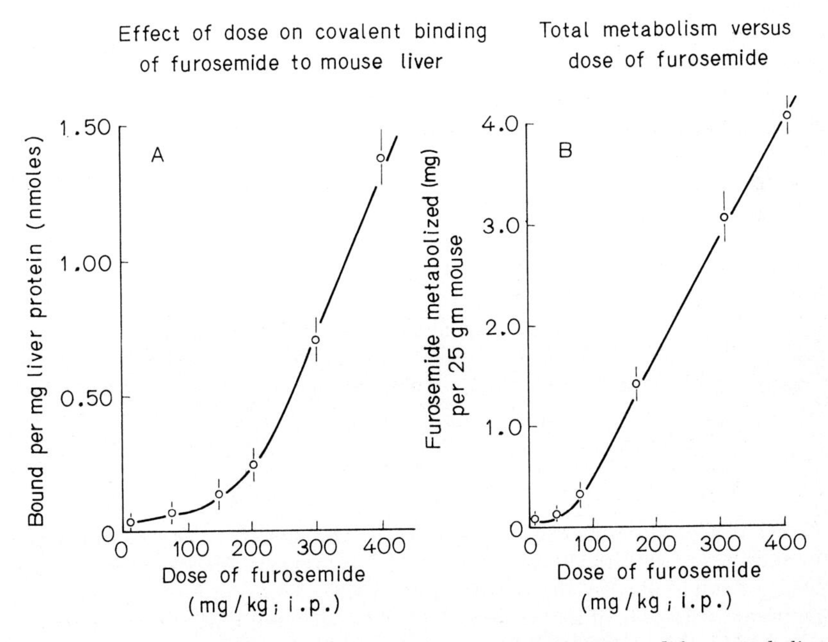

Fig. 6A and B. Dose response curves for hepatic covalent binding and for metabolism of ^{3}H-furosemide determined 3 hr after intraperitoneal administration of drug to mice. Points represent means ± S.E. of determinations on 4 mice. A. Covalent binding to liver versus dose, demonstrating dose threshold for arylation. B. Total amount of metabolites formed versus dose, demonstrating dose threshold for metabolism. Amount of metabolites determined by measuring disappearance of ^{3}H-furosemide from a homogenate of each whole mouse plus its excreta. (Unpublished data from POTTER, JOLLOW, and MITCHELL)

emide-induced hepatic necrosis. The covalent binding occurred a few hours before the onset of histologically recognizable necrosis and prior to biochemical changes in the hepatocytes [e.g., decreased cytochrome P-450-dependent drug metabolism and decreased protein synthesis (THORGEIRSSON et al., 1973b)]. Thus, the formation of a reactive furosemide metabolite may be causally related to the development of furosemide-induced hepatic necrosis.

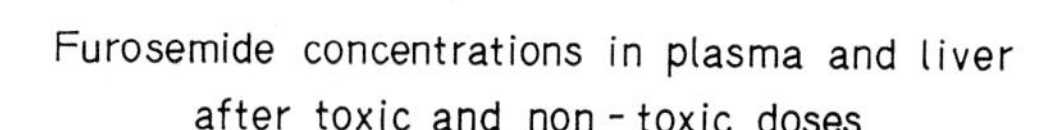

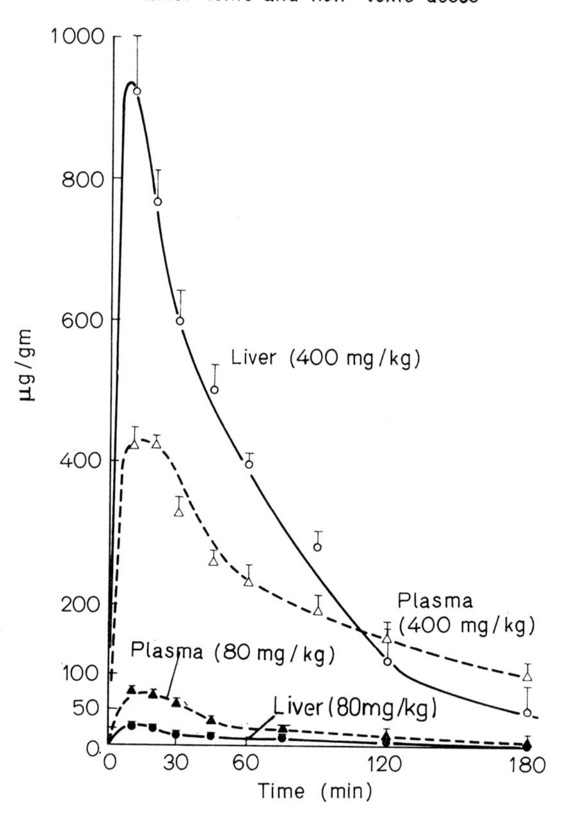

Fig. 7. Change in tissue distribution of furosemide after toxic and nontoxic doses to mice. (Unpublished data from POTTER, JOLLOW, and MITCHELL)

As with acetaminophen and phenacetin, the hepatic necrosis produced by furosemide also exhibited a dose threshold for toxicity (MITCHELL et al., 1973d, 1974b; POTTER et al., 1973b; WEIHE et al., 1974). No arylation (Fig. 6a), necrosis or significant amounts of metabolism (Fig. 6b) occurred until a dose of 100 mg/kg was exceeded. Unlike the dose threshold for acetaminophen and phenacetin hepatotoxicity, however, the furosemide threshold is not due to a protective role of glutathione since furosemide did not deplete hepatic glutathione. Studies of the metabolism, distribution and reversible plasma protein binding of furosemide after nontoxic and toxic doses indicate that the dose threshold for toxicity results from a saturation of the organic anion binding sites on plasma proteins after toxic doses of furosemide. More free furosemide is then available to the liver for metabolism (Fig. 7), i.e., the average extraction ratio is increased to a greater extent in the liver than in the kidney as the dose is increased.

The hepatic injury produced by furosemide [*N*-(2′-furylmethyl)-4-chloro-5-sulfamoylanthranilic acid] apparently results from the metabolic activation of the furan ring, possibly by an epoxidation (Fig. 8) similar to that proposed for the hepatocarcinogenic dihydrofuran aflatoxins (SWENSON et al., 1973). For

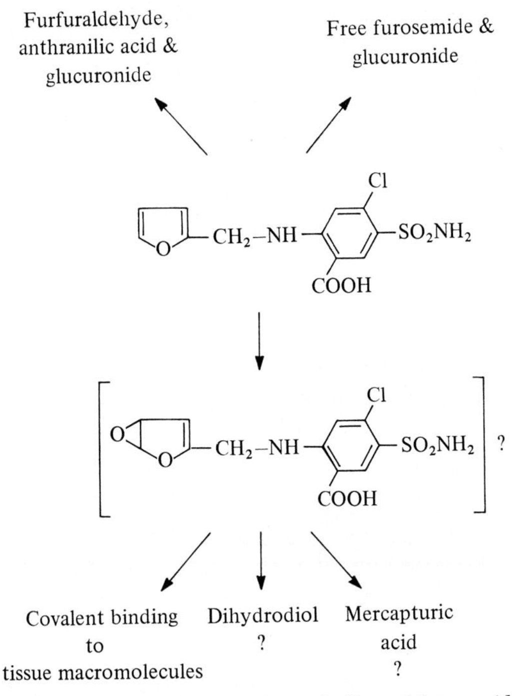

Fig. 8. Proposed pathways of metabolism of furosemide

example, furosemide radiolabeled with tritium in its furan moiety was bound covalently to hepatic microsomes in the presence of oxygen and NADPH to the same extent as furosemide radiolabeled specifically with ^{35}S in its sulfonamide moiety, demonstrating that the bound metabolite contained both parts of the furosemide molecule. To determine where the binding occurred on the molecule, the metabolite-protein conjugates isolated from the liver were hydrolyzed under mild acid conditions (pH 4.5) that split furosemide into its methylfuran and sulfamoylanthranilic acid portions. The binding of furan-radiolabeled furosemide to protein was unchanged, whereas the binding of ^{35}S-labeled furosemide was lost. Thus, the metabolic activation of furosemide must have occurred on the furan ring.

Hepatotoxicity also is caused by structural analogues of furosemide such as 2-hydroxymethylfuran, 2-acetylfuran and even furan itself (Table 2) (MITCHELL et al., 1974b). Moreover, pretreatment with phenobarbital shifts the zone of necrosis produced by furosemide and by other furan-containing compounds, such as ngaione (SEAWRIGHT and HRDLICKA, 1972) and 3-hydroxymethyl furan (*N*,*N*-

diethyl)-carbamate (SEAWRIGHT and MATTOCKS, 1973), from a centrilobular to midzonal distribution in mice. Furosemide, furan, 2,3-benzofuran and certain other furano compounds cause acute tubular necrosis in the kidney (Table 2) in addition to the liver, whereas ipomeanol and other furan analogues selectively produce lung damage and pulmonary edema (WILSON et al., 1970, 1971). Thus, a variety of tissue lesions seen after furan-containing compounds may result from a metabolic activation similar to that proposed for furosemide.

Table 2. *Liver, kidney, and lung toxicity in mice and rats caused by furan analogues*

Liver[a]	Dose[b] mmole/kg i.p.	Kidney[a]	Dose[b] mmole/kg i.p.	Lung[a]
Furan	3.5	Furan	3.5	Ipomeanol[d]
Furosemide	1.1	2-Ethyl furan	2.4	
2-Furamide	0.7	2,3-Benzofuran	1.3	
2-Acetyl furan	2.0	2-Furoic acid	1.9	
2-Furfurol	3.9			
2-Ethyl furoate	3.6			
2-Methoxy furan	8.4			
Dibenzofuran	5.8			
Ngaione[c]	—			

[a] Primary site of lesion. However, the other organs frequently manifested damage but to a lesser degree. Moreover, the site of the lesion often was shifted from one organ to another by pretreatments that altered drug metabolism, e.g., administration of phenobarbital or piperonyl butoxide (unpublished results).

[b] Dose producing necrosis of 6 to 25% of parenchymal hepatocytes or 6 to 25% of cells of renal convoluted tubules. (Data from MITCHELL et al., 1974b.)

[c] Data from SEAWRIGHT and HRDLICKA, 1972.

[d] Data from WILSON et al., 1970, 1971.

The implications of the present findings for the clinical use of furosemide are uncertain. Since extensive clinical experience with furosemide has not resulted in reports of liver damage, the drug presumably is not hepatotoxic at the usual therapeutic doses (40 to 600 mg/day). However, the enzyme pathway responsible for the formation of the toxic metabolite is apparently present in humans because human cadaver microsomes, similar to mouse microsomes, can convert furosemide to an arylating metabolite (POTTER and MITCHELL, 1974). The possibility arises, therefore, that the remarkable safety of furosemide after low therapeutic doses may result from the presence in humans of a dose-threshold phenomenon similar to that found in mice. If this is so, then tissue concentrations of furosemide in patients with renal failure probably are much greater in relation to the furosemide dose, since furosemide is eliminated from the body primarily by renal excretion of the unchanged drug (KINDT and SCHMIDT, 1970). In addition, patients with chronic renal failure often have decreased concentrations of plasma proteins, making more free (nonbound) furosemide available to the liver for metabolism. Since huge doses of furosemide (e.g., 2000 mg by rapid i.v. injection) are currently used to treat patients with acute and chronic renal failure (ELLIOTT et al., 1971), the potential hepatotoxicity of this therapeutic regimen should be carefully evaluated. The possible consequences of the formation in humans of an arylating metabolite from furosemide should also be considered, because many compounds that arylate tissue macromolecules *in vivo* produce neoplasia (MILLER, 1970).

4. Cephaloridine

Extension of the furan studies to thiophene, an analogue of furan in which the oxygen atom in the five-membered ring is replaced with sulfur, has shown that several thiophene-containing compounds also produce massive hepatic or renal necrosis in animals (POTTER and MITCHELL, 1974). Since cephaloridine-induced renal necrosis is prevented, but not reversed, by treatment of mice with two inhibitors of drug metabolism, piperonyl butoxide and cobaltous chloride, it seems likely that the necrosis is mediated by an active metabolite. But whether the nephrotoxicity produced in humans by cephaloridine and cephalothin results from the metabolic activation of the thiophene nucleus in these antibiotic drugs remains to be determined.

The renal tubular necrosis seen after the thiophene and furan analogues is especially interesting because most of them, but not cephaloridine or furosemide, markedly deplete renal glutathione and demonstrate dose thresholds for renal toxicity. Preliminary studies with ^{3}H-furamide have revealed a direct correlation between depletion of renal glutathione, arylation of renal proteins and renal tubular necrosis after various pretreatment regimens that alter renal necrosis, just as seen previously in the liver after acetaminophen and phenacetin. The lack of effect of cephaloridine and furosemide on renal glutathione levels is unknown but perhaps the reactive metabolites of these substances are poor substrates for the glutathione-*S*-epoxide transferases in kidney.

5. Isoniazid and Iproniazid

The cephaloridine studies in experimental animals suggest that the concept of drug metabolism as a cause of toxicity may apply not only to drugs given in huge overdoses (e.g., acetaminophen) but also to important drug reactions seen clinically after therapeutic doses of drugs. Three of our recent studies with isoniazid and iproniazid further implicate metabolic activation as a cause of hepatic injury after therapeutic doses of a drug.

First was a prospective study in which 250 patients receiving isoniazid were examined monthly for serum glutamic-oxaloacetic transaminase (SGOT) and bilirubin elevations (MITCHELL et al., 1974c). About 20% showed abnormal values which subsided while the patients continued to take isoniazid. Thus, isoniazid was hepatotoxic in a large proportion of individuals, most of whom adapted to the insult and did not progress to severe hepatic injury. Measurement of the 6 hr plasma concentrations of isoniazid in these patients failed to show a correlation between slow rates of metabolism (acetylation) and abnormal liver function tests. In other studies some investigators have reported such a correlation in a small number of patients (LAS et al., 1972), whereas others also have failed to find a correlation between isoniazid hepatic injury and slow acetylation of isoniazid (SMITH et al., 1972; RAISFELD et al., 1973). In our study no anti-isoniazid antibodies were found and no correlation was seen between hepatic injury and antinuclear antibodies measured at the end of the study (MITCHELL et al., 1974c).

The second study was a retrospective analysis of 114 patients with possible isoniazid-related hepatic injury (BLACK et al., 1973). These patients were part of a special surveillance program carried out by the U. S. Public Health Service on 13000 recipients of isoniazid to determine the actual incidence of isoniazid-related hepatic injury. Some of the important findings of this study were: a) isoniazid-related liver injury was indistinguishable biochemically (SGOT, bilirubin, alkaline phosphatase) and morphologically from iproniazid-induced liver damage or from other causes of acute hepatocellular injury such as viral hepatitis; b) no clinical evidence for a hypersensitivity mechanism was apparent; c) about 40% of the patients with hepatic reactions were residents of Honolulu and of Oriental ancestry; this was important because 90% of the Oriental population is known to be fast acetylators of isoniazid (KALOW, 1962).

These observations led to the third study. Non-Oriental patients who had had isoniazid hepatitis were phenotyped after recovery into rapid and slow acetylators of isoniazid, by the sulfamethazine method (RAO et al., 1970). Of the 21 individuals, 18 (86%) of them were rapid acetylators (MITCHELL et al., 1974d), whereas the expected frequency of rapid acetylators among black and white Americans was 45% (KALOW, 1962).

For these reasons, acetylisoniazid and isopropylisoniazid (iproniazid) were given to rats, mice and hamsters to see if they would produce hepatic necrosis (SNODGRASS et al., 1974). Rare, scattered single cell necrosis was seen occasionally in normal rats and mice, but after pretreatment with phenobarbital, a marked hepatic necrosis occurred with doses of these isoniazid analogs above 100 mg/kg. This necrosis was prevented by simultaneous treatment with inhibitors of drug metabolism such as cobaltous chloride, aminotriazole and piperonyl butoxide. Hepatic necrosis was not produced by isoniazid in normal or phenobarbital-pretreated rats even after lethal doses.

When the metabolism of isoniazid and its acetyl and isopropyl analogues was examined in rats, very little isoniazid was hydrolyzed to isonicotinic acid, whereas over 30% of the administered acetylisoniazid or isopropylisoniazid was hydrolyzed to isonicotinic acid and presumably to acetyl hydrazine or isopropylhydrazine, respectively. Since many hydrazines are known to be potent hepatotoxins, carcinogens and mutagens (BACK and THOMAS, 1970; DRUCKREY, 1973), the hepatotoxicity of various hydrazines was examined in animals. After phenobarbital pretreatment, acetylhydrazine and isopropylhydrazine produced a striking mid-zonal and centrilobular necrosis at doses as low as 15 mg/kg (SNODGRASS et al., 1974). The necrosis produced by acetylhydrazine and isopropylhydrazine was prevented by pretreatment of animals with inhibitors of drug metabolism such as cobaltous chloride, aminotriazole and piperonyl butoxide. Thus, the necrosis caused by acetyl hydrazine and isopropyl hydrazine is mediated by active metabolites of these substances. By contrast, only rare single cell necrosis was seen even after LD_{50} doses of hydrazine, methylhydrazine, ethylhydrazine and *n*-propylhydrazine, presumably because these hydrazines are not significantly converted to active metabolites in the liver.

As further proof that acetylisoniazid was converted in the body to a chemically reactive metabolite, ^{14}C-acetyl-labeled acetylisoniazid was administered to rats. A large amount of covalent binding of the radiolabeled material (about 0.5 nmole/mg protein) was found in liver, the target organ. This binding was proportional to dose (no dose threshold), was increased by pretreatment with phenobarbital and was markedly decreased by pretreatment with cobaltous chloride or aminotriazole. However, no covalent binding occurred after acetylisoniazid when the radiolabel was in the pyridine ring, indicating that the reactive metabolite came only from the acetylhydrazine moiety.

Finally, isoniazid itself was shown to produce acute hepatic necrosis in phenobarbital-pretreated rats. Since the proportion acetylated decreases markedly as the dose is increased above 100 mg/kg (TIMBRELL et al., 1974), presumably because the acetylase becomes saturated, isoniazid was administered in divided rather than large single doses (e.g., 100 mg/kg hourly for 7 doses). With this dosage schedule it was possible to produce acute hepatic necrosis in phenobarbital-pretreated rats.

Thus, metabolic activation of the liberated hydrazines satisfactorily accounts for the hepatic necrosis produced by isoniazid, acetylisoniazid and isopropylisoniazid in animals (Fig. 9). A possible explanation, therefore, is provided for the increased incidence of isoniazid hepatitis in humans with fast acetylator phenotype, since human fast acetylators of isoniazid produce more acetylhydrazine than do slow acetylators (Table 3).

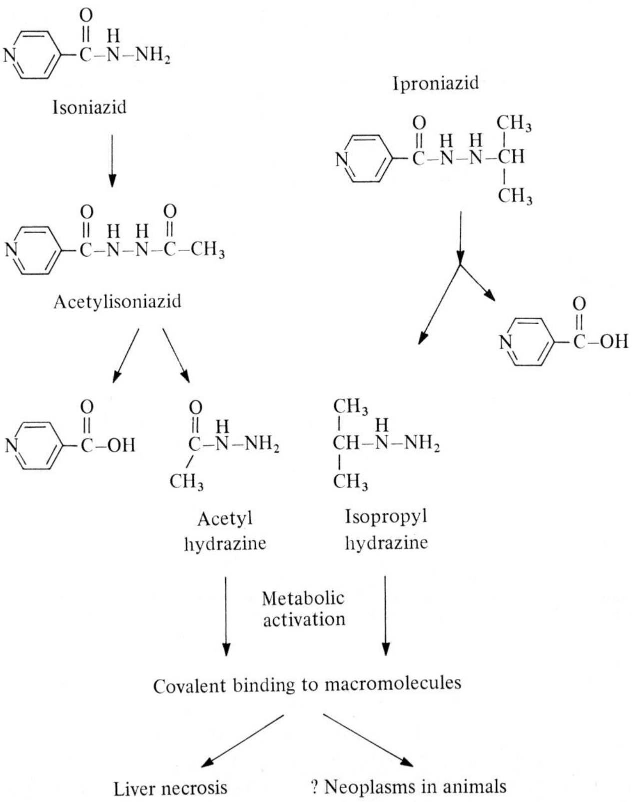

Fig. 9. Proposed metabolic activation pathways for isoniazid, acetylisoniazid and isopropylisoniazid (iproniazid)

Other clinical observations also make this hypothesis an attractive one. For example, the increased incidence of isoniazid injury in patients concomitantly receiving rifampin (LEES et al., 1971, 1972), an excellent inducer of human drug metabolizing enzymes (REMMER et al., 1973), might be due to an increased activation of acetylhydrazine to its toxic form. Similarly, the allegedly decreased incidence of isoniazid hepatitis in patients receiving isoniazid in combination with *p*-aminosalicylic acid (PAS) might be due to inhibition of the acetylation of isoniazid by PAS (TIITINEN, 1969).

6. Fluroxene

This anesthetic agent occasionally causes massive hepatic necrosis in humans (REYNOLDS et al., 1972). We previously reported that it does not cause liver

Table 3

Acetylator Phenotype	INH[a]	INH Hydrazones	AcINH[b]	INA Derivatives[c]	Estimated acetyl hydrazine[d] mmoles/70 kg man
	% of dose				
(A) Twenty-four hour urinary excretion of ^{3}H-isoniazid and metabolites in 7 humans receiving isoniazid (5 mg/kg, p.o.)					
Fast (3)	2.8 ± 0.4	3.6 ± 0.4	49.2 ± 1.9	44.4 ± 3.9	0.783 ± 0.11
Slow (4)	10.9 ± 0.8	26.5 ± 4.8	32.1 ± 1.2	30.5 ± 3.5	0.537 ± 0.09
(B) Twenty-four hour urinary excretion of ^{3}H-acetylisoniazid and metabolites in 5 humans receiving acetylisoniazid (5 mg/kg, p.o.)					
Fast (2)	—	—	54.9 ± 2.2	45.1 ± 2.7	0.88 ± 0.11
Slow (3)	—	—	53.8 ± 1.2	46.2 ± 1.1	0.90 ± 0.12

[a] INH — isoniazid.
[b] AcINH — acetylisoniazid.
[c] Isonicotinic acid and isonicotinyl glycine.
[d] The validity of this estimation is documented in Table 3B (unpublished results) and by previous investigations (PETERS et al., 1965). The recovery of AcINH after administration of AcINH to fast or slow acetylators was 54.5% (Table 3B). Accordingly, the AcINH formed after administration of INH to fast or slow acetylators must have equaled the % recovered plus the % hydrolyzed [e.g., for fast acetylators in Table 3A, 54.5% × total AcINH formed = amount recovered (49.2%); therefore total AcINH formed = 90.2% of INH dose]. Thus, almost all the INA derivatives must have arisen from hydrolysis of AcINH (at least 41.0% of the 44.4% of recovered INA derivatives in Table 3A, namely, 90.2% — 49.2%) and equal the amount of acetyl hydrazine liberated *in vivo*.

necrosis in experimental animals after intraperitoneal injection (MITCHELL et al., 1973a). A recent investigation by HARRISON and SMITH (1973) has demonstrated that fluroxene produces massive hepatic necrosis in phenobarbital-pretreated rats when the drug is given over several hours as an anesthetic gas. This route permits much larger doses of fluroxene to be given than after intraperitoneal administration. The potentiation of fluroxene-induced hepatic necrosis by prior treatment with phenobarbital suggests that again a toxic metabolite is responsible for the hepatic injury caused by therapeutic doses of a drug, although careful metabolic studies are necessary to exclude the possibility that the phenobarbital effect could be related to increased hepatic blood flow with increased hepatic clearance of fluroxene (SHAND et al., 1974).

7. Acetanilide

Another example of metabolic activation is drug-induced methemoglobinemia and hemolysis. Aromatic amines including aniline are known to cause methemoglobinemia by being converted to phenylhydroxylamines which on oxidation to their nitroso derivatives and subsequent reduction back to the phenylhydroxylamines promote the formation of methemoglobin (KIESE, 1966; UEHLEKE, 1973). In addition, certain aniline-derivatives used as drugs (acetanilide, phenacetin) also cause hemolysis, especially in patients having a genetic deficiency in erythrocyte glucose-6-phosphate dehydrogenase (BEUTLER, 1969). ^{51}Cr-labeled rat erythrocytes injected intravenously into rats have been used to measure the effects of drugs on hemolysis (MITCHELL et al., 1973a; DI CHIARA et al., 1972, 1973). After the intraperitoneal administration of aromatic amines, such as aniline, 4-phenetidine and 4-chloroaniline, the rate at which ^{51}Cr declined in blood was greatly increased, confirming that these compounds induced hemolysis in rats. The hemolysis is apparently caused by metabolites rather than by the parent amines, since these compounds in incubation mixtures had no effect on erythrocyte survival. Moreover, pretreatment of rats with carbon tetrachloride, which decreased the metabolism of the aromatic amines, prevented the hemolysis induced *in vivo*.

The corresponding *N*-acetylated drugs, namely acetanilide and phenacetin, also caused hemolysis in this system, although to a lesser degree than did the free amines. The hemolysis produced by the acetylated drugs was prevented by pretreatment of the rats with *bis-p*-nitrophenyl phosphate, which inhibits deacetylation (HEYMANN et al., 1969). In contrast, the administration of the deacetylase inhibitor did not prevent the hemolysis induced by the free amines. Thus, these analgesic drugs apparently require metabolic transformation in two steps in order to cause hemolysis, presumably deacetylation followed by *N*- or *C*-hydroxylation.

In the past the same active intermediate was thought to cause both hemolysis and methemoglobinemia. However, phenobarbital pretreatment of rats increases the methemoglobinemia but decreases the hemolysis caused by acetanilide and aniline, whereas pretreatment with piperonyl butoxide, an inhibitor of liver cytochrome P-450 enzymes, evokes exactly the opposite effects (DI CHIARA et al., 1973). These results clearly demonstrate that methemoglobinemia and drug-induced hemolysis are mediated by different active metabolites and suggest that methemoglobinemia is not a prerequisite for hemolysis.

8. Halobenzenes

Studies with bromobenzene and other halobenzene hepatotoxins also demonstrate relationships between hepatic necrosis (BRODIE et al., 1971; MITCHELL et al., 1971; REID et al., 1971), hepatic glutathione (JOLLOW et al., 1974b) and arylation of hepatic macromolecules by reactive metabolites (REID, 1973; JOLLOW et al., 1974b). Studies *in vivo* and *in vitro* of the metabolism and covalent binding of bromobenzene by rat liver permitted a direct assessment of the relative roles of the known metabolic pathways (Fig. 10) (ZAMPAGLIONE et al., 1973; JOLLOW et al., 1974b). The results provided direct evidence that virtually all of the metabolism of bromobenzene is through the formation of 3,4-bromobenzene oxide. Depending on its steady-state concentration, this epoxide then undergoes rearrangement nonenzymically to 4-bromophenol, hydration by an epoxide hydrase to 3,4-bromophenyldihydrodiol or conjugation with glutathione by a glutathione-*S*-epoxide transferase to form a glutathione-conjugate that ultimately is converted to a mercapturic acid. At low doses of bromobenzene, the 3,4-bromobenzene oxide normally was conjugated preferentially with glutathione (Fig. 11). However, when glutathione became rate-limiting for the glutathione transferase-catalyzed reaction, the steady-state concentrations of 3,4-bromobenzene oxide rose, as reflected by the concomitant increase in both the nonenzymic rearrangement to 4-bromophenol and in the nonenzymic arylation of the microsomal protein (Fig. 11). In normal rats, the epoxide hydrase activity apparently was low and relatively unimportant for detoxification of 3,4-bromobenzene oxide, since only insignificant amounts of 3,4-bromophenyldihydrodiol and 3,4-bromocatechol were formed even when glutathione concentrations were low.

To understand the importance of these studies with model compounds to drug-induced hepatotoxicity, it must be remembered that many aromatic drug hydroxylations in the body occur via epoxidation reactions as demonstrated above for bromobenzene (DALY et al., 1972). Thus, it is not surprising that several commonly used drugs, such as diphenylhydantoin (HORNING et al., 1971), phenobarbital (HARVEY et al., 1972), mephobarbital (HARVEY et al., 1972), and methsuximide (HORNING et al., 1973), are metabolized in humans to dihydrodiol derivatives which indicate the formation of epoxide intermediates of the drugs. However,

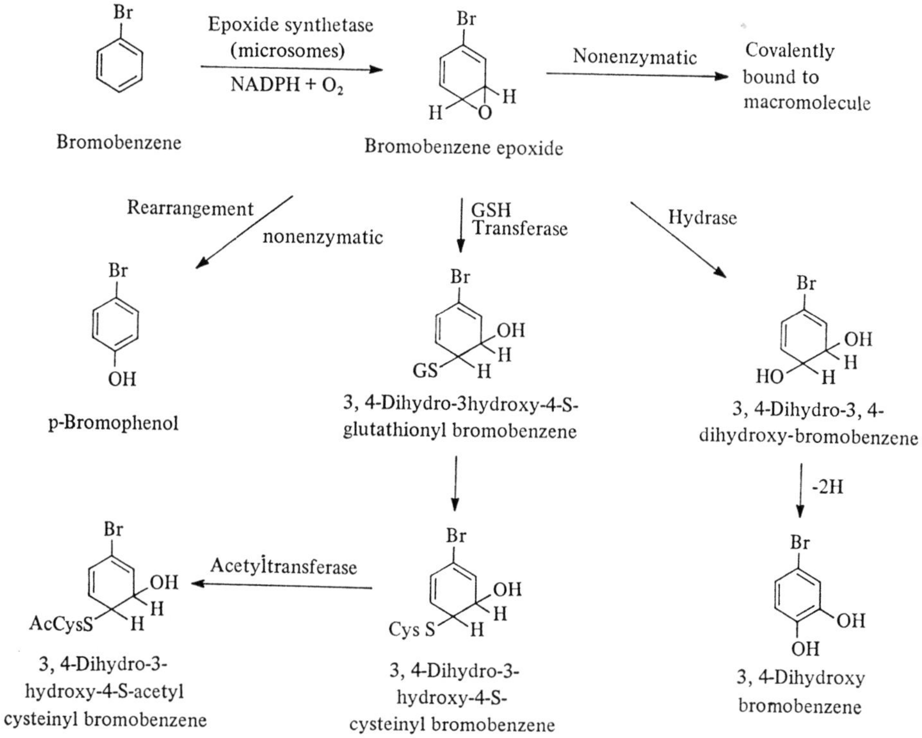

Bromobenzene
Epoxide synthetase (microsomes)
$NADPH + O_2$
Bromobenzene epoxide
Nonenzymatic
Covalently bound to macromolecule
Rearrangement nonenzymatic
GSH Transferase
Hydrase
p-Bromophenol
3, 4-Dihydro-3hydroxy-4-S-glutathionyl bromobenzene
3, 4-Dihydro-3, 4-dihydroxy-bromobenzene
-2H
Acetyltransferase
3, 4-Dihydro-3-hydroxy-4-S-acetyl cysteinyl bromobenzene
3, 4-Dihydro-3-hydroxy-4-S-cysteinyl bromobenzene
3, 4-Dihydroxy bromobenzene

Fig. 10. Pathways of metabolism of bromobenzene

the finding that epoxides are formed from drugs does not necessarily mean that they are toxic. As clearly illustrated by the bromobenzene studies, the toxicity of an epoxide metabolite is dependent on many factors, including the rate of formation, the inherent chemical stability and the rate of detoxification by other metabolic pathways. All these factors must be considered when assessing the potential toxicity of a particular drug.

9. Spironolactone

Although spironolactone induces cytochrome P-450 enzymes in liver, it causes a selective destruction of testicular cytochrome P-450 and thereby decreases the synthesis of testosterone (STRIPP et al., 1973). Since NADPH is required for the destruction of testicular cytochrome P-450 by spironolactone *in vitro*, the effect probably is mediated by an active metabolite (MENARD et al., 1974a). But neither the identity of the active metabolite nor its mechanism of action is known. The impairment in testicular cytochrome P-450 is not due to cellular necrosis and is fully reversible within 5 days after withdrawal of the drug (STRIPP et al., 1973). Experiments *in vivo* in male dogs demonstrate that spironolactone not only decreases testicular cytochrome P-450 but also decreases the release of both testosterone and estradiol into the plasma (MENARD et al., 1974b). As in animals, spironolactone may decrease the plasma levels of testosterone in patients (PENTIKAINEN et al., 1973; DYMLING et al., 1972). But whether the decrease is due

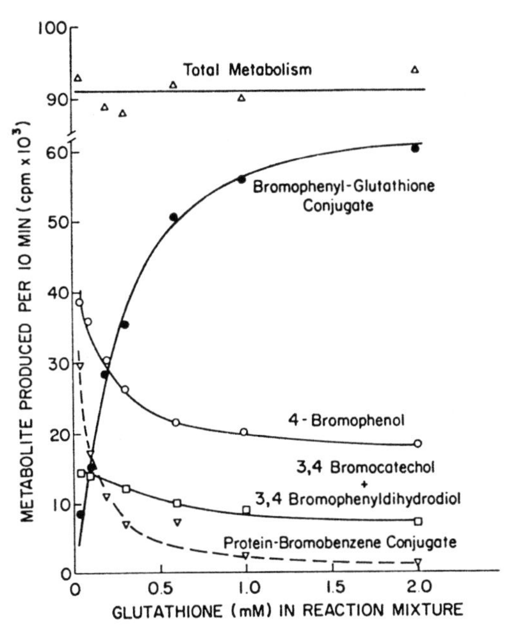

Fig. 11. Metabolism of bromobenzene by the 9000 × g supernatant fraction of liver from normal rats showing the effect of concentration of glutathione in the reaction medium on the formation of the hepatic protein-bromobenzene conjugate and on the formation of metabolites known to arise from 3,4-bromobenzene oxide (Fig. 10). The toxic metabolite was conjugated preferentially with glutathione unless the concentration of glutathione was rate-limiting. The steady-state concentrations of 3,4-bromobenzene oxide then rose, as reflected by the concomitant increase in both the nonenzymic rearrangement to 4-bromophenol and in the nonenzymic arylation of microsomal protein. Epoxide hydrase activity was low and relatively unimportant for the detoxification of 3,4-bromobenzene oxide, since only insignificant amounts of 3,4-bromophenyl-dihydrodiol and 3,4-bromocatechol were formed even when glutathione concentrations were low. (Data from JOLLOW et al., 1974b)

primarily to a decrease in synthesis in testis or an increase in catabolism in liver is not known. It is also not known whether these effects of spironolactone on the metabolism of sex steroids account for the gynecomastia and decreased libido observed in humans receiving the drug.

10. Porphyria

At least 3 different types of experimental porphyria can be induced by drugs and other foreign compounds: 1) That produced by allyl ureide and allyl acetamide compounds, such as sedormid, allylisopropylacetamide (AIA), and secobarbital; 2) that caused by hexachlorobenzene; 3) that caused by dicarbethoxydihydrocollidine and related compounds. In addition to causing porphyria and increasing the synthesis of δ-aminolevulinic acid, the allyl compounds such as AIA (DEMATTEIS, 1973; LEVIN et al., 1973), secobarbital, allobarbital and aprobarbital (LEVIN et al., 1973) also catalyze the destruction of the various heme compounds in liver, especially cytochrome P-450. Recently, it was shown that a number of other allyl compounds, including allyl alcohol, acrylamide and the anesthetic gas, fluroxene, also cause a selective destruction of liver microsomal cytochrome P-450 (GILLETTE, 1973). The finding that these effects are markedly increased after pretreatment of rats with phenobarbital suggest that they are caused by active metabolites of the foreign compounds. In support of this view, secobarbital decreases the cytochrome P-450 in liver microsomes only when NADPH is present in the incubation system (LEVIN et al., 1973). In contrast to the destructive effects of compounds such as CCl_4 (REINER et al., 1972) on cytochrome P-450 *in vitro*, however, the effects of secobarbital are not blocked by EDTA (LEVIN et al., 1973). Thus, the effects of the allyl compounds probably are not mediated by lipid peroxidation. Since allyl compounds are converted to epoxides, it seems possible that these might be the active metabolites. However, this possibility has not been confirmed nor is it known whether the active metabolites evoke their action by becoming covalently bound to cytochrome P-450 or to other macromolecules.

The destruction of heme compounds in liver may be an initial event in the development of porphyria (DEMATTEIS, 1973). Since heme is known to exert a negative feed-back control on the conversion of succinyl-CoA to δ-aminolevulinic acid, the rate-limiting step in heme synthesis, any mechanism that decreases the level of the intracellular heme pool that exerts this control would be expected to increase the rate of porphyrin synthesis. However, the rate-controlling pool of heme probably is not cytochrome P-450, because the heme in cytochrome P-450 does not exchange with free heme (MAINES and ANDERS, 1973). The finding that heme in the mitochondrial and the soluble fractions of liver are also decreased after administration of AIA (DEMATTEIS, 1973) raises the possibility that a heme pool in these fractions may serve to control porphyrin synthesis.

B. Drug-Induced Neoplasia

Since the pioneering work of the MILLERS in Wisconsin and of MAGEE and co-workers in England, it has become increasingly evident that most if not all chemical carcinogens bring about their effects by combining covalently with DNA and other tissue macromolecules or by being transformed to chemically reactive metabolites that in turn combine covalently with tissue macromolecules (MILLER and MILLER, 1966; MILLER, 1970; MAGEE and BARNES, 1967).

The many studies on the mechanism of formation of carcinogenic metabolites in the body have revealed that chemically inert substances can be converted to chemically reactive metabolites by a variety of different reactions (Fig. 12). For example, secondary amines, such as *N*-methyl-4-aminoazobenzene, primary amines including β-naphthylamine and aminobiphenyl, and acetylated primary amines including 2-acetylaminofluorene are *N*-hydroxylated by either cytochrome P-450 enzymes or amine *N*-oxidase. In some instances, the metabolites are further activated by being converted to *N*-0-sulfate esters (MILLER, 1970). Dialkylnitrosamines are *N*-demethylated by cytochrome P-450 enzymes to monoalkylnitrosamines, which in turn spontaneously rearrange to unknown active metabolites, possibly alkyl carbonium ions (MAGEE and BARNES, 1967). Cycasin is hydrolyzed in the intestine by bacterial β-glucosidase to methylazoxymethanol which acts as a methylating agent (LAQUEUR, 1964). Pyrrolizidine alkaloids are thought to be dehydrogenated to chemically reactive pyrrole derivatives (MATTOCKS, 1973; MC LEAN, 1970). Nitroaryl compounds, such as 4-nitroquinoline *N*-oxide, may act either as arylating agents *per se* or be converted to chemically reactive hydroxylamine derivatives (ENDO and KUME, 1965).

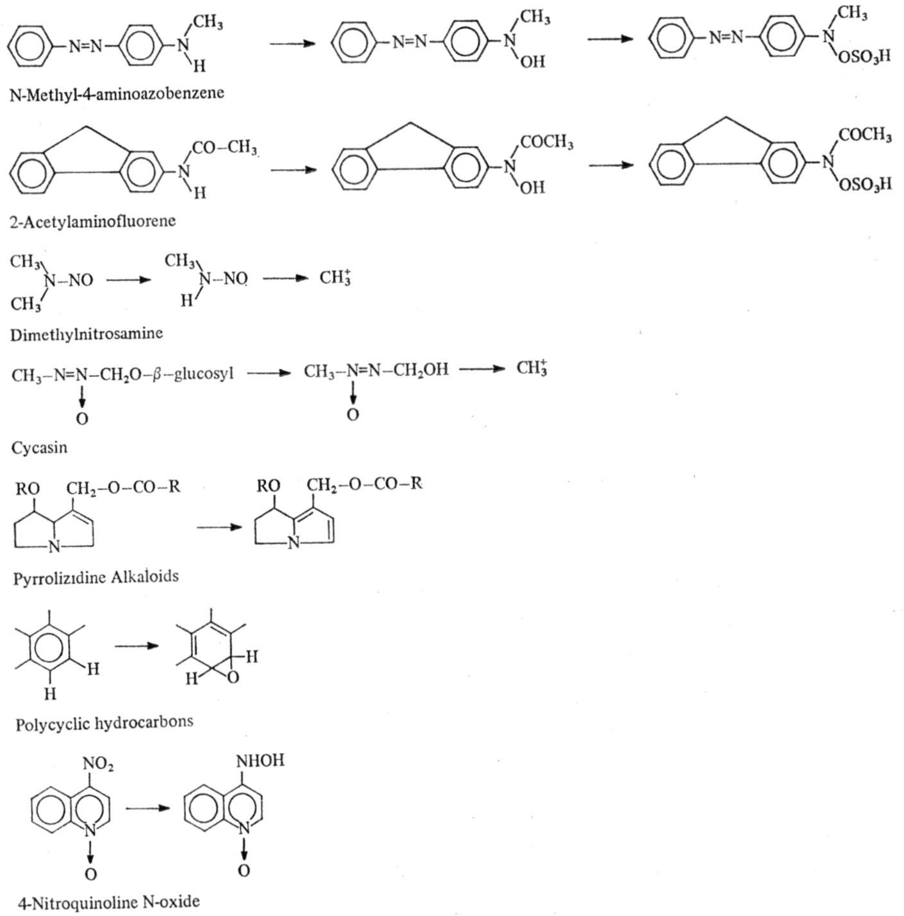

N-Methyl-4-aminoazobenzene

2-Acetylaminofluorene

Dimethylnitrosamine

Cycasin

Pyrrolizidine Alkaloids

Polycyclic hydrocarbons

4-Nitroquinoline N-oxide

Fig. 12. Biochemical mechanisms for the formation of chemically reactive metabolites of carcinogens

From these studies it is apparent that drugs which might be carcinogenic in humans would most likely be reactive alkylating or arylating agents when given *in vivo*.

1. Antineoplastic Drugs

The validity of the latter assumption is confirmed by the known carcinogenic activity of the biological alkylating agents, such as cyclophosphamide (STEINBERG et al., 1972; CLAYSON, 1972). It seems desirable to identify the most carcinogenic of the antineoplastic agents and to restrict their use to those patients in whom the possible benefit outweighs the risk of inducing further tumors.

2. Estrogens

Administration of natural or synthetic estrogens to many species of laboratory animals may lead to tumors at one or more sites. The sites affected are most commonly in endocrine tissues, such as the ovaries or testes, or in hormone-responsive tissues such as the breast. However, diethylstilbestrol and other estrogens lead to adenocarcinomas in the kidney of hamsters and this tissue is not hormone-responsive (MATTHEWS et al., 1947). Moreover, a large scale trial of estrogens in the treatment of men with coronary artery disease was recently terminated not only for a lack of efficacy but also because of an apparent increase in the number of patients dying from cancer (Coronary Drug Project, 1973). Another tragic example of carcinogenesis in humans caused by estrogens was the appearance of vaginal adenocarcinoma in young women aged 15 to 19 (GREENWALD et al., 1971; HERBST et al., 1972). In every case, the mothers of the victims had taken diethylstilbestrol during pregnancy as possible prophylaxis against abortion.

Estrogens, such as ethinyl estradiol and mestranol, have been shown to be converted to chemically reactive metabolites that covalently bind to tissue macromolecules (REMMER and BOCK, 1974). Since the metabolites apparently did not combine with DNA, the metabolites were considered to be innocuous. But this interpretation may be overly optimistic in view of the emerging evidence of the carcinogenicity of many estrogens. Diethylstilbestrol also has been shown to be converted to a potent arylating agent (CORSINI et al., 1972).

3. Phenacetin

Recently, HULTENGREN et al. (1965) and BENGTSSON et al. (1968, 1970) have drawn attention to the development of renal pelvic carcinoma as a possible further complication of the abuse of phenacetin-containing analgesics. The annual incidence of this form of cancer in Sweden is 1 case in 183000 inhabitants. Yet among 103 patients with renal papillary necrosis probably secondary to analgesic abuse, HULTENGREN et al. (1965) found 4 with renal pelvic carcinoma. Similarly, BENGTSSON et al. (1968) discovered 9 cases of renal pelvic carcinoma among 104 patients with nonobstructive pyelonephritis who had abused analgesics. In addition, ANGERVALL et al. (1969) noted 15 instances of this carcinoma presenting in one district hospital, of which 10 patients and possibly 12, had consumed excessive amounts of phenacetin analgesics.

The nature of the carcinogenic agent is unknown and the role of phenacetin has been questioned (Editorial, 1969). In view of the arylating capability of phenacetin discussed above, it seems possible that phenacetin is carcinogenic in patients who consume analgesics to such an excessive extent that they overwhelm the capacity of glutathione to detoxify the reactive metabolite.

4. Drugs Used Clinically Not Known to be Carcinogenic in Humans but Known to Induce Cancer in Animals

There are unresolved problems in applying the results of carcinogenicity tests in animals to potential risk for humans (CLAYSON, 1972). Some of these are concerned with differences in the transport, storage, binding, metabolism and excretion of drugs in man and experimental animals. Others reflect variations between one species and another in tissue response. Nevertheless, it seems prudent to regard all animal carcinogens as potentially carcinogenic in man.

a) *Isoniazid*

This drug exemplifies the most difficult problem of assessing benefit-risk ratios in humans. The drug has undoubtedly been responsible for the great reduction in morbidity due to tuberculosis. The fact that it offers relief from an incapacitating and potentially fatal disease must be set against the strongly supported experimental observation that it induces pulmonary adenomas in mice (JUHASZ et al., 1957; MORI et al., 1960; SCHWAN, 1961, 1962; TOTH and SHUBIK, 1966). The consensus of informed opinion is that its use is justified but there must be considerable reservations about its indiscriminate use for chemoprophylaxis, especially in children who have their entire life span in which to develop tumors (Editorial, 1966).

The mechanisms by which isoniazid is converted to a potent alkylating agent in animals and presumably in humans have been discussed under the section on cell necrosis.

b) *Nitrofuran Derivatives*

Nitrofurazone, a compound used as a topical antibacterial agent, and several related nitrofuran compounds cause mammary and bladder tumors in rats (COHEN and BRYAN, 1973; ERTUK et al., 1967, 1969). Moreover, it also produced a decrease in testicular cytochrome P-450 and impaired testosterone synthesis in mice (STRIPP et al., 1973). A metabolite of the drug was covalently bound to proteins of the liver, kidney, testis and mammary gland in mice (STRIPP et al., 1973, 1974), perhaps as a result of nitrofurazone reduction to its reactive hydroxylamino derivative by xanthine oxidase (MORITA et al., 1971) and aldehyde oxidase (WOLPERT et al., 1973) in those tissues as well as by liver microsomal NADPH cytochrome *c* reductase (FELLER et al., 1971).

c) *Miscellaneous Drugs*

Drugs such as griseofulvin (EPSTEIN et al., 1967), parenteral iron (ROE, 1967) and penicillin G (DICKENS et al., 1967a and b) have been reported to cause neoplasia in various animals. The mechanism and significance of these findings are unknown. If the studies with penicillin withstand criticism (CLAYSON, 1962), the acylating potential of this compound should be noted.

C. Drug Allergy

The mechanisms by which drugs cause allergic responses remain largely unexplored. Since the classic work of LANDSTEINER and his colleagues (1945), many investigators have shown that small molecules can serve as antigens only after they become covalently bound to macromolecules such as plasma albumin (ERLANGER, 1973). Consequently, the finding that drugs can be converted to chemically reactive metabolites which combine covalently to macromolecules raised the possibility that the formation of chemically reactive metabolites might be an initial step in drug-induced hypersensitivity reactions. Indeed the failure to demonstrate covalent binding of drugs to macromolecules either *in vivo* or *in vitro* might suggest that the drug would not evoke hypersensitivity reactions in man. On the other hand, it has become evident that covalent binding of the drug does not always lead to antibody formation or an immunologic response. Indeed, investigators usually fail to find antibodies to a given drug either in patients manifesting hypersensitivity reactions (HOIGNE, 1972) or in animals receiving a drug that is known to be covalently bound to macromolecules after activation *in vivo* (KRISHNA and GILLETTE, 1974).

The area is further complicated by a lack of universal agreement about the criteria for classification of a drug reaction as allergic. Often there is a mere coincidence in time between the administration of the drug and onset of the disease, without rigorous proof of an allergic basis. Yet the presence of an antigen-antibody reaction is the only valid criterion for acceptance of an allergy. By this criterion it has been possible to demonstrate clearly that drug allergy accounts for the destruction of blood cells by demonstrating the presence of an antibody that reacts specifically with the suspected drug (see NIEWEG, 1972). In some instances, this may come about through antigen attachment to the cells (as shown for penicilloyl derivatives in penicillin allergy) followed by antibody attachment and cell agglutination and the subsequent removal of affected cells in the reticulo-endothelial system. Alternatively, an antibody-antigen complex may attach to cells (e.g. platelets) and with the participation of complement lead to cell lysis without covalent attachment of drug to the target cell having occurred (SHULMAN, 1963). In addition to the hematologic syndromes, sensitization with substances having hapten characteristics also plays a prominent role in allergic contact dermatitis, the skin being the most frequently affected organ in drug allergies (ACKROYD and ROOK, 1963; HOIGNE, 1965).

Proof of the involvement of other organs in allergic drug reactions is not available. The problem of antibody identification is fraught with great technical difficulties, especially concerning sensitization with haptens. Moreover, like the infectious diseases, the mere identification of an antibody by no means proves that a certain disease manifestation has been caused by the antibody (LEVINE, 1966). The diagnosis of a drug allergy is therefore usually derived from a combination and correlation of anamnestic facts, the type of clinical reaction and results of various testing methods. Although these clinical observations and results of re-exposure (challenge) to the drug have been helpful in the diagnosis of drug allergy in the hematologic syndromes, the complexities and questionable specificity of such diagnostic criteria are well illustrated by two drugs generally thought to be hepatotoxic on an allergic basis, isoniazid and halothane.

MADDREY and BOITNOTT (1973) have reviewed the evidence supporting the view that hypersensitivity or allergy to isoniazid is responsible for its hepatotoxicity. In a small proportion of patients with isoniazid hepatitis the illness includes features that suggest an allergic mechanism. This syndrome consists of fever, chills and deterioration of hepatic function within a few hours after re-challenge of the patients with isoniazid (HABER and OSBORNE, 1959; MERRIT and FETTER, 1959; MADDREY and BOITNOTT, 1973; ADRIANY, 1960). However, other individuals fail to show constitutional features that suggest an allergic response and often demonstrate hepatic injury only after lengthy reexposure to isoniazid (MADDREY and BOITNOTT, 1973). Circulating antibodies to isoniazid, moreover, have not been reported (MITCHELL et al., 1974c) and although anti-nuclear antibodies may appear during isoniazid therapy (ALARCON-SEGOVIA et al., 1971; MITCHELL et al., 1974c) these do not correlate with evidence of hepatic damage (MITCHELL et al., 1974c). Isolated instances of lymphocyte transformation in the presence of isoniazid and serum from suspect patients have been construed as evidence for an allergic basis for isoniazid-related hepatic disease (ASSEM et al., 1969; MATTHEWS et al., 1971). Negative results with the lymphocyte transformation test also have been reported (DOVE et al., 1972), however, and the relevance of the positive results remains to be established.

Our study of 114 patients with isoniazid hepatitis also failed to support the view that isoniazid liver disease is usually due to allergy (BLACK et al., 1973). Accepted hallmarks of allergy such as skin rash and peripheral eosinophilia were notably uncommon in these subjects and a febrile onset of the disease was absent in most of the patients. Furthermore, the variable and often prolonged period of exposure to isoniazid before the appearance of hepatic injury provided an additional argument against hypersensitivity as the main cause of isoniazid hepatic injury. Drug-induced hepatic damage attributed to an allergic mechanism is usually characterized by a fixed and fairly brief (1 to 4 week) period of exposure to the agent (ZIMMERMAN, 1972; KLATSKIN, 1968). Finally, several studies (SCHARER and SMITH, 1969; BYRD et al., 1972; BAILEY et al., 1973; MITCHELL et al., 1974c) have shown that isoniazid produces mild hepatic injury in 12 to 20% of recipients and is not restricted to rare idiosyncratic or allergic subjects.

Thus, the combined data provide strong support for our view that isoniazid related hepatic injury in humans results from the formation in the liver of a chemically reactive metabolite of the drug (see above Section II A 5). They also are consistent with the evidence suggesting an allergic mechanism in the pathogenesis of isoniazid injury in occasional patients. In such individuals the alkylation of tissue macromolecules by acetylhydrazine may have produced an antigen, leading to a true drug allergic reaction. However, the possibility also exists that the allergic syndrome in these patients did not result at all from a direct sensitization to the drug itself. The readministration of isoniazid might have produced direct hepatic injury with the release of sequestered, and therefore antigenic, intracellular proteins and other constituents. Since the patient could have been sensitized by the release of these cellular substances during the previous episode of isoniazid hepatitis, the observed allergic syndrome might have resulted entirely from an anamnestic response to the released intracellular antigens.

Halothane-induced liver injury similarly demonstrates the difficulties in proving a drug allergy pathogenesis and has been well discussed in an articulate, objective review by CONN (1974). For example, a case of supposed allergy to halothane in an anesthetist was carefully documented, with recurrent attacks of hepatitis at every exposure to halothane but not to other anesthetic agents (KLATSKIN and KIMBERG, 1970). Moreover, the known metabolism of halothane is compatible with the formation of allergenic protein conjugates. However, during the same process of metabolic activation, the halothane metabolite can destroy hepatic cytochrome P-450 enzymes, increase lipid diene conjugation and apparently produce hepatic necrosis (REYNOLDS et al., 1973). Thus one is left with a chicken or the egg argument—which comes first or is the more important for the development of halothane hepatitis, the direct injury of hepatocytes by the chemically reactive metabolite or the subsequent immunologic response triggered by the reaction? Moreover, how does a large molecular weight antibody penetrate the plasma membrane of viable hepatocytes to react with the altered intracellular drug-protein antigens? If this passage is not necessary (i.e., the antibody reacts with antigens on the plasma surface of the hepatocytes), why is the liver the only organ affected? Is it possible that the allergic phenomena observed clinically are purely epiphenomena unrelated to the pathogenesis of the hepatic injury and reflect only a common initiator, namely the formation of a chemically reactive metabolite from halothane? Clearly, answers to these questions must await the development of new methodology for measuring antibody formation and for delineating the role of antigen-antibody complexes in the pathogenesis of hepatic injury.

III. Drug Reactions of Unknown Etiology

In order to promote scientific investigation and communication, we feel that all adverse drug reactions which cannot be documented to correspond to Section

I or II should be included in Category III. Accordingly, many drug reactions currently categorized as idiosyncratic have been shown to be due to pharmacogenetic susceptibility of particular individuals to known actions of the drug and should now be classified under Sections I or II. Thus, patients who experience paralysis after normal doses of succinylcholine are essentially receiving overdoses of the drug because they metabolize it abnormally slowly. Similarly, acetanilide drugs that produce hemolytic anemia in glucose-6-phosphate dehydrogenase deficient subjects will cause hemolytic anemia in normal individuals if given in higher doses.

IV. Perspective

Although this paper has dealt only with toxic drug reactions, we anticipate that the metabolic activation of compounds to toxic substances may eventually be implicated in the pathogenesis of a wide variety of tissue lesions induced by environmental pollutants as well as drugs. For example, repeated injections of benzene produce bone marrow aplasia in rats. The lesion is prevented when the metabolism of benzene is altered by a variety of pretreatments including SKF 525-A, piperonyl butoxide and phenobarbital (MITCHELL, 1971). Although the finding that phenobarbital pretreatment protects against benzene-induced bone marrow damage has led investigators to propose subsequently that benzene itself is the toxic agent (IKEDA and OHTSUJI, 1971), this seems unlikely since pretreatments such as cobaltous chloride and aminotriazole, which inhibit the synthesis of cytochrome P-450 in rats, block the metabolism of benzene and markedly reduce the bone marrow damage (POTTER and MITCHELL, 1974). Thus, it seems likely the bone marrow aplasia caused by benzene also is mediated by an unknown active metabolite.

Many of the initial concepts of metabolic activation were developed during studies of chemical carcinogenesis; the work of the MILLERS in the United States and of MAGEE and co-workers in England has been especially illuminating. The realization that the enzyme pathways responsible for the metabolic activation of carcinogens are the same microsomal mixed-function oxidases that metabolize most drugs led BRODIE (1967) to speculate that drug-induced tissue lesions might also be mediated through the covalent binding of reactive metabolites. The lack of reactivity of most chemically stable drugs and the frequent localization of tissue damage only in those organs or to those animal species having the necessary drug-metabolizing enzymes supported this view.

Evaluation of this possibility, however, has been difficult. Potential therapeutic agents that reproducibly cause pathological lesions in animal toxicity tests are rarely marketed for clinical use. Moreover, most active drug metabolites produce their effects by combining reversibly with receptor sites. Thus, their pharmacologic activity frequently can be evaluated simply by measuring the concentration of the metabolite in plasma (BRODIE and MITCHELL, 1973). But when the response is tissue damage caused by the covalent binding of chemically reactive metabolites to tissue macromolecules, rarely can the relationship between the plasma level of the metabolite and the severity of the lesion be determined. Indeed, with highly reactive metabolites, little or none ever reaches the plasma.

How then can one readily determine the formation of such chemically unstable and reactive metabolites? It seemed possible that there might be a relationship between the severity of the lesion and the amount of covalently bound metabolite for any particular drug. The covalent binding of the reactive metabolite could then be used as an index of the formation of the metabolite. Furthermore, this parameter might well be the most reliable estimate of the availability of the metabolite *in situ* for causing tissue damage, since much of the metabolite often decomposes or is further metabolized before it can be isolated in body fluids or urine. Thus, our approach to the problem has been to determine whether radiolabeled drugs administered to animals over a wide dose range are covalently bound to macromolecules in target tissues that subsequently become necrotic.

It is now apparent that many drugs and foreign substances can be converted to chemically reactive metabolites since they become covalently bound to microsomal protein in incubation systems containing liver microsomes and NADPH or are mutagenic in such systems when appropriate test bacteria are included. However, such studies should be interpreted carefully. Although they can be very useful in biochemical studies of the enzyme pathway leading to a chemically reactive metabolite (Figs. 11 and 13), they do not necessarily predict the occurrence of toxicity *in vivo*. As discussed above, the manifestation of toxicity caused by reactive drug metabolites depends on several critical kinetic factors, namely, the rate of formation, the inherent reactivity and the rate of detoxification of the toxic metabolites at their site of action in the target organ.

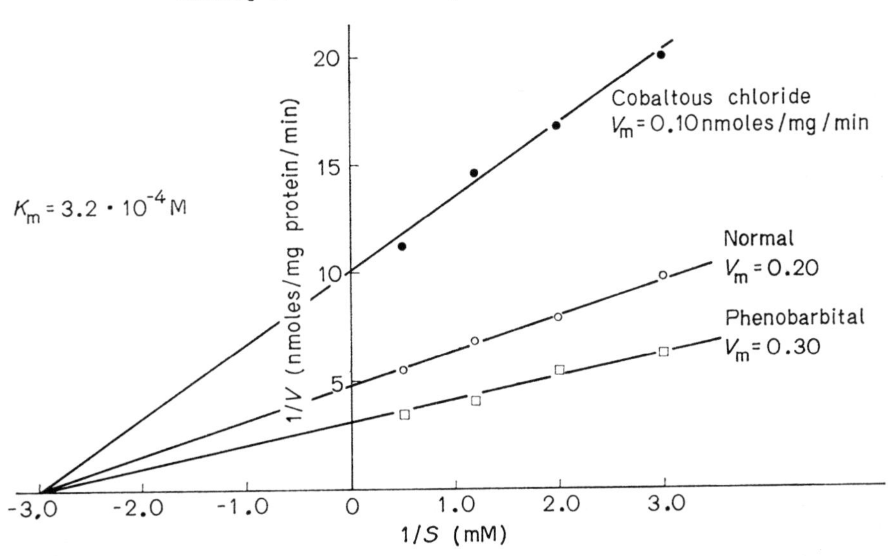

Fig. 13. Double reciprocal plots of mixed-function oxidase-dependent ^{3}H-acetaminophen covalent binding to microsomal protein *in vitro*. Plot for acetaminophen incubated for 12 min with cofactors and microsomes from normal mice or from mice pretreated with cobaltous chloride or phenobarbital. Values are means of at least 4 incubations; S.E.M. were less than $\pm$ 6%. Note that the alterations in the rate of acetaminophen binding after pretreatments correlate with the changes in acetaminophen hepatotoxicity, demonstrating that the enzyme activating acetaminophen is a hepatic microsomal mixed function oxidase. (Data from POTTER et al., 1973a)

The implications of covalent binding *in vivo* by reactive drug metabolites are unknown but are obviously of great concern. It is clear that covalent binding can be used as a valuable index of the formation in the body of alkylating or arylating metabolites from chemically stable parent drugs. However, knowledge of the biochemical nature of the reactive metabolite does not explain why a particular reaction such as *N*-oxidation leads to tissue necrosis after one compound but to neoplasia after another. The chemical-tissue interactions critical to particular necrotic or neoplastic processes must be defined and the mechanisms by which these critical interactions produce necrosis or neoplasia must be determined before a full assessment of the meaning of the covalent binding of drug metabolites to tissue macromolecules can be made.

In attempting to relate the formation of chemically reactive metabolites with their toxicities, we have found it useful to view the incidence and severity of any given toxicity as the product of four mathematical functions, i.e. incidence or severity = ABCD. In this equation, A is the proportion of the dose of the toxicant that is converted to a chemically reactive metabolite; B is the proportion of the chemically reactive metabolite that becomes covalently bound to various kinds of cellular components including macromolecules such as protein, DNA and RNA; C is the proportion of the covalently bound metabolite that is attached to vitally important cellular components; and D is the proportion of C that cannot be replaced or repaired or that leads to the formation of genetic abnormalities. Obviously, the values of C and D will vary with the toxicant, the tissue in which the covalent binding occurs, and the mechanism of toxicity. When the product of $C \times D$ is small, toxicities may not occur even when considerable amounts of covalently bound metabolite are found in tissues. But when the product of $C \times D$ is high for a given kind of toxicity, the toxicity may occur when small amounts of covalently bound metabolite are found in tissues. Since little is known of the mechanisms of most toxicities, the importance of numerical values of C and D are virtually impossible to estimate at the present time and therefore measurements of covalent binding alone cannot be used to predict whether a given foreign compound will cause a given kind of toxicity. Nevertheless, when it has been established that a drug causes a toxicity, we can determine whether the toxicity is mediated through the formation of a chemically reactive metabolite by correlating *changes* in the amount of covalently bound metabolite (that is the product of $A \times B$) with *changes* in the incidence and severity of the toxicity. In addition, this correlation will also determine whether treatments that alter the rates of drug metabolism also alter either A or B.

In conclusion, it is practical to consider three major questions when studying the mechanisms of drug-induced toxicity. Is the toxicity caused by the drug itself or is it caused by a metabolite? Does the toxic drug or its metabolite exert its effect by combining reversibly or irreversibly (covalently) with tissue components? What factors affect the severity of the lesion after the toxic drug or its metabolite reacts with intracellular sites? Although little is known about the last set of factors, considerable progress has been made in elucidating those factors that control the formation and fate of toxic drug metabolites, which has been the subject of this review. A better understanding of these factors will hopefully lead to more rational approaches to the design of nontoxic therapeutic agents in the future.

References

ABEL, J. A.: Analgesic nephropathy — A review of the literature. Clin. Pharmacol. Ther. **12**, 583—598 (1970).

ACKROYD, J. F., ROOK, A. J.: Drug reactions. In: GELL, P. G. H., COOMBS, R. R. A. (Eds.): Clinical aspects of immunology, pp. 448—459. Oxford: Blackwell 1963.

ADRIANY, J.: Toxicity to para-amino salicylic acid and isoniazid. Dis. Chest **38**, 107—108 (1960).

Alarcon-Segovia, D., Fishbein, E., Alcala, H.: Isoniazid acetylation rate and development of antinuclear antibodies upon isoniazid treatment. Arthritis Rheum. **14**, 748—752 (1971).
Angervall, L., Bengtsson, U., Zetterlund, C. G., Zsigmond, M.: Renal pelvic carcinoma in a Swedish district with abuse of a phenacetin-containing drug. Brit. J. Urol. **41**, 401—405 (1969).
Assem, E. S. K., Ndoping, N., Nicholson, H. et al.: Liver damage and isoniazid allergy. Clin. exp. Immunol. **5**, 439—442 (1969).
Back, K. C., Thomas, A. A.: Aerospace problems in pharmacology and toxicology. Ann. Rev. Pharmacol. **10**, 395—412 (1970).
Bailey, W. C., Taylor, S. L., Dascomb, H. E., Greenberg, H. B., Ziskind, M. M.: Disturbed hepatic function during isoniazid chemoprophylaxis for tuberculosis. Amer. Rev. resp. Dis. **107**, 523—529 (1973).
Bengtsson, U., Angervall, L., Ekman, H., Lemann, L.: Transitional cell tumors of the renal pelvis in analgesic abusers. Scand. J. Urol. Nephrol. **2**, 145—151 (1968).
Bengtsson, U., Angervall, L.: Analgesic abuse and tumors of renal pelvis. Lancet **1970 I**, 305—311.
Beutler, E.: Drug-induced hemolytic anemia. Pharmacol. Rev. **21**, 73—103 (1969).
Black, M., Mitchell, J. R., Zimmerman, H. J., Ishak, K., Epler, G. R.: Isoniazid-associated hepatic disease (Abstr.). Gastroenterology **65**, A-4/528 (1973).
Boyd, E. M., Bereczky, G. M.: Liver necrosis from paracetamol. Brit. J. Pharmacol. **26**, 606—614 (1966).
Boyland, E., Chasseaud, L. F.: Enzyme-catalyzed conjugations of glutathione with unsaturated compounds. Biochem. J. **104**, 95—102 (1967).
Boyland, E., Chasseaud, L. F.: The role of glutathione and glutathione-transferases in mercapturic acid biosynthesis. Advanc. Enzymol. **32**, 173—211 (1969).
Brodie, B. B.: Idiosyncrasy and intolerance. In: Wolstenholme, G., Porter, R. (Eds.): Drug responses in man, pp. 188—201. London: Churchill 1967.
Brodie, B. B., Reid, W. D., Cho, A. K., Sipes, G., Krishna, G., Gillette, J. R.: Possible mechanism of liver necrosis caused by aromatic organic compounds. Proc. nat. Acad. Sci. (Wash.) **68**, 160—164 (1971).
Brodie, B. B., Mitchell, J. R.: The value of correlating biological effects of drugs with plasma concentrations. In: Davies, D. C., Prichard, B. N. C. (Eds.): Biological effects of drugs in relation to their plasma concentration, pp. 1—12. London: MacMillan 1973.
Büch, H., Gerhards, W., Pfleger, K., Rudiger, W., Rummel, W.: Enhancement of phenacetin metabolism by phenobarbital. Biochem. Pharmacol. **16**, 1585—1593 (1967).
Byrd, R. B., Nelson, R., Elliott, R. C.: Isoniazid toxicity. J. Amer. Med. Assn. **220**, 1471—1473 (1972).
Calder, I. C., Creek, M. J., Williams, P. J.: *N*-Hydroxylation of *p*-acetophenetidide as a factor in nephrotoxicity. J. med. Chem. **16**, 499—502 (1973).
Calder, I. C., Creek, M. J., Williams, P. J.: *N*-Hydroxyphenacetin as a precursor of 3-substituted 4-hydroxyacetanilide metabolites of phenacetin. Chem. Biol. Interact. **8**, 87—90 (1974).
Clark, R., Borirakchanyavat, V., Davidson, A. R., Thompson, R. P. H., Widdop, B., Gouldine, R., Williams, R.: Hepatic damage and death from overdose of paracetamol. Lancet **1973 I**, 66—69.
Clayson, D. B.: Chemical carcinogenesis. London: Churchill 1962.
Clayson, D. B.: Carcinogenic hazards due to drugs. In: Meyler, L., Peck, H. M. (Eds.): Drug-induced diseases, Vol. 4, pp. 91—109. Amsterdam: Excerpta Medica 1972.
Cohen, S. M., Bryan, G. T.: Carcinogenesis caused by nitrofuran derivatives. Proc. Int. Congr. Pharmacol., 5th, **2**, 164—170. Basel: Karger 1973.
Conn, H. J.: Halothane-associated hepatitis: A disease of medical progress. Israel J. med. Sci. **10**, 404—415 (1974).
Coronary Drug Project: Findings leading to discontinuation of the 2.5 mg/day estrogen group. J. Amer. med. Ass. **226**, 652—657 (1973).
Corsini, G. U., Krishna, G., Brodie, B. B., Gillette, J. R.: Studies on covalent binding of estrogens to microsomes. 1972, Proc. 5th Int. Congr. Pharmacol. Abstr. Vol. Papers, p. 47.
Daly, J. W., Jerina, D. M., Witkop, B.: Arene oxides and the NIH shift: The metabolism, toxicity and carcinogenicity of aromatic compounds. Experientia (Basel) **28**, 1129—1149 (1972).
Davis, D. C., Potter, W. Z., Jollow, D. J., Mitchell, J. R.: Species differences in hepatic glutathione depletion, covalent binding and hepatic necrosis after acetaminophen. Life Sci. **14**, 2099—2109 (1974).
DeMatteis, F.: Drug-induced destruction of cytochrome P-450. Drug Metab. Disp. **1**, 267—274 (1973).

di Chiara, G., Potter, W. Z., Jollow, D., Mitchell, J. R., Gillette, J. R., Brodie, B. B.: Acetaminophen-induced hepatic necrosis. IV. Comparison with methemoglobin formation by acetanilide analogs (Abstr.). Fifth Intl. Cong. on Pharmacol., p. 57 (1972).
di Chiara, G., Hinson, J. A., Potter, W. Z., Jollow, D. J., Mitchell, J. R.: Hemolysis produced by acetanilide and analogues (Abstr.). Fed. Proc. **32**, 305 (1973).
Dickens, F.: Drugs with lactone groups as potential carcinogens. In: Truhaut, R. (Ed.): Potential carcinogenic hazards from drugs, pp. 144—160. U. I. C. C. Monographs 7. Berlin-Heidelberg-New York: Springer 1967a.
Dickens, F.: Mold products, including antibiotics, as carcinogens. In: Staff of the Anderson, M.D. Hospital and Tumor Institute (Ed.): A broad critique, pp. 447—463. Baltimore: Williams and Wilkins 1967b.
Dingell, J. V., Sulser, F., Gillette, J. R.: Species differences in the metabolism of imipramine and desmethylimipramine (DMI). J. Pharmacol. exp. Ther. **143**, 14—22 (1964).
Dodds, M. G., Foord, R. D.: Enhancement by potent diuretics of renal tubular necrosis induced by cephaloridine. Brit. J. Pharmacol. **40**, 227—236 (1970).
Dove, J. T., Chaparas, S. D., Hedrick, S. R.: Failure to demonstrate transformation of lymphocytes of patients with isoniazid associated hepatitis. Amer. Rev. resp. Dis. **106**, 485—487 (1972).
Druckrey, H.: Specific carcinogenic and teratogenic effects of indirect alkylating methyl and ethyl compounds, and their dependency on stages of ontogenic development. Xenobiotica **3**, 271—303 (1973).
Dubach, U. D., Levy, P. S., Minder, F.: Epidemological study of analgesic intake and its relationship to urinary tract disorders in Switzerland. Helv. med. Acta **34**, 297—307 (1968).
Dymling, J. F., Nilsson, K. O., Hokfelt, B.: The effect of soldactona (Canrenoate-potassium) on plasma testosterone and androstenedione and urinary 17-ketosteroids and 17-hydroxycorticosteroids. Acta Endocr. (Kbh.) **70**, 104—112 (1972).
Editorial: Isoniazid: how much a carcinogen? Lancet **1966 II**, 1452—1453.
Editorial: Analgesic abuse and tumors of the renal pelvis. Lancet **1969 II**, 1233—1234.
Elliot, R. W., Kerrand, D. N. S., Lewis, A. A. G.: Frusemide in renal failure. Postgrad. med. J. Suppl. **47**, 5—57 (1971).
Endo, H., Kume, F.: Introduction of sarcoma in rats by a single inection of 4-hydroxyaminoquinoline-1-oxide. Gann **56**, 261—266 (1965).
Epstein, S. S., Andrea, J., Joshi, S., Mantel, N.: Hepatocarcinogenicity of griseofulvin following parenteral administration to infant mice. Cancer Res. **27**, 1900—1913 (1967).
Erlanger, F. B.: Principles and methods for the preparation of drug protein conjugates for immunological studies. Pharmacol. Rev. **25**, 271—280 (1973).
Erturk, E., Price, J. M., Morris, J. E., Cohen, S., Leith, R. S., von Esch, A. M., Crovetti, A. J.: The production of carcinoma of the urinary bladder in rats by feeding *N*-[4-(*S*-nitro-2-furyl)-2-thiazolyl] foramide. Cancer Res. **27**, 1998—2018 (1967).
Erturk, E., Cohen, S. M., Price, J. M., Bryan, G. T.: Pathogenesis, histology and transplantability of urinary bladder carcinomas induced in albino rats by oral administration of *N*-[4-(5-nitro-2-furyl)-2-thiazolyl] foramide. Cancer Res. **29**, 2219—2235 (1969).
Feller, D. R., Morita, M., Gillette, J. R.: Reduction of heterocyclic nitro compounds in the rat liver. Proc. Soc. exp. Biol. (N. Y.) **137**, 433—442 (1971).
Foord, R. D.: Cephaloridine and the kidney. VI. International Congress of Chemotherapy, Tokyo, Proceedings **1**, 597—604 (1969).
Gault, M. H., Rudwal, T. C., Engees, W. D., Dossetor, J. B.: Syndrome associated with the abuse of analgesics. Ann. intern. Med. **4**, 906—908 (1968).
Gillette, J. R.: Factors that affect the covalent binding and toxicity of drugs. In: Loomis, T. A. (Ed.): Pharmacology and the future of man. Proc. Fifth Intl. Cong. on Pharmacol., San Francisco 1972, pp. 187—202. Basel: S. Karger 1973.
Gillette, J. R., Mitchell, J. R., Brodie, B. B.: Biochemical basis for drug toxicity. Ann. Rev. Pharmacol. **14**, 271—288 (1974).
Greenwald, P., Barlow, J. J., Nasca, P. C., Burnet, W. S.: Vaginal cancer after maternal treatment with synthetic oestrogens. New Engl. J. Med. **285**, 390—391 (1971).
Grover, P. L., Sims, P.: Conjugation with glutathione *S*-aryl-transferase in vertebrate species. Biochem. J. **90**, 603—606, 1964.
Haber, E., Osborne, R. K.: Icterus and febrile reactions in response to isonicotinic acid hydrazine. New Engl. J. Med. **260**, 417—420 (1959).
Harvey, D. J., Glazener, L., Stratton, C., Nowlin, J., Hill, R. M., Horning, E. C.: Detection of a 5-(3,4-dihydroxy-1,5-cyclohexadien-1-yl)-metabolite of phenobarbital and mephobarbital in rat, guinea pig and human. Res. Commun. Chem. Path. Pharmacol. **3**, 557—565 (1972).

HARRISON, G.G., SMITH, J.S.: Massive lethal hepatic necrosis in rats anesthetized with fluroxene after microsomal enzyme induction. Anesthesiology **39**, 619—625 (1973).
HEYMANN, E., KRISCH, K., BÜCH, H., BUZELLO, W.: Inhibition of phenacetin- and acetanilide-induced methemoglobinemia in the rat by the carboxylesterase inhibitor *bis*-(*p*-nitrophenyl) phosphate. Biochem. Pharmacol. **18**, 801—811 (1969).
HERBST, A.L., KURMAN, R.J., SCULLY, R.E., POSKANZER, D.C.: Clear-cell adenocarcinoma of the genital tract in young females. New Engl. J. Med. **287**, 1259—1265 (1972).
HINSON, J.A., MITCHELL, J.R., JOLLOW, D.J.: Evidence for *N*-oxidation of acetanilide derivatives (Abstr.). Fed. Proc. **33**, 573 (1974a).
HINSON, J.A., JOLLOW, D.J., MITCHELL, J.R.: Evidence for *N*-oxidation *in vivo* of acetanilide derivatives (Abstr.). Pharmacologist **16**, 496 (1974b).
HINSON, J.H., MITCHELL, J.R.: Unpublished results (1974).
HOIGNE, R.: Arzneimittel-Allergien, Klinische und serologisch-experimentelle Untersuchungen. Berne: Huber 1965.
HOIGNE, R.: Hypersensitivity to drugs. In: MEYLER, L., PEEK, H.M. (Eds.): Drug-induced diseases, **4**, pp. 220—233. Amsterdam: Excerpta Medica 1972.
HORNING, M.G., STRATTON, C., WILSON, A., HORNING, E.C., HILL, R.M.: Detection of 5-(3,4-dihydroxyl-1,5-cyclohexadien-1-yl)-5-phenylhydantoin as a major metabolite of 5,5-diphenylhydantoin (dilantin) in the newborn human. Anal. Letters **4**, 537—545 (1971).
HORNING, M.G., BUTLER, C., HARVEY, D.J., HILL, R.M., ZION, T.C.: Metabolism of *N*,2-methyl-2-phenylsuccinimide (methsyximide) by the epoxide diol pathway in rat, guinea pig and human. Res. Commun. Chem. Path. Pharmacol. **6**, 565—578 (1973).
HULTENGREN, N., LAGERGREN, C., LJUNGQVIST, A.: Carcinoma of the renal pelvis in renal papillary necrosis. Acta chir. scand. **130**, 314—327 (1965).
IKEDA, M., OTSUJI, H.: Phenobarbital-induced protection against toxicity of toluene and benzene in the rat. Toxicol. appl. Pharmacol. **20**, 30—43 (1971).
JAGENBURG, O.R., TOCZKO, K.: The metabolism of acetophenetidine, isolation and characterization of *S* (1-acetamido-4-hydroxy-phenyl)-cysteine metabolite of acetophenetidine. Biochem. J. **92**, 639—643 (1964).
JOLLOW, D.J., MITCHELL, J.R., POTTER, W.Z., DAVIS, D.C., GILLETTE, J.R., BRODIE, B.B.: Acetaminophen-induced hepatic necrosis. II. Role of covalent binding *in vivo*. J. Pharmacol. exp. Ther. **187**, 175—202 (1973).
JOLLOW, D.J., THORGEIRSSON, S.S., POTTER, W.Z., HASHIMOTO, M., MITCHELL, J.R.: Acetaminophen-induced hepatic necrosis. VI. Metabolic disposition of toxic and nontoxic doses of acetaminophen. Pharmacology, 1974a (in press).
JOLLOW, D.J., MITCHELL, J.R., ZAMPAGLIONE, N., GILLETTE, J.R.: Bromobenzene-induced hepatic necrosis: Protective role of glutathione and evidence for 3,4-bromobenzene oxide as the hepatic metabolite. Pharmacology **11**, 151—169 (1974b).
JUHASZ, J., BALO, J., KENDREY, G.: Über die Geschwulsterzeugung. Wirkung der Isonicotinsäure Hydrazid (INH). Z. Krebsforsch. **62**, 188—194 (1957).
KALOW, W.P.: Pharmacogenetics, heredity and the response to drugs, p. 99. Philadelphia: W. B. Saunders Co. 1962.
KASANEN, A., FORSSTROM, J., JACMI, H.A.: On the commonness, cause and detrimental effects of the use of analgesics by factory workers. Acta med. scand. **172**, 15—27 (1962).
KIESE, M.: The biochemical production of ferrihemoglobin-forming derivatives from aromatic amines, and mechanisms of ferrihemoglobin formation. Pharmacol. Rev. **18**, 1091—1161 (1966).
KINDT, H., SCHMIDT, E.: Urinary excretion of furosemide in healthy subjects and patients with liver cirrhosis. Pharmacol. Clin. **2**, 221—227 (1970).
KLATSKIN, G.: Toxic and drug-induced hepatitis. In: SCHIFF, L. (Ed.): Disease of the liver, 2nd ed., pp. 498—601. Philadelphia: J. B. Lippincott Co. 1968.
KLATSKIN, G., KIMBERG, D.V.: Recurrent hepatitis attributable to halothane sensitization in an anesthetist. New Engl. J. med. **280**, 515—523 (1970).
KRISHNA, G., GILLETTE, J.R.: Unpublished results (1974).
LANDSTEINER, K.: The specificity of serological reactions. Cambridge: Harvard Univ. Press 1945.
LAQUEUR, G.L.: Carcinogenic effects of cacad meal and cycasin, methylazomethanol glycoside, in rats and effects of cycasin in germ free rats. Fed. Proc. **23**, 1386—1388 (1964).
LARSEN, K., MOLLER, C.E.: A renal lesion caused by abuse of phenacetin. Acta med. scand. **164**, 53 (1959).
LAS, S., SINGHEL, S.N., BURLEY, D.M., CROSSLEY, G.: Effect of rifampicin and isoniazid on liver function. Brit. med. J. **1972 I**, 148—150.

LAVAN, J. N., BENSON, W. J., GATENBY, A. H., POSEN, S.: The consumption of analgesics by Australian Hospital patients. Med. J. Aust. 2, 694—695 (1966).

LEES, A. W., ALLAN, G. W., SMITH, J., TYRRELL, W. F., FALLON, R. J.: Toxicity from rifampicin plus isoniazid and rifamidicin plus ethambutol therapy. Tubercle 52, 182—190 (1971).

LEES, A. W., ALLAN, G. W., SMITH, J., TYRRELL, W. F., FALLON, R. J.: Rifampicin plus isoniazid in initial therapy of pulmonary tuberculosis and rifampicin and ethambutol in retreatment cases. Chest 61, 579—582 (1972).

LEVIN, W., JACOBSON, M., SERNATINGER, E., KUNTZMAN, R.: Breakdown of cytochrome P-450 heme by secobarbital and other allyl-containing barbiturates. Drug Metab. Disp. 1, 275—285 (1973).

LEVINE, B. B.: Immunochemical mechanisms of drug allergy. Ann. Rev. Med. 17, 23—38 (1966).

MADDREY, W. C., BOITNOTT, J. R.: Isoniazid hepatitis. Ann. intern. Med. 70, 1—12 (1973).

MAGEE, P. H., BARNES, J. M.: Carcinogenic nitroso compounds. Advan. Cancer Res. 10, 163—246 (1967).

MAINES, M. D., ANDERS, M. W.: The possible implication of heme transfer from cytochrome P-420 to albumin in the metabolism of cytochrome P-450. Drug Met. Disp. 1, 293—313 (1973).

MARKS, P. A., BANKS, J.: Drug-induced hemolytic anemias associated with glucose-6-phosphate dehydrogenase deficiency: A genetically heterogeneous trait. Ann. N. Y. Acad. Sci. 123, 198—208 (1965).

MATTHEWS, V. S., KIRKMAN, H., BACON, R. L.: Kidney damage in the golden hamster following chronic administration of diethylstilbestrol and sesame oil. Proc. Soc. exp. Biol. (N. Y.) 66, 195—198 (1947).

MATHEWS, K. D., PAN, P. M., WELLS, J. H.: Experience with lymphocyte transformation tests in the evaluation of allergy to aminosalicylic acid, isoniazid, and streptomycin (Abstr.). J. Allergy 47, 105 (1971).

MATTOCKS, A. R.: Mechanisms of pyrrolizidine alkaloid toxicity. Proc. Int. Congr. Pharmacol. 5th, 2, 114—123. Basel: Karger 1973.

MCLEAN, E. K.: The toxic action of pyrrolizidine (senecio) alkaloids. Pharmacol. Rev. 22, 429—471 (1970).

MENARD, R. H., STRIPP, B., GILLETTE, J. R.: Spironolactone and testicular cytochrome P-450: Decreased testosterone formation in several species and changes in hepatic drug metabolism. Endocrinology 94, 1628—1636 (1974a).

MENARD, R. H., STRIPP, B., LORIAUX, D. L., BARTTER, F. C., GILLETTE, J. R.: Spironolactone (I) and testicular cytochrome P-450: Mechanism of inhibition of androgen formation in testis. Presented at the 56th Annual Meeting of the Endocrine Society, Abstr. No. 98, p. A-104 (1974b).

MERRITT, A. D., FETTER, B. F.: Toxic hepatic necrosis (hepatitis) due to isoniazid: report of a case with cirrhosis and death due to hemorrhage from esohageal varrices. Ann. intern. Med. 50, 804—810 (1959).

MICHIELSEN, P., DE SCHEPPER, P. J., DE BROE, M., VALKENBORGH, G., TRICOT, J. P.: Renal diseases due to drugs. In: MEYLER, L., PECK, H. M. (Eds.): Drug-induced diseases, Vol. 4, pp. 261—324. Amsterdam: Excerpta Medica 1972.

MILLER, E. C., MILLER, J. A., ENOMOTO, M.: The comparative carcinogenicities of 2-acetylaminofluorene and its *N*-hydroxymetabolite in mice, hamsters and guinea pigs. Cancer Res. 24, 2018—2026 (1964).

MILLER, E. C., MILLER, J. A.: Mechanisms of chemical carcinogenesis: Nature of proximate carcinogens and interactions with macromolecules. Pharmacol. Rev. 18, 805—838 (1966).

MILLER, J. A.: Carcinogenesis by chemicals: An overview — G. H. A. Clowes Memorial Lecture. Cancer Res. 30, 559—576 (1970).

MITCHELL, J. R.: Mechanism of benzene-induced aplastic anemia (Abstr.). Fed. Proc. 30, 2044 (1971).

MITCHELL, J. R., CAVANAUGH, J. H., DINGELL, J. V., OATES, J. A.: Guanethidine and related agents. II. Metabolism by hepatic microsomes and its inhibition by drugs. J. Pharmacol. exp. Ther. 172, 105—114 (1970).

MITCHELL, J. R., REID, W. D., CHRISTIE, B., MOSKOWITZ, J., KRISHNA, G., BRODIE, B. B.: Bromobenzene-induced hepatic necrosis. Species differences and protection by SKF 525-A. Res. Commun. Chem. Path. Pharmacol. 2, 877—888 (1971).

MITCHELL, J. R., JOLLOW, D. J., GILLETTE, J. R., BRODIE, B. B.: Drug metabolism as a cause of drug toxicity. Drug Metab. Disp. 1, 418—423 (1973a).

MITCHELL, J. R., JOLLOW, D. J., POTTER, W. Z., DAVIS, D. C., GILLETTE, J. R., BRODIE, B. B.: Acetaminophen-induced hepatic necrosis. I. Role of drug metabolism. J. Pharmacol. exp. Ther. 187, 185—194 (1973b).

MITCHELL, J. R., JOLLOW, D. J., POTTER, W. Z., GILLETTE, J. R., BRODIE, B. B.: Acetaminophen-induced hepatic necrosis. IV. Protective role of glutathione. J. Pharmacol. exp. Ther. **187**, 211—217 (1973c).

MITCHELL, J. R., POTTER, W. Z., JOLLOW, D. J.: Furosemide-induced hepatic and renal tubular necrosis. I. Effects of treatments which alter drug-metabolizing enzymes (Abstr.). Fed. Proc. **32**, 395 (1973d).

MITCHELL, J. R., THORGEIRSSON, S. S., POTTER, W. Z., JOLLOW, D. J., KEISER, H.: Acetaminophen-induced hepatic injury: Protective role of glutathione and rationale for therapy. Clin. Pharmacol. Ther. **16**, 676—684 (1974a).

MITCHELL, J. R., POTTER, W. Z., HINSON, J. A., JOLLOW, D. J.: Massive hepatic necrosis caused by furosemide, a furan-containing diuretic. Nature (Lond.) New Biol. **251**, 508—511 (1974b).

MITCHELL, J. R., LONG, M. W., THORGEIRSSON, U. P., JOLLOW, D. J.: Acetylation rates and monthly liver function tests during one year of isoniazid preventive therapy. Chest 1974c (in press).

MITCHELL, J. R., THORGEIRSSON, U. P., BLACK, M., TIMBRELL, J. A., POTTER, W. Z., SNODGRASS, W. R., JOLLOW, D. J., KEISER, H.: Correlation of rapid acetylator phenotype and isoniazid hepatitis. Possible explanation by analysis of urinary metabolites. Clin. Pharmacol. Ther. 1974d (in press).

MITCHELL, J. R., JOLLOW, D. J.: Metabolic activation of drugs to toxic substances. Gastroenterology 1974 (in press).

MORI, K., YASUNO, A., MATSUMOTO, K.: Induction of pulmonary tumors in mice with isonicotinic acid hydrazide. Gann **51**, 83—91 (1960).

MORITA, M., FELLER, D. R., GILLETTE, J. R.: Reduction of niridazole by rat liver xanthine oxidase. Biochem. Pharmacol. **20**, 217—226 (1971).

NERY, R.: Some new aspects of the metabolism of phenacetin in the rat. Biochem. J. **122**, 317—326 (1971).

NIEWEG, H. O.: Drug-induced diseases as seen by a haematologist. In: MEYLER, L., PECK, H. M. (Eds.): Drug-induced diseases, Vol. 4, pp. 325—344. Amsterdam: Excerpta Medica 1972.

PENTIKAINEN, P. J., PENTIKAINEN, L. A., HUFFMAN, H., AZARNOFF, D. L.: The effect of spironolactone on sexual hormones in males (Abstr.). Clin. Res. **21**, 472 (1973).

PETERS, J. H., MILLER, K. S., BROWN, P.: Studies on the metabolic basis for the genetically determined capacities for isoniazid inactivation in man. J. Pharmacol. exp. Ther. **150**, 298—304 (1965).

Physicians Desk Reference: Medical Economics Co. Oradell, N. J., USA, 1973, p. 774.

POTTER, W. Z., DAVIS, D. C., MITCHELL, J. R., JOLLOW, D. J., GILLETTE, J. R., BRODIE, B. B.: Acetaminophen-induced hepatic necrosis. III. Cytochrome P-450-mediated covalent binding *in vitro*. J. Pharmacol. exp. Ther. **187**, 203—210 (1973a).

POTTER, W. Z., NELSON, W. L., THORGEIRSSON, S. S., SASAME, H., JOLLOW, D. J., MITCHELL, J. R.: Furosemide-induced hepatic necrosis. II. Comparison of necrosis with covalent binding of furosemide (Abstr.). Fed. Proc. **32**, 305 (1973b).

POTTER, W. Z., THORGEIRSSON, S. S., JOLLOW, D. J., MITCHELL, J. R.: Acetaminophen-induced hepatic necrosis. V. Correlation of hepatic necrosis, covalent binding and glutathione depletion in hamsters. Pharmacology (1974) (in press).

POTTER, W. Z., MITCHELL, J. R.: Unpublished results (1974).

PRESCOTT, L. F., WRIGHT, N., ROSCOE, P., BROWN, S. S.: Plasma paracetamol half-life and hepatic necrosis in patients with paracetamol overdosage. Lancet **1971 I**, 519—522.

PRESCOTT, L. F., NEWTON, R. W., SWAINSON, C. P., WRIGHT, N., FORREST, A. R. W., MATTHEW, H.: Successful treatment of severe paracetamol overdosage with cysteamine. Lancet **1974 I**, 588—592.

RAISFELD, I. H., PFISTER, L., FEINGOLD, M.: The effect of isoniazid on hepatic endoplasmic reticulum in man and rat (Abstr.). Gastroenterology **65**, A-42/566 (1973).

RAO, K. V. N., MITCHISON, D. A., NAIR, N. G. K., PRENA, K., TRIPATHY, S. P.: Sulphadimidine acetylation test for classification of patients as slow or rapid inactivators of isoniazid. Brit. med. J. **1970 III**, 495—497.

REID, W. D., CHRISTIE, B., KRISHNA, G., MITCHELL, J. R., MOSKOWITZ, J., BRODIE, B. B.: Bromobenzene metabolism and hepatic necrosis. Pharmacology **6**, 41—55 (1971).

REID, W. D.: Relationship between tissue necrosis and covalent binding of toxic metabolites of halogenated aromatic hydrocarbons. Proc. 5th Int. Congr. Pharmacology, Vol. 2, pp. 187—202. Basel: Karger 1973.

REINER, O., ATHANASSOPOULOS, S., HELLMER, K. H., MURRAY, R. E., UEHLEKE, H.: Bildung von Chlroroform aus Tetrachlorokohlenstoff in Lebermikrosomen, Lipid peroxidation und Zerstörung von Cytochrome P-450. Arch. Toxicol. **29**, 219—233 (1972).

REMMER, H., SCHOENE, B., FLEISCHMANN, R. A.: Induction of unspecific microsomal hydroxylase in human liver. Drug Metab. Disp. **1**, 224—230 (1973).

REMMER, H., BOCK, K. W.: The role of the liver in drug metabolism. In: SCHAFFNER, F., SHERLOCK, S., LEEVY, C. (Eds.): The liver and its disease, pp. 34—42. New York: Intercontinental Medical Book Corporation 1974.

REYNOLDS, E. S., BROWN, B. R., VANDAM, L. D.: Massive hepatic necrosis after fluroxene anesthesia. New Engl. J. Med. **286**, 530—531 (1972).

REYNOLDS, E. S., MOSLEN, M. T.: Liver injury following halothane anesthesia in rats (Abstr.). Fed. Proc. **32**, 306 (1973).

ROE, F. J. C.: On potential carcinogenicity of the iron macromolecular complexes. In: TRUHAUT, R. (Ed.): Potential carcinogenetic hazards from drugs, pp. 105—130, U. I. C. C. Monographs **7**, Berlin-Heidelberg-New York: Springer 1967.

SCHARER, L., SMITH, J. P.: Serum transaminase elevations and other hepatic abnormalities in patients receiving isoniazid. Ann. intern. Med. **71**, 1113—1120 (1969).

SCHWAN, S.: Isonicotinic acid hydrazide as a carcinogenic factor in mice. I. Pat. Pol. **12**, 53—60 (1961).

SCHWAN, S.: Isonicotinic acid hydrazide as a carcinogenic agent in mice. II. Pat. Pol. **13**, 42—48 (1962).

SEAWRIGHT, A. A., HRDLICKA, J.: The effect of prior dosing with phenobarbitone and diethylaminoethyl diphenylpropyl acetate (SKF 525-A) on the toxicity and liver lesion caused by ngaione in the mouse. Brit. J. exp. Path. **53**, 242—252 (1972).

SEAWRIGHT, A. A., MATTOCKS, A. R.: The toxicity of two synthetic 3-substituted furan carbamates. Experientia (Basel) **29**, 1197—1200 (1973).

SHAHIDI, N. J.: Acetophenetidin-induced methemoglobinemia. Ann. N. Y. Acad. Sci. **151**, 822—832 (1968).

SHAND, D., MITCHELL, J. R., OATES, J. A.: Pharmacokinetic drug interactions. In: GILLETTE, J. R., MITCHELL, J. R. (Eds.): Handbuch der experimentellen Pharmakologie, this volume. Berlin-Heidelberg-New York: Springer 1974.

SHELLEY, J. H.: Phenacetin, through the looking glass. Clin. Pharm. Ther. 8, 427—471 (1967)-

SHULMAN, N. R.: Mechanism of blood cell damage by adsorption of antigen-antibody complexes. In: Immunology, pp. 338—351. Basel: Schwabe 1963.

SMITH, J., TYRRELL, W. F., GOW, A., ALLAN, G. W., LEES, A. W.: Hepatotoxicity in rifampin isoniazid treated patients related to their rate of isoniazid inactivation. Chest **61**, 587—588 (1972).

SNODGRASS, W., POTTER, W. Z., TIMBRELL, J., JOLLOW, D. J., MITCHELL, J. R.: Possible mechanism of isoniazid-related hepatic injury (Abstr.). Clin. Res. **22**, 323A (1974).

SNODGRASS, W. R., MITCHELL, J. R.: Unpublished results (1974).

SPUHLER, O., ZOLLINGER, H. V.: Die chronisch-interstitielle Nephritis. Z. Klin. Med. **151**, 1—9 (1953).

STEINBERG, A. D., PLOTZ, P. H., WOLFF, S. M., WONG, V. G., AGUS, S. G., DECKER, J. L.: Cytotoxic drugs in treatment of nonmalignant diseases. Ann. Inst. Med. **76**, 619—642 (1972).

STRIPP, B., MENARD, R. H., ZAMPAGLIONE, N. G., HAMRICK, M. E., GILLETTE, J. R.: Effect of steroids on drug metabolism in male and female rats. Drug Metab. Disp. **1**, 216—223 (1973).

STRIPP, B., MENARD, R. H., GILLETTE, J. R.: Interaction of nitrofurazone with hepatic and extrahepatic tissues from mice and rats (Abstr.). Toxicol. Appl. Pharmacol. **29**, 91 (1974).

SWENSON, D. H., MILLER, J. A., MILLER, E. C.: 2,3-Dihydro-2,3-dihydroxy-aflatoxin B1: An acid hydrolysis product of an RNA-aflatoxin B1 adduct formed by hamster and rat liver microsomes *in vitro*. Biochem. Biophys. Res. Commun. **53**, 1260—1267 (1973).

THORGEIRSSON, S. S., JOLLOW, D. J., SASAME, H. A., GREEN, I., MITCHELL, J. R.: The role of cytochrome P-450 in *N*-hydroxylation of 2-acetylaminofluorene. Mol. Pharmacol. **9**, 398—404 (1973a).

THORGEIRSSON, S. S., SASAME, H., POTTER, W. Z., NELSON, W. L., JOLLOW, D. J., MITCHELL, J. R.: Biochemical changes after acetaminophen and furosemide-induced liver injury. Fed. Proc. **32**, 305 (1973b).

TIITINEN, H.: Modification by para-aminosalicylic acid and sulfamethazine of the isoniazid inactivation in man. Scand. J. resp. Dis. **50**, 281—290 (1969).

TIMBRELL, J. A., JOLLOW, D. J., MITCHELL, J. R.: Unpublished results (1974).

TOTH, B., SHUBIK, P.: Carcinogenesis in Swiss mice by isonicotinic acid hydrazide. Cancer Res. **26**, 1473—1479 (1966).

UEHLEKE, H.: Mechanisms of methemoglobin formation by therapeutic and environmental agents. In: LOOMIS, T. A. (Ed.): Pharmacology and the future of man, pp. 124—136. Basel: S. Karger 1973.

VESELL, E. S., PASSANANTI, G. T., GREENE, F. E.: Impairment of drug metabolism in man by allopurinol and nortriptyline. New Engl. J. Med. **283**, 1484—1488 (1972).

WEIHE, M., POTTER, W. Z., NELSON, W. L., JOLLOW, D. J., MITCHELL, J. R.: Mechanism of dose threshold for furosemide hepatotoxicity (Abstr.). Toxicol. Appl. Pharmacol. **29**, 90 (1974).

WEISBURGER, J. H., WEISBURGER, E. K.: Biochemical formation and properties of hydroxylamines and hydroxamic acids. Pharmacol. Rev. **25**, 1—66 (1973).

WILSON, B. J., YANG, D. T. C., BOYD, M. R.: Toxicity of mould damaged sweet potatoes (*Ipomoea Batatas*). Nature (Lond.) **227**, 521—522 (1970).

WILSON, B. J., BOYD, M. R., HARRIS, T. M., YANG, D. T. C.: A lung oedema factor from mouldy sweet potatoes (*Ipomoea Batatas*). Nature (Lond.) **231**, 52—53 (1971).

WOLPERT, M. K., ALTHAUS, J. R., JOHNS, D. G.: Nitroreductase activity of mammalian liver aldehyde oxidase. J. Pharmacol. exp. Ther. **185**, 202—213 (1973).

WRIGHT, N., PRESCOTT, L. F.: Potentiation by previous drug therapy of hepatotoxicity following paracetamol overdosage. Scot. Med. J. **18**, 56—58 (1973).

ZAMPAGLIONE, N., JOLLOW, D. J., MITCHELL, J. R., STRIPP, B., HAMRICK, M., GILLETTE, J. R.: Role of detoxifying enzymes in bromobenzene-induced liver necrosis. J. Pharmacol. exp. Ther. **187**, 218—227 (1973).

ZIMMERMAN, H. J.: Drug-induced hepatic injury. In: SAMTER, M., PETERS, G. (Eds.): Hypersensitivity to drugs, pp. 299—366. I. E. P. T. Oxford: Pergamon Press 1972.

Author Index

Numbers in brackets refer to the numbers of the reference within the current text and the references

Page numbers in *italics* refer to the bibliography

Abdou, I. A., see Eisenbrand, L. L. 135, *146*
Abel, J. A. 391, *412*
Abraham, E. P., see Sabath, L. D. 352, *357*
Abrams, W. B., see Solomon, H. M. 286, 288, *313*, *314*
Abt, A. F., see Kagan, B. M. 261, *270*
Ackermann, S. J., see Brook, S. M. 286, *309*
Ackroyd, J. F., Rook, A. J. 408, *412*
Adams, B. J., see Shideman, F. E. 242, *256*
Adams, H. J., see Boyes, R. N. 163, *166*
Adams, H. J., see Caldwell, H. C. 95, 101, 108, 109, *109*
Adams, H. J., see Dittert, L. W. 88, 95, 102, 103, 104, *109*, 110
Adams, H. J., see Misher, A. 94, 95, *111*
Adams, H. J., see Swintosky, J. V. 95, 101, *112*
Adams, W. P., see Reidenberg, M. M. 260 264, *270*
Adamson, R. H., 338, *338*
Adir, J., see Barr, W. H. 279, *308*
Adjepon-Yamoah, K. K., Scott, D. B., Prescott, L. F. 237, 238, *252*
Adler, T. K. 115, *126*
Adler, T. K., see Eisenbrandt, L. L. 135, *146*
Adler, T. K., see Way, E. L. 134, *149*
Adolph, E. F. 223, *230*
Adriany, J. 408, *412*
Aebi, H., Baggiolini, M., Dewald, B., Lauber, E., Suter, H., Micheli, A., Frei, H. 174, *203*
Aebi, H., Heiniger, J. P., Bütler, R., Hässig, A. 174, *203*
Aebi, H., see Michelli, A. 174, *208*
Aebi, H. E. 174, *203*
Affrime, M., see Reidenberg, M. M. 307, *313*
Agersborg, H. P. K., Batchelor, A., Cambridge, G. W., Rule, A. W. 95, 96, *109*
Agersborg, H. P. K., see Glassman, J. M. 143, *147*
Aggeler, P. M., O'Reilly, R. A. 279, 296, 297, 301, *307*
Aggeler, P. M., O'Reilly, R. A., Leong, L., Kowitz, P. E. 184, *203*
Aggeler, P. M., O'Reilly, R. A., Leong, L., Kowitz, P. M. 296, *307*
Aggeler, P. M., see Levy, G. *312*
Aggeler, P. M., see O'Reilly, R. A. 29, *34*, 184, 185, 186, 197, *209*, 301, *312*
Agus, S. G., see Steinberg, A. D. 405, *418*
Ahlquist, R. P. 317, *322*
Akgun, S., see Steiner, A. 134, *149*
Alarcón-Segovia, D., Fishbein, E., Alcala, H. 178, *203*, 408, *413*
Alcala, H., see Alarcón-Segovia, D. 178, *203*, 408, *413*
Albert, A. 86, *109*
Alexander, F., see Dittert, L. W. 95, 103, *110*
Alexander, W. D., see Papapetrou, P. D. 145, *148*
Alexanderson, B. 298, *307*
Alexanderson, B., Borga, O. 298, *308*
Alexanderson, B., Borga, O., Alvan, G. [1] 76, *84*, 363, *381*
Alexanderson, B., Price Evans, D. A., Sjöquist, F. 198, *203*, 286, *308*
Allan, G. W., see Lees, A. W. 299, *416*
Allan, G. W., see Smith, J. 397, *418*
Allansmith., M, McClellan, B. H., Butterwordt, M., Maloney, J. R. 220, *230*
Allen, J. C., see Besch H. R., Jr. 316, *322*
Allen, J. E., see Jusko, W. J. 216, *232*
Allen, L. M., see Creaven, P. J. 249, *253*
Alleva, F. R., see Vernier, V. G. 121, *129*
Altenmeier, W. A., see Scheiner, J. 236, *256*, 280, *313*
Althaus, J. R. see Wolpert, M. K. 407, *419*
Altland, K., see Goedde, H. W. 179, 181, *206*
Alvan, G., see Alexander, B. 363, *381*
Alvan, G., see Alexanderson, B. [1] 76, *84*
Alving, A. S., see Dern, R. J. 190, *205*
Alway, C. D., see Wagner, J. G. *34*
Ambre, J. J., Fischer, L. J. 141, *145*, 280, *308*
Amsel, L. P., see Levy, G. 26, *34*
Anders, M. W. 283, *308*
Andersen, D. H., see Silverman, W. A. 297, *313*
Andersson, S. 236, *252*
Andersson, S., Thorstrand, C. *313*
Andrea, J., see Epstein, S. S. 407, *414*
Andres, M. W., see Maines, M. D. 404, *416*
Andrew, W. 223, *230*
Andrew, W., Pruett, D. 223, *230*
Andrews, R. H., see Devadatta, S. 176, 177, *205*
Andrioli, V. T., see Bell, N. H. 337, *338*
Angervall, L., Bengtsson, U., Zetterlund, C. G., Zsigmond, M. 406, *413*
Angervall, L., see Bengtsson, U. 406, *413*

Anthony,E.M., see Carlson, H.B. 176, *204*
Anton,A.H., Corey,W.T. 241, 242, *252*, 260, *268*
Anton,A.H., see Tjandramaga,T.B. 319, *324*
Apajalahti,A., see Manninen, V. 280, *312*
Apter,J.T., see Hardison, W.G.M. 145, *147*
Arai,K., see Layne,D.S. 139, *148*
Arai,L., Golab,T., Layne, D.S., Pincus,G. 139,*145*
Arataki,M. 223, *230*
Archer,S., see Rosi,D. 126, *128*
Arcieri,G.M., see Falco,F.G. 337, *340*
Ardeman,S., see Maxwell, J.D. 286, *312*
Arias,I.M., Gartner,L.M., Cohen,B., Ben Ezzer,J., J., Levi,A.J. 287, *308*
Arias,I.M., see Biempica,L. 261, *268*
Arias,I.M., see Black,M. 261, *268*
Arias,I.M., see Levi,A.J. 222, *232*
Arias,I.M., see Lichter,M. 259, 262, 265, *270*
Arias,I.M., see Litwack,G. 222, *232*
Arias,I.M., see Mishkin,S. 222, *233*, 301, *313*
Arias,I.M., see Reyes,H. 222, *233*
Arias,L., see Mitchell,J.R. 299, 300, *312*
Ariens,E.J. 93, 95, *109*
Ariens,E.J., Simanis,A.M., van Rossum,J.M. 315, 322
Armaly,M.F. 193, 194, *203*
Armstrong,A.R., Peart, H.E. 177, *203*
Armstrong,B.K., Ulkich, A.W., Goatcher,P.M. 235, *252*
Arnold,A., see Lands,A.M. 317, *323*
Aronow,L., see Goldstein,A. 218, 219, 220, 223, 225, 226, *231*, 259, *269*, *310*
Arora,S.C., see McMullin, G.P. *312*
Arvela,P., see Sotaniemi,E. 246, *256*
Åsberg,M., Price-Evans, D.A., Sjöqvist,F. 198, *204*
Ashbrook,J.D., see Spector, A.A. 221, *233*
Ashley,J.J., see Perrier,D. 28, *34*
Assem,E.S.K., Ndoping,N., Nicholson,H. 408, *413*
Athanassopoulos,S., see Reiner,O. 404, 418
Atkinson,R.M., see Busfield, D. 278, *309*
Auker,S., see Venho,V.M.K. 239, *257*
Aur,R., see Pinkel,D. 332, *341*
Avellaneda,M. 285, *308*
Avioli,L.V., Birge,S., Won Lee,S., Slatopolsky, E. 259, 261, *268*
Avioli,L.V., see Hahn,T.J. 286, *310*
Axelrod,J., see Brodie,B.B. 117, *127*, 262, *268*
Axelrod,J., see Davis,J.M. 291, *310*
Axelrod,J. see Sobermann,R. 197, *210*
Axelrod,J., see Weiner,M. 197, *211*
Axline,S.G., Simon,H.J. 224, *230*
Axline,S.G., Yaffe,S.J. Simon,H.J. 225, *230*
Ayres,P., see Greaves,J.H. 184, *206*
Azarnoff,D.L., see Borga,O. 291, 298, 300, *308*
Azarnoff,D.L., see Hunninghake,D.B. 297, *311*
Azarnoff,D.L., see Kolmodin, B. 288, *311*
Azarnoff,D.L., see Pentikainen,P.J. 403 *417*
Azouz,W.M., Parke,D.V., Williams,R.T. 378, *381*

Back,G., see Siegers,C.-P. 263, *256*
Back,K.C., Thomas,A.A. 398,*413*
Back,N., see Sandberg, A.A. 130, *148*
Bacon,G.E., Kenny,F.M., Murdaugh, H.V., Richards,C. 262, 265, *268*
Bacon,R.L., see Matthews, V.S. 406, *416*
Bagdade,J.D. 262, *268*
Baggiolini,M., see Aebi,H. 174, *203*
Bagley,C.M.,Jr., Devita,V.T., Berard,C.W., Canellos,G.P. 325, 335, *338*
Baglioni,C. 191, *204*
Bailey,G.L., see Hampers,C.L. 262, *269*
Bailey,W.C., Taylor,S.L., Dascomb,H.E., Greenberg,H.B., Ziskind,M.M. 409, *413*
Baker,D.R., see Rapoport,H. 95, 99, *111*
Baker,S.J., Kumar,S., Swaminathan,S.P. 130, *145*
Balasubramaniam,K., Mawer,G.E., Pohl,J.E.P., Simons, P.J.G. 250, *252*
Balasubramaniam,K., Mawer,G.E., Simons, P.J. 241, *252*
Bale,W.F., see Izzo,J.L. 262, *269*
Baliah,T., see Yaffe,S.J. 203, *212*, 287, *314*
Ballingall,D.L.K., see Jick,H. 196, *207*
Ballinger,B., Browning,M., O'Malley,K., Stevenson, I.H. 246, *253*
Balo,J., see Juhasz,J. 407, *415*
Baltzer,G., see Gossmann, H.H. 259, *269*
Banks,J., see Marks,P.A. *416*
Banziger,R.F., see Randall,L.O. 119, 121, *128*
Baraka,A., Gabali,F. 220, *230*, 242, *253*
Barker,T.L., see Bennett, L.L., Jr. 329, *338*
Barlow,J.J., see Greenwald,P. 406, *414*
Barnden,R.L., Evans,R.M., Hamlet,J.C., Hems,B.A., Jansen,A.B.A., Trevett, M.E., Webb,G.B. 95, *109*
Barner,H.D., see Cohen,S.S. 328, *339*
Barnes,J.M., see Magee,P.H. 404, 410, *416*
Barnes,M., see Ikeda,M. 184, *207*
Barnett,H.L. 213, *230*
Barnett,H.L., McNamara, H., Schultz,S., Tompsett,R. 225, *230*
Barnicot,N.A. 192, *204*
Bar-Or,R., see Szeinberg,A. 177, *210*

Barowsky, H., Schwartz, S. A. 116, *126*
Barquet, A., see Davies, J. E. 288, *310*
Barr, W. H., Adir, J., Garrettson, L. 279, *308*
Barron, E. S. G., Singer, T. P. 189, *204*
Barrowman, J., see Thompson, G. R. 281, *314*
Barrowman, J. A., D'Mello, A., Herxheimer, A. 281, *308*
Barry, H., see Jaffe, J. M. 235, *254*
Barry, V. C. 325, *338*
Bartett, J. W., see Izzo, J. L. 262, *269*
Bartter, F. C., see Henkin, R. I. 193, *206*
Bartter, F. C., see Menard, R. H. 403, *416*
Baselt, R. C., Casarett, L. J. 135, *145*
Bass, M. 158, *166*
Batchelor, A. see Agersborg, H. P. K. 95, 96, *109*
Batchelor, F. R., see Sutherland, R. 351, *358*
Batt, E. R., Schachter, D. 216, *230*
Bauer, W., see Hodgkin, W. E. 180, 181, *206*
Baur, E. W. 174, *204*
Baxter, C., see Jick, H. 196, *207*
Baylis, E. M., Crowley, J. M., Preece, J. M., Sylvester, P. E., Marks, V. 286, *308*
Bazzano, G., Bazzano, G. S. 281, *308*
Bazzano, G. S., see Bazzano, G. 281, *308*
Beard, O. W., see Dougherty, J. E. 137, *146*
Beard, W. J., Free, S. M., Jr. 95, 101, *109*
Beaty, H. N., Turck, M., Petersdorf, R. G. 354, *355*
Becak, W., see Saldanha, P. H. 192, *209*
Becker, B., Morton, W. R. 192, *204*
Becker, B. A., Hindmann, K. L., Gibson, J. E. 130, *146*
Becker, B. A., see Gibson, J. E. 130, *147*
Becker, C. E., see Blum, M. R. 260, *268*
Becker, C. E., see Cavalieri, R. R. 286, *309*
Beckers, T., see Prins, H. K. 189, *209*
Beckett, A. H., Brookes, L. G. 122, *126*
Beckett, A. H., Hossie, R. D. 152, *166*
Beckett, A. H., Shenoy, E. V. B., Salmon, J. A. 122, *126*
Beermann, B., Hellström, K., Rosen, A. 235, 239, *253*
Behrman, R. E., see Gotoff, S. P. 344, *356*
Bell, J. C., see Mitchell, R. S. 177, *208*
Bell, N. H., Andrioli, V. T., Sabesin, S. M., Utz, J. P. 337, *338*
Bellet, S., Kershbaum, A., Fink, E. M. 221, *230*
Bellet, S., Roman, L. R., Boza, A. 248, *253*
Bender, A. D. 217, *230*
Bender, M. L. 92, *109*
Benet, L. Z., see Rowland, M. [14] 24, *34*, 49, *85*, 283, 300, *313*
Ben Ezzer, J., see Arias, I. M. 287, *308*
Bengtsson, U., Angervall, L. 406, *413*
Bengtsson, U., Angervall, L., Ekman, H., Lemann, L. 406, *413*
Bengtsson, U., see Angervall, L. 406, *413*
Benjamin, D. M., see Robinson, D. S. 281, *313*
Ben-Maimon, Rabbi Moses 355, *355*
Benner, E. J. 266, *268*
Bennett, L. L., Jr., Simpson, L., Golden, J., Barker, T. L. 329, *338*
Bennett, W. G., see Gatz, E. E. 195, *205*
Bennett, W. M., Singer, I., Coggins, C. H. 259, 261, 266, *268*
Benoit, M., see McDermott, W. 153, *167*
Benowitz, N., Rowland, M., Forsyth, R., Melmon, K. L. 293, *308*
Benson, H., see Place, V. A. 235, *255*, 279, *313*
Benson, W. J., see Lavan, J. N. 391, *416*
Berard, C. W., see Bagley, C. M., Jr. 325, 335, *338*
Berberian, D. A., see Rosi, D. 126, *128*
Bereczky, G. M., see Boyd, E. M. 384, 391, *413*
Beresford, C. H., Neale, R. J., Brooks, O. G. 238, *253*
Berger, F. M., Riley, R. F. 95, 104, *109*
Berger, J., see Dowling, R. H. 131, *146*
Berger, L., see Burns, J. J. 115, *127*
Berglund, F., see Sjöqvist, F. 300, *313*
Bergstrom, S. 131, *146*
Berkowitz, W. D., see Rosenblum, R. 321, *324*
Berlin, A., see Lund, L. 240, *254*
Berlin, C. M., see Chignell, C. F. 197, *204*, 241, *253*
Berman, L. B., Katz, S. 337, *338*
Berman, M., Weiss, M. F. 33, *33*
Berneis, K., Kofler, M., Bollag, W., Kaiser, A., Langemann, A. 329, *338*
Bernhard, K., see Keberle, H. 130, 131, 140, *147*
Bernstein, E., see Brodie, B. B. [2] 61, *85*
Bernstein, J. G., see Sellers, E. M. 290, 302, *313*
Bers, L. 6, *33*
Berthelot, P., Erlinger, S., Dhumeaux, D., Preauz, A.-M. 144, *146*
Bertino, J. R. 329, 330, *338*
Bertino, J. R., see Capizzi, R. L. 338, *339*
Bertino, J. R., see Levitt, M. 335, *341*
Besch, H. R., Jr., Allen, J. C., Glick, G., Schwartz, A. 316, *322*
Betke, K., see Frick, P. G. 191, *205*
Betke, K., see Hitzig, W. H. 191, *206*
Betke, K., see Huisman, T. H. J. 191, *207*
Beutler, E. 190, *204*, 400, *413*
Beutler, E., see Dern, R. J. 190. *205*
Beyer, K. H., Russo, H. F., Tillson, E. K., Miller, A. K., Verwey, W. F., Gass, S. R. 361, *381*
Bianchine, J. R., Calimlim, L. R., Morgan, J. P., Dujuvne, C. A., Lasagna, L. 152, *166*, 238, *253*, 281, *308*
Bieber, S., see Elion, G. B. 328, 337, *339*
Biehl, J. P. 177, *204*

Biehl, J.P., see Hughes, H.B. 176, *207*
Biempica, L., Goldfischer, S., Black, M., Biempica, S.L., Grossman, S.B., Arias, I.M. 261, *268*
Biempica, S.L., see Biempica, L. 261, *268*
Bigland, B., Goetzee, B., MacLagan, J., Zaimis, E. 252, *253*
Billups, N.F., see Doluisio, J.T. 236, *253*
Bing, R.J., Maraist, F.M., Damman, J.F., Jr., Draper, A., Jr., Heimbecker, R., Daley, R., Gerard, R., Calazel, P. 315, *322*
Binns, T.B. 278, 279, *308*
Bird, O.D., see Thompson, R.Q. 130, *149*
Birge, S., see Avioli, L.V. 259, 261, *268*
Birge, S.G., see Hahn, T.J. 286, *310*
Bischoff, K.B., Dedrick, R.L. 24, *33*
Bjorksten, K., see Neuvonen, P.J. 280, *312*
Black, M., Grossman, S.B., Arias, I.M. 261, *268*
Black, M., Mitchell, J.R., Zimmerman, H.J., Ishak, K., Epler, G.R. 397, 408, *413*
Black, M., Sherlock, S. 287, *308*
Black, M., see Biempica, L. 261, *268*
Black, M., see Lichter, M. 259, 262, 265, *270*
Black, M., see Mitchell, J.R. 398, *417*
Blackburn, E.K., see Guyer, R.J. 333, 334, *340*
Blackwell, see Briant, 161,
Blair, A.H., Vallee, B.L. 195, *204*
Blake, D.A., see Cascorbi, H.F. 198, *204*
Blakeslee, A.F. 192, *204*
Blanc, W.A., see Johnson, L. 297, *311*
Blanc, W.A., see Silverman, W.A. 297, *313*
Blanchard, A.W., see Gwilt, J.R. 238, *254*
Blaszkowski, T.P., see Bogdanski, D.F. 286, *308*
Block, J.B., see Mendelson, D. 335, *341*
Bloomfield, A.L., Polland, W.S. 216, *231*
Blount, J.G. 347, *355*
Blount, R.E. 325, *338*
Bluemle, L.W., Goldberg, M.J. 240, *253*
Blum, M.R., Riegelman, S., Becker, C.E. 260, *268*
Bock, K.W., see Remmer, H. 406, *418*
Boddy, K., see Lawson, D.H. 259, *270*
Bodey, G.P., see Freireich, E.J. 333, *340*
Bönicke, R., Lisboa, B.P. 177, 198, *204*
Bönicke, R., Reif, W. 176, *204*
Boesman, M., see Gitlin, D. 220, *231*
Bogafrt, M., see McNay, J.L. 321, *324*
Bogdanski, D.F., Blaszkowski, T.P., Brodie, B.B. 286, *308*
Bogoch, A., Davis, T.W., Jow, E., Wrenshall, G.A., 250, *253*
Boitnott, J.R., see Maddrey, W.C. 408, *416*
Bollag, W., see Berneis, K. 329, *338*
Boman, G. 281, 282, *308*
Borella, L., see Pinkel, D. 332, *341*
Boreus, L.O. 230, *231*
Borgó, O. *308*
Borgó, O., Azarnoff, D.L., Forshell, G.P., Sjöqvist, F. 291, 298, 300, *308*
Borgó, O., see Alexanderson, B. [1] 76, *84*, 298, *308* 363, *381*
Borgó, O., see Reidenberg, M.M. 242, *256*, 260, 264, *270*
Borgó, O., see Sjöqvist, F. 300, *313*
Borgstrom *et al.* 1957 131, *146*
Borirakchanyavat, V., see Clark, R. 389, *413*
Borman, A., see Kessler, W.B. 95, 99, *111*
Borowitz, J.L., Moore, P.F., Yim, G.K.W., Miya, T.S. 280, *308*
Bostwick, L., see Evans, J.S. 333, *339*
Boulton, T.B., see Gilbertson, A.A. 251, *253*
Bourne, J.G., Collier, H.O.J., Somers, G.F. 178, *204*
Bovet, D., Bovet-Nitti, F., Guarino, S. Longo, V.G., Marotta, M. 179, *204*
Bovet-Nitti, F. 179, *204*
Bovet-Nitti, F., see Bovet, D. 179, *204*
Boyd, E.M., Bereczky, G.M. 384, 391, *413*
Boyd, M.R., see Wilson, B.J. 396, *419*
Boyer, P.D., Lardy, H., Myrbäck, K. 261, *268*
Boyes, R.N., Adams, H.J., Duce, B.P. 163, *166*
Boyes, R.N., Keenaghan, J.B. 163, *166*
Boyes, R.N., Scott, D.B., Jebson, P.J., Goodman, M.J., Julian, D.G. 163, 164, *166*
Boyes, R.N., see Gibaldi, M. 24, *33*, 284, *310*
Boyland, E., Chasseaud, L.F. 387, 389, *413*
Boyle, C., see Field, J.B. 292, *310*
Boza, A., see Bellet, S. 248, *253*
Bradshaw, M., see Wachstein, M. 223, *233*
Brainerd, H.D., see Jawetz, E. 353, 354, *356*
Branch, R.A., Herbert, C.M., Read, A.E. 288, 293, 307, *308*
Branch, R.A., Nies, A.S., Shand, D.G. 283, 293, 307, *308*
Branch, R.A., Shand, D.G., Nies, A.S. 307, *308*
Branch, R.A., Shand, D.G., Wilkinson, G.R., Nies, A.S. 300, 306, *308*, 363, *381*
Brand, L., see Kuntzman, R.G. 248, *254*
Braunwald, E., see Kaiser, G.A. 321, *323*
Braunwald, E., see Mason, D.T. 251, *255*
Braverman, L.E., see Vagenakis, A.G. 251, *257*
Brazeau, P. 153, 157, *166*
Breckenridge, A., Orme, M. 203, *204*, 285, 288, *308*
Breckenridge, A., Orme, M.L.E., Davies, L., Thorgeirsson, S.S., Davies, D.S. 305, *308*
Breckenridge, A., see Ohnhaus, E.E. 300, *312*, 363, *382*
Bremer, J. 132, *146*
Brennan, R., see Kutt, H. 289, *311*

Brennan, R. W., Dehejia, H., Kutt, H., McDowell, F. 183, *204*
Brennan, R. W., Dehejia, H., Kutt, H., Verebely, K., McDowell, F. 289, *308*
Brent, T. P., see Roberts, J. J. 329, *341*
Brest, A. N., see Seller, R. H. 316, *324*
Briant, Blackwell 161,
Bridges, M. T., see Huisman, T. H. J. 191, *207*
Britt, B. A., Kalow, W. 194, *204*
Broadie, L. L., see Wenzel, D. C. 200, *211*
Brock, N., Hohorst, H. J. 126, *126*
Brod, L. H., see Fife, T. H. 94, *110*
Brodie, B. B. 236, *253*, 278, 279, *308*, 410, *413*
Brodie, B. B., Axelrod, J. 117, *127*
Brodie, B. B., Axelrod, J., Soberman, R., Levy, B. B. 262, *268*
Brodie, B. B., Chang, C. C., Costa, E. 299, *308*
Brodie, B. B., Lowman, E. W., Burns, J. J., Lee, P. R., Chenkin, T., Goldman, A., Weiner, M., Steele, J. M. 157, *166*
Brodie, B. B., Mark, L. C., Papper, E. M., Lief, P. A., Bernstein, E., Rovenstine, E. A. [2] 61, *85*
Brodie, B. B. Mitchell, J. R. 410, *413*
Brodie, B. B., Reid, W. D., Cho, A. K., Sipes, G., Krishna, G., Gillette, J. R. 401, *413*
Brodie, B. B., see Bogdanski, D. F. 286, *308*
Brodie, B. B., see Burns, J. J. 117, 118, *127*, 197, *204*
Brodie, B. B., see Corsini, G. U. 406, *413*
Brodie, B. B., see DiChiara, G. 400, *414*
Brodie, B. B., see Gillette, J. R. 121, *127*, 384, *414*
Brodie, B. B., see Jollow, D. J. 245, *254*, 384, 386, *415*
Brodie, B. B., see Mitchell, J. R. 242, 245, *255*, 378, *382*, 384, 385, 386, 387, 389, 390, 391, 392, 394, 400, 401, *416*, *417*
Brodie, B. B., see Potter, W. Z. 245, *255*, 377, *382*, 390, 411, *417*
Brodie, B. B., see Reid, W. D. 401, *417*
Brodie, B. B., see Soberman, R. 197, *210*
Brodie, B. B., see Weiner, M. 197, *211*
Brodie, D. A., see Hucker, H. B. 131, 139, 140, *147*
Brodwall, E. K., see Myhre, E. 249, *255*
Brodwall, E. K., see Stenbaek, Ø. 249, *256*
Brookes, L. G., see Beckett, A. H. 122, *126*
Brooks, O. G., see Beresford, C. H. 238, *253*
Brooks, S. M., Werk, E. E., Ackerman, S. J., Sullivan, I., Trasher, K. 286, *309*
Broome, J. D. 329, *338*
Brossi, A., see Schwartz, D. E. 118, *128*
Broughton, C., see Cattell, W. R. 353, *355*
Brown, B. R., see Reynolds, E. S. 399, 409, *418*
Brown, H., see Englert, E., Jr. 262, *269*
Brown, P., see Peters, J. H. 176, 177, *209*, 400, *417*
Brown, S. S., see Prescott, L. F. 244, *256*, 389, 417
Brown, T. G., Jr., see Lands, A. M. 317, *323*
Browning, M., see Ballinger, B. 246, *253*
Browning, M., see O'Malley, K. 286, *312*
Bruce, W. R., Meeker, B. E., Valeriote, F. A. 329, *338*
Bruderer, H., see Schwartz, D. E. 118, *128*
Bruff, J. A., see Gunnison, J. B. 350, *356*
Brunetti, P., see Larizza, P. 190, *208*
Brunner, H., Hedwall, P. R., Maitre, L., Meier, M. 123, *127*
Bryan, G. T., see Cohen, S. M. 407, *413*
Bryan, G. T., see Erturk, E. 407, *414*
Bryant, R. E., Hood, A. F., Hood, C. E., Koenig, M. G. 344, *355*
Büch, H., Gerhards, W., Pfleger, K., Rudiger, W., Rummel, W. 392, *413*
Büch, H., Karachristianidis, G., Rudiger, W. 362, *381*
Büch, H., Pfleger, K., Rudiger, W. 362, *381*
Büch, H., see Heymann, E. 401, *415*
Bütler, R., see Aebi, H. 174, *203*
Bulger, R. J. 353, *355*
Bulger, R. J., Kirby, W. M. M. 353, *355*
Bulger, R. J., Roosen-Runge, U. 353, *355*
Bullard, J. C., see Sessions, J. T. 243, *256*
Bulloch, R. T., see Dougherty, J. E. 137, *146*
Bullock, W. E., see Hoffman, T. A. 265, *269*
Bunn, P. A., see McDermott, W. 153, *167*
Burchall, J. J., see Hitchings, G. H. 328, *340*
Burchell, H. B. 158, *166*
Burchenal, J. H., Dollinger, M. R. 333, *339*
Burgert, E. O., Jr., Glidewell, O. 335, *339*
Burke, C. W., see Lewis, G. P. 242, *254*
Burkitt, D. 335, *339*
Burley, D. M., see Las, S. 397, *416*
Burnet, W. S., see Greenwald, P. 406, *414*
Burns, J. J., Rose, R. K., Chenkin, T., Goldman, A., Schulert, A., Brodie, B. B. 197, *204*
Burns, J. J., Salvador, R. A., Lemberger, L. 122, *127*
Burns, J. J., Yu, T. F., Berger, L., Gutman, A. B. 115, *127*
Burns, J. J., Yu, T. F., Dayton, P. G., Gutman, A. B., Brodie, B. B. 116, 116, *127*
Burns, J. J., Yu T. F., Ritterbrand, A., Perel, J. M., Gutman, A. B., Brodie, B. B. 118, *127*
Burns, J. J., see Brodie, B. B. 157, *166*
Burns, J. J., see Chen, W. 288, 301, *309*
Burns, J. J., see Conney, A. H. 117, *127*
Burns, J. J., see Cucinell, S. A. 285, *309*, 364, *381*
Burns, J. J., see Welch, R. M. 301, *314*
Burstein, S., Klaiber, E. L. 286, *309*
Burzynski, S. 258, *268*
Burzynski, S., see Czerniak, Z. 258, *269*

Busfield, D., Child, K. J., Atkinson, R. M., Tomich, E. G. 278, *309*
Bush, C. A., see Caldwell, J. H. 131, 137, *146*, 281, *309*
Bush, G. H., Roth, F. 180, *204*
Butcher, R. W., see Robison, G. A. 317, *324*
Butler, A. M., Richie, R. H. 214, *231*
Butler, C., see Horning, M. G. 402, *415*
Butler, T. 217, *231*
Butler, T. C. [3] 57, *85*, 115, *127*, 155, *166*, 182, 183, *204*
Butler, T. C., Mahaffee, C., Waddell, W. J. 266, *268*
Butler, T. C., see Waddell, W. J. 240, *257*, 361, *382*
Butler, U. P., Jr., see Schmidt, D. H. 317, *324*
Butler, V. P., Jr. see Smith, T. W. 317, *324*
Butterwordt, M., see Allansmith, M. 220, *230*
Buzello, W., see Heymann, E. 401, *415*
Bye, A., see Conney, A. H. 165, 200, *205*, 243, *253*
Byrd, R. B., Nelson, R., Elliot, R. C. 409, *413*

Caesar, J., Shaldon, S., Chiandussi, L., Guevara, L., Sherlock, S. 301, *309*
Calazel, P., see Bing, R. J. 315, *322*
Calcagno, P. L., Rubin, M. I. 225, *231*
Calder, I. C., Creek, M. J., Williams, P. J. 392, *413*
Caldwell, H. C., Adams, H. J., Jones, R. G., Mann, W. A., Dittert, L. W., Chong, C. W., Swintosky, J. V. 95, *109*
Caldwell, H. C., Adams, H. J., Rivard, D. E., Swintosky, J. V. 95, 101, *109*
Caldwell, H. C., see Dittert, L. W. 88, 93, 94, 95, 102, *109*, *110*
Caldwell, H. C., see Swintosky J. V. 95, 101, 102, *112*
Caldwell, J. H., Bush, C. A., Greenbergen, J. J. 131, 137, *146*
Caldwell, J. H., Bush, C. A., Greenberger, N. J. 281, *309*
Caldwell, J. H., Dring, L. G., Williams, R. T. 141, *146*
Caldwell, J. H., Greenberger, N. J. 137, *146*, 281, *309*
Caldwell, R. W., Goldberg, L. I. 321, *322*
Calimlim, L. R., see Beckett, A. H. 152, *166*
Calimlim, L. R., see Bianchine J. R. 238, *253*, 281, *308*
Callahan, M., see Yesair, D. W. 140, *149*
Callahan, S., see Elion, G. B. 328, 337, *339*
Callender, S. T. E., see Dubach, R. 259, *269*
Cambridge, G. W., see Agersborg, H. P. K. 95, 96, *109*
Campbell, D. E. S., see Schröder, H. 235, 239, *256*
Canellos, G. P., Devita, V. T., Whang-Peng, J., Carbone, P. P. 334, *339*
Canellos, G. P., see Bagley, C. M., Jr. 325, 335, *338*
Canellos, G. P., see Devita, V. T., Jr. 335, *339*
Cangiano, J., see Seller, R. H. 316, *324*
Cantwell, N. H. R., see Hucker, H. B. 131, 139, 140, *147*
Capizzi, R. L., Nichols, R., Mullins, J. 338, *339*
Capizzi, R. L., Summers, W. P., Bertino, J. R. 338, *339*
Carbone, J. J., see Schwartz, M. A. 120, *128*
Carbone, P. P., Spurr, C. 334, *339*
Carbone, P. P., see Canellos, G. P. 334, *239*
Carbone, P. P., see Devita, V. T., Jr. 325, 334, *339*
Carey, R. W. 333, *339*
Carliner, N. A., Denune, D. P., Finch, C. S., Jr., Goldberg, L. I. 320, *322*
Carliner, N. H., Gilbert, C. A., Pruitt, A. W., Goldberg, L. I. 317, *322*
Carlson, H. B., Anthony, E. M., Russell, W. F., Jr., Middlebrook, G. 176, *204*
Carlsson, A., Corrodi, H., Fuxe, K., Hökfelt, T. 121, *127*
Carlsson, A., Fuxe, K., Hamberger, B., Lindqvist, M. 121, *127*
Carmody, S., see Perry, H. M., Jr. 178, *209*
Carnes, H. E., see Glazko, A. J. 95, 98, 100, *110*
Carpenter, F. H. 95, 96, *109*
Carpenter, O. S., see Wagner, J. G. *34*
Carr, G., see Sellers, E. M. 290, 302, *313*
Carrara, C., see Jori, A. 121, *128*
Carrella, M., see Hunter, J. 202, *207*, 286, *311*
Carroll, D. I., see Horning, E. C. 230, *231*
Carson, S. N., see Nishimura, E. T. 174, *208*
Carter, C. H., see Davies, J. E. 288, *310*
Carter, G. G., see Whelton, A. 240, *257*, 260, *271*
Carter, S., see Yahr, M. D. 182, *212*
Carulli, N., Manenti, F., Gallo, M., Salvioli, G. F. 288, 302, *309*
Casarett, L. J., see Baselt, R. C. 135, *145*
Cascorbi, H. F., Vesell, E. S., Blake, D. A., Helrich, M. 198, *204*
Casey, J. I., see Sabath, L. D. 266, *271*
Castle, M. C., Lage, G. L. 137, *146*
Casy, A. F., see Kupchan, S. M. 95, 99, 100, *111*
Cavalieri, R. R., Sung, L. C., Becker, C. E. 286, *309*
Cavanaugh, J. H., see Mitchell, J. R. 299, 300, *312*, 389, *416*
Celmer, W. D. 95, 97, *109*
Celmer, W. D., Els, H., Murai, K. 95, 97, *109*
Ceriotti, G., de Franceshi, A., de Carneri, I., Zamboni, V. 95, 100, *109*
Cerletty, J. M., Engbring, N. H. 262, *269*
Cestero, R., see Hoffman, T. A. 265, *269*
Cattell, W. R., Chamberlain, D. A., Fry, I. K., McSherry M. A., Broughton, C., O'Grady, F. 353, *355*
Chamberlain, D. A., see Cattell, W. R. 353, *355*
Chamberlain, D. A., see White, R. J. 225, *257*
Chang, C. C., see Brodie, B. B. 299, *308*
Chang, R., see Conney, A. H. 169, 200, *205*, 243, *253*
Chang, T. W., Weinstein, L. 350, *355*
Chaparas, S. D., see Dove, J. T. 408, *414*
Chaplin, H., Jr., see Perry, H. M., Jr. 178, *209*

Chaput de Saintonge, D. M. Herxheimer, A. 279, *309*
Charytan, C. 131, 140, 141, *146*
Chasis, H., see West, J. R. *233*
Chasseaud, L. F., see Boyland, E. 387, 389, *413*
Chavelier, L., Glidewell, O. 332, *339*
Chen, W., Vrindten, P. A., Dayton, P. G., Burns, J. J. 288, 301, *309*
Chenkin, T., see Brodie, B. B. 157, *166*
Chenkin, T., see Burns, J. J. 197, *204*
Chenkin, T., see Dayton, P. G. 285, *310*
Cherrick, G. R., Stein, S. W., Leevy, C. M., Davidson, C. S. 301, *309*
Chiandussi, L., see Caesar, J. 301, *309*
Chidsey, C., see Gilmore, E. 320, *332*
Chignell, C. F. 230, *231*
Chignell, C. F., Vesell, E. S., Starkweather, D. K., Berlin, C. M. 197, *204*, 241, *253*
Child, K. J., see Busfield, D. 278, *309*
Cho, A. K., see Brodie, B. B. 401, *413*
Chodos, R. B., see Webb, D. I. 239, *257*, *314*
Choi, Y., Thrasher, K., Werk, E. E., Jr., Sholiton, L. J., Olinger, C. 286, 287, *309*
Choi, Y., see Werk, E. E., Jr. *314*
Chong, C. W., see Caldwell, H. C. 95, 108, 109, *109*
Chong, C. W., see Dittert, L. W. 88, 95, 102, 103, 104, *109*, *110*
Chong, C. W., see Swintosky, J. V. 95, 101, 102, *112*
Chong, C. W. S. 92, 95, 97, *109*
Chow, A. W., see Winters, R. E. 350, *358*
Christensen, F. 184, *204*
Christensen, L. K., Hansen, J. M., Kristensen, M. 289, 297, 302, *309*
Christensen, L. K., Skovsted, L. 289, *309*
Christensen, L. K., see Hansen, J. M. 288, 289, *310*
Christie, B., see Reid, W. D. 401, *417*
Christie, B. B., see Mitchell, J. R. 401, *416*
Christopher, T. G., see Cutler, R. E. 262, *269*
Chu, M.-Y., Fischer, G. A. 329, *339*
Clark, A. G., Fischer, L. J., Millburn, P., Smith, R. L., Williams, R. T. 131, 138, *146*
Clark, B. B., see Levine, R. M. 88, 89, *111*
Clark, D. W., Goldberg, L. I. 320, *323*
Clark, R., Borirakchanyavat, V., Davidson, A. R., Thompson, R. P. H. 389, *413*
Clark, S. W., Glaubiger, G. A., Ladu, B. N. 179, *204*
Clarke, C. A., see Kitchin, F. D. 192, *207*
Clarke, D. A. 326, *339*
Clarke, R. A., Julian, D. G., Nimmo, J., Prescott, L. F., Talbot, R. 245, *253*
Clarkson, B., Ohkita, T., Ota, K., Fried, J. 333, *339*
Clarkson, B. D., see Gee, T. S. 333, *340*
Clayson, D. B. 405, 406, 407, *413*
Clements, F. W., Wishart, J. W. 192, *204*
Clifton, J. A., see Schedl, H. P. 239, *256*
Clisby, K. H., see Richter, C. P. 192, *209*
Clubb, R. J., see Tobey, R. E. 251, *256*
Cluff, L. E., Thornston, G., Seidl, L., Smith, J. 173, *205*
Cluff, L. E., Thornston, G. F., Seidl, L. G. 343, *355*
Cockburn, T., see Harris, F. W. 181, *206*
Coggins, C. H., see Bennett, W. M. 259, 261, 266, *268*
Cohen, B., see Arias, I. M. 287, *308*
Cohen, B. D., Spritz, N., Lubash, G. D., Rubin, A. L. 316, *323*
Cohen, J. L., see Jao, J. W. 290, *311*
Cohen, S., see Erturk, E. 407, *414*
Cohen, S., see Kabins, S. A. 264, *270*
Cohen, S. M., Bryan, G. T. 407, *413*
Cohen, S. M., see Erturk, E. 407, *414*
Cohen, S. N., Weber, W. W. 169, *205*
Cohen, S. N., see Weber, W. W. 177, 178, *211*
Cohen, S. S., Flaks, J. G., Barner, H. D., Loeb, M. R., Lichtenstein, J. 328, *339*
Cohn, M. I., Middlebrook, G., Russell, W. R., Jr. 344, *355*
Colaizzi, J. L., see Jaffe, J. M. 235, *254*
Coleman, N., see Eyles, D. E. 353, *356*
Coleman, V. R., see Gunnison, J. B. 349, *356*
Coleman, V. R., see Jawetz, E. 346, 347, 348, 352, 354, *356*
Coleman, V. R., see Miles, C. P. 352, *357*
Coles, G. A., Peters, D. K., Jones, J. H. 260, *269*
Coles, F. K., see Huf, E. G. 95, 104, *111*
Collier, H. O. J., see Bourne, J. G. 178, *204*
Coltart, D. J., Shand, D. G. 164, 165, *166*
Comfort, M., see Medoff, G. 353, *357*
Conn, H. J. 409, *413*
Conney, A. H. 203, *205*, 283, *309*, 362, *381*
Conney, A. H., Kuntzman, R. 258, *269*
Conney, A. H., Sansur, M., Soroko, F., Koster, R., Burns, J. J. 117, *127*
Conney, A. H., Welch, R., Kuntzman, R., Chang, R., Jacobson, M., Munrofaure, A. D., Peck, A. W., Bye, A. Poland, A., Poppers, P. J., Finster, M., Wolff, J. A. 169, 200, *205*, 243, *253*
Conney, A. H., see Cucinell, S. A. 285, *309*, 364, *381*
Conney, A. H., see Gilman, A. G. 363, *381*
Conney, A. H., see Kuntzman, R. 202, *207*
Conney, A. H., see Pantuck, E. J. 363, *382*
Conney, A. H., see Welch, R. M. 288, 301, *314*
Connors, K. A., see Shah, A. A. 95, 102, *111*
Connors, T. A., Cox, P. J., Farmer, P. D., Forster, A. D., Jarman, M. 290, *309*
Connors, T. A., Jones, M. 338, *339*
Conolly, M. E., Davies, D. S., Dollery, C. T., Morgan, C. D., Paterson, J. W., Sandler, M. 159, *166*

Conolly, M.E., see Dollery, C. T. 124, *127*, 159, 164, 165, *166*, 318, *323*
Conolly, M.E., see George, C. F. 293, *310*
Conolly, M.E., see Paterson, J.W. 162, 164, *167*
Constantino, R.T., see Stenson, R.E. 163, *168*, 307, *314*, 363, *382*
Cook, D.A., Hagerman, L.M., Schneider, D.L. 133, *146*
Cook, J.D., see Eschbach, J. W. 259, *269*
Cooke, A.R., Hunt, J.N. 279, *309*
Cooper, J.R., see Mackay, F. J. 114, *128*
Cooper, J.R., see Weiner, M. 197, *211*
Cooper, P.D., see Rowley, D. 353, *357*
Cooper, R.G. 325, 329, 336, *339*
Cooper, R.G., see Hayes, A. 164, *167*
Cooper, R.G., see Paterson, J.W. 164, *167*
Corey, W.T., see Anton, A.H. 241, 242, *252*, 260, *268*
Corn, M. 285, 287, 306, *309*
Corn, M., Rockett, J.F. 285, *309*
Corrodi, H., see Carlsson, A. 121, *127*
Corsini, G.U., Krishna, G., Brodie, B.B., Gillette, J. R. 406, *413*
Costa, E., see Brodie, B.B. 299, *308*
Costa, J., see Hayes, D.M. 330, *340*
Costio, J.R., see Wilbur, J.R. 336, *342*
Cote, R., see Vagenakis, A.G. 251, *257*
Cotten, M.DeV. 230, *231*
Court, J.M., Dunlop, M.E., Leonard, R.F. 221, *231*
Covino, B.G. 234, *253*
Cox, P.J., see Connors, T.A. 290, *309*
Cox, S.V., see Huckes, H.B. 131, 139, 140, *147*
Craig, W.A., see Kunin, C.M. 222, *232*
Cram, D.J., Hammond, G.S. 95, 101, *109*
Crane, C.A., see Moore, R.B. 133, *148*
Cranley, J.J., Krause, R.J., Strasser, E.S., Hafner, C. D. 320, *323*
Crathorn, A.R., see Roberts, J.J. 329, *341*
Creasey, W.A. 329, 336, *339*
Creaven, P.J., Allen, L.M. 249, *253*
Credner, K., see Holtz, P. 321, *323*
Creek, M.J., see Calder, I.C. 392, *413*
Crew, M.C., Gala, E.L., Haynes, L.J., Dicarlo, F.J. 142, *146*
Crigler, J.F., Jr., Gold, N.I. 287, *309*
Crooks, J., see O'Malley, K. 200, *208*, 243, 244, *255*
Crossley, G., see Las, S. 397, *416*
Crounse, R.G. 236, *253*, 279, *309*
Crovetti, A.J., see Erturk, E. 407, *414*
Crowley, J.M., see Baylis, E. M. 286, *308*
Crozier, D.N., see Silverman, W.A. 297, *313*
Cuatrecasas, P. 230, *231*
Cucinell, E.A., see Finlay, G. D. 319, *323*
Cucinell, S.A., Conney, A.H., Sansur, M., Burns, J.J. 285, *309*, 364, *381*
Culp, H.W., see McMahon, R. E. 116, 125, *128*
Cunningham, M.D., Mace, J. W., Peters, E.R. 287, *309*
Curry, Marshall 197
Curry, J.H., Jr., see Okita, G. T. 136, *148*
Curry, S.H. 162, *166*, 298, *309*
Curry, S.H., Davis, J.M., Janowsky, D.S., Marshall, J. H.L. 286, *309*
Curry, S.H., D'Mello, A., Mould, G.P. 162, *166*
Curry, S.H., see Hollister, L. E. 162, *167*
Curtis, J.R., Marshall, M.J. 249, *253*
Cutler, R., see Gyselynck, A. M. 266, *269*
Cutler, R.E., Forrey, A.W., Christopher, T.G., Kimpel, B.M. 262, *269*
Cutler, R.E., Orme, B.M. 248, *253*
Cutler, R.E., see Orme, B.M. 260, *270*
Czerniak, Z., Burzynski, S. 258, *269*

Dahm, P.A., see Orgell, W.H. 181, *208*
Daley, R., see Bing, R.J. 315, *322*
Dalrymphe, G.L., see Dougherty, J.E. 137, *146*
Daly, J.W., Jerina, D.M., Witkop, B. 378, 379, *381*, 390, 402, *413*
Dammann, J.F., Jr., see Bing, R.J. 315, *322*
Dans, P.E., see Smith, C.B. 344, *357*
Danysz, A., Wisniewski, K. 282, *309*
D'Arconte, L., see Koechlin, B.A. 118, *128*
Darby, C.W., see Hardisty, R.M. 331, *340*
Darken, M.A., see MacDonald H. 235, *255*
Das, K.M., Eastwood, M.A. 239, *253*
D'Ascensio, I.-L., see Grunberg, E. 176, *206*
Dascomb, H.E., see Bailey, W.C. 409, *413*
Dasdia, T., see diMarco, A. 329, *339*
Dash, C.H., see Gower, P.E. 235, *254*
Daskalakis, E.G., see Hanahan, D.J. 138, *147*
Dauben, J.H., Jr., see Hanahan, D.J. 138, *147*
Davenport, H.W. 94, *109*
Davidov, M., see Mroczek, W. J. 154, *167*
Davidson, A.R., see Clark, R. 389, *413*
Davidson, C.S., see Cherrick, G.R. 301, *309*
Davidson, I.W.F., Rollins, F. O., DiCarlo, F.J., Miller, H.S., Jr. 152, *166*
Davidson, M. 216, *231*
Davies, see Thorgeirsson 151
Davies, D., see Kattamis, C. 180, *207*
Davies, D.S., Thorgeirsson, S. S. 202, *205*
Davies, D.S., see Conolly, M. E. 159, *166*
Davies, D.S., see Dollery, C. T. 124, *127*, 159, 164, 165, *166*, 318, *323*
Davies, D.S., see Ohnhaus, E. E. 300, *312*, 363, *382*
Davies, D.S., see Paterson, J. W. 162, *167*
Davies, J.E., Edmundson, W. F., Carter, C.H., Barquet, A. 288, *310*
Davies, L., see Breckenridge, A. 305, *308*
Davies, R.O., Marton, A.V., Kalow, W. 179, *205*
Davies, R.O., see Kalow, W. 179, *207*

Davis, A. E., Pirola, R. C. 239, *253*
Davis, B. D., see Plotz, P. H. 352, *357*
Davis, D. C., Potter, W. Z., Jollow, D. J., Mitchell, J. R. 384, 391, *413*
Davis, D. C., see Jollow, D. J. 245, *254*, 384, 386, *415*
Davis, D. C., see Mitchell, J. R. 242, 245, *255*, 378, *382*, 384, 386, 390, 391, 392, 394, *416*
Davis, D. C., see Potter, W. J. 245, *255*, 377, *382*, 390, 411, *417*
Davis, D. S., see Breckenridge, A. 305, *308*
Davis, J. M., Kopin, I. J., Lemberger, L., Axelrod, J. 291, *310*
Davis, J. M., see Curry, S. H. 286, *309*
Davis, J. M., see Fann, W. E. 299, 300, *310*
Davis, T. W., see Bogoch, A. 250, *253*
Davison, A. N. 122, *127*
Day, A. R., see Gordon, M. 89, *110*
Day, R., see Johnson, L. 297, *311*
Dayton, P. G., Perel, J. M. 29, *33*
Dayton, P. G., Tarcan, Y., Chenkin, T., Weiner, M. 285, *310*
Dayton, P. G., see Burns, J. J. 117, *127*
Dayton, P. G., see Chen, W. 288, 301, *309*
Dayton, P. G., see Garrettson, L. K. 289, *310*
Dean, M. A., see Tobey, R. E. 251, *256*
deBroe, M., see Michielsen, P. 391, *416*
deCarli, L. M., see Lieber, C. S. 196, *208*
deCarneri, I., see Ceriotti, G. 95, 100, *109*
Decker, J. L., see Steinberg, A. D. 405, *418*
Deconti, R. C., see Levitt, M. 335, *341*
Dedrick, R. L., see Bischoff, K. B. 24, *33*
Defendi, V., Manson, L. A. 333, *339*
deFranceshi, A., see Ceriotti, G. 95, 100, *109*
deFreudenreich, J., see Fabre J. 266, *269*
deFreudenreich, J., see Richet, G. 261, 267, *271*
DeGoodman, W. S., see Smith, F. R. 261, *271*
deGuzman, N. T., see Yeh, B. K. 317, *324*
Dehejia, H., see Brennan, R. W. 183, *204*, 289, *308*
Dehejia, H., see Kutt, H. 289, *311*
Dehn, F., see Mitoma, C. 200, *208*
Deigh, R. A., see Hewitt, W. L. 353, 354, *356*
DeLuca, H. F. 261, *269*
deMaine, J. B., see Kirby, W. M. M. 264, 266, *270*
deMaine, J. B., see Standiford, H. D. 352, 354, *358*
deMarco, S., see diMarco, A. 329, *339*
DeMatteis, F. 404, *413*
Denbesten, L. 131, *146*
Denborough, M. A., Forster, J. F. A., Lovell, R. R. H., Maplestone, P. A., Villiers, J. D. 194, *205*
Denborough, M. A., see King, J. O. 252, *254*
Denning, G. S., Jr., see Pals, D. T. 320, *324*
Dennis, E. W., see Rosi, D. 126, *128*
Denune, D. P., see Carliner, N. A. 320, *322*
Deren, J. J., see Toskes, P. P. 281, *314*
Dern, R. J., Beutler, E., Alving, A. S. 190, *205*
Derr, J. E., see Hollister, L. E. 162, *167*
Desbaillets, L., see Menguy, R. 246, 247, *255*
deSchepper, P. J., see Michielsen, P. 391, *416*
DeSilva, J. A. F., Puglisi, C. V. 120, *127*
deTaveau, R. M., see Hunt, R. 178, *207*
Dettli, L. 9, *33*, 266, 267, *269*
Dettli, L., Spring, P. 265, *269*
Dettli, L., Spring, P., Habersang, R. 248, *253*, 265, *269*
Dettli, L., Spring, P., Ryter, S. 265, 266, *269*
Devadatta, S., Gangadharam, P. R. J., Andrews, R. H., Fox, W., Ramakrishnan, C. V., Selkon, J. B., Velu, S. 176, 177, *205*
Devita, V. T., see Bagley, C. M., Jr. 325, *338*
Devita, V. T., see Canellos, G. P. 334, *339*
Devita, V. T., see Lowenbraun, S. 335, *341*
Devita, V. T., Jr., Canellos, G. P. 335, *339*
Devita, V. T., Jr., Serpick, A. A., Carbone, P. P. 325, 334, *339*
Dewald, B., see Aebi, H. 174, *203*
Dhumeaux, D., see Berthelot, P. 144, *146*
Diamond, J. A., see Hampers, C. L. 262, *269*
Diamond, L., see Doluisio, J. T. 236, *253*
Diamond, L., see Wiebel, F. J. 363, *382*
DiCarlo, F. J., Viau, J. P. 121, *127*
DiCarlo, F. J., Viau, J. P., Epps, J. E., Haynes, L. J. 121, *127*
DiCarlo, F. J., see Crew, M. C. 142, *146*
DiCarlo, F. J., see Davidson, I. W. F. 152, *166*
DiChiara, G., Hinson, J. A., Potter, W. Z., Jollow, D. J. Mitchell, J. R. 400, 401, *414*
DiChiara, G., Potter, W. Z., Jollow, D., Mitchell, J. R., Gillette, J. R., Brodie, B. B. 400, *414*
Dickens, F. 407, *414*
Dicker, S. E. *231*
Dickhaus, D. W., see Goss, J. E. 285, *310*
Dienst, J., see Heaton, A. D. 344, *356*
Dietschy, J. M. 131, 132, 133, *146*
Digenis, G., see Swintowsky, J. V. [18] 39, *85*
Dill, W. A., see Glazko, A. J. 95, 98, *110*
diMarco, A., Silvestrini, R., deMarco, S., Dasdia, T. 329, *339*
Dinan, B., see Jick, H. 196, *207*
Dingell, Gillette 40
Dingell, J. V., Sulser, F., Gillette, J. R. 390, *414*
Dingell, J. V., see Gillette, J. R. 121, *127*
Dingell, J. V., see Mitchell, J. R. 389, *416*
Dittert, L. W., Adams, H. J., Alexander, F., Chong, C. W., Ellison, T., Swintosky, J. V. 95, 103, *110*
Dittert, L. W., Adams, H. J., Chong, C. W., Swintosky, J. V. 95, 104, *110*

Dittert, L. W., Caldwell, H. C., Adams, H. J., Irwin, G. M., Swintosky, J. V. 88, 95, 102, *109*
Dittert, L. W., Caldwell, H. C., Ellison, T., Irwin, G. M., Rivard, D. E., Swintosky, J. V. 93, 94, 95
Dittert, L. W., Irwin, G. M., Chong, C. W., Swintosky, J. V. 88, 95, 102, *109*
Dittert, L. W., Irwin, G. M., Rattie, E. S., Chong, C. W., Swintosky, J. V. 95, 102, *110*
Dittert, L. W., see Caldwell, H. C. 95, 108, 109, *109*
Dittert, L. W., see Rattie, E. S. 95, 102, *111*
Dittert, L. W., see Swintosky, J. V. 95, 101, 102, *112*
Dixon, J. E., see Venter, J. C. 230, *233*
D'Mello, A., see Barrowman, J. A. 281, *308*
D'Mello, A., see Curry, S. H. 162, *166*
Dobbs, H. E., Hall, J. M. 131, 135, *146*
Dobbs, H. E., Hall, J. M., Steiger, B. 131, 136, *146*
Dodds, M. G., Foord, R. D. 393, *414*
Doherty, J. E. 317, *323*
Doi, K., see Nishimura, E. T. 174, *208*
Doi, K., see Takahara, S. 173, *210*
Doi, M., see Takahara, S. 173, *210*
Doll, R., see Speizer, F. E. 162, *168*
Dollery, C. T. 154, *166*
Dollery, C. T., Davies, D. S., Conolly, M. E. 124, *127*, 159, 164, 165, *166*, 318, *323*
Dollery, C. T., see Conolly, M. E. 159, *166*
Dollery, C. T., see George, C. F. 293, *310*
Dollery, G. T., see Paterson, J. W. 162, 164, *167*
Dollinger, M. R., see Burchenal, J. H. 333, *339*
Doluisio, J. T., Tan, G. H., Billups, N. F., Diamond, L. 236, *253*
Doluisio, J. T., see Kojima, S. 236, *254*, 279, *311*
Donaldson, R. M., Jr. 215, *231*
Done, A. K. 116, *127*, 214, *231*
Donelli, M. G., Garattini, S. 338, *339*
Donno, L., Sanguineti, V., Ricciardi, M. L., Soldati, M. 353, *356*
Dossetor, J. B., see Gault, M. H. 391, *414*
Dost, F. H. 14, 15, *33*
Dougherty, J. E., Flanagan, W. J., Murphy, M. L., Bulloch, R. T., Dalrymphe G. L., Beard, O. W., Perkins, W. H. 137, *146*
Dougherty, T. F. 329, *339*
Dove, J. T., Chaparas, S. D., Hedrick, S. R. 408, *414*
Dowling, H. F. 354, *356*
Dowling, H. F., see Lepper, M. H. 345, 350, *357*
Dowling, R. H. 131, 132, *146*
Dowling, R. H., Mack, E., Picott, J., Berger, J., Small, D. M. 131, *146*
Draper, A., Jr., see Bing, R. J. 315, *322*
Dreifus, L. S., Durate, C., Kodama, R., Moyer, J. H. 250, *253*
Dresel, P. E., Slater, I. H. 95, 104, 106, *110*
Dressler, S. H., see Russell, W. F., Jr. 344, *357*
Dreyer, G., Walker, E. W. A. 214, *231*
Dring, L. G., see Caldwell, J. H. 141, *146*
Dring, L. G., see Reidenberg, M. M. 246, 247, *256*
Druckrey, H. 398, *414*
Drummond, K. H., Hellman, D. A., Marchessault. J. H. V., Feldman, W. 290, *310*
Dubach, R., Callender, S. T. E. Moore, C. V. 259, *269*
Dubach, U. C. 264, *269*
Dubach, U. D., Levy, P. S., Minder, F. 391, *414*
Dubois, K. P., see Tardiff, R. G. 338, *341*
Dubois, R., see McDermott, W. 153, *167*
Duce, B. P., see Boyes, R. N. 163, *166*
Duce, B. R., see Smith, E. R. 123, *129*
Duckert, A., see Fabre, J. 266, *269*
Dujuvne, C. A., see Bianchine J. R. 152, *166*, 238, *253*, 218, *308*
Duke, E., see O'Malley, K. 200, *208*, 243, 244, *255*
Dulin, W. E., Gerritsen, G. C. 125, *127*
Dulin, W. E., see Smith, D. L. 125, *129*
Dumas, K., see Shubin, H. 95, 97, *111*
Dundee, J. W. 252, *253*
Dundee, J. W., Richards, R. K. 242, *253*, 264, *269*
Dunlop, D., Shanks, R. G. 317, *323*
Dunlop, M. E., see Court, J. M. 221, *231*
Durant, J. R., see Lacher, M. J. 334, *341*
Durante, P., see Letteri, J. M. 249, *254*, 264, *270*
Durate, C., see Dreifus, L. S. 250, *253*
Dvorak, G., see Schmid, E. 259, *271*
Dymling, J. F., Nilsson, K. O., Hokfelt, B. 403, *414*

Eastwood, M. A., see Das, K. M. 239, *253*
Ebadi, M. S., Kugel, R. B. 242, 245, *253*
Eble, J. N. 321, *323*
Eckstein, J. W., see Schuelke, D. M. 321, *324*
Edelstein, D., see Salatka, K. 220, *233*
Edgerton, W. H., see Glatzko, A. J. 95, 98, *110*
Edmundson, W. F., see Davies, J. E. 288, *310*
Edrada, L. S., see O'Connor, W. J. 216, *233*
Edwards, J. A., Price Evans, D. A. 195, *205*
Edwards, K. D. G. 262, *269*
Edwards, L. J. 91, *110*
Edwards, T., see Hanahan, D. J. 138, *147*
Eger, E. I. II., Larson, C. P., Jr. 158, *167*
Ehrenberg, L. 123, *127*
Eickhoff, T. C. 353, *356*
Eiden, W., see Hahn, K. J. 281, *310*
Eisenbrandt, L. L., Adler, T. K., Elliott, H. W., Abdou, I. A. 135, *146*
Ekman, H., see Bengtsson, U. 406, *413*
Elder, H. A., see Sabath, L. D. 352, 354, *357*
Elias, R. A. 185, *205*
Elion, G. B., Callahan, S., Nathan, H., Bieber, S., Rundles, R. W., Hitchings, G. H. 328, 337, *339*
Elion, G. B., Kovensky, A., Hitchings, G. H., Metz, E., Rundles, R. W. 118, *127*

Elion, G. B., Singer, S., Hitchings, G. H. *339*
Ellard, G. A., Gammon, P. T., Wallace, S. M. 177, *205*
Ellard, G. A., Garrod, J. M. B., Scales, B., Snow, G. A. 153, *167*
Elliot, R. W., Kerrand, D. N. S., Lewis, A. A. G. 396, *414*
Elliott, H. W., see Eisenbrandt, L. L. 135, *146*
Elliott, H. W., see March, C. H. 135, *148*
Elliott, R. C., see Byrd, R. B. 409, *413*
Ellison, T., see Dittert, L. W. 93, 94, 95, 103, *110*
Ellison, T., see Swintosky, J. V. 95, 101, *112*
Ellsworth, R. M., see Hyman, G. A. 336, *340*
Els. H., see Celmers, W. D. 95, 97, *109*
Elslager, E. F., Gavrilis, Z. B. Phillips, A. A., Worth, D. F. 95, 107, *110*
Elslager, E. F., Phillips, A. A. Worth, D. F. 95, 107, 108, *110*
Elslager, E. F., see Worth, D. F. 95, 107, *112*
Embden-Myerhof 189
Emery, J. L., see MacDonald, M. S. 223, *232*
Endo, H., Kume, F. 405, *414*
Engbring, N. H., see Cerletty, J. M. 262, *269*
Engers, W. D., see Gault, M. H. 391, *414*
Englert, E., Jr., Brown, H., Willardson, D. G., Wallach, S., Simons, E. L. 262, *269*
Enna, S. J., Shanker, L. S. 155, *167*
Enomoto, M., see Miller, E. C. 391, *416*
Epler, G. R., see Black, M. 397, 408, *413*
Epps, J. E., see DiCarlo, F. J. 121, *127*
Epstein, S. E., see Levey, G. S. 318, *323*
Epstein, S. S., Andrea, J., Joshi, S., Mantel, N. 407, *414*
Epstein, W. L., see Riegelman, S. 278, 301, *313*
Eriksson, H. 131, 138, *146*
Erlanger, F. B. 407, *414*
Erlinger, S., see Berthelot, P. 144, *146*
Erturk, E., Cohen, S. M., Price, J. M., Bryan, G. T. 407, *414*
Erturk, E., Price, J. M., Morris, J. E., Cohen, S., Leith, R. S., von Esch, A. M., Crovetti, A. J. 407, *414*
Eschbach, J. W., Cook, J. D., Finch, C. A. 259, *269*
Esplin, D. W., see Woodbury, D. M. 182, *212*
Estrich, D., see Wood, P. 133, *149*
Eubank, L. L., see Huf, E. G. 95, 104, *111*
Evans, A. E., Heyn, R. M., Newton, W. A., Jr., Leikin, S. L. 333, 336, *339*
Evans, F. T., Gray, P. W. S., Lehmann, H., Silk, E. 178, *205*
Evans, G. H., Nies, A. S., Shand, D. G. 276, 297, *310*
Evans, G. H., see Nies, A. S. 293, *312*
Evans, G. H., see Shand, D. G. 164, *168*
Evans, J. S., Musser, E. A., Bostwick, L., Mengel, G. D. 333, *339*
Evans, M. E., see Walker, S. R. 155, *168*
Evans, R. M., see Barnden, R. L. 95, *109*
Evard, E., see Eyssen, H. 134, *146*
Eviator, L., see Szeinberg, A. 181, *210*
Eyles, D. E., Coleman, N. 353, *356*
Eyssen, H., Evard, E., Vanderhaeghe, H. 134, *146*

Fabre, J., de Freudenreich, J., Duckert, A., Pitton, J. S., Rudhart, M., Virieux, C. 266, *269*
Fabre, J., see Richet, G. 267, *271*
Facklam, R. R., see Wilkowske, C. J. 353, *358*
Falco, F. G., Smith, H. M., Arcieri, G. M. 337, *340*
Falk, H., see MacDonald 235, *255*
Falkson, G., see van Eden, E. B. 330, *342*
Falkson, H. C., see van Eden, E. B. 330, *342*
Fallon, R. J., see Lees, A. W. 399, *416*
Faloon, W. W. 281, *310*
Faloon, W. W., see Webb, D. I. 239, *257*, *314*
Fanelli, R., see Kvetina, J. 119, *128*
Fanelli, R., see Marcucci, F. 119, *128*
Fann, W. E., Davis, J. M., Janowsky, D. S., Oates, J. A. 299, *310*
Fann, W. E., Kaufmann, J. S. Griffith, J. D., Davis, J. M. Janowsky, D. S., Oates, J. A. 299, 300, *310*
Farah, A., Witt, P. N. 316, *323*
Farah, A., see Stanburry, J. B. 316, *324*
Farber, L. R., see Levitt, M. 335, *341*
Farmer, P. D., see Connors, T. A. 290, *309*
Feagin, O. T., see Oates, J. A. 300, *312*
Feind, C. R., see Hyman, G. A. 336, *340*
Feingold, D. S., see Greenfield, S. 353, *356*
Feingold, M., see Raisfeld, I. H. 397, *417*
Feiwel, M., see James, V. H. T. 155, *167*
Feldman, H. A. 353, *356*
Feldman, S., see Gibaldi, M. 24, *33*, 284, *310*
Feldman, W., see Drummond, K. H. 290, *310*
Feller, D. R., Morita, M., Gillette, J. R. 407, *414*
Feller, D. R., see Morita, M. 407, *417*
Felts, J. M., see Gousios, A. G. 315, *323*
Fengler, H., see Palm, D. 124, *128*
Fenna, D., Mix, L., Schaefer, O., Gilbert, J. A. L. 196, *205*
Fenyvesi, T., see George, C. F. 293, *310*
Ferrier, W. R., see Prescott, L. F. 235, *256*
Ferry, D. G., see McQueen, E. G. 288, *312*
Fersht, A. R., Kirby, A. J. 91, *110*
Fetter, B. F., see Merritt, A. D. 408, *416*
Feuer, G., Sosa-Lucero, J. C., Lumb, G., Moddel, G. 363, *381*
Field, C. E., see Yeung, C. Y. 287, *314*
Field, H., Jr., see Swell, L. 221, *233*

Field, J.B., Ohta, M., Boyle, C., Remer, A. 292, *310*
Fife, T.H., Brod, L.H. 94, *110*
Fife, T.H., Jao, L.D. 94, 95, *110*
Figueroa, W.G., Thompson, J.H. 145, *146*
Finch, C.A., see Eschbach, J. W. 259, *269*
Finch, C.S., Jr., see Carliner, N.A. 320, *322*
Finck, E.M., see Bellet, S. 221, *230*
Finkelstein, S., see Fogelman, A, M, 317, *323*
Finland, M., Meads, M., Ory, E.M. 216, *231*
Finland, M., see McCarthy, C.G. 153, *167*
Finland, M., see Kunin, C.M. 264, 265, 266, *270*
Finland, M., see Peterson, O. L. 238, *255*
Finland, M., see Sabath, L.D. 266, *271*, 352, 354, *357*
Finland, M., see Smith, C.B. 344, *357*
Finlay, G.D., Whitsett, T.L., Cucinell, E.A., Goldberg, L.I. 319, *323*
Finnerty, F.A., see Mroczek, W.J. 154, *167*
Finster, M., see Conney, A.H. 169, 200, *205*, 243, *253*
Finster, M., see Welch, R.M. 288, *314*
Fisch, C., Martz, B.L., Priebe, F.H. 316, *323*
Fischer, E. 242, *253*
Fischer, G.A., see Chu, M.-Y. 329, *339*
Fischer, L.J., Kent, T.H., Weissinger, J.L. 138, *146*
Fischer, L.J., Millburn, P. 138, *146*
Fischer, L.J., Millburn, P., Smith, R.L., Williams, R. T. 131, 138, *146*
Fischer, L.J., Weissinger, J.L. Kent, T.H. 138, *147*
Fischer, L.J., see Ambre, J.J. 141, *145*, 280, *308*
Fischer, L.J., see Clark, A.G. 131, 138, *146*
Fischer, L.J., see Hanasono, G.K. 338, *340*
Fishbein, E., see Alarcon-Segovia, D. 178, *203*, 408 *413*
Fishler, J.J., see Misher, A. 94, 95, *111*
Flaks, J.G., see Cohen, S.S. 328, *339*
Flanagan, T.L., see Van Loon, E.J. 143, *149*
Flanagan, W.J., see Dougherty, J.E. 137, *146*
Flatz, G., see Vogel, F. 196, *211*
Fleischer *et al.* 38
Fleischmann, R.A., see Remmer, H. 399, *418*
Fleischmann, R.A., see Schoene, B. 243, *256*
Fleming, I.D., see Pratt, C.B. 336, *341*
Fletcher, J.E., see Spector, A. A. 221, *233*
Fodor, G., Kovacs, O. 90, *110*
Fodor, G., Nador, K. 90, *110*
Fogelman, A.M., Lamont, J. T., Finkelstein, S., Rado, E., Pearce, M.L. 317, *323*
Foldes, F.F. 180, *205*
Foord, R.D. 393, *414*
Foord, R.D., see Dodds, M.G. 393, *414*
Forbat, A., Lehmann, H., Silk, E. 179, *205*
Forestner, J.E., see Forman, D.T. 133, *147*
Forist, A.A., see Smith, D.L. 125, *129*
Forman, D.T., Garvin, J.E., Forestner, J.E., Taylor, C. B. 133, *147*
Forrest, A.R.W., see Prescott L.F. 390, *417*
Forrest, F.M., Forrest, I.S., Serra, M.T. 286, *310*
Forrest, I.S., see Forrest, F. M. 286, *310*
Forrest, J.A.H., see Prescott, L.F. 246, 250, *256*
Forrey, A., see Gyselynck, A. M. 266, *269*
Forrey, A.W., see Cutler, R. E. 262, *269*
Forshell, G.P., see Borgå, O. 291, 298, 300, *308*
Forsstrom, J., see Kasangen, A. 391, *415*
Forster, A.D., see Connors, T. A. 290, *309*
Forster, J.F.A., see Denborough, M.A. 194, *205*
Forsyth, R., see Benowitz, N. 293, *308*
Foulkes, D.M., see Thompson, F.D. 249, *256*
Fox, A.L. 192, *205*
Fox, W., see Devadatta, S. 176, 177, *205*
Fraher, M.A., Jawetz, E. 352, *356*
Franch, R.H. 156, *167*
Franco, A.P., see Kattamis, C. 180, *207*
Frantz, I.D., Jr., see Moore, R.B. 133, *148*
Fraser, G.R. 192, *205*
Fraser, I.M., Tilton, B.E., Vesell, E.S. 190, *205*
Fraser, I.M., Vesell, E.S. 190, *205*
Frederiksen, E., see von Daehne, W. 95, 96, *109*
Fredrickson, D.S., see Glueck, C.J. 220, *231*
Free, S.M., Jr., see Beard, W. J. 95, 101, *109*
Freedman, L., see Shapiro, S. L. 95, 106, *111*
Freele, H., see Rosi, D. 126, *128*
Frei, C., see Perry, H.M., Jr. 178, *209*
Frei III, E., Freireich, E.J. 331, *340*
Frei III, E., see Freireich, E. J. 333, *340*
Frei, J., see Aebi, H. 174, *203*
Freireich, E.J., Bodey, G.P., Hart, S., Rodriguez, V., Whitecar, J.P., Frei III, E. 333, *340*
Freireich, E.J., see Frei III, E. 331, *340*
Frenkel, E.P., Skinner, W.N., Smiley, J.D. 329, *340*
Frick, P.G., Hitzig, W.H., Betke, K. 191, *205*
Frick, P.G., see Hitzig, W.H. 191, *206*
Frieck, J., see Clarkson, B. 333, *339*
Friis-Hansen, B. 218, 219, *231*
Friman, A., see Mattila, M.J. 239, *255*
Frisby, S.A., see Zager, R.F. 337, *342*
Fry, I.K., see Cattell, W.R. 353, *355*
Fujimoto, J.M., see Smith, D.S. 135, *148*
Fuxe, K., see Carlsson, A. 121, *127*
Fuxe, K., see Lidbrink, P. 121, *128*

Gabali, F., see Baraka, A. 220, *230*, 242, *253*
Gaetano, L.F., see Jick, H. 196, *207*
Gaetjens, E., Morawetz, H. 92, *110*
Gaetjens, E., see Morawetz, H. 92, *111*

Gaffney, T. E., see Knapp, D.R. 230, *232*
Gala, E. L., see Crew, M. C. 142, *146*
Gallo, M., see Carulli, N. 288, 302, *309*
Galton, F. 198, *205*
Gammon, P. T., see Ellard, G.A. 177, *205*
Gangadharam, P. R. J., Selkon, J. B. 177, *205*
Gangadharam, P. R. J., see Devadatta, S. 176, 177, *205*
Garattini, S., Marcucci, F., Mussini, E. [4] 39, *85*
Garattini, S., Mussini, E. Marcucci, F., Guaitani, A. 120, *127*
Garattini, S., see Donelli, M.G. 338, *339*
Garattini, S., see Jori, A. 121, *128*
Garattini, S., see Marcucci, F. 119, *128*
Garattini, S., see Morselli, P.L. 289, 302, *312*
Garattini, S., see Samanin, R. 121, *128*
Garattini, S., see Tognoni, G. 122, *129*
Garcia, M., see Wallace, J. F. 349, 350, *358*
Garrett, E. R. 91, *110*
Garrettson, L., see Barr, W. H. 279, *308*
Garrettson, L. K., Perel, J. M., Dayton, P. G. 289, *310*
Garrod, J. M. B., see Ellard, G.A. 153, *167*
Garrod, L. P., Waterworth, P.M. 346, *356*
Gartler, S. M., see Shepard, T. H., II 192, *210*
Gartner, L. M., see Arias, I. M. 287, *308*
Garvin, J. E., see Forman, D.T. 133, *147*
Gass, S. R., see Beyer, K. H. 361, *381*
Gatenby, A. H., see Lavan, J.N. 391, *416*
Gatmaitan, Z., see Levi, A. J. 222, *232*
Gatmaitan, Z., see Mishkin, S. 222, *233*
Gatmaitan, Z., see Reyes, H. 222, *233*, 301, *313*
Gatz, E. E., Hull, M. J., Bennett, W. G., Jones, J.R. 195, *205*
Gault, M. H., Rudwal, T. C., Engers, W. D., Dossetor, J.B. 391, *414*
Gavan, T. L., see McHenry, M.C. 248, *255*, 266, *270*
Gavras, H., see Papapetrou, P.D. 145, *148*
Gavrilis, Z. B., see Elslager, E.F. 95, 107, *110*
Gazes, P. C., see Walton, R. P. 315, *324*
Gee, T. S., Tu, K.-P., Clarkson, B. D. 333, *340*
Geer, T., see Sachs, J. 259, 266, *271*
Geilling, E. M. K., see Okita, G.T. 136, *148*
Gelb, N. A., see Song, C. S. 191, *210*
Gelboin, H. V., see Wiebel, F.J. 363, *382*
Genest, K., see Kalow, W. 179, 180, 182, *207*
George, C. F., Fenyvesi, T., Conolly, M. E., Dollery, C.T. 293, *310*
Geraci, J. E., see Wilkowske, C.J. 353, *358*
Gerard, R., see Bing, R. J. 315, *322*
Gerhard, R. E., Knouss, R. P., Thyrum, P. T., Luchi, R. J., Morris, J. J. 291, *310*
Gerhards, W., see Büch, H. 392, *413*
Gerritsen, G. C., see Dulin, W. E. 125, *127*
Geurkink, N. A., see McHenry, M. C. 248, *255*
Geurnick, N. A., see McHenry, M. C. 266, *270*
Ghezzi, D., see Samanin, R. 121, *128*
Giacola, G. P., see Krasner, J. 221, *232*
Gibaldi, M. 22, *33*
Gibaldi, M., Boyes, R. N., Feldman, S. 24, *33*, 284, *310*
Gibaldi, M., Grundhofer, B. 281, *310*
Gibaldi, M., Levy, G., Hayton, W. 24, 31, *33*
Gibaldi, M., Levy, G., Weintraub, H. 31, *33*
Gibaldi, M., Nagashima, R., Levy, G. [5] 20, *33*, 63, *85*
Gibaldi, M., Schwartz, M. A. 95, 96, 103, *110*
Gibaldi, M., Weintraub, H. *33*
Gibaldi, M., see Levy, G. 32, *34*, 35, 63, 67, 68, 238, *254*, 280, *312*
Gibaldi, M., see Mayersohn, M. 19, *34*
Giblett, E. R., see Hodgkin, W.E. 180, 181, *206*
Gibson, J. E., Becker, B. A. 130, *147*
Gibson, J. E., see Becker, B. A. 130, *146*
Gifford, R. W., Jr., see McHenry, M. C. 248, *255*, 266, *270*
Gilbert, C. A., see Carliner, C.A., 317, *322*
Gilbert, J. A. L., see Fenna, D. 196, *205*
Gilbertson, A. A., Boulton, T.B. 251, *253*
Gilfillan, J. L., see Huff, J. W. 133, *147*
Gill, J. R., Jr., see Henkin, R.I. 193, *206*
Gillette, see Dingell 40,
Gillette, C. A. M. [6] 38, 69, 74, *85*
Gillette, J. R. 169, 182, 198, *205*, *206*, 245, *253*, 294, 299, *310*, 379, 380, *381*, 404, *414*
Gillette, J. R., Dingell, J.V., Sulser, F., Kuntzman, R., Brodie, B. B. 121, *127*
Gillette, J. R., Mitchell, J. R. 290, *310*
Gillette, J. R., Mitchell, J. R., Brodie, B. B. 384, *414*
Gillette, J. R., see Brodie, B.B. 401, *413*
Gillette, J. R., see Corsini, G.U. 406, *413*
Gillette, J. R., see Dichiara, G. 400, *414*
Gillette, J. R., see Dingell, J.V. 390, *414*
Gillette, J. R., see Feller, D.R. 407, *414*
Gillette, J. R., see Jollow, D.J. 245, *254*, 376, 378, 379, 380, *381*, 384, 386, 401, 403, *415*
Gillette, J. R., see Krishna, G. 407, *415*
Gillette, J. R., see Menard, R.H. 402, 403, *416*
Gillette, J. R., see Mitchell, J.R. 242, 245, *255*, 290, *312*, 362, 375, 376, 377, 378, *382*, 384, 385, 386, 387, 389, 390, 391, 392, 394, 400, 407, *416*, *417*
Gillette, J. R., see Potter, W.Z. 245, *255*, 377, *382*, 390, 411, *417*
Gillette, J. R., see Sjöquist, F. 367, *382*
Gillette, J. R., see Stripp, B. 402, 403, 407, *418*

Gillette, J. R., see Zampaglione, N. 378, 380, *382*, 401, *417*
Gilman, A., see Goodman, L. S. 351, *356*
Gilman, A. G., Conney, A. H. 363, *381*
Gilmore, E., Weil, J., Chidsey, C. 320, *323*
Gitlin, D. 220, *231*
Gitlin, D., Boesman, M. 220, *231*
Gladtke, E. 248, *253*
Glasser, J. E., Weiner, I. M., Lack, L. 133, *147*
Glassman, J. M., Warren, G. H., Rosenman, S. B., Agersborg, H. P. K. 143, *147*
Glaubiger, G. A., see Clark, S. W. 179, *204*
Glazener, L., see Harvey, D. J. 402, *414*
Glazko 143
Glazko, A. J., Carnes, H. E., Kazenko, A., Wolf, L. M., Reutner, T. F. 95, 98, 100, *110*
Glazko, A. J., Dill, W. A., Kazenko, A., Wolf, L. M., Carnes, H. E. 95, 98, *110*
Glazko, A. J., Edgerton, W. H., Dill, W. A., Lenz, W. R. 95, 98, *110*
Glazko, A. J., see Kunin, C. M. 264, 265, 266, *270*
Glazko, A. J., see Letteri, J. M. 249, *254*, 264, *270*
Glazko, A. J., see Thompson, R. Q. 130, *149*
Glazko, A. J., see Winder, C. V. 142, *149*
Gleason, R. E., see Hampers, C. L. 262, *269*
Glick, D. 95, 97, *110*
Glick, G. 318, *323*
Glick, G., Parmley, W. W., Wechsler, A. S., Sonnenblick, E. H. 318, *323*
Glick, G., see Besch, H. R., Jr. 316, *322*
Glick, J. M., see Reid, W. D. 381, *382*
Glidewell, O., see Burgert, E. O., Jr. 335, *339*
Glidewell, O., see Chavelier, L. 332, *339*
Glidewell, O. J., see Hoogstraten, B. 335, *340*
Glueck, C. J., Levy, R. I., Glueck, H. I., Gralnick, H. R., Greten, H., Fredrickson, D. S. 220, *231*
Glueck, H. I., see Glueck, C. J. 220, *231*
Goatcher, P. M., see Armstrong, B. K. 235, *252*
Godtfredsen, W. O., see von Daehne, W. 95, 96, 97, *109*
Goedde, H. W., Altland, K. 181, *206*
Goedde, H. W., Altland, K., Schloot, W. 179, *206*
Goetzee, B., see Bigland, B. 252, *253*
Goff, J. B., Schlegel, J. U., O'Dell, R. M. 259, *269*
Goff, J. B., see Schlegel, J. U. 259, 266, *271*
Golab, T., see Arai, K. 139, *145*
Golab, T., see Layne, D. S. 139, *148*
Golbey, R., see Li, M. C. 336, *341*
Gold, N. I., see Crigler, J. F., Jr. 287, *309*
Goldberg, L. I. 316, 319, 321, *323*
Goldberg, L. I., Musgrave, G. E. 322, *323*
Goldberg, L. I., Sonneville, P. F., McNay, J. L. 321, 322, *323*
Goldberg, L. I., Talley, R. C. 318, 320, *323*
Goldberg, L. I., see Caldwell, R. W. 321, *322*
Goldberg, L. I., see Carliner, N. A. 317, 320, *322*
Goldberg, L. I., see Clark, D. W. 320, *323*
Goldberg, L. I., see Finlay, G. D. 319, *323*
Goldberg, L. I., see McNay, J. L. 321, *324*
Goldberg, L. I., see Meyer, J. F. 299, *312*
Goldberg, L. I., see Simanis, J. 318, *324*
Goldberg, L. I., see Yeh, B. K. 321, *324*
Goldberg, L. T., see Tjandramaga, T. B. 319, *324*
Goldberg, M. J., see Bluemle, L. W. 240, *253*
Golden, J., see Bennett, L. L., Jr. 329, *338*
Goldfinger, S. E., see Heizer, W. D. 239, *254*, 258, *269*
Goldfischer, S., see Biempica, L. 261, *268*
Goldin, A., Mantel, N. 325, *340*
Goldin, A., see Kline, I. 328, *340*
Goldin, A., see Vadlamudi, S. 336, *341*
Goldin, A., see Venditti, J. M. 326, *342*
Goldman, A., see Brodie, B. B. 157, *166*
Goldman, A., see Burns, J. J. 197, *204*
Goldman, L., see Gwilt, J. R. 238, *254*
Goldsmith, G. A., see Miller, O. N. 124, *128*
Goldstein, A., Aronow, L., Kalman, S. M. 218, 219, 220, 223, 225, 226, *231*, 259, *269*, *310*
Goldstein, J. A., Taurog, A. 144, *147*
Gommi, B. W., see Welch, R. M. 288, *314*
Goodman, D. S. 221, *231*
Goodman, L. S., Gilman, A. 351, *356*
Goodman, M. J., see Boyes, R. N. 163, 164, *166*
Gorbach, S. L., Plaut, A. G., Nahas, L., Weinstein, L., Spanknebel, G., Levitan, R. 215, *231*
Gordon, G. R., see Peters, J. H. 176, *209*
Gordon, M., Miller, J. G., Day, A. R. 89, *110*
Goss, J. E., Dickhaus, D. W. 285, *310*
Gossmann, H. H., Baltzer, G., Helms, H. 259, *269*
Gothoni, G., see Neuvonen, P. J. 280, *312*
Gotoff, S. P., Behrman, R. E. 344, *356*
Gousios, A. G., Felts, J. M., Havel, R. J. 315, *323*
Gow, A., see Smith, J. 397, *418*
Gow, J. G., Evans, D. A. P. 176, *206*
Gower, P. E., Dash, C. H. 235, *254*
Grabstald, H., see Li, M. C. 336, *341*
Graham, G. G., see Rowland, M. [14] 49, *85*, 283, 300, *313*
Gralnick, H. R., see Glueck, C. J. 220, *231*
Gram, L. F., Overo, K. F. 289, *310*
Grant, W. M. 194, *206*
Grasbeck, R., Nyberg, W., Reizenstein, P. 130, *147*

Graul, E. H., Schaumlöffel, E., Hundeshagen, H., Willmanns, H., Simon, G. 337, *340*
Graves, L., see Lewis, G. P. 242, *254*
Gray, P. W. S., see Evans, F.T. 178, *205*
Greaves, J. A., Ayres, P. 184, *206*
Green, I., see Thorgeirsson, S.S. 390, 391, *418*
Greenberg, H. B., see Bailey, W.C. 409, *413*
Greenberger, J. J., see Caldwell, J. H. 131, 137, *146*
Greenberger, N. J., Thomas, F.B. 137, *147*
Greenberger, N. J., see Caldwell, J. H. 281, *309*
Greene, F. E., see Vesell, E. S. 203, *211*, 289, *314*, 390, *417*
Greenfield, R. E., see Price, V.E. 174, *209*
Greenfield, S., Feingold, D. S. 353, *356*
Greenwald, P., Barlow, J. J., Nasca, P. C., Burnet, W. S. 406, *414*
Greer, M. A. 192, *206*
Greten, H., see Glueck, C. J. 220, *231*
Griffith, J. D., see Fann, W. E. 299, 300, *310*
Grignani, F., see Larizza, P. 190, *208*
Griner, P. F., Raisz, L. G., Rickles, F. R., Wiesner, P.J., Odoroff, C. L. 296, *310*
Grobecker, H., see Palm, D. 124, *128*
Gross, see Marks 190
Grossman, M., Ticknor, W. 216, *231*
Grossman, M. I., see Ivy, A. A. 216, *232*
Grossman, S. B., see Biempica, L. 261, *268*
Grossman, S. B., see Black, M. 261, *268*
Grover, P. L., Sims, P. 388, *414*
Grubb, R. 220, *231*
Gruber, J. W., see Luchi, R. J. 246, *254*
Grunberg, E., Leiwant, B., D'Ascensio, I.-L., Schnitzer, R. J. 176, *206*
Grundbacher, F. J., Shreffler, D.C. 220, *231*
Grundhofer, B., see Gibaldi, M. 281, *310*
Guaitani, A., see Garattini, S. 120, *127*
Guarino, S., see Bovet, D. 179, *204*
Guevara, L., see Caesar, J. 301, *309*
Guichard, A., see Rowland, M. 163, *167*
Gulyassy, P. F., Peters, J. H., Lin, S. C., Ryan, P. M. 258, *269*
Gulyassy, P. F., Peters, J. H., Lin, S. C., Ryan, P. M., Schoenfeld, P. 258, 260, *269*
Gundersen, E., see von Daehne, W. 95, 96, *109*
Gunn, D. R., see Kalow, W. 179, 180, *207*
Gunnison, J. B., Kunishige, E., Coleman, V. R., Jawetz, E. 349, *356*
Gunnison, J. B., Speck, R. S., Jawetz, E., Bruff, J. A. 350, *356*
Gunnison, J. B., see Jawetz, E. 343, 346, 347, 348, 350, 352, 354, *356*
Gunnison, J. B., see Miles, C.P. 352, *357*
Gutierrez, L., see Thompson, G.R. 281, *314*
Gutman, A. B., see Burns, J.J. 115, 117, 118, *127*
Gutsche, B. B., Scott, E. M., Wright, R. C. 181, *206*
Guyer, R. J., Winfield, D. A., Shahani, R. T., Blackburn, E. K. 333, 334, *340*
Gwilt, J. R., Robertson, A., Goldman, L., Blanchard, A.W. 238, *254*
Gyselynck, A. M., Forrey, A., Cutler, R. 266, *269*

Haass, A., Lüllmann, H., Peters, T. 239, *254*
Haber, E., Osborne, R. K. 408, *414*
Haber, E., see Lefkowitz, R. J. 317, *323*
Haber, E., see Smith, T. W. 317, *324*
Habersang, R., see Dettli, L. 248, *253*, 265, *269*
Hackman, R., see Neuvonen, P.J. 280, *312*
Hackney, R. L., Jr., see Payne, H.M. 95, 101, *111*
Hässig, A., see Aebi, H. 174, *203*
Hafner, C. D., see Cranley, J. J. 320, *323*
Hagerman, L. M., see Cook, D.A. 133, *146*
Hague, D. E., Smith, M. E., Ryan, J. R., McMahon, F. G. 289, *310*
Hahn, F. E., Wisseman, C. L., Jr. 351, *356*
Hahn, K.-J., Eiden, W., Schettle, M., Hahn, M., Walter, E., Weber, E. 281, *310*
Hahn, M., see Hahn, K.-J. 281, *310*
Hahn, M. A., see Liegler, D. G. 337, *341*
Hahn, T. J. 286, *310*
Hahn, T. J., Birge, S. J., Scharp, C. R., Avioli, L. V. 286, *310*
Hall, J. M., see Dobbs, H. E. 131, 135, 136, *146*
Haltalin, K. C., see Nelson, J. D. 235, 238, 245, *255*
Hamberger, B., see Carlsson, A. 121, *127*
Hamburger, M., see Tan, J. S. 354, *358*
Hamilton, H. B., Neel, J. V. 173, *206*
Hamilton, H. B., see Nishimura, E. T. 174, *208*
Hamilton, H. B., see Shibata, Y. 174, *210*
Hamilton, H. B., see Takahara, S. 174, *210*
Hamilton, J. G., see Miller, O.N. 124, *128*
Hamlet, J. C., see Barnden, R. L. 95, *109*
Hammar, C.-H., Prellwitz, W. 243, *254*
Hammer, W., Martens, S., Sjöqvist, F. 202, *206*
Hammer, W., see Sjöqvist, F. 300, *313*, 367, *382*
Hammond, G. S., see Cram, D.J. 95, 101, *109*
Hampers, C. L., Soeldner, J. S. Gleason, R. E., Bailey, G. L., Diamond, J. A., Merrill, J.P. 262, *269*
Hamrick, M., see Zampaglione, N. 378, 380, *382*, 401, *417*
Hamrick, M. E., see Stripp, B. 402, 403, 407, *418*
Hanahan, D. J., Daskalakis, E.G., Edwards, T., Dauben, H. J., Jr. 138, *147*
Hanasono, G. K., Fischer, L. J. 338, *340*
Hansen, J. M., Kristensen, M., Skovsted, L. 289, *310*

Hansen, J.M., Kristensen, M., Skovsted, L., Christensen, L.K. 288, *310*
Hansen, J.M., Siersbaek-Nielsen, K., Skovsted, L. 286, *310*
Hansen, J.M., see Christensen, L.K. 289, 297, 302, *309*
Hansen, J.M., see Kristensen, J. 285, 288, *311*
Hansen, T., see Myhre, E. 249, *255*
Hansen, T., see Stenbaek, Ø. 249, *256*
Hanson, H.M., see Vernier, V.G. 121, *129*
Hanzlik, P.J., Presho, N.E. 93, 95, *110*
Harding, R.S., see Powell, R. C. 134, *148*
Hardison, W.G.M., Apter, J. T. 145, *147*
Hardisty, R.M., McElwain, T. J. Darby, C.W. 331, *340*
Harley, J.B., see Hayes, D. M. 330, *340*
Harley, J.B., see Hoogstraten, B. 335, *340*
Harper, N.J. 86, 98, 99, 100, *110*
Harris, F.W., Cockburn, T. 181, *206*
Harris, H., Hopkinson, D.A., Robson, E.B., Whittaker, M. 181, 182, *206*
Harris, H., Kalmus, H. 192, *206*
Harris, H., Kalmus, H., Trotter, W.H. 192, *206*
Harris, H., Whittaker, M. 179, 180, 181, *206*
Harris, H., Whittaker, M., Lehmann, H., Silk, E. 180, *206*
Harris, H.W. 176, 177, *206*
Harris, H.W., Knight, R.A., Selin, M.J. 177, *206*
Harris, H.W.M., see Knight, R.A. 177, *207*
Harris, L., Jr., see Salatka, K. 220, *233*
Harris, R.C., Lucey, J.F., MacLean, J.R. 297, *310*
Harris, T.M., see Wilson, B. J. 396, *419*
Harrison, D.C. 157, *167*
Harrison, D.C., see Stenson, R.E. 163, *168*, 307, *314*, 363, *382*
Harrison, G.G., Smith, J.S. 400, *415*
Harrison, P.M., see Stewart, G.T. 143, *149*
Harrison, T.S. 251, *254*
Harrison, Y.E., see Welch, R. M. 288, 301, *314*
Hart, S., see Freireich, E.J. 333, *340*
Hartiala, K., Kasanen, A., Raussi, M. 236, *254*
Hartz, S., see Jick, H. 196, *207*
Harvey, D.J., Glazener, L., Stratton, C., Nowlin, J., Hill, R.M., Horning, E.C. 402, *414*
Harvey, D.J., see Horning, M.G. 230, *232*, 402, *415*
Hashim, S.A., Van Itallie, T. B. 133, *147*
Hashimoto, M., see Jollow, D. J. 376, *381*, 385, 388, 390, *415*
Hashizume, A., see Hollingsworth, J.W. 213, *231*
Havel, R.J., see Gousios, A.G. 315, *323*
Hayakawa, T., Kanai, N., Yamada, R., Kuroda, R., Highashi, H., Mogami, H., Jannai, D. 338, *340*
Hayes, A., Cooper, R.G. 164, *167*
Hayes, A., see Paterson, J.W. 164, *167*
Hayes, A.H. 240, *254*
Hayes, D.M., Costa, J., Moon, J.H., Hoogstraten, B., Harley, J.B. 330, *340*
Hayes, E.R., see Keys, A. 222, *232*
Hayes, M.F., see McMullin, G. P. *312*
Haynes, C., see Perry, H.M., Jr. 178, *209*
Haynes, J., see Kutt, H. 285, 306, *311*
Haynes, L.J., see Crew, M.C. 142, *146*
Haynes, L.J., see DiCarlo, F. J. 121, *127*
Hayton, W., see Gibaldi, M. 24, 31, *33*
Heading, R.C., Nimmo, J., Prescott, L.F., Tothill, P. 236, *254*, 280, *310*
Heading, R.C., see Nimmo, J. 237, 239, *255*
Heaf, P., see Speizer, F.E. 162, *168*
Heathcote, A.G.S., Nassau, E. 95, 96, *110*
Heaton, A.D., Russell, W.F., Jr., Dienst, J., Middlebrook, G. 344, *356*
Heaton, A.D., see Mandel, W. 344, *357*
Heaton, A.D., see Russell, W. F., Jr. 344, *357*
Hecht, R.H., see Winters, R. E. 350, *358*
Hedrick, S.R., see Dove, J.T. 408, *414*
Hedwall, P.R., see Brunner, H. 123, *127*
Heidelberger, C. 329, 330, *340*
Heimbecker, R., see Bing, R. J. 315, *322*
Heiniger, J.P., see Aebi, H. 174, *203*
Heinonen, O. P., see Jick, H. 196, *207*
Heinz 190
Heizer, W.D., Smith, T.W., Goldfinger, S.E. 239, *254*, 258, *269*
Held, H., Olderhausen, H.F. 243, *254*
Hellman, D.A., see Drummond, K.H. 290, *310*
Hellmer, K.H., see Reiner, O. 404, *418*
Hellström, K., see Beermann, B. 235, 239, *253*
Helm, J.D., Jr., see Rammelkamp, C.H. 216, *233*
Helms, H., see Gossmann, H. H. 259, *269*
Helrich, M., see Cascorbi, H. F. 198, *204*
Hems, B.A., see Barnden, R.-L. 95, *109*
Henderson, E.S., Samaha, R. J. 331, 332, *340*
Henderson, E.S., see Liegler, D.G. 292, *312*, 337, *341*
Henderson, J.F., Junga, I.G. 334, *340*
Henkin, R.I., Bartter, F.C. 193, *206*
Henkin, R.I., Gill, J.R., Jr., Bartter, F.C. 193, *206*
Herbert, C.M., see Branch, R. A. 288, 293, 307, *308*
Herbert, V. 130, *147*, 150, *167*
Herbst, A.L., Kurman, R.J., Scully, R.E., Poskanzer, D.C. 406, *415*
Hernandez, K., see Pinkel, D. 332, *341*
Herr, M.E., see Heyl, F.W. 95, 108, *111*
Hertz, R., Lewis, J., Jr., Lipsett, M.B. 335, *340*
Herxheimer, A., see Barrowman, J.A. 281, *308*
Herxheimer, A., see Chaput de Saintonge, D.M. 279, *309*
Hewitt, W.L., Seligman, S.J., Deigh, R.A. 353, 354, *356*

Hewitt, W. L., see Winters, R. E. 350, *358*
Heyl, F. W., Herr, M. E. 95, 108, *111*
Heymann, E., Krisch, K., Buch, H., Buzello, W. 401, *415*
Heyn, R. M., see Evans, A. E. 333, 336, *339*
Hicks, J. D., see Nanra, R. S. 240, *255*
Higashi, H., see Hayakawa, T. 338, *340*
Higashi, T., see Shibata, Y. 174, *210*
High, R. H., see Huang, N. N. 216, *232*
Higuchi, T., see Hussain, A. 91, *111*
Hill, D. L., Laster, W. R., Jr., Struck, R. F. 126, *127*, 337, *340*
Hill, R. M., see Harvey, D. J. 402, *414*
Hill, R. M., see Horning, M. G. 230, *232*, 402, *415*
Hindman, K. L., see Becker, B. A. 130, *146*
Hines, J. D., see Wolinsky, E. 267, *271*, 337, *342*
Hinson, J. A., Jollow, D. J., Mitchell, J. R. 390, 391, 392, *415*
Hinson, J. A., Mitchell, J. R., Jollow, D. J. 390, 391, *415*
Hinson, J. A., see DiChiara, G. 400, 401, *414*
Hinson, J. A., see Mitchell, J. R. 290, *312*, 362, 375, 376, 377, 378, *382*, 393, 395, 396, *417*
Hinson, J. H., Mitchell, J. R. 390, 391, *415*
Hirai, H., see Shibata, Y. 174, *210*
Hirsch, G. H., Hook, J. B. 225, *231*
Hitchings, G. H. 351, 353, *356*
Hitchings, G. H., Burchall, J. J. 238, *340*
Hitchings, G. H., see Elion, G. B. 118, *127*, 328, 337, *339*
Hitzig, W. H., Frick, P. G., Betke, K., Huisman, T. H. J. 191, *206*
Hitzig, W. H., see Frick, P. G. 191, *205*
Hitzig, W. H., see Huisman, T. H. J. 191, *207*
Ho, W. K. L., see Ockner, R. K. 222, *233*
Hoag, M. S., see O'Reilly, R. A. 185, 186, *209*
Hoch, J. H. 315, *323*
Hodgin, U. G., Sanford, J. P. 344, *356*
Hodgkin, W. E., Giblett, E. R. Levine, H., Bauer, W., Motulsky, A. G. 180, 181, *206*
Hökfelt, T., see Carlsson, A. 121, *127*
Hoffman, T. A., Cestero, R., Bullock, W. E. 265, *269*
Hofmann, A. F. 131, 133, *147*
Hofmann, A. F., Small, D. M. 131, *147*
Hoffmann, K., see Keberle, H. 130, 131, 140, *147*
Hofstee, B. H. J. 88, *111*
Hogben, C. A. M. [7] 35, *85*
Hohorst, H. J., see Brock, N. 126, *126*
Hoigne, R. 407, 408, *415*
Hokfelt, B., see Dymling, J. F. 403, *414*
Holder, G. M., Ryan, A. J., Watson, T. R., Wiebe, L. I. 142, *147*
Holger-Madsen, T. 220, *231*
Holland, J. F. 325, 331, 332, *340*
Holland, N. H., see West, C. D. 220, 223, 224, *233*
Hollander, V., see Soberman, R. 197, *210*
Hollingsworth, J. W., Hashizume, A., Jablon, S. 213, *231*
Hollister, L., Levy, G. 289, *311*
Hollister, L. E., Curry, S. H., Derr, J. E., Kanter, S. L. 162, *167*
Holmes, E. L., see Winder, C. V. 142, *149*
Holmes, R. S., Masters, C. J. 174, *207*
Holt, P. R., see Pierson, R. N. 94, *111*
Holton, C., see Pinkel, D. 332, *341*
Holtz, P., Credner, K. 321, *323*
Holtzman, N. A., see Rieder, R. F. 191, *209*
Hong, R., see West, C. D. 220, 223, 224, *233*
Hood, A. F., see Bryant, R. E. 344, *355*
Hood, C. E., see Bryant, R. E. 344, *355*
Hoogstraten, B., Owens, A. H. Lenhard, R. E., Glidewell, O. J., Leone, L. A., Olson, K. B., Harley, J. B., Townsend, S. R., Miller, S. P., Spurr, C. L. 335, *340*
Hoogstraten, B., see Hayes, D. M. 330, *340*
Hook, E. W., see Kaye, D. 353, *356*
Hook, E. W., see Mandell, G. L. 353, *357*
Hook, J. B., see Hirsch, G. H. 225, *231*
Hopkinson, D. A., see Harris, H. 181, 182, *206*
Horita, A. 121, *127*
Horning, E. C., Horning, M. G., Carroll, D. I., Stillwell, R. N. 230, *231*
Horning, E. C., see Harvey, D. J. 402, *414*
Horning, E. C., see Horning, M. G. 402, *415*
Horning, M. G., Butler, C., Harvey, D. J., Hill, R. M., Zion, T. C. 402, *415*
Horning, M. G., Harvey, D. J., Nowlin, J., Stillwell, W. G. Hill, R. M. 230, *232*
Horning, M. G., Stratton, C., Wilson, A., Horning, E. C., Hill, R. M. 402, *415*
Horning, M. G., see Horning, E. C. 230, *231*
Horton, B., see Huisman, T. H. J. 191, *207*
Horton, E. S., Johnson, C., Lebovitz, H. E. 262, *269*
Hossie, R. D., see Beckett, A. H. 152, *166*
Howard, E., see Steiner, A. 134, *149*
Howard, M., see White, R. J. 235, *257*
Howel-Evans, W., see Kitchin, F. D. 192, *207*
Howell, A., Sutherland, R., Robinson, G. N. 240, *254*
Howell, R. B., see Wyngaarden, J. B. 173, *212*
Hoyt, R., see Lippman, R. W. 259, *270*
Hrdlicka, J., see Seawright, A. A. 395, 396, *418*
Hsio, K. C., see Pantuck, E. J. 363, *382*
Hsu, K.-P., see Li, M. C. 336, *341*
Huang, N. N., High, R. H. 216, *232*
Hubbard, S. J., see Kalmus, H. 192, *207*

Hucker, H.B., Zacchei, A.G., Cox, S.V., Brodie, D.A., Cantwell, N.H.R. 131, 139, 140, *147*
Huf, E.G., Coles, F.K., Eubank, L.L. 95, 104, *111*
Huff, J.W., Gilfillan, J.L., Hunt, V.M. 133, *147*
Huffman, H., see Pentikainen, P.J. 403, *417*
Hughes, H.B. 176, *207*
Hughes, H.B., Biehl, J.P., Jones, A.P., Schmidt, L.H. 176, *207*
Hughes, H.B., Schmidt, L.H., Biehl, J.P. 176, *207*
Hughes, J., Vane, J.R. 321, *323*
Huhti, E., see Sotaniemi, E. 246, *256*
Huisman, T.H.J., Horton, B., Bridges, M.T., Betke, K., Hitzig, W.H. 191, *207*
Huisman, T.H.J., see Hitzig, W.H. 191, *206*
Hull, M.J., see Gatz, E.E. 195, *205*
Hultengren, N., Lagergren, C., Ljungqvist, A. 406, *415*
Humphreys, S.R., see Venditti, J,M. 326, *342*
Hundeshagen, H., see Graul, E.H. 337, *340*
Hunninghake, D.B., Azarnoff, D.L. 297, *311*
Hunninghake, D.B., see Kupferberg, H.J. 289, *311*
Hunt, J.N., see Cooke, A.R. 279, *309*
Hunt, R., de Taveau, R.M. 178, *207*
Hunt, V.M., see Huff, J.W. 133, *147*
Hunter, J., Maxwell, J.D., Carrella, M., Stewart, D.A., Williams, R. 202, *207*, 286, *311*
Hunter, J., see Maxwell, J.D. 286, *312*
Hunter, T.H., Paterson, P.Y. 354, *356*
Hurwitz, A. 279, 280, 281, 282, *311*
Hussain, A., Higuchi, T., Stella, V. 91, *111*
Hussain, A., Yamasaki, M., Truelove, J.E. 94, 95, *111*
Hustu, H.O., see Pratt, C.B. 336, *341*
Hustu, O., see James, D.H., Jr. 325, 336, *340*
Hyman, A.I., see Ryan, J.F. 252, *256*
Hyman, G.A., Ellsworth, R.M., Feind, C.R., Tretter, P. 336, *340*
Hyvarinen, M., Zeltzer, P., Oh, W., Stiehm, E.R. 220, *232*

Iber, F.L., see Kater, R.M.H. 246, *254*, 288, 311
Ibrahim, G.W., Watson, C.J. 130, *147*
Ikeda, M., Otsuji, H. 410, *415*
Ikeda, M., Sezesny, B., Barnes, M. 184, *207*
Ilett, K.F., see Reid, W.D. 381, *382*
Ingelfinger, F.J., see Sessions, J.T. 243, *256*
Ingold, C.K. 88, *111*
Inman, W.H.W., see Jick, H. 196, *207*
Irwin, G.M., see Dittert, L.W. 88, 93, 94, 95, 102, *109*, *110*
Irwin, G.M., see Swintosky, J.V. 95, 101, 102, *112*
Ishak, K., see Black, M. 397, 408, *413*
Issekutz, B., Jr., see Rodahl, K. 221, *233*
Isselbacher, K.J., see Playoust, M.R. 133, *148*
Ivler, D., see Mathies, A.W., Jr. 350, *357*
Ivy, A.A., Grossman, M.I. 216, *232*
Izzo, J.L., Bartlett, J.W., Roncone, A., Izzo, M.J., Bale, W.F. 262, *269*
Izzo, M.J., see Izzo, J.L. 262, *269*

Jablon, S., see Hollingsworth, J.W. 213, *231*
Jackson, B., see Nanra, R.S. 240, *255*
Jackson, E.H., see Welt, S.I. 191, *211*
Jackson, G.G., see Lepper, M.H. 345, *357*
Jackson, G.G., see McCabe, W.R. 353, 354, *357*
Jackson, G.G., see Riff, L.J. 266, *271*
Jacmi, H.A., see Kasangen, A. 391, *415*
Jacob, H.S., Jandl, J.H. 190, *207*
Jacobsen, P.M., see Tobey, R.E. 251, *256*
Jacobson, M., see Conney, A.H. 169, 200, *205*, 243, *253*
Jacobson, M., see Kuntzman, R. 202, *207*
Jacobson, M., see Levin, W. 404, *416*
Jaffe, J.J., see MacDonald, M.G. 285, 287, *312*
Jaffe, J.M., Colaizzi, J.L., Barry, H. 235, *254*
Jagenburg, O.R., see Toczko, K. 388, *415*
James, D.H., Jr., Hustu, O., Wrenn, E.L., Jr., Pinkel, D. 325, 336, *340*
James, M., see Reidenberg, M.M. 246, 247, *256*
James, V.H.T., Munro, D.D., Feiwel, M. 155, *167*
Jandel, J.H., see Jacob, H.S. 190, *207*
Janowsky, D.S., see Curry, S.H. 286, *309*
Janowsky, D.S., see Fann, W.E. 299, 300, *310*
Jansen, A.B.A., Russell, T.J. 95, 96, *111*
Jansen, A.B.A., see Barnden, R.L. 95, *109*
Jansen, G.R., Zanetti, M.E. 133, *147*
Janssen, P.A.J., see Niemegeers, C.J.E. 116, *128*
Jantzen, R., see Ritz, E. 261, *271*
Jao, J.W., Jusko, W.J., Cohen, J.L. 290, *311*
Jao, L.D., see Fife, T.H. 94, 95, *110*
Jarman, M., see Connors, T.A. 290, *309*
Jawetz, E. 343, 344, 345, 347, 352, *356*
Jawetz, E., Brainerd, H.D. 353, 354, *356*
Jawetz, E., Gunnison, J.B. 343, 346, 350, 352, *356*
Jawetz, E., Gunnison, J.B., Coleman, V.R. 347, 348, 352, 354, *356*
Jawetz, E., Gunnison, J.B., Coleman, V.R., Kempe, H.C. 346, *356*
Jawetz, E., Gunnison, J.B., Speck, R.S. 350, *356*
Jawetz, E., Sonne, M. 354, *356*
Jawetz, E., see Fraher, M.A. 352, *356*
Jawetz, E., see Gunnison, J.B. 349, 350, *356*
Jawetz, E., see Miles, C.P. 352, *357*

Jawetz, E., see Sonne, M. 353, *358*
Jebson, P. J., see Boyes, R. N. 163, 164, *166*
Jeffery, W., see Kupferberg, H. J. 289, *311*
Jencks, W. P. 91, 95, 107, *111*
Jenne, J. W. 177, 178, *207*
Jenne, J. W., Macdonald, F. M., Mendoza, E. 176, *207*
Jerina, D. M., see Daly, J. W. 378, 379, *381*, 390, 402, *413*
Jewitt, D. E., Kishon, Y., Thomas, M. 163, *167*
Jick, H., Slone, D., Shapiro, S., Heinonen, O. P., Lawson, D. H., Lewis, G. P., Jusko, W., Ballingall, D. L. K., Siskind, V., Hartz, S., Gaetano, L. F., MacLaughlin, D. S., Parker, W. J., Wizwer, P., Dinan, B., Baxter, C., Miettinen, O. S. 196, *207*
Jick, H., Slone, D., Westerholm, B., Inman, W. H. W., Vessey, M. P., Shapiro, S., Lewis, G. P., Worcester, J. 196, *207*
Jick, H., see Lewis, G. P. 196, 197, *208*
Jinnai, D., see Hayakawa, T. 338, *340*
Joekes, A. M., see Thompson, F. D. 249, *256*
Johns, D. G., see Wolpert, M. K. 407, *419*
Johnson, C., see Horton, E. S. 262, *269*
Johnson, L., Sarmiento, F., Blanc, W. A., Day, R. 297, *311*
Jollow, D. J., Mitchell, J. R., Potter, W. Z., Davis, D. C., Gillette, J. R., Brodie, B. B. 384, 386, *415*
Jollow, D. J., Mitchell, J. R., Potter, W. Z., Davis, W. Z. Gillette, J. R., Brodie, B. B. 245, *254*
Jollow, D. J., Mitchell, J. R., Zampaglione, N., Gillette, J. R. 376, 378, 379, 380, *381*, 401, 403, *415*
Jollow, D. J., Thorgeirsson, S. S., Potter, W. Z., Hashimoto, M., Mitchell, J. R. 376, *381*, 385, 388, 390, *415*
Jollow, D. J., see Davis, D. C. 384, 391, *413*
Jollow, D. J., see DiChiara, G. 400, 401, *414*
Jollow, D. J., see Hinson, J. A. 390, 391, 392, *415*
Jollow, D. J., see Mitchell, J. R. 242, 245, *255*, 378, *382*, 384, 385, 386, 387, 388, 389, 390, 391, 392, 393, 394, 395, 396, 397, 398, 400, 408, 409, *416*, *417*
Jollow, D. J., see Potter, W. Z. 245, *255*, 377, *382*. 384, 385, 386, 388, 389, 390, 392, 393, 394, 411, *417*
Jollow, D. J., see Snodgrass, W. 398, *418*
Jollow, D. J., see Thorgeirsson S. S. 390, 391, 394, *418*
Jollow, D. J., see Timbrell, J. A. 398, *419*
Jollow, D. J., see Weihe, M. 393, 394, *419*
Jollow, D. J., see Zampaglione, N. 378, 380, *382*, 401, *417*
Jones, A. P., see Hughes, H. B. 176, *207*
Jones, D. P., Perman, E. S., Lieber, C. S. 221, *232*
Jones, J. H., see Coles, G. A. 260, *269*
Jones, J. R., see Gatz, E. E. 195, *205*
Jones, L. G., see McCracken, G. H., Jr. 224, *232*
Jones, M., see Connors, T. A. 338, *339*
Jones, N. F., see McGregor, G. A. 261, *270*
Jones, R. G., see Caldwell, H. C. 95, 108, 109, *109*
Jones, R. G., see Misher, A. 94, 95, *111*
Jones, R. J. 158, *167*
Jonsson, G., see Lidbrink, P. 121, *128*
Jordan, D. A., see Kagan, B. M. 261, *270*
Jori, A., Carrara, C., Paglialunga, S., Garattini, S. 121, *128*
Joseph, H. W. 261, *269*
Jow, E., see Bogoch, A. 250, *253*
Joyce, T. H., see Tolomie, J. D. 252, *256*
Jubiz, W., Levinson, R. A., Meikle, A. W., West, C. D., Tyler, F. H. 287, *311*
Jubiz, W., see Meikle, A. W. 287, *312*
Juhasz, J., Balo, J., Kendrey, G. 407, *415*
Julian, D. G., see Boyes, R. N. 163, 164, *166*
Julian, D. G., see Clarke, R. A. 245, *253*
Junga, I. G., see Henderson, J. F. 334, *340*
Jusko, W., see Jick, H. 196, *207*
Jusko, W. J., Khanna, N., Levy, G., Stern, L., Yaffe, S. J. 216, 217, *232*
Jusko, W. J., Levy, G. 28, *33*
Jusko, W. J., Levy, G., Yaffe, S. 216, 217, *232*
Jusko, W. J., Levy, G., Yaffe, S. J., Allen, J. E. 216, *232*
Jusko, W. J., see Jao, J. W. 290, *311*
Jusko, W. J., see Lewis, G. P. 242, *254*
Jussila, J., see Kekki, M. 236, *254*, 279, *311*
Jussila, J., see Siurala, M. 236, *256*, 279, *313*
Jussila, J., see Venho, V. M. K. 239, *257*

Kabins, S. A., Cohen, S. 264, *270*
Kagan, B. M., Kaiser, E. 261, *270*
Kagan, B. M., Thomas, E. M. Jordan, D. A., Abt, A. F. 261, *270*
Kagen, L. J., see Ryan, J. F. 252, *256*
Kahle, C. T., see Tobey, R. E. 251, *256*
Kaiser, A., see Berneis, K. 329, *338*
Kaiser, E., see Kagan, B. M. 261, *270*
Kaiser, G. A., Ross, J., Jr., Braunwald, E. 321, *323*
Kalant, H., see Sellers, E. M. 289, 302, *313*
Kalman, S. M., see Goldstein, A. 218, 219, 220, 223, 225, 226, *231*, 259, *269*, *310*
Kalmus, H., Hubbard, S. J. 192, *207*
Kalmus, H., see Harris, H. 192, *206*
Kalow, W. 169, 170, 176, 179, 180, 194, 195, *207*
Kalow, W., Davies, R. O. 179, *207*
Kalow, W., Genest, K. 179, 180, 182, *207*
Kalow, W., Gunn, D. R. 179, 180, *207*
Kalow, W., Staron, N. 179, 180, 182, *207*
Kalow, W., see Britt, B. A. 194, *204*
Kalow, W., see Davies, R. O. 179, *205*

Kalow, W., see LaDu, B.N. 170, *208*
Kalow, W.P. 398, *415*
Kamat, S.A. 353, *356*
Kamii, K., see Ueda, H. 243, *256*
Kanai, N., see Hayakawa, T. 338, *340*
Kanter, S.L., see Hollister, L. E. 162, *167*
Kaplan, N.O., see Venter, J.C. 230, *233*
Kaplan, S., see Tan, J.S. 354, *358*
Karachristianidis, G., see see Büch, H. 362, *381*
Karesoja, M., see Manninen, V. 280, *312*
Karnofsky, D. 330, *340*
Karo, W., see Sandler, S.R. 95, 107, *111*
Kasanen, A., see Hartiala, K. 236, *254*
Kasangen, A., Forsstrom, J., Jacmi, H.A. 391, *415*
Kass, I., see Russel, W.F., Jr. 344, *357*
Kater, R.M.H., Roggin, G., Tobon, F., Zieve, P., Iber, F.L. 246, *254*, 288, *311*
Kattamis, C., Zannos-Mariolea, L., Franco, A.P., Liddell, J., Lehmann, H., Davies, D. 180, *207*
Katz, M., Poulsen, B.J. 157, *167*
Katz, S., see Berman, L.B. 337, *338*
Katzung, B.G., see Meyers, F.H. 136, *147*
Kaufmann, J.S., see Fann, W.E. 299, 300, *310*
Kaufmann, J.S., see Oates, J.A. 300, *312*
Kaump, D.H., see Winder, C. V. 142, *149*
Kaye, D., Koenig, M.G., Hook, E.W. 353, *356*
Kaye, D., see Koenig, M.G. 354, *357*
Kazenko, A., see Glatzko, A. J. 95, 98, 100, *110*
Keating, R.P., see Pierson, R. N. 94, *111*
Keberle, H., Hoffmann, K., Bernhard, K. 130, 131, 140, *147*
Keech, G.M., see Levy, G. *312*
Keen, P. [8] 28, *33*, 68, *85*, 218, 226, *232*, 260, *270*, 294, *311*
Keenaghan, J.B., see Boyes, R.N. 163, *166*
Keiding, S., see Winkler, K. 283, *314*
Keiser, H., see Mitchell, J.R. 388, 389, 390, 398, *417*
Kekki, M., Pyörälä, K., Mustala, O., Salmi, H., Jussila, J., Siurala, M. 236, *254*, 279, *311*
Kellerman, L., Posner, A. 194, *207*
Kelly, A.R., see Shideman, F. E. 242, *256*
Kempe, H.C., see Jawetz, E. 246, *356*
Kendrey, G., see Juhasz, J. 407, *415*
Kennedy, T.J., see Zubrod, C.G. 299, *314*
Kenny, F.M., see Bacon, G.E. 262, 265, *268*
Kensler, C.J., see Wodinski, I. 333, *342*
Kensler, C.J., see Yesair, D. W. 140, *149*
Kent, T.H., see Fischer, L.J. 138, *146*, *147*
Kenward, D.H., see Losowsky, M.S. 262, *270*
Kerrand, D.N.S., see Elliot, R.W. 396, *414*
Kershbaum, A., see Bellet, S. 221, *230*
Kessler, W.B., Borman, A. 95, 99, *111*
Ketterer, B., see Litwack, G. 222, *232*
Keys, A., Mickelson, A., Miller, E.V.O., Hayes, E. R., Todd, R.L. 222, *232*
Khanna, N., see Jusko, W.J. 216, 217, *232*
Kiese, M. 400, *415*
Kim, K.E., see Seller, R.H. 316, *324*
Kimball, A.P., Wilson, M.J. 329, *340*
Kimberg, D.V. see Klatskin, G, 409, *415*
Kimmel, H.B., Walkenstein, S.S. 118, *128*
Kimmelman, L.J., see Klein, M. 344, *357*
Kimpel, B.M., see Cutler, R. E. 262, *269*
Kincaid-Smith, P., see Nanra, R.S. 240, *255*
Kindt, H., Schmidt, E. 396, *415*
King, J.O., Denborough, M. A., Zapf, P.W. 252, *254*
King, P.C., see Lawson, D.H. 259, *270*
Kingma, S., see Muller, C.J. 191, *208*
Kiørboe, E. 289, *311*
Kirby, A.J., see Fersht, A.R. 91, *110*
Kirby, J.C., see Olsson, R.A. 350, *357*
Kirby, W.M.M., deMaine, J. B., Serrill, W.S. 264, 266, *270*
Kirby, W.M.M., see Bulger, R.J. 353, *355*
Kirby, W.M.M., see Standiford, H.D. 352, 354, *358*
Kirdani, R.Y., see Sandberg, A.A. 130, *148*
Kirkman, H., see Matthews, V.S. 406, *416*
Kirkman, H.N., see Welt, S.I. 191, *211*
Kishon, Y., see Jewitt, D.E. 163, *167*
Kitchin, F.D., Howel-Evans, W., Clarke, C.A., McConnell, R.B., Shepard, P.M. 192, *207*
Kitchin, F.D., see Price Evans, D.A. 193, *209*
Klaassen, C.D. 130, 143, 144, *147*
Klaassen, C.D., Plaa, G.L. 143, *147*
Klaiber, E., see Burstein, S. 286, *309*
Klatskin, G., Kimberg, D.V. 409, *415*
Klatskin, G. 409, *415*
Klaus, W., see Lee, K.S. 316, *323*
Klein, M., Kimmelman, L.J. 344, *357*
Klein, S.W., see Koch-Weser, J. 247, *254*
Kline, I., Venditti, J.M., Mead, J.A.R., Tyrer, D. D., Goldin, A. 328, *340*
Knapp, D.R., Gaffney, T.E. 230, *232*
Knight, R.A., Selin, M.J., Harris, H.W.M. 177, *207*
Knight, R.A., see Harris, H. W. 177, *206*
Knight, R.N., Young, L. 378, *381*
Knill-Jones, R.P., see Murray-Lyon, I.M. 252, *255*
Knouss, R.P., see Gerhardt, R.E. 291, *310*
Kobara, T.Y., see Nishimura, E.T. 174, *208*
Kobara, T.Y., see Takahara, S. 174, *210*
Kobayashi, G.S., see Medoff, G. 353, *357*
Koch-Weser, J. 197, 202, *207*, 235, 241, *254*
Koch-Weser, J., Klein, S.W. 247, *254*

Koch-Weser, J., Sellers, E.M. 285, *311*
Koch-Weser, J., see Sellers, E. M. 154, 155, *167*, 289, 290, 296, 302, *313*
Kodama, R., see Dreifus, L.S. 250, *253*
Koechlin, B.A., D'Arconte, L. 118, *128*
Koenig, M.G., Kaye, D. 354, *357*
Koenig, M.G., see Bryant, R. E. 344, *355*
Koenig, M.G., see Kaye, D. 353, *356*
Kofler, M., see Berneis, K. 329, *338*
Koivisto, O., see Sotaniemi, E. 246, *256*
Kojima, S., Smith, R.B., Doluisio, J.T. 236, *254*, 279, *311*
Kokenge, R., see Kutt, H. 182, 183, 197, *208*
Kolmodin, B., Azarnoff, D.L., Sjöquist, F. 288, *311*
Kolmodin-Hedman, B. 288
Kopin, I.J. 123, *128*
Kopin, I.J., see Davis, J.M. 291, *310*
Kornguth, M., see Kunin, C. M. 222, *232*
Koskiner, R., see Mattila, M. J. 239, *255*
Kosower, E.M. 107, *111*
Kostenbauder, H., see Reidenberg, M.M. 260, 264, *270*
Koster, R., see Conney, A.H. 117, *127*
Kovacs, O., see Fodor, G. 90, *110*
Kovensky, A., see Elion, G.B. 118, *127*
Kowitz, P.E., see Aggeler, P. M. 184, *203*
Kowitz, P.M., see Aggeler, P. M. 296, *307*
Kozulitzina, T.I., see Smirnov, G.A. 178, *210*
Kramer, J., see Whelton, A. 240, *257*
Kramer, K.-P., see Rauen, H. M. 337, *341*
Krasner, J., Giacola, G.P., Yaffe, S.J. 221, *232*
Krause, R.J., see Cranley, J. J. 320, *323*
Kreis, W., Yen, W. 329, *340*
Kresge, D., see Salatka, K. 220, *233*
Krisch, K., see Heymann, E. 401, *415*
Krishna, G., Gillette, J.R. 407, *415*
Krishna, G., see Brodie, B.B. 401, *413*
Krishna, G., see Corsini, G.U. 406, *413*
Krishna, G., see Mitchell, J.R. 401, *416*
Krishna, G., see Reid, W.D. 379, 380, 381, *382*, 401, *417*
Kristensen, J., Hansen, J.M. 288, *311*
Kristensen, M., Hansen, J.M., Skovsted, L. 285, *311*
Kristensen, M., see Christensen, L.K. 289, 297, 302, *309*
Kristensen, M., see Hansen, J.M. 288, *310*
Kropatkin, M.L., see O'Reilly R.A. 185, 186, *209*
Krüger-Thiemer, E., Levine, R.R. 27, *33*
Kruger, J., see Vogel, F. 196, *211*
Kueter, K.E., see Richards, R.K. 264, *271*
Kugel, R.B., see Ebadi, M. 242, 245, *253*
Kumar, S., see Baker, S.J. 130, *145*
Kume, F., see Endo, H. 405, *414*
Kunin, C.M. 259, 261, 266, *270*, 297, *311*
Kunin, C.M., Craig, W.A., Kornguth, M., Monson, R. 222, *232*
Kunin, C.M., Finland, M. 264, 265, *270*
Kunin, C.M., Glazko, A.J. Finland, M. 264, 265, 266, *270*
Kunin, C.M., see Sachs, J. 259, 266, *271*
Kunishige, E., see Gunnison, J.B. 349, *356*
Kuntzman, R. 283, *311*
Kuntzman, R., Jacobson, M., Levin, W., Conney, A.H. 202, *207*
Kuntzman, R., see Conney, A. H. 169, 200, *205*, 243, *253*, 258, *269*
Kuntzman, R., see Gillette, J.R. 121, *127*
Kuntzman, R., see Levin, W. 404, *416*
Kuntzman, R., see Pantuck, E.J. 363, *382*
Kuntzman, R.G., Tsai, I., Brand, L., Mark, L.C. 248, *254*
Kupchan, S.M., Casy, A.F., Swintosky, J.V. 95, 99, 100, *111*
Kupferberg, H.J., Jeffery, W. Hunninghake, D.B. 289, *311*
Kurman, R.J., see Herbst, A. L. 406, *415*
Kuroda, R., see Hayakawa, T. 338, *340*
Kusmiesz, H.T., see Nelson, J.D. 235, 238, 245, *255*
Kutt, H., Brennan, R., Dehejia, H., Verebely, K. 289, *311*
Kutt, H., Haynes, J., Verebely, K., McDowell, F. 285, 306, *311*
Kutt, H., Verebely, K., McDowell, F. 183, *208*
Kutt, H., Winters, W., Kokenge, R., McDowell, F. 182, 183, 197, *208*
Kutt, H., Winters, W., McDowell, F.H. 289, *311*
Kutt, H., Winters, W., Scherman, R., McDowell, F. 243, *254*
Kutt, H., Wolk, M., Scherman, R., McDowell, F. 182, 183, 197, *208*
Kutt, H., see Brennan, R.W. 183, *204*, 289, *308*
Kutt, H., see Letteri, J.M. 249, *254*, 264, *270*
Kvetina, J., Marcucci, F., Fanelli, R. 119, *128*

Lacher, M.J., Durant, J.R. 334, *341*
Lack, L., Weiner, I.M. 131, 132, 133, *147*, *148*
Lack, L., see Glasser, J.E. 133, *147*
Ladomery, L.G., Ryan, A.J., Wright, S.E. 131, 141, *148*
La Du, B.N. 178, 180, 182, *208*
La Du, B.N., Kalow, W. 170, *208*
La Du, B.N., see Clark, S.W. 179, *204*
Lage, G.L., see Castle, M.C. 137, *146*
Lagergren, C., see Hultengren, N. 406, *415*
Lahey, M.E., see Lanzkowsky, P. 217, *232*
Laidlaw, J., Read, A.E., Sherlock, S. 252, *254*
Lamont, J.T., see Fogelman, A.M. 317, *323*
Lands, A.M., Arnold, A., McAuliff, J.P., Luduena, F.P., Brown, T.G., Jr. 317, *323*

Lands, A. M., see Minatoya, H. 158, *167*
Lands, A. M., see Portmann, G. A. 158, *167*
Landsteiner, K. 407, *415*
Lang, M., see Sellers, E. M. 289, 302, *313*
Langemann, A., see Berneis, K. 329, *338*
Lanzkowsky, P., Lloyd, E. A., Lahey, M. E. 217, *232*
Lanzkowsky, P., Madenlioglu, M., Wilson, J. F., Lahey, M. E. 217, *232*
Lardy, H., see Boyer, P. D. 261, *268*
Larizza, P., Brunetti, P., Grignani, F., Ventura, S. 190, *208*
Larmi, T. K. I., see Mattila, M. J. 239, *255*
Larsen, K., Moller, C. E. 391, *415*
Larson, C. P., Jr., see Eger, E. I. II 158, *167*
Las, S., Singhel, S. N., Burley, D. M., Crossley, G. 397, *416*
Lasagna, L., see Bianchine, J. R. 152, *166*, 238, *253*, 281, *308*
Laster, W. R., Jr., see Hill, D. L. 126, *127*, 337, *340*
Latham, A. M., Millbank, L., Richens, A., Rowe, D. J. F. 286, *311*
Lauber, E., see Aebi, H. 174, *203*
Laurence, D. R., Rosenheim, M. L. 300, *311*
Lavan, J. N., Benson, W. J., Gatenby, A. H., Posen, S. 391, *416*
Lawson, D., see Rosenblum, R. 321, *324*
Lawson, D. H., Boddy, K., King, P. C., Linton, A. L., Will, G. 259, *270*
Lawson, D. H., see Jick, H. 196, *207*
Layne, D. S., Golab, T., Arai, K., Pincus, G. 139, *148*
Layne, D. S., see Arai, K. 139, *145*
LeBlanc, E., see Sellers, E. M. 289, 302, *313*
Lebovitz, H. E., see Horton, E. S. 262, *269*
Lee, K. S., Klaus, W. 316, *323*
Lee, L. E., see Shideman, F. E. 242, *256*
Lee, P. R., see Brodie, B. B. 157, *166*
Leedom, J. M., see Mathies, A. W., Jr. 350, *357*
Lees, A. W., Allan, G. W., Smith, J., Tyrrell, W. F., Fallon, R. J. 399, *416*
Lees, A. W., see Smith, J. 397, *418*
Leevy, C. M., see Cherrick, G. R. 301, *309*
Lefkowitz, R. J., Haber, E. 317, *323*
Lehmann, H., Liddell, J. 180, *208*
Lehmann, H., Ryan, E. 179, *208*
Lehmann, H., Silk, E. 179, *208*
Lehmann, H., see Evans, F. T. 178, *205*
Lehmann, H., see Forbat, A. 179, *205*
Lehmann, H., see Harris, H. 180, *206*
Lehmann, H., see Kattamis, C. 180, *207*
Leibel, B. A., see Mroczek, W. J. 154, *167*
Leikin, S. L., see Evans, A. E. 333, 336, *339*
Leishman, A. W. D., Matthews, H. L., Smith, A. J. 299, *311*
Leith, R. S., see Erturk, E. 407, *414*
Leiwant, B., see Grunberg, E. 176, *206*
Lemann, L., see Bengtsson, U. 406, *413*
Lemberger, L., see Burns, J. J. 122, *127*
Lemberger, L., see Davies, J. M. 291, *310*
Lenaerts, F. M., see Niemegeers, C. J. E. 116, *128*
Lenhard, R. E., see Hoogstraten, B. 335, *340*
Lenz, W. R., see Glazko, A. J. 95, 98, *110*
Leonard, R. F., see Court, J. M. 211, *231*
Leonards, J. R., Levy, G. 94, *111*
Leone, L. A., see Hoogstraten, B. 335, *340*
Leong, J. L., see Wattenberg, L. W. 362, *382*
Leong, L., see Aggelier, P. M. 184, *203*, 296, *307*
Leong, L. S., see O'Reilly, R. A. 29, *34*, 185, 186, *209*
Le Page, G. A. 329, *341*
Le Page, G. A., White, S. C. 334, *341*
Le Page, G. A., see Sartorelli, A. C. 329, *341*
Lepper, M. H., Dowling, H. F. 350, *357*
Lepper, M. H., Dowling, H. F., Jackson, G. G., Spies, H. W., Mellody, M. 345, *357*
Laqueur, G. L. 404, *415*
Lerner, P. I., Weinstein, L. 325, *341*
Leslie, D. W., see Nanra, R. S. 240, *255*
Leslie, L. G., see Wagner, J. G. 21, *34*
Lester, R., Schumer, W., Schmid, R. 130, *148*
Letteri, J. M., Mellk, H., Louis, S., Kutt, H., Durante, P., Glazko, A. 249, *254*, 264, *270*
Leutz, J. C., see Wiebel, F. J. 363, *382*
Le Valley, S. E., see Mitoma, C. 200, *208*
Levey, G. S., Prindle, K. H., Jr., Epstein, S. E. 318, *323*
Levi, A., Sherlock, S., Walker, D. 288, 307, *311*
Levi, A. J., Gatmaitanz, Z., Arias, I. M. 222, *232*
Levi, A. J., Sherlock, S., Walker, D. 243, *254*
Levi, A. J., see Arias, I. M. 287, *308*
Levi, J. A., see Reyes, H. 222, *233*, 301, *313*
Levin, W., Jacobson, M., Sernatinger, E., Kuntzman, R. 404, *416*
Levin, W., see Kuntzman, R. 202, *207*
Levine, B. B. 408, *416*
Levine, H., see Hodgkin, W. E. 180, 181, *206*
Levine, R., see Reyes, H. 301, *313*
Levine, R. M., Clark, B. B. 88, 89, *111*
Levine, R. R. 235, 236, *254*, 278, 279. 280, *311*
Levine, R. R., see Krüger-Thiemer, E. 27, *33*
Levine, R. R., see Walsh, C. T. 215, *233*
Levine, W. G. 144, *148*
Levine, W. G., Millburn, P., Smith, R. L., Williams, R. T. 144, *148*
Levinson, R. A., see Jubiz, W. 287, *311*
Levitan, R., see Gorbach, S. L. 215, *231*
Levitt, M., Marsh, J. C., Deconti, R. C., Mitchell, M. S., Skeel, R. T., Farber, L. R., Bertino, J. R. 335, *341*

Levy, B. 317, *324*
Levy, B.B., see Brodie, B.B. 262, *268*
Levy, B.B., see Soberman, R. 197, *210*
Levy, G. 26, 28, 29, 30, 31, 32, *34*, 154, *167*, 362, *382*
Levy, G., Gibaldi, M. 32, *34*, 35, 63, 67, 68
Levy, G., Gibaldi, M., Procknal, J.A. 238, *254*, 280, *312*
Levy, G., O'Reilly, R.A., Aggeler, P.M., Keech, G. M. *312*
Levy, G., Rao, B.K. 281, *311*
Levy, G., Tsuchiya, T. 29, *34*
Levy, G., Tsuchiya, T., Amsel, L.P. 26, *34*
Levy, G., Yamada, H. 363, 364, *382*
Levy, G., see Gibaldi, M. [5] 20, 24, 31, *33*, 63, *85*
Levy, G., see Hollister, L. 289, *311*
Levy, G., see Jusko, W.J. 28, *33*, 216, 217, *232*
Levy, G., see Leonards, J.R. 94, *111*
Levy, G., see Nagashima, R. 24, 33, *34*
Levy, G., see Perrier, D. 28, *34*
Levy, G., see Tsuchiya, T. 29, *34*
Levy, G., see Yaffe, S.J. 203, *212*, 287, *314*
Levy, P.S., see Dubach, U.D. 391, *414*
Levy, R.I., see Glueck, C.J. 220, *231*
Lewis, A.A.G., see Elliot, R. W. 396, *414*
Lewis, G.P., Jick, H., Slone, D., Shapiro, S. 196, 197, *208*
Lewis, G.P., Jusko, W.J., Burke, C.W., Graves, L. 242, *254*
Lewis, G.P., see Jick, H. 196, *207*
Lewis, J., Jr., see Hertz, R. 335, *340*
Li, M.C., Hsu, K.-P. 336, *341*
Li, M.C., Whitmore, W.F., Jr., Golbey, R., Grabstald, H. 336, *341*
Libow, L.S. 220, *232*
Lichtenstein, J., see Cohen, S. S. 328, *339*
Lichter, M., Black, M., Arias, I.M. 259, 262, 265, *270*
Lidbrink, P., Jonsson, G., Fuxe, K. 121, *128*
Liddell, J., see Kattamis, C. 180, *207*
Liddell, J., see Lehmann, H. 180, *208*
Lie, J.T., see Nanra, R.S. 240, *255*
Lieber, C.S., de Carli, L.M. 196, *208*
Lieber, C.S., see Jones, D.P. 221, *232*
Lief, P.A., see Brodie, B.B. [2] 61, *85*
Liegler, D.G., Henderson, E. S., Hahn, M.A., Oliviero, V.T. 292, *312*, 337, *341*
Lin, S.C., see Gulyassy, P.F. 258, 260, *269*
Lindenberg et al. (1966) 264
Lindqvist, M., see Carlsson, A. 121, *127*
Lindsey, E., see Mandell, G.L. 353, *357*
Lindstedt, S. 131, *148*
Lindup, W.E., see Parke, V. D. 220, *233*
Link, K.P. 184, *208*
Linton, A.L., see Lawson, D. H. 259, *270*
Lippman, R.W., Wrobel, C.J., Rees, R., Hoyt, R. 259, *270*
Lipsett, M.B. 287, *312*
Lipsett, M.B., see Hertz, R. 335, *340*
Lisboa, B.P., see Bönicke, R. 177, 198, *204*
Lischner, H.W. 130, *148*
Litwack et al. (1971) 38
Litwack, see Morey 38
Litwack, G., Ketterer, B., Arias, I.M. 222, *232*
Ljungqvist, A., see Hultengren, N. 406, *415*
Lloyd, E.A., see Lanzkowsky, P. 217, *232*
Loeb, M.R., see Cohen, S.S. 328, *339*
Loewe, S. 326, *341*
Lombroza, L., see Mitoma, C. 200, *208*
London, I., Poet, R.B. 95, 104, 106, *111*
Long, M.W., see Mitchell, J. R. 397, 408, 409, *417*
Longo, V.G., see Bovet, D. 179, *204*
Loo, J., see Riegelman, S. [12] 63, *85*
Loo, J.C.K., Riegelman, S. 21, 22, *34*
Loos, J.A., see Oort, M. 189, *208*
Loos, J.A., see Prins, H.K. 189, *209*
Lopez de Novales, E., see Richet, G. 261, 267, *271*
Loriaux, D.L., see Menard, R. H. 403, *416*
Losowsky, M.S., Kenward, D. H. 262, *270*
Louis, S., see Letteri, J.M. 249, *254*, 264, *270*
Lovell, R.R.H., see Denborough, M.A. 194, *205*
Lowell, V.F., see Shideman, F.E. 242, *256*
Lowenbraun, S., Devita, V. T., Serpick, A.A. 335, *341*
Lowman, E.W., see Brodie, B.B. 157, *166*
Lubash, G.D., see Cohen, B. D. 316, *323*
Lucchesi, B.R. 318, *324*
Lucey, J.F., see Harris, R.C. 297, *310*
Luchi, R.J., Gruber, J.W. 246, *254*
Luchi, R.J., see Gerhard, R. E. 291, *310*
Luduena, F.P., see Lands, A. M. 317, *323*
Lüllmann, H., see Haass, A. 239, *254*
Lumb, G., see Feuer, G. 363, *381*
Lumb, G.A., Mawer, E.B., Stanbury, S.W. 259, *270*
Lumb, G.A., see Stanbury, S. W. 259, *271*
Lund, F., see von Daehne, W. 95, 96, *109*
Lund, L., Berlin, A., Lunde, P.K.M. 240, *254*
Lunde, P., see Odar-Cederlöf, I. 242, *255*, 260, 268, *270*
Lunde, P.K.M., see Lund, L. 240, *254*
Lundquist, F., Wolthers, H. 26, 27, *34*

Maas, A.R., see Van Loon, E.J. 143, *149*
Macdonald, F.M., see Jenne, J.W. 176, *207*
MacDonald, H., Place, V.A., Falk, H., Darken, M.A. 235, *255*
MacDonald, M.G., Robinson, D.S. 285, *312*
MacDonald, M.G., Robinson, D.S., Sylwester, D., Jaffe, J.J. 285, 287, *312*
MacDonald, M.G., see Robinson, D.S. 285, *313*
MacDonald, M.S., Emery, J.L. 223, *232*
Mace, J.W., see Cunningham, M.D. 287, *309*
MacGee, J., see Werk, E.E., Jr. 287, *314*

MacGregor, G. A., Poole-Wilson, P. A., Jones, N. F. 261, *270*
Mack, E., see Dowling, R. H. 131, *146*
Mackay, F. J., Cooper, J. R. 114, *128*
Mackenzie, A. R. 337, *341*
MacLagan, J., see Bigland, B. 252, *253*
MacLaughlin, D. S., see Jick, H. 196, *207*
MacLean, H., see Schlichter, J. G. 347, *357*
MacLean, J. R., see Harris, R. C. 297, *310*
Maddrey, W. C., Boitnott, J. R. 408, *416*
Madenlioglu, M., see Lanzkowsky, P. 217, *232*
Maezawa, H., see Ueda, H. 243, *256*
Magee, P. H., Barnes, J. M. 404, 410, *416*
Maggio, A., see Pantuck, E. J. 363, *382*
Magnussen, M. P. 279, *312*
Mahaffee, C., see Butler, T. C. 266, *268*
Mahar, C. Q., see Webb, D. I. *314*
Mahar, C. R., see Webb, D. I. 239, *257*
Mahler, J. F., see Schreiner, G. E. 261, *271*
Maines, M. D., Andres, M. W. 404, *416*
Maitre, L., see Brunner, H. 123, *127*
Mally, J., see Meyer, H. 176, *208*
Maloney, J. R., see Allansmith, M. 220, *230*
Mandala, P. S., see O'Connor, W. J. 216, *233*
Mandel, W., Heaton, A. D., Russell, W. F., Jr., Middlebrook, G. 344, *357*
Mandell, G. L., Lindsey, E., Hook, E. W. 353, *357*
Manenti, F., see Carulli, N. 288, 302, *309*
Manley, K. A., see Price Evans, D. A. 177, *209*
Mann, W. A., see Caldwell, H. C. 95, 108, 109, *109*
Manninen, V., Apajalahti, A., Melin, J., Karesoja, M. 280, *312*
Manning, J. A., see Ockner, R. K. 222, *233*
Manson, L. A., see Defendi, V. 333, *339*
Mantel, N. 327, 328, *341*
Mantel, N., see Epstein, S. S. 407, *414*
Mantel, N., see Goldin, A. 325, *340*
Mantel, N., see Venditti, J. M. 326, *342*
Mapleson, W. W. [9] 57, *85*
Maplestone, P. A., see Denborough, M. A. 194, *205*
March, C. H., Elliott, H. W. 135, *148*
Marchand, B., see Papapetrou, P. D. 145, *148*
Marchessault, J. H. V., see Drummond, K. H. 290, *310*
Marcucci, F., Mussini, E., Fanelli, R., Garratini, S. 119, *128*
Marcucci, F., see Garattini, S. [4] 39, *85*, 120, *127*
Marcucci, F., see Kvetina, J. 119, *128*
Mariast, F. M., see Bing, R. J. 315, *322*
Mark, A. L., see Schuelke, D. M. 321, *324*
Mark, L. C. 114, *128*
Mark, L. C., see Brodie, B. B. [2] 61, *85*
Mark, L. C., see Kuntzmann, R. G. 248, *254*
Markors, P. R., see Venter, J. C. 230, *233*
Marks, Gross 190
Marks, P. A., Banks, J. *416*
Marks, V., see Baylis, E. M. 286, *308*
Markus, G. 260, *270*
Marotta, M., see Bovet, D. 179, *204*
Marquardt, D. W. 33, *34*
Mars, G., see Sereni, F. 241, *256*
Marsh, J. C., see Levitt, M. 335, *341*
Marshall, see Curry 197
Marshall, F. J., see McMahon, R. E. 116, 125, *128*
Marshall, J. H. L., see Curry, S. H. 286, *309*
Marshall, M. J., see Curtis, J. R. 249, *253*
Martens, S., see Hammer, W. 202, *206*
Marton, A. V., see Davies, R. O. 179, *205*
Martz, B. L., see Fisch, C. 316, *323*
Marubini, E., see Sereni, F. 241, *256*
Mason, D. T., Braunwald, E. 251, *255*
Masters, C. J., see Holmes, R. S. 174, *207*
Masters, Y. F., see Menguy, R. 246, 247, *255*
Masucci, F. D., see Pals, D. T. 320, *324*
Mathews, K. D., Pan, P. M., Wells, J. H. 408, *416*
Mathies, A. W., Jr., Leedom, J. M., Ivler, D., Wehrle, P. F., Portnoy, B. 350, *357*
Matsukura, S., see Meikle, A. W. 287, *312*
Matsumoto, K., see Mori, K. 407, *417*
Matsuzawa, T., see Yaffe, S. J. 203, *212*, 287, *314*
Matthew, H., see Prescott, L. F. 390, *417*
Matthews, H. L., see Leisman, A. W. D. 299, *311*
Matthews, V. S., Kirkman, H., Bacon, R. L. 406, *416*
Mattila, M. J., Friman, A., Larmi, T. K. I., Koskiner, R. 239, *255*
Mattok, G. L., see McGilveray, I. J. 236, *255*
Mattocks, A. R. 405, *416*
Mattocks, A. R., see Seawright, A. A. 396, *418*
Mawer, E. B., see Lumb, G. A. 259, *270*
Mawer, E. B., see Stanbury, S. W. 259, *271*
Mawer, G. E., Miller, N. E., Turnberg, L. A. 242, 244, 249, *255*
Mawer, G. E., see Balasubramanian, K. 241, 250, *252*
Maxwell, J. D., Hunter, J., Stewart, D. A., Ardeman, S., Williams, R. 286, *312*
Maxwell, J. D., see Hunter, J. 202, *207*, 286, *311*
Mayer, S. E. 317, 318, *324*
Mayersohn, M., Gibaldi, M. 19, *34*
Maynert, E, W. 182, *208*
McAllister, C. K., see Meyer, J. F. 299, *312*
McAuliff, J. P., see Lands, A. M. 317, *323*
McCabe, W. R. 351, *357*
McCabe, W. R., Jackson, G. G. 353, 354, *357*
McCall, C. E., see Sabath, L. D. 352, 354, *357*
McCarthy, C. G., Finland, M. 153, *167*
McClellan, B. H., see Allansmith, B. H. 220, *230*
McConnell, R. B., see Kitchin, F. D. 192. *207*

McCormack, J. J., see Robinson, D. S. 281, *313*
McCracken, G. H., Jr., *232*
McCracken, G. H., Jr., Jones, L. G. 224, *232*
McDermott, W., Bunn, P. A., Benoit, M., Dubois, R., Reynolds, M. E. 153, *167*
McDonald, R. K., see Miller, J. H. 224, *232*
McDowell, F., see Brennan, R. W. 183, *204*, 289, *308*
McDowell, F., see Kutt, H. 183, 197, *208*, 243, *254*, 285, 289, 306, *311*
McElwain, T. J., see Hardisty, R. M. 331, *340*
McGilveray, I. J., see Mattok, G. L. 236, *255*
McHenry, M. C., Gavan, T. L., Gifford, R. W., Jr., Geurkink, N. A., van Ommen, R. A., Town, M. A., Wagner, J. G. 248, *255*, 266, *270*
McKusick, V. A. 170, *208*
McKusick, V. A., see Price Evans, D. A. 177, *209*
McLean, E. K. 405, *416*
McMahon, F. G., see Hague, D. E. 289, *310*
McMahon, R. E., Culp, H. W. Marshall, F. J. 116, *128*
McMahon, R. E., Marshall, F. J., Culp, H. W. 125, *128*
McMahon, T. 214, *232*
McMenamy, R. H., Oncley, J. L. 221, *232*
McMullin, G. P. 287, *312*
McMullin, G. P., Hayes, M. F., Arora, S. C. *312*
McNamara, H., see Barnett, H. L. 225, *230*
McNamara, J. H., see Nanra, R. S. 240, *255*
McNay, J. L., Bogafrt, M., Goldberg, L. I. 321, *324*
McNay, J. L., Goldberg, L. I. 321, *324*
McNay, J. L., see Goldberg, L. I. 321, 322, *323*
McNay, J. L., see Yeh, B. K. 321, *324*
McQueen, E. G., Owen, D., Ferry, D. G. 288, *312*
McSherry, M. A., see Cattell, W. R. 353, *355*
Mead, J. A. R., see Kline, I. 328, *340*
Meads, M., see Finland, M. 216, *231*
Medoff, G., Comfort, M., Kobayashi, G. S. 353, *357*
Meeker, B. E., see Bruce, W. R. 329, *338*
Meier, M., see Brunner, H. 123, *127*
Meikle, A. W., Jubiz, W., Matsukura, S., West, C. D., Tyler, F. H. 287, *312*
Meikle, A. W., see Jubiz, W. 287, *311*
Melin, J., see Manninen, V. 280, *312*
Mellett, L. B. 126, *128*
Mellk, H., see Letteri, J. M. 249, *254*, 264, *270*
Mellody, M., see Lepper, M. H. 345, *357*
Melmon, K. L., see Benowitz, N. 293, *308*
Melmon, K. L., see Rowland, M. 163, *167*
Melmon, K. L., see Thomson, P. D. 156, *168*, 241, 243, 245, *256*, 307, *314*
Menard, R. H., Stripp, B., Gillette, J. R. 402, *416*
Menard, R. H., Stripp, B., Loriaux, D. L., Bartter, F. C., Gillette, J. R. 403, *416*
Menard, R. H., see Stripp, B. 402, 403, 407, *418*
Mendelson, D., Block, J. B., Sherpick, A. A. 335, *341*
Mendelssohn, S., see Seller, R. H. 316, *324*
Mendoza, E., see Jenne, J. W. 176, *207*
Mengel, G. D., see Evans, J. S. 333, *339*
Menguy, R., Desbaillets, L., Okabe, S., Masters, Y. F. 246, 247, *255*
Merril, J. P., see Hampers, C. L. 262, *269*
Merrit, A. D., Fetter, B. F. 408, *416*
Merritt, H. H., see Yahr, M. D. 182, *212*
Metys, J., see Metysova, J. 121, *128*
Metysova, J., Metys, J. Votava, Z. 121, *128*
Metz, E., see Elion, G. B. 118, *127*
Metzler, C. M. 33, *34*
Metzler, C. M., see Wagner, J. G. 21, *34*
Meyer, H., Mally, J. 176, *208*
Meyer, J., Sorter, H., Oliver, J., Necheles, H. 217, *232*
Meyer, J. F., McAllister, C. K., Goldberg, L. I. 299, *312*
Meyer, M. B., Zelechowski, K. 234, *255*
Meyers, B. R., see Rahal, J. J., Jr. 353, 354, *357*
Meyers, F. H., see Katzung, B. G. 136, *147*
Meythaler, C., see Schmid, E. 259, *271*
Micheli, A., see Aebi, H. 174, *203*
Michelli, A., Aebi, H. 174, *208*
Michielsen, P., de Schepper, P. J., de Broe, M., Valkenborgh, G., Tricot, J. P. 391, *416*
Mickelson, O., see Keys, A. 222, *232*
Middlebrook, G., see Carlson, H. B. 176, *204*
Middlebrook, G., see Cohn, M. I. 344, *355*
Middlebrook, G., see Heaton, A. D. 344, *356*
Middlebrook, G., see Mandel, W. 344, *357*
Middlebrook, G., see Russell, W. F., Jr. 344, *357*
Miettinen, O. S., see Jick, H. 196, *207*
Mihara, S., see Takahara, S. 173, *210*
Miles, C. P., Coleman, V. R., Gunnison, J. B., Jawetz, E. 352, *357*
Millbank, L., see Latham, A. M. 286, *311*
Millburn, P., see Clark, A. G. 131, 138, *146*
Millburn, P., see Fischer, L. J. 131, 138, *146*
Millburn, P., see Levine, W. G. 144, *148*
Millburn, P., see Williams, R. T. 130, 143, *149*
Miller, A. K., see Beyer, K. H. 361, *381*
Miller, E., Rock, H. J., Moore, M. L. 95, 106, *111*
Miller, E. C., Miller, J. A., Enomoto, M. 391, *416*
Miller, E. C., see Swenson, D. H. 395, *418*
Miller, E. V. O., see Keys, A. 222, *232*
Miller, H. I., see Rodahl, K. 221, *233*
Miller, H. S., Jr., see Davidson, I. W. F. 152, *166*
Miller, J. A. 396, 404, *416*

Miller, J. A., see Miller, E. C. 391, *416*
Miller, J. A., see Miller, R. C. 404, *416*
Miller, J. A., see Swenson, D. H. 395, *418*
Miller, J. G., see Gordon, M. 89, *110*
Miller, J. H., McDonald, R. K., Shock, N. W. 224, *232*
Miller, K. S., see Peters, J. H. 176, 177, *209*, 400, *417*
Miller, M. E., see Vagenakis, A. G. 251, *257*
Miller, N. E., see Mawer, G. E. 242, 243, 244, 249, *255*
Miller, O. N., Hamilton, J. G., Goldsmith, G. A. 124, *128*
Miller, R. A. 215, *232*
Miller, R. C., Miller, J. A. 404, *416*
Miller, S. P., see Hoogstraten, B. 335, *340*
Minatoya, H., Lands, A. M., Portmann, G. A. 158, *167*
Minatoya, H., see Portmann, G. A. 158, *167*
Minder, F., see Dubach, U. D. 391, *414*
Minkel, H. P., see Sessions, J. T. 243, *256*
Mirkin, B. L. 230, *232*
Misher, A., Adams, H. J., Fishler, J. J., Jones, R. G. 94, 95, *111*
Mishkin, S., Stein, L., Gatmaitan, Z., Arias, I. M. 222, *233*
Mitchell, G. D., see Tolomie, J. D. 252, *256*
Mitchell, J. R. 410, *416*
Mitchell, J. R., Arias, L., Oates, J. A. 299, *312*
Mitchell, J. R., Cavanaugh, J. H., Arias, L., Oates, J. A. 299, 300, *312*
Mitchell, J. R., Cavanaugh, J. H., Dingell, J. V., Oates, J. A. 389, *416*
Mitchell, J. R., Jollow, D. J. 384, 386, *417*
Mitchell, J. R., Jollow, D. J., Gillette, J. R., Brodie, B. B. 384, 385, 390, 400, *416*
Mitchell, J. R., Jollow, D. J., Potter, W. Z., Davis, D. C., Gillette, J. R., Brodie, B. B. 242, 245, *255*, 378, *382*, 384, 386, 390, 391, 392, 394, *416*
Mitchell, J. R., Jollow, D. J., Potter, W. Z., Gillette, J. R., Brodie, B. B. *382*, 385, 386, 387, 389, 390, 393, *417*
Mitchell, J. R., Long, M. W., Thorgeiersson, U. P., Jollow, D. J. 397, 408, 409, *417*
Mitchell, J. R., Oates, J. A. 299, *312*
Mitchell, J. R., Potter, W. Z., Hinson, J. A., Jollow, D. J. 393, 395, 396, *417*
Mitchell, J. R., Potter, W. Z., Hinson, J. A., Snodgrass, W. R., Timbrell, J. A., Gillette, J. R. 290, *312*, 362, 375, 376, 377, 378, *382*
Mitchell, J. R., Potter, W. Z., Jollow, D. J. 393, *417*
Mitchell, J. R., Reid, W. D., Christie, B., Moskowitz, J., Krishna, G., Brodie, B. B. 401, *416*
Mitchell, J. R., Thorgeirsson, U. P., Black, M., Timbrell, J. A., Potter, W. Z., Snodgrass, W. R., Jollow, D. J., Kreiser, H. 398, *417*
Mitchell, J. R., Thorgeirsson, S. S., Potter, W. Z., Jollow, D. J., Keiser, H. 388, 389, 390, *417*
Mitchell, J. R., see Black, M. 397, 408, *413*
Mitchell, J. R., see Brodie, B. B. 410, *413*
Mitchell, J. R., see Dates J., A. 300, *312*
Mitchell, J. R., see Davis, D. C. 384, 391, *413*
Mitchell, J. R., see DiChiara, G. 400, 401, *414*
Mitchell, J. R., see Gillette, J. R. 290, *310*, 384, *414*
Mitchell, J. R., see Hinson, J. A. 390, 391, 392, *415*
Mitchell, J. R., see Jollow, D. J. 245, *254*, 376, 378, 379, 380, *381*, 384, 385, 386, 388, 390, 401, 403, *415*
Mitchell, J. R., see Potter, W. Z. 245, *255*, 377, *382*, 384, 385, 386, 388, 389, 390, 392, 393, 394, 396, 397, 410, 411, *417*
Mitchell, J. R., see Reid, W. D. 401, *417*
Mitchell, J. R., see Shand, D. G. [16] 68, 69, 74, *85*, *418*
Mitchell, J. R., see Snodgrass, W. 391, 392, 398, *418*
Mitchell, J. R., see Thorgeirsson, S. S. 390, 391, 394, *418*
Mitchell, J. R., see Timbrell, J. A. 398, *419*
Mitchell, J. R., see Weihe, M. 393, 394, *419*
Mitchell, J. R., see Zampaglione, N. 378, 380, *382*, 401, *417*
Mitchell, M. S., see Levitt, M. 335, *341*
Mitchell, R. S., Bell, J. C., Riemensnider, D. K. 177, *208*
Mitchison, D. A., see Rao, K. V. N. 398, *417*
Mitoma, C., Lombroza, L., LeValley, S. E., Dehn, F. 200, *208*
Mix, L., see Fenna, D. 196, *205*
Miya, T. S., see Borowitz, J. L. 280, *308*
Moddel, G., see Feuer, G. 363, *381*
Moellering, R. C., Jr., Weinberg, A. N. 353, *357*
Moellering, C. R., Jr., Wennersten, C., Weinberg, A. N. 352, 354, *357*
Moellering, R. C., Jr., see Zimmermann, R. A. 352, 353, *358*
Mogami, H., see Hayakawa, T. 338, *340*
Moller, C. E., see Larsen, K. 391, *415*
Monson, R., see Kunin, C. M. 222, *232*
Montgomery, J. 329, *341*
Moon, J. H., see Hayes, D. M. 330, *340*
Moore, C. V., see Dubach, R. 259, *269*
Moore, M. L., see Miller, E. 95, 106, *111*
Moore, P. F., see Borowitz, J. L. 280, *308*
Moore, R. B., Crane, C. A., Frantz, I. D., Jr. 133, *148*
Morawetz, H., Gaetjens, E. 92, *111*
Morawetz, H., Westhead, E. W., Jr. 92, *111*
Morawetz, H., see Gaetjens, E. 92, *110*
Morch, P., see von Daehne, W. 95, 96, *109*
Morey, Litwack 38
Morgan, C. D., see Conolly, M. E. 159, *166*

Morgan, J. P., see Bianchine, J. R. 152, *166*, 238, *253*, 281, *308*
Mori, K., Yasuno, A., Matsumoto, K. 407, *417*
Morita, M., Feller, D. R., Gillette, J. R. 407, *417*
Morita, M., see Feller, D. R. 407, *414*
Morris, J. E., see Erturk, E. 407, *414*
Morris, J. J., see Gerhard, 291, *310*
Morselli, P. L., Rizzo, M., Garattini, S. 289, 302, *312*
Morselli, P. L., see Tognoni, G. 122, *129*
Morton, W. R., see Becker, B. 192, *204*
Moskowitz, J., see Mitchell, J. R. 401, *416*
Moslen, M. T., see Reynolds, E. S. 399, 409, *418*
Motulsky, A. G. 169, 170, 177, 181, 187, 198, *208*
Motulsky, A. G., Yoshida, A., Stamatoyannopoulos, G. 187, 188, 189, *208*
Motulsky, A. G., see Hodgkin, W. E. 180, 181, *206*
Motulsky, A. G., see Yoshida, A. 182, *212*
Mould, G. P., see Curry, S. H. 162, *166*
Mountain, I. M., see Tarnowski, G. S. 330, *341*
Moyer, J. H., see Dreifus, L. S. 250, *253*
Mroczek, W. J., Leibel, B. A., Davidov, M., Finnerty, F. A. 154, *167*
Muggleton, P. W., see Ungar, J. 95, 96, *112*
Muller, C. J., Kingma, S. 191, *208*
Mullins, J., see Capizzi, R. L. 338, *339*
Munro, D. D., see James, V. H. T. 155, *167*
Munro, H. N. 260, *270*
Munrofaure, A. D., see Conney, A. H. 169, 200, *205*, 243, *253*
Murai, K., see Celmer, W. D. 95, 97, *109*
Murdaugh, H. V., see Bacon, G. E. 262, 265, *268*
Murphy, M. L., see Dougherty, J. E, 137, *146*
Murray, R. E., see Reiner, O. 404, *418*
Murray-Lyon, I. M., Young, J., Parkes, J. D., Knill-Jones, R. P., Williams, R. 252, *255*
Musgrave, G. E., see Goldberg, L. I. 322, *323*
Musser, E. A., see Evans, J. S. 333, *339*
Mussini, E., see Garattini, S. [4] 39, *85*, 120, *127*
Mussini, E., see Marcucci, F., 119, *128*
Mustala, O., see Kekki, M. 236, *254*, 279, *311*
Mustala, O., see Siurala, M. 236, *256*, 279, *313*
Myhre, E., Brodwall, E. K., Stenbaek, Ø., Hansen, I. 249, *255*
Myhre, E., see Stenbaek, Ø. 249, *256*
Myrbäck, K., see Boyer, P. D. 261, *268*

Nador, K., see Fodor, G. 90, *110*
Nagashima, R., Levy, G., O'Reilly, R. A. 24, *34*
Nagashima, R., O'Reilly, R. A., Levy, G. 32, *34*
Nagashima, R., see Gibaldi, M. [5] 20, *33*, 63, *85*
Nahas, L., see Gorbach, S. L. 215, *231*
Nair, N. G. K., see Rao, K. V. N. 398, *417*
Nakajima, A., see Ueda, H. 243, *256*
Nakamura, K., see Pantuck, E. J. 363, *382*
Nanra, R. S., Hicks, J. D., McNamara, J. H., Lie, J. T., Leslie, D. W., Jackson, B., Kincaid-Smith, P. 240, *255*
Narahara, K., see Shapiro, W. 317, *324*
Nasca, P. C., see Greenwald, P. 406, *414*
Nassau, E., see Heathcote, A. G. S. 95, 96, *110*
Nathan, H., see Elion, G. B. 328, 337, *339*
Nathanielsz, P. W. *233*
Ndoping, N., see Assem, E. S. K. 408, *413*
Neale, R. J., see Beresford, C. H. 238, *253*
Nebert, D. W. 230, *233*
Necheles, H., see Meyer, J. 217, *232*
Neel, J. V., Schull, W. J. 198, *208*
Neel, J. V., see Hamilton, H. B. 173, *206*
Neel, J. V., see Takahara, S. 174, *210*
Neitlich, H. W. 181, 182, *208*
Nelson, E. 243, *255*
Nelson, E., O'Reilly, I. 8, *34*
Nelson, E., see Wagner, J. G. 13, *34*
Nelson, J. D., Shelton, S., Kusmiesz, H. T., Haltalin, K. C. 235, 238, 245, *255*
Nelson, R., see Byrd, R. B. 409, *413*
Nelson, W. L., see Potter, W. Z. 393, 394, *417*
Nelson, W. L., see Thorgeirsson, S. S. 394, *418*
Nelson, W. L., see Weihe, M. 393, 394, *419*
Nery, R. 392, *417*
Neuvonen, P. J., Gothoni, G., Hackman, R., Bjorksten, K. 280, *312*
Newton, R. W., see Prescott, L. F. 390, *417*
Newton, W. A., Jr., see Evans, A. E. 333, 336, *339*
Ng, L., see Vesell, E. S. 203, *211*
Nicholls, P. J., Orton, T. C. 115, *128*
Nichols, R., see Capizzi, R. L. 338, *339*
Nicholson, H., see Assem, E. S. K. 408, *413*
Niemegeers, C. J. E., Lenaerts, F. M., Janssen, P. A. J. 116, *128*
Nies, A. S., Evans, G. H., Shand, D. G. 293, *312*
Nies, A. S., Oates, J. A. 289, *312*
Nies, A. S., see Branch, R. A. 283, 293, 300, 306, 307, *308*, 363, *381*
Nies, A. S., see Evans, G. H. 276, 297, *310*
Nies, N. S., see Shand, D. G. 164, *168*
Nieweg, H. O. 408, *417*
Nilsson, K. O., see Dymling, J. F. 403, *414*
Nimmo, J. 241
Nimmo, J., Heading, R. C., Tothill, P., Prescott, L. F. 237, 239, *255*
Nimmo, J., see Clarke, R. A. 245, *253*
Nimmo, J., see Heading, R. C. 236, *254*, 280, *310*
Nimmo, J., see Prescott, L. F. 235, 236, 245, *255*
Nishimura, E. T., Carson, S. N., Kobara, T. Y. 174, *208*

Nishimura, E.T., Hamilton, H.B., Kobara, T.Y., Takahara, S., Ogura, Y., Doi, K. 174, *208*
Nishimura, E.T., see Takahara, S. 174, *210*
Noell, P., see Sachs, J. 259, 266, *271*
Noone, P., Pattison, J.R. 350, *357*
Norman, A., Sjovall, J. 131, *148*
Norris, A.H., Shock, N.W. 213, *233*
North, H.B., see Pfiffner, J.J. 99, *111*
Northam, J.I., see Wagner, J.G. *34*
Novak, E., see Wagner, J.G. 21, *34*
Novick, W.J., see van Loon, E.J. 143, *149*
Nowlin, J., see Harvey, D.J. 402, *414*
Nowlin, J., see Horning, M.G. 230, *232*
Nuckolls, E.M., see Shand, D.G. 197, *210*, 318, *324*
Nunes, W.T., see Powell, R.C. 134, *148*
Nyberg, W., see Grasbeck, R. 130, *147*

Oates, J.A., Mitchell, J.R., Feagin, O.T., Kaufmann, J.S., Shand, D.G. 300, *312*
Oates, J.A., see Fann, W.E. 299, 300, *310*
Oates, J.A., see Mitchell, J.R. 299, 300, *312*, 389, *416*
Oates, J.A., see Nies, A.S. 289, *312*
Oates, J.A., see Pettinger, W.A. 301, *313*
Oates, J.A., see Shand, D.G. [16] 68, 69, 74, *85*, 197, *210*, 318, *324*, *418*
O'Brien, J.P., Sharpe, A.R., Jr. 262, *270*
Ockner, R.K., Manning, J.A., Poppenhauser, B.B., Ho, W.K.L. 222, *233*
O'Connor, W.J., Warren, G.H., Edrada, L.S., Mandala, P.S., Rosenman, S.B. 216, *233*
Odar-Cederlöf, I., Lunde, P., Sjöqvist, F. 242, *255*, 260, 268, *270*
Odar-Cederlöf, I., see Reidenberg, M.M. 242, *256*, 260, 264, *270*
Odell, G.B. 297, *312*
O'Dell, R.M., see Goff, J.B. 259, *269*
O'Dell, R.M., see Schlegel, J.U. 259, 266, *271*
Odoroff, C.L., see Griner, P.F. 296, *310*
Ogg, C.S. 259, *270*
O'Grady, F., see Cattell, W.R. 353, *355*
O'Grady, F., see Sales, J.E.L. 239, 240, *256*
Ogura, Y., see Nishimura, E.T. 174, *208*
Ogura, Y., see Takahara, S. 174, *210*
Oh, W., see Hyvarinen, M. 220, *232*
Ohkita, T., see Clarkson, B. 333, *339*
Ohnhaus, E.E. 249, *255*
Ohnhaus, E.E., Thorgeirsson, S.S., Davies, D.S., Breckenridge, A. 300, *312*, 363, *382*
Ohta, M., see Field, J.B. 292, *310*
Okabe, S., see Mengay, R. 246, 247, *255*
Okita, G.T. 136, 137, *148*
Okita, G.T., Talso, P.J., Curry, J.H., Jr., Smith, F.D., Jr., Geilling, E.M.K. 136, *148*
Olderhausen, H.F., see Held, H. 243, *254*
Olesen, O.V. 289, *312*
Olinger, C., see Choi, Y. 286, 287, *309*
Olinger, C., see Werk, E.E., Jr. *314*
Oliver, J., see Meyer, J. 217, *232*
Oliviero, V.T., see Liegler, D.G. 292, *312*, 337, *341*
Oliviero, V.T., see Zager, R.F. 337, *342*
Olson, B.M., see Schroeder, W.A. 174, *210*
Olson, K.B., see Hoogstraten, B. 335, *340*
Olsson, R.A., Kirby, J.C., Romansky, M.J. 350, *357*
O'Malley, K., Browning, M., Stevenson, I., Turnbull, M.J. 286, *312*
O'Malley, K., Crooks, J., Duke, E., Stevenson, I.H. 200, *208*, 243, 244, *255*
O'Malley, K., see Ballinger, B. 246, *253*
Oncley, J.L., see McMenamy, R.H. 221, *232*
Oort, M., Loos, J.A., Prins, H.K. 189, *208*
Oort, M., see Prins, H.K. 189, *209*
O'Reilly, I., see Nelson, E. 8, *34*
O'Reilly, R.A. 186, *208*, 289, *312*
O'Reilly, R.A., Aggeler, P.M. 184, 185, 197, *209*, 301, *312*
O'Reilly, R.A., Aggeler, P.M., Hoag, M.S., Leong, L.S., Kropatkin, M.L. 185, 186, *209*
O'Reilly, R.A., Aggeler, P.M., Leong, L.S. 29, *34*
O'Reilly, R.A., Pool, J.G., Aggeler, P.M. 184, 185, 186, *209*
O'Reilly, R.A., Sahud, M.A., Robinson, A.J. 297, *313*
O'Reilly, R.A., see Aggeler, P.M. 184, *203*, 279, 296, 297, 301, *307*
O'Reilly, R.A., see Levy, G. *312*
O'Reilly, R.A., see Nagashima, R. 24, 32, *34*
Orgell, W.H., Vaidya, K.A., Dahm, P.A. 181, *208*
Orme, B.M., Cutler, R.E. 260, *270*
Orme, B.M., see Cutler, R.E. 248, *253*
Orme, M., see Breckenridge, A. 203, *204*
Orme, M.L.E., see Breckenridge, A. 305, *308*
Orme, R., see Breckenridge, A. 285, *288*, *308*
Ornstein, G.G., see Robitzek, E.H. 176, *209*
Orton, T.C., see Nicholls, P.J. 115, *128*
Ory, E.M., see Finland, M. 216, *231*
Osborne, R.K., see Haber, E. 408, *414*
Ostfeld, E., see Szeinberg, A. 181, *210*
Ota, K., see Clarkson, B. 333, *339*
Ota, M., see Ueda, H. 243, *256*
Otsuji, H., see Ikeda, M. 410, *415*
Ove, P., see Salatka, K. 220, *233*
Overo, K.F., see Gram, L.F. 289, *310*
Owen, D., see Mc Queen, E.G. 288, *312*
Owens, A.H., see Hoogstraten, B. 335, *340*

Padarathsingh, M., see Vadlamudi, S. 336, *341*
Page, J. G., see Vesell, E. S. 265, *271*, 285, 287, 288, 289, 306, *314*
Paglialunga, S., see Jori, A. 121, *128*
Palm, D., Grobecker, H., Fengler, H. 124, *128*
Pals, D. T., Masucci, F. D., Sipos, F., Denning, G. S., Jr. 320, *324*
Pan, P. M., see Mathews, K. D. 408, *416*
Pannacciulli, I., see Salvidio, E. 190, *210*
Pantuck, E. J., Hsiao, K. C., Maggio, A., Nakamura, K., Kuntzman, R., Conney, A. H. 363, *382*
Papapetrou, P. D., Marchand, B., Gavras, H., Alexander, W. D. 145, *148*
Papper, E. M. 158, *167*
Papper, E. M., see Brodie, B. B. [2] 61, *85*
Parke, D. V., Lindup, W. E. 220, *233*
Parke, D. V., see Azouz, W. M. 378, *381*
Parker, J. C., see Wel t, S. I. 191, *211*
Parker, W. J., see Jick, H. 196, *207*
Parkes, J. D., see Murray-Lyon, I. M. 252, *255*
Parmley, W. W., see Glick, G. 318, *323*
Passananti, G. T., see Vesell, E. S. 287, 289, *314*, 390, *419*
Paterson, J. W., Conolly, M. E., Davies, D. S., Dollery, C. T. 162, *167*
Paterson, J. W., Conolly, M. E., Dollery, C. T., Hayes, A., Cooper, R. G. 164, *167*
Paterson, J. W., see Conolly, M. E. 159, *166*
Paterson, J. W., see Walker, S. R. 155, *168*
Paterson, P. Y., see Hunter, T. H. 354, *356*
Payne, H. M., Hackney, R. L., Jr. 95, 101, *111*
Payoust, M. R., Isselbacher, K. J. 133, *148*
Pearce, M. L., see Fogelman, A. M. 317, *323*
Peart, H. E., see Armstrong, A. R. 177, *203*
Peck, A. W., see Conney, A. H. 169, 200, *205*, 243, *253*
Pentikainen, L. A., see Pentikainen, P. J. 403, *417*
Pentikainen, P. J., Pentikainen, L. A., Huffman, H., Azaroff, D. L. 403, *417*
Perel, J. M., see Burns, J. J. 118, *127*
Perel, J. M., see Dayton, P. G. 29, *33*
Perel, J. M., see Garrettson, L. K. 289, *310*
Perletti, L., see Sereni, F. 241, *256*
Perman, E. S., see Jones, D. P. 221, *232*
Perrier, D., Ashley, J. J., Levy, G. 28, *34*
Perry, H. M., Jr., Chaplin, H., Jr., Carmody, S., Haynes, C., Frei, C. 178, *209*
Perry, H. M., Jr., Sakamoto, A., Tan, E. M. 178, *209*
Perry, H. M., Jr., Tan, E. M., Carmody, S., Sakamoto, A. 178, *209*
Perry, R. P. 329, *341*
Perry, S., see Skipper, H. E. 331, *341*
Peruzzotti, G., see Rosi, D. 126, *128*
Peters, D. K., see Coles, G. A. 260, *269*
Peters, E. R., see Cunningham, M. D. 287, *309*
Peters, J. H. 176, *209*
Peters, J. H., Gordon, C. R., Brown, P. 176, *209*
Peters, J. H., Miller, K. S., Brown, P. 176, 177, *209*, 400, *417*
Peters, J. H., see Gulyassy, P. F. 258, 260, *269*
Peters, T., see Haass, A. 239, *254*
Petersdorf, R. G., see Beaty, H. N. 354, *355*
Petersdorf, R. G., see Wallace, J. F. 349, 350, *358*
Petersen, H. J., see von Daehne, W. 95, 96, *109*
Petersen, R. E., see Wyngaarden, J. B. 139, *149*
Peterson, O. L., Finland, M. 238, *255*
Peterson, R. J., see Smith, D. S. 135, *148*
Petruscak, J., see Smith, R. B. 251, *256*
Pettinger, W. A., Oates, J. A. 301, *313*
Pfiffner, J. J., North, H. B. 99, *111*
Pfister, L., see Raisfeld, I. H. 397, *417*
Pfleger, K., see Büch, H. 362, *381*, 392, *413*
Phillips, A. A., see Elslager, E. F. 95, 107, 108, *110*
Phillips, A. A., see Worth, D. F. 95, 107, *112*
Phillips, A. P. 92, *111*
Picott, J., see Dowling, R. H. 131, *146*
Pierson, R. N., Holt, P. R., Watson, R. M., Keating, R. P. 94, *111*
Pietersen, J. H., see Stubbe, L. 94, *112*
Pinakatt, T., see Yeh, B. K. 317, *324*
Pincus, G., see Arai, K. 139, *145*
Pincus, G., see Layne, D. S. 139, *148*
Pinkel, D., Hernandez, K., Borella, L., Holton, C., Aur, R., Samoy, G., Pratt, C. 332, *341*
Pinkel, D., see James, D. H., Jr. 325, 336, *340*
Pipano, S., see Szeinberg, A. 181, *210*
Pirola, R. C., see Davis, A. E. 239, *253*
Pitton, J. S., see Fabre, J. 266, *269*
Plaa, G. L. [10] 40, *85*, 130, *148*
Plaa, G. L., see Klaassen, C. D. 143, *147*
Plaa, G. L., see Roberts, R. J. 143, *148*
Plaa, G. L., see Stowe, C. M. 130, 143, *149*
Place, V. A., Benson, H. 235, *255*, 279, *313*
Place, V. A., see Mac Donald, H. 235, *255*
Plaut, A. G., see Gorbach, S. L. 215, *231*
Plotz, P. H., Davis, B. D. 352, *357*
Plotz, P. H., see Steinberg, A. D. 405, *418*
Podevin, R., see Richet, G. 261, 267, *271*
Poet, R. B., see London, I. 95, 104, 106, *111*
Pohl, J. E. P., see Balasubramanian, K. 250, *252*
Poland, A., see Conney, A. H. 169, 200, *205*, 243, *253*

Polland, W.S., see Bloomfield, A.L. 216, *231*
Pool, J.G., see O'Reilly, R.A. 184, 185, 186, *209*
Poole-Wilson, P.A., see Mac Gregor, G.A. 261, *270*
Poppenhausen, B.B., see Ockner, R.K. 222, *233*
Poppers, P.J., see Conney, A.H. 169, 200, *205*, 243, *253*
Poppers, P.J., see Welch, R.M. 288, *314*
Portmann, G.A., Minatoya, H., Lands, A.M. 158, *167*
Portmann, G.A., see Minatoya, H. 158, *167*
Portnoy, B., see Mathies, A.W., Jr. 350, *357*
Posen, S., see Lavan, J.N. 391, *416*
Poskanzer, D.C., see Herbst, A.L. 406, *415*
Posner, A., see Kellerman, L. 194, *207*
Postma, E., see Schwartz, M.A. 118, 119, *128*
Potter, V.R. 328, *341*
Potter, W.Z., Davis, D.C., Mitchell, J.R., Jollow, D.J., Gillette, J.R., Brodie, B.B. 245, *255*, 377, *382*, 390, 411, *417*
Potter, W.Z., Jollow, D.J., Mitchell, J.R. 393, 394,
Potter, W.Z., Mitchell, J.R. 390, 392, 396, 397, 410, *417*
Potter, W.Z., Nelson, W.L., Thorgeirsson, S.S., Sasame, H., Jollow, D.J., Mitchell, J.R. 393, 394, *417*
Potter, W.Z., Thorgeirsson, S.S., Jollow, D.J., Mitchell, J.R. 384, 385, 386, 388, 389, 390, 392, *417*
Potter, W.Z., see Davis, D.C. 384, 391, *413*
Potter, W.Z., see Di Chiara, G. 400, 401, *414*
Potter, W.Z., see Jollow, D.J. 245, *254*, 376, *381*, 384, 385, 386, 388, 390, *415*
Potter, W.Z., see Mitchell, J.R. 242, 245, *255*, 290, *312*, 362, 375, 376, 377, 378, *382*, 384, 385, 386, 387, 388, 389, 390, 391, 392, 393, 394, 395, 396, 398, *416*, *417*
Potter, W.Z., see Snodgrass, W. 398, *418*
Potter, W.Z., see Thorgeirsson, S.S. 394, *418*
Potter, W.Z., see Weihe, M. 393, 394, *419*
Pottison, J.R., see Noone, P. 350, *357*
Poucher, R.L., Vecchio, T.J. 301, *313*
Poulsen, B.J., see Katz, M. 157, *167*
Powell, R.C., Nunes, W.T., Harding, R.S., Vacca, J.B. 134, *148*
Pratt, C., see Pinkel, D. 332, *341*
Pratt, C.B., Fleming, I.D., Huster, H.O. 336, *341*
Preauz, A.M., see Berthelot, P. 144, *146*
Preece, J.M., see Baylis, E.M. 286, *308*
Prellwitz, W., see Hammar, C.-H. 243, *254*
Prena, K., see Rao, K.V.N. 398, *417*
Prescott, L.F., 247, *255*
Prescott, L.F., Newton, R.W., Swainson, C.P., Wright, N., Forrest, A.R.W., Matthew, H. 390, *417*
Prescott, L.F., Nimmo, J. 235, 236, 245, *255*
Prescott, L.F., Roscoe, P., Forrest, J.A.H. 246, 250, *256*
Prescott, L.F., Roscoe, P., Wright, N., Brown, S.S. 244, *256*
Prescott, L.F., Steel, R.F., Ferrier, W.R., 235, *256*
Prescott, L.F., Stevenson, I.H. 243, 244, 245, *256*
Prescott, L.F., Wright, N., Roscoe, P., Brown, S.S. 389, *417*
Prescott, L.F., see Adjepon-Yamoah, K.K. 237, 238, *252*
Prescott, L.F., see Clarke, R.A. 245, *253*
Prescott, L.F., see Heading, R.C. 236, *254*, 280, *310*
Prescott, L.F., see Nimmo, J. 237, 239, *255*
Prescott, L.F., see Wright, N. 245, *257*, 389, *419*
Presho, N.E., see Hanzlik, P.J. 93, 95, *110*
Price, H.L. 156, *167*
Price, J.M., see Erturk, E. 407, *414*
Price, V.E., Greenfield, R.E. 174, *209*
Price Evans, D.A. 176, 177, *209*, 243, *256*
Price Evans, D.A., Kitchin, F.D., Riding, J.E. 193, *209*
Price Evans, D.A., Manley, K.A. McKusick, V.A. *177*, *209*
Price Evans, D.A., White, T.A. 176, 177, *209*
Price Evans, D.A., see Alexanderson, B. 198, *203*, 286, *308*
Price Evans, D.A., see Åsberg, M. 198, *204*
Price Evans, D.A., see Edwards, J.A. 195, *205*
Price Evans, D.A., see Gow, J.G. 176, *206*
Price Evans, D.A., see White, T.A. 177, *211*
Price Evans, D.A., see Whittaker, J.A. 198, 201, *211*
Priebe, F.H., see Fisch, C. 316, *323*
Prindle, K.H., Jr., see Levey, G.S. 318, *323*
Prins, H.K., Oort, M., Loos, J.A., Zurcher, C., Beckers, T. 189, *209*
Prins, H.K., see Oort, M. 189, *208*
Procknal, J.A., see Levy, G. 238, *254*, 280, *312*
Prox, A. 230, *233*
Pruett, D., see Andrew, W. 223, *230*
Pruitt, A.W., see Carliner, N.H. 317, *322*
Puglisi, C.V., see De Silva, J.A.F. 120, *127*
Puig, J.R., see Ross, S. 95, 100, *111*
Pyörälä, K., see Kekki, M. 236, *254*, 279, *311*
Pyörälä, K., see Ristola, P. 245, *256*

Rado, E., see Fogelman, A.M. 317, *323*
Rahai, J.J., Jr., Meyers, B.R., Weinstein, L. 353, 354, *357*
Raisfeld, I.H., Pfister, L., Feingold, M. 397, *417*
Raisz, L.G., see Griner, P.F. 296, *310*
Rall, D.P. [11] 35, 40, 61, *85*
Ramakrishnan, C.V., see Devadatta, S. 176, 177, *205*
Ramboer, C., Thompson, R.P.H., Williams, R. 287, *313*

Rammelkamp, C. H., Helm, J. D., Jr. 216, *233*
Rand, M. J., Trinker, F. R. 301, *313*
Rand, M. J., see Trinker, F. R. 301, *314*
Randall, L. O., Schallek, W. 118, *128*
Randall, L. O., Scheckel, C. L., Banziger, R. F. 119, 121, *128*
Randall, L. O., see Zbinden, G. 118, *129*
Rao, B. K., see Levy, G. 281, *311*
Rao, K. V. N., Mitchison, D. A., Nair, N. G. K., Prena, K., Tripathy, S. P. 398, *417*
Rapoport, H., Baker, D. R., Reist, H. N. 95, 99, *111*
Rattie, E. S., Shami, E. G., Dittert, L. W., Swintosky, J. V. 95, 102, *111*
Rattie, E. S., see Dittert, L. W. 95, 102, *110*
Rauen, H. M., Kramer, K.-P. 337, *341*
Raussi, M., see Hartiala, K. 236, *254*
Read, A. E., see Branch, R. A. 288, 293, 307, *308*
Read, A. E., see Laidlaw, J. 252, *254*
Redinger, R. N., Small, D. M. 144, *148*
Reed, C. F., see Weed, R. I. 190, *211*
Rees, R., see Lippman, R. W. 259, *270*
Reich, S., see Solomon, H. M. 288, *314*
Reid, W. D. 401, *418*
Reid, W. D., Ilett, K. F., Glick, J. M., Krishna, G. 381, *382*
Reid, W. D., Christie, B., Krishna, G., Mitchell, J. R., Moskowitz, J. R., Brodie, B. B. 401, *417*
Reid, W. D., Krishna, G. 379, 380, *382*
Reid, W. D., see Brodie, B. B. 401, *413*
Reid, W. D., see Mitchell, J. R. 401, *416*
Reidenberg, M. M. 242, 247, *256*, 260, *270*
Reidenberg, M. M., Affrime, M. 307, *313*
Reidenberg, M. M., James, M., Dring, L. G. 246, 247, *256*
Reidenberg, M. M., Kostenbauder, H., Adams, W. P. 264, *270*
Reidenberg, M. M., Kostenbauder, H., Adams, W. P., Odar-Cederlöf, I., von Bahr, C., Borgå, O., Sjöqvist, F. 260, 264, *270*
Reidenberg, M. M., Odar-Cederlöf, I., von Bahr, C., Borgå, O., Sjöqvist, F. 242, *256*
Reif, W., see Bönicke, R. 176, *204*
Reiner, O., Athanassopoulos, S., Hellmer, K. H., Murray R. E., Uehleke, H. 404, *418*
Reist, H. N., see Rapoport, H. 95, 99, *111*
Reizenstein, P., see Grasbeck, R. 130, *147*
Remer, A., see Field, J. B. 292, *310*
Remington, L., see Yesair, D. W. 140, *149*
Remmer, H., Bock, K. W. 406, *418*
Remmer, H., Schoene, B., Fleischmann, R. A. 399, *418*
Remmer, H., see Schoene, B. 243, *256*
Renyi, A. L., see Ross, S. B. 121, *128*
Reutner, T. F., see Glazko, A. J. 95, 98, 100, *110*
Reyes, H., Levi, J. A., Levine, R., Gatmaitan, Z., Arias, I. M. 301, *313*
Reyes, H., Levi, J. A., Gatmaitan, Z., Arias, I. M. 222, *233*
Reynolds, E. H. 286, *313*
Reynolds, E. S., Brown, B. R., Vandam, L. D. 399, 409, *418*
Reynolds, E. S., Moslen, M. T. 399, 409, *418*
Reynolds, M. E., see McDermott, W. 153, *167*
Ricciardi, M. L., see Donno, L. 353, *356*
Richards, A. J., see Walker, S. R. 155, *168*
Richards, C., see Bacon, G. E. 262, 265, *268*
Richards, R. K., Taylor, J. D., Kueter, K. E. 264, *271*
Richards, R. K., see Dundee, J. W. 242, *253*, 264, *269*
Richens, A., see Latham, A. M. 286, *311*
Richet, G., Fabre, J., de Freudenreich, J., Podevin, R. 267, *271*
Richet, G., Fabre, J., de Freudenreich, J., Podevin, R., Lopez de Novales, E., Verroust, P. 267, *271*
Richie, R. H., see Butler, A. M. 214, *231*
Richter, C. P., Clisby, K. H. 192, *209*
Rickles, F. R., see Griner, P. F. 296, *310*
Riding, J. E., see Price Evans, D. A. 193, *209*
Rieder, J., see Schwartz, D. E. 118, *128*
Rieder, R. F., Zinkham, W. H., Holtzman, N. A. 191, *209*
Riegelman, S., Loo, J., Rowland, M. [12] 63, *85*
Riegelman, S., Rowland, M., Epstein, W. L. 278, 301, *313*
Riegelman, S., see Blum, M. R. 260, *268*
Riegelman, S., see Loo, J. C. K. 21, 22, *34*
Riegelman, S., see Rowland, M. 24, *34*
Riegelman, S., see Wan, S. H. 261, *271*
Riemensnider, D. K., see Mitchell, R. S. 177, *208*
Riff, L. J., Jackson, G. G. 266, *271*
Riggs, D. S. [13] 63, *85*
Riley, R. F., see Berger, F. M. 95, 104, *109*
Ristola, P., Pyörälä, K. 245, *256*
Ritterbrand, A., see Burns, J. J. 118, *127*
Ritz, E., Jantzen, R. 261, *271*
Rivard, D. E., see Caldwell, H. C. 95, 101, *109*
Rivard, D. E., see Dittert, L. W. 93, 94, 95, *110*
Rivard, D. E., see Swintosky, J. V. 95, 101, *112*
Rivera-Calimlim, L. 282, *313*
Rizzo, M., see Morselli, P. L. 289, 302, *312*
Robbins, W. C., Tompsett, R. 354, *357*
Roberts, J. A. F. 170, *209*
Roberts, J. J., Brent, T. P., Crathorn, A. R. 329, *341*
Roberts, P. W., see Rowley, D. 353, *357*
Roberts, R. J., Plaa, G. L. 143, *148*
Robertson, A., see Gwilt, J. R. 238, *254*
Robertson, D. N. 95, 99, *111*

Robinson, A.J., see O'Reilly, R.A. 297, *313*
Robinson, D.S., Benjamin, D.M., McCormack, J.J. 281, *313*
Robinson, D.S., MacDonald, M.G. 285, *313*
Robinson, D.S., see MacDonald, M.G. 285, 287, *312*
Robinson, G.N., see Howell, A. 240, *254*
Robison, G.A., Butcher, R.W., Sutherland, E.W. 317, *324*
Robitzek, E.H., Selikoff, I.J., Ornstein, G.G. 176, *209*
Robson, E.B., see Harris, H. 181, 182, *206*
Robson, J.M., Sullivan, F.M. 176, *209*
Rock, H.J., see Miller, E. 95, 106, *111*
Rockett, J.F., see Corn, M. 285, *309*
Rodahl, K., Miller, H.I., Issekutz, B., Jr. 221, *233*
Rodriguez, V., see Freireich, E.J. 333, *340*
Roe, D.A. 133, *148*
Roe, F.J.C. 407, *418*
Roels, O.A. 261, *271*
Roggin, G., see Kater, R.M.H. 246, *254*, 288, *311*
Roholt, K., see von Daehne, W. 95, 96, 97, *109*
Rollins, F.O., see Davidson, I.W.F. 152, *166*
Roman, L.R., see Bellet, S. 248, *253*
Romansky, M.J., see Olsson, R.A. 350, *357*
Roncone, A., see Izzo, J.L. 262, *269*
Rongey, K.A., Weisman, H. 322, *324*
Rook, A.J., see Ackroyd, J.F. 408, *412*
Roosen-Runge, U., see Bulger, R.J. 353, *355*
Roscoe, P., see Prescott, L.F. 244, 246, 250, *256*, 389, *417*
Rose, I.M., see Shapiro, S.L. 95, 106, *111*
Rose, R.K., see Burns, J.J. 197, *204*
Rosen, A., see Beermann, B. 235, 239, *253*
Rosenblum, R., Berkowitz, W.D., Lawson, D. 321, *324*
Rosenheim, M.L., see Laurence, D.R. 300, *311*
Rosenman, S.B., see Glassman, J.M. 143, *147*
Rosenman, S.B., see O'Connor, W.J. 216, *233*
Rosi, D., Peruzzotti, G., Dennis, E.W., Berberian, D.A., Freele, H., Tullar, B.F., Archer, S. 126, *128*
Ross, J., Jr., see Kaiser, G.A. 321, *323*
Ross, R.R. 176, *209*
Ross, S., Puig, J.R., Zaremba, E.A. 95, 100, *111*
Ross, S.B., Renyi, A.L. 121, *128*
Roth, F., see Bush, G.H. 180, *204*
Roth, F.E., Tabachnick, I.A. 322, *324*
Rothstein, E. 289, *313*
Rovenstine, E.A., see Brodie, B.B. [2] 61, *85*
Rowe, D.J.F., see Latham, A.M. 286, *311*
Rowland, M. 277, 284, 285, *313*
Rowland, M., Benet, L.Z., Graham, G.G. [14] 49, *85*, 283, 300, *313*
Rowland, M., Benet, L.Z., Riegelman, S. 24, *34*
Rowland, M., Thomson, P.D., Guichard, A., Melmon, K.L. 163, *167*
Rowland, M., see Benowitz, N. 293, *308*
Rowland, M., see Riegelman, S. [12] 63, *85*, 278, 301, *313*
Rowland, M., see Thomson, P.D. 156, *168*, 241, 243, 245, *256*, 307, *314*
Rowley, D., Cooper, P.D., Roberts, P.W., Smith, E.L. 353, *357*
Rubin, A.L., see Cohen, B.D. 316, *323*
Rubin, H.S. 93, 95, *111*
Rubin, M.I., see Calcagno, P.L. 225, *231*
Ruch, P.A., see Sabath, L.D. 266, *271*
Rudhart, M., see Fabre, J. 266, *269*
Rudiger, W., see Büch, H. 362, *381*, 392, *413*
Rudwal, T.C., see Gault, M.H. 391, *414*
Rule, A.W., see Agersborg, H.P.K. 95, 96, *109*
Rummel, W., see Büch, H. 392, *413*
Rundles, R.W., see Elion, G.B. 118, *127*, 328, 337, *339*
Russell, T.J., see Jansen, A.B.A. 95, 96, *111*
Russell, W.F., Jr., Kass, I., Heaton, A.D., Dressler, S.H., Middlebrook, G. 344, *357*
Russell, W.F., Jr., see Carlson, H.B. 176, *204*
Russell, W.F., Jr., see Heaton, A.D. 344, *356*
Russell, W.F., Jr., see Mandel, W. 344, *357*
Russell, W.R., Jr., see Cohn, M.I. 344, *355*
Russo, H.F., see Beyer, K.H. 361, *381*
Ryan, A.J., see Holder, G.M. 142, *147*
Ryan, A.J., see Ladomery, L.G. 131, 141, *148*
Ryan, E., see Lehmann, H. 179, *208*
Ryan, J.F., Kagen, L.J., Hyman, A.I. 252, *256*
Ryan, J.R., see Hague, D.E. 289, *310*
Ryan, P.M., see Gulyassy, P.F. 258, 260, *269*
Ryter, S., see Dettli, L. 265, 266, *269*

Sabath, L.D., Abraham, E.P. 352, *357*
Sabath, L.D., Casey, J.I., Ruch, P.A., Stumpf, L.L., Finland, M. 266, *271*
Sabath, L.D., Elder, H.A., McCall, C.E., Finland, M. 352, 354, *357*
Sabesin, S.M., see Bell, N.H. 337, *338*
Sachs, J., Geer, T., Noell, P., Kunin, C.M. 259, 266, *271*
Sahud, M.A., see O'Reilly, R.A. 297, *313*
Sakamoto, A., see Perry, H.M., Jr. 178, *209*
Sakurai, T., see Ueda, H. 243, *256*
Salatka, K., Kresge, D., Harris, L., Jr., Edelstein, D., Ove, P. 220, *233*
Saldanha, P.H., Becak, W. 192, *209*
Sales, J.E.L., Sutcliffe, M., O'Grady, F. 239, 240, *256*
Salman, J.A., see Beckett, A.H. 122, *126*
Salmi, H., see Kekki, M. 236, *254*, 279, *311*
Salvador, R.A., see Burns, J.J. 122, *127*
Salvidio, E., Pannacciulli, I., Tizianello, A. 190, *210*

Salvioli, G.F., see Carulli, N. 288, 302, *309*
Samaha, R.J., see Henderson, E.S. 331, 332, *340*
Samanin, R., Ghezzi, D., Garattini, S. 121, *128*
Samoy, G., see Pinkel, D. 332, *341*
Samuel, P., Waithe, W.I. 134, *148*, 281, *313*
Samuel, P., see Steiner, A. 134, *148*
Sandberg, A.A., Kirdani, R. Y., Back, N., Weyman, P., Slaunwhite, W.R., Jr. 130, *148*
Sandler, M., see Conolly, M.E. 159, *166*
Sandler, S.R., Karo, W. 95, 107, *111*
Sanford, J.P., see Hodgin, U. G. 344, *356*
Sanguineti, V., see Donno, L. 353, *356*
Sansur, M., see Conney, A.H. 117, *127*
Sansur, M., see Cucinell, S.A. 285, *309*, 364, *381*
Santos, E.C., see Spector, A. A. 221, *233*
Sapir, D.G., see Whelton, A. 240, *257*, 260, *271*
Sarmiento, F., see Jonson, L. 297, *311*
Sartorelli, A.C. 325, 326, 329, *341*
Sartorelli, A.C., Lepage, G.A. 329, *341*
Sasame, H., see Potter, W.Z. 393, 394, *417*
Sasame, H., see Thorgeirsson, S.S. 390, 391, 394, *418*
Sato, H., see Takahara, S. 173, *210*
Satterlee, W., see Schimpff, S. 344, *357*
Scales, B., see Ellard, G.A. 153, *167*
Schachter, D., see Batt, E.R. 216, *230*
Schaefer, O., see Fenna, D. 196, *205*
Schallek, W., see Randall, L. O. 118, *128*
Schanker, L.S. 215, *233*
Scharer, L., Smith, J.P. 409, *418*
Scharp, C.R., see Hahn, T.J. 286, *310*
Schaudig, H., see Schmid, E. 259, *271*
Schaumlöffel, E., see Graul, E.H. 337, *340*
Scheckel, C.L., see Randall, L.O. 119, 121, *128*
Schedl, H.P., Clifton, J.A. 239, *256*
Scheiner, J., Altenmeier, W. A. 236, *256*, 280, *313*
Scheline, R.R. 281, *313*
Scherman, R., see Kutt, H. 182, 183, 197, *208*, 243, *254*
Schettle, M., see Hahn, K.-J. 281, *310*
Schimpff, S., Satterlee, W., Young, V.M., Serpick, A. 344, *357*
Schlegel, J.U., Goff, J.B., O'Dell, R.M. 259, 266, *271*
Schlegel, J.U., see Goff, J.B. 259, *269*
Schlichter, J.G., MacLean, H. 347, *357*
Schloot, W., see Goedde, H. W. 179, *206*
Schmid, E., Meythaler, C., Dvorak, G., Schaudig, H. 259, *271*
Schmid, P.G., see Schuelke, D.M. 321, *324*
Schmid, R., see Lester, R. 130, *148*
Schmidt, D.H., Butler, V.P., Jr. 317, *324*
Schmidt, E., see Kindt, H. 396, *415*
Schmidt, L.H., see Hughes, H.B. 176, *207*
Schmiedel, A. 177, *210*
Schmock, C.L., see Schumacher, E.E., Jr. 158, *167*
Schneider, D.L., see Cook, D. A. 133, *146*
Schnitzer, R.J., see Grunberg, E. 176, *206*
Schoenbach, E.B. 100, *111*
Schoene, B., Fleischmann, R. A., Remmer, H. 243, *256*
Schoene, B., see Remmer, H. 399, *418*
Schoenfeld, P., see Gulyassy, P.F. 258, 260, *269*
Schou, J. [15] 36, 39, *85*, 157, *167*
Schreiner, G.E. 248, *256*
Schreiner, G.E., Mahler, J.F. 261, *271*
Schröder, H., Campbell, D.E. S. 235, 239, *256*
Schroeder, W.A., Shelton, J. R., Shelton, J.B., Olson, B.M. 174, *210*
Schrogie, J.J., Solomon, H.M. 297, *313*
Schrogie, J.J., see Solomon, H.M. 184, *210*, 288, 297, 306, *313*, *314*
Schuelke, D.M., Mark, A.L., Schmid, P.G., Eckstein, J. W. 321, *324*
Schürch, P.M., see von Wartburg, J.P. 195, *211*
Schulert, A., see Burns, J.J. 197, *204*
Schull, W.J., see Neel, J.V. 198, *208*
Schultz, S., see Barnett, H.L. 225, *230*
Schumacher, E.E., Jr., Schmock, C.L. 158, *167*
Schumacher, H., see Sjöqvist, F. 367, *382*
Schumer, W., see Lester, R. 130, *148*
Schwan, S. 407, *418*
Schwartz, A., see Besch, H.R., Jr. 316, *322*
Schwartz, D.E., Bruderer, H., Rieder, J., Brossi, A. 118, *128*
Schwartz, M.A., Carbone, J. J. 120, *128*
Schwartz, M.A., Postma, E. 118, 119, *128*
Schwartz, M.A., see Gibaldi, M. 95, 96, 103, *110*
Schwartz, S.A., see Barowsky, H. 116, *126*
Sciarra, D., see Yahr, M.D. 182, *212*
Scott, D.B., see Adjepon-Yamoah, K.K. 237, 238, *252*
Scott, D.B., see Boyes, R.N. 163, 164, *166*
Scott, E.M., see Gutsche, B. B. 181, *206*
Scully, R.E., see Herbst, A.L. 406, *415*
Seawright, A.A., Hrdlicka, J. 395, 396, *418*
Seawright, A.A., Mattocks, A. R. 396, *418*
Seidl, L.G., see Cluff, L.E. 173, *205*, 343, *355*
Seligman, S.J., see Hewitt, W.L. 353, 354, *356*
Selikoff, I.J., see Robitzek, E.H. 176, *209*
Selin, M.J., see Harris, H.W. 177, *206*
Selin, M.J., see Knight, R.A. 177, *207*
Selkon, J.B., see Devadatta, S. 176, 177, *205*
Selkon, J.B., see Gangadharam, P.R.J. 177, *205*

Seller, R. H., Cangiano, J., Kim, K. E., Mendelssohn, S., Brest, A. N., Swartz, C. 316, *324*
Sellers, E.M., Carr, G., Bernstein, J. G., Sellers, S., Koch-Weser, J. 290, 302, *313*
Sellers, E.M., Koch-Weser, J. 154, 155, *167*, 296, *313*
Sellers, E.M., Lang, M., Koch-Weser, J., Le Blanc, E., Kalant, H. 289, 302, *313*
Sellers, E.M., see Koch-Weser, J. 285, *311*
Sellers, S., see Sellers, E.M. 290, 302, *313*
Sereni, F., Perletti, L., Marubini, E., Mars, G. 241, *256*
Sernatinger, E., see Levin, W. 404, *416*
Serpick, A. A., see Devita, V. T., Jr. 325, 334, *339*
Serpick, A. A., see Lowenbraun, S. 335, *341*
Serpick, A. A., see Mendelson, D. 335, *341*
Serpick, A., see Schimpff, S. 344, *357*
Serra, M. T., see Forrest, F. M. 286, *310*
Serrill, W. S., see Kirby, W. M. M. 264, 266, *270*
Sessions, J. T., Minkel, H. P., Bullard, J. C., Ingelfinger, F. J. 243, *256*
Sezesny, B., see Ikeda, M. 184, *207*
Shah, A. A., Connors, K. A. 95, 102, *111*
Shahani, R. T., see Guyer, R. J. 333, 334, *340*
Shahidi, N. J. 392, *418*
Shahidi, N. T. 184, 185, *210*
Shaldon, S., see Caesar, J. 301, *309*
Shami, E. G., see Rattie, E. S. 95, 102, *111*
Shand, D., Mitchell, J. R., Oates, J. A. *418*
Shand, D. G., Evans, G. H., Nies, N. S. 164, *168*
Shand, D. G., Mitchell, J. R., Oates, J. A. [16] 68, 69, 74, *85*
Shand, D. G., Nucholls, E. M., Oates, J. A. 318, *324*
Shand, D. G., Nuckolis, E. M., Oates, J. A. 197, *210*
Shand, D. G., see Branch, R. A. 283, 293, 300, 306, 307, *308*, 363, *381*
Shand, D. G., see Coltart, D. J. 164, 165, *166*
Shand, D. G., see Evans, G. H. 276, 297, *310*
Shand, D. G., see Nies, A. S. 293, *312*
Shand, D. G., see Oates, J. A. 300, *312*
Shanker, L. S. 153, *168*
Shanker, L. S., see Enna, S. J. 155, *167*
Shanks, R. G., see Dunlop, D. 317, *323*
Shannon, J. A., see Zubrod, C. G. 299, *314*
Shapiro, S., see Jick, H. 196, *207*
Shapiro, S., see Lewis, G. P. 196, 197, *208*
Shapiro, S., see Weiner, M. 197, *211*
Shapiro, S. L., Rose, I. M., Freedman, L. 95, 106, *111*
Shapiro, W., Narahara, K., Taubert, K. 317, *324*
Sharpe, A. R., Jr., see O'Brien, J. P. 262, *270*
Sheba, C., see Szeinberg, A. 177, *210*
Shelley, J. H. 391, *418*
Shelton, J. B., see Schroeder, W. A. 174, *210*
Shelton, J. R., see Schroeder, W. A. 174, *210*
Shelton, S., see Nelson, J. D. 235, 238, 245, *255*
Shenoy, E. V. B., see Beckett, A. H. 122, *126*
Shepard, P. M., see Kitchin, F. D. 192, *207*
Shepard, T. H., II, Gartler, S. M. 192, *210*
Sherlock, S., see Black, M. 287, *308*
Sherlock, S., see Caesar, J. 301, *309*
Sherlock, S., see Laidlaw, J. 252, *254*
Sherlock, S., see Levi, A. 288, 307, *311*
Sherlock, S., see Levi, A. J. 243, *254*
Shibata, Y., Higashi, T., Hirai, H., Hamilton, H. B. 174, *210*
Shideman, F. E., Kelly, A. R., Lee, L. E., Lowell, V. F., Adams, B. J. 242, *256*
Shioda, R., see Wood, P. 133, *149*
Shock, N. W. 223, *233*
Shock, N. W., see Miller, J. H. 224, *232*
Shock, N. W., see Norris, A. H. 213, *233*
Sholiton, L. J., see Choi, Y. 286, 287, *309*
Sholiton, L. J., see Werk, E. E., Jr. 287, *314*
Shreffler, D. C., see Grundbacher, F. J. 220, *231*
Shubik, P., see Toth, B. 407, *419*
Shubin, H., Dumas, K., Sokmensuer, A. 95, 97, *111*
Shulman, N. R. 408, *418*
Siegers, C.-P., Strubelt, O., Back, G. 236, *256*
Siersbaek-Nielsen, K., see Hansen, J. M. 286, *310*
Silk, E., see Evans, F. T. 178, *205*
Silk, E., see Forbat, A. 179, *205*
Silk, E., see Harris, H. 180, *206*
Silk, E., see Lehmann, H. 179, *208*
Silverman, W. A., Andersen, D. H., Blanc, W. A., Crozier, D. N. 297, *313*
Silvestrini, R., see di Marco, A. 329, *339*
Simanis, A. M., see Ariens, E. J. 315, *322*
Simanis, J., Goldberg, L. I. 318, *324*
Simon, G., see Graul, E. H. 337, *340*
Simon, H. J., see Axline, S. G. 224, 225, *230*
Simons, E. L., see Englert, E., Jr. 262, *269*
Simons, P. J. G., see Balasubramaniam, K. 241, 250, *252*
Simpson, L., see Bennett, L. L., Jr. 329, *338*
Sims, P., see Grover, P. L. 388, *414*
Singer, I., see Bennett, W. M. 259, 261, 266, *268*
Singer, S., see Elion, G. B. *339*
Singer, T. P., see Barron, E. S. G. 189, *204*
Singhel, S. N., see Las, S. 397, *416*
Sipes, G., see Brodie, B. B. 401, *413*
Sipos, F., see Pals, D. T. 320, *324*
Siskind, V., see Jick, H. 196, *207*
Siurala, M., Mustala, O., Jussila, J. 236, *256*, 279, *313*
Siurala, M., see Kekki, M. 236, *254*, 279, *311*

Sjöqvist, F., Berglund, F., Borga, O., Hammer, W., Andersson, S., Thorstrand, C. 300, *313*
Sjöqvist, F., Hammer, W., Schumacher, H., Gillette, J.R. 367, *382*
Sjöqvist, F., see Alexanderson, B. 198, *203*, 286, *308*
Sjöqvist, F., see Åsberg, M. 198, *204*
Sjöqvist, F., see Borgá, O. 291, 298, 300, *308*
Sjöqvist, F., see Hammer, W. 202, *206*
Sjöqvist, F., see Kolmodin, B. 288, *311*
Sjöqvist, F., see Odar-Cederlöf, I. 242, *255*, 260, *270*
Sjöqvist, F., see Reidenberg, M.M. 242, *256*, 260, 264, *270*
Sjovall, J., see Norman, A. 131, *148*
Skeel, R.T., see Levitt, M. 335, *341*
Skinner, W.N., see Frenkel, E.P. 329, *340*
Skipper, H.E., Perry, S. 331, *341*
Skovsted, L., see Christensen, L.K. 289, *309*
Skovsted, L., see Hansen, J. M. 286, 288, 289, *310*
Skovsted, L., see Kristensen, M. 285, *311*
Sladek, N.E. 337, *341*
Slater, I.H., see Dresel, P.E. 95, 104, 106, *110*
Slatopolsky, E., see Avioli, L. V. 259, 261, *268*
Slaunwhite, W.R., Jr., see Sandberg, A.A. 130, *148*
Slone, D., see Jick, H. 196, *207*
Slone, D., see Lewis, G.P. 196, 197, *208*
Small, D.M. 132, *148*
Small, D.M., see Dowling, R. H. 131, *146*
Small, D.M., see Hofmann, A.F. 131, *147*
Small, D.M., see Redinger, R. N. 144, *148*
Smih, C.B. 301, *313*
Smiley, J.D., see Frenkel, E. P. 329, *340*
Smirnov, G.A., Kozulitzina, T.I. 178, *210*
Smith, A.J., see Leishman, A.W.D. 299, *311*
Smith, C.A. 215, 216, *233*
Smith, C.B., Dans, P.E., Wilfert, J.N., Finland, M. 344, *357*
Smith, D.L., Forist, A.A., Dulin, W.E. 125, *129*
Smith, D.S., Peterson, R.E., Fujimoto, J.M. 135, *148*
Smith, E.L., see Rowley, D. 353, *357*
Smith, E.R., Duce, B.R. 123, *129*
Smith, F.D., Jr., see Okita, G.T. 136, *148*
Smith, F.R., De Goodman, W. S. 261, *271*
Smith, H.M., see Falco, F.G. 337, *340*
Smith, H.W., see West, J.R. *233*
Smith, I.M., Soderstrom, W. H. 95, 97, *111*
Smith, J., Tyrrell, W.F., Gow, A., Allan, G.W., Lees, A.W. 397, *418*
Smith, J., see Cluff, L.E. 173, *205*
Smith, J., see Lees, A.W. 399, *416*
Smith, J.P., see Scharrer, L. 409, *418*
Smith, J.S., see Harrison, G. G. 400, *415*
Smith, M.E., see Hague, D.E. 289, *310*
Smith, R.B., Petruscak, J. 251, *256*
Smith, R.B., see Kojima, S. 236, *254*, 279, *311*
Smith, R.H., see Wallace, J. F. 349, 350, *358*
Smith, R.L. [17] 39, 40, *85*, 130, *148*
Smith, R.L., see Clark, A.G. 131, 138, *146*
Smith, R.L., see Fischer, L.J. 131, 138, *146*
Smith, R.L., see Levine, W.G. 144, *148*
Smith, R.L., see Williams, R.T. 130, 143, *149*
Smith, T.W., Butler, V.P., Jr. Haber, E. 317, *324*
Smith, T.W., see Heizer, W. D. 239, *254*, 258, *269*
Smith, T.W., see White, R.J. 235, *257*
Smith, W.K. 184, *210*
Snodgrass, W., Potter, W.Z., Timbrell, J., Jollow, D.J., Mitchell, J.R. 398, *418*
Snodgrass, W.R., Mitchell, J. R. 391, 392, *418*
Snodgrass, W.R., see Mitchell J.R. 290, *312*, 362, 375, 376, 377, 378, *382*, 398, *417*
Snow, G.A., see Ellard, G.A. 153, *167*
Snyder, L.H. 192, *210*
Soberman, R., Brodie, B.B., Levy, B.B., Axelrod, J., Hollander, V., Steele, J.M. 197, *210*
Soberman, R., see Brodie, B. B. 262, *268*
Soderstrom, W.H., see Smith, I.M. 95, 97, *111*
Soeldner, J.S., see Hampers, C.L. 262, *269*
Sokmensuer, A., see Shubin, H. 95, 97, *111*
Soldati, M., see Donno, L. 353, *356*
Solomon, H.M. 184, *210*
Solomon, H.M., Abrams, W. B. 286, *313*
Solomon, H.M., Reich, S., Spirt, N., Abrams, W.B. 288, *314*
Solomon, H.M., Schrogie, J. J. 184, *210*, 288, *313*, *314*
Solomon, H.M., Schrogie, J. J., Williams, D. 297, 306, *314*
Solomon, H.M., see Schrogie, J.J. 297, *313*
Somers, G.F., see Bourne, J. G. 178, *204*
Somlyo, A.P., Somlyo, A.V. 319, *324*
Somlyo, A.V., see Somlyo, A. P. 319, *324*
Song, C.S., Gelb, N.A., Wolff, S.M. 191, *210*
Song, Y.K., see Vogel, F. 196, *211*
Sonne, M., Jawetz, E. 353, *358*
Sonne, M., see Jawetz, E. 354, *356*
Sonnenblick, E.H., see Glick, G. 318, *323*
Sonneville, P.F., see Goldberg, L.I. 321, 322, *323*
Soroko, F., see Conney, A.H. 117, *127*
Sorter, H., see Meyer, J. 217, *232*
Sosa-Lucero, J.C., see Feuer, G. 363, *381*
Sotaniemi, E., Huhti, E., Arvela, P., Koivisto, O. 246, *256*
Spanknebel, G., see Gorbach, S.L. 215, *231*
Speck, R.S., see Gunnison, J. B. 350, *356*
Speck, R.S., see Jawetz, E. 350, *356*
Spector, A.A. *233*
Spector, A.A., Santos, E.C., Ashbrook, J.D., Fletcher, J.E. 221, *233*

Speizer, F. E., Doll, R., Heaf, P. 162, *168*
Spencer, R. P. 215, 217, *233*
Spies, H. W., see Lepper, M. H. 345, *357*
Sping, P., see Dettli, L. 248, *253*
Spinks, A., Waring, W. S. 114, 115, *129*
Spirt, N., see Solomon, H. M. 288, *314*
Splitter, S., see Wood, P. 133, *149*
Spring, P., see Dettli, L. 265, 266, *269*
Spritz, N., see Cohen, B. D. 316, *323*
Spuhler, O., Zollinger, H. V. 391, *418*
Spurr, C., see Carbone, P. P. 334, *339*
Spurr, C. L., see Hoogstraten, B. 335, *340*
Stahl, W. R. 214, *233*
Stamatoyannopoulos, G., see Motulsky, A. G. 187, 188, 189, *208*
Stanbury, J. B., Farah, A. 316, *324*
Stanburry, S. W., Lumb, G. A. Mawer, E. B. 259, *271*
Stanbury, S. W., see Lumb, G. A. 259, *270*
Standiford, H. D., de Maine, J. B., Kirby, W. M. M. 352, 354, *358*
Starkweather, D. K., see Chignell, C. F. 241, *253*
Staron, N., see Kalow, W. 179, 180, 182, *207*
Starweather, D. K., see Chignell, C. F. 197, *204*
Stathers, G. M. 130, *149*
Steel, R. F., see Prescott, L. F. 235, *256*
Steele, J. M., see Brodie, B. B. 157, *166*
Steele, J. M., see Soberman, R. 197, *210*
Steiger, B., see Dobbs, H. E. 131, 136, *146*
Stein, L., see Mishkin, S. 222, *233*
Stein, S. W., see Cherrick, G. R. 301, *309*
Steinberg, A. D., Plotz, P. H., Wolff, S. M., Wong, V. G., Agus, S. G., Decker, J. L. 405, *418*
Steinberg, M. S., see Weber, W. W. 177, 178, *211*
Steiner, A., Howard, E., Akgun, S. 134, *149*
Steiner, A., see Samuel, P. 134, *148*
Stella, V., see Hussain, A. 91, *111*
Stenbaek, Ø., Myhre, E., Brodwall, E. K., Hansen, T. 249, *256*
Stenbaek, Ø., see Myhre, E. 249, *255*
Stenson, R. E., Constantino, R. T., Harrison, D. C. 163, *168*, 307, *314*, 363, *382*
Stern, L., see Jusko, W. J. 216, 217, *232*
Stevenson, I. H., see Ballinger, B. 246, *253*
Stevenson, I. H., see O'Malley K. 200, *208*, 243, 244, *255*, 286, *312*
Stevenson, I. H., see Prescott, L. F. 243, 244, 245, *256*
Stewart, D. A., see Hunter, J. 202, *207*, 286, *311*
Stewart, D. A., see Maxwell, J. D. 286, *312*
Stewart, G. T., Harrison, P. M. 143, *149*
Stiehm, E. R., see Hyvarinen, M. 220, *232*
Stillwell, R. N., see Horning, E. C. 230, *231*
Stillwell, W. G., see Horning, M. G. 230, *232*
Stock, C. C., see Tarnowski, G. S. 330, *341*
Stohlman, F., Jr., see Vagenakis, A. G. 251, *257*
Stone, C. A., see Vernier, V. G. 121, *129*
Stowe, C. M., Plaa, G. L. 130, 143, *149*
Strasser, E. S., see Cranley, J. J. 320, *323*
Stratton, C., see Harvey, D. J. 402, *414*
Stratton, C., see Horning, M. G. 402, *415*
Stripp, B., Menard, R. H., Gillette, J. R. 407, *418*
Stripp, B., Menard, R. H., Zampaglione, N. G., Hamrick, M. E., Gillette, J. R. 402, 403, 407, *418*
Stripp, B., see Menard, R. H. 402, 403, *416*
Stripp, B., see Zampaglione, N. 378, 380, *382*, 401, *417*
Ström, J. 350, *358*
Strubelt, O., see Siegers, C.-P. 236, *256*
Struck, R. F., see Hill, D. L. 126, *127*, 337, *340*
Stubbe, L., Pietersen, J. H., van Heulen, C. 94, *112*
Stumpf, L. L., see Sabath, L. D. 266, *271*
Sturge, L. M., Whittaker, V. P. 92, 95, 97, *112*
Sturtevant, M., see Thompson R. Q. 130, *149*
Sudborough, J. J. 95, 99, *112*
Suhrland, L. G., Weisberger, A. S. 261, 264, *271*
Sullivan, F. M., see Robson, J. M. 176, *209*
Sullivan, I., see Brooks, S. M. 286, *309*
Sullivan, M. P., see Wilbur, J. R. 336, *342*
Sulser, F., see Dingell, J. V. 390, *414*
Sulser, F., see Gillette, J. R. 121, *127*
Summers, W. P., see Capizzi, R. L. 338, *339*
Sunahara, S. 177, *210*
Sung, L. C., see Cavalieri, R. R. 286, *309*
Sutcliffe, M., see Sales, J. E. L. 239, 240, *256*
Suter, H., see Aebi, H. 174, *203*
Sutherland, E. W., see Robison, G. A. 317, *324*
Sutherland, R., Batchelor, F. R. 351, *358*
Sutherland, R., see Howell, A. 240, *254*
Sutow, W. W., see Wilbur, J. R. 336, *342*
Swainson, C. P., see Prescott, L. F. 390, *417*
Swaminathan, S., P., see Baker, S. J. 130, *145*
Swartz, C., see Seller, R. H. 316, *324*
Swell, L., Field, H., Jr., Treadwell, C. R. 221, *233*
Swenson, D. H., Miller, J. A., Miller, E. C. 395, *418*
Swintosky, J. V. 95, 104, 106, *112*
Swintosky, J. V., Adams, H. J., Caldwell, H. C., Dittert, L. W., Ellison, T., Rivard, D. E. 95, 101, *112*
Swintosky, J. V., Caldwell, H. C., Chong, C. W., Irwin G. M., Dittert, L. W. 95, 101, 102, *112*
Swintowsky, J. V., Digenis, G. [18] 39, *85*
Swintosky, J. V., see Caldwell, H. C. 95, 101, 108, 109, *109*
Swintosky, J. V., see Dittert, L. W. 88, 93, 94, 95, 102, 103, 104, *109*, *110*

Swintosky, J.V., see Kupchan, S.M. 95, 99, 100, *111*
Swintosky, J.V., see Rattie, E.S. 95, 102, *111*
Sylvester, P.E., see Baylis, E.M. 286, *308*
Sylwester, D., see MacDonald M.G. 285, 287, *312*
Szeinberg, A., Bar-Or, R., Sheba, C. 177, *210*
Szeinberg, A., Pipano, S., Ostfeld, E., Eviator, L. 181, *210*
Szmuszkovicz, J. 95, 108, *112*

Tabachnick, I.A., see Roth, F.E. 322, *324*
Tabachnick, M. 221, *233*
Takahara, S. 173, 175, *210*
Takahara, S., Doi, K. 173, *210*
Takahara, S., Hamilton, H.B., Neel, J.V., Kobara, T.Y., Ogura, Y., Nishimura, E.T. 174, *210*
Takahara, S., Sato, H., Doi, M., Mihara, S. 173, *210*
Takahara, S., see Nishimura, E.T. 174, *208*
Talbot, R., see Clarke, R.A. 245, *253*
Talley, R.C., see Goldberg, L.I. 318, 320, *323*
Talso, P.J., see Okita, G.T. 136, *148*
Tan, E.M., see Perry, H.M., Jr. 178, *209*
Tan, G.H., see Doluisio, J.T. 236, *253*
Tan, J.S., Terhune, C.A., Jr., Kaplan, S., Hamburger, M. 354, *358*
Tanford, C., Lovrien, R. 174, *210*
Tarcan, Y., see Dayton, P.G. 285, *310*
Tardiff, R.G., Dubois, K.P. 338, *341*
Tarnowski, G.S., Mountain, I.M., Stock, C.C. 330, *341*
Tatsumi, K., see Tsukamoto, H. 106, *112*
Taubert, K., see Shapiro, W. 317, *324*
Taurog, A., see Goldstein, J.A. 144, *147*
Taylor, C.B., see Forman, D.T. 133, *147*
Taylor, H.G., see Wilbur, J.R. 336, *342*
Taylor, J.D., see Richards, R.K. 264, *271*
Taylor, P.W. [19] 37, *85*
Taylor, S.L., see Bailey, W.C. 409, *413*
Terhune, C.A., Jr., see Tan, J.S. 354, *358*
Thomas, A.A., see Back, K.C. 398, *413*
Thomas, E.M., see Kagan, B.M. 261, 270
Thomas, F.B., see Greenberger, N.J. 137, *147*
Thomas, M., see Jewitt, D.E. 163, *167*
Thompson, G.R. 281, *314*
Thompson, G.R., Barrowman, J., Gutierrez, L. 281, *314*
Thompson, J.H., see Figueroa, W.G. 145, *146*
Thompson, R.P.H., see Clark, R. 389, *413*
Thompson, R.P.H., see Ramboer, C. 287, *313*
Thompson, R.Q., Sturtevant, M., Bird, O.D., Glazko, A.J. 130, *149*
Thomson, P.D., Rowland, M., Melmon, K.L. 156, *168*, 241, 243, 245, *256*, 307, *314*
Thomson, P.D., see Rowland, M. 163, *167*
Thorgeirsson, Davies 151
Thorgeirsson, S.S., Jollow, D.J., Sasame, H.A., Green, I., Mitchell, J.R. 390, 391, *418*
Thorgeirsson, S.S., Sasame, H., Potter, W.Z., Nelson, W.L., Jollow, D.J., Mitchell, J.R. 394, *418*
Thorgeirsson, S.S., see Breckenridge, A. 305, *308*
Thorgeirsson, S.S., see Davies, D.S. 202, *205*
Thorgeirsson, S.S., see Jollow, D.J. 376, *381*, 385, 388, 390, *415*
Thorgeirsson, S.S., see Mitchell, J.R. 388, 389, 390, *417*
Thorgeirsson, S.S., see Ohnhaus, E.E. 300, *312*, 363, *382*
Thorgeirsson, S.S., see Potter, W.Z. 384, 385, 386, 389, 390, 392, 393, 394, *417*
Thorgeirsson, U.P., see Mitchell, J.R. 397, 398, 408, 409, *417*
Thornston, G.F., see Cluff, L.E. 343, *355*
Thornton, G., see Cluff, L.E. 173, *205*
Thorp, J.M. 125, *129*
Thorstrand, C., see Sjöqvist, F. *313*
Thorup, O.A., Jr., Carpenter, J.T., Howard, P. 174, *210*
Thrasher, K., see Choi, Y. 286, 287, *309*
Thrasher, K., see Werk, E.E., Jr. *314*
Thyrum, P.T., see Gerhard, R.E. 291, *310*
Ticknor, W., see Grossman, M. 216, *231*
Tillson, E.K., see Beyer, K.H. 361, *381*
Tilton, B.E., see Fraser, I.M. 190, *205*
Timbrell, J., see Snodgrass, W. 398, *418*
Timbrell, J.A., Jallow, D.J., Mitchell, J.R. 398, *419*
Timbrell, J.A., see Mitchell, J.R. 290, *312*, 362, 375, 376, 377, 378, *382*, 398, *417*
Tintinen, H. 399, *419*
Tizianello, A., see Salvidio, E. 190, *210*
Tjandramaga, T.B., Anton, A.H., Goldberg, L.I. 319, *324*
Tobey, R.E., Jacobsen, P.M., Kahle, C.T., Clubb, R.J., Dean, M.A. 251, *256*
Tobon, F., see Kater, R.M.H. 246, *254*, 288, *311*
Toczko, K., see Jagenburg, O.R. 388, *415*
Todd, R.L., see Keys, A. 222, *232*
Tognoni, G., Morselli, P.L., Garattini, S. 122, *129*
Tolomie, J.D., Joyce, T.H., Mitchell, G.D. 252, *256*
Tomich, E.G., see Busfield, D. 278, *309*
Tompsett, R., see Barnett, H.L. 225, *230*
Tompsett, R., see Robbins, W.C. 354, *357*
Tompson, F.D., Joekes, A.M., Foulkes, D.M. 249, *256*
Tornquist, R. 194, *210*
Toskes, P.P., Deren, J.J. 281, *314*
Toth, B., Shubik, P. 407, *419*
Tothill, P., see Heading, R.C. 236, *254*, 280, *310*

Tothill, P., see Nimmo, J. 237, 239, *255*
Town, M. A., see McHenry, M.C. 248, *255*, 266, *270*
Townsend, S. R., see Hoogstraten, B. 335, *340*
Trasher, K., see Brooks, S.M. 286, *309*
Treadwell, C. R., see Swell, L. 221, *233*
Tretter, P., see Hyman, G. A. 336, *340*
Trevett, M. E., see Barnden, R. L. 95, *109*
Tricot, J. P., see Michielson, P. 391, *416*
Trinker, F. R., Rand, M. J. 301, *314*
Trinker, F. R., see Rand, M. J. 301, *313*
Tripathy, S. P., see Rao, K. V. N. 398, *417*
Trolle, D. 287, *314*
Trotter, W. H., see Harris, H. 192, *206*
Truelove, J. E., see Hussain, A. 94, 95, *111*
Tsai, I., see Kuntzman, R. G. 248, *254*
Tseng, L. T., Walaczek, E. J. 321, *324*
Tsuchiya, T., Levy, G. 29, *34*
Tsuchiya, T., see Levy, G. 26, 29, *34*
Tsukamoto, H., Yoshimura, H., Tatsumi, K. 106, *112*
Tsukamoto, H., see Yamamoto, A. 95, 106, *112*
Tu, K.-P., see Gee, T. S. 333, *340*
Tullar, B. F., see Rosi, D. 126, *128*
Turck, M., see Beaty, N. H. 354, *355*
Turnberg, L. A., see Mawer, G.E. 242, 243, 244, 249, *255*
Turnbull, M. J., see O'Malley, K. 286, *312*
Tybring, L., see von Daehne, W. 95, 96, 97, *109*
Tygstrup, N., see Winkler, K. 283, *314*
Tyler, F. H., see Jubiz, W. 287, *311*
Tyler, F. H., see Meikle, A. W. 287, *312*
Tyrer, D. D., see Kline, I. 328, *340*
Tyrrell, W. F., see Lees, A. W. 399, *416*
Tyrrell, W. F., see Smith, J. 397, *418*

Udall, J. A. 296, *314*
Ueda, H., Sakurai, T., Ota, M., Nakajima, A., Kamii, K., Maezawa, H. 243, *256*
Uehleke, H. 400, *419*
Uehleke, H., see Reiner, O. 404, *418*
Ukich, A. W., see Armstrong, B. K. 235, *252*
Ungar, J., Muggleton, P. W. 95, 96, *112*
Utz, J. P., see Bell, N. H. 337, *338*

Vacca, J. B., see Powell, R. C. 134, *148*
Vadlamudi, S., Padarath-singh, M., Waravdekar, V. S., Goldin, A. 336, *341*
Vagenakis, A. G., Cote, R., Miller, M. E., Braverman, L. E., Stohlman, F., Jr. 251, *257*
Vaidya, K. A., see Orgell, W.H. 181, *208*
Valeriote, F. A., see Bruce, W. R. 329, *338*
Valkenborgh, G., see Michielson, P. 391, *416*
Vallee, B. L., see Blair, A. H. 195, *204*
Vanderhaeghe, H., see Eyssen, H. 134, *146*
Vandam, L. D., see Reynolds, E.S. 399, 409, *418*
Vane, J. R., see Hughes, J. 321, *323*
van Eden, E. B., Falkson, H. C., Falkson, G. 330, *342*
van Heulen, C., see Stubbe, L. 94, *112*
Van Itallie, T. B., see Hashim, S. A. 133, *147*
Van Loon, E. J., Flanagan, T. L., Novick, W. J., Maass, A.R. 143, *149*
van Ommen, R. A., see McHenry, M. C. 248, *255*, 266, *270*
van Rossum, J. M., see Ariens, E. J. 315, *322*
Vecchio, T. J., see Poucher, R. L. 301, *313*
Velu, S., see Devadatta, S. 176, 177, *205*
Venditti, J. M., Humphreys, S. R., Mantel, N., Goldin, A. 326, *342*
Venditti, J. M., see Kline, I. 328, *340*
Venho, V. M. K., Jussila, J., Aukee, S. 239, *257*
Venter, J. C., Dixon, J. E., Markors, P. R., Kaplan, N. O. 230, *233*
Ventura, S., see Larizza, P. 190, *208*
Verebely, K., see Brennan, R. W. 289, *308*
Verebely, K., see Kutt, H. 183, *208*, 285, 289, 306, *311*
Vernier, V. G., Alleva, F. R., Hanson, H. M., Stone, C.A. 121, *129*
Verroust, P., see Richet, G. 261, 267, *271*
Verwey, W. F., Williams, H. R., Jr. 260, *271*
Verwey, W. F., see Beyer, K. H. 361, *381*
Vesell, E. S. 169, 170, 187, 198, *210*, *211*
Vesell, E. S., Ng, L., Passananti, G. T., Chase, T. N. 203, *211*
Vesell, E. S., Page, J. G. 198, 202, 203, *211*, 265, *271*, 285, 288, 306, *314*
Vesell, E. S., Page, J. G., Passananti, G. T. 198, *211*, 287, 289, *314*
Vesell, E. S., Passananti, G. T. 202, *211*
Vesell, E. S., Passananti, G.T., Greene, F. E. 289, *314*, 390, *419*
Vesell, E. S., Passananti, G.T., Lee, C. H. 203, *211*
Vesell, E. S., Passananti, G.T., Greene, F. E. 203, *211*
Vesell, E. S., see Cascorbi, H. F. 198, *204*
Vesell, E. S., see Chignell, C. F. 197, *204*, 241, *253*
Vesell, E .S., see Fraser, I. M. 190, *205*
Vessey, M. P., see Jick, H. 196, *207*
Viau, J. P., see DiCarlo, F. J. 121, *127*
Villiers, J. D., see Denborough, M. A. 194, *205*
Virieux, C., see Fabre, J. 266, *269*
Vogel, F. 169, *211*
Vogel, F., Kruger, J. 196, *211*
Vogel, F., Kruger, J., Song, Y. K., Flatz, G. 196, *211*
Vogler, W. R. 330, *342*
von Bahr, C., see Reidenberg, M.M. 242, *256*, 260, 264, *270*

von Daehne, W., Frederiksen, E., Gundersen, E., Lund, F., Morch, P., Petersen, H.J., Roholt, K., Tybring, L., Godtfredsen, W.O. 95, 96, *109*
von Daehne, W., Godtfredsen, W.O., Roholt, K., Tybring, L. 95, 97, *109*
von Esch, A.M., see Erturk, E. 407, *414*
von Wartburg, J.P., Schürch, P.M. 195, *211*
Votava, Z., see Metysoda, J. 121, *128*
Vrindten, P.A., see Chen, W. 288, 301, *309*

Wachstein, M., Bradshaw, M. 223, *233*
Waddell, W.J., Buttler, T.C. 240, *257*, 361, *382*
Waddell, W.J., see Buttler, T.C. 266, *268*
Wagner, J. 214, *233*
Wagner, J.G. [20] 24, 27, 28, 31, *34*, 78, *85*
Wagner, J.G., Nelson, E. 13, *34*
Wagner, J.G., Northam, J.I., Alway, C.D., Carpenter, O.S. *34*
Wagner, J.G., Novak, E., Leslie, L.G., Metzler, C.M. 21, *34*
Wagner, J.G., see McHenry, M.C. 248, *255*, 266, *270*
Waithe, W.I., see Samuel, P. 134, *148*, 218, *313*
Walaczek, E.J., see Tseng, L.T. 321, *324*
Walkenstein, S.S., see Kimmel, H.B. 118, *128*
Walker, D., see Levi, A.J. 243, *254*, 288, 307, *311*
Walker, E.W.A., see Dreyer, G. 214, *231*
Walker, S.R., Evans, M.E., Richards, A.J., Paterson, J.W. 155, *168*
Walker, W.G., see Whelton, A. 240, *257*
Wallace, J.F., Smith, R.H., Garcia, M., Petersdorf, R.G., 349, 350, *358*
Wallace, S.M., see Ellard, G.A. 177, *205*
Wallach, S., see Englert, E., Jr. 262, *269*
Walsh, C.T., Levine, R.R. 215, *233*
Walter, E., see Hahn, K.-J. 281, *310*
Walton, R.P., Gazes, P.C. 315, *324*
Wan, S.H., Riegelman, S. 261, *271*
Waravdekar, V.S., see Vadlamudi, S. 336, *341*
Wardell, W. 294, *314*
Waring, W.S., see Spinks, A. 114, 115, *129*
Warren, G.H., see Glassman, J.M. 143, *147*
Warren, G.H., see O'Connor, W.J. 216, *233*
Washington II, J.A., see Wilkowske, C.J. 353, *358*
Watanakunakorn, C. 352, 354, *358*
Waterworth, P.M., see Garrod, L.P. 346, *356*
Watson, C.J., see Ibrahim, G.W. 130, *147*
Watson, R.M., see Pierson, R.N. 94, *111*
Watson, T.R., see Holder, G.M. 142, *147*
Wattenberg, L.W., Leong, J. L. 362, *382*
Way, E.L. 116, *129*
Way, E.L., Adler, T.K. 134, *149*
Webb, D.I., Chodos, R.B., Mahar, C.R., Faloon, W.W. 239, *257*
Webb, D.I., Chodos, R.B., Mahar, C.Q., Faloon, W.W. *314*
Webb, G.B., see Barnden, R.L. 95, *109*
Weber, E., see Hahn, K.-J. 281, *310*
Weber, W.W. 178, *211*
Weber, W.W., Cohen, S.N., Steinberg, M.S. 177, 178, *211*
Weber, W.W., see Cohen, S.N. 169, *205*
Wechselberg, K. 234, *257*
Wechsler, A.S., see Glick, G. 318, *323*
Weed, R.I., Reed, C.F. 190, *211*
Weeks, J.R. 320, *324*
Wehrle, P.F., see Mathies, A.W., Jr. 350, *357*
Weihe, M., Potter, W.Z., Nelson, W.L., Jollow, D.J., Mitchell, J.R. 393, 394, *419*
Weil, J., see Gilmore, E. 320, *323*
Weil, W.B., Jr. 223, *233*
Weinberg, A.N., see Moellering, R.C., Jr. 352, 353, 354, *357*
Weinberg, A.N., see Zimmermann, R.A. 352, 353, *358*
Weiner, I.M. [21] 39, *85*
Weiner, I.M., see Glasser, J.E. 133, *147*
Weiner, I.M., see Lack, L. 131, 132, 133, *147*, *148*
Weiner, M., Shapiro, S., Axelrod, J., Cooper, J.R., Brodie, B.B. 197, *211*
Weiner, M., see Brodie, B.B. 157, *166*
Weiner, M., see Dayton, P.G. 285, *310*
Weinstein, L., see Chang, T.W. 350, *355*
Weinstein, L., see Gorbach, S.L. 215, *231*
Weinstein, L., see Lerner, P.I. 325, *341*
Weinstein, L., see Rahal, J.J., Jr. 353, 354, *357*
Weintraub, H., see Gibaldi, M. 31, *33*
Weisberger, A.S., see Suhrland, L.G. 261, 264, *271*
Weisburger, E.K., see Weisburger, J.H. 373, *382*, 392, *419*
Weisburger, J.H., Weisburger, E.K. 373, *382*, 392, *419*
Weisman, H., see Rongey, K.A. 322, *324*
Weiss, M.F., see Berman, M. 33, *33*
Weissinger, J.L., see Fischer, L.J. 138, *146*, *147*
Welch, R., see Conney, A.H. 169, 200, *205*, 243, *253*
Welch, R.M., Harrison, Y.E., Conney, A.H., Burns, J.J. 301, *314*
Welch, R.M., Harrison. Y.E., Gommi, B.W., Poppers, P.J., Finster, M., Conney, A.H. 288, *314*
Wells, J.H., see Mathews, K.D. 408, *416*
Welt, S.I., Jackson, E.H., Kirkman, H.N., Parker, J.C. 191, *211*
Wennersten, C., see Moellering, R.C., Jr. 352, 354, *357*
Wenzel, D.C., Broadie, L.L. 200, *211*
Werk, E.E., see Brooks, S.M. 286, *309*
Werk, E.E., Jr., MacGee, J., Sholiton, L.J. 287, *314*
Werk, E.E., Jr., Trasher, K., Sholiton, L.J., Olinger, C., Choi, Y. *314*

Werk,E.E.,Jr., see Choi,Y. 286, 287, *309*
West,C.D., Hong,R., Holland,N.H. 220, 223, 224, *233*
West,C.D., see Jubiz,W. 287, *311*
West,C.D., see Meikle,A.W. 287, *312*
West,J.R., Smith,H.W., Chasis,H. 233
Westerholm,B., see Jick,H. 196, *207*
Westhead,E.W.,Jr., see Morawetz,H. 92, *111*
Weyman,P., see Sandberg, A.A. 130, *148*
Whang-Peng,J., see Canellos, G.P. 334, *339*
Whelton,A., Sapir,D.G., Carter,G.G. 260, *271*
Whelton,A., Sapir,D.G., Carter,G.G., Kramer,J., Walker,W.G. 240, *257*
White,P.D. 169, *211*
White,R.J., Chamberlain, D.A., Howard,M., Smith, T.W. 235, *257*
White,S.C., see Lepage,G.A. 334, *341*
White,T.A., Price Evans, D.A. 177, *211*
White,T.A., see Price Evans, D.A. 176, 177, *209*
Whitecar,J.P., see Freireich, E.J. 333, *340*
Whitmore,W.F.,Jr., see Li,M.C. 336, *341*
Whitsett,T.L., see Finlay, G.D. 319, *323*
Whittaker,J.A., Price Evans, D.A. 198, 201, *211*
Wittaker,M., see Harris,H. 179, 180, 181, 182, *206*
Whittaker,V.P., Wijesundera,S. 179, *211*
Whittaker,V.P., see Sturge, L.M. 92, 95, 97, *112*
Wiebe,L.I., see Holder,G.M. 142, *147*
Wiebel,F.J., Leutz,J.C., Diamond,L., Gelboin, H.V. 363, *382*
Wiesner,P.J., see Griner, P.F. 296, *310*
Wijesundera,S., see Whittaker,V.P. 179, *211*
Wilbur,J.R., Sutow,W.W., Sullivan,M.P., Costio, J.R., Taylor,H.G. 336, *342*
Wilfert,J.N., see Smith,C.B. 344, *357*
Wilkerson,G.R., see Branch, R.A. 363, *381*
Wilkinson,G.R., see Branch, R.A. 300, 306, *308*
Wilkowske,C.J., Facklam, R.R., Washington II, J.A., Geraci,J.E. 353, *358*
Will,G., see Lawson,D.H. 259, *270*
Willardson,D.G., see Englert,E.,Jr. 262, *269*
Williams,D., see Solomon, H.M. 297, 306, *314*
Williams,H.R.,Jr., see Verwey,W.F. 260, *271*
Williams,P.J., see Calder, I.C. 392, *413*
Williams,R., see Hunter,J. 202, *207*, 286, *311*
Williams,R., see Maxwell, J.D. 286, *312*
Williams,R., see Murray-Lyon,I.M. 252, *255*
Williams,R., see Ramboer,C. 287, *313*
Williams,R.T. 86, 99, *112*
Williams,R.T., Millburn,P., Smith,R.L. 130, 143, *149*
Williams,R.T., see Azouz, W.M. 378, *381*
Williams,R.T., see Caldwell, J.H. 141, *146*
Williams,R.T., see Clark, A.G. 131, 138, *146*
Williams,R.T., see Fischer, L.J. 131, 138, *146*
Williams,R.T., see Levine, W.G. 144, *148*
Willimott,S.G. 181, *211*
Willmanns,H., see Graul, E.H. 337, *340*
Wilson,A., see Horning,M.G. 402, *415*
Wilson,B.J., Boyd,M.R., Harris,T.M., Yang, D.T.C. 396, *419*
Wilson,B.J., Yang,D.T.C., Boyd,M.R. 396, *419*
Wilson,G.S. 181, *212*
Wilson,I.B. 179, *212*
Wilson,J.F., see Lanzkowsky,P. 217, *232*
Wilson,J.T. 287, *314*
Wilson,M.J., see Kimball, A.P. 329, *340*
Winder,C.V., Kaump,D.H., Glazko,A.J., Holmes, E.L. 142, *149*
Winfield,D.A., see Guyer, R.J. 333, 334, *340*
Winkler,K., Keiding,S., Tygstrup,N. 283, *314*
Winters,R.E., Chow,A.W., Hecht,R.H., Hewitt, W.L. 350, *358*
Winters,W., see Kutt,H. 182, 183, 197, *208*, 243, *254*, 289, *311*
Wishart,J.W., see Clements, F.W. 192, *204*
Wisniewski,K., see Danysz, A. 282, *309*
Wisseman,C.L.,Jr., see Hahn,F.E. 351, *356*
Witkop,B., see Daly,J.W. 378, 379, *381*, 390, 402, *413*
Witt,P.N., see Farah,A. 316, *323*
Wizwer,P., see Jick,H. 196, *207*
Wodinski,I., Kensler,C.J. 333, *342*
Wolf,A.R., see Wyngaarden, J.B. 139, *149*
Wolf,L.M., see Glazko,A.J. 95, 98, 100, *110*
Wolff,J.A., see Conney,A.H. 169, 200, *205*, 243, *253*
Wolff,P.H. 196, *212*
Wolff,S.M., see Song,C.S. 191, *210*
Wolff,S.M., see Steinberg, A.D. 405, *418*
Wolinsky,E., Hines,J.D. 267, *271*, 337, *342*
Wolk,M., see Kutt,H. 182, 183, 197, *208*
Wolpert,M.K., Althaus,J.R., Johns,D.G. 407, *419*
Wolthers,H., see Lundquist, F. 26, 27, *34*
Wong,V.G., see Steinberg, A.D. 405, *418*
Won Lee,S., see Avioli,L.V. 259, 261, *268*
Wood,P., Shioda,R., Estrich, D., Splitter,S. 133, *149*
Woodbury,D.M., Esplin, D.W. 182, *212*
Woods,L.A. 131, 134, *149*
Worcester,J., see Jick,H. 196, *207*
Worth,D.F., Elslager,E.F., Phillips,A.A. 95, 107, *112*
Worth,D.F., see Elslager, E.F. 95, 107, 108, *110*
Wrenn,E.L.,Jr., see James, D.H.,Jr. 325, 336, *340*
Wrenshall,G.A., see Bogoch, A. 250, *253*
Wright,N., Prescott,L.F. 245, *257*, 389, *419*

Wright, N., see Prescott, L. F. 244, *256*, 389, *417*
Wright, R. C., see Gutsche, B. B. 181, *206*
Wright, S. E., see Ladomery, L. G. 131, 141, *148*
Wrobel, C. J., see Lippman, R. W. 259, *270*
Wyngaarden, J. B., Howell, R. B. 173, *212*
Wyngaarden, J. B., Petersen, R. E., Wolf, A. R. 139, *149*

Yaffe, S. J., Levy, G., Matsuzawa, T., Baliah, T. 203, *212*, 278, *314*
Yaffe, S. J., see Axline, S. G. 225, *230*
Yaffe, S. J., see Krasner, J. 221, *232*
Yaffe, S., see Jusko, W. J. 216, 217, *232*
Yahr, M. D., Merritt, H. H. 182, *212*
Yahr, M. D., Sciarra, D., Carter, S., Merritt, H. H. 182, *212*
Yamada, H., see Levy, G. 363, 364, *382*
Yamada, R., see Hayakawa, T. 338, *340*
Yamamoto, A., Yoshimura, H., Tsukamoto, H. 95, 106, *112*
Yamasaki, M., see Hussain, A. 94, 95, *111*
Yang, D. T. C., see Wilson, B. J. 396, *419*
Yasuno, A., see Mori, K. 407, *417*
Yata, H. 174, *212*
Yeh, B. K., de Guzman, N. T., Pinakatt, T. 317, *324*
Yeh, B. K., McNay, J. L., Goldberg, L. I. 321, *324*
Yen, W., see Kreis, W. 329, *340*
Yesair, D. W., Callahan, M., Remington, L., Kensler, C. J. 140, *149*
Yeung, C. Y., Field, C. E. 287, *314*
Yim, G. K. W., see Borowitz, J. L. 280, *308*
Yoshida, A. 187, *212*
Yoshida, A., Motulsky, A. G. 182, *212*
Yoshida, A., see Motulsky, A. G. 187, 188, 189, *208*
Yoshimura, H., see Tsukamoto, H. 106, *112*
Yoshimura, H., see Yamamoto, A. 95, 106, *112*
Young, J., see Murray-Lyon, I. M. 252, *255*
Young, L., see Knight, R. N. 378, *381*
Young, V. M., see Schimpff, S. 344, *357*
Yu, T. F., see Burns, J. J. 115, 117, 118, *127*

Zacchei, A. G., see Hucker, H. B. 131, 139, 140, *147*
Zager, R. F., Frisby, S. A., Oliverio, V. T. 337, *342*
Zaimis, E., see Bigland, B. 252, *253*
Zamboni, V., see Ceriotti, G. 95, 100, *109*
Zampaglione, N., Jollow, D. J., Mitchell, J. R., Stripp, B., Hamrick, M., Gillette, J. R. 378, 380, *382*, 401, *417*
Zampaglione, N., see Jollow, D. J. 376, 378, 379, 380, *381*, 401, 403, *415*
Zampaglione, N. G., see Stripp, B. 402, 403, 407, *418*
Zanetti, M. E., see Jansen, G. R. 133, *147*
Zannos-Mariolea, L., see Kattamis, C. 180, *207*
Zapf, P. W., see King, J. O. 252, *254*
Zaremba, E. A., see Ross, S. 95, 100, *111*
Zbinden, G., Randall, L. O. 118, *129*
Zelechowski, K., see Meyer, M. B. 234, *255*
Zeltzer, P., see Hyvarinen, M. 220, *232*
Zetterlund, C. G., see Angervall, L. 406, *413*
Zieve, P., see Kater, R. M. 246, *254*, 288, *311*
Zimmerman, H. J. 409, *419*
Zimmerman, H. J., see Black, M. 397, 408, *413*
Zimmermann, R. A., Moellering, R. C., Jr., Weinberg, A. N. 352, 353, *358*
Zinkham, W. H., see Rieder, R. F. 191, *209*
Zins, G. R. 124, *129*
Zion, T. C., see Horning, M. G. 402, *415*
Ziskind, M. M., see Bailey, W. C. 409, *413*
Zollinger, H. V., see Spuhler, O. 391, *418*
Zubrod, C. G., Kennedy, T. J., Shannon, J. A. 299, *314*
Zurcher, C., see Prins, H. K. 189, *209*
Zsigmond, M., see Angervall, L. 406, *413*

Subject Index

Absorption of drugs
  and drug interactions 278, 279, 281, 282
  effects of aging on 214—217
  effect of various factors on 234—240
  in chronic renal failure 258, 259
  kinetics 240—242
    in linear one-compartment systems 3—17
    in linear multicompartment systems 17—24
    in non-linear systems 29
Acatalasemia, see acatalasia
Acatalasia 171, 173—175
Acedapsone, see 4′,4‴-sulfonylbisacetanilide
Acetaldehyde
  drug interactions 290
p-Acetamidophenyltrichloroethyl-carbonate 103
4-Acet-amido-phenyl-2,2,2-trichloroethyl
  synthesis of carbonate 101
Acetaminophen
  carbonate prodrug esters of 88, 89, 102, 105
  drug interactions 362, 363, 373
  from phenacetin 116
  metabolic pathways 387
  toxicity 244, 245, 375—381, 384—390, 411
  variations in absorption of 235—241, 280
  with trichloroethanol 101
Acetanilide 390
  drug interactions 373
  in G6PD deficiency 172, 410
  liver disease and metabolism of 243
  toxicity 400, 401, 410
Acetazolamide
  renal excretion of 248
Acetohexamide 125
  drug interactions 292
  renal excretion of 248
Acetophenetidin
  and methemoglobinemia 184, 185
  in G6PD deficiency 172
N-Acetyl-4-amino-antipyrine 117
2-Acetylaminofluorene 404, 405
N-Acetyl-*p*-aminophenol
  from phenacetin 184
  in chronic renal failure 264
Acetylation and hepatotoxicity 397—399
Acetylcholine
  and acid-base, electrolyte balance 251, 252, 280
  drug interactions 322
Acetylisoniazid 176, 397—399
1-[(L-Acetylphenyl)sulfonyl]-3-cyclohexyl-urea 125
Acetylmethadol
  active metabolites of 116
O-Acetylsalicylamide
  hydrolysis of 91
Acetylsalicylic acid 172
  drug interactions 337
  from salicylates 116
  in G6PD deficiency 172
  liver disease and serum levels 246
N_2-Acetylsulfanilamide
  in G6PD deficiency 172
Acetyl transferase
  and isoniazid metabolism 176
Acid ash
  in drug interactions 291
Acid-base and electrolyte balance
  effect on drug action 251, 252
Actinomycin D
  in drug combinations 330, 336
  mechanism of action 329
Active drug transport
  effect on rate of distribution 39
Active metabolites
  and drug interactions 290
  biotransformation of drugs to 113—129
  definition 113, 114
Acyloxymethyl ester 95
Adams-Stokes syndrome
  use of isoproterenol 158
ADH, see alchohol dehydrogenase 195
  see also vasopressin
Adrenergic blocking agents 123
  α-methyldopa 123
  α-methyl-*m*-tyrosine 123
  and drug interactions 293, 299, 317—322
  and glaucoma 194
  enhanced response to 250, 251
Adrenocortical steroids
  and diabetes mellitus 170
Aging effects and drugs 211—230
AIA, see Allylisopropylacetamide
Albumin
  and drug binding 241—243
    in uremia 260
  and reaction to prednisone 197
  binding to sulfaphenazole 197
  in drug interactions 295—298
  levels at various ages 220, 221, 229
Alcohol dehydrogenase
  atypical 195
Alcohol prodrug derivatives 97—106
Alcoholism
  relation to ADH 195

Alkylating agents
in treatment of neoplastic diseases 126, 334—338
Allergic drug reactions 407—409
Allobarbital 404
Allocatalasia 175
Allopurinol
active metabolites of 117
effect on drug metabolism 203
renal excretion of 248
drug interactions 289, 337
Alloxanthine
from allopurinol 117
Allyl acetamide compounds
and porphyria 403, 404
Allyl ureide compounds
and porphyria 403, 404
Allylisopropylacetamide (AIA)
and porphyria 403, 404
Aluminium hydroxide
effect on drug absorption 280, 282
Amaranth
biliary excretion 144
Amiloridine
renal excretion of 248
Amine prodrug derivatives 106—109
Amino acid
levels in chronic renal failure 259, 260
4-Aminoantipyrine
from aminopyrine 117
p-Aminobenozic acid 90
2-Amino-5-chlorobenzoxazole 115
Aminoglycosides
drug interactions 350, 352—354
renal excretion of 248
p-Aminohippurate
drug interactions 302
extraction related to age 225
δ-Aminolevulinic acid 404
Aminophylline
drug interactions 318
Aminopyrine
active metabolites of 117
in chronic renal failure 261
in G6PD deficiency 172
Aminosalicylic acid
see also *p*-aminosalicylic acid
drug interactions 344
p-Aminosalicylic acid
absorption of 239
acetylation of 176
and diphenylhydantoin 183
and malabsorption syndromes 239, 281, 282
from salazopyrin 239
in G6PD deficiency 172
liver disease and metabolism of 243
renal excretion of 248, 293
Amitriptyline 121
drug interactions 300
Amobarbital
binding 241, 249
volume of distribution 241
with acetaminophen 245
Amphetamine
absorption of 152
enterohepatic circulation of 141
from fenproporex 122
from prenylamine 124
in drug interactions 291, 300, 301
renal excretion of 248
Amphotericin B
drug interactions 337, 353
Ampicillin
absorption 96, 238, 240
age-related changes in excretion 225
drug interactions 344, 350, 353, 354
enterohepatic circulation of 143
in chronic renal failure 260, 267
to sterilize intestinal flora 136
Amyl nitrite
effect on volume of distribution 241
Analgesics
absorption 278
abuse 391
in G6PD deficiency 172
Analgesic-antipyretics 116—118
allopurinol 117
aminopyrine 117
diminished response to 250
phenacetin 116, 117
phenylbutazone 117
4-phenylthioethyl 118
salicylates 116
Analgesics, narcotic 115, 116
acetylmethadol 116
codeine 115
diphenoxylate 116
meperidine 116
tolerance to 250
Anectine, see succinylcholine
Anemia
and thiopental anesthesia 252
pernicious and drug absorption 236
Anesthetics
and malignant hyperthermia 194, 252
influences on absorption 234, 252
inhalation of 158
Angiotensin 320
Aniline 400
ANIT, see α-naphthylisothiocyanate
Anorexia nervosa
effect on drug absorption 236
Antacids
effect on drug absorption 279—282
Antagonism
of combined drug effect 347—351
Antiarrhythmic drugs
2′,6′-dimethyl-phenoxy-2-aminopropane 245
drug interactions 305
effects of acid-base and electrolyte balance 251, 252
lidocaine 123
propranolol 123
Antibacterial agents
in G6PD deficiency 172

Antibiotics
absorption 154
drug interactions 281
enterohepatic circulation of 143
local action 155
prodrugs of 97, 98
Anticholinesterase drugs 122
parathion 122
Anticoagulant drugs
drug interactions 285
resistance to 185, 186
sensitivity to 184
Anticonvulsant drugs 114, 115
N-demethyldiazepam 119
diphenylhydantoin 182, 183
drug interactions 286, 305
effect of drug metabolism 245
mephobarbital 114
methsuximide 115
N-methyloxazepam 119
oxazepam 119
primidone 114
trimethadione 115
Antidepressant drugs 286, 289
drug interactions 293, 298, 299
Antidiarrheal drugs
diphenoxylate 116
Antifungal drugs
absorption 278
local action 155
Antihypertensive drugs
diazoxide 154
drug interactions 305
effects of acid-base, electrolyte balance 251, 252
Anti-inflammatory drugs 116—118
see also analgesic-antipyretics
mefenamic acid 142
Antileprotic drugs
ditophal 153
sulfone 107, 108
Antimalarial drugs
drug interactions 299
in G6PD deficiency 172
sulfone 107, 108
Antipyrine
drug interactions 285, 286, 288, 289, 293, 300—302, 306, 307
genetic variations in clearance 197—203
in G6PD deficiency 172
in uremic patients 259, 262, 263, 265
liver necrosis and metabolism of 244
routes of administration 151, 152
Antistreptococcal drugs
sulfanilamide 106
Anti-tuberculosis drugs 176—178
drug interactions 289, 344
APAP, see acetaminophen
Apnea
in atypical pseudocholinesterase 179
Apomorphine
drug interactions 322
Apotryptophanase
and isoniazid 176
Aprobarbital 404
Arabinosyl-cytosine (Ara-C)
drug combinations 330, 333—335, 338
mechanism of actions 329
Area under curve
applied to metabolites 67, 68
calculation of 64, 65
estimation of 6
total drug concentration 79, 80, 275—277
Arrhythmias
after digoxin 196, 316, 317, 319
and drug interactions 283
and drug sensitivity 251
Arylcarboxyl tetrahydropans
hydrolysis of 94
L-Asparaginase
mechanism of action 329, 338
Aspirin
absorption of 236
and gastric bleeding 94
drug interactions 279, 291, 297
hydrolysis of 91, 94, 247
Asthmatics
drug interactions 286
Atabrine, see quinacrine
in G6PD deficiency 172
Ataxia
from diphenylhydantoin 182
ATPase
and cardiac glycosides 316
Atropine
and glaucoma 194
drug interactions 322
effect on drug absorption 237, 238
renal excretion of 248
Atypical alcohol dehydrogenase 195
Atypical pseudocholinesterase 171, 178—182
AUC, see area under curve 6
Autosomal recessive genes
and pharmacogenetic conditions 171—172
Availability of drug
factors influencing 274
Azaserine (AZA)
in drug combinations 327—330
Azathioprine
drug interactions 289
Azotemic patients
response to drugs 267, 268
Azulfidine, see salicylazosulfapyridine
in G6PD deficiency 172

Bacitracin
drug interactions 344, 350, 352, 353
Bacteremia
use of combined chemotherapy 343
Bacteriuria
combined chemotherapy 350, 351
BAL
in G6PD deficiency 172
Barbital
renal excretion of 248, 361

Barbiturates
absorption of 236, 261
and porphyria 170
drug interactions 278, 279, 285—287
in chronic renal failure 264
liver disease and metabolism of 243, 244
Benzene toxicity 410
Benzoate ester
as a prodrug 88
7,8-Benzoflavone
drug interactions 363
Benzoic acid derivatives
in chronic renal failure 261
Benzphetamine hydroxylase
in chronic renal failure 261
3,4-Benzpyrene
drug interaction 362, 363
effect on biliary flow 144
effect on drug metabolism 200, 202
Benzprene hydroxylase
drug interactions 288
Beta, in multicompartment systems 20
Bethanidine
drug interactions 299, 300, 305
BHT, see butylated hydroxytoluene
Bile acids
drug interactions 281
Bile salts
absorption 132—134
enterohepatic circulation 130—134
excretion 132
Biliary excretion of drugs 143—145, 239, 240
Biliary lipids 130
Biliary phospholipids 130
Bilirubin
and drug metabolism 243
binding changes with age 221, 222
drug interactions 286, 287, 297
Bioavailability and absorption kinetics
in linear one-compartment systems 11—14
in linear multicompartment systems 20, 21
Biotransformation of drugs
to active metabolites 113—129
Biphenyl
biliary excretion 144
2-[Bis(2-chloroethyl)amino]-tetrahydro-2*H*-1,3,2-oxazo-phosphorine-2-oxide 126
1,3-Bis-(2-chloroethyl)-1-nitrosourea (BCNU)
in drug combinations 330, 331, 332, 334
mechanism of action 329
Bishydroxycoumarin
genetic variations in clearance 197, 198, 202, 203
sensitivity 171, 184
Bis-*p*-nitrophenyl phosphate 401
Body water compartments
and age-related changes 218, 219, 224—230
Bretylium
absorption 154
Bromobenzene
metabolic pathways 402
toxicity 378—381, 401—403
Bromsulphophthalein
age-related changes in binding 222
and drug metabolism 243
drug interactions 302
Brucellosis
combined chemotherapy 353
BSP, see sulfobromophthalein
Bulbocapnine 322
Burns
acid-base and electrolyte balance 251
Butylated hydroxytoluene (BHT)
enterohepatic circulation of 141, 142
4-Butyl-3,5-dioxo-1,2-diphenyl-pyrazolidine 117
Butyrophenones 319, 321

Caffeine
and free fatty acid levels 221
and malignant hyperthermia 194, 195
effect on drug metabolism 200, 202
Calcium
absorption in uremic patients 259
and digitalis 316
and malignant hyperthermia 194
effect on drug absorption 280, 282
Camptothecin
renal clearance of 249
Capillary compartment I 43—80
see also multicompartment models
Carbamate esters
as prodrugs 95
Carbamazepine
drug interactions 285
Carbenicillin
in chronic renal failure 265—267
in drug interactions 344, 353
4-Carbethoxy-1-methyl-4-phenyl-piperdine 116
Carbimazol
biliary excretion 145
Carbon tetrachloride 401
Carbonate esters
as prodrugs 95, 101
of acetaminophen 102, 105
of phenol 105
Carbonic anhydrase
in drug-protein complexes 37
inhibitors, and cardiac glycosides 316
Carboxylic acid drugs
as prodrugs 92—97
selection of alcohol moiety 92
5-Carboxy-3-methyl-isoxazole 125
Carcinogens 404—407
biochemical mechanisms 405
Carcinomatous myopathy
drug sensitivity 251
Cardiac glycosides
drug interactions 315—319
Cardiovascular drugs
absorption of 239, 240, 251, 252
active metabolites 123—125
drug interactions 305, 315—322

Carotene 281
drug interactions 281
Catalase 174
Catecholamine
effect of prenylamine 124
Caternary model 43—80
Central nervous system
drugs affecting 114—121
sensitivity to 252
tolerance to 250
Cephalexin
biliary excretion of 239, 240
renal excretion of 267
Cephaloridine
drug interaction 393
renal excretion of 249, 250, 267
toxicity 397
Cephalosporins
drug interactions 351, 352, 354
renal excretion of 248
Cephalothin
absorption of 240
drug interaction 337
kinetics in uremic patients 264, 266, 267
toxicity 397
Cerebrospinal fluid flow
influence on pharmacokinetics 40
Charcoal activated
drug binding 281
Chenodeoxycholic acid 132
Chloral hydrate
active metabolites of 114
drug interactions 289, 302
Chlorambucil
in combination chemotherapy 334, 336
Chloramphenicol
age-related changes in binding 222
biliary excretion of 130
drug interactions 289, 350, 353
effects of ester derivatives 98, 100, 101
in chronic renal failure 261, 264, 266, 267
in G6PD deficiency 172
liver disease and metabolism of 243
Chlorcyclizine 144
Chlordane
effect on biliary flow 144
Chlordiazepoxide
active metabolites of 118
Chlorethate 101
4-Chloroaniline 400
5-Chloro-2-benzoxazolinone 115
7-Chloro-1 (cyclopropylmethyl)-1,3-dihydro-5-phenyl-2*H*-1,4-benzodiazepine-2-one 121
7-Chloro-2,3-dihydro-1-methyl-5-phenyl-1*H*-1,4-benzodiazepine 120
7-Chloro-1,3-dihydro-1-methyl-5-phenyl-2*H*-1,4-benzodiazepine-2-one 119
Chloroformate esters 95
Chloromethylacetate 95
7-Chloro-2-methylamino-5-phenyl-1,4-benzodiazepine-4-oxide 118
Chloromethylmethyl ether 95
Chloromethylpivalate 95
Chloromycetin
drug interactions 348, 349
Chlorophenoxyisobutyrate
age-related changes in binding 221
p-Chlorophenyl-(*m*-trifluoromethylphenoxy)-acetate
age-related changes in binding 221
Chlorotetracycline
see also chlortetracycline
drug interactions 350
Chlorothiazide
absorption 281
Chloroxazone 115
Chlorpromazine
absorption of 239, 282
biliary excretion of 143
drug interactions 286, 293, 298, 299
genetic variations in clearance 197
in liver disease 252
routes of administration 162, 163
Chlorpropamide
drug interactions 288
renal excretion of 248
Chlortetracycline 134
renal excretion 267
Cholanolic acid 132
Cholecystokinin
effect on drug absorption 239
Cholestasis
produced by phenothiazine 184
Cholesterol
and bile excretion 145
drug interactions 281
levels at various ages 221, 222, 229, 230
serum concentrations and neomycin 134
serum concentrations and nicotinic acid 124
Cholesterol 7-α-hydroxylase
and bile acid synthesis 132
Cholestipol
and drug absorption 281
Cholestyramine
and bile salt absorption 130, 133, 137
drug interactions 281, 282
Cholic acid 132
Cholinergic drugs 322
see also acetylcholine
Cholinergic nerves
blockage and glaucoma 194
^{51}Chromium
gastric emptying 280
Chronic renal failure
effect on drug absorption 258, 259
effect on drug distribution 259, 260
effect on drug elimination 265—267
effect on drug metabolism 261—265
effect on drug response 267, 268
Chymotrypsin
in hydrolysis of carbonate diesters 102
inactivation of oxytocin 153
Ciba-Geigy 36278
renal excretion 267
Cinchophen
effect of prothrombinopenic response 184

Cleft lip 170
Cleft palate 170
Clofibrate 124
 in drug interactions 297
Closed model, pharmacokinetics 41—44
Cloxacillin
 drug interaction 352
 tissue distribution 240
Club foot 170
CNS depressants, see central nervous system
Cobaltous chloride 378
 and acetaminophen 384, 386
 and cephaloridine 397
 and furosemide 393
 and isoniazid 398
 and phenacetin 391, 392
Cocaine
 and glaucoma 194
 drug interactions 337
Codeine
 active metabolites of 115
Colchicine
 absorption of 239
 and biliary secretion 130
 drug interactions 281
Colistin
 drug interactions 337
 renal excretion of 248, 267
Computer programs
 in pharmacokinetics 5, 21
Concussion
 increased pseudocholinesterase 181
Corticosteroids
 drug interactions 337
 enhanced response to 251
 formulation and effectiveness 154
 in glaucoma 172, 193
 in G6PD deficiency 172
Corticosterone
 as a transport carrier 79
 enterohepatic circulation 138
Cortisol
 absorption of 239
 drug interactions 286, 287
Coumarin anticoagulants
 drug interactions 281, 297, 305, 306
 sensitivity to 184, 251
CPIB, see chlorophenoxyisobutyrate 221
Cranberry juice
 in drug interactions 291
Creatinine
 clearance in renal disease 248, 249, 260, 266
 in determining drug clearance 9
CRF, see chronic renal failure
Crigler-Najjar syndrome
 pharmacogenetics 170
Curare
 and malignant hyperthermia 194
N-2-Cyanoethylamphetamine 122
Cycasin 404, 405
Cyclamate
 absorption of 239
Cyclic AMP
 and sympathomimetic amines 317, 318
Cyclophosphamide 126, 405
 drug interactions 290
 in drug combinations 330—338
Cyclopropane
 and malignant hyperthermia 172, 194
 in G6PD deficiency 172
Cycloserine
 renal excretion of 248
Cynthiana variant
 atypical pseudocholinesterase 181, 182
Cysteamine 390
Cysteine 379, 387, 390
Cytochrome P-450
 and spironolactone 402, 403
 in chronic renal failure 261
 in drug interactions 282, 359, 363
 in liver disease 243, 244, 378—381, 389, 393, 394
 in porphyria 404

Dapsone
 in chronic renal failure 260
 protein binding 242
 slow acetylators 171
 see also diamino diphenylsulfone
Daunomycin
 mechanism of action 329—333. 338
DBSP, see phenol-3,6-dibromophthalein disulphonate
DDS, see diamino diphenylsulfone
DDT
 drug interactions 288
Debrisoquin
 drug interactions 299
Decamethonium
 and acid-base, electrolyte balance 251, 252
7,8-Dehydro-4,5-epoxy-6-hydroxy-3-methoxy-*N*-methylmorphinan 115
N-Demethyl-chlordiazepoxide 118
N-Demethyldiazepam 119—121
Demethyldoxepine 121
Deoxycholic acid 132
Deoxycorticosterone sulphate 138
Desacetylcephalothin
 renal clearance 264
Deschlorobenzoylindomethacin 140
Desferrioxamine
 effect on biliary excretion 145
Desipramine
 drug interactions 299, 300, 367, 389, 390
Desmethylimipramine
 genetic differences in metabolism 202
 in chronic renal failure 260
 in drug interactions 291
 protein binding 242
Desmethylindomethacin 140
2-Desoxy-5-ethyl-5-phenyl barbituric acid
 active metabolites of 114
Dexamethasone
 drug interaction 286, 287
Dexamethasone-21-phosphate
 effect on ocular pressure 193, 194

Diabetes mellitus
and adrenocortical steroids 170
blood type factor 196
multigenic effects 170
Dialkylnitrosamines 404
Diallylmelamine 124
2,4-Diamino-6-(diallylamino)-*s*-triazine 124
Diamino diphenylsulfone 107
Diazepam
active metabolites of 119, 120
in liver disease 252
Diazoxide
by intravenous injection 154
renal excretion of 248
Dibucaine number 180
Dicarbethoxydihydrocollidine
and porphyria 403, 404
Dicoumarol
drug interactions 278, 280, 285, 288, 289, 301, 364
formulation and effectiveness 154
1-[(2-Diethylaminoethyl)amino]-4-methyl-thioxanten-9-one 125
Diethylaminoethyl ester 95
Diethyl dithiolisophthalate 153
N,N-Diethylethanolamine 95
Diethyl maleate 379, 387
O,O-Diethyl-*O*-(*p*-nitrophenyl) phosphorothionate 122
Diethylstilbestrol
as a carcinogen 406
biliary excretion 130, 144
enterohepatic circulation of 138
pretreatment with kanamycin 138
pretreatment with neomycin 138
Difenoxine 116
Diffusion barriers in extracellular space
effect on rate of distribution 39
Digitalis 315—317, 319
factors affecting action 251
Digitoxin
absorption 281
drug interaction 286, 288, 305
enterohepatic circulation of 136, 137
renal excretion 267
Digoxin
absorption of 239, 246, 258, 280, 282
and biliary secretion 130
and renal excretion 247, 248, 267
development of arrhythmias 196
Dihydrocholate
effect on biliary excretion 145
10,11-Dihydro-*N,N*-dimethyl-5*H*-dibenzo [*a,d*]cycloheptene-Δ^5,γ-propylamine 121
Dihydropyran 95
4,5-Dihydroxy-3-(*p*-nitrophenylazo)-2,7-naphthalene disulfonic acid 37
Dimercaprol 390
in G6PD deficiency 172
6-Dimethylamino-4-diphenyl-3-acetoxy-heptane 116
4-Dimethylamino-2,3-dimethyl-1-phenyl-3-pyrazoline-5-one 117
Dimethylaminoethyl ester 95
5-(3-Dimethylaminopropyl)-10,11-dihydro-5*H*-dibenz [*b,f*]azepine 121
N,N-Dimethyl-dibenz[*b,e*]oxepin-Δ^{11}-(6*H*)γ-propylamine 121
N,N-Dimethyldibenzo[*b,e*]thiepin-Δ^{11}-(6*H*)γ-propylamine 121
N,N-Dimethylethanolamine 95
3,5-Dimethyl-5-ethyloxazolidine-2,4-dione 115
3,5-Dimethylisoxazole 125
Dimethylnitrosamine 405
2′,6′-Dimethyl-phenoxy-2-aminopropane
metabolism of 245
N-α-dimethyl-α-phenylsuccinimide 115
3,5-Dimethylpyrazole 125
Dimidione
active metabolites of 115
2,4-Dinitrolysine 37
Dinitrophenol
and malignant hyperthermia 195
Diphenoxylate
active metabolites of 116
2,2-Diphenyl-4-(4-carbethoxy-4-phenyl-1-piperidino) butyronitrile 116
Diphenylhydantoin 402
deficient parahydroxylation of 171, 182, 183
drug interactions 281, 285—287, 290, 293, 294, 298, 302, 305, 306, 364
generated by ester prodrug 91
genetic variations in clearance 197
in chronic renal failure 260, 264, 268
in hypokalemia 251
liver disease and metabolism of 243, 244
protein binding 242, 249
tissue distribution 240
Diplosal® (see salicylsalicylic acid) 92
Disodium cromoglycate 155
Disposition rate constants 20
Dissociation rate constants
effect of reversible binding 37, 38
Distribution of drugs
age-related changes in 217—228
in chronic renal failure 259, 260
kinetics of
effect of reversible binding 36—38
factors affecting 35—40
in linear multicompartment systems 17—24
Disulfiram
drug interactions 289
effect on drug metabolism 203
4-[(5,7-Disulfonic acid naphth-2-ol)-1-azo] benzene sulfonamide 37
Ditophal 153
Diuretics 251, 252
see also thiazide diuretics
L-Dopa
absorption of 238, 239, 281, 282
after oral dosing 152
drug interactions 319
effect on drug metabolism 203
Dopamine
drug interactions 318, 319, 321, 322

Down's syndrome
and drug binding 242, 245
Doxepine 121
drug interactions 299, 300
Doxycycline
renal excretion 267
Drowsiness
from diphenylhydantoin 182
Drug absorption
age-related changes 214—217
Drug binding
in plasma
and drug interactions 294—298
in tissue
and drug interactions 299, 300
Drug clearance
and drug interactions 282—294, 360—362, 368
and kinetic parameters 276
effect of doubling intrinsic clearance 284
effect of reversible binding 69, 70
genetic control of variations in 197
Drug interactions 343—355
clinical relevance 304—307
mechanisms 378—382
methods of study 345—347, 302—304, 326—328
Drug metabolite levels
in linear one-compartment system 9—11, 364—368
Duodenal ulcer
blood type factor 196
Dysarthria
from diphenylhydantoin 182

EDDP 135
EDTA 352
Electroconvulsive therapy
with succinylcholine 179
Elimination of drugs
age-related changes in 217—228
in drug interactions 282—294
kinetics of
in linear one-compartment systems 3—17
in linear multicompartment systems 17—24
in non-linear systems 24—27
Emotional disorders
increased pseudocholinesterase 181
Enamines
use in prodrug design 95, 108
Endocarditis
treatment with combined chemotherapy 354
Enteric fevers
combined chemotherapy 353
Enterococci
treatment with combined chemotherapy 352—355
Enterohepatic circulation 130—149
methods of studying 131
of bile salts 131—134
of drugs 134—143
Ephedrine
and glaucoma 194
drug interactions 337
Epilepsy
and drug interactions 283, 285
Epinephrine
and glaucoma 194
drug interactions 320, 322
Ergot derivatives 320
ergotamine 152
Erythrocytes
and G6PD deficiency 187—191
Erythromycin
effects of ester derivatives 97, 98, 100
renal excretion 267
Estopen 96
Estradiol 406
effects of ester derivatives 98
Estriol 130
Estrogens
as a transport carrier 79
as carcinogens 406
Estrone 130
Ethacrynic acid 137
Ethambutol
absorption of 239
drug interactions 344
renal excretion of 248
Ethanol
absorption of 236, 279
and free fatty acid levels 221
drug interactions 288, 289
genetic differences in metabolism 198
metabolized by ADH 195
pharmacokinetics 26
racial differences in metabolism 195, 196
Ether
and malignant hyperthermia 172, 194
in G6PD deficiency 172
solubility and clearance 158
Ethionamide
absorption of 239
p-Ethoxy-acetanilide 116
(1-Ethoxy)ethyl-(2-acetoxy)benzoate 94
Ethyl alcohol
drug interactions 289
2-Ethylamino-2′,6′-acetoxylidide 123
Ethyl biscoumacetate
drug interactions 289
Ethyl-(*p*-chlorophenoxy)isobutyrate 125
3-Ethyl-5,5-dimethyl-oxazolidine-2,4-dione 115
2-Ethylidine-1,5-dimethyl-3,3-diphenyl-pyrrolidine 135
Ethyl mercaptan 153
Ethylvinyl ether
in prodrugs 95
17α-Ethynyl-3 β-dihydroxy-5(10)-estrene 139
Etisul® 153
Etorphine
enterohepatic circulation of 135, 136
Extracellular compartment II 43—80
see also multicompartment models

Extracellular water
relation to age 224—229
Extrahepatic drug metabolism
effects of disease on 246, 247

FAA, see N-2-Fluorenylacetamide
Familial periodic paralysis
drug sensitivity 251
Favism 172, 187—191
see also glucose-6-phosphate dehydrogenase deficiency
Feathering, see Residuals
Fenamates
enterohepatic circulation of 142
Fenfluramine 122
overdosage 248
Fenproporex 122
Fick's Law of Diffusion 41, 45
First-Pass effect
after oral administration 154
in drug interactions 284
in linear multicompartment systems 23, 24
Flip-flop model
in pharmacokinetics 13
Flufenamic acid 143
N-2-Fluorenylacetamide
biliary excretion 145
drug interaction 373
Fluorescein
in chronic renal failure 260
Fluorocarbon aerosol propellants
hazard of 158
5-Fluorocytosine
drug interaction 353
renal excretion 267
5-Fluorouracil (5-FU)
in drug combinations 328, 330, 336, 338
mechanism of action 329
Fluroxene
toxicity 399, 400, 404
Folate
drug interactions 286, 353
Folic acid 130
deficiency 286
Food intake and drug absorption 235, 279
Free fatty acids
changes with age 221, 222
Frusemide, see Furosemide
Furadantin®, see nitrofurantoin
in G6PD deficiency 172
Furan analogues
toxicity 396
Furazolidone
drug interactions 301
in G6PD deficiency 172
Furosemide
dosage and metabolism 362
in chronic renal failure 262
metabolic pathways 395
toxicity 393—396, 397
with cephaloridine 250

G6PD deficiency 172, 173, 187—191, 400, 401
see also glucose-6-phosphate dehydrogenase
G-fast tissues 54—65
G-slow tissues 54—56
Galactose
absorption, age-related 217
Galactosemia
pharmacogenetics 170
Gallamine
receptor sensitivity 252
renal excretion of 248
Gamma-globulin
in drug-protein complexes 37
Ganglion-blocking drugs
absorption 154
Gantrisin®, see sulfisoxazole
in G6PD deficiency 172
Gas-liquid-chromatographic method
drug samples in uremic patients 263
Gastric acidity
age-related changes 215, 229
and drug interactions 279
effect on drug absorption 236
Gastric bleeding
from aspirin 94
Gastric carcinoma
blood type factor 196
Gastric disease and drug absorption 236, 248
Gastric emptying rate
and drug interactions 279, 280
effect on drug absorption 236—239
Gastrointestinal function
in uremia 258
Gastrointestinal tract
and drug interactions 278—282
drug absorption from 235—240, 259
Genetic variations
in drug disposition 169—203
in response to drugs 169—203
Gentamicin
age-related changes in excretion 223, 224
drug interactions 337, 344, 352, 353
in chronic renal failure 266, 267
Gilbert's syndrome
drug interactions 287
Glaucoma
intraocular pressure and steroids 193, 194
pharmacogenetic 172
relation to PTC tasting 192
Glomerular filtration
in drug interactions 290—294
in renal disease 248
relation to age 222—225, 227—229
Glomerulonephritis
and drug distribution 260
Glucagon 315, 318
D-Glucaric acid
drug interaction 286

Glucose levels in blood
drugs that lower 124, 125
Glucose-6-phosphate dehydrogenase
deficiency 172, 187—191, 400, 401
see also G6PD
Glutaric acid
in prodrugs 95, 99
Glutathione 367, 378—381
and acetaminophen 385—390
and halobenzenes 401, 402
and phenacetin 391, 392
Glutethimide
absorption 281
and hemodialysis 265
drug interactions 281, 287, 306
enterohepatic circulation of 140, 141
metabolites 130
Glyceryl trinitrate
receptor sensitivity to 250
Glycine 367
and bile salt excretion 132
Glycocholic acid 133
Gout
and thiazide diuretics 170
Gram negative bacteria
treatment with combined chemotherapy 344, 352—355
Griseofulvin
toxicity 407
variations in absorption 236, 278, 279, 301
Guanethidine
drug interactions 299, 300, 305

Half-life
and kinetic parameters 275
Halobenzenes
toxicity 401, 402
Halofenate
age-related changes in binding 221
Halothane
allergy to 408, 409
and malignant hyperthermia 172, 194, 195
genetic differences in metabolism 198
in G6PD deficiency 172
solubility and effectiveness 158
Helminthiasis
lucanthone in treatment 125
Hemodialysis
and drug metabolism 248, 249, 260, 261, 263, 265
Hemodynamic drug interactions 292—294
Hemoglobin H
drug sensitive 172, 191, 192
Hemoglobin Zurich
drug sensitive 172, 191, 192
Hemolysis 400, 401
Hemolytic anemia 172, 191, 192, 400, 401, 410
Henderson-Hasselbach equation 291
Hepatic drug metabolizing enzymes
influences on activity 243, 362, 363
Hepatic necrosis
and drug metobolism 243, 244
induced by
acetaminophen 244, 245
fluoroxene 399, 400
furosemide 393—396
halobenzenes 401, 402
hydrazines 384—390, 398
isoniazid 397—399
phenacetin 391—393
Heptabarbital
drug absorption 278
drug interaction 301
Hexachlorobenzene
and porphyria 403, 404
Hexamethonium 150
absorption 154
renal excretion of 248
Hexobarbital
genetic variations in activity 198
Histamine 322
Hodgkin's disease
combination chemotherapy 334, 335
HPPH, see 5-Phenyl-5'-parahydroxyphenyl-hydantoin
Hyaluronic acid
in tissue barriers 39
Hyaluronidase
effect on rate of diffusion 39
Hycanthone 126
Hydralazine
slow acetylators 171, 176, 178
Hydrazines
hepatoxicity 398
Hydrocortisone 139
Hydrogen peroxide
and acatalasia 171
Hydroxocobalamin 150
N-Hydroxy metabolites
toxicity of 390, 391, 392
4-Hydroxy-acetanilide 116
p-Hydroxy-amphetamine 124
4-Hydroxyamphetamine glucuronide 140
N-Hydroxy-2-fluorenylacetamide
biliary excretion 145
Hydroxyhexamide 125
4-Hydroxymethamphetamine glucuronide 141
4-Hydroxynorephedrine glucuronide 141
2-Hydroxyphenacetin
and methemoglobinemia 184, 185
2-Hydroxyphenetidin
and methemoglobinemia 184, 185
γ-Hydroxy-phenylbutazone 117
4-Hydroxypropranolol 123
Hydroxyurea
mechanism of action 329
Hyperbilirubinemia
drug interactions 287
genetic factors in treatment 203
Hypercalcemia
effect on drug action 251, 252
Hypersensitivity
to drugs 407—409

Hypertension
and drug interactions 283, 320, 321
increased pseudocholinesterase 181
Hyperthermia, see malignant hyperthermia
Hyperthyroidism
enhanced response to drugs 251
Hypnotics
chloral hydrate 114
Hypocalcemia
drug interactions 286
Hypoglycemic agents
drug interactions 305
Hypokalemia
effect on drug action 251, 252
Hypothyroidism
drug sensitivity 252

IgG antiglobulin
and slow acetylators 178
IgG globulin
levels at various ages 220
Imidoquinone 390, 329
Imipramine 121
Indocyanine green
biliary excretion 130, 144
drug interactions 300—302
Indomethacin
enterohepatic circulation 139—140
INH, see isoniazid
Inhalation of drugs 157—158
Insecticides
drug interactions 288
effect on drug metabolism 200, 202
Insulin
drug absorption 282
in chronic renal failure 262
loss in gut lumen 153
resistence to 250
route of administration 150
Intal, see disodium cromoglycate
Intestinal flora
and age changes 215
Intracellular compartment III 43—80
see also multicompartment models
Intramuscular drug injection 156, 157
effect on drug absorption 234
Intravenous drug administration
effect on drug absorption 234
elimination 155, 156
in linear one-compartment systems 3—5
in linear multicompartment systems 17—19
in non-linear systems 24—26
Intravenous infusion of drug
in linear one-compartment systems 14, 15
in linear multicompartment systems 21, 22
in studying drug interactions 303, 304
Inulin
age-related changes 223
in determining drug clearance 8, 9
Ion exchange resins
effect on drug absorption 281
Iopanoic acid 130
Iproniazid 397—399
metabolic pathway 399
Iron
absorption of 238, 259, 280, 281
biliary excretion 145
toxicity 407
Isoniazid
absorption of 239
allergy to 408, 409
and diphenylhydantoin toxicity 183
drug interactions 280, 282, 289, 306, 344
genetic differences in metabolism 198, 201
metabolic pathway 399
slow acetylation of 171, 176—178, 397—399
toxicity 243, 397—399, 408
Isonicotinic acid 176
Isoprenaline, see isoproterenol
Isoprenaline sulphate, see isoproterenol
2-(*N*-Isopropylamino)-1-(2,5-dimethoxyphenyl)-1-propanol 122
1-Isopropylamino-3-(1-naphthyloxy)-2-propanol-HCl 123
N-Isopropylmethoxamine 122
Isoproterenol
absorption of 239
drug interactions 293, 318, 319, 321
influence of route of administration 158—162
routes of metabolism 155, 160

Kanamycin
age-related changes in excretion 223, 224
and enterohepatic circulation 134, 138
drug interactions 337, 344, 352—354
in chronic renal failure 260, 267
Koalin-pectin mixtures
effect on drug absorption 281
Kernicterus 297
Ketamine
narcosis time in CRF 261
Kinetic models, *in vivo* 52—61
caternary 3-pool system, Model I 52—54
mammillary, caternary system, Model II 56—58
mammillary, degenerate diffusion limited system, Model IA 54—56
mammillary, 3-pool, flowed-limited system, Model IV 60, 61
2-pool system, Model III 58, 59
Kinetic parameters of drug interaction
biologic determinants 274—278
Kynex®, see sulfamethoxypyridazine
in G6PD deficiency 172

Latentiation, drug 86—112
Lesch-Nyhan syndrome
pharmacogenetics 170
Leukemia
acute childhood
combination chemotherapy 331, 332
acute myeloblastic
combination chemotherapy 332, 333

Lidocaine
absorption of 234, 237, 238, 240, 241
clearance in heart patients 245, 247
drug interactions 292, 293, 302, 305, 307
in hypokalemia 251
intravenous infusion 156
liver disease and metabolism of 243
metabolism of 123
routes of administration 163, 164
Ligandin
drug interactions 300
in drug binding 38, 222
Lignocaine, see lidocaine
Lincomycin
drug interactions 281
renal excretion of 248, 267
Lindane
drug interactions 288
Linear systems
effects of reversible binding 71—78
pharmacokinetics 3—24
Lineweaver-Burk equation 27
Linked animals
in study of enterohepatic circulation 131, 138, 139, 142, 143
Lipid levels in blood
drugs that lower 124, 125
acetohexamide 125
clofibrate 125
3,5-dimethylisoxazole 125
3,5-dimethylpyrazole 125
neomycin 134
nicotinic acid 125
in uremia 262
Lipid solubility and drug ionization 36
Lithocholic acid 132, 134
Liver disease
sensitivity to anticoagulants 184
Lucanthone 125
Lymphatic flow
influence on pharmacokinetics 40
Lymphomas
combination chemotherapy 334, 335
Lymphosarcomas
combination chemotherapy 335

Macrolides
drug interactions 350
Magnesium
hydroxide
effect on drug absorption 279, 282
ions
and digitalis 316
Malabsorption syndromes 238, 239, 258, 281, 282
Malaria
combined chemotherapy 353
Malignant hyperthermia
after anesthetic agnets 194, 195, 252
pharmacogenetic 172
Mandrax, see methaqualone
Mannitol
age-related changes 223
in hepatocytes 26
Mecamylamine
renal excretion of 248
Mechanisms of drug interactions 278—302
Medazepam
active metabolites of 120
Meningitis
combined chemotherapy 350
Meperidine
active metabolites of 116
Mephenesin
prodrugs of 104
Mephentermine 321
Mephobarbital
active metabolites of 114
Meprobamate
biliary secretion 130
carbonate esters of 101, 106
liver disease and metabolism of 243
6-Mercaptopurine (6-MP)
drug interactions 289
in drug combinations 327, 328, 330—333, 337
mechanism of action 329
Mercapturic acid 376, 385—389
Mercurials
and cardiac glycosides 316
Mesenteric blood flow
effect on drug absorption 239
Mestranol 406
Metabolism of drugs
in chronic renal failure 261—265
in drug interactions 282—290
kinetics 51—80
Metaraminol 123
Methadone
enterohepatic circulation of 135
Methamphetamine
enterohepatic circulation of 141
Methaqualone
absorption of 239
plus diphenhydramine 246
Methemoglobinemia
induced by acetanilide 400, 401
induced by acetophenetidin 184, 185
pharmacogenetic 171
Methicillin
age-related changes in excretion 225
drug interactions 352, 353
renal excretion 267
Methimazol
biliary excretion 145
Methotrexate (MTX)
drug interactions 292
in drug combinations 328, 330—338
mechanism of action 329
renal excretion of 248
Methoxamine 122
Methoxyflurane
and malignant hyperthermia 172, 194
and renal function 250
in G6PD deficiency 172
Methoxymethyl ethers
in prodrugs 95

Methsuximide
  active metabolites of 115, 402
2-Methylamino-2',6'-acetoxylidide 123
N-Methyl-4-aminozobenzene 404, 405
3-Methylcholanthrene
  and acetaminophen 384, 390
  and phenacetin 291
  effect on biliary flow 144
  effect on drug metabolism 200, 363
Methyldigoxin
  renal excretion 267
L-α-Methyl-3,4-dihydroxyphenylalanine 123
Methyldopa
  acetylation 178
  metabolites of 123
  renal excretion of 248, 249
Methylene blue
  in G6PD deficiency 172
L-γ-Methyl-3-hydroxyphenylalanine 123
3-O-Methylisoprenaline 160
D-Methylnorepinephrine 123
N-Methyloxazepam 119, 120
D-α-Methylphenethylamine 106
Methylphenidate
  drug interactions 289
α-Methyl-α-phenylsuccinimide 115
2-Methyl-3-piperidino-pyrazine 121
Methylthiouracil
  metabolism by PTC tasters 193
α-Methyl-*m*-tyrosine 123
Methylxanthine derivatives 315, 318, 319
Metoclopramide
  effect on drug absorption 237, 280
Metrazol 120
Metyrapone
  drug interactions 287, 288
Michaelis-Menten equation 24—29
Monoaminoxidase inhibitors 121
  drug interactions 301
MOPP
  in Hodgkin's disease 334, 335
Morphine
  enhanced response to 250—252
  enterohepatic circulation of 134, 135
  ethereal sulfate 135
  from codeine 115
  glucuronide 135, 363
Multicompartment models 17—24
  absorption kinetics 17—24
  bioavailability and absorption 20, 21
  deriving equations 43—80
  diagram 2
  distribution kinetics 17—24
  distribution of single i. v. doses 17—19
  elimination kinetics 17—24
  elimination of single i. v. doses 17—19
  first-pass effect 23, 24
  intravenous infusion of drug 21, 22
  repetitive drug administration 22, 23
  reversible pharmacologic effects 31, 32
  volume of distribution 20
Muscle relaxants
  centrally acting 115
  effect of acid-base, electrolyte balance 251, 252
  mephenesin 104
  succinylcholine 178—182
  zoxazolamine 115
Muscular rigidity
  in malignant hyperthermia 194, 195, 252
Myelosuppressive drugs
  see procarbazine
  see nitrogen mustard
Myelotoxic drugs 331
Myasthenia gravis
  drug sensitivity 251
Myocardial contractility
  drugs that increase 315—319
  see also digitalis
  see also sympathomimetic amines
Myocardial ischemia 319
Myoglobinemia
  from succinylcholine 252
Myotonia dystrophica
  drug sensitivity 251

Nafcillin 143
Nalidixic acid
  absorption of 238, 245, 266
Naphtalene
  in G6PD deficiency 172
1-Naphthol-2-sulfonic acid-4-[4-(4'-azobenzene-azo)]phenylarsonic acid 37
α-Naphthylisothiocyanate (ANIT) 384, 386, 393
Neomycin
  absorption of 154, 239
  and enterohepatic circulation 134, 138
  drug interactions 281, 344
Neoplastic diseases
  cyclophosphamide in treatment 126
Neostigmine
  renal excretion of 248
Nephropathy
  from analgesics 391—393
Neural injuries
  and sensitivity to succinylcholine 251
Neuritis
  after isoniazid 176
Neuropsychiatric disturbance
  drug-induced
    in patients with chronic renal failure 261
Nicotine
  drug interaction 337
  effect on drug metabolism 200, 202
  route of administration 152
Nicotinic acid 124
Nicotinuric acid 124
Nikethamide 144
Nitrates
  tolerance to 250
Nitrofuran derivatives
  toxicity 407

Nitrofurantoin
in chronic renal failure 259, 266
in G6PD deficiency 172
Nitrofurazone
toxicity 407
Nitrogen mustard
drug combinations 330, 334, 335, 338
intravenous infusion 156
mechanism of action 329
4-Nitroquinoline-*N*-oxide 405
Nitrous oxide
and malignant hyperthermia 194
Non-linear systems
binding 78
pharmacokinetics 24—33
Noracetylmethadol 116
Noradrenaline, see also norepinephrine 121
Norepinephrine 123
and RO-1-9571 118
drug interactions 293, 299—301, 318
Norethynodrel 139
Norfenfluramine 122
Normeperidine 116
Norprothiadene 121
Nortriptyline
as active metabolite 121
clearance 76
drug interactions 286, 289, 298, 363
genetic differences in metabolism 198, 201—203
Nor-ψ-tropine bicyclic ring system 90
Nystagmus
from diphenylhydantoin 182

Oleandomycin 97
effect of acetylation 97
Omega volume of distribution 63
One-compartment models 3—17
absorption kinetics 3—17
bioavailability and absorption 11—14
diagram 3—17
drug metabolite levels 9—11
elimination kinetics 3—17
elimination of single i. v. doses 3—5
intravenous infusion of drug 14, 15
repetitive drug administration 15—17
reversible pharmacologic effects 30, 31, 364—368
urinary excretion of drugs 7—9
volume of distribution 6
Oral contraceptives
blood group and thromboembolism 196
Oral dosing 151—155, 235
for studying drug interactions 302, 303
Ouabain
biliary excretion 130, 144
drug interactions 282
Oxacillin
age-related changes in excretion 225
renal excretion 267
Oxazepam 118—121
2-Oxo-3-isobutyl-9,10-dimethoxy-1,2,3,4,6,7-hexahydro-11*bH*-benzo[*a*]quinolizine 118
Oxotremorine
drug interactions 367
Oxyphenbutazone
in drug interactions 293
Oxyphenylbutazone 117
genetic differences in metabolism 202
Oxytocin
loss in gut lumen 153

PAH
extraction related to age 225
Pamaquine
drug interactions 299
in G6PD deficiency 172
Paracetamol, see acetaminophen
Parahydroxylation
of diphenylhydantoin 171, 182, 183
Paramethadione
active metabolites of 115
Paraoxon 122
Parasitic diseases
drugs used in treatment 125, 126
Parathion
metabolites of 122
Pargyline
drug interactions 301
PAS, see *p*-aminosalicylic acid
Penamecillin 96
Penicillin
absorption 96, 103, 153, 239, 240
age-related changes in binding 222
age-related changes in excretion 225
drug interactions 279, 292, 297, 298, 348—354, 361
esters of 95, 96
renal excretion of 248, 267, 294
Penicillin G, see also penicillin 407
Pentaerythritol trinitrate
absorption of 152
enterohepatic circulation of 142
Pentaquine
in G6PD deficiency 172
Pentetrazole 120
Pentobarbital
effect on biliary excretion 145
sodium, absorption 279
Pentolinium
absorption 154
Pentothal
in chronic renal failure 264
Percutaneous drug administration 157
Perfusion systems (nonrecirculatory) 44—51
emptying of tissue compartments 50, 51
Model 1 45—49, 52—54
Model 2 49
Model 3 49, 50
Model 4 50
Permeability of capillaries and cells
effect on rate of distribution 35, 36
Pernicious anemia
blood type factor 196
Perphenazine 130
drug interactions 289
Persistin 92
Peruvoside
renal excretion 267

Pethidine
absorption 282
pH differences between cells and blood 36
Pharmacodynamic drug reactions 384
Pharmacogenetics 169—203
in drug interactions 306, 307
Pharmacokinetic models
closed model 41—44
development of 40—80
diagram 2
of covalent binding 368—370
of drugs 360, 365
perfusion systems 44—85
see multicompartment model
see one-compartment model
Pharmacokinetics
and drug interactions 305—307, 360
drug reactions 383
Chapter 59 1—34
Chapter 60 35—85
Pharmacokinetics of drug distribution and metabolism after i. v. drug administration 51—80
central compartment 62
in vivo kinetic models 52—61
volume of distribution, concepts 62—67
Phenacetin
active metabolites of 116
and methemoglobinemia 184, 185
in chronic renal failure 264
in G6PD deficiency 172
toxicity 391—393, 400, 401, 406
variations in absorption 235
o-Phenanthroline
ADH inhibitor 195
Phenelzine
drug interactions 301
slow acetylators 171, 176, 178
4-Phenetidine 400
Phenformin
diminished response to 250
in chronic renal failure 261
renal excretion of 248
Phenobarbital
and acetaminophen 384, 386, 389
and drug absorption 278, 285—288
and drug distribution 240
and fluroxene 400
and furosemide 395
and isoniazid 398
and phenacetin 185, 391, 392, 402
compared with chlorethate 101
drug interactions 290, 291, 300—302, 306, 337, 361—364, 378—381
effect on bile acid synthesis 132
effect on biliary excretion 137, 143, 144
from mephobarbital 114
genetic variations in activity 203
in chronic renal failure 266
intoxication 248
parahydroxylation of 182, 183
Phenol
carbamate esters of 104
Phenol red
biliary excretion 144
Phenol-3,6-dibromophthalein disulphonate 143
Phenolphthalein
biliary excretion 144
Phenothiazine
active metabolites of 121
α-adrenergic blocker 319—321
effect on prothrombinopenic response 184
enterohepatic circulation of 143
Phenoxybenzamine
drug interactions 319, 321
Phentolamine
drug interactions 319
Phenyl acid glutarate
hydrolysis of 92
Phenylalanine
drug interactions 281
parahydroxylation 182
Phenylbutazone
active metabolites of 117
drug interactions 281, 288, 292, 296, 297, 301, 302, 307
effect on biliary flow 144
effect on prothrombinopenic response 184
enhanced response to 251
genetic variations in clearance 197—203
importance of formulation 154
intramuscular injection 157
liver disease and metabolism of 243, 250
Phenylephrine
and glaucoma 194
Phenylethylmalondiamide 114
5-Phenyl-5-ethyl-1-methyl-barbituric acid
active metabolites of 114
Phenylhydrazine
in G6PD deficiency 172
Phenylhydroxylamines
and methemoglobinemia 400
Phenylketonuria
pharmacogenetics 170
5-Phenyl-5'-parahydroxyphenylhydantoin 182
Phenylpropanolamine 108
N-(3-Phenyl-2-propyl)-1,1-diphenyl-3-propylamine 124
Phenylthiocarbamide
inability to taste 172, 192, 193
4-Phenylthioethyl-3,5-dioxo-1,2-diphenyl-pyrazolidine 118
Phenylthiourea
inability to taste 172, 192, 193
Phenyramidol
drug interactions 288
effect on thrombinopenic response 184
Phlorridizin 132
Phosphodiesterase
drug interactions 318
Phospholipids
biliary excretion 145
Phtivazid
acetylation 178

Pindolol
renal excretion of 249
Piperonyl butoxide
and acetaminophen 384, 386
and acetanilide 401
and cephaloridine 397
and furosemide 393
and isoniazed 398
and phenacetin 391
drug interactions 378
effect on biliary excretion 145
Pivampicillin 96
Polycyclic hydrocarbons 405
Polymixin
B, renal excretion of 248, 267
drug interactions 344, 350, 353
Porphyria
and barbiturates 170
drug-induced 403, 404
Porphyrins
enterohepatic circulation 130
Potassium
and cardiac glycosides 316, 317
Practolol
renal excretion of 248
Prazepam
active metabolites of 121
Prednisone
adverse reactions and albumin levels 197
drug combinations 330—336
drug interactions 286
mechanism of action 329
protein binding 242
Pregnenolone 138
sulphate 138
Prenylamine 124
Primaquine
in G6PD deficiency 172, 187—191
in hemolytic anemia 191, 192
Primidone
active metabolites of 114
Probenecid
drug interactions 292, 361
in G6PD deficiency 172
Procainamide
absorption 278
and malignant hyperthermia 194
drug interactions 305
effects of renal disease on 247, 248
genetic variations in clearance 197
hydrolysis of 246
volume of distribution 241
Procaine-amide ethobromide
biliary excretion 144
Procarbazine
combination chemotherapy 334, 335, 338
mechanism of action 329
Prochlorperazine 143
Prodrugs 86—112
conversion to active form 39
Progesterone
effects of ester derivatives 99
Promazine 130
Propranolol
absorption of 239, 249
drug interaction 292, 293, 297, 298, 300, 302, 318, 363
enhanced response to 251
genetic variations in clearance 197
paths of metabolism 123, 164
routes of administration 154, 164, 165
Propantheline
biliary excretion of 239
drug absorption 236, 237, 239, 279, 280
Propoxyphene
in drug interactions 293
Propylthiouracil
biliary excretion 145
in G6PD deficiency 172
Prostaglandins 320
Proteolytic enzymes
in hydrolysis of carbonate diesters 102
Prothiadene 121
Protoporphyrin 130
Protriptyline
drug interactions 299, 300
Pseudocholinesterase
and enzymatic cleavage 88, 102
and liver disease 246
and succinylcholine sensitivity 171, 178—182
Pseudoephedrine
renal excretion of 248
Psychiatric chemotherapeutic agents 118—121
chlordiazepoxide 118
diazepam 119
MAO inhibitors 121
medazepam 120
phenothiazines 121
prazepam 121
PTC, see phenylthiourea
Pyloric stenosis 170
Pyramidon®, see aminopyrine
in G6PD deficiency 172
Pyrazole
ADH inhibitor 195
1*H*-Pyrazolo[3,4-*d*]pyrimidin-4-ol 117
Pyridoxine
and isoniazid 176
Pyrimethamine
drug interactions 351, 353
Pyrrolizidine alkaloids 404, 405
Pyruvate esters 95
Pyruvic acid
in prodrugs 95, 99

Quinacrine
drug interactions 299
in G6PD deficiency 172
Quinalbarbital
drug interactions 305, 306

Quinidine
absorption of 239
in chronic renal failure 260
in drug interactions 291, 305
in G6PD deficiency 172
in hypokalemia 251
protein binding 242
renal excretion of 248
Quinine
drug interactions 281
in G6PD deficiency 172

Racial differences
in ethanol metabolism 195, 196
in G6PD deficiency 187—190
in isoniazid acetylation 177, 397, 398
in pseudocholinesterase activity 181
Reactive metabolites 374—381
Receptor sensitivity 250—252
Rectal drug administration 155
Renal clearance of drugs
in drug interactions 290—292
in renal failure 247—250, 258—267
relation to age 225—230
Renal necrosis
from cephaloridine 397
from furosemide 393
Repetitive drug administration
in linear multicompartment systems 22, 23
in linear one-compartment systems 15—17
in non-linear systems 29, 30
Reserpine 124
and RO-1-9751 118
Residuals, method of
in pharmacokinetics 12, 13, 18, 22
Reversible binding of drugs to proteins
diffusion into extracellar space 38
dissociation rate constants 37, 38
effect on drug clearance 69, 70
effect on rate of distribution 36—38
effect on volume of distribution 70, 71
general considerations 36, 37, 68, 69
reversible binding and drug clearance 38
Reversible pharmacologic effects
kinetics of
one-compartment systems 30, 31
multicompartment systems 31, 32
Rheumatic heart disease
blood type factor 196
resistance to digoxin 246
Riboflavin
absorption 216, 217, 238, 280
Rifampin
absorption 282
drug interactions 281
RO-1-9571 118
RO-5-0883/1 118
RO-5-2092 118
Rolitetracycline
renal excretion 267
Routes of drug administration 150—158
influence on drug response 158—166

Salazopyrin
absorption of 239
Salicylamide 116
absorption of 239
drug interaction 363, 364
Salicylate
active metabolites of 116
and hemodialysis 248
binding differences with age 221
dosage and metabolism 362
drug interaction 281, 282, 291, 292
ester 95
from salicylsalicylic acid 92
intoxication 248
pharmacokinetics 26
protein binding 242, 245
sodium salicylate 92
Salicylazosulfapyridine
in G6PD deficiency 172
Salicylic acid, see also salicylate
from aspirin 247
hexyl carbonate derivative 94
in prodrugs 95
Salicylsalicylic acid
synthesis of 92, 93
Salysal®
see also salicylsalicylic acid 92
Schiff bases
use in design of prodrugs 95, 106
Schistomicidal drug 126
Schizophrenia
increased pseudocholinesterase 181
Scoline, see succinylcholine
Scopolamine
and glaucoma 194
Secobarbital
and porphyria 403, 404
Sedatives
chloral hydrate 114
chlorethate 101
phenobarbital 101
trichloroethanol 101
Sedormid
and porphyria 403, 404
Serotonin 121
Serum albumin
in drug-protein complexes 37
Serum globulins
levels at various ages 220
Serum proteins
in chronic renal failure 259, 262
levels at various ages 220, 221
Shigellosis
and drug absorption 238, 245
Sodium bicarbonate
and drug absorption 279, 291
Sodium nitroprusside
and ergotism 320
Sodium salicylate 116
Solanine poisoning 181
Spectrophotometric method
drug sample in uremic patients 263
Spironolactone 137
and cardiac glycosides 316, 317
toxicity 402, 403

Steroids
drug interactions 286, 287
enterohepatic circulation of 138, 139
local action 155
Streptomycin
age-related changes in excretion 223, 224
drug interactions 344, 348, 349, 352—354
renal excretion 267
Strophantin G
renal excretion 267
Strophantin K
-acetyldigoxin
renal excretion 267
renal excretion 267
Subcutaneous drug administration 157
Sublingual drug administration 155
Succinic acid
in prodrugs 95, 99
Succinylcholine
and acid-base, electrolyte balance 251, 252
and liver disease 246
and malignant hyperthermia 172, 194
hydrolysis of 92
in G6PD deficiency 172
sensitivity to 171, 178—182, 410
Succinylmonocholine 179
Succinylsulphathiazole
absorption 154
biliary excretion 144
Sulfacetamide
in G6PD deficiency 172
Sulfadiazine
absorption 279
drug interaction 353
renal excretion 267
Sulfadimethoxine
and hemolytic anemia 191, 192
Sulfamaprine
slow acetylators 171
Sulfamethazine
in chronic renal failure 260
slow acetylators 171, 176, 177
Sulfamethiazole
in chronic renal failure 266
Sulfamethonium, see succinylcholine
Sulfamethoxazole
drug interaction 353
renal excretion 267
Sulfamethoxypyridazine
acetylation 177
in G6PD deficiency 172
Sulfanilamide
acetylation 177
N-acyl derivatives 106
antistreptococcic agent 106
in G6PD deficiency 172
Sulfaphenazole
binding to plasma albumin 197
drug interactions 289, 297, 302
Sulfapyridine
absorption of 239
in G6PD deficiency 172
Sulfate
drug interactions 362, 367, 373
Sulfazoxazole
in drug interactions 297
Sulfinpyrazone
renal excretion of 248
Sulfisomidine
renal excretion 267
Sulfisoxazole
absorption of 239
in chronic renal failure 264
in G6PD deficiency 172
Sulfobromophthalein 143
and ligandin 38
Sulfonamides
absorption of 154, 238, 241
and hemolytic anemia 191, 192
in chronic renal failure 260, 264
in drug interactions 297, 298, 350, 351, 353
in G6PD deficiency 172
protein binding 242
renal excretion of 248
Sulfones
in G6PD deficiency 172
see also 4'4'''-sulfonylbisacetanilide 107
4',4'''-Sulfonylbisacetanilide 107
4,4'-Sulfonyldianiline (see diamino diphenyl-sulfone)
Sulfonylureas
receptor sensitivity to 250
Sulfoxone
in G6PD deficiency 172
Sulthiamine
drug interactions 289
Superposition, principle of
in pharmacokinetics 3, 26
Suramin 150
Suxethonium, see succinylcholine
Sympathomimetic drugs 122
as vasoconstrictors 319, 320
as vasodilators 321
drug interactions 301, 317—319
enhanced response to 251
fenfluramine 122
fenproporex 122
N-isopropylmethoxamine 122
isoproteronol 158—162
tolerance to 250
Synergism
of combined drug effect 351—355

Takahara's disease 175
Taurine
and bile salt excretion 132, 133
Taurocholate
effect on biliary excretion 145
Taurocholic acid 133
Testosterone
and spironolactone 402, 403
as a transport carrier 79
effects of ester derivatives 99, 100
Tetanus
treatment with succinylcholine 179

Tetrabenazine
  active metabolites of 118
Tetracaine
  and pseudocholinesterase 180
Tetracycline
  age-related changes in binding 222
  drug interactions 344, 350, 351, 353
  oral dosing 153
  renal excretion of 248, 250, 267
  variations in absorption of 235, 239, 279, 280, 282
Tetrahydropyran 95
Tetrahydropyranyl ethers 95
Theophylline 153
Thiamphenicol
  renal excretion 267
Thiazide diuretics
  and cardiac glycosides 316
  and gout 170
  renal excretion of 248
  resistance to 250
Thiazolsulfone
  in G6PD deficiency 172
Thiobarbituric acid
  drug interactions 337
6-Thioguanine (6-TG)
  in drug combinations 329, 330, 333, 334, 338
  mechanism of action 329
Thiopental
  distribution kinetics 61
  drug binding 242
  in anemic patients 252
Thiopentone, see also thiopental
  intravenous infusion 156
  metabolism by PTC tasters 193
Thromboembolism
  after oral contraceptives 196
Thyroid disease
  relation to PTC 192, 193
Thyrotoxicosis
  increased pseudocholinesterase 181
Thyroxine 125
  absorption 281, 282
  age-related changes in binding 221
  as a transport carrier 79
  biliary excretion of glucuronide 144
  drug interactions 286
Time-delay factors
  effect on rate of distribution 39, 40
Tolbutamide
  drug interactions 288, 289, 297, 301, 302, 305, 306
  metabolism 243, 246
Toxoplasmosis
  treatment with combined chemotherapy 353
Tranquilizers
  meprobamate 101, 106
Tremorine
  drug interactions 367
Triacetyloleandomycin 97
Triamterene
  and cardiac glycosides 316
Trichloroacetate
  drug interaction 281
Trichloroethanol 114
  carbonate ester of 101
  drug interactions 289, 290, 302
  prodrug of 93
  with acetaminophen 102
2,2,2-Trichloroethyl chloroformate 101
Trifluoperazine
  biliary excretion of 143
1-(3-Trifluoromethylphenyl)-2-aminopropane 122
1-(3-Trifluoromethylphenyl)-2-ethyl aminopropane 122
Triglyceride 125
Trihydroxycholanoic acid 133
Triiodothyronine
  absorption 282
  age-related changes in binding 221
Trimethadione
  active metabolites of 115
Trimethoprim
  drug interactions 351, 353
  renal excretion 267
3,5,5-Trimethyloxazolidine-2-4-dione 115
Trinitrotoluene
  in G6PD deficiency 172
Trypanosomiasis
  treatment with suramin 150
Trypsin
  in hydrolysis of carbonate diesters 102
Tryptophan
  age-related changes in binding 221
  levels in uremic patients 259, 260
Tuberculosis
  patients on isoniazid 176—178
  see also anti-tuberculosis drugs
d-Tubocurarnie
  and acid-base, electrolyte balance 251, 252
  protein binding 242
  renal excretion of 248
Tubular secretion
  in drug interactions 290—292
  in renal disease 248
  relation to age 229
Tumors, solid
  combination chemotherapy 335—337
Tyramine
  drug interactions 299, 300, 301

Urea 95
Uremia
  drug absorption 258, 259
  drug distribution 259, 260
  drug elimination 265—267
  drug metabolism 261—265
  drug sensitivity 249, 251
  response to drugs 267, 268
Urinary excretion of drug 247—250
  in linear one-compartment system 7—9
Urinary tract infection
  treatment with combined chemotherapy 344, 352—355

Urine flow and pH changes
effect on drug clearance 148, 149, 361
in drug interactions 290—292
Urobilinogen 130

Vancomycin
drug interactions 350, 352
renal excretion of 248, 267
Vasoconstricting drugs 319, 320
see also angiotensin
see also ergot derivatives
see also sympathomimetic amine
Vasodilation
effect on drug absorption 234
Vasodilator drugs
diallylmelamine 124
drug interactions 320, 321
prenylamine 124
Vasopression
intramuscular injection 157
Vinblastine
in drug combinations 330, 334, 336
mechanism of action 329
Vinca alkaloids
in combination chemotherapy 333, 334, 336
Vincristine
in drug combinations 331—337
mechanism of action 329
1,5-Vinyl-2-thiooxazolidone
goitrogenic compound 192
Vitamin A
drug interactions 281
in chronic renal failure 261
Vitamin B_{12} 130, 239
drug interactions 281
Vitamin C
in drug interactions 291
Vitamin D
absorption 259
drug interactions 286
in chronic renal failure 261, 262
Vitamin D_3
drug interactions 286
Vitamin K
deficiency and sensitivity to anticoagulants 184
drug interactions 281
in G6PD deficiency 172
resistance to warfarin 185, 186, 251
Volume of distribution 240, 241
concepts and equations 62—67
effect of reversible drug binding 70, 71
in drug interactions 277, 283, 294—296, 298, 299, 364, 365
in one-compartment systems 6
in multicompartment systems 20
in renal failure 248, 259, 260, 262
omega 63
variations relating to age 225—230

Warfarin
absorption of 236, 245
drug interactions 279, 281, 282, 285, 288, 289, 294, 296, 297, 301, 305—307
genetic variations in clearance 197
resistance to 171, 184—186
Wilson's disease
pharmacogenetics 170

X-linked dominant genes
and pharmacogenetic conditions 171, 172
Xylocaine, see lidocaine
Xylose
absorption, age-related 217
D-Xylose
drug interactions 281

Y protein
age-related changes 222

Z protein
age-related changes 222
Zoxazolamine
active metabolites of 115
paralysis time in CRF 261

Handbuch der experimentellen Pharmakologie/
Handbook of Experimental Pharmacology
Heffter — Heubner. New Series

Vol. IV:	**General Pharmacology** ISBN 3-540-04845-6	DM 86,—	US $ 35.10
Vol. X:	**Die Pharmakologie anorganischer Anionen** ISBN 3-540-01465-9	DM 255,—	US $ 104.10
Vol. XI:	**Lobelin und Lobeliaalkaloide** ISBN 3-540-01910-3	DM 20,—	US $ 8.20
Vol. XII:	**Morphin und morphinähnlich wirkende Verbindungen** ISBN 3-540-02158-2	DM 95,—	US $ 38.80
Vol. XIII:	**The Alkali Metal Ions in Biology.** In preparation		
Vol. XIV:			
Part 1	**The Adrenocortical Hormones I** ISBN 3-540-02830-7	DM 290,—	US $ 118.40
Part 2	**The Adrenocortical Hormones II** ISBN 3-540-03146-4	DM 90,—	US $ 36.80
Part 3	**The Adrenocortical Hormones III** ISBN 3-540-04147-8	DM 170,—	US $ 69.40
Vol. XV:	**Cholinesterases and Anticholinesterase Agents** ISBN 3-540-02988-5	DM 360,—	US $ 146.90
Vol. XVI:	**Erzeugung von Krankheitszuständen durch das Experiment**		
Part 1	**Blut/Blood.** In preparation		
Part 2	**Atemwege** ISBN 3-540-04517-1	DM 150,—	US $ 61.20
Part 3	**Heart and Circulation.** In preparation		
Part 4	**Niere, Nierenbecken, Blase** ISBN 3-540-03305-X	DM 180,—	US $ 73.50
Part 5	**Liver.** In preparation		
Part 6	**Schilddrüse.** In preparation		
Part 7	**Zentralnervensystem** ISBN 3-540-02831-5	DM 150,—	US $ 61.20
Part 8	**Stütz- und Hartgewebe** ISBN 3-540-04518-X	DM 120,—	US $ 49.00
Part 9	**Infektionen I** ISBN 3-540-03147-2	DM 180,—	US $ 73.50
Part 10	**Infektionen II** ISBN 3-540-03531-1	DM 200,—	US $ 81.60
Part 11A	**Infektionen III** ISBN 3-540-03840-X	DM 180,—	US $ 73.50
Part 11B	**Infektionen IV.** In preparation		
Part 12	**Tumoren I** ISBN 3-540-03532-X	DM 180,—	US $ 73.50
Part 13	**Tumoren II** ISBN 3-540-03533-8	DM 110,—	US $ 44.90
Part 14	**Tumoren III.** In preparation		
Part 15	**Kohlenhydratstoffwechsel, Fieber/Carbohydrate Metabolism, Fever** ISBN 3-540-03534-6	DM 180,—	US $ 73.50
Vol. XVII:			
Part 1	**Ions, alcalino-terreux I. Systèmes isolés** ISBN 3-540-02989-3	DM 190,—	US $ 77.60
Part 2	**Ions, alcalino-terreux II. Organismes entiers** ISBN 3-540-03148-0	DM 240,—	US $ 98.00
Vol. XVIII:			
Part 1	**Histamine** ISBN 3-540-03535-4	DM 230,—	US $ 93.90
Part 2	**Anti-Histaminics.** In preparation		
Vol. XIX:	**5-Hydroxytryptamine and Related Indolealkylamines** ISBN 3-540-03536-2	DM 230,—	US $ 93.90

Vol. XX:			
Part 1	**Pharmacology of Fluorides I** ISBN 3-540-03537-0	DM 175,—	US $ 71.40
Part 2	**Pharmacology of Fluorides II** ISBN 3-540-04846-4	DM 165,—	US $ 67.40
Vol. XXI:	**Beryllium** ISBN 3-540-03538-9	DM 48,—	US $ 19.60
Vol. XXII:			
Part 1	**Die Gestagene I** ISBN 3-540-04148-6	DM 370,—	US $ 151.00
Part 2	**Die Gestagene II** ISBN 3-540-04519-8	DM 370,—	US $ 151.00
Vol. XXIII:	**Neurohypophysial Hormones and Similar Polypeptides** ISBN 3-540-04149-4	DM 230,—	US $ 93.90
Vol. XXIV:	**Diuretica** ISBN 3-540-04520-1	DM 290,—	US $ 118.40
Vol. XXV:	**Bradykinin, Kallidin and Kallikrein** ISBN 3-540-04847-2	DM 275,—	US $ 112.20
Vol. XXVI:	**Vergleichende Pharmakologie von Überträgersubstanzen in tiersystematischer Darstellung** ISBN 3-540-05132-5	DM 180,—	US $ 73.50
Vol. XXVII:	**Anticoagulantien** ISBN 3-540-05133-3	DM 225,—	US $ 91.80
Vol. XXVIII:			
Part 1	**Concepts in Biochemical Pharmacology I** ISBN 3-540-05134-1	DM 190,—	US $ 77.60
Part 2	**Concepts in Biochemical Pharmacology II** ISBN 3-540-05389-1	DM 275,—	US $ 112.20
Part 3	**Concepts in Biochemical Pharmacology III** ISBN 3-540-07001-X.	DM 248.—	US $ 101.20
Vol. XXIX:	**Oral wirksame Antidiabetika** ISBN 3-540-05554-1	DM 340,—	US $ 138.80
Vol. XXX:	**Modern Inhalation Anesthetics** ISBN 3-540-05135-X	DM 220,—	US $ 89.80
Vol. XXXI:	**Antianginal Drugs** ISBN 3-540-05365-4	DM 190,—	US $ 77.60
Vol. XXXII:			
Part 1	**Insulin I** ISBN 3-540-05470-7	DM 275,—	US $ 112.20
Part 2	**Insulin II.** ISBN 3-540-07006-0.		
Vol. XXXIII:	**Catecholamines** Out of print		
Vol. XXXIV:	**Secretin, Cholecystokinin — Pancreozymin and Gastrin** ISBN 3-540-05952-0	DM 165,—	US $ 67.40
Vol. XXXV:			
Part 1	**Androgene I** ISBN 3-540-05706-4	DM 320,—	US $ 130.60
Part 2	**Androgens II and Antiandrogens. Androgene II und Antiandrogene** ISBN 3-540-06883-X	DM 398,—	US $ 162.40
Vol. XXXVI:	**Uranium-Plutonium-Transplutonic Elements** ISBN 3-540-06168-1	DM 374,—	US $ 152.60
Vol. XXXVII:	**Angiotensin** ISBN 3-540-06276-9	DM 224,—	US $ 91.40
Vol. XXXVIII:			
Part 1	**Antineoplastic and Immunosuppressive Agents I** ISBN 3-540-06402-8	DM 258,—	US $ 105.30
Part 2	**Antineoplastic and Immunosuppressive Agents II** ISBN 3-540-06633-0	DM 348,—	US $ 142.00